Green Energy and Technology

Climate change, environmental impact and the limited natural resources urge scientific research and novel technical solutions. The monograph series Green Energy and Technology serves as a publishing platform for scientific and technological approaches to "green"—i.e. environmentally friendly and sustainable—technologies. While a focus lies on energy and power supply, it also covers "green" solutions in industrial engineering and engineering design. Green Energy and Technology addresses researchers, advanced students, technical consultants as well as decision makers in industries and politics. Hence, the level of presentation spans from instructional to highly technical.

Indexed in Scopus.

Indexed in Ei Compendex.

Mika Järvinen · Hanna Paulomäki
Editors

Designing Renewable Energy Systems within Planetary Boundaries

A Textbook for Energy Engineers

Springer

Editors
Mika Järvinen
Department of Energy and Mechanical Engineering
Aalto University
Espoo, Finland

Hanna Paulomäki
Department of Energy and Mechanical Engineering
Aalto University
Espoo, Finland

ISSN 1865-3529 ISSN 1865-3537 (electronic)
Green Energy and Technology
ISBN 978-3-031-69858-3 ISBN 978-3-031-69856-9 (eBook)
https://doi.org/10.1007/978-3-031-69856-9

The book has been funded by Tiina and Antti Herlin Foundation and Aalto University, School of Engineering.

This Springer imprint is published by the registered company Springer Nature Switzerland AG
The registered company address is: Gewerbestrasse 11, 6330 Cham, Switzerland

Preface

The journey for producing this book started as "living room couch"-discussions between an engineer and an ecologist. We were intrigued by the notion that the way we perceive the current ecological crises—including climate change, loss of nature, pollution, or social injustice—and the solutions to these crises, is affected by our prior education, our accustomed methods of problem-solving, and the terminology we use, which is largely influenced by the scientific traditions we have been taught. Most importantly, we recognized the significant potential for mutual learning and collaboration across disciplines in addressing the challenges at hand. To effectively tackle these issues, we must foster greater interdisciplinary cooperation and build shared understanding.

We were born in 327.5 and 333.7 ppm of CO_2 in the atmosphere. We met at 399.1 ppm and began drafting the concept for this book at 414.6 ppm. We are now in the final stages of manuscript preparation at 421.6 ppm. Scientists have agreed that carbon dioxide in the atmosphere should remain within the range of 390–430 ppm to ensure the continued existence of life on Earth as we know it and to limit global warming to 1.5 °C. So, it is fair to say that we are in urgency to bring greenhouse gas emissions back within the safe limits.

Unfortunately, climate change is not our only worry. The overconsumption of natural resources, and unsustainable way of living has led us into the intersection of multiple, interlinked crises—a phenomenon called the poly-crisis. Unlike isolated challenges, a poly-crisis consists of a range of crises that interact in ways that exacerbate each other, creating a situation where the entity is far more hazardous than the sum of its parts. The concurrent climate change, loss of nature, pollution, geopolitical tensions, and social inequities represent just a few strands in this complex web.

Addressing these crises one by one is not enough; there is a need for action and cooperation across disciplines and sectors. Energy transition is linked to many of the challenges faced by humanity and therefore also provides an opportunity to address many of them. Our current energy systems, heavily reliant on fossil fuels, are a major driver of climate change, pollution, geopolitical conflicts, and nature loss. Therefore, transitioning to renewable energy sources is not merely a climate action. The adoption of renewable energy technologies offers a pathway to achieving multiple objectives, including the reduction of greenhouse gas emissions, protecting and restoring biodiversity, enhancing energy security and economic resilience, and building social equity.

Engineers play a pivotal role in our societies; their impact can be witnessed everywhere in man-modified environments. It is, therefore, crucial to recognize that the perception of the world and the value placed on different aspects of it by engineers can have a significant impact on the direction of technological development. For this reason, we have expanded the focus of this textbook beyond the traditional techno-economic solutions to climate change. We will explore the opportunities and potential trade-offs large-scale renewable energy projects can have, considering specifically ecological and social impacts, both positive and negative, and ways to mitigate negative impacts.

This textbook has been co-authored by professionals from various fields, including engineering, biology, forestry, agrology, social sciences, and economics. The objective is to discuss solution pathways in the world of finite resources, in the middle of the climate change, and biodiversity loss. By bringing together insights and perspectives from diverse fields, we aim to foster a deeper understanding of the various ways energy transition can provide solutions towards a more sustainable, equitable, and resilient future. The topic is broad, and it was not feasible to cover every single aspect of this theme. However, though we are not able to provide all the answers, our objective is to provide enough information for asking the right questions when initiating a large-scale energy project. We want the future generations of energy engineers to be real game changers, bringing solutions to the climate change while avoiding harmful lockdowns or creating new problems when developing energy systems that are technically and economically feasible and build on ecological and social sustainability.

The journey ahead is undoubtedly challenging. But the good news is that we already have enough knowledge, the tools, and technologies required to adjust human life with the planetary boundaries. Sustainable energy transition could be a major vehicle in enabling this transformation.

This is an exciting time to be an energy engineer. The current energy transition away from fossil fuels provides a unique transformative potential that should not be missed. With this in mind, we are inviting readers to join us in exploring this book and envisioning and realizing an energy transition that supports achieving a better future for all.

In the end, we will conserve only what we love,
we will love only what we understand,
and we will understand only what we are taught.
—Baba Dioum, 1968

Espoo, Finland
June 2024

Mika Järvinen
Hanna Paulomäki

Acknowledgements

We would like to express our sincerest gratitude to all those who have joined us on this journey. We are particularly grateful to our co-authors, who have taken on the challenge of drafting the book chapters and case studies with us. We would like to thank the following individuals for their contributions to this project: Alicja Dańkowska, Annukka Santasalo-Aarnio, Antti Arasto, Anu Akujärvi, Ari Seppälä, Ayesha Sadiqa, Benjamin Wilson, Christian Pansch, Dilip Khatiwada, Edris Pouresmaeil, Eetu Laitila, Eeva-Lotta Apajalahti, Elina Oksanen, Emma Luoma, Georgi Popov, Han van Kasteren, Heini Kujala, Iryna Herzon, Janne Hirvonen, Janne S Kotiaho, Jonna Kangasoja, Konsta Turunen, Maiju Peura, Mari Lundström, Maria Holmberg, Mariana Galvão Lyra, Martin Forsius, Matti Lehtonen, Minna Pekkonen, Neha Garg, Paul Thompson, Rogier Ralph Floors, Roosa Talala, Saeed Rahimpour Golroudbary, Sam Cross, Sami El Geneidy, Shouzhuang Li, Torsti Schulz and Virpi Junttila.

We would also like to express our appreciation to all the reviewers who provided valuable feedback, comments, and questions, which greatly helped us improve the quality of the text. In particular, we sincerely thank the following esteemed reviewers: Katariina Buure (LUT University) and Toldy Árpád (Aalto University) for Chaps. 1 and 2; Antti Arasto (VTT) and Mohamed Magdeldin (Sumitomo Heavy Industries – Foster Wheeler) for Chap. 3; Antti Tolvanen (Endeas Oy) for Chap. 4; Magnus Hellström (Åbo Akademi), Hannele Holttinen (Aalto University), Thaysa Portela de Carvalho (Åbo Akademi), and Duncan Heathfield (World in a Box) for Chap. 5; Timo Laukkanen and Muddasser Inayat (Aalto University) for Chap. 6; Toldy Árpád for Chap. 7; Hannele Holttinen for Chap. 8; and Alicja Dańkowska (LUT University) for Chap. 9. We also extend our thanks to Marko Hinkkanen for his detailed and valuable comments on Chap. 8.

We would like to thank Maija Karala for her illustrations and excellent collaboration. Our thanks also go to Chris Stewart (for Chapters 1–2, 4, 6–9) and Kate Ravilious (for Chapters 3 and 5) for their outstanding work with language editing, and, at times, even at content itself.

We would also like to acknowledge and thank all those who participated in the initial stages of commenting on the book contents in the workshops. We would like to specifically thank Magnus Hellström (Åbo Akademi) for his support throughout the process, and Massimo Santarelli (Politecnico di Torino) for his inspirational conversations and for taking the time to visit us at Aalto. We would also like to thank Duarte de Mesquita e Sousa (Técnico Lisboa) and Cesar Valderrama Angel (UPC Barcelona) for their valuable input regarding the structure of the book.

Inge de Waard (EIT InnoEnergy), your enthusiasm and helpful tips were invaluable. Thank you for bringing your unique skill set to the project.

We would also like to thank Leo Stranius, Veikko Sajaniemi and Henni Savolainen from Third Rock Finland for organizing a special workshop for industry actors with us. We want to thank Haide Szilvia (Flexens), Kati Vilonen (UPM), Niklas Hokka (OX2), Piia Heikkinen (Nordi), Sanna Häsä (Siemens), Taina Kokkarinen (Helen), Thomas Kohl (Neste) and Ville Reinikainen (Granlund) for helping us gain a deeper understanding of the challenges and bottlenecks faced by the companies. Many thanks also to Seppo Hakola (Ilmatar Service) for hosting us during several visits on your production sites for wind and solar power.

Thank you Hannu Tikkanen (Metsähallitus) for discussions and approving the use of your work in case study 5.1. We are also grateful for Teemu Lehtiniemi (Birdlife) for insightful discussions on the impacts of wind power on birds.

Hanna wishes to express her gratitude to the Puistokatu 4 community, which has been a valuable source of support and inspiration throughout this project. Puistokatu 4 serves as a prime example of the advantages of interdisciplinary and cross-disciplinary collaboration. By uniting individuals from diverse fields, it has fostered novel ideas and enhanced comprehension of sustainability challenges and solutions.

We are grateful to the Tiina and Antti Herlin Foundation for their financial support, which made this project possible. We would like to thank in particular Harri Lammi for his invaluable support and for engaging in constructive and insightful conversations throughout the project.

And most importantly, we want to thank all those Mika's students over several years, who initially inspired us to embark on this project with their insightful questions and ideas. The course on Renewable Energy Engineering has served as a living lab and test bench for many of the topics and exercises that ended-up in this book. Thank you for all the discussions and active participation; it has been a pleasure to work with you all! Looking forward to continue engaging with new students, see you on lectures!

How to Cite This Book

The whole book should be cited as: Järvinen MP, Paulomäki H (Eds.) 2025. *Designing Renewable Energy Systems within Planetary Boundaries*—A Textbook for Energy Engineers. Green Energy and Technology-series, Springer Publishing, Switzerland.

Individual chapters should be cited with section author names.

Contents

About the Editors

Mika Järvinen is Professor in Aalto University. He is currently Head of Energy Conversion and Systems research group and Director of Aalto's Hydrogen Innovation Center. His research is focused on different CCS technologies, P2X solutions, and thermal utilization of waste-based materials. Mika is also Enthusiastic Teacher on renewable energy technologies and related sustainability aspects.

Hanna Paulomäki is Ecologist who has worked with biodiversity, nature conservation and sustainability questions for twenty years as Expert and in leadership positions in international and national organizations including HELCOM, Oceana, Greenpeace Norden, and the Finnish Nature Panel. She has also been Member of expert groups, such as national IPBES working group. At current she is Doctoral Researcher at LUT University studying land use and socio-ecological impacts resulting from renewable energy expansion in Finland.

Chapter 1
Sustainable Energy Systems in a Changing World—An Introduction

Hanna Paulomäki and Mika Järvinen

Nature is not a sector of our society, it's a prerequisite for it.
—Carl Folke

The Sun is the enabler of almost all life on Earth, and the source of most of the resources and materials we humans consume. Photosynthesis of plants and some other organisms converts and stores solar energy into chemical form. An important raw material for photosynthesis is CO_2, which is captured from the atmosphere and turned into oxygen and sugars consumed by humans and other animals. Our planet's ideal distance from the Sun has created the perfect conditions for diverse life to develop, and a system to circulate materials endlessly and form flows of energy (Fig. 1.1). When living organisms die, pile up and form sediments over millions of years, a small part of the biomass residues are slowly converted into peat, different types of coal, oil and methane-rich gas. Collectively we call these fossil fuels, but in fact, together with the CO_2 bound to rock minerals, they form a safe and stable storage of carbon dioxide, captured from the atmosphere and locked away, making the planet Earth habitable as we know it.

H. Paulomäki (✉) · M. Järvinen
Aalto University, Espoo, Finland
e-mail: hanna.paulomaki@lut.fi

M. Järvinen
e-mail: mika.jarvinen@aalto.fi

M. Järvinen and H. Paulomäki (eds.), *Designing Renewable Energy Systems within Planetary Boundaries*, Green Energy and Technology,
https://doi.org/10.1007/978-3-031-69856-9_1

Fig. 1.1 The Earth and the Sun—an optimal 150 million km apart [27]

The Sun's electromagnetic energy has a natural heating effect. Temperature differences resulting from the solar heating of the Earth's surface and atmosphere, and the resulting density gradients, are the driving forces of winds, and hence the energy that can be harvested from the wind. We have also learned to convert the energy of the photons into electrical current and power with the use of photovoltaic solar panels.

In addition to these solar-based renewable energy sources, the rotation of the Earth on its axis and the gravity field of the Moon creates tidal water currents that can also be utilized as an energy source. Moreover, there is an almost infinite source of geothermal heat inside the Earth's crust that can be used in heating or generating electricity.

Modern human, *Homo sapiens*—"the wise man"—emerged about 200,000 years ago, and began engineering their environment in ways never seen before in the history of Earth. For the first hundreds of thousands of years human influence on the environment was relatively minor, but the extent and intensity of the impact started to increase after the development of agricultural communities around 10,000 years ago. At first humans exploited fire and burning biomass, but the discovery that the mechanical power of flowing water and blowing winds can be exploited to carry-out tasks, accelerated the development of societies and their impact on the environment.

The real upheaval occurred when Thomas Newcomen unveiled his steam engine in 1712, which was followed by improved designs by James Watt from 1776 onwards [7] to achieve higher efficiencies. During the era of industrialization beginning in the 1850s, steam power and use of fossil fuels enabled the scaling-up of many industrial activities and transportation. Further inventions enabled even larger-scale power production, such as the Rankine cycle invented by William Rankine in 1859, and the steam turbine invented by Charles Parsons in 1884. These and other modern inventions powered by fossil fuels heralded a new era that profoundly changed society's way of life, intensified land use, and boosted population growth. In just a little over 150 years human actions have interfered with many natural system cycles, altered the climate, and pushed many non-human species towards or into extinction—all to the extent that our own way of life is severely threatened. Our footprint is so visible across the world that it has been proposed that we have entered the epoch of the Anthropocene: the era of mankind.

The development of the fossil fuel industry has created markets, stimulated economic growth, and increased welfare—all things that we consider progress. However, since the late 1800s scientists have raised concerns about the excess burning of fossil fuels (see for instance Phillips [33], Arrhenius [1]), over-consumption of natural resources [26], and unexpected consequences related to the loss of biodiversity [29]. They have warned us not to exceed planetary boundaries [39], and showed how our economies, livelihoods and wellbeing all depend on the health of nature [8].

Many believe that new technologies will help us in overcoming these problems—and while we do need new technologies, we also require a systemic change in our behavior, and more caution about how our actions impact the living and non-living environment. In any case, new technologies take decades mature, and so far technological development has mostly led to increased consumption, emissions, and land use. This is due to the so-called "rebound effect" [9], where efficiency improvements or energy-saving measures lead to increased consumption (often via reduced cost or other systemic responses), offsetting any initial savings.

At present we are heading towards a 2.8 °C warmer planet [23, 52] and sixth mass extinction [4, 16]. While these warnings have not been totally ignored like the mythical prophecies of Cassandra, and several international, regional, and national agreements exist to tackle climate change and loss of biodiversity, progress on implementing sufficient measures is slow and deadlines for action have been postponed decade after decade.

Our collective major challenge is that our human ability to comprehend the passage of time is limited. We tend to set baselines on how things are now, and treat the present moment as the norm. This "shifting baselines" concept [25, 32] refers to the slow, gradual drift that happens over time in climate, in nature and in our environment, and how failing to identify the correct baseline can result in underestimating the actual depletion that has occurred—a loss of perception of change occurs as each generation redefines what is "natural". Also, our seemingly endless capacity to adapt to prevailing conditions has made us vulnerable to climate change, ongoing environmental degradation and loss of biodiversity, sometimes a change in the natural environment is hidden from us until it is too late. This is referred to as extinction debt [49]: populations of species can persist for long periods of time, at least in human timescales, and yet still face extinction unless effective restoration and conservation measures are applied [22, 31].

Climate change and loss of biodiversity have dominated the risk landscape in recent years [56]. The world's foremost risk experts and leaders in business, government, and civil society ranked *extreme weather*, *critical change to Earth systems*, *biodiversity loss*, and *natural resource shortages* as the top four most severe global risks over the next ten years. These are interlinked with many other social and economic risks, and so it is comforting that solutions to many of these risks are also interlinked—finding solutions to one issue can help in solving many others. For example, a transition to clean energy production is needed to curb climate change, and this in turn helps overcoming: in loss of biodiversity and climate change-induced economic and social risks. We must take care, however, to ensure that proposed solutions do not contradict targets for ecological and social sustainability. As an example, while climate change is one of the major drivers of biodiversity loss, the loss of biodiversity also exacerbates climate change [34]. Therefore, climate action alone is not enough, it is important to protect and restore nature, secure the function and resilience of natural systems and recover and increase natural carbon sinks and stores. Otherwise, we not only risk losing thousands of species, but we may also be ambushed by biologically-driven fluxes of carbon from the ocean, tundra, wetlands and soil, trapped for millennia [16] and then released, worsening climate change—which in turn deepens the loss of biodiversity further.

To remain within safe operating bounds for climate change, scientists argue it is critical to limit global warming to 1.5 °C [18]. Since pre-industrial times, the mean temperature of our planet has risen already by around 1.5 °C—so to avoid dangerous warming and unpredictable events, greenhouse gas emissions must be halved every decade to reach carbon neutrality by 2050, and carbon negativity after that. This poses both a huge challenge and an opportunity for the energy sector.

With this book we aim to shake up the traditional techno-economical approach to engineering education. We want to plant a seed, to grow a new generation of engineers able to master energy projects that help to keep humanity within planetary boundaries, acknowledging that the Earth is home to at least 8.7 million species (including us) that are mutually dependent on each other. Current and future generations of engineers can become real game-changers, helping to redesign the world, prevent harmful lockdowns, and avoid creating new problems, by redesigning energy systems in an economically profitable and ecologically sustainable way. Figure 1.2 presents this strong sustainability approach, where renewable energy projects are embedded into a broader ecosystem (see also Fig. 1.5 showing how the United Nations' Sustainable Development Goal (SDG) 7 on affordable and clean energy is placed within a strong sustainability concept). We want to hit the "bull's eye" in this development.

Alongside providing an introduction to the technical design of renewable energy projects, we also want to demonstrate that energy projects do not happen in isolation from society and nature, and that poorly planned and executed projects can have detrimental, sometimes irreversible, effects. To achieve this, we want to familiarize the reader with the basic theories, terminologies and causalities occurring in nature and in society in a way that is useful for an energy engineer. We will present a wide range of case examples from existing scientific literature and projects, and though the responses in nature are always case-dependent, we will draw out lessons to be learned, and mitigation strategies that can be adapted for specific cases. We introduce some of the main ways renewable energy projects can contribute to ecosystem change and biodiversity loss, and how to mitigate these. We will see that measures that are too narrowly focused on climate mitigation and adaptation can create new problems and induce unwanted direct or indirect impacts on nature—and as a result, on the

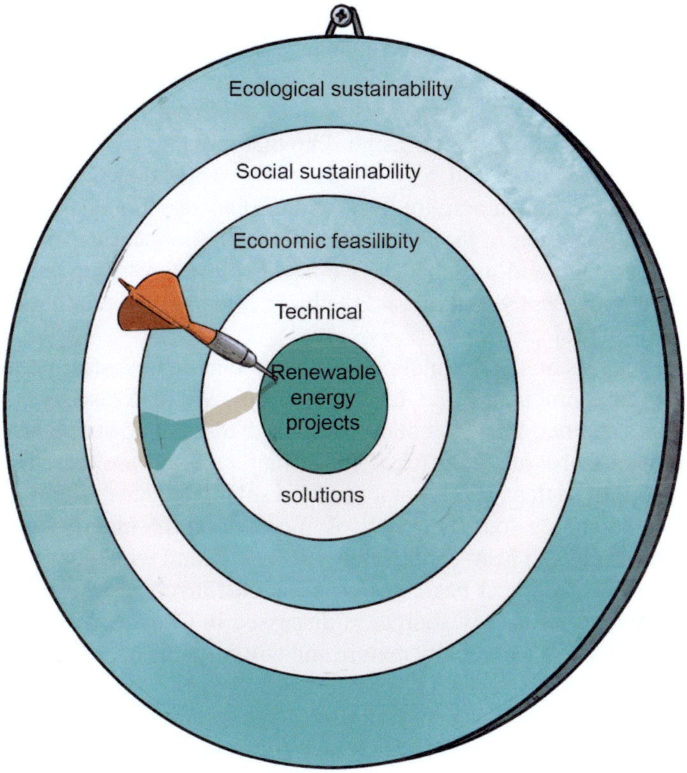

Fig. 1.2 Renewable energy engineering in a strong sustainability concept (© Maija Karala)

'services' that nature provides, and therefore on societies dependent on these services. Importantly, where synergies and spin-offs do exist between climate and biodiversity actions, these should be actively cultivated and favored.

Since the focus of the book is on renewable energy sources, we do not discuss nuclear power. However, we believe that nuclear energy, especially small modular reactors (SMRs), can help to lower CO_2 emissions, and could be used as a power-production control method to balance the variability of renewables and reduce the need for large scale storage. SMR technology is already available, but challenges remain around legal frameworks and social acceptability. The development of fusion technology is also continuously progressing, with indications that commercial fusion reactors could be available within 15–30 years [3]. While nuclear technologies may be important in the future, right now we are dependent on existing and mature technologies in realizing the energy transition at the pace required by the Paris Agreement.

In this first chapter we introduce and briefly discuss some sustainability concepts, such as principles of planetary boundaries, social foundation, and the safe operating zone for the planet; we then connect these to the techno-economical design of energy systems based on renewable resources. We examine the history of industrialization to understand the pace of human-induced change, and the time-frame in which we need to act if we are to avoid the adverse effects of global warming and loss of biodiversity. We then discuss briefly how this ongoing transition challenges the competence profile of an energy engineer and give our insight into that topic.

Ecologically sustainable and socially just energy transition is one of the major vehicles in bringing human activity back within planetary boundaries. In Chap. 2 we focus on the principles of three particular planetary boundaries: climate change, biodiversity loss, and land-use change. We also explore how large energy projects are interlinked with society, and discuss the basic principles of economic design for renewable energy projects.

Chapter 3 focuses on the thermal power plant technologies: steam power plants, gas turbine cycle, combined cycle, and heat pumps. We also discuss concentrated solar power and geothermal power plants. Although thermal power plants were originally developed to be run with fossil fuels, they can and are already used with renewable energies such as geothermal heat and solar heat.

On renewable energy sources, this book focusses on the main ways to produce renewable electricity and heat—solar energy (Chap. 4) and wind power (Chap. 5)—and we discuss the technical basic design principles involved in each. The use of waste and bioenergy as energy sources is discussed in Chap. 6. We focus on these technologies as they are technically mature and will form the backbone of the energy transition.

Chapter 7 gives an overview of the main available solutions for energy storage, and Chap. 8 outlines the requirements for electrical power systems and the electricity grid when using intermittent and variable sources. These include, for example, pumped hydro (this forms the connection with hydro-power, which is not discussed as a separate chapter) and chemical energy storage. Each section will also provide an overview on the different impacts of renewable energy production on nature, and discusses social acceptability. Finally, in Chap. 9 we summarize the implications of the topics introduced in this book for the transition to a green economy.

1.1 Sustainable Living Between Ecological Ceiling and Social Foundation

Human life could not exist without well-functioning, healthy ecosystems—Nature is invaluable to us. We are part of a living system that has evolved on Earth over billions of years. No living organism can survive in isolation; all require the support of the network of biotic (living things) and abiotic (physical components, like minerals and water) elements. Well-functioning ecosystems are also the prerequisite for climate stability, food production, material and energy production, clean water, and flood control, just to mention a few of the benefits (or ‘ecosystem services’) that nature provides for us [16].

Based on the National Footprint [28] on the Earth as a whole, today in 2024 we are consuming natural resources on average at a rate that requires 1.75 planet Earths to provide. The Earth Overshoot Day (EOD) concept was defined to make the limits and overuse of natural resources more comprehensible [28]. The EOD is computed by dividing the planet's biocapacity (the annual amount of ecological resources the planet can generate each year) by humanity's ecological footprint (our annual demand for resources), multiplied by 365, the number of days in a year. The historical development of the EOD from 1971 onwards is presented in Fig. 1.3.

To better understand the differences between countries, Fig. 1.4 presents individual countries' overshoot days that considers national biocapacities along with per capita consumption. Biocapacity is an estimate of an ecosystem's biological production, and its ability to take in and recycle waste. For example, Namibia's biocapacity is much smaller than Spain's—and so, even though Namibia consumes much less than Spain and European countries in general, they still reach their overshoot day on 19 August. With the warming climate, rainfall and biocapacity in the African region are forecast to decrease further [24], diminishing the region's biocapacity even more.

The wellbeing of nature and society are interlinked [38]. Therefore, when we talk about saving nature, besides taking care of the diversity of life, we talk also about the wellbeing and health of people, their economic prosperity, and their food, energy, and global security [8, 16, 17].

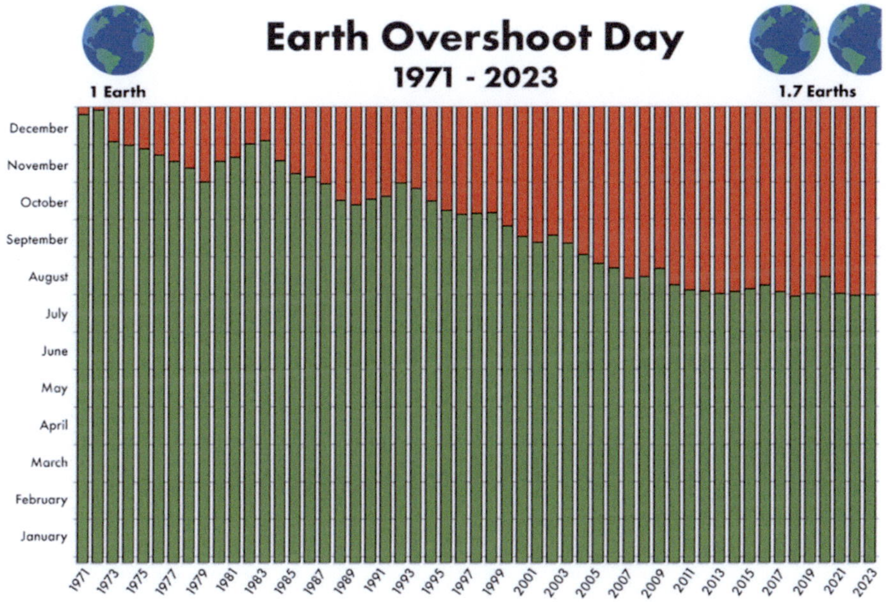

Fig. 1.3 Historical development of the Earth Overshoot Day. Red indicates the time that our ecological footprint is exceeding Earth's biocapacity ([28], © Global Footprint Network 2024, www.overshootday.org)

Fig. 1.4 Earth Overshoot Days for different countries in 2023. The overshoot days are calculated from the beginning of the year, and tell us how long the Earth's biocapacity could support us if all countries in the world consumed at the same rate as the country named. © Global Footprint Network 2024, www.overshootday.org

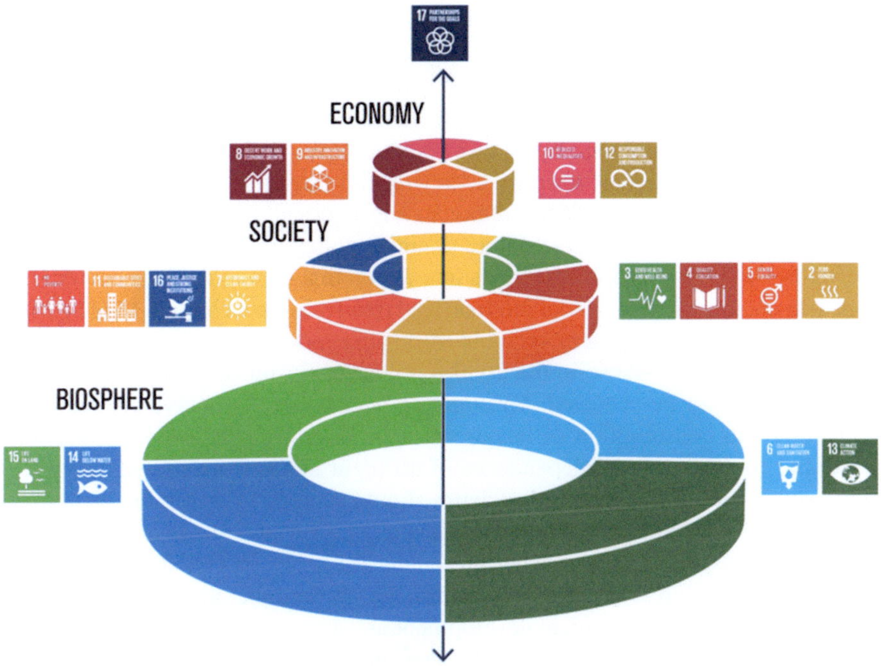

Fig. 1.5 The United Nations' sustainable development goals presented within the concept of strong sustainability (© Azote images for Stockholm Resilience Centre, CC BY-ND 3.0)

The world's nations have agreed on several measures to halt climate change and the loss of biodiversity, and presented a blueprint for sustainable development. The United Nations Framework Convention on Climate Change (UNFCCC) and the Convention on Biological Diversity (CBD), together with the United Nations Convention to Combat Desertification (UNCCD), were signed in 1992 at the Earth Summit in Rio de Janeiro, Brazil. The aim of the UNFCCC is to solve the "dangerous human interference with the climate system". The Paris agreement [51], a legally binding treaty under UNFCCC, sets a target to keep global temperature rise this century well below 2 °C (compared to pre-industrial levels), and to strive to limit the temperature increase to 1.5 °C by 2050.

The CBD aims to conserve biological diversity, secure the sustainable use of its components, and ensure fair and equitable sharing of benefits arising from genetic resources. The UN 2030 Agenda for Sustainable Development [50] proposed 17 Sustainable Development Goals (SDGs, see also Fig. 1.5), creating a blueprint for social, economic, and environmental sustainability to be achieved by 2030. Noteworthy within the SDGs is the recognition of the interconnectedness of individual goals: action in one area will affect the outcome of other goals. Therefore, implementation of any SDG requires a balance so that achievement of one goal cannot undermine opportunities to achieve the rest.

Over the years, several strategic plans have been agreed under the CBD, such as the Aichi targets [5] to mitigate biodiversity loss across the globe, and the Kunming-Montreal Global Biodiversity Framework [6], setting a pathway to implement the CBD and support the achievement of the Sustainable Development Goals. Although these agreements are not legally binding (except for the Paris agreement), countries need to report on progress towards achieving the targets via national and global reviews. These help to set global, regional, and national sustainability targets and guidelines to protect and restore nature, identify policy gaps, and provide valuable information to investors to direct funds into action that creates the best outcome.

At present these actions under international treaties are not enough. Rapid decline in biodiversity and ecosystem services mean that most international societal and environmental goals—those in the 2030 Agenda for Sustainable Development, for example—will not be achieved. We urgently need a transformative change in the production and consumption of resources, including minerals, food, animal feed, water, and energy. The interdependence of different SDGs is illustrated in Fig. 1.5, which shows how economies and societies should be seen as embedded parts of the biosphere. Thriving economies and societies can only be built on top of a healthy and functioning biosphere. This concept presented in Fig. 1.5 is called strong sustainability, and it requires moving away from the current sectorial approach where social, economic, and ecological development are seen as separate, sometimes even competing, interests. SDG 7 on access to clean and affordable energy lies at the nexus of biosphere, society, and economy.

While we will not discuss international, regional, or national policies any further in this book, it is however important to be acquainted with these few international treaties, and the ways they steer policy and legislation, inform markets and investors of the course of action, and via these guides also the energy transition.

In addition to the international agreements, different science-based frameworks and concepts have been developed to assess and describe sustainable existence on the planet that secures long-term survival of both human and non-human species. These frameworks include planetary boundaries [37, 39, 46], planetary well-being ([19], see also Sect. 2.3), the doughnut economy model [35, 36], and safe and just earth system boundaries [38].

We will now take a closer look at the planetary boundaries concept, first introduced in 2009 [39] to describe the system processes and environmental boundaries critical for sustaining the current state of life on Earth. For example, there is a maximum limit that we can utilize biomass and other natural resources, or increase the CO_2 content in the atmosphere, without jeopardizing the functionality of the whole system. The concept defines nine boundaries (Fig. 1.6): climate change; biosphere integrity; land-system change; freshwater use; biochemical flows; ocean acidification; atmospheric aerosol loading; stratosphere ozone depletion; and novel entities.

The framework was developed by a group of scientists with the understanding that, since industrialization, humans have become the main driver of global environmental change. According to the framework [39]:

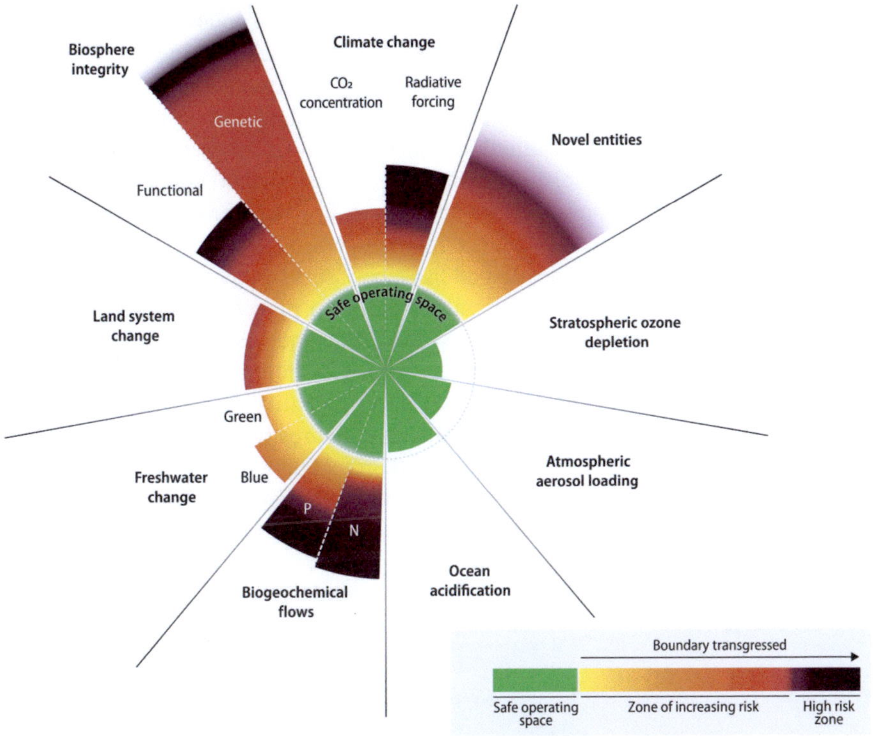

Fig. 1.6 The planetary boundaries' framework [37, 39, 46]. © Azote for Stockholm Resilience Centre, Stockholm University, licenced under CC BY-NC-ND 3.0

> *Transgressing one or more planetary boundaries may be deleterious or even catastrophic due to the risk of crossing thresholds that will trigger non-linear, abrupt environmental change within continental-scale to planetary-scale systems.*

This means that crossing even one of the boundaries puts the stability of the whole planetary system at risk through a domino effect of interlinked and cascading consequences. The boundaries represent critical thresholds beyond which risks to the ecosystem, and hence risks to people, increase. The connectedness of the boundaries also underlines the importance of addressing all boundaries together, and not just focusing on one such as climate change. The journey towards a sustainable and resilient future requires focusing more on the connections at the boundaries, particularly on the interplay between climate change and loss of biodiversity [34, 38]. The latest assessment, shown in Fig. 1.6, Richardson et al. [37] shows that we have already overshot six of the nine boundaries.

The doughnut economy model for sustainable development develops these ideas further by adding a social foundation at the inner circle of the planetary boundary framework, as shown in Fig. 1.7 [35, 36]. The twelve social foundations are derived from the UN SDGs, which can be considered the minimum social standards essential for a good life. These include food security, health, education, income and work, peace and justice, political voice, social equity, gender equality, housing, networks, energy, and water. Based on this framework, human life can thrive sustainably when all social foundations are met without overshooting any of the planetary boundaries. A study looking into data from 150 countries concluded that all countries are falling short of these requirements and are not able to satisfy its citizens' basic needs while maintaining a globally sustainable level of resource use [30].

Rockström et al. [38] have developed these concepts further to discuss and quantify the complex processes and outcomes related to planetary boundaries, social foundations, and their interactions, by assessing Safe and Just Earth System Boundaries (ESBs). This assessment builds on the planetary boundaries' framework [39], doughnut economics [35, 36], and the SDGs [50], identifying safe global and sub-global boundaries for maintaining and enhancing the resilience of the Earth system over time, and whether these boundaries could protect people from significant harm. It also incorporates concepts of justice from multiple viewpoints, such as between species, generations, countries, communities, and individuals. The assessment looks

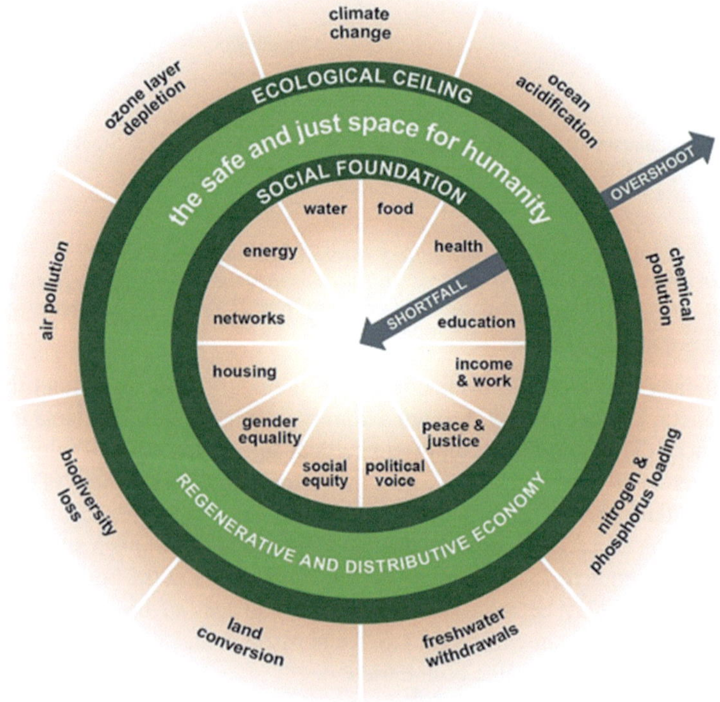

Fig. 1.7 The doughnut economy model [35, 36] (Licensed under CC BY-SA 4.0 Deed)

not only into the harm that human activities inflict on Earth systems and processes, but also the harm these changes in Earth systems does to non-human species, and to people in different parts of the world.

The proposed eight ESBs are climate, the biosphere (functional integrity and natural ecosystem area), fresh water (surface water and ground water), nutrients (nitrogen and phosphorus) and air pollution (aerosols) at global and sub-global scales. Currently seven out of eight globally assessed limits are exceeded [38].

The assessment distinguishes between "safe" and "just" limits, which result in differences in outcomes. This is shown for example with the limit on global warming. Global average temperature increases of more than 1.0 °C from pre-industrial levels (already exceeded) can trigger tipping points, such as melting of the Greenland ice sheet, while global warming of 1.5 or 2.0 °C increases the likelihood of such events to high or very high, respectively. To avoid the most harmful impacts on humans and nature, the safe Earth system boundary is set to 1.5 °C, which is also in line with the Paris Agreement. However, an assessment of significant harm revealed that even with 1.0 °C warming, tens of millions of people have been impacted by extreme temperatures. Further warming would put even more people, mostly those already in vulnerable positions, at risk for impacts from extreme temperatures, weather events, and sea level rise. Therefore, to avoid significant harm, the boundary of a limit that is both safe and just should be set lower at 1.0 °C increase in average global temperature [38].

Let's also take a closer look at the two biosphere boundaries: natural ecosystem area, and functional integrity. These two, together with climate, are topical in the context of the energy transition and increased development of renewable energy, and interlinked to each other via land use. For biodiversity, the Safe and Just Earth System Boundaries assessment considered two aspects: the area of largely intact natural ecosystem, and the functional integrity of human-modified ecosystems. Functional integrity refers to maintaining ecosystem functions and processes, such as nutrient recycling, pollination, carbon sinks and stores, soil protection, water quality, and halting species extinction—these are often called ecosystem services, or nature's contributions to people. These two are interlinked, since maintaining intact natural ecosystems also secures the maintenance of ecosystem functions [38].

Around 75% of land area and over 65% of the oceans are substantially impacted by human uses [16]. A safe Earth system boundary for largely intact natural ecosystem would be achieved by ensuring that 50–60% of global land and sea surface remains largely intact. Though the exact figure varies between ecosystems and regions, restoration efforts are needed to maintain natural systems and their functions, including the vital services on which we humans depend. Note that this target is not limited to nature conservation areas, nor does it require stopping all human activities in these areas—sustainable use and living are not excluded in this model. For example, development of renewable energy on degraded lands could provide combined energy production and restoration opportunities [38].

For functional integrity of human-modified land, the boundary has been assessed as the requirement for 20–25% of diverse, seminatural habitat in each square kilometer to maintain local ecosystem services. Functional integrity is closely related to the overall system's resilience—the capacity of an ecosystem to withstand natural hazards and climate change, or food production—and sustainable agricultural practices, such as pest control, nutrient leakages, and pollution [38].

Though the boundary on climate is well-established at a high confidence level, the confidence for other boundaries varies from high to low. The overall confidence of the biosphere boundaries is medium meaning that there is a need to enhance our understanding of the ways natural ecosystems maintain and support their many functions and services, like soil protection and buffering against natural hazards [38]. Regardless, this concept provides a framework to conceptualize and assess how human-induced changes impact the planetary system, nature and its processes, and our ability to live sustainably on this planet.

Hopefully, the frameworks and concepts outlined above have helped you to understand better how all human activity, our wellbeing, our economies, natural systems, and all non-human species are interconnected. Every action we take has an impact, either negative or positive, on the entire system. Sometimes the impacts on the environment can only be seen after a time lag, or the impacts of different projects can accumulate over time—which underlines the importance of keeping a focus on ecological and social sustainability when carrying out extensive projects, like those related to the ongoing energy transition. In large projects like the development of renewable energy technologies, a mitigation hierarchy (see Sect. 2.3, Fig. 2.13) should be applied, from early planning to decommissioning of the project, to assess and mitigate any unnecessary, harmful impact on nature, and to take action to compensate harm that cannot be mitigated. In this book we will focus on the impacts of selected renewable energy technologies on climate change, biodiversity, land use, and societies. These issues are addressed through available information based on scientific literature and case examples.

In the next section we discuss how humans have pushed the boundaries of our planet to the point of destabilizing the natural system. The root causes and drivers are linked to energy production, the use of natural resources, and increases in consumption.

1.2 From Industrialization to Green Energy Transition

Looking back at our history, examining how technology and society as a whole operated in the past, helps us to understand the state of the world today—and, perhaps, helps us to avoid making the same mistakes again.

Figure 1.8 presents the global primary energy consumption by source between 1800 and 2022. The horizontal axis also indicates the invention of key thermal power processes and other milestones, such as Einstein's patent on photoelectric effect in 1921, the first large wind turbine, and the first application of solar photovoltaics (PV) in space.

Before 1850, combustion of traditional biomass was the dominating source of energy, especially for heating, cooking, and metallurgy. A significant share of work was done by humans with the help of animal labor, such as horses and bulls. In addition, hydropower and wind power were used locally to do mechanical work, though these are not shown in Fig. 1.8. This development path reached its limits with decreasing sources of biomass in many parts of Europe, especially in Great Britain [43]. For example, the available land area was no longer able to support the increasing need of biomass for metallurgical purposes.

When people learned to utilize fossil coal resources efficiently, a fast lane to industrial development and economic growth opened up. This development was coupled with the discovery of new processes to produce iron, steel and many other metals, and new machinery to do work powered by fossil fuels. The invention of the steam engine, and James Watt's significantly improved design with an additional condenser in 1769, enabled an exponential increase in energy production and consumption. The market entrance of this new machine followed the classical S-curve of product development [41]. After the original invention of Watt's condenser, it still took decades for this new technology to be adopted at commercial scale. The expansion accelerated after

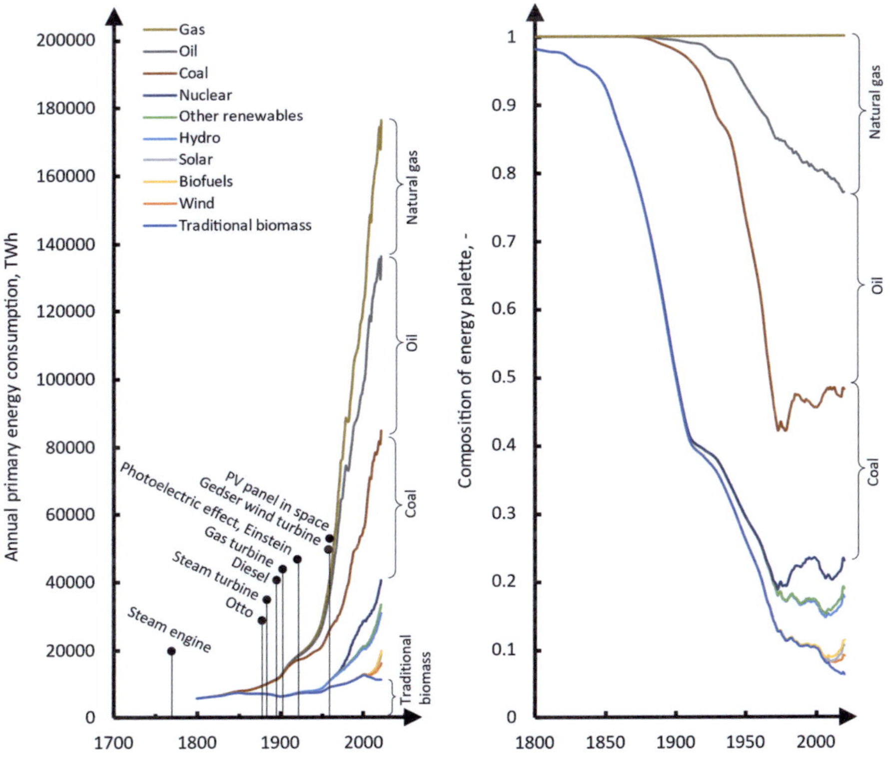

Fig. 1.8 Global annual primary energy consumption by source, 1800–2022 (left), and development of the composition of the energy palette (right). Data before 1965 is obtained from Smil [43], and data from 1965 onwards is based on BP [2]

Watt's patent expired in 1800, and this opened up the technology for further development and innovation. The most efficient steam engines reached thermal efficiencies of around 25% [7, 43].

The invention of the internal combustion engine, the diesel engine, the steam turbine, and the gas turbine finally unleashed the rapid growth of fossil energy use. The operational principle of the internal combustion engine was presented in 1862 by Alphonse Beau, and in 1877 Nicolaus Otto patented an engine based on spark-ignition-based technology. The first efficient prototype of this was built by Daimler and Maybach in 1883, with other significant additional improvements. At the same time, Benz had been developing his own designs, and after Otto's patent expired, he patented his own engine in 1886.

Soon after, Rudolph Diesel developed an engine design where spark ignition energy was not required—instead the system was based on a very high-pressure compression stage, leading to spontaneous ignition of the fuel. Diesel's engine was patented in 1895, and the thermal efficiency of his system exceeded 30% in the fall of 1897.

Charles Parsons developed the steam turbine in 1884 and patented it that same year. The largest steam turbine built before the Second World War, installed in Chicago in 1912, achieved a remarkable capacity of 25 MW. Turbine technologies in general had the benefit of continuous steady rotation, efficient scalability, and high rpm, that together enabled high generator efficiency. Steam turbine technology is currently the standard for industrial-scale power production, and single-unit capacities can reach over 1000 MW, using pressures over 340 bar and live steam temperatures of over 600 °C [7, 43].

Another important invention was the gas turbine process developed by Stolze in 1872, using a multi-stage turbine section and a flow compressor. Unfortunately, his design was never able to generate net power. The first successful gas turbine was built in 1903, producing excess power using both rotary compressors and turbines.

These major inventions are presented on the timeline in Fig. 1.8, illustrating their relation to the increase in energy consumption of first coal in steam engines and steam turbines, then oil in internal combustion engines and gas turbines, and finally the use of natural gas in gas turbines. The precondition for the development and spread of these technologies was the availability of necessary metals (such as steel and copper) and mature manufacturing technologies. As we will discuss in Chap. 3, the improvement of the thermal efficiency is still today limited by constraints in the development of high-temperature and high-strength materials.

Another important energy transition took place during the 1970s when the oil crisis hit most of the western world [7]—and with it the realization that oil reserves would not last forever, and the drive to develop alternative energy forms.

Scientific evidence is clear that climate change is caused by the excessive anthropogenic emissions of greenhouse gasses to the atmosphere from the use of fossil fuels, land- and sea-use change, and over-consumption [18]. At the time of writing this book in 2024, we are in the middle of the largest energy transition since the invention of steam power, often labelled as a "green energy transition". The use of wind and solar power has grown steadily, and though their combined share was a modest 4.3% of primary energy consumption in 2022 (Fig. 1.8), wind and solar power are predicted to supply 30–60% of the total electricity demand by 2050 [40]. Note that "green transition" here refers here to the process of shifting away from fossil fuels to renewable energy sources, but realizing a sustainable green transition requires expanding the focus beyond technological solutions, building a system that does not push planetary boundaries and is socially just. For that reason, a concept of "systemic change" is often referred to when discussing the green transition: to be successful in bringing about sustainable solutions, the change has to be applied throughout society and the economy—in much the same way that earlier large-scale transitions reshaped our societies. The question is, how do we ensure that we do not commit ourselves to an energy production solution that, while freeing us from fossil fuel dependency, creates new, severe problems of its own?

One could argue that the invention of ways to derive energy from coal, while enabling the development of societies, innovations, and our current quality of life, was also an unfortunate breakthrough because it locked us into a harmful system that is difficult to change. Bioenergy has been proposed as one solution to replace fossil resources—and if we observe the period prior to the widespread use of coal, before 1850, biomass was the main source of thermal energy used. Further, biomass provided a significant share of the energy used in metallurgy for reduction and melting purposes. Yet even at that time, the available land area was not able to sustain the bioenergy needs of the growing population, and thereby to sustain economic growth. From this perspective, reinventing the use of biomass as a solution to substitute fossil fuels seems implausible; the limits of its use were already reached 170 years ago. Even with increased efficiency and recycling the potential for sustainable biomass is limited, especially considering that the current human population is significantly larger, and the standard of living much higher, than in the mid-1800s. Furthermore, the present-day level of biomass use has already resulted in a global loss of biodiversity, and created conflicts over land-use [16]. This topic will be discussed in detail the Chap. 6.

Globally, the top five countries for energy use are China, United States, India, Russian Federation and Japan, as shown in Fig. 1.9 (data from 2019)—the European Union is also shown, treated as one entity. We see large differences between the country profiles—for example, the energy use for industry in India and China is about half of their total energy consumption, while this fraction is significantly smaller in other countries. This means that much of global production is done in these countries, and therefore the carbon emissions related to goods and services used globally are outsourced to these countries. We see also that the energy used in for transport in the US, comprising almost half of their total energy consumption, is more than the total energy consumption of Russia or Japan. This also explains the high proportion of oil use in the US.

There are huge challenges in the front of us to phase out highly-mature fossil energy technologies, and to transition to renewable energy solutions in ecologically sustainable and socially just ways. Successful transition should engage all sectors of society, and be based on sustainable economic growth where natural resources

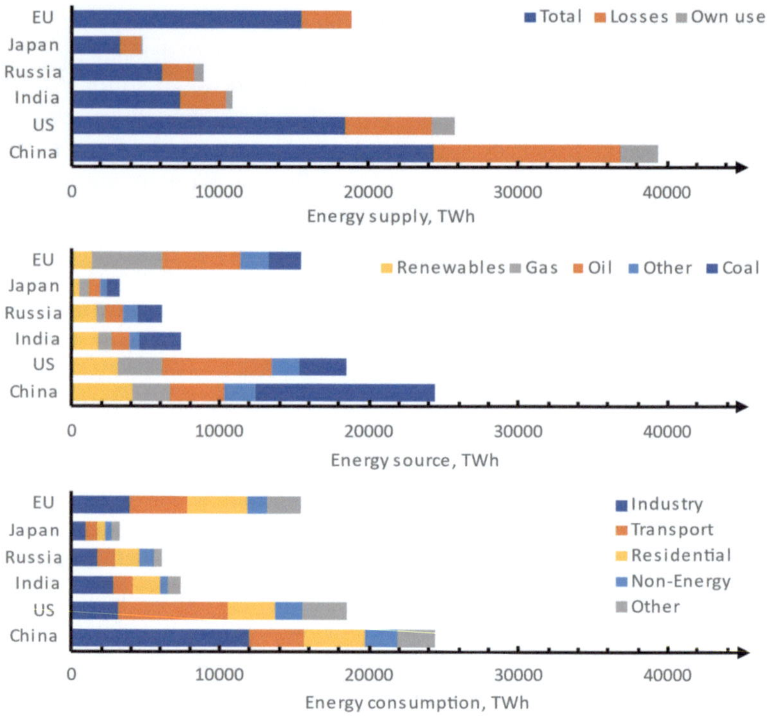

Fig. 1.9 Annual energy supply and losses (top), energy sources (middle) and consumption (bottom) in the top five countries and EU [13]. Peat and oil shale are combined with coal. "Other" in sources (middle) includes nuclear, electricity trade, heat, and non-renewable waste. "Other" in consumption (bottom) includes commercial and public services, agriculture/forestry, fishing, and non-specified. In the upper figure of supply in EU, "Losses and Own use" are combined in "Losses"

are not overused. Identifying externalities—the costs or benefits of economic activities not reflected in market prices—and making them visible is an important step towards building an equitable economic system (see an example, of calculating the biodiversity footprint of district heating in the University of Jyväskylä, Sect. 2.3, Case example 2.1).

At present, high-income countries have externalized the manufacturing of many products to mid- or low-income countries, who also bear the cost of pollution, greenhouse-gas emissions, and other negative externalities. Implementing a sustainable green transition requires working simultaneously on reducing greenhouse-gas emissions, improving resource and materials efficiency, and ensuring that the transition is socially just, that the loss of nature is halted, and that ecosystems' processes and resources are protected and restored.

Apart from adopting new technologies, initiatives and investments, a green energy transition also requires changes in policy, infrastructure, and the behavior of people. Renewable energy development, energy efficiency measures, sustainable transport, and an enhanced circular economy are all at the core of this transition to securing a sustainable energy future. Phasing out the fossil-based energy system requires policies that promote renewable energy sources, such as carbon pricing, tight regulations on emissions, or the elimination of subsidies for polluting industries. Policies should also direct investments for infrastructure renewal to incorporate renewable energy, such as requiring suitable grid connections, building decentralized solutions such as using electric vehicles as temporary energy storage, and maximizing the use of demand-side response solutions.

Successful transition also requires bringing people on board through awareness campaigns and education, and by involving local groups and communities in the development of decentralized energy systems. A green energy transition is not only critical in addressing climate change; due to the strong interconnections between the various solutions to climate change and the loss of nature, this is also an opportunity to create an ecologically sustainable and socially just future for all.

1.3 Potentials and Challenges of Variable Renewable Energy Technologies

Renewable energy is energy derived from natural sources that replenish at a higher rate than they are consumed. Sunlight and wind are such sources, available as long as the Sun is shining. From a human perspective, the stars are an infinite source of energy due to their long lifetime—the Sun is expected to shine for approximately 5 to 8 billion years [20]. Similarly, heat is continuously produced inside the Earth, making geothermal energy also a renewable energy source. And, as discussed earlier, there are limits for the sustainable use of biomass: bioenergy is renewable only as long as the feedstock replenishes at higher rates than it is consumed. Sustainable use of biomass also requires that the negative impacts on land-use and land-use change, food production, and biodiversity are minimized. Also, as we learned earlier, natural ecosystems have a crucial role in producing and maintaining provisions that support all life on Earth, including humans and non-human species. In modified landscapes these provisions are often reduced or impaired, further limiting the sustainable use of biomass. Transitioning from fossil fuels, which currently account for the main share of emissions, to renewable energy is a key solution to addressing the climate crisis.

The Earth's journey around the Sun takes 365 days, and it rotates around its own axis in 24 h. Since the Earth's rotational axis is tilted by 23.5°, we have four seasons that are stronger the further we are from the equator. At the poles the seasons are more dramatic: the polar night lasts for six months, after which the sun does not set for another six months. By contrast, close to the equator the seasons are not as clear, and the length of the day is fairly constant. These annual, daily, and locational variations greatly affect the ability to utilize renewable energy sources coming from the Sun. In addition, cloudiness and precipitation vary from one location to another and between seasons. For example, in South and South-East Asia, monsoon season brings thick cloud cover between May and September, decreasing solar irradiation;

conversely, outside monsoon season the possibilities for utilizing solar energy are great. This variability can be addressed through designing proper energy storage systems, for example sufficient battery energy storage or a back-up power system, such as an internal combustion engine running on biodiesel, or an e-Fuel produced from CO_2 and green hydrogen.

The variability of solar irradiation, wind speed and wind direction are key parameters in harnessing the available energy. This variable availability is typically described by the capacity factor, C_f, defined as the ratio of actual energy (e.g. in kWh, MWh) produced during an observation period, divided by the energy that would be produced if, for the entire observation period, the Sun shone at the local maximum value (standard value 1 kW/m^2) or wind blew at the rated speed of the wind turbine (the speed that gives the rated power production, for example $P_r = 10$ MW @ 12 m/s) for 8760 h during one year (or any other period of observation). Mathematically this can be expressed as follows.

$$C_f = \frac{\int_{t=0}^{\tau} P(t)dt}{P_r \tau}$$

The time range of observation, τ, is typically one year (8760 h). For example, if the capacity factor is equal to 0.5, it means that the system can produce the rated power for 50% of the hours across the year. This variable energy production is spread over the whole observation period. As will be shown in the coming chapters, capacity factors vary significantly for different locations, and there are also great differences in C_f between solar (Chap. 4) and wind power (Chap. 5). As an example of this variation, Fig. 1.10 presents the onshore and offshore wind energy production 2021 for different months of the year, in European wind power plants.

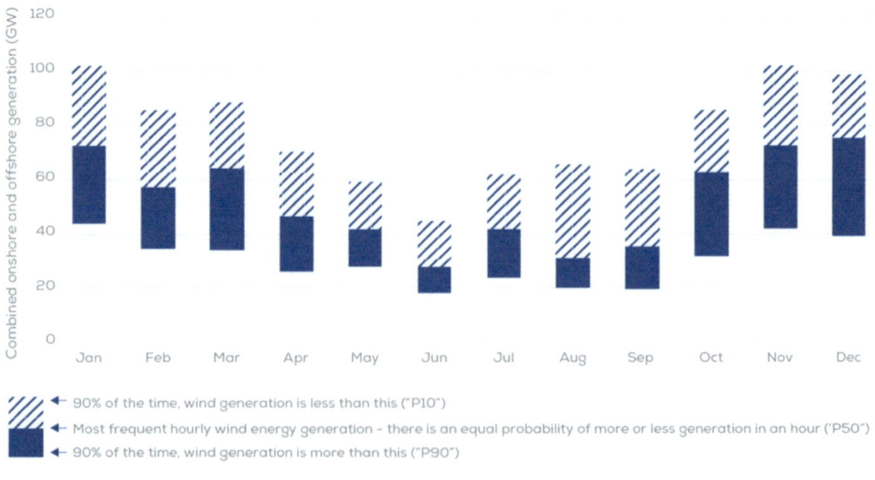

Fig. 1.10 Spread of hourly wind energy generation in 2023 [55]. Total installed onshore and offshore capacity was 207 GW and 28 GW, respectively

Based on the data from Wind Europe (2022), capacity factors for new onshore and offshore power plants are 30–35% and 42–55%, respectively. Due to the variability of renewable energy availability and production, there is a need either to adapt consumption patterns, or use an appropriate energy storage system to balance the energy supply. Variability and the lack of large-scale storage systems are the major challenges when transitioning from conventional energy technologies to renewables. Capacity factor describes how much of the maximum potential can be harnessed, but it does not indicate what size of energy storage is required to maintain constant energy supply, or what changes may be required to the electricity grid. For this, we need to analyze the production over several years to first find out the maximum storage capacity (MWh) and power needed to charge the storage (MW). The theoretically required maximum discharge rate of power to empty the storage is equal to maximum rate of consumption, and the maximum storage capacity is defined by the duration of power outages and required charging times. Often, several kinds of storage systems are needed to cover variations over different time scales. One option is to use biomass combustion or nuclear energy-based power and heat generation to stabilize the variability. Different energy storage concepts will be discussed in Chap. 7, and in Chap. 8 we will briefly examine the principles of electric power systems.

Bioenergy, which refers to energy produced from any organic material, is also seen as one of the key alternatives replacing fossil fuels. Biomass can be utilized by existing, mature technologies developed for fossil fuels, making it a lucrative alternative. Moreover, some plants and algae grow quickly and can generate several harvests per year under optimal circumstances. However, the sustainability of using biomass as an energy source depends on several factors including the type and source of the biomass, quantities of biomass used, production methods, and overall impacts on nature and local communities. Biomass production often competes with other land-use purposes such as food production, and intensive production of energy crops can increase the demand for agricultural land, leading to deforestation, soil erosion, and loss of biodiversity [16]. In addition, combustion of biomass creates emissions of greenhouse gases and other harmful particles, contributing to air pollution and impacting human health. Therefore, we need to carefully consider these factors to ensure that the production and use of biomass is sustainable. Biomass and waste as energy sources will be discussed in more detail in Chap. 6.

One challenge, and a bottleneck in the expansion of renewable energy systems, is the need for advanced materials that are often available at very low concentrations in the Earth's crust, or available only at certain locations. The increase in renewable energy means increased demand for necessary materials, like neodymium for permanent magnets used in wind turbines, silver used in solar panels, and cobalt, lithium and graphite needed for energy storage. Mining of these materials can have significant impacts on biodiversity and local communities [12]. Mining and the associated infrastructure cause direct degradation of land and nature, and fragmentation and loss of habitats [44]. Previously inaccessible and remote areas may also be exposed to new human settlements, threatening indigenous and local communities [12]. Poorly handled mining wastes cause pollution and threaten both native ecosystems and local livelihoods, which can lead to conflicts. Though mining activities are often welcomed because of the economic benefits they provide to society, strategic planning and careful implementation of all phases of the mining process are needed to prevent and manage conflicts with nature and local communities at early stages of development. Deep-sea mining is a growing industry as the need for minerals grow. However, this industry is still not fully developed, and our understanding of associated environmental risks is so limited that advancing such plans should be coupled with strong precautions [45].

One way to improve the ecological and social sustainability of mining activities is through continued development of sustainability standards for industry, and promotion of certification schemes that guarantee that materials are sustainably acquired. It is also important to improve recycling and re-use practices, since most of the materials used in wind and solar power plants can be recycled and reused. Note though that the recycling of some materials, like copper, lithium, and some rare earth metals such as dysprosium and neodymium, is still a challenge that needs to be solved. In this book we are not covering mining operations and their impacts on nature and local communities in detail, but it is good to keep in mind that the sustainability of the "green energy transition" depends to large extent on how the growing need for critical minerals is solved: how the extraction of virgin materials versus recycling will be prioritized, for instance, and how the technologies requiring these materials evolve. Several studies have been done on the available resources (e.g., Gielen [11]) and example of recycling of photovoltaic solar panels is given in Sect. 4.6.1. We also return to the availability of critical materials in Chap. 9. We emphasize the need to transition away from fossil-fuel energy and use the maximum amount of renewable energy sources that can be sustainably harnessed.

Next, we will examine how on the one hand, a transition to sustainable energy challenges the skills and competencies of an energy engineer, but on the other hand, creates new opportunities.

1.4 Competence Profile of a Game-Changing Engineer in Current and Future Energy Technologies

An energy transition driven by ecological crisis creates the need for a critical look at the skill sets and competencies of professional engineers. With climate change, natural catastrophes become more frequent and more violent. Do we have enough know-how to design and build infrastructure able to withstand natural disasters and extreme weather? Do we have the expertise to protect current infrastructure, often decades old, from these hazards? How do we incorporate biodiversity into energy planning so that, while we reduce greenhouse gas emissions, we also prevent further deterioration of nature? Future energy systems, with a greater share of renewables and increased demand fluctuations as electrification proceeds, will also require better adaptation to production demand—are current engineer candidates up to the task?

Sustainability has become a megatrend, and sustainability issues have permanently changed the operating environment of the energy industry. Stakeholders, customers, investors, employees, the public, and suppliers all expect action from companies. Policy and regulations are constantly evolving, bringing new obligations to businesses. Today, environmental, social, and governance (ESG) criteria need to sit at the core of every organization's strategy and leadership.

This ongoing green transition also brings a major structural change to labor markets, and we are witnessing a massive employment transition involving new kinds of technologies, expertise, policies, and regulations [14, 53]. ILO [15] defines "green jobs" broadly as those that contribute to preserving or restoring the environment, either in traditional sectors such as manufacturing and construction, or in new sectors such as renewable energy and energy efficiency. Green jobs contribute to improving energy and material efficiency, reducing greenhouse-gas emissions, minimizing waste and pollution, supporting climate-change adaptation, and protecting and restoring ecosystems [15].

Renewable energy is one of the most significant growth sectors for new job opportunities and the development of new occupational profiles [14]. At the same time, the transition to an environmentally sustainable and socially just future creates a growing need for expertise that is different from the traditional techno-economically focused competence profile—for example, environmental engineers needing to gain proficiency with restoration or industrial ecology, where industrial systems are designed to operate like natural ecosystems, in closed loop [14]. Non-technical skills and competencies, such as communication and team-work skills, are increasingly needed to excel in today's workplace, and in society in general.

This has created a need to evolve energy engineering education, to train students better able to address the current ecological crises and social issues related to the green transition [42]. Not only must engineers be able to tackle technical problems, they must also balance local and global sustainability issues, including ecological and social sustainability, as well as economic feasibility. Balancing these sometimes-contradictory viewpoints requires astute use of technology coupled with a broader understanding of planetary boundaries and just transition. It is also important to widen the scope beyond carbon reduction, to avoid developing a "carbon tunnel vision" (see Fig. 1.11)—net-zero emissions cannot be achieved at the expense of nature or human rights. Companies and individual experts focusing only on carbon emissions might fail to recognize and address the far-reaching environmental and social impacts, risking both their business and the wellbeing of people and our planet.

And so we need to discuss what it means to be an energy expert, now and in the future, as we progress through the green energy transition. On one hand there is a plethora of information available, and technologies are developing at an accelerating pace. On the other hand, deepening ecological crises force us to ask how this information and technology can be best harnessed to bring sustainable solutions. What strengths and responsibilities are demanded of experts these days? What kind of toolkit will an engineer need in the future? How does an engineer today and into the future need to think differently from a "traditional" engineer? What kinds of skills

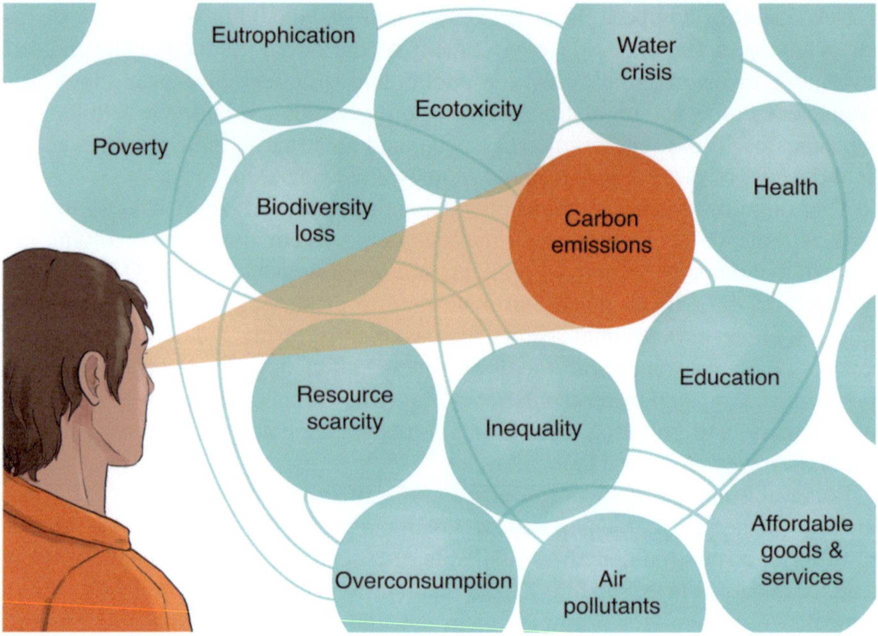

Fig. 1.11 Carbon tunnel vision in the ongoing sustainability transition. Redrawn after [21] © Maija Karala

are required if technology is to be used actively in achieving ecological and societal goals?

To gain more insight into these questions we, the editors of this textbook, hosted a workshop for main national and international industry actors in Espoo, Finland on 23rd September 2022, which we followed with a survey questionnaire. The workshop followed Chatman House Rule[1] to encourage openness and engagement and was facilitated by Third Rock Finland. The results of the workshop are outlined in the remainder of this section.

Workshop participants felt that there is a sense of urgency for the energy transition, and more resources and human capital is moving into the development of sustainable energy systems regardless that the future holds also uncertainties. We asked the industry participants how they perceive this change and how the ongoing ecological crisis is affecting the operating environments of the energy sector. Next, we asked about the skills and mindset required to solve the linear over-consumption of natural resources in the energy industry, and to ensure adaptive and resilient energy infrastructure.

[1] Chatman House Rule states: "When a meeting, or part thereof, is held under the Chatham House Rule, participants are free to use the information received, but neither the identity nor the affiliation of the speaker(s), nor that of any other participant, may be revealed." https://www.chathamhouse.org/about-us/chatham-house-rule.

The ongoing and upcoming changes in the operating environment were divided into several categories: political, economic, societal, technological, environmental and values in society. Political decision making is regulating businesses via rules and regulations. These have become stricter over time as environmental degradation has deepened. During the workshop, industry participants identified challenges relating to the discrepancy between the speed of technological development and political decision making: technology evolves quickly while regulations change slowly, permit applications are seen as complex, and overall the perspective of political decision making is bound to election cycles that are much shorter than the development time-frames of business and industry.

Shifts in the economic operating environment included those related to energy prices and the cost of technologies. On one hand, industry participants reckoned that energy prices will become more realistic, yet there will be higher uncertainty and fluctuations. Cost-parity with fossil fuels, internalization of external costs, and decreases in renewable energy technology prices have all increased investor interest in renewable energy. Participants also agreed that societies and companies that have become rich from fossil fuels should remodel their economies, and invest in and enable pilot renewable energy projects.

Sustainability principles have crossed over to society at large, bringing both opportunities and challenges for business. Opportunities arise when broader sustainability is included in all aspects of society. At the same time, greenwashing was seen as a potential risk, since it might reduce the public's ability to identify truly impactful and sustainable choices. Citizen engagement and education were recognized as key strategies in the fight against misinformation, to minimize fears of new technology and reduce possible local opposition to renewable energy projects.

Renewable energy production requires novel technical solutions, and increased electrification brings greater need for grid flexibility and energy storage. Our workshop participants identified policy regulation as lagging behind the rapid technical development of renewable energy technologies, hampering the implementation of the best available solutions.

Participants' environmental concerns covered challenges related to the sustainability of mining, sufficiency of raw materials and recycling which were identified as key questions within the growing green technology industry. They recognize a growing need to understand entire lifecycle impacts of renewable energy projects, and raised the problem of environmental "unknown unknowns"—risks that industry has not yet identified—as well as the need for ways to incorporate biodiversity and nature into energy planning more organically and holistically. In the area of sustainable mining, the potential impacts on biodiversity of the expansion of renewable technologies were recognized; different renewable energy systems have also direct and indirect impacts on nature and may drive biodiversity loss if poorly planned and executed (see for instance Gasparatos et al. [10]).

The sustainability megatrend is readily identified in shifts in societal values. Environmentally friendly lifestyles, recycling and re-use, saving energy, household or community energy production, climate friendly diets, and minimizing environmental footprint are all becoming normalized. Consumers increasingly demand low-carbon products and services, and company reputation has become increasingly important. The challenge is to find a balance between human needs, environmental requirements, available resources, and technical possibilities.

Changes in both operating field of energy industry and in society demand changes in the skills and competencies required by experts, and defining the exact set of competencies needed for the future is challenging. We asked the industry participants how they see the future skill profile of an energy engineer: will the skillset become broader, or more specialized? There was no single consensus, but overall, they predicted that the ongoing transition will require the ability to understand the big picture, and that there is a need to broaden experts' knowledge base outside the technical field of energy engineering. While technical expertise will remain as a core-competence, this is a growing need for non-technical skills, such as the ability to communicate complex topics to both internal and external stakeholders. Industry also recognizes a growing need to actively engage with experts from other fields and to build teams across many disciplines. In the future, optimal energy solutions will involve multidisciplinary teams of engineers, economists, biologists, and social scientists. Each team member will have their own T-profile (see Fig. 1.12): they will be true experts on their own field, for example energy conversion processes or environmental law, but will also have a solid broad understanding of the field as a whole. When we combine a group of people with different T-profiles, we create a team with the required overall competence field.

A positive mindset, together with the courage to fail, were also considered important characteristics. Complex challenges require creativity and curiosity, the ability to seek solutions in new directions, and to examine each situation case-by-case to find an optimal solution. The industry participants invest hope in younger generations to

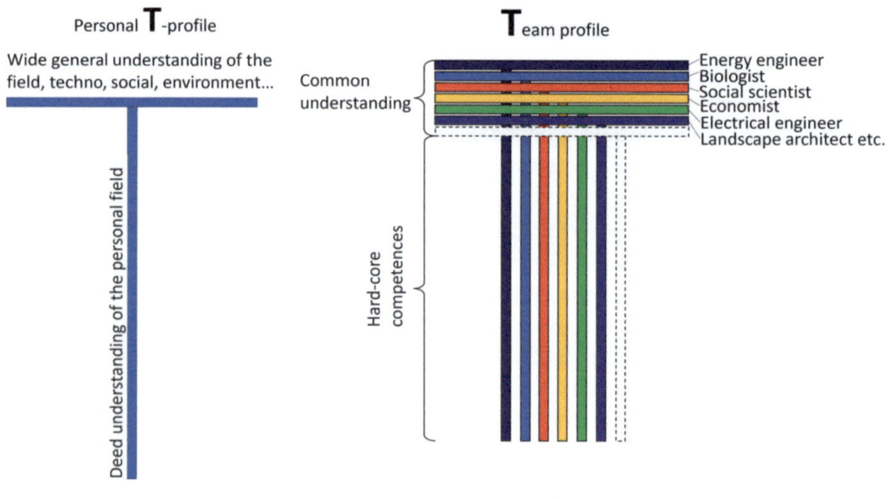

Fig. 1.12 Competence profile of an individual team member, and the profile of a renewable energy "dream-team"

master sustainability considerations in the energy sector and the circular economy, and to pass that knowledge to the older generation.

Similar results emerged also from a study commissioned by Engineers in Finland [48] that studied the effects of the green transition to employment and educational requirements. Though the study focused only on needs arising from carbon reduction targets, the complexity of the whole field increases the need for teamwork, stressing the importance of non-technical skills. Along with core technical skills, to be successful in the labor market candidates need good communication skills, resilience, and a willingness to learn continuously [14, 47, 48, 54]. However, the mathematical-technical core remains as the basis for engineers, and the green transition will not reduce this demand.

The ongoing energy transition will have a profound effect on expertise, education, employment—in fact, all aspects of our lives: the need to bring human activities within the boundaries of our planet will eventually impact everything that we use and do. This transition needs its champions in all fields, but without a doubt energy engineers will have a key role to play. Therefore, it is important to stay curious, be open-minded, and look for bold solutions to harness technology for the benefit of nature, society, and our planet as a whole.

1.5 How to Use the Book Effectively?

There is a growing need for interdisciplinary sustainability education for energy engineering students: they are this book's primary audience. It is intended to be used as a textbook, and therefore will follow a structure that is suitable for one or several courses on renewable energy utilization. The first two chapters and the last chapter of the book provide an overview of and introduction to ecological and social sustainability in energy engineering, and can be used as an introduction to sustainability challenges of renewable energy project design.

The book focuses mostly on basic phenomena and principles that do not become quickly outdated, such as the technical fluid mechanics design of effective wind turbines, electrical principles of solar energy systems, thermodynamics of thermal power plants, heat pumps and energy storage, and chemical principles of hydrogen applications in energy storage. We also discuss the importance of biodiversity, the main ways large scale energy projects impact nature and wildlife, and how to avoid these negative impacts. We also consider the social sustainability of energy projects, and examine the economic design of such systems.

Most of the chapters include exercises and project-work descriptions that are useful in practicing new skills and linking them with real world examples. We also provide some MS Excel© and Matlab©-based resources, which can be downloaded from the publisher's portal that can be used during courses and as a starting point in real-life projects. Please feel free to use these in your own exercises and projects and modify them if needed.

In this book we want to broaden the standard syllabus of energy engineering studies. This book discusses proposed solution pathways in a world of finite resources in the midst of climate change and biodiversity loss. While many uncertainties remain, established and mature key solutions are introduced, together with caveats and open questions.

The motivation behind this book was to contribute to building a new education toolkit for energy engineering teaching, one that is better fit for bringing solutions to ecological crises.

References

1. Arrhenius S (1896) On the influence of carbonic acid in the air upon the temperature of the ground. Philos Mag J Sci Ser 5(41):237–276. https://www.rsc.org/images/Arrhenius1896_tcm18-173546.pdf
2. BP (2022) BP statistical review of world energy. 71st ed. https://www.bp.com/content/dam/bp/business-sites/en/global/corporate/pdfs/energy-economics/statistical-review/bp-stats-review-2022-full-report.pdf
3. Ball P (2021) The race to fusion energy. Nature 599(18):362–366
4. Barnosky A, Matzke N, Tomiya S, Wogan GOU, Swartz B, Quental TB, Marshall C, McGuire JL, Lindsey EL, Maguire KC, Mersey B, Ferre EA (2011) Has the Earth's sixth mass extinction already arrived? Nature 471:51–57. https://doi.org/10.1038/nature09678
5. CBD (2011) Strategic plan for biodiversity 2011-2020. Including Aichi Biodiversity Targets. https://www.cbd.int/sp/
6. CBD (2022) CBD/COP/DEC/15/5. Decision adopted by the conference of the parties to convention on biological diversity. Monitoring framework for the Kunming-Montreal Global Biodiversity Framework. https://www.cbd.int/doc/decisions/cop-15/cop-15-dec-05-en.pdf
7. Cleveland CJ (2009) Concise encyclopedia of the history of energy. Academic, New York. ISBN: 9780123751171.
8. Dasgupta P (2021) The economics of biodiversity: the Dasgupta review. HM Treasury, London, 610pp. https://www.gov.uk/government/publications/final-report-the-economics-of-biodiversity-the-dasgupta-review
9. Freeman R (2018) A theory on the future of the rebound effect in a resource-constrained world. Front Energy Res 6:81. https://doi.org/10.3389/fenrg.2018.00081
10. Gasparatos A, Doll CNH, Esteban M, Ahmed A, Olang TA (2017) Renewable energy and biodiversity: implications for transitioning to a green economy. Renew Sust Energ Rev 70:161–184. https://doi.org/10.1016/j.rser.2016.08.030
11. Gielen D (2021) Critical minerals for the energy transition. International Renewable Energy Agency, Abu Dhabi
12. Haddaway NR, Cooke SJ, Lesser P, Macura B, Nilsson AE, Taylor JJ, Raito K (2019) Evidence of the impacts of metal mining and the effectiveness of mining mitigation measures on social–ecological systems in Arctic and boreal regions: a systematic map protocol. Environ Evid 8(9). https://doi.org/10.1186/s13750-019-0152-8
13. IEA (2019) Top five countries by total energy supply by sector. IEA, Paris. https://www.iea.org/data-and-statistics/charts/top-five-countries-by-total-energy-supply-by-sector-2019. IEA. License: CC BY 4.0
14. ILO (2019) Skills for a greener future. Key findings. International Labor Office, Geneva
15. ILO (2016) Green jobs: progress report 2014–2015. International Labour Office, Geneva
16. IPBES (2019) Global assessment report on biodiversity and ecosystem services of the Intergovernmental Science-Policy Platform on Biodiversity and Ecosystem Services. In: Brondizio

ES, Settele J, Díaz S, Ngo HT (eds). IPBES secretariat, Bonn, Germany. 1148p. https://doi.org/10.5281/zenodo.3831673
17. IPBES (2020) Workshop report on biodiversity and pandemics of the intergovernmental platform on biodiversity and ecosystem services. Daszak P, Amuasi J, das Neves CG, Hayman D, Kuiken T, Roche B, Zambrana-Torrelio C, Buss P, Dundarova H, Feferholtz Y, Földvári G, Igbinosa E, Junglen S, Liu Q, Suzan G, Uhart M, Wannous C, Woolaston K, Mosig Reidl P, O'Brien K, Pascual U, Stoett P, Li H, Ngo HT, IPBES secretariat, Bonn, Germany. https://doi.org/10.5281/zenodo.4147317
18. IPCC (2018) Global Warming of 1.5 °C. An IPCC special report on the impacts of global warming of 1.5°C above pre-industrial levels and related global greenhouse gas emission pathways
19. JYU Wisdom community (2021) Planetary well-being. Humanit Soc Sci Commun 8:258. https://doi.org/10.1057/s41599-021-00899-3
20. Karttunen H, Kröger P, Oja H, Poutanen M, Donner K-J (1996) Fundamental astronomy. Springer, Berlin. ISBN: 978-3-662-03215-2
21. Konietzko J (2022) Moving beyond carbon tunnel vision with a sustainability data strategy. Forbes. https://www.forbes.com/sites/cognizant/2022/04/07/moving-beyond-carbon-tunnel-vision-with-a-sustainability-data-strategy/
22. Kuussaari M, Bommarco R, Heikkinen R, Helm A, Krauss J, Lindborg R, Öckinger E, Pärtel M, Pino J, Rodà F, Stefanescu C, Teder T, Zobel M, Steffan-Dewenter I (2009) Extinction debt: a challenge for biodiversity conservation. Trends Ecol Evol 24:564–571. https://doi.org/10.1016/j.tree.2009.04.011
23. Lee H et al. (2023) AR6 Synthesis report: Climate changen 2023, IPCC.
24. Mamalakis A et al (2021) Nature climate change, 11 Feb, pp 143–151, https://doi.org/10.1038/s41558-020-00963-x
25. McHarg I (1969) Design with nature. The Natural History Press, pp 67–70
26. Meadows DH, Meadows DL, Randers J, Behrens III WW (1972) The limits to growth; a report for the Club of Rome's project on the predicament of mankind. Universe Books, New York. ISBN: 0876631650
27. NASA (2022) https://www.nasa.gov/image-article/sun-shines-above-earths-horizon/
28. National Footprint (2022) National footprint and biocapacity accounts, 2022 edition. data.footprintnetwork.org
29. National Research Council (1988) Biodiversity. The National Academies Press, Washington, 538pp. https://doi.org/10.17226/989
30. O'Neill DW, Fanning AL, Lamb WF, Steinberger JK (2018) A good life for all within planetary boundaries. Nat Sustain 1:88–95. https://doi.org/10.1038/s41893-018-0021-4
31. Ovaskainen O, Hanski I (2002) Transient dynamics in metapopulation response to perturbation. Theor Popul Biol 61:285–295
32. Pauly D (1995) Anecdotes and the shifting baseline syndrome of fisheries. Trends Ecol Evol 10(10):430
33. Phillips HA (1882) Pollution of the atmosphere. Nature 127. https://www.nature.com/articles/027127c0.pdf
34. Pörtner HO, Scholes RJ, Agard J, Archer E, Arneth A, Bai X, Barnes D, Burrows M, Chan L, Cheung WL, Diamond S, Donatti C, Duarte C, Eisenhauer N, Foden W, Gasalla MA, Handa C, Hickler T, Hoegh-Guldberg O, Ichii K, Jacob U, Insarov G, Kiessling W, Leadley P, Leemans R, Levin L, Lim M, Maharaj S, Managi S, Marquet PA, McElwee P, Midgley G, Oberdorff T, Obura D, Osman E, Pandit R, Pascual U, Pires APF, Popp A, Reyes-García V, Sankaran M, Settele J, Shin YJ, Sintayehu DW, Smith P, Steiner N, Strassburg B, Sukumar R, Trisos C, Val AL, Wu J, Aldrian E, Parmesan C, Pichs-Madruga R, Roberts DC, Rogers AD, Díaz S, Fischer M, Hashimoto S, Lavorel S, Wu N, Ngo HT (2021) IPBES-IPCC co-sponsored workshop report on biodiversity and climate change. IPBES IPCC. https://doi.org/10.5281/zenodo.4782538
35. Raworth K (2012) A safe and just space for humanity. Oxfam Discussion Paper, p. 26, Feb 2012

36. Raworth K (2017) Doughnut economics: seven ways to think like a 21st-century economist. Random House UK, 384 pp. ISBN: 9781847941398
37. Richardson K, Steffen W, Lucht W, Bendtsen J, Cornell SE, Donges JF, Drüke M., Fetzer I, Bala G, von Bloh W, Feulner G, Fiedler S, Gerten D, Gleeson T, Hofmann M, Huiskamp W, Kummu M, Mohan C, Nogués-Bravo D, Petri S, Porkka M, Rahmstorf S, Schaphoff S, Thonicke K, Tobian A, Virkki V, Wang-Erlandsson L, Weber L, Rockström J (2023) Earth beyond six of nine planetary boundaries. Sci Adv 9:eadh2458. https://doi.org/10.1126/sciadv.adh2458
38. Rockström J, Gupta J, Qin D et al (2023) Safe and just Earth system boundaries. Nature 619:102–111. https://doi.org/10.1038/s41586-023-06083-8
39. Rockström J, Steffen W, Noone K, Persson Å, Chapin III FS, Lambin E, Lenton TM, Scheffer M, Folke C, Schellnhuber H, Nykvist B, De Wit CA, Hughes T, van der Leeuw S, Rodhe H, Sörlin S, Snyder PK, Costanza R, Svedin U, Falkenmark M, Karlberg L, Corell RW, Fabry VJ, Hansen J, Walker B, Liverman D, Richardson K, Crutzen P, Foley J (2009) Planetary boundaries: exploring the safe operating space for humanity. Ecol Soc 14(2): 32. http://www.ecologyandsociety.org/vol14/iss2/art32/
40. Rogelj J, Shindell D, Jiang K, Fifita S, Forster P, Ginzburg V, Handa C, Kheshgi H, Kobayashi S, Kriegler E, Mundaca L, Séférian R, Vilariño MV (2018) Mitigation pathways compatible with 1.5°C in the context of sustainable development. In: Global Warming of 1.5°C. An IPCC Special Report on the impacts of global warming of 1.5°C above pre-industrial levels and related global greenhouse gas emission pathways, in the context of strengthening the global response to the threat of climate change, sustainable development, and efforts to eradicate poverty. IPCC
41. Rogers EM (1962) Diffusion of innovations, 1st edn. Free Press of Glencoe, New York
42. Ruiz-Rivas U, Martinez-Crespo J, Venegas M, Chinchilla-Sanchez M (2020) Energy engineering curricula for sustainable development, considering underserved areas. J Clean Prod 258:120960
43. Smil V (2017) Energy and civilization—a history. The MIT Press, Cambridge
44. Sonter LJ, Dade MC, Watson JEM, Valenta RK (2020) Renewable energy production will exacerbate mining threats to biodiversity. Nat Commun 11:4174. https://doi.org/10.1038/s41467-020-17928-5
45. Spearman J, Taylor J, Crossouard N et al (2020) Measurement and modelling of deep-sea sediment plumes and implications for deep sea mining. Sci Rep 10:5075. https://doi.org/10.1038/s41598-020-61837-y
46. Steffen W, Richardson K, Rockström J, Cornell SE, Fetzer I, Bennett EM, Biggs R, Carpenter SR, de Vries W, de Wit CA, Folke C, Gerten D, Heinke J, Mace GM, Persson LM, Ramanathan V, Reyers B, Sörlin S (2015) Planetary boundaries: guiding human development on a changing planet. Science 347:1259855. https://doi.org/10.1126/science.1259855
47. Strietska-Ilina O, Hofmann C, Haro MD, Jeon, S (2016) Skills for green jobs: a global view. Synthesis report based on 21 country studies, Geneva, ILO
48. TEK (2022) TEK graduate survey 2022, Results. https://www.tek.fi/sites/default/files/attachments/tek_graduate_survey_2022.pdf
49. Tilman D, May RM, Lehman CL, Nowak MA (1994) Habitat destruction and the extinction debt. Nature 371:65–66
50. United Nations (2015) Transforming our world: the 2030 Agenda for Sustainable Development. Resolution adopted by the General Assembly on 25 Sept 2015. A/RES/70/1
51. United Nations (2016) The Paris agreement, United Nations framework convention on climate change (UNFCCC)
52. United Nations Environment Programme (2022) Emissions Gap Report 2022: the closing window—climate crisis calls for rapid transformation of societies. Nairobi. https://www.unep.org/emissions-gap-report-2022
53. Vona F, Marin G, Consoli D and Popp D (2018) Environmental regulation and green skills: an empirical exploration. J Assoc Environ Resour Econ 5(4). https://doi.org/10.1086/698859

54. Wikman M, Nyrhilä L, Roschier S (2022) The effects of the green transition to the employment and educational requirements of engineers in Finland. A study commissioned by Engineers Finland from Gaia Consulting
55. Wind Europe (2023) Wind energy in Europe, 2023 Statistics and the outlook for 2024–2030
56. World Economic Forum (2024) The global risk report, 19th ed. https://www3.weforum.org/docs/WEF_The_Global_Risks_Report_2024.pdf

Chapter 2
A Sustainable Path for Renewable Energy Technologies

Mika Järvinen, Hanna Paulomäki, Elina Oksanen, Dilip Khatiwada, Sami El Geneidy, Maiju Peura, Janne S. Kotiaho, Iryna Herzon, Eeva-Lotta Apajalahti, Alicja Dańkowska, Mariana Galvão Lyra, Ayesha Sadiqa, Roosa Talala, Samuel Cross, and Antti Arasto

Therein is the tragedy. Each man is locked into a system that compels him to increase his herd without limit—in a world that is limited..
—Elinor Ostrom

The original version of the chapter has been revised. A correction to this chapter can be found at https://doi.org/10.1007/978-3-031-69856-9_10

Supplementary Information The online version contains supplementary material available at https://doi.org/10.1007/978-3-031-69856-9_2.

M. Järvinen (✉) · H. Paulomäki · S. Cross
Aalto University, Espoo, Finland
e-mail: mika.jarvinen@aalto.fi

E. Oksanen
University of Eastern Finland, Kuopio, Finland

D. Khatiwada
KTH Royal Institute of Technology, Stockholm, Sweden

M. Järvinen and H. Paulomäki (eds.), *Designing Renewable Energy Systems within Planetary Boundaries*, Green Energy and Technology,
https://doi.org/10.1007/978-3-031-69856-9_2

2.1 Introduction

Securing intergenerational well-being on a planet with limited resources means that we cannot use natural resources faster than they are replenished. We are now living in a new era, "the Anthropocene", that is strongly affected by human-induced changes in the environment and the planetary system. Over-consumption, the excessive use of goods and services (energy, land, water or materials), is the root cause of depletion of natural resources. Use of fossil fuels—coal, oil and natural gas—has led to air pollution, greenhouse gas emissions and climate change. Extraction of minerals and metals, like iron, copper, gold, and rare earth elements needed for various industries, including renewable energy technologies, have led to habitat destruction, soil and water pollution, and other environmental impacts. Unsustainable land use has resulted both in loss of arable land and deforestation. Freshwater resources are depleted in many regions because of increased demand, pollution, and inefficient water management. Overfishing and unsustainable fishing practices are threatening fish populations and marine ecosystems, while also affecting the livelihoods of communities dependent on fishing. All these human activities have impacted nature and its species and habitats so severely that today more species are threatened with global extinction than ever before [102]. Maintaining the functionality of natural ecosystems plays a key role in securing ecosystem services, also called "nature's contribution to people", such as carbon fixation and storage, pollination, air quality, climate stability, water quality, soil functionality, food, bioenergy and materials production.

The current exploitation rates of many limited natural resources raise concerns about their sustainability and the long-term consequences of their depletion. Moreover, many natural resources are distributed among people in an unjust manner, and as these resources become scarcer, their availability can lead to geopolitical tensions and conflicts. Changes in our environment, in particular rapid decline in biodiversity and ecosystem services, mean that it will be hard to meet most of the 17 international societal and environmental goals defined in the 2030 Agenda for Sustainable Development, the Sustainable Development Goals (SDGs) (Chap. 1.1). To address these challenges, sustainable resource management, conservation efforts and technological advancements are needed. A shift towards sustainable and clean renewable energy production is a crucial step towards a resilient and sustainable future.

S. El Geneidy · M. Peura · J. S. Kotiaho
University of Jyväskylä, Jyväskylä, Finland

I. Herzon
University of Helsinki, Helsinki, Finland

E.-L. Apajalahti · A. Dańkowska · M. Galvão Lyra · A. Sadiqa · R. Talala
LUT University, Lappeenranta, Finland

A. Arasto
VTT Technical Research Centre of Finland, Espoo, Finland

Many of the environmental risks faced by today's societies are not new. For example, the first sign of air pollution impacts on vegetation and human health was detected in London in 1661 [61]. Today the threat that air pollutants cause to plant production, forestry, agriculture, and the vitality of ecosystems is well-understood, and strict emission-control measures have been applied in many parts of the world, in particular Europe and the United States, leading to a decrease in the amount of air pollutants [171]. At the same time the health of the atmospheric environment has worsened, especially in rapidly developing regions such as Asia [171].

One success story in the history of humankind is the restoration of one of the critical planetary boundaries, the ozone layer. Humanity came close to destroying this shield against the Sun's ultraviolet radiation. Strict bans on the use of ozone-depleting chemicals (chlorine-fluorine-carbons, CFCs) turned the tide, and today though the hole in the ozone layer still exists, a full recovery is expected by around 2060 if no new CFCs are released into the atmosphere.

Planetary boundaries, a concept that defines critical Earth system processes such as climate regulation and biodiversity [190], is one way to discuss and measure large-scale changes in our environment (see also Sect. 1.1). Six out nine planetary boundaries—climate change, novel entities, biochemical flows, land system change, biosphere integrity and freshwater change—have already been crossed [184]. To bring our lives back within sustainable limits, a transformative change is needed in the production and consumption of all resources. Technological innovations, such as digital and AI solutions, together with the development of alternatives to critical metals and enhancing the circular economy, are integral parts of the solution. But more importantly there is a need to have a critical look at levels of consumption, particularly in high-income countries, and to rethink how economic profit is measured. Introducing, for example, the idea of natural capital—the value of nature—into the accounting systems of companies and nations (see Sect. 2.3.5) would help bring human activities towards a sustainable path.

An energy transition should be used as an opportunity to strengthen the sustainability transformation, through the development of significant capacities of large-scale renewable energy from wind and solar power plants, or from sustainable harvest of biomass, to replace the heat and power produced by the fossil fuels. Sustainably-executed energy systems, including sustainable land use measures, could bring solutions to both climate change and the loss of biodiversity, and help also strengthen societies and local communities.

The objective of this chapter is to discuss in more detail three planetary boundaries that are strongly interlinked, and that also relate closely to renewable energy production: climate change, biosphere integrity (biodiversity loss) and land system change. We will also discuss social sustainability, and introduce basic economic tools to evaluate the economic feasibility of renewable energy projects. Energy engineers should have a solid foundational understanding of the different aspects of energy projects in their everyday work in designing and building the sustainable energy systems for a de-carbonized, sustainable and just future.

2.2 Climate Change and the Impacts of Anthropogenic Carbon Emissions on Nature and Society

The Earth's climate has changed several times throughout its history [191, 236]. Climate change refers to long-term changes in global climate and weather patterns, and over time the climate has fluctuated from warmer periods to ice ages. At the moment we live in a warm, post-glacial epoch that began about 11 000 years ago.

Changes in climate are natural, but since the 1800s human activities, namely the burning of fossil fuels, have been the major driver of the warming global climate. The current average global temperature is the highest ever observed across measurement history [103]. The crucial question is whether the new equilibrium we are heading towards is suitable for supporting the life of the 9 to 11 billion humans predicted to live on Earth by 2060–2100 [219].

Climate change does not simply refer to the warming of the global climate or the increase in atmospheric CO_2 concentrations. It has far larger impacts on many Earth systems. The warming climate leads to increased evaporation, and this results in more precipitation. Rising temperature increases the frequency of storms and extreme weather events such as hurricanes and typhoons, flooding, and drought. At higher temperatures more ice will melt from polar regions and glaciers, causing sea level rise and threatening the habitability of coastal areas. Warming climate also has impacts on living organisms by changing their established habitats and forcing them to adapt or vanish, or move location where that is possible. This also includes us humans: for instance, areas close to the equator may become too hot for habitation or food production; in other regions drought and floods will get more intense, also changing the extent of habitable area for humans and agriculture. This creates a massive social, humanitarian and safety concern that cannot be overlooked. For example, Australia has already experienced seasons of severe high temperatures and dryness, bringing forest fires that cause massive damage to nature and communities [34]. At the other extreme, countries in the hurricane area in the Atlantic Ocean and the north-eastern Pacific Ocean are experiencing increasingly disastrous storm events, causing severe damage to societies and infrastructure [236].

Climate change is driving and enforcing many transformations seen in the environment. The land- and seascapes are currently changing by such magnitudes that the risk of "flips"—rapid changes from one state to another—is increasing. Biological systems are characterized by resilience; they tend to resist change and stay in the equilibrium by slightly adjusting to changes in the environment. When a flip occurs, cascading and long-lasting effects push the system over a tipping point into a new state of balance. Researchers have warned that such processes might be happening already, for example in the Amazon rainforest, and could exacerbate deforestation and transform large areas into a dry, savanna-like system [25, 241].

The probability of a flip from one state to another increases if the feedback loops become self-amplifying, further enforcing the negative trend. These are called "vicious cycles" because they make the situation worse, accelerating problems like climate change and the loss of biodiversity. For example, the Amazonian rainforest has long been a carbon sink of global relevance, and also one of the world's biodiversity hotspots.[1] Changes in a system like this can have transformative consequences across the globe, and so it is worrying that the Amazon rainforest has turned into a net carbon emitter [86]. Examples of such feedback loops are presented in the Fig. 2.1, and reviewed in the detail below. Note that these examples are rough simplifications: in real life everything is interconnected, and the feedback loops presented here would be coupled with wider ecosystem effects at local level that can be either positive (resisting the change) or negative (further accelerating the change).

These feedback loops can also reverse direction, and it is possible to create a self-amplifying process that decreases the negative trend—for instance, by restoring degraded lands and rewilding environments with locally occurring species to capture carbon rather than releasing greenhouse gasses.

To better understand climate change and climate warming, we first need to briefly discuss the Earth's energy balance and the greenhouse effect that has made the temperature of the Earth suitable for living. Then we focus on anthropogenic carbon emissions and carbon cycles that define the atmospheric CO_2 concentration. This is again tightly connected to radiative properties of the atmosphere and atmospheric energy balance. We also study the interconnected influence of fossil emissions, population growth, economic growth, and sources of energy on CO_2 emissions. What is required to turn the curve of CO_2 emissions negative?

2.2.1 Earth's Energy Balance and Greenhouse Effect

The surface of the Sun corresponds well to a black-body emitter at a temperature of ~ 5800 K. When the emitted energy travels ~ 150 million km from the Sun to the Earth, its intensity decreases as the energy is spread over a spherical surface area that increases with the square of the distance r from the Sun: $A = 4\pi r^2$. Just outside the Earth's atmosphere, the rate of irradiation is approximately $S_0 = 1367\ \mathrm{W/m^2}$, known as the solar constant (Fig. 2.2).

Based on the solar constant and the Earth's cross-sectional area, through which the incident solar energy crosses, the total continuous energy flow rate to Earth's atmosphere can be calculated as $\phi = S_0 A_0 = 174315\ \mathrm{TW}$.

Let's first discuss what would happen if there was no atmosphere (Fig. 2.3, upper image). If we assume the planet is in radiative equilibrium and has reached a constant temperature, then the absorbed energy must be equal to the emitted energy. Approximately 30% of the incoming radiation of visible sunlight is reflected without any other interaction; the remaining 70% is absorbed as thermal energy at Earth's surface. The

[1] Biodiversity hotspots are areas with global significance in terms of maintaining biodiversity. UNEP defines a biodiversity hotspot as "a region with significant levels of biodiversity that is threatened with destruction. To qualify as a hotspot, a region must meet two criteria: it must contain at least 1,500 species of vascular plants (> 0.5 percent of the world's total) as endemics, and it must have lost at least 70% of its primary vegetation. These sites support nearly 60% of the world's plant, bird, mammal, reptile, and amphibian species, with a very high share of endemic species."

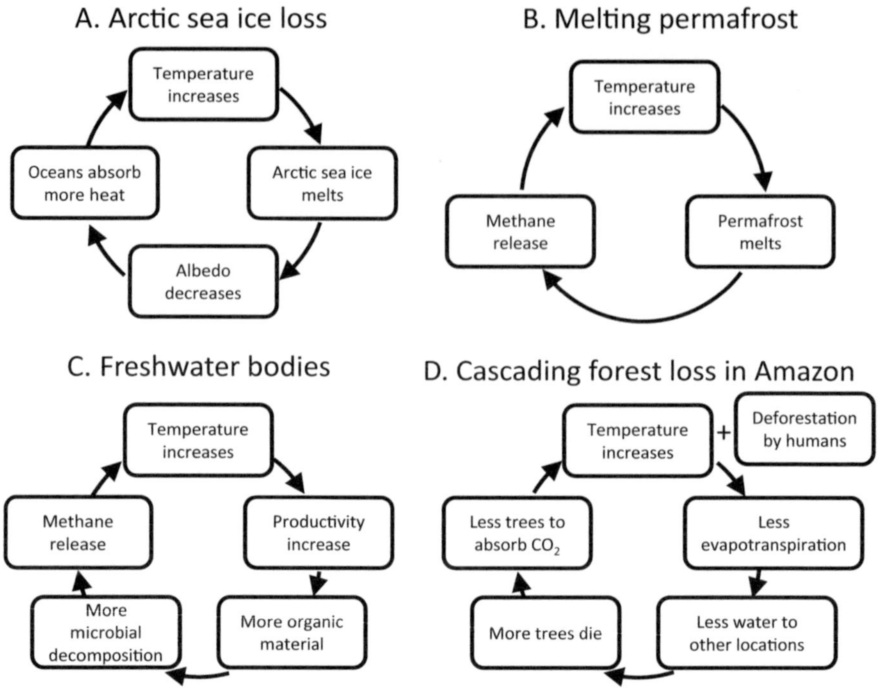

Fig. 2.1 Simplified examples of self-enforcing feedback loops due to increasing temperature. **a** Arctic sea ice loss. The Arctic region has warmed by nearly 2 °C since the 1970s, resulting in reduction in sea ice cover. Shrinking ice cover has resulted in decrease of Arctic planetary albedo from 0.52 to 0.48 between 1979 and 2011, corresponding to an additional 6.4 ± 0.9 W/m^2 of solar energy input to the Arctic Ocean region since 1979 [177]. Darkening of the surface heats the Arctic region resulting in an even faster melt of the sea ice. **b** Melting permafrost. Organic material in the permafrost region stores approximately twice as much carbon as is in the atmosphere. Climate warming promotes the melting of permafrost and decomposition of the organic matter, which can release excessive amounts of carbon into the atmosphere, further warming the climate and worsening the thaw [165]. **c** Methane releases from freshwater bodies. Freshwater ecosystems are major sources of methane, accounting to about 6–16% of natural CH_4 emissions as compared to 1% from the oceans. Methane emissions are particularly high in lakes with high productivity (i.e. with a number of aquatic plants). Warming temperatures accelerate accumulation and decomposition of organic matter, leading to higher methane emissions. [55]. **d** Loss of Amazon rainforest. Climate-induced drought and deforestation, and forest degradation, together decrease evapotranspiration locally, and lower moisture transport as a result. This decreases rainfall in other locations, leading to further loss of forest and reduced moisture transport [25, 241], weakening the resilience of the system and deepening the cycle

heated surface then emits energy in the infrared region of the spectrum. At infrared wavelengths the Earth acts as an efficient emitter-absorber, so its emissivity can be approximated as $\varepsilon = 1$. Radiative energy balance states that the incoming radiation through the cross section A_0 is equal to the emitted radiation from the overall surface of the planet, since the Earth is continuously rotating and cooling from the whole surface.

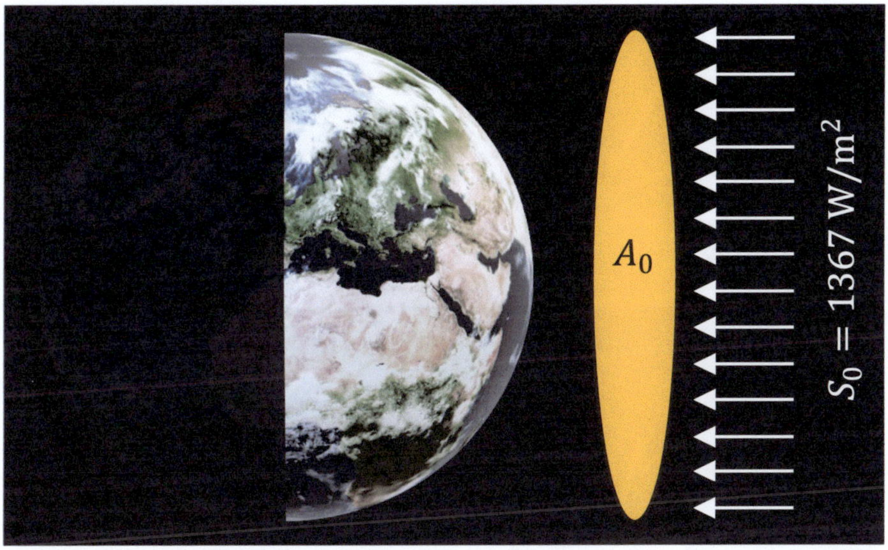

Fig. 2.2 Solar energy input just outside the Earth's atmosphere. Diameter of the Earth is 12742 km

$$0.7 \times \phi = A\varepsilon\sigma\left(T_s^4 - T_\infty^4\right), \text{ or } 0.7 \times \phi/A = \sigma T_s^4$$

where $\sigma = 5.67 \times 10^{-8} W/m^2K^4$ is the Stefan-Boltzmann constant, T_s is the temperature of Earth's surface, and T_∞ is the temperature of space ~ 0 K (2.7 K to be more accurate), so it can be neglected from the equation. With the above numerical values, we have

$$\frac{0.7 \times 174315 \times 10^{12}\mathrm{W}}{5.1 \times 10^{14} m^2} = 5.67 \times 10^{-8} \frac{W}{m^2K^4} T_s^4$$

Solving for the surface temperature, we get $T_s = 254.9\mathrm{K}$ or $t_s = -18.3$ °C. This is clearly not correct, since we know we have liquid water on Earth, for example. The true average temperature of the Earth is higher, around 288 K or 15 ºC. The temperature difference between a freezing cold third rock from the Sun with no atmosphere and our habitable warm Earth is 33.3 ºC.

Next, we study our planet as it is today, with an absorbing and emitting atmosphere. This system is presented in terms of total energy flows in Fig. 2.3 (lower image). As already discussed above, the energy of the incoming irradiation is partly in the short-wavelength visible regime that mostly penetrates the atmosphere. Approximately 7% (11 727 TW) of the incoming radiation is reflected back by dust and aerosol particles in the outer layer of the atmosphere. Of the remaining 93% (162 586 TW) of transmitted radiation, 24% (39 482 TW) is absorbed by the atmosphere, heating the air, and 76% (123 104 TW) reaches the Earth's surface. Of this, 68% (83 675 TW) is absorbed and 32% (39 431 TW) is reflected.

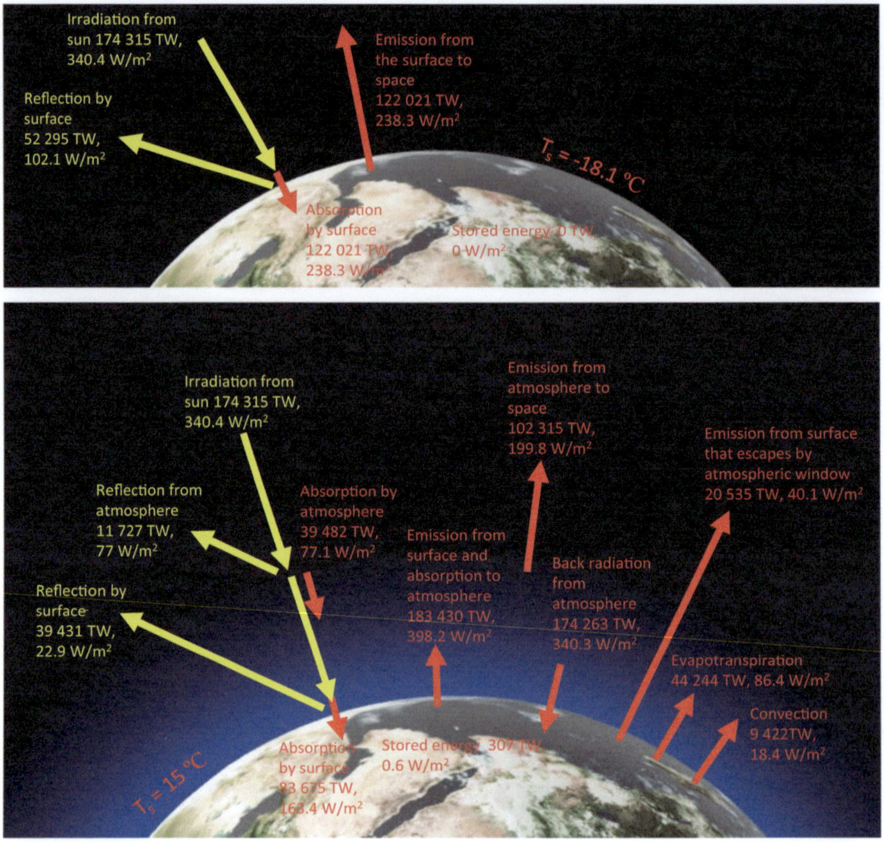

Fig. 2.3 Earth's energy balance with no atmosphere (upper) and with atmosphere (lower). Yellow arrows indicate transmission in visible wavelengths without absorption, red indicates thermal radiation in the infra-red region, and the blue region represents the control volume defined by the atmosphere and clouds. Based on [236]

The part that is absorbed at the ground heats the Earth's surface. The surface then emits thermal radiation at longer-wavelengths, now in the infrared region of the spectrum. This re-emitted IR radiation is not able to penetrate the atmosphere effectively; it is absorbed, heats the air, which then re-emits the energy partly back to Earth and partly outwards to space. This mechanism of heating the atmosphere is called the *greenhouse effect*.

As can be seen from Fig. 2.3, the fraction of energy absorbed in the Earth's system that is used for evaporation is 44,244/83,675 = 0.529. This same amount of energy is transported and released in the atmosphere when the vapor condenses at higher altitudes into liquid, which then falls as rain. This forms a continuous water cycle that transports energy—this is also the driving force for the hydropower.

The fraction of incoming solar energy converted to wind and ocean currents is ~ 0.01; in terms of energy this corresponds to $0.01 \times 174\,315 = 1743$ TW. This energy transforms eventually back to heat by viscous friction and turbulence. Based on [236], photosynthesis consumes only a small fraction, around 0.0008 (0.08%), from the incoming energy at the surface, corresponding to $0.0008 \times 174\,315 = 139$ TW. Annually, this mechanism stores 139×8760 TWh = 1 221 600 TWh of solar energy in biomass. In 2022 our annual primary energy consumption was ~ 180 000 TWh, which is around 15% of the solar energy stored annually in biomass—which suggests that there should be plenty of opportunities to increase our use of bioenergy. However, this same amount of carbon bound up by photosynthesis is released via plant and bacterial respiration and decomposition; only a negligible fraction is actually stored in the system, eventually forming coal, oil, gas and peat. This will be discussed in more detail in Chap. 6.

The overall rate of excess energy stored as heat is 307 TW, and this is the heat flow rate responsible for heating the Earth and pushing the climate system into its new equilibrium. Dividing by the total surface area of the Earth, this corresponds only ~ 0.6 W/m^2. To stop the climate from warming, this should be equal to zero.

The absorption of the thermal radiation emitted from the Earth's surface by the atmosphere is a function of the atmosphere's composition of gasses, for example the concentrations of H_2O, CO_2 and CH_4. The relatively rapid release of fossil carbon by combustion over the past 100–150 years has changed the concentration of CO_2 in the atmosphere. This affects the radiative properties of the atmosphere and therefore, it also has an impact on the mean temperature. This is a very complex mechanism and is interlinked to many other systems on our planet. The combined effect of fossil carbon emissions, changes in land use, and reduction of forests has been shown to cause the climate warming ([103], see also Fig. 2.1).

2.2.2 *Finite Fossil Energy Resources and Carbon Budget*

Peat, coal, oil, and natural gas are all chemically-stored solar energy. The carbon dioxide captured from the atmosphere via photosynthesis is the natural Bio-Energy Carbon Capture and Storage process, BECCS, developed during the evolution of the life on Earth. Solid coal was formed from biomass that grew on the ground, died, piled up, and accumulated into thick layers. Under high temperature and pressure, this material converted into different forms of coal, from brown coal, to lignite, and eventually into anthracite, which is almost pure carbon. Oil and natural gas were formed from plants and animals that lived in marine environments, died and accumulated on the seabed. These formed very stable storages of carbon and resulted in a decrease in atmospheric CO_2. It is therefore easy to understand that, while it took millions of years to store the carbon in coal, oil and gas, releasing this carbon over just a few hundred years will throw the climate system off the balance. We are now running and observing a planet-scale experiment on this very idea.

When we contemplate solutions for this energy resource problem, we need to remember that the possibilities for sustainable use of bioenergy are very limited. The carbon cycles already existing in nature, the flow of carbon between the atmosphere and the organisms on Earth, should be handled with extreme care. Changes or disturbances in these flows create cascading impacts throughout the system, like atmospheric warming, ocean acidification, loss of biodiversity, and further releases of carbon. It is very difficult to replace this enormous and natural CO_2 capture mechanism with any kind of man-made system.

Our use of fossil fuels has increased the CO_2 content of the Earth's atmosphere in just around 170 years, from a pre-industrialization level of 285 ppm in 1850 to the current value of 417 ppm, recorded on 5th January 2023 [33]. Next, we will take a brief look at the carbon budget remaining to keep within the Paris Agreement, and also consider how much fossil fuels we have left in the ground at the current rate of use. Though new mining and oil extraction techniques are still being developed, at some point the cost of extracting these diminishing resources will increase significantly. From a climate change point of view, seeking to prevent further disturbance to the balance of nature, the most rational and cost-efficient decision would be to leave the rest of the fossil fuel reserves in the ground, and accelerate the transformative change towards sustainability, including a transition to clean energy systems.

It has been estimated that to keep the atmospheric temperature increase below 1.5 °C from pre-industrial times, we can release a total amount of CO_2 that corresponds to nine years of emissions at the 2022 rate, equal to 380 billion tons [68]. Figure 2.4 shows a graphical representation of different scenarios for 1.5, 1.7 and 2.0 °C temperature increase [68].

Decarbonization and transformation of the current system can be achieved, but it requires that the task is taken seriously by policy makers and industrial actors. The sooner the work is started, the more time there is to change the system in a coordinated and equitable manner. This transformation can also be presented as a simple equation illustrating the impact of delay on the total time left to stop the use

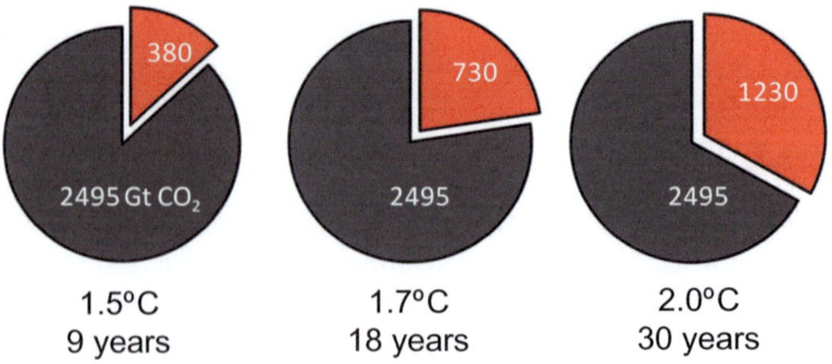

Fig. 2.4 Released fossil fuels and land-use change CO_2 emission before 2022, 2495 Gt, and the remaining carbon budget in Gt of CO_2 for different allowed atmospheric temperature increases, if we assume that the emissions remain at 2022 level. All results are given on 50% confidence interval. The countdown started at 2022. Data from [68]

of fossil energy. In this example we assume that the emissions are linearly reduced from annual emission value in 2022, e_{2022} (in Gt per year), to zero after a delaying time of t_d (continuing at the same level of emission as year 2022). The total time, t, that we would have for the de-carbonization can then be calculated:

$$t = \frac{2B_{CO_2}}{e_{2022}} - t_d$$

Here B_{CO_2} refers to the available CO_2 budget (Gt). For example, for the 1.5 °C scenario, if we start the linear decarbonization immediately in 2022 ($t_d = 0$), and the annual emissions were 42.2 Gt in 2022, we will have $2 \times 380/42.2 = 18$ years to decarbonize the energy system globally. The longer we wait, the shorter the time available for reaching the net zero emissions target becomes. This linear model is a simplified model of reality, but it gives a rough idea of the mechanism and underlines the urgency for action.

Even if we disregarded climate change and other negative impacts created by the use and extraction of fossil fuels, we would still need to find alternative energy sources since fossil fuel resources are finite. Based on data from [14], the current reserves of coal, oil and gas will last 139, 54 and 49 years, respectively, with the current production and consumption rate—just a few human generations.

If we burn all the known fossil sources, how much CO_2 would be released? Alternatively, instead of continuing to use all three fossil energy sources in the proportions that we do today, how long would each fuel last if used one at the time to cover current energy consumption, using first all available natural gas, then all oil, and finally all of the coal? Table 2.1 presents these calculations.

Table 2.1 Proven global fossil fuel resources 2022, their lifetimes [14] with current production, and potential CO_2 release

	Coal	Oil	Natural gas
Proved resources (Gt)	1074	244	138
Lifetime with current production (years)	139	54	49
Average carbon content (mass fraction)	0.75	0.85	0.75
CO_2 emissions if burned (Gt)	2954	762	380
Lower heating value, LHV (MJ/kg*)	29	39	47
Energy content (TWh)	8 653 000	2 648 000	1 809 000
Duration of energy as a single source, years[1]	64	19	13
Annual emissions as a single source, Gt[1]	46.4	39.1	28.6
Time to exhaust 1.5 °C budget as single source, years[2]	8	10	13
Time to exhaust 2 °C budget as single source, years[3]	27	31	43

[1]Annual consumption of fossil fuels energy is assumed constant of 136 000 TWh [14], [2]Carbon budget 380 Gt, [3]Carbon budget 1230 Gt.* Lower heating value, assuming that water vapor exits the processes as gas

Looking at the Table 2.1, if we consider only using natural gas as an energy source, the known reserves will last 13 years, and the carbon budget of 1.5 °C scenario would also be reached after the same period. This would require a four-fold increase in the production of natural gas, which in the real world is highly unlikely. If only coal is used, known reserves would last for 64 years, but we would reach the 1.5 °C limit in just 8 years. This underlines the argument that we should leave most of the coal to the ground.

With the 2 °C scenario, we could use all known gas and oil resources, but most of the coal should be left as a stable carbon storage. We would have around 32 years for this transition (gas and oil would last 13 and 19 as single sources, respectively); in this case it would take 74 years to reach 2 °C (though the resources would be exhausted earlier). Finally, burning coal alone would give us 27 years to implement the energy transition and stay below 2 °C. Although these are very simplified back-of-the-envelope calculations, they nonetheless give an idea of the magnitude of the challenge ahead of us.

If we continue business-as-usual and consume all fossil resources, we enter a scenario that is difficult to forecast. Based on the estimates, by 2300 this would mean an average global temperature increase of between 6.4 and 9.5 °C, the Arctic regions would warm by between 14.7 and 19.5 °C, and average regional precipitation would increase by more than a factor of four [210]. As this is not really a viable option from human perspective, we urgently need to implement a transition to sustainable and renewable energy, to develop technologies to reduce and/or capture CO_2 emissions directly from emission sources, to develop Direct Air Capture (DAC) technologies to further decrease the atmospheric CO_2 concentration, and—most importantly—to focus on reducing our consumption to sustainable levels.

2.2.3 Carbon Emissions and Carbon Cycles

We discussed earlier how the utilization of coal opened the tap for the use of oil and natural gas resources. This can be clearly seen when observing the annual global emissions by fuel type in Fig. 2.5, which shows anthropogenic carbon dioxide emissions since 1850s. Until the 1940s the main source of CO_2 emissions was land-use change. Flaring refers to the burning of unused gases in refineries, and this energy is totally wasted, releasing CO_2 and other pollutants into the atmosphere.

Combustion of coal has historically been the biggest source of fossil carbon emission, and still today its share is around 40%. From 1967 to 2003, emissions from the use of oil were greater those from coal; at present the share of oil-sourced CO_2 emissions is ~ 32%, and ~ 21% of emissions originate from natural gas. A significant share of CO_2 is also released from the cement industry (4.5%) due to the use of limestone in the calcination process. The impact of the Covid-19 pandemic can be clearly seen as a sudden drop in the emissions of oil combustion in transportation—unfortunately, as we know, emissions soon jumped back to "normal" levels after the pandemic [41].

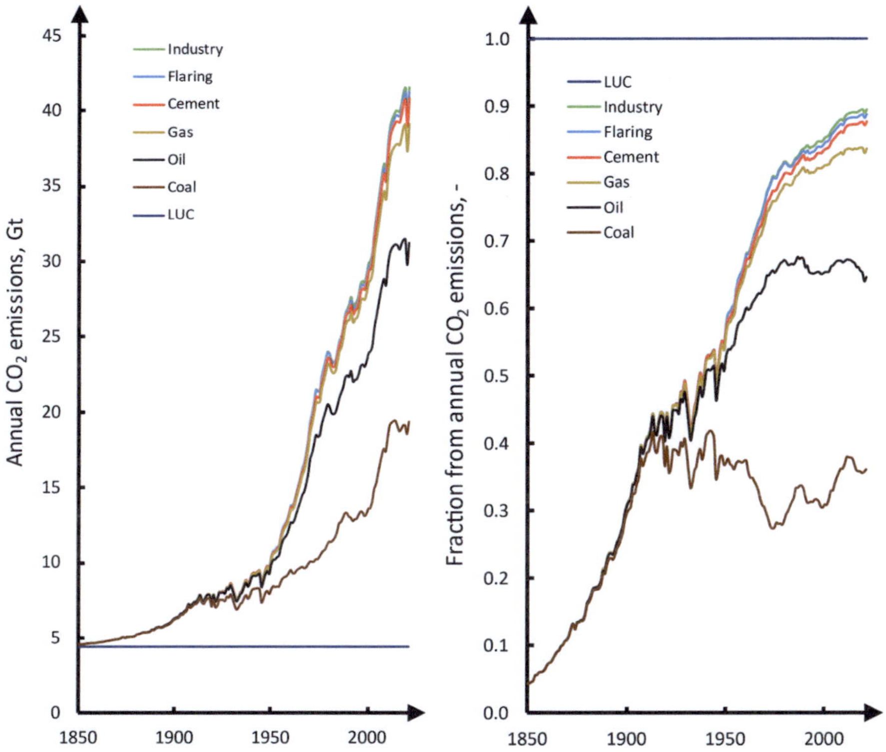

Fig. 2.5 Stacked carbon dioxide emissions by fossil fuels and land-use change since 1850, and their share of all emissions [76]. Please note that the share of different sources is added on the top of previous emission sources and fractions below and the topmost value is the total emission and therefore, the fractions add up to 100% for all times

The anthropogenic carbon dioxide emissions described above are known with quite good accuracy. Next, we discuss what happens to the CO_2 once it is in the atmosphere. Figure 2.6 presents the average surface temperature and CO_2 concentration between 1850 and 2022, and the correlation between the temperature and CO_2 concentration.

Figure 2.6 shows that there are much larger oscillations in the temperature than in the CO_2 concentration, but nonetheless the trend is clear. Based on the linear trendline (red line, right figure), if the CO_2 concentration increases from the current value 421 ppm (2023) to 476 and 525 ppm, atmospheric temperature increases to 1.5–2 °C, respectively. This, again, is a simplified model that does not consider any other effects such as the land-use change.

The detailed climate models, which incorporate carbon cycle mechanisms and detailed atmospheric phenomena, demonstrate significant variation in the results between the different climate models. For example, results from [23] (supplementary material for CO_2 only) show that the average CO_2 concentrations predicted by 13

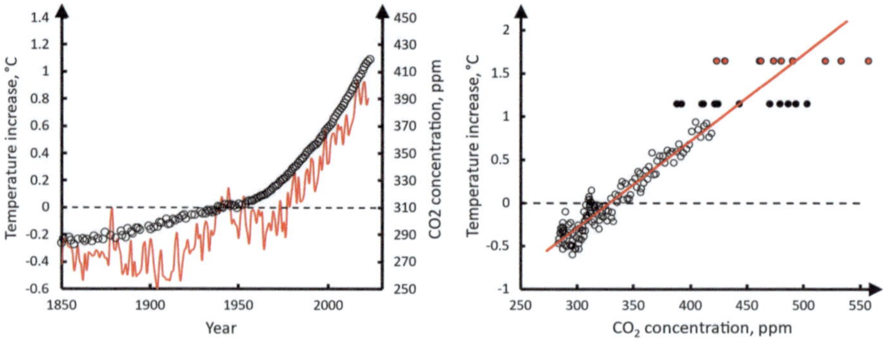

Fig. 2.6 Atmospheric CO_2 concentration (open circles) and temperature increase (solid red line) between 1850 and 2022 (left figure), and correlation between CO_2 concentration and temperature increase (right). Reference temperature is the average temperature from 1960 to 1990. Filled dots present (red 1.7 and black 1.15 °C compared with reference range of 1861–1880, see below) predictions from 13 detailed climate models [23]

climate models resulting in 1.15 and 1.7 °C temperature increases were 442 ppm and 505 ppm respectively, based on a reference time range of 1861–1880. These results are shown as filled dots on the right-side graph of Fig. 2.6, and as can be seen there is significant variation in the model predictions.

At least in principle, the amount of CO_2 accumulated in the atmosphere could be estimated from the CO_2 concentration (as molar fraction), x_{CO_2}, temperature, T, and pressure profiles, p, in the atmosphere. In theory, these could be systematically measured for example using weather balloons or from airplanes, and then integrated over the total volume of the atmosphere, V, to calculate the average concentration and carbon dioxide amount as follows—note that all variables in the integral are not constant, but vary with height, longitude and latitude, and time. The latter expression is presented in terms of averaged values of mole fraction, pressure and temperature over the total volume of the atmosphere, $R_u = 8.314\mathrm{J/molK}$ is the universal gas constant.

$$m_{\mathrm{CO_2}} = \frac{M_{\mathrm{CO_2}}}{R_u} \int_0^V \frac{x_{\mathrm{CO_2}} p}{T} dV = \frac{M_{\mathrm{CO_2}}}{R_u} \frac{\bar{x}_{\mathrm{CO_2}} \bar{p}\ V}{\bar{T}}$$

We will use this equation below to do some simple calculations. The released fossil CO_2 emissions (Fig. 2.5) and the atmospheric CO_2 concentration (Fig. 2.6 left) can also be plotted against each other (independent of time) to see the correlation between released CO_2 and its accumulation in the atmosphere. If we assume that the volume of the atmosphere, pressure, and temperature of the air have been constant, we can also estimate what the CO_2 concentration would be if all released CO_2 stays in the atmosphere without any capture mechanism, such as photosynthesis or dissolution in the ocean (Fig. 2.7). This can be written also as.

$$x_{CO_2} = x^0_{CO_2}\left(m_{CO_2}/m^0_{CO_2}\right),$$

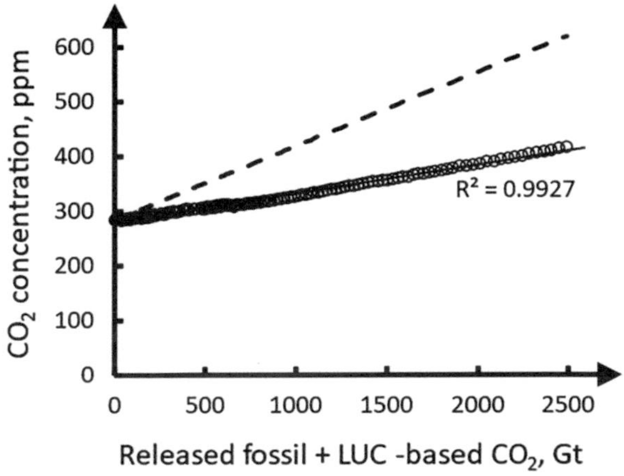

Fig. 2.7 Observed carbon dioxide concentration as a function of total cumulative (fossil and land use change based emission) released CO_2 since 1850. Measured concentration (open circles and thin, solid trend-line), and model concentration if we assume all CO_2 stays in the atmosphere (dashed line)

Where $x^0_{CO_2}$ and $m^0_{CO_2}$ refer to some known concentration (molar fraction basis) and mass of CO_2 in the atmosphere, respectively (e.g., year 1850, 284 ppm and 2106 Gt as CO_2).

As can be seen from Fig. 2.7, the correlation is linear ($R^2 = 0.9927$) between the analyzed concentration and released fossil CO_2. The observed concentration at maximum release is more than 30% lower than the value if all CO_2 is assumed to stay in the atmosphere (dashed line). This means that almost 40% of the released fossil carbon since year 1850 has been removed from the atmosphere by various mechanisms.

Carbon is a key element in all living organisms, including us humans—life on Earth is all carbon based. Our main solid and liquid fuels are carbon-based compounds, our diet is comprised of various carbon species, and our societies and cities are built from carbon-based raw materials such as wood, plastics, and concrete. Plants, algae, and some other organisms use atmospheric carbon as raw material to produce oxygen, sugars, and more complex products via a process called photosynthesis. Though the total amount of carbon on the ground and in the atmosphere stays constant, it is continuously transported between different reservoirs (Fig. 2.8).

There is currently around 850 Gt of carbon stored in the atmosphere, corresponding to 3117 Gt of CO_2 (~421 ppm). As a reference, CO_2 concentration was as high as 4000 ppm 500 million years ago, corresponding to 8184 Gt of atmospheric carbon and 30,000 Gt of CO_2 in the atmosphere.

Today, living organisms hold 550 Gt of carbon, mostly in plants [236]. The soil contains 2300 Gt of carbon. In the oceans, 1000 Gt of carbon is stored in the surface ocean and living biota, 37,000 Gt in the deep ocean, and deep-ocean sediments are estimated to store 6000 Gt. The amount of carbon stored as coal, oil, and gas is difficult to approximate accurately, and estimates vary between 2000 and 10,000 Gt. The largest storage of carbon, by far, is in rocks and minerals, and we will discuss this in more detail later.

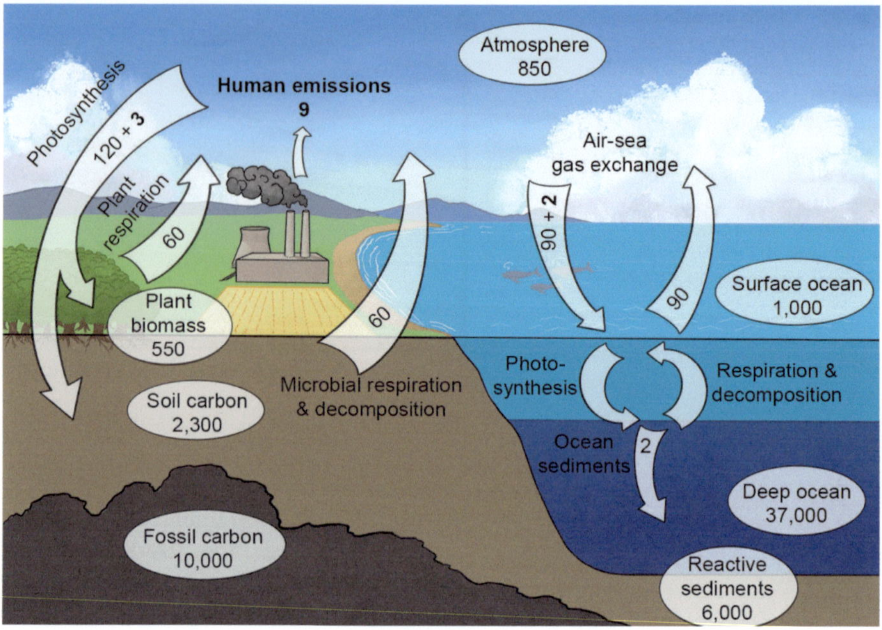

Fig. 2.8 Terrestrial and ocean carbon cycles (flows are presented in Gt per year and stored amounts in Gt unit). Note that carbon bound to rock minerals is not presented (data from [163]). © Maija Rajala

Carbon is transported between these reservoirs by several mechanisms. Emissions from human actions and combustion of fossil fuels release annually about 10 Gt of carbon, or 37 Gt of CO_2. Photosynthesis removes approximately 123 Gt of carbon annually from the atmosphere, including 3 Gt from the fossil emissions. Nature's mechanisms—animal, microbial and plant respiration and decomposition, and forest fires—release 120 Gt carbon back to atmosphere. The fraction of organic matter that enters the process of turning biomass into fossil fuels is very small: this terrestrial cycle captures 3 Gt of extra carbon from anthropogenic sources per year.

In the oceans, approximately 90 Gt of carbon is bound annually by ocean surface photosynthesis, and released via respiration and decomposition. Organisms that die then sink to the seabed, while physical processes such as diffusion and upwelling move carbon towards the surface. The net flow of carbon is approximated as 2 Gt downwards and, therefore, oceans can capture 2 Gt of carbon annually.

In addition to the relatively fast carbon cycle outlined above, another carbon cycle stores even greater amounts of carbon in rock material, but this process operates on much longer time scales. Approximately 10–100 Mt of carbon moves through this slow cycle annually, a rate of carbon transportation much lower than found in the fast carbon cycle.

The slow cycle begins with CO_2 dissolved into rainwater, forming a weak carbonic acid. When this slightly acidic water falls on rocks, it slowly dissolves calcium, magnesium and other earth alkali metals. The dissolved metal ions—Ca^{+2}, Mg^{+2} etc.—are then transported to oceans by rivers. The oceans contain huge amounts of dissolved CO_2 as bicarbonate ions, HCO_3^-, and these combine with the metal ions to form carbonates, such as $CaCO_3$ and $MgCO_3$. Most of this chemical processing is performed by shell-building animals such as corals, oysters, mussels, urchins, and starfish, and when these organisms die, they fall to the seabed. Over long periods of time these remains cement together and turn into rock, storing the carbon as limestone and its derivatives.

The remaining 20% of the carbon in the slow cycle originates from living organisms in the oceans. As they die, these pile up in layers of mud and, over millions of years under high pressure and temperature, form sediments. If these dead plant sediments form faster than they decompose, the layers of organic matter turn into coal, oil and natural gas.

From the slow cycle, carbon is released to the atmosphere by erupting volcanoes. These are currently less than one percent of the scale of anthropogenic fossil-fuel carbon emissions.

If the CO_2 concentration in the atmosphere increases, leading to higher atmospheric temperatures, this will result in increased precipitation. More precipitation dissolves more metal ions that combine with dissolved CO_2 as carbonates, which fall to the seabed, removing carbon from the air and storing it. This cycle is very slow, taking several hundreds or thousands of years to balance the carbon cycle [9].

In summary, properly functioning natural terrestrial and ocean systems can take in 5 Gt of human fossil carbon emissions, leaving an additional 5 Gt of carbon—or more than 18 Gt of CO_2—in the atmosphere every year, resulting in the increasing CO_2 concentration and temperature. Both mechanisms are strongly affected by climate conditions, and the well-being and healthy function of natural ecosystems.

2.2.4 Connection Between Total Emissions, Population, GDP, Energy Intensity and Energy Sources

Economic growth and modern human society require significant amounts of primary energy. If the focus is on the fast economic improvement, the cheaper the energy is, the faster growth can be achieved in the short-term. There is a strong connection between gross domestic product (GDP) and energy consumption per capita. Countries with high living standards have, both historically and in the present day, had high per capita emissions from cheap fossil fuels combustion, which have led to environmental impacts and global warming. At a macro-level analysis, environmental impact is determined by three factors—population, wealth or affluence, and technology—which together form the "IPAT" model for the relationship between human population, resource consumption, and the impact on ecosystems:

$$I = P \times A \times T$$

The environmental impact (I) corresponds to the total fossil primary energy consumption, and hence the resulting emissions. The right side combines the main factors or drivers of CO_2 emissions: population growth or population density (P), economic development measured by GDP (A), and technological advancements measured by energy intensity, resource efficiency and carbon intensity (T). The model was developed by Ehrlich and Holdre [51] and it is widely used to understand the rebound effect from the perspective of technological progress in development contexts (e.g., [200, 226]). Countries purchase raw materials and primary energy to produce power and heat for society and for industries that manufacture products and services. Gross domestic product is the market value of all of these final goods and services produced and sold in a year (or within any other time range used) by a country or countries. Primary energy refers to an energy form found in nature that has not been subject to any human-engineered conversion process: it is energy contained in fuels, and other forms of energy, including waste, received as an input to a system. For example, in a pulverized combustion power plant, the primary energy source is the coal, and its chemical energy is converted into thermal energy. Thermal energy is then converted into mechanical work in the steam turbine, and finally to electricity in the generator. The efficiency of converting this primary energy to electricity is around 30–40%, depending on the specific technology, which means that 60–70% of the primary energy is converted to process heat or waste heat. When a coal fired power plant is built, the necessary materials require energy for mining, refining and producing the power plant components. This creates additional, indirect emissions.

For comparison, renewable energy sources such as wind power, solar power, hydropower and geothermal power do not need any fuel to be purchased during their operation, as their primary energy comes from the Sun, wind or geothermal sources. Even though these processes do not have direct emissions, there are several sources of indirect emissions during the manufacture of wind turbines and solar PV panels, as well as land-use change emissions, and other direct and indirect impacts on nature.

The correlation between total CO_2 emissions m_{CO_2}, primary energy emission factor e_{CO_2}, gross domestic product G, primary energy consumption E^*, and population P, is captured by the Kaya identity [115]:

$$m_{CO_2} = \frac{m_{CO_2}}{E^*}\frac{E^*}{G}\frac{G}{P}P = e_{CO_2}e_E e_G P$$

where $e_{CO_2} = m_{CO_2}/E^*$, $e_E = E^*/G$ and $e_G = G/P$ are the primary energy emission factor (Mton CO_2/MWh), energy intensity (MWh/M\$) and per capita *GDP* (M\$/million persons), respectively. Dividing total emissions by population gives the emissions per capita, also known as the average carbon footprint (ton CO_2/person),

$$\frac{m_{CO_2}}{P} = e_{CO_2}e_E e_G$$

Table 2.2 presents these Kaya parameters for selected countries, and shows that both population P and gross domestic product per capita e_G have been increasing since 1965 for all selected countries (shown in red). For China in particular this increase in e_G has been very large, by more than a factor of ten, and there has been a dramatic increase in the Chinese economy. Much of the industrial production of the westerns countries has been externalized to China, and a large share of industrial CO_2 emissions are therefore also externalized.

Since the first two terms in Table 2.2 have increased for all selected countries, one or both remaining terms must therefore decrease to lower total emissions. From these selected countries, the third term e_E, energy intensity, has continuously decreased for Denmark, Germany, Spain, and the US from 1965 to 2018. For all other countries listed energy intensity peaked during the 1990s, which means that less energy is now needed to generate wealth. Energy intensity in South Africa and China is the highest, at 2115–2125 MWh/M$ (or 2.12–2.23 kWh/$) respectively. The emission factors of all countries are also already decreasing, which indicates that they are increasing energy efficiencies and transitioning to cleaner energy sources. In spite of these reductions and efficiencies, the total emissions of the most countries are still increasing.

Historically, environmental damage (e.g., resource use and greenhouse gas emissions) has strongly correlated with economic growth (GDP/capita) or national wealth [224, 236]. The richer we are or wish to become, the more resources we tend to use, and consequently the more emissions we produce due to the combustion of fossil oil and coal. From historical observations, we see that GDP increase first leads to increasing emissions. At the early stages of development, pollution and emissions increase due to increased use of fossil energy resources (which, historically, were much cheaper than renewable energy alternatives, reducing environmental quality. This trend reverses at the high-income level, leading to environmental improvement, at this point, economic growth decouples from emissions.

This phenomenon is typically presented graphically as the Environmental Kuznets Curve (EKC), shown in Fig. 2.9. The trajectory of the curve depends on various indicators such as technological advancements, policy instruments, and the market economy. Many countries have managed to achieve economic growth while reducing emissions—for example, since 1990 the United Kingdom has increased GDP by more than 40% while also reducing emissions by 40%.

As Fig. 2.9 shows, several possible pathways to economic wealth exist. The choices made by the UK and Germany have led to around 50% lower peak emissions during their economic development, and they are also reaching lower per capita emissions much sooner in their development trajectories than the US. It will be very interesting to observe the progress of development in China and India in the future—there is great hope that they could expand their economies with lower per capita peak emissions by relying on renewable energy from an earlier stage in their growth curve. This may be feasible, since the prices of solar and wind energy are already at the level of fossil energy sources.

The Environmental Kuznets Curve outlined above is based on a pattern of increasing, peaking, then decreasing levels of emissions/environmental harm versus gross domestic product. As shown in Fig. 2.9, different countries can have significantly different trajectories in their development. The EKC theory was created in

Table 2.2 Kaya identity parameters for selected countries. Red = increasing, orange = peak, and green = decreasing [173]

		P	$e_G = G/P$	$e_E = E/G$	$e_{CO_2} = m_{CO_2}/E$	m_{CO_2}
	Year	M	M$/M	MWh/M$	ton/MWh	Mton
	1965	84.6	3617	911	0.205	57
	1991	153.3	7888	1309	0.145	229
Brazil	2018	210.2	14034	1152	0.140	475
	1965	723.8	1253	1714	0.326	506
	1991	1170.6	3051	2376	0.313	2653
China	2018	1417.1	13102	2125	0.268	10590
	1965	4.8	17459	2001	0.266	44
	1991	5.2	29915	1510	0.276	64
Denmark	2018	5.8	46312	743	0.174	35
	1965	4.6	12226	2068	0.219	25
	1991	5.0	25175	2617	0.167	55
Finland	2018	5.5	38897	1558	0.137	46
	1965	76.0	14642	2681	0.323	965
	1991	79.8	26770	1911	0.248	1012
Germany	2018	82.9	46178	976	0.199	743
	1965	500.1	1229	1039	0.248	158
	1991	888.9	2062	1377	0.253	640
India	2018	1369.0	6806	1048	0.281	2742
	1965	32.1	7420	1417	0.210	71
	1991	39.0	19958	1405	0.218	238
Spain	2018	46.8	31497	1097	0.165	267
	1965	7.8	17239	2947	0.159	63
	1991	8.6	27614	2719	0.090	58
Sweden	2018	10.2	45542	1348	0.067	42
	1965	189.7	21390	3474	0.235	3319
	1991	251.6	36464	2442	0.224	5026
United States	2018	332.1	55335	1475	0.201	5447
	1965	19.2	5673	3138	0.362	124
	1991	40.9	6035	4310	0.319	339
South Africa	2018	57.3	12166	2115	0.306	451

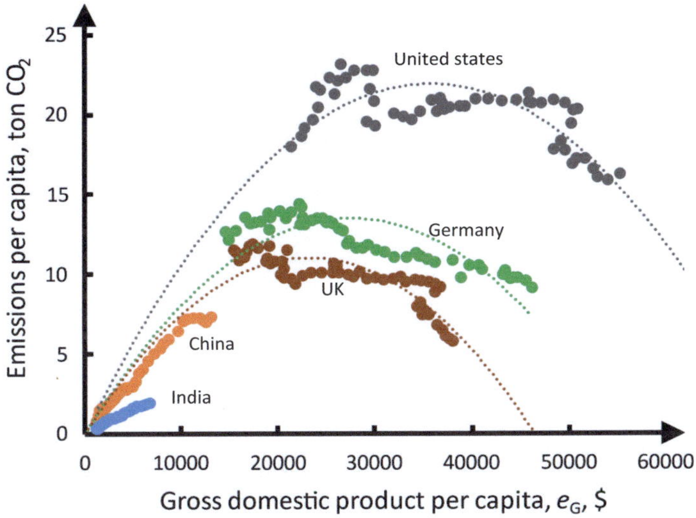

Fig. 2.9 Environmental Kuznets Curve for India, China, UK, Germany and United States. Dashed lines are 2nd order trend-lines set to start from the emission value 0 to demonstrate the typical behavior of the EKC [173]

a world dominated by fossil fuels, and where those were seen as the easiest path towards economic growth. But is this still a valid assumption? In the early 2020s we have already seen renewable energy prices significantly lower than fossil fuels, and so even developing countries could use these to increase their standard of living without harming the environment. In 2021 for example, the price of onshore wind (0.033 $/kWh) and solar PV (0.048 $/kWh) were already below the cost of fossil energies (0.05–0.17 $/kWh) [105].

One challenge for many developing countries and emerging economies in transitioning to renewables is that they often have significant, easily available fossil resources that offer a fast-track for improving the economy. Not only are the infrastructure and markets available and mature, but trade agreements and contracts with third parties are often signed for decades into the future, making it difficult pull away from fossil fuels.

As discussed in Sect. 2.2, a significant share of anthropogenic CO_2 emissions come from the combustion of fossil fuels. However, renewable energy sources also create emissions, for example by requiring the mining of rare-earth minerals such as lithium or neodymium. Table 2.3 presents these impacts from fossil fuels, nuclear energy, and renewable energy technologies in *electrical power generation.*

UN conventions on energy statistics counts the electricity made from wind and solar as primary energy sources [217]. To compare wind and solar energy to fossil fuels and nuclear energy, the installed rated capacities (MW) of wind turbines and PV panels are counted as primary energy sources, and then capacity factors and system efficiencies are used to calculate how much of these sources can be turned

Table 2.3 Impact factors $a_{i,j}$ for energy technologies j on environmental indicators i: Emission, freshwater eutrophication, carcinogenic effects, ionizing radiation, direct and indirect (raw material mining activities) land use, dissipated water, and minerals and metals coefficients for electrical power generation [217, 218]

	Indicators of impact A_i						
Impact factors $a_{i,j}$	Climate change emission	Freshwater eutrophication	Carcinogenic effects	Ionizing radiation	Land use	Dissipated water	Minerals and metals
[Unit]/MWh	tCO_2-e	gP-e	mCTUh	kgU_{235}-e	m^2	m^3	mgSb-e
$j=$	$i=1$	2	3	4	5	6	7
1 Hard coal PC	1.000	490	7.3	9.1	10.3–26.6	2.9	520
2 Hard coal SC	0.950	460	6.9	8.2	9.7–24.9	2.6	500
3 N-Gas NGCC	0.430	20	1.3	9.2	0.7–5.8	1.2	240
4 Hard coal PC, CCS	0.370	690	10	13	14.9–38.0	5.1	780
5 Hard coal SC, CCS	0.330	640	9.7	12	13.8–34.9	4.6	740
6 N-Gas NGCC, CCS	0.130	24	1.7	11	0.9–7.0	2	310
7 Hydro 660 MW	0.150	13	2.6	12	8.7–15.2	0.37	610
8 Hydro 360 MW	0.011	1.3	0.35	0.84	24.1–42.0	0.039	61
9 Nuclear average	0.005	5.8	0.51	14	0.52–1.02	2.4	330
10 CSP tower	0.022	11	2.1	4.5	12.6–58.7	0.18	340
11 CSP trough	0.042	14	6.3	6.1	11.6–45	0.34	650
12 PV poly-Si, ground	0.037	28	4.1	9.1	13.7–40.0	0.58	4500
13 PV poly-Si, roof	0.037	39	1.6	9.8	7.6–21.7	0.63	7200

(continued)

Table 2.3 (continued)

	Indicators of impact A_i						
Impact factors $a_{i,j}$	Climate change emission	Freshwater eutrophication	Carcinogenic effects	Ionizing radiation	Land use	Dissipated water	Minerals and metals
14 PV CdTe, ground	0.012	8.8	3.4	1.9	7.9–26.4	0.13	1500
15 PV CdTe, roof	0.015	14	1.1	1.9	1.1–3.2	0.16	2600
16 PV CIGS, ground	0.011	8.8	3.4	1.8	7.6–25.5	0.13	1700
17 PV CIGS, roof	0.014	14	1.1	1.8	1.1–3.1	0.16	2800
18 Wind onshore	0.012	6.7	6.6	1	0.6–3.1	0.18	680
19 Offshore, concrete	0.014	7	5.5	1.2	0.9–1.5	0.16	980
20 Offshore, steel	0.013	6.8	7	1.2	0.8–1	0.16	990

PC = pulverized combustion, SC = super-critical steam power plant, NGCC = Natural gas combined cycle, CSP = concentrated solar power, ground = ground-mounted PV panels, roof = roof-mounted PV panels, concrete/steel = foundation type, CTUh = Human Toxicity Cancer-indicator, Land-use impact of different kinds of biomass is presented under Sect. 2.4

into electrical energy at the load. Nonetheless, as Table 2.3 shows, the emissions of CO_2 in particular are significantly lower than from the fossil energy sources. Since the capacity factors for, for example, solar PV and onshore wind are only 12% and 35% respectively, proportionally larger systems are required to produce same amount of energy than if the rated power was continuously available (for solar the scale-up factor is 1/0.12 = 8.3 times larger, and for on-shore wind 1/0.35 = 2.9 times larger—plus suitably large energy storage systems). This challenges the development of variable renewable energy in areas where the capacity factor is low during some seasons, because it means that at certain times the system is running at the rated power and the production of electricity is high, while at low season there is a power gap that must be solved either via demand-side responses or large storage systems (See more on storages in Chap. 7).

To compare different technologies in Table 2.3, all values are based on production of 1 MWh of electrical power. Using this data and a country's profile of power generation, E_j (MWh), for different technologies j, we can estimate that country's total emissions, land use impact, etc., from power generation. Table 2.3 shows $a_{i,j}$, the impact factor of technology j on studied indicator i. The columns are the indicators A_i: total equivalent CO_2 emissions A_1 (ton), total freshwater eutrophication impact A_2 (g P-e), … and total minerals and metals consumption A_7. These are calculated with the following equation, where i and j refer to indicator and technology, respectively:

$$A_i = \overline{a}_i E = \sum_{j=1}^{20} \left(a_{i,j} E_j\right) = E \sum_{j=1}^{20} \left(a_{i,j} f_j\right)$$

where E is the total electrical energy generation, $\overline{a}_i$ is the average impact coefficient of a country, and f_j is the energy fraction of technology j,

$$\overline{a}_i = \sum_{j=1}^{20} \left(a_{i,j} f_j\right)$$

For example, if a country is currently using pulverized coal (PC) combustion power plant and wants to reduce their emissions by either investing in carbon capture and storage (CCS) with the PC combustion, or switching to natural gas-based combined cycle (NGCC). Based on the Table 2.3, emissions would decrease from 1000 kg/MWh of CO_2 to 370 and 430 kg/MWh CO_2 for the PC combustion with CCS and the natural gas cases, respectively. However, all other impacts such as dissipated water, land use, or minerals and materials use are significantly higher when using coal with CCS than using NGCC. Further, if the country's power needs could be delivered by renewables, for example wind power, all other impacts would drop to negligible fractions except minerals and materials use, which would increase by 30–90%, depending on the technology used.

The indicators included in Table 2.3 do not include biodiversity and nature-related indicators, and these are discussed in detail in Sect. 2.3. Also, while the land-use indicator incorporates both direct and indirect impacts, the indirect use here refers to impacts from materials extraction and does not consider, for example, the area between the turbines. Land-use impacts of different renewable energy technologies will be discussed in Sect. 2.4.

2.2.5 Conclusions

The Earth's climate has undergone natural fluctuations throughout its history, but human activities, particularly the burning of fossil fuels since the 1800s, have accelerated global warming and climate change. Climate change results in various impacts including extreme weather events, sea-level rise, habitat shifts, and biodiversity loss. Moreover, the effects of climate change on nature, sometimes coupled with other human activities, can trigger carbon fluxes to the atmosphere. Climate change also poses significant societal and humanitarian risks that can escalate if unmanaged—therefore, climate change is also a security challenge.

We have explored the relationship between economic growth, energy consumption, and carbon emissions with the use of the Environmental Kuznets Curve. This concept suggests that economic growth initially leads to increased emissions, but advanced economies can achieve environmental improvement with technological advancements and policy interventions. While decoupling economic growth from emissions is essential for sustainable development, it is highly likely that transformative change towards sustainability requires us to critically assess and modify our consumption levels. Moreover, developing countries face challenges in transitioning to renewable energy due to existing infrastructure, agreements and investments centered around fossil fuels. Ideally, investments would direct developing countries to sustainable and renewable energy systems without going through the fossil fuel phase.

Transitioning to renewable energy sources is imperative to reduce greenhouse gas emissions and mitigate climate change and its impacts. But while solving for climate change, there is also a dire need to halt the loss of biodiversity, and these priorities should not compete with each other. Next, we will have a look at biodiversity, its loss and its importance to humanity, with the aim of highlighting the need to find solutions that can tackle both climate change and biodiversity loss.

2.3 Biodiversity Loss

2.3.1 What is Biodiversity?

According to the Convention on Biological Diversity (CBD), the shorthand term "biodiversity" means the variability among living organisms from all sources including, inter alia, terrestrial, marine, and other aquatic ecosystems, and the ecological complexes of which they are part; this includes diversity within species, between species, and of ecosystems. Basically, biodiversity can be understood simply as the variability of all life on Earth. Nature is dependent on biodiversity, as biodiversity underpins the functioning and processes of ecosystems. Importantly, biodiversity supports the resistance (capacity to resist change) and resilience (capacity to recover from change) of nature against different disturbances [82]. For example, ecosystems having more diverse species' communities are more likely to recover from a disturbance compared to ecosystems having less diverse species' communities. The human pressure on nature has increased rapidly in recent decades, which means that more resistance and resilience are required for nature to function normally.

2.3.2 How to Measure Biodiversity?

Biodiversity is a multidimensional concept and thus it is not easily described by a single number or measure [178, 228]. Biodiversity can be expressed as the variability of genes, organisms, or ecosystems at different spatial scales. Multiple different sampling techniques and ways of counting exist, and the final method needs to be selected depending on the context [142, 143]. Methodologies range from microscopy to remote sensing with satellites, and from sampling and sequencing of DNA to monitoring of global ecosystems. Genetic diversity can be measured, for example, by measuring the variation of genetic factors within a species population [143, 228]. Species diversity can be measured, for example, by counting the number of species in a certain area. The diversity of ecosystems can be measured, for example, by estimating the extent and condition of different ecosystems. The state of ecosystems can be estimated by measuring the structural characteristics of ecosystems, the structure of species communities, and the functioning of ecosystems (see info box 2.1 Planetary well-being). There are numerous measures of biodiversity, and the use of a suitable measure depends on the context and the question [167].

In general, the state of biodiversity is often measured using species-related metrics [31]. Local biodiversity can be expressed as the number of species (i.e., species richness, alpha diversity) in the specific area or ecosystem. For example, we can measure the number of different bird species in a specific forest area (Fig. 2.10). On a larger spatial scale, biodiversity can be measured by comparing the diversity between ecosystems, usually measured as the magnitude of species change between the ecosystems (beta diversity). For example, two different ecosystems may have the same number of species but the identity and/or the relative abundance of the species, that is the structure of the community of the species, differs. At a larger scale, gamma diversity expresses the overall number of species by combining alpha and beta diversity (Fig. 2.11).

It should be noted that a higher number of species does not always mean that the state of biodiversity is better. For example, a human-driven disturbance can increase the number of species in a specific area, but with new species that are not native for the original ecosystem [155]. Therefore, the natural state of biodiversity, measured for example as the community composition of species without any human impact, should be used as reference state [101, 121, 122].

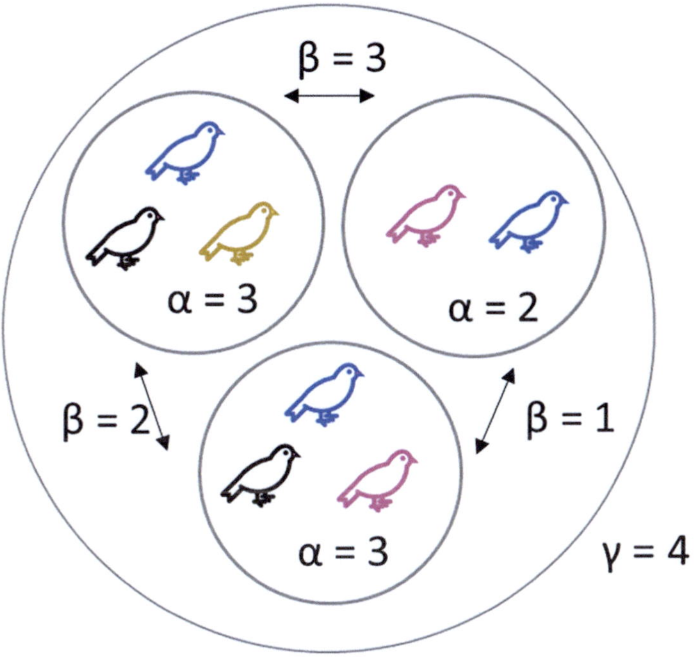

Fig. 2.10 Different scales of diversity. In the example, colours indicate different species and small circles indicate local species communities. Biodiversity can be expressed, for example, by the local species number (α = alpha diversity), by the difference between local communities (β = beta) or by the overall species number (γ = gamma)

There are species groups, such as birds, mammals and vascular plants, whose diversity is well known. On the other hand, there are also poorly-known species groups, and species that are not even found yet. In practice, when measuring biodiversity, it is impossible to measure all its dimensions [142]. Therefore, for example, well-known species or species groups and the extent and condition of natural ecosystems can be used as indices for biodiversity.

The IUCN (International Union for Conservation of Nature) Red List of Threatened Species is the world's most comprehensive inventory of the global status of plant and animal species [107]. It uses several quantitative criteria to evaluate the extinction risk of species. These criteria are relevant to most species and all regions of the world. The extinction risks are evaluated based on species' population sizes and geographic ranges. Globally the status of over 150,000 species has been evaluated. In addition to species, this data is integrated at the ecosystem level as the IUCN Red List for Ecosystems [117].

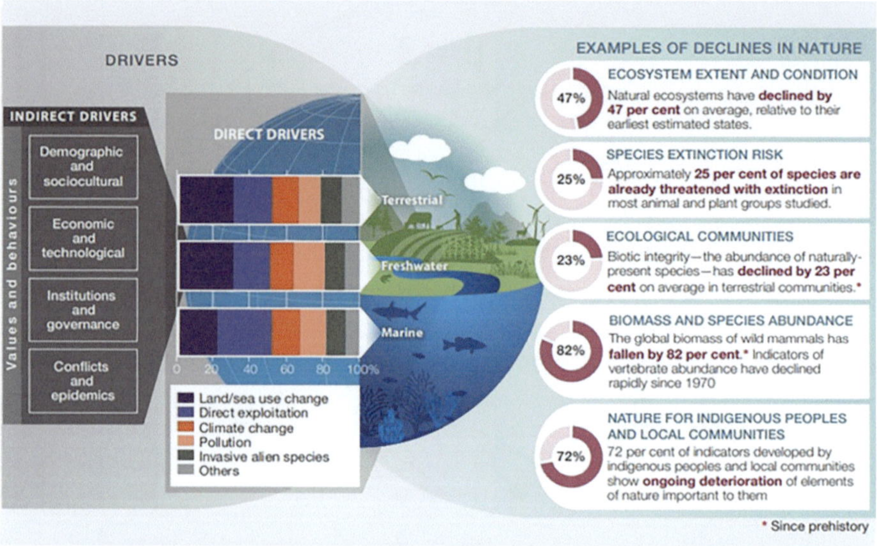

Fig. 2.11 Examples of global declines in nature, emphasizing declines in biodiversity, that have been and are being caused by direct and indirect drivers of change [102]

Info Box 2.1. Planetary Well-Being

Human actions threaten to cause irreversible changes to the Earth system, with several critical safety limits, or so-called planetary boundaries, already exceeded as outlined in Chap. 1 [207]. Transformative changes to social, economic, and technological systems are increasingly called for to change course towards a more sustainable future [45, 233].

A challenge with many existing frameworks around sustainability or sustainable development is that they seldom focus on the systems and processes that support life, well-being, and biodiversity at different spatial scales. [119] recognized the need to conceptualize wellbeing in a way that is non-anthropocentric, and that encourages a systems-oriented, multi-scalar outlook. For this purpose, they coined the concept of *planetary well-being*, and explained it as follows: Planetary well-being is a state in which the integrity of Earth system and ecosystem processes remains unimpaired to a degree that lineages can persist to the future as parts of ecosystems, and organisms (human and nonhuman) can realize their typical characteristics and capacities.

Essential for the proper understanding this concept is to note that planetary well-being emphasizes the integrity of Earth-system and ecosystem processes, on systems orientation, and on needs-based, objective accounts of wellbeing. Integrity of Earth-system and ecosystem processes is fundamental for the survival and evolutionary potential of species and lineages (a group of organisms with a shared genetic ancestry that is distinct from other such groups), and thus for the existence and well-being of organisms and the ecosystems they inhabit.

A *system* is an entity that is comprised of its components, that can be impacted by the environment, has characteristic relations and interactions between its components, and has system-specific characteristics and capacities that stem from the system processes [28, 119]. Every cell is a system, as is every organ, individual, population, species, ecosystem, and so on.

Well-being is the functional integrity of the system. In other words, the system has well-being when the integrity of the system processes remains intact to a degree that allows the system to continue its existence and realize its system-specific characteristics and capacities. To achieve and maintain well-being—that is, for the system processes to function—specific objective *needs* (conditions of dependence on inputs) must be satisfied.

Instead of focusing on the needs themselves, however, the planetary well-being framework focuses on the systems and processes that are necessary for the satisfaction of the needs of diverse life forms on Earth. In other words, the focus of planetary well-being is on life-supporting systems. Life on Earth can be conceptualized as a set of interdependent systems, and well-being at any level as the integrity of that particular system. From the planetary well-being viewpoint, human exploitation of ecosystems, while necessary to fulfil the objective needs of the species *Homo sapiens*, must not compromise the ability of other species and lineages to persist in their ecosystems into the future. Therefore, as concluded by the original authors [119], planetary well-being can be understood as the opportunity for both humans and nonhumans to satisfy their objective needs now and in the future, and this should be the ultimate goal of humanity's actions.

2.3.3 State of Nature and Drivers of Biodiversity Loss

The current state of biodiversity is not good, and biodiversity loss is now faster than ever in human history [101, 102]. Globally, 75% of the land surface is significantly altered. The average abundance of native species in most major terrestrial biomes has fallen by at least 20%. Over 65% of the ocean area is experiencing increasing cumulative impacts, and over 85% of wetland area has been lost. An average of around 25% of species in assessed animal and plant groups are threatened. Overall, around 1 million of the estimated 8 million species already face extinction, many within decades, unless action is taken to stop biodiversity loss. The global rate of species extinction is tens to hundreds of times higher than it has averaged over the past 10 million years [102, 157].

There are many reasons behind global biodiversity loss and deterioration of ecosystem processes (Fig. 2.11, [102]). The causes, or drivers, of biodiversity loss can be divided into direct and indirect drivers. Direct drivers (natural and anthropogenic) are those that directly influence biodiversity and ecosystem processes (also referred to as 'pressures') [102]. Main direct drivers of biodiversity and ecosystem change are land-use change, natural resource use and exploitation, climate change, pollution, and invasive species. These impact biodiversity and ecosystem change at a more proximate level, often involving synergies with other direct drivers, and finally feeding back into indirect drivers. Indirect drivers alter and influence direct drivers as well as other indirect drivers (sometimes also referred to as underlying or ultimate causes) [102]. Impacts of indirect drivers often stem from values and human behavior, for example through consumption. Next, we briefly explain each of the direct drivers in more detail.

Land use refers to both land-use change or land transformation, and land occupation [225]. Land transformation means modifying land to make it suitable for intended use, such as cutting forest and converting it to field for agricultural use. Land occupation, on the other hand, means using land in the intended productive way (e.g., agriculture) (see also [101]). In energy production, land is used for the harvesting of renewable and nonrenewable fuel types, but also energy plant itself including mining of materials for the construction, decommissioning, and the handling of waste ([215], see also Sect. 2.4 Renewable energy and land use change). For example, the use of wood- and peat-based energy requires land use, which has negative impacts on nature [222]. From the species perspective, land use results in habitat loss and degradation, which decreases the size of species' populations, reduces their viability, and eventually can drive species to extinction [85]. The direct exploitation of species and natural resources leads to small population sizes and extinctions. For example, overexploitation of fishes has strongly disturbed marine habitats and biodiversity [35]. Climate change drives environmental changes that disturb natural ecosystems and species [18]. Due to climate change, for example, the spatial location of habitats suitable for any given species may change, also causing habitat loss and degradation, and finally extinctions. Pollution drives biodiversity loss especially in freshwater and marine habitats [44]. The most severe pollutants are nitrogen and phosphorus fertilizers that leak into the environment and cause eutrophication, strongly disturbing aquatic life and biodiversity. Invasive species disturb natural functioning of ecosystems, often by outcompeting local and native species for resources [160].

2.3.4 *Value of Biodiversity*

People have known for decades that the main causes of the loss of biodiversity are human driven. But why should we try to alleviate the loss and safeguard biodiversity? First, biodiversity, which is life in all its different shapes and forms, has an intrinsic value. Intrinsic value means that biodiversity has value that is independent of its' potential usefulness for humans [74]. All organisms and species are valuable in themselves, and their value does not need to be justified in any way.

We humans are heterotrophic animals, which means we cannot survive without other organisms. Thus, from a human perspective, biodiversity has invaluable utilitarian value. We are simply completely dependent on other organisms and their functioning, i.e., on nature [16, 83]. Benefits people get from nature are also called ecosystem services (sometimes also nature's contributions to people) [49, 157]. For example, production of oxygen, carbon sequestration and climate regulation, pollination, purification of water and air, and food, fodder, fibers, and other raw materials that people get from nature are all ecosystem services. Without ecosystem services there are no people. Thus, safeguarding biodiversity is safeguarding all life on earth, including our human lives. Biodiversity loss and climate change are severe existential risks for humanity, and they should be tackled together [179]. Importantly, protecting biodiversity can help us mitigate and adapt to climate change.

2.3.5 Assessing Biodiversity Footprints in Different Contexts

To better understand, communicate and visualize our impact on nature, we need indicators and methodologies that can quantify the biodiversity loss associated with different human activities. While we have been able to quantify and communicate the direct drivers of biodiversity loss, the impacts of the consumption of individuals and organizations has remained less recognized. Even though nature would be directly exploited, say, by a mining company or an energy producer, the production of goods and services (such as energy) is driven by demand from organizations and individuals further along the supply chain. Thus, we need tools to quantify the biodiversity impacts of consumption.

Biodiversity footprint assessments provide an interesting possibility in this regard. Biodiversity footprint refers to the negative biodiversity impacts caused by the production and consumption of products and services [145]. Biodiversity footprints can be calculated to assess the direct and indirect biodiversity impacts of products and services [11], organizations [27, 52, 53], as well as nations and regions [24, 38, 133]. For example, an energy-producing company could assess the biodiversity footprint of the harvesting of the different fuels it consumes, giving it an option to assess how to avoid and mitigate negative biodiversity impacts along its supply chain, as we will later show in this chapter. Furthermore, organizations have used biodiversity footprints to assess which operations are most detrimental to biodiversity, to allow them to focus their mitigation efforts on those [27, 53]. In a completely different sector, banks have assessed the biodiversity footprints of the funds and loans they manage to allow and encourage projects to consider their impact on nature, and biodiversity footprints can also be used in funding decisions of investors [114].

To measure the biodiversity footprint of consumption, it is possible, and even necessary, to utilize existing research and knowledge about how different activities

Fig. 2.12 The four key pieces of information needed in biodiversity footprint assessment © Maija Karala

influence nature. To assess the biodiversity footprint of any activity, four key pieces of information are needed (Fig. 2.12). Next, we will elaborate the key pieces of information step by step.

STEP 1. Type and amount of consumption

First, we need to know what was consumed, and how much. For example, to assess the biodiversity footprint of energy consumption, we need to collect information about the amount of energy consumption (e.g., in MWh) by fuel type. Or, to assess the biodiversity footprint of food consumption, we need to know what kind of food products were consumed, and how much. This type of information is often collected by organizations but sometimes it is challenging to obtain, especially when the level of desired detail increases. One possibility is to utilize one of the most common information systems of organizations: financial accounting. While financial consumption information might not be as accurate as physical consumption information, it can be accurate enough to help decision-making purposes. We will later explain how financial consumption can be used to calculate biodiversity footprints.

STEP 2. Type and amount of environmental impact caused by the consumption

Next, the type and amount of environmental impact caused by the consumption needs to be identified. We can utilize the direct drivers of biodiversity loss mentioned in Fig. 2.11 in identifying relevant environmental impacts. In essence, this means, for example, assessing how much and what kinds of land (ecosystems) are being used when we produce and consume a certain fuel type for energy. It can also mean measuring the climate-change impacts caused by the consumption. Currently, biodiversity footprint assessments are most commonly done for land use, but they can be conducted also for the other direct drivers.

There are numerous ways to gain information on the environmental impacts of different activities. Apart from measuring the environmental impact on-site, which can be time consuming and costly, the information can often be found from already existing research. Nevertheless, more on-site research is needed to improve existing information. For example, the United Nations Economic Commission for Europe [215] has conducted a life-cycle assessment (LCA) of electricity sources. The environmental impact information of different electricity sources can be used to assess biodiversity footprints in the energy sector, and later we will elaborate how these environmental impacts are connected to biodiversity. However, in cases where information in the form of research articles or reports is non-existent, conducting LCAs from scratch or utilizing environmentally-extended input–output (EEIO) assessments may become necessary.

LCA is a tool for in-depth assessment of environmental impacts of products and services. It can be used to compare the environmental impacts of different products, for example two different wind turbine materials, or produced wind energy versus produced natural gas, or larger systems, such as two different wind parks. In addition, with LCA it is possible to assess how different steps of the life cycle contribute to the total environmental impacts of a given product. For example, in the LCA of a wind-power plant, we might be interested in knowing the size of impacts originating from producing materials for the plant's infrastructure, from establishing and operating the plant, or from the dismantling and recycling of the infrastructure. There are numerous ways to conduct LCAs, and readymade software (such as openLCA and SimaPro) and databases (such as ecoinvent and ELCD) are available for use.

Collecting information for an LCA can be very time consuming and difficult. Supply chains of products (and services) can be highly complex, and tracing the origin of the environmental impacts remains difficult (see Step 3). This is where environmentally extended input–output (EEIO) analyses can help. EEIO databases are used to assess the environmental impacts of entire supply chains. In essence, the input–output methodology assesses the inputs that an economic sector needs to produce its goods and services, and the outputs an economic sector provides to other sectors or to final consumption [118, 134, 195]. EEIO databases, such as EXIOBASE, Eora, GTAP and WIOD, connect environmental impacts, such as land use, greenhouse gas emissions, and water pollution, with economic activities and transactions. Their aim is to reveal the environmental impacts associated with consumption of products and services by organizations, public sector, households, and individual consumers [118, 134, 195]. However, EEIO analyses cannot usually produce the level of detail that can be reached with LCA. For example, comparing two different products from a similar sector (e.g., two different types of wind power generators) remains difficult with EEIO analysis.

LCA and EEIO analyses can also complement each other. So-called hybrid-LCA or EEIO-LCA combine the strengths of both approaches [36, 164, 208]. Organizations have used both LCA and EEIO methodologies to assess their biodiversity footprints [27, 53, 145].

STEP 3. Location of the environmental impact

One of the key features that separate biodiversity footprint assessments from carbon footprint assessments is the need for location-specific data. Nature is local, while climate is global. Characteristics of biodiversity—that is, the identity and community composition of species—are different when we compare, for example, Finland and Egypt, or China and Mexico. Biodiversity varies also at a smaller scale across few hundred kilometers within a certain country. Thus, it is vital to have at least approximate information about the location of the environmental impact that was determined in Step 2.

STEP 4. Biodiversity loss caused by the environmental impact in a given location

Now that we have determined the type and amount of the consumption, the type and amount of environmental impact caused by that consumption, and the location of the environmental impact, the final step is to assess the biodiversity loss caused by the environmental impact. Even if we knew that a wind-power plant used, for example, around 99 m^2 of land per MWh of produced electricity [188] and that the manufacturing of all the parts needed for a functioning windmill produced 11.9 kg of CO_2 equivalents per MWh produced electricity around the world [215], we still don't know how much biodiversity loss these activities caused.

Currently there are few emerging tools and databases that allow the assessment of the biodiversity loss by certain activities in certain locations. One such database is the LC-IMPACT database [225]. With LC-IMPACT it is possible to estimate the biodiversity footprint of different environmental drivers, such as land use, climate change and pollution, in different countries of the world [225]. Environmental impacts can also be converted to biodiversity loss with the ReCiPe framework [96]. Basically, these databases make use of impact factors for biodiversity loss, for example, in terms of biodiversity loss per square meter of land used or per ton of nitrogen discharge in a specific location. A few other tools for somewhat coarser or local biodiversity footprint analyses include the Species Threat Abatement and Restoration Metric (STAR), ENCORE and Bioscope. As mentioned, assessments of biodiversity footprints have already been conducted for products [11], organizations [27, 53], regions [38] and global consumption [24, 133].

Case study 2.1 presents an example on how biodiversity footprint assessment can be completed for a district heating system that produces biomass-based energy (peat and wood). The example helps to understand the basic principles of conducting a detailed process-based biodiversity footprint assessment. But before diving deep into the case study, we will explain what kind of biodiversity footprint indicators exist and how they should be interpreted.

2.3.6 Biodiversity Footprint Indicators

Biodiversity footprint indicators express the (often negative) change in nature caused by the direct drivers of biodiversity loss. Several different biodiversity footprint indicators exist [145]. The simplest indices report the area of land-use types, such as agriculture or forestry. Other indices consider the viability of species and/or the condition of ecosystems. Biodiversity footprints are generally expressed in the form of a single indicator, such as the potentially disappeared fraction of species, mean species abundance, or habitat hectares [37, 129].

The potentially disappeared fraction of species (PDF) metric accounts for a fraction of species that are potentially lost due to an environmental impact [77, 225]. It indicates the biodiversity impacts relative to a counterfactual natural state without any human impact. PDF ranges from 0 to 1, where 1 means that all estimated species are at risk of extinction, and 0 means that no species is at risk of extinction. For instance, a PDF of 0.01 means that 1% of the species pool is committed to go extinct if the pressure causing the impact (e.g., land use) continues to happen. The metric is based on species area relationship models, geographic ranges of species, the vulnerability of species to different human impacts, and the extinction risk classification of species. Global PDF indicates an irreversible extinction risk of species on a global level, while regional PDF indicates the potential biodiversity loss at the regional scale. Openly available characterization factors (e.g., PDF/m^2 or PDF/kg; also known as impact factors) exist at least for land use, water stress, several pollutants, and climate change [225].

Mean species abundance (MSA) estimates the average abundance, richness, or geographic extent of species relative to the expectation in a pristine site [5]. In other words, the reference state that MSA utilizes is again the counterfactual natural state of the biodiversity. MSA is derived by combining separate meta-analyses of how species abundance responds to difference impacts, such as land use and infrastructure. MSA ranges from 0 to 1, where 1 means that the species assemblage is fully intact (i.e., in its natural state), and 0 means that all original species are lost. MSA is a local measure, and it does not consider the threatened status of species; thus, unlike global PDF, MSA cannot capture global biodiversity loss [144].

Potentially disappeared fraction of species and mean species abundance are indicators for biodiversity loss based on species richness and abundance. Habitat hectares, on the other hand, are based on ecosystem quality or condition [176]. The habitat hectare of a given hectare ranges from 0 to 1, where 1 means that the ecosystem is in natural state, and 0 means that the ecosystem is destroyed. A value of 0.2 habitat hectares in a given hectare means that 80% of the natural condition of the area has been lost. The impact of any activity is derived from the difference between the pre-existing condition of the ecosystem and condition after the impact. Habitat hectare can be used to evaluate how much natural habitat is lost due to any activity. It can also be used to balance the biodiversity losses and gains in a biodiversity offsetting project [158, 159], Info Box 2.2).

Box 2.2: Mitigation Hierarchy

A mitigation hierarchy is a framework that can be used as a guide to limit negative impacts on nature and biodiversity [10, 13, 158, 159]. Applying full mitigation hierarchy typically sets a goal of either no net loss (NNL) of biodiversity or net positive impacts (NPI) on. The mitigation hierarchy includes four steps that are designed to be implemented sequentially [10]: Avoidance, Minimization; Restoration; and Offsetting (Fig. 2.13). The order of the last two steps may differ depending on the goal [159], which is briefly explained at the end of box.

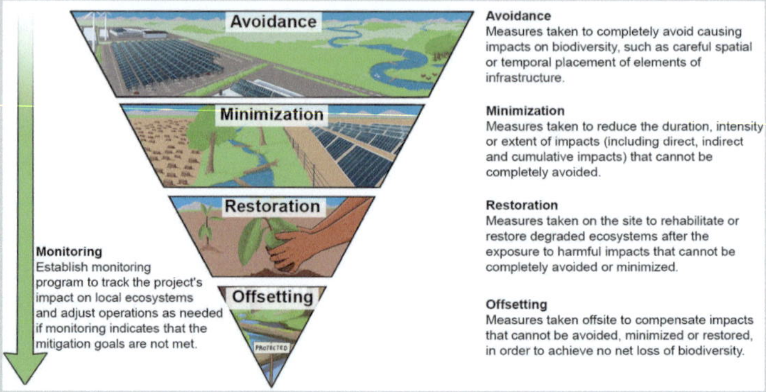

Fig. 2.13 Mitigation hierarchy definitions © Maija Karala

1. **Avoidance**: Measures taken to completely avoid impacts on certain components of biodiversity [13], such as careful spatial or temporal placement of elements of infrastructure. In practice this can mean, for example, screening potential risks prior to the project and selecting a project design and development site with the least damage to biodiversity [10].
2. **Minimization**: Measures taken to reduce the duration, intensity and/or extent of impacts (including direct, indirect, and cumulative impacts, as appropriate) that cannot be completely avoided at the first step [13]. In practice, for example, this can mean the use of more environmentally friendly energy sources or construction methods [10].
3. **Restoration**: Local measures taken to rehabilitate or restore degraded ecosystems after exposure to harmful impacts that cannot be completely avoided or minimized [13]. In practice this can mean, for example, reseeding affected land [10].

4. **Offset**: Offsite measures taken to compensate for impacts that cannot be avoided, minimized or restored, in order to achieve no net loss of biodiversity [13] or net positive impacts. Offsetting means balancing ecological losses with ecological gains generated through actions, such as ecological restoration, establishment of new protected areas, or some form of habitat management [159].

It has also been proposed that restoration should become the last step in the hierarchy, and that its role should be recast as a time-delayed source of net positive impact rather than counting toward the reduction of the original residual impacts of the project that need to be offset [159]. This would lead to a situation where more offset action would be required to produce the gains needed for no net loss, but at the same time conservation actions would be brought forward in time, uncertainty would be reduced, and net gain for biodiversity would be delivered, all of which would benefit nature.

Case Study 2.1: Biodiversity footprint of district heating in the University of Jyväskylä

Here we will demonstrate how the four steps of biodiversity footprint assessment (Fig. 2.12) was applied to a district heating system that produces biomass-based energy (peat and wood). The example is derived from the biodiversity footprint assessment project of the University of Jyväskylä in Finland [53, 221, 222].

STEP 1. Type and Amount of Consumption

First, we needed to estimate how much energy University of Jyväskylä uses for heating, and determine the source of the energy being consumed. The information was gained directly from the local energy company, ALVA. The total energy consumption in 2021 was 19 689 MWh, and the fuels used were mostly wood (62%) and peat (35%), with a small amount of coal, oil or biogas as backup fuel (3%). In this exercise we focus on wood fuels and peat. There are different sources for wood fuels, each with different environmental impacts, and thus this fuel was further divided into two categories: forest residue and roundwood chips, accounting for 34% of the total fuel consumption; and industry by-products, accounting for 28%. Here we leave out the latter and focus on the former for the remainder of this example biodiversity footprint assessment.

STEP 2. Type and Amount of Environmental Impact

For this example, we only estimated the extent of land use required for extracting the different fuels, leaving out other direct drivers of biodiversity loss. To investigate land use for energy production from wood fuels and peat, we calculated how much energy can be obtained from one ha, E_A, which is given by:

$$E_A = E_V V_A$$

where E_V is the energy yield per volumetric unit of each fuel type (MWh/m^3), and V_A is the volumetric yield of each fuel type per unit of extraction area (m^3/ha). We derived values for E_V and V_A from the energy company, from reports, and from scientific literature. With these we calculated the annual energy yield per area, which was 230 MWh/ha for forest residue chips, 90 MWh/ha for roundwood chips, and 400 MWh/ha for peat.

STEP 3. Location of the Environmental Impact

In Finland, biomass-based district heating systems generally aim to harvest their fuels within a relatively small distance from the power plant, and therefore the location analysis in this case was rather straightforward: we simply assumed that most of the peat and wood fuels were derived from the region of Central Finland. When searching for the required information from reports and scientific literature we emphasized results that were specific to the region, however where these were not available we used average national figures for Finland. In international supply chain analyses location analysis can be much more complex, but this process can be aided by EEIO databases as discussed earlier in this chapter.

STEP 4. Biodiversity Loss Caused by the Environmental Impact

Finally, we needed to determine how much biodiversity loss is caused by the land use. In this case, we decided to use habitat hectare as an indicator of biodiversity footprint (for details about the indicator, see the section *Biodiversity footprint indicators* above). From the literature we derived an estimate for how much each of the ecosystems will be degraded, per unit area, by the harvesting of the different fuel types. Degree of ecosystem degradation caused by each of the fuel types in terms of lost habitat hectares can be interpreted as the biodiversity footprint of each of the fuel types.

We found that the biodiversity footprint, $B_{A.i}$, for forest residue chips was 0.023 habitat hectares per harvested hectare (hha/ha), 0.125 hha/ha for roundwood chips, and 0.020 hha/ha for peat [222]. In other words, one hectare of peat harvesting for energy use causes biodiversity loss equivalent to the complete destruction of 0.020 hectares of a pristine peatland. By dividing biodiversity footprint values, $B_{A.i}$, (habitat hectares per harvested hectare) with the energy yield per hectare, $E_{A.i}$, we get biodiversity footprint per unit of energy for each fuel type, $B_{E.i}$:

$$B_{E.i} = B_{A.i}/E_{A.i}$$

The fuel-type specific biodiversity footprint is then calculated by multiplying the biodiversity footprint per unit of energy with the total energy consumption attributable to each of the fuel types, E_i:

$$B_i = B_{E.i}E_i$$

The total biodiversity footprint is then simply the sum of all fuel contributions:

$$B = \sum_{i}^{N} B_i = B_{E.i}E_i$$

The total biodiversity footprint, B_i, was 0.29 hha for forest residue chips, 4.87 hha for roundwood chips, and 0.40 hha for peat [222]. This means that the consumption of district heating of the University of Jyväskylä in 2021 caused biodiversity loss that is equivalent to complete destruction of $0.29 + 4.87 = 5.16$ ha of pristine woodlands, and 0.40 ha of peatlands.

It is worth recalling that in this example we only concentrated on land use and ignored other potentially significant impacts. For example, if we consider the biodiversity footprint per energy unit for each of the fuel types from the above equations, we will see that wood fuels have 2–28 times greater biodiversity footprint per energy unit compared to peat fuel. However, if we account for other direct drivers of biodiversity loss such as climate change and pollution, the result would change, since the net climate impact of peat firing is significantly greater than that of wood firing in the long term (e.g., 100 years) due to peat being practically non-renewable, and the effect of carbon sequestration in growing forests [206]. In the short term (e.g., 10 years), however, emissions might be comparable in size [168].

2.3.7 Conclusions

Biodiversity loss is an existential risk for humanity. All of the main drivers of the biodiversity loss are anthropogenic, i.e., resulting from the influence of human beings. In this chapter we described a few methods of how to quantify biodiversity loss—that is, how to measure biodiversity footprints—and gave a practical example. The next step is to use this knowledge to implement measures with which biodiversity loss can be avoided and mitigated. For example, we should strive to use less energy, and the energy we use should be produced with less damaging methods for the environment. We should design our societies such that we can achieve planetary well-being (see info box 2.1).

Humans are heterotrophic animals, which means that it is very difficult to avoid all negative impacts on nature. Therefore, to achieve no net loss of biodiversity, biodiversity offsetting is needed. In biodiversity offsetting the lost nature values are compensated by providing new nature values in another area by restoring already damaged ecosystems, or by protecting ecosystems under threat of becoming degraded (see e.g., [158]). Finally, it is important to realize that every action takes us either towards strong sustainability, or away from it. Therefore, every individual and all sectors of the society are needed in stopping biodiversity loss.

2.4 Renewable Energy and Land-Use Change

For millennia humans have altered the planet by modifying natural areas into built environments, cropland, pastures, managed forests, areas used for energy production, mining, and other uses. Growing human population and affluence create growing needs and competition over land resources. Over 70% of the land and over 65% of the ocean is now substantially altered by human use, and this has had a negative impact on many species, ecosystems, and ecosystem services such as those maintained by functional, healthy soil [101–103, 157].

Land is altered for reasons related to human livelihoods and economic activities—that is, to create benefits for society. But these activities also often result in land degradation and increase in land-based greenhouse gas emissions [69]. Land-use activities can result in emissions of carbon dioxide (CO_2), methane (CH_4) and nitrous oxide (N_2O) to the atmosphere; conversely, restorative land-use practices can help to remove these greenhouse gases from the atmosphere (see also Sect. 2.2.3: Carbon emissions and carbon cycles).

With access to energy-dense fossil fuels, human economic activity has expanded hugely in the past century, with associated environmental impacts and release of CO_2 previously captured in fossil-energy sources [54]. Land conversion and use of resources also drastically change biodiversity and, in most cases, lead to dramatic losses of species and genetic diversity, and even the loss of entire biomes [150]. In this section, we take a brief look at what is meant by land-use change and land degradation, their extent today, and how renewable energy production relates to land use.

Land-use change measures the conversion of land from one state to another. Land-use change can happen via direct or indirect human activity. Clearing a forest for biofuel crops, field or transforming an existing cropland into a bioenergy plantation, are examples of the direct land-use change. Direct land-use change can be observed and measured, for example, from satellite images.

Indirect land-use change happens when, for example, agricultural land in one location is used for biofuel production, but the demand for food and livestock feed remains the same. Maintaining supply to meet these demands leads to higher crop prices, and often results in clearing new areas for food production elsewhere. In the end, the amount of agricultural land remains the same in one place, but creates change in land use in another place. This phenomenon is called "leakage", and, depending on the circumstances, can lead to an increase in net greenhouse gas emissions from land use, and loss of biodiversity. For example, replacing carbon-rich environments, such as forests, permanent grasslands or wetlands, with agricultural land increases net greenhouse gas emissions [69]. Often, carbon-rich environments also host high biodiversity [203]. Usually, direct land-use change is easier to observe than indirect change—indirect change is driven by international markets, and may vary depending on socio-economic contexts in all affected region, which may be far apart. Indirect land-use change is most often measured through modelling. Note though that changes in ecosystems are usually very complex, and therefore long-term studies are needed for modelling purpose. Moreover, invisible changes (e.g., soil microbiota) are more difficult to study compared to visible ones (e.g., on vegetation, insects or mammals). Also, land-use change is not necessarily negative, improving the status of degraded land can result in positive land-use change, and this should be our target direction.

Land degradation is a result of unsustainable land-use practices, and a major problem across the globe. It is defined (after [101]) as a human-induced process that results in loss of biodiversity, weakens ecosystem functions or ecosystem services such as productivity of the soil, carbon sequestration, nutrient recycling or water regulation (see Fig. 2.14), and thereby also creating economic damage. Land degradation and related adverse effects like loss of biodiversity typically occur gradually, over decades rather than years [57], and this can reduce the sense of urgency to act. As we discussed earlier, we humans seem to be better equipped to react to sudden changes than to gradual ones.

Degraded land is prone to soil erosion, where the removal of vegetation allows water or wind to erode the productive, nutrient-rich topsoil. Desertification is an extreme form of land degradation in dry areas. Climate change worsens land degradation and aggravates biodiversity loss, while both degraded land and biodiversity loss exacerbate climate change. For example, deforestation and fires are pushing the

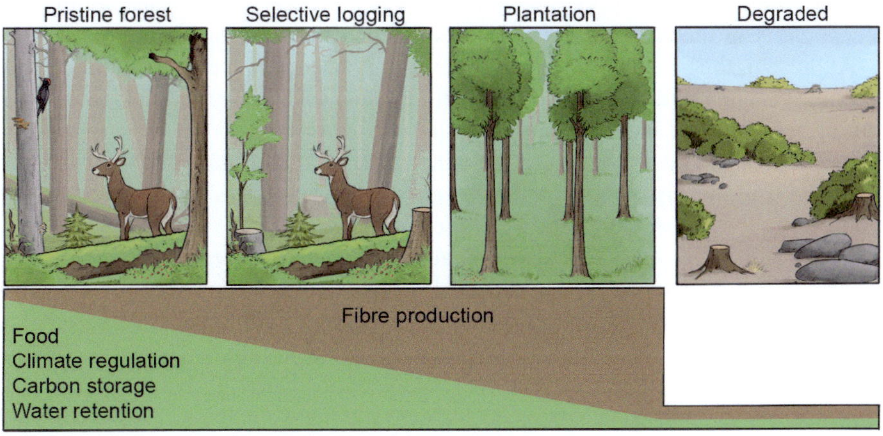

Fig. 2.14 Stages of land degradation © Maija Karala

Amazon rainforest towards a “tipping point”, after which the existing forest community may lose the ability to restore itself, and start turning into a sparsely forested savannah (see also Sect. 2.2, and [25]).

It has been estimated that by year 2050, land degradation and climate change together will reduce crop yields by an average of 10% globally, and up to 50% in the most vulnerable regions, such as drylands [101]. Estimates of total degraded area varies between 10 and 60 million km^2 [70, 75], with another estimated 27 million km^2 for marginal land (including drylands) [3]. Restoration (returning an ecosystem to its natural condition) and rehabilitation (when restoration is not possible, returning an ecosystem into a self-sustaining state as close to the natural state as possible) are needed to reverse land degradation. When full restoration is not possible, even partial restoration or rehabilitation of degraded land areas has many positive impacts on climate, well-being, and society, including increased food, energy crop and material production, carbon storage, and reduced greenhouse-gas emissions [101].

Land use is also a matter of social justice. Increasing demand for resources, including land, raises issues of land-use rights and the threat of increased conflict over resources. The impact is particularly severe for those already in vulnerable positions: people who are directly dependent on natural resources and have little ability to influence the distribution of land [57]. Along with increased pressures for using land for human needs, there is also a growing need to set aside land for nature conservation and protection of natural carbon stores. The world’s governments have agreed to protect 30% of the planet’s land and oceans by 2030 [32]. Energy development and land use are linked to almost all UN Sustainable Development Goals, giving an opportunity to promote sustainable living while simultaneously addressing sustainable land use and energy development [69]. The transition to renewable energy should be carried out in a responsible and just manner, balancing multiple demands for the same land resources, and accounting for the needs of many different social groups, and present and future generations. While there is huge potential for technological innovations in every sector, we require a serious assessment of humanity’s overall levels of consumption and waste production. The social sustainability and acceptability of renewable energy projects will be discussed further in Sect. 2.5.

At present, the production of renewable energy has a relatively small direct land-use impact when compared with many other uses, such as agriculture, forestry or mining; demand for land for energy systems is additional to the already existing uses, and this land-use pressure is growing [57]. Increased human use of land, in particular for bioenergy, could increase ecological costs such as soil degradation and biodiversity loss, even with improved efficiency [123].

Renewable energy projects also change the physical landscape, and can lead to concerns about how much land is occupied by these technologies in contrast to other uses, such as food production and housing [69]. The visual impact can be great in scenic or rural areas. Such negative impacts can create conflicts with local communities, particularly if they feel that their interests or values are being disregarded. This can lead to opposition to projects, delays, and legal challenges. An example of a collaborative platform to solve such problems is presented in Chap. 5, Case study 5.3: Wind power meets reindeer husbandry in northern Finland. However, renewable energy can be combined with other uses of land, such as pastures [152], [132] or vegetable production (i.e., agrivoltaic farming), and agroforestry can provide options for combined food and bioenergy production [201]. Solar power can also easily be installed in urban settlements and attached to buildings [95, 140]. Siting decisions of renewable energy power plants should favor degraded lands, renewable energy development can be used to rehabilitate the land if the construction is coupled with actions to increase plant cover to enhance and protect the soil [87, 90, 95].

Today, energy production accounts for 2% of total land use globally [69]. Renewable energy has many benefits over fossil fuels, but the rapid development and expansion of these technologies will increase their direct and indirect land footprints, which can lead to increased land-use conflicts [69]. Compared with fossil fuel acquisition, renewable energy is not constrained by abundance of or access to a resource, but by availability of suitable land areas for renewable energy development. However, availability of critical materials is a challenge, and building renewable energy power plants and infrastructure, and mining of critical materials can alter, destroy, or fragment natural environments on land, at sea, and in inland waterbodies, posing a threat

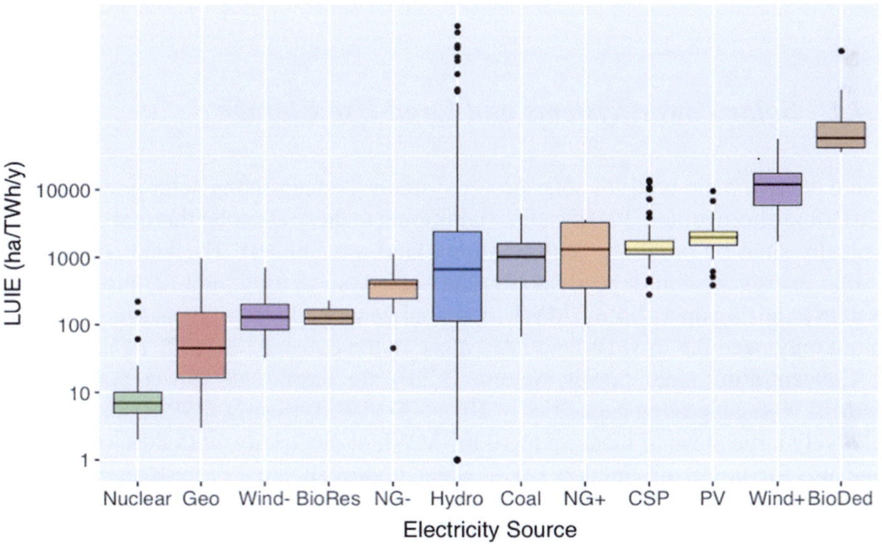

Fig. 2.15 Land use intensity of electricity production (LUIE), 100 ha/TWh corresponds to 1 km^2/TWh and 1 m^2/MWh [140]. Boxes represent the inter-quartile range with the median as the middle bar. Whiskers extend to the highest or lowest data point that is within 1.5 times the inter-quartile range, points outside this range represent outliers. Electricity sources: nuclear energy (Nuclear), geothermal energy (Geo), wind energy with footprint only (Wind−), natural gas footprint only (NG−) and including spacing (NG +), hydroelectric power for single purpose dams (Hydro), coal (Coal), concentrating solar power (CSP), ground-mounted photovoltaic solar energy (PV), wind energy with footprint (Wind-) and spacing (Wind +), and residual biomass (BioRes) and dedicated biomass (BioDed)

to species living in those habitats (we discuss the impacts on nature of the various energy technologies in their respective chapters in this book).

The land-use impact of different energy systems varies between energy sources, and the total land-use impact per amount of energy produced (called land use intensity) depends also on the location and local conditions, such as availability of wind and solar irradiance, see Fig. 2.15 for one example of an analysis on this.

Fossil fuel production typically has larger land-use footprints, with impacts on soil, water, ecosystems and their functions, than renewable energy systems or nuclear power (Fig. 2.15). The direct land footprint of wind and solar power (photovoltaic) production is small, but the entire power plant can spread over large areas—the indirectly impacted area can be significantly larger. In particular, in areas of concentrated wind power production, the "free" space between the turbines can be vast, but the area has suffered a major change, and this will have impacts on the ecosystem and species. In the case of biomass, removing all residues from the forest removes carbon and nutrients that would otherwise be returned to the soil and used by growing plants.

Next, we will briefly examine the main impacts on land use of solar power, wind power, biomass and hydropower. We will discuss the impacts on nature of these technologies in more detail in their respective chapters.

2.4.1 Solar Power Systems and Land Use Change

There are two main types of solar power systems: concentrating solar power system (CSP) and photovoltaic (PV) systems (additional to these are solar thermal collectors, typically sited on buildings with no direct land-use impact). The land footprint of a solar energy system depends on (1) the solar system used, and (2) insolation. A system requiring up to 1.5 m^2/MWh of annual generated electricity at high latitudes, might only need 0.5 m^2/MWh situated close to the equator [69, 137, 147].

Concentrating solar power systems (CSP) are large-scale power plants using mirrors, with generating capacities of 10 to 500 MW. Typically, CSP systems require relatively large areas of land, up to 10 m^2/MWh of electricity, with 80% of the total land use for power production [217]. Solar photovoltaic (PV) systems vary from small (a single panel generating around 50 W) to large-scale utility systems with capacities up to the GW level. Direct land-use impact of PV varies greatly on siting, but in general the impact of PV systems on land use is quite marginal, varying from 0 (roof- and wall-mounted systems) to 1.5 m^2/MWh of electricity [87, 172]. For ground-mounted panels 60% of the total land use footprint is due to the installation itself, the rest of the land-use impact is linked to the energy use and acquisition of the various elements, and critical minerals and metals [217]. PV systems can be easily integrated with buildings and other urban structures, such as highway noise barriers, and in these cases the direct land impact is close to zero [69, 217].

In recent decades, construction of centralized, large-scale, ground-mounted solar power plants has increased, and consequently the land-use pressure and impacts on nature have grown. Compared with wind-power plants, the impacts of solar-power plants on nature are not as well studied. The power plant area, including the arrays of panels, access roads, substation and powerlines, is typically heavily modified. This changes the land cover and can cause habitat loss and fragmentation (including soil disturbance and degradation of aquatic ecosystems), affect abiotic conditions resulting in changes in seed germination and species composition under solar panels, increases mortality in some local populations, creates barriers for animal movement, and may enable introductions of invasive alien species [73, 81, 90, 128, 161, 212, 230]. The exact impacts on nature are site-specific and depend on the impacted total area and the status of the site's natural environment prior to construction. [127] present a systematic map of the available knowledge on the effects of photovoltaic and solar thermal installations on terrestrial and semi-aquatic species and natural/semi-natural habitats and ecosystems in order to support decision-making.

Construction of solar-power plants can also result in *land displacement*, in which one form of land use is replaced by another, leading to the occupation of new land elsewhere, sometimes far away, to continue the original activity. For example, [91] estimated that in California, US, about 30% of solar power development occurs on croplands and pastures. This can result in clearing of new agricultural land elsewhere, and opening up previously remote areas to access for other human uses. This is an example of indirect land use change.

PV systems can also be co-located with other kinds of land uses, increasing the overall land use efficiency, such as so-called agrivoltaic systems that co-locate PV with agriculture [47, 69]. Degraded lands, such as landfills and abandoned open mines, can be rehabilitated and used for solar energy production. In the United States alone nearly 20 000 km^2 of degraded land could be used for renewable energy production to generate 114 TWh of electricity (3% of the US total in 2016) with no new direct land intake [166]. Concentrating solar power systems are discussed further in Chap. 3, and photovoltaic systems in Chap. 4.

2.4.2 Wind Power and Land-Use Change

Wind-power plants typically have a relatively small direct impact on land- and seascape use in electricity production, in the order of 1 m^2/MWh (100 ha/TWh), depending on the wind conditions and topography [69]. The directly impacted area is found to be equivalent for both on- and offshore wind power plants [217]. Most of the area occupied by a wind power plant is not directly used or changed by the turbines, roads, power lines, power station and other structures, leaving this space available for other uses like grazing or cropping [132, 152]. However, land use can be restricted due to noise disturbance [116], and both on- and offshore wind power developments have impacts that can conflict with nature [73, 231, 239].

Life-cycle assessments often account for the directly-impacted area only [217] and this may result in underestimates for the indirect land-use impacts of wind power. For instance, Denholm [43] has proposed distinguishing between direct, indirect, and total impact areas. For example, the indirect impact area of a wind-power plant depends to a large extent on the location, but the total impacted area can be six times greater than for ground-mounted photovoltaic panels [81].

The most documented indirect impact of wind power on wildlife is the mortality of birds and bats from collisions with turbines or power transmission lines [48, 88]. Many species–breeding and migrating birds, mammals and fish–have been shown to change their behavior or avoid locations near wind turbines, though the responses are often case dependent [19, 136, 138, 146, 149, 198].

Clearance of forests for wind production can be particularly detrimental to forest species. Previous research has shown that tree clearance creates fragmentation of the pre-existing continuous habitat and can lead to an "edge effect", creating disruption for some forest species for hundreds of meters in all directions [85, 151]. In offshore locations, the wind-wake effect can result in impacts on marine life over tens of kilometers away from the turbines [39, 65, 180]. These concerns will be discussed in more detail in Chap. 5.

In general, proper siting is key to avoiding and minimizing negative direct and indirect land- and sea-use impacts [116, 124, 169]. Sometimes this results in lower capacity factors and, therefore, an increase in the total land footprint needed, but balancing energy production and land-use impacts may nonetheless give the best possible total outcome when considering all factors, from societal impacts and impacts on nature to efficacy of energy production. Land use efficiency can also be improved with co-locating different uses in the same area where this is possible [69], like pastures. At sea, for example, offshore wind-power plants have been used to restore degraded oyster beds in the North Sea [112].

2.4.3 Bioenergy and Land Use Change

The land-use impact of bioenergy from dedicated feedstocks is the largest of all energy sources, renewable or otherwise (Table 2.4), although the total impacted area depends on the type of biomass used, the specific methods of cultivation and harvesting, and the area where the biomass crop has been planted. Most of the global variation comes from differences in growth location. Impacts of bioenergy on nature will be discussed in more detail in Chap. 6.

The demand for bioenergy, including wood pellets, is expected to increase due to high energy prices and policies to phase out fossil fuels [57, 213]. Modern bioenergy production is more efficient in terms of land use than traditional biomass, which often has low end-use efficiency and can lead to overharvesting of resources [126]—an estimated 30% of the wood fuel harvests in tropical regions are unsustainable [15]. Unsustainable use of land can create a negative cycle: the land becomes further degraded, and new areas need to be cleared to produce food, feed and fuels [69].

Availability of arable land is one of the most difficult challenges in sustaining the growing human population. Efficient plant production is essential for maintaining the increasing need for food, animal feed, raw materials (such as wood and fiber), and energy. There are approximately 400 000 vascular plant species known to science, of

Table 2.4 Land use intensity of different biomass fuels, on a heating value basis

Product	m^2/MWh	ha/TWh	Source
Sugarcane (residue)	0.1	10	a
Cellulose (residue)	0.1	10	a
Fossil oil	0.4	40	a
Sugarcane	75–82	7500–8200	b
Wheat	130–141	13,000–14,100	b
Corn (maize)	220–237	22,000–23,700	a
Sugarcane (juice-based)	239–274	23,900–27,400	a
Reed canary grass	255–837	25,500–83,700	b
Soybean	296–479	29,600–47,900	a
Rape seed	333–360	33,300–36,000	b
Cellulose (short rotation)	410–565	41,000–56,500	a

Sources **a** Fritche et al. [69], **b** Hakala et al. [84]—data taken from the source b includes land use for spacing and upstream life-cycles (mining etc.). Here m^2/MWh = km^2/TWh = 100 ha/TWh

which around 31 000 species have documented uses. Roughly 5500 plant species are known to be used for human food, while more than 11 000 plant species are used for material production, and over 1600 for bioenergy production. The basis of human nutrition is built on surprisingly few plant species, and just three main crops—rice, wheat, and maize—produce over 50% of our calorie needs. Plant sources for energy use can be categorized as woody biomass, food crops, and perennial energy crops (see more on this categorization in Chap. 6). Use of a wider selection of crops, and local varieties and breeds, for agriculture and bioenergy helps maintain biodiversity, including genetic diversity, and build the resilience of agricultural systems against potential threats such as pathogens, pests, drought, and other extreme climatic conditions. The development of agricultural practices such as agro-ecological cropping systems can further facilitate sustainable land use, enhancing biodiversity, carbon sequestration, and restoration of soil fertility.

There is increasing competition over cropland as a limiting resource. The land is needed to grow crops either for food, fiber or bioenergy; meanwhile, preventing the conversion of natural ecosystems into cultivated land. There is also an increasing need for restoration of natural ecosystems for climate change mitigation and preventing further biodiversity loss. Based on a review looking into competition over limited resources, such as land, water, labor, and capital, [162] suggest that, as a general strategy, the direct use of cropland for human food should be prioritized over feed for animals or bioenergy. Direct consumption of crops by humans is more resource-efficient than the consumption of animal-source food produced using crops as feed [72].

There are also cases where biomass production can provide positive impacts when compared to the previous state of the land. For example, cultivation of some plants, such as willow (*Salix*) or giant reed (*Arundo donax*), can be used as vegetation filters that capture nutrients in runoff from farmlands [63, 78], and permanent bioenergy crops can help prevent soil erosion, reduce evaporation, stabilize sediment, and enhance water infiltration into the soil [69]. Some biomass feedstocks can also be cultivated on land in danger of becoming degraded, and degraded land can be rehabilitated for crop production. Moreover, not all biomass has a direct land-use impact: the land footprint of biomass residues and waste is close to zero, though there are also competing ways of using them (e.g. as soil fertilizers or livestock feed, or other industrial inputs). However, since this technology relies on combustion, it will generate emissions of CO_2, particulate matter, and nitrogen oxides. These issues and other impacts of bioenergy will be discussed in more detail in Chap. 6.

2.4.4 Hydropower and Land Use Change

Hydropower (technical aspects are discussed under Chap. 7 on energy storage) has usually large total land-use impacts, and in some cases, totally changes the local landscape as flooding excludes all other uses. The impacts vary from place to place, but hydropower can have significant effects on nature and wildlife by destroying habitats, and impacting the quantity and quality of waterbodies; dams introduce migration barriers for fish [69, 79, 238]. Many new hydropower projects underestimate their direct and indirect impacts on biodiversity [234]. The indirect impacts result from changes in hydrology of catchments, microclimate, and extinctions of numerous species decades after the construction due to the disrupted interactions among species and lower population sizes (extinction debts, [111]). For example, a study in South America revealed widespread species extinctions across plants, vertebrates and invertebrates in the area of fragmented by a hydroelectric reservoir [175]. Particularly in tropical countries anaerobic digestion of submerged vegetation can produce leakages of methane gas in amounts equivalent to the greenhouse gas emissions from using natural gas [42, 67].

The total land footprint varies greatly from place to place, from 2.3 m^2/MWh of electricity (230 ha/TWh) in Sysenvatnet in Norway, to 156 m^2/MWh (15 600 ha/TWh) in Churchill Falls in Canada [69], and even up to 100–300 m^2/MWh (10 000–30 000 ha/TWh) in Argentina, Brazil and Venezuela [20]. Hydropower can also have large impacts on local communities. Globally at least 100 million people have been relocated due to hydro-power development, and the subsequent resettlement of people impacts land use elsewhere. In some cases, relocation has also resulted in social disruption of, in particular, indigenous communities, whose social structures and livelihoods can rarely be restored [196]. The great environmental and social impacts resulting from hydropower mean that the opportunities to construct much more hydropower are limited. Nonetheless, the existing power plants form an important mean to produce and store electricity.

2.4.5 Conclusions

Land-use change, driven by humanity's increasing demand for land and resources, provides livelihoods and economic benefits, while also causing direct and indirect environmental and socio-economic damage. Land degradation, one of the drivers of biodiversity loss, also increases carbon emissions, threatens food security, and creates conflicts and competition over land. The land-use impact of renewable energy production is still relatively small compared to many other land uses, but increased demand for cleaner energy means that this form of land use is growing.

Renewable energy production affects land use both directly and indirectly. Direct impacts are created by the power facility itself, power lines, maintenance buildings, and access roads. For example, ground-mounted solar PV plants cover large areas with panel arrays. Wind power plants, on the other hand, use relatively little space directly, with only about 10% of the power plant area occupied by the turbines, grid connection and maintenance roads but the entire power plant impacts a relatively large area with its potential effects on species and local communities. Bioenergy typically has large direct and indirect impacts on land use. The indirect effects can be difficult to measure as the change may occur far from the bioenergy production site—for example, when food production is moved elsewhere. Growth in renewable energy production can create habitat change and fragmentation, affecting species composition, behavior and ecosystem function.

In some cases, renewable energy projects, such as ground-mounted solar photovoltaic farms or bioenergy crops, can help to revegetate and rehabilitate degraded lands. Careful land-use planning in renewable energy projects is key to avoiding further land degradation and halting habitat loss and fragmentation. Renewable energy power plants should be developed and implemented with a broad ambition: beyond just climate mitigation targets and a transition to clean energy, towards solutions that also support ecological and social sustainability. Sustainable energy development and land use requires finding a fair balance between environmental and societal impacts, and this topic will be discussed next.

2.5 Societal Impacts of Renewable Energy

Social sustainability is one of the three fundamental pillars of sustainability; however, it is not as commonly discussed as environmental sustainability or the economic benefits of energy systems. Often, considerations of the sustainability of energy systems focus on greenhouse gas emissions, and how to reduce them to alleviate climate change. While there is a clear need to tackle climate change, decarbonize societies, and curb the emissions of energy systems, the implementation of a sustainable energy transition also requires that we equally take on board societal impacts and social sustainability, together with impacts on nature. One reason for neglecting social sustainability has been its wide definition and lack of consensus on how social sustainability should be measured [1, 194]. However, social sustainability in energy has recently gained more attention—for example, through emerging discussions on energy justice, equality, and local impacts of energy projects.

Societal and environmental impacts are inseparably linked to each other [189]. Evidence shows how changes in planetary climatic and ecological systems are inevitably changing our societies. For example, extreme weather conditions, loss of biodiversity, lack of water, and entire regions becoming unlivable due to drought, are all threatening basic human needs and leading to climate immigration. Furthermore, in the process of phasing-out fossil fuels, we need to consider local employment impacts, changes to local economies, and how installations of renewable energy technologies impact local societies, communities, and livelihoods that are often tightly linked to land use (see also Sect. 2.4). These are part of the justice perspective of energy systems, which include questions of who has the access to clean energy, and who can afford that energy.

This section sheds light on social sustainability to ensure just energy transitions, and introduces the kinds of societal impacts inherent to renewable energy systems. We also discuss the influence and power of major players, such as large energy companies, and how they shape the energy system through influencing policy and political lobbying, as well as the societal consequences of too-narrow decision-making. Finally, we look into societal perceptions of energy systems and their benefits for local communities, and how democratic participation is crucial for social acceptability, thus determining whether or not a given technology is successfully taken up. We'll also explain how a change in an energy system is also a socio-technological change, involving the participation of a large number of different actors. Energy engineers, being experts in their field, have a key role in shaping future energy policies and societies.

2.5.1 Social Sustainability of Energy Systems

Social sustainability incorporates a set of different aspects that influence the potential of a person or a society to achieve a decent quality of life. The Bruntland Commission's well-known definition of sustainability highlights human needs by stating that sustainability is "meeting the needs of the present without compromising the ability of future generations to meet their own needs" [26]. Humans use of fossil energy in ways that are directly tied to growing gross domestic product has on the one hand, increased the welfare of industrial economies, but on the other undermined future generations' ability to fulfil their basic needs. Overconsumption of natural resources has led to degradation of nature and its ecosystems, which produce and maintain many vital ecosystem services that we humans depend on [102].

In addition to basic needs such as access to clean water, nutrition, and safety, social sustainability includes well-being and the right to pursue good life, social participation, and gender equality. In addition, it encompasses health care, and the possibility of developing social capacity via learning, education, and work. In energy systems, social sustainability includes the right to access electricity and clean water, the ability to influence one's own energy production, consumption, and source of energy, the participation in energy decision-making and energy communities, and the possibility to build the skills and capacity required to fulfill the needs listed above. Social sustainability is thus tightly linked with democratic principles and the notion of energy justice, an enhanced social inclusion in transition agendas to ensure a fair distribution of benefits and prioritize previously marginalized social groups [220].

Energy systems are linked to all spheres of well-being, sustainable economies, energy-basis for food production, healthy ecosystems, cities and rural areas, and access to energy, which is one of the basic human rights. Agenda 2030 forms "a plan of action for people, planet and prosperity" with 17 Sustainable Development Goals (SDGs, see also Sect. 1.1) and six grand transformations that are necessary to reach a more sustainable future [216]—energy decarbonization and universal energy access is one of these six grand transformations.

One way to measure the impact of energy investments is to employ the broadly-used Social Impact Assessment (SIA) framework. In general, the SIA is "the process of identifying the future consequences of current or proposed actions, which are related to individuals, organizations and social macro-systems" [17]. Therefore, social impacts have influence on different levels (individual, community, national impacts), and also vary in time—for example, immediate impacts on people, intended long-term impacts for communities, and unintended consequences [232]. Several definitions of social impact exist. The widest definition [29] includes: "All social and cultural consequences to human populations of any public or private actions that alter the ways in which people live, work, play, relate to one another, organize to meet their needs, and generally cope as members of society." As can be seen from these examples, the definitions of social impacts are broadly in line with each other, but vary in scale, and which definition is the most useful depends on what impacts are assessed. Next, we turn towards (positive) social impacts in renewable energy systems.

The keyword for beneficial and positive social impact is *participation*. Participatory processes of social impact assessment support affected populations (particularly the most vulnerable social groups), proponents and regulatory agencies, enhance understanding of the changes taking place, as well as equally distribute the benefits and mitigate the negative impacts of change [58]. This applies to all energy planning projects.

Afshari et al [1] studied social sustainability indicators in the energy supply chain and summarized different social impacts in different phases of energy production (Fig. 2.16). Enhancing social sustainability is the responsibility of several different actors in the energy supply chain, from the responsible sourcing of materials, to the management of waste from renewable energy technologies and how it affects people in nearby communities, to the treatment of workers in each stage of the supply chain (safety, salaries, conditions).

An extensive literature review on social impacts of energy supply systems by [1] further categorized the above-mentioned indicators and phases into four themes: (1) human rights and social life, (2) occupational indicators, (3) business-related

Supply	Production	Distribution	Customers	Reverse logistics
• Fair sourcing • Avoid child labor • Relationship with suppliers • Sourcing from marginalized groups • Compliant with culture of sourcing origin • Investing in workplace improvement of suppliers	• Health and safety of employees • Equity in hiring • Fair wages • Commitment to the stakeholders needs • Average distance travelled by employees to the company • Displacing local communities • Minimize noise annoyance/expos ure • Safety of animals/birds/sp ecies	• Energy coverage • Logistics traceability • Network with local companies • Reliability of distribution system • Provision abroad for the crew	• Societal impact of technology • Social mixing, cohesion/ identity • Quality of life • Community • Engagement/ involvement • Complaints or feedback system • Community improvement • Local preference • Affordability	• Appropriate land use for landfill sites • End-of-life responsibility • Legislation for recycling • Waste consumption rate • Legal waste disposal • Legal demolition

Fig. 2.16 Social sustainability indicators in the energy supply chain [1]

indicators, and (4) legal, political, and government-related indicators. Social impacts of energy projects in the first theme of *human rights and social life* include the basic human rights of access to energy, reliability of energy provision, local concerns of noise, annoyance and safety near power production, and the consequences of displacing communities for energy-related projects.

The second theme on *occupational indicators* in renewable energy projects include basic work-related indicators such as a safe working environment—particularly relevant to some renewable energy fields, which carry specific risks such as working at height, exposure to chemicals, nuclear technologies, and offshore sites.

The third theme on *business-related indicators* concerns the wider societal impacts of company activities on local communities and stakeholders, and includes corporate social responsibility (CSR) measures. Business-related impacts in the energy sector range from sustainable sourcing of raw materials and related social impacts (e.g., no child labor, working environment in mines), to local impacts on livelihoods, to the responsibility to provide reliable energy and to manage the waste products of energy production [1].

The last set of social sustainability indicators in the energy sector is *legal, political, and government-related indicators*, which according to [1] includes mainly beneficial social impacts of regulating the amount of waste located near residential areas, support for innovation, and recycling decommissioned material from renewable energy technology.

It is important to note that, similar to impacts on nature, social impacts can be positive, negative or neutral, can affect different groups of people in different ways, and can also be in contradiction with other pillars of sustainability. In the following section, we examine the macro social impacts of energy systems, before exploring the benefits and challenges of renewable energy technologies.

2.5.2 *Energy Consumption's Contribution to Welfare*

The energy intensity of an economy describes how much energy is required for the economic activities of that country, often measured by Gross Domestic Product (GDP). Historical development figures show how energy consumption in western industrialized countries is tightly coupled with economic growth, which is then assumed to increase welfare. This is true to some extent—development can be seen to happen with the help of energy consumption—but it is also a consequence of industrialization. Furthermore, GDP is solely a measure of the amount of economic throughput, and does not take into account the quality of consumption, nor the quality of economic transactions. Therefore, using GDP as a signal for welfare is problematic.

Energy intensity varies between different regions and countries. Some differences can be explained by climate conditions—for example, the application of energy use differs from one region to another: colder climates require heating, while in warmer regions cooling becomes more important. A country in a hot climate needs a significant amount of power for air conditioning to keep room temperatures at healthy levels; in developing countries, however, resources are limited and living conditions are already unhealthy, or even lethal, due to extremely high temperatures. Infrastructure, logistic solutions and population density impacts the total energy use, and the share of different uses, in different countries. Often GDP is used to indicate welfare impact, however it is only designed to depict the value of products and services that flow through the economy of a country.

A well-known and more accurate key performance indicator for the measure of quality of life is the Human Development Index (HDI) proposed by the United Nations. The HDI is a summary measure of average achievement in essential dimensions of human development: living a long and healthy life, attaining knowledge and education, and having a decent standard of living [214].

The HDI has been used to depict energy consumption and its impact on quality of life in different stages of a country's development. At first, when countries start to develop their societies and economies, an increase in the per capita energy consumption increases the quality of life. However, this increase eventually slows down and stops, and additional energy consumption does not give any additional measurable benefits. On the contrary, in several countries average standard of living has reached a point where it is no longer sustainable. As an example, [236] discusses the impact of per capita energy consumption on life expectancy, one important term in the HDI. Below consumption per capita of 5 kW, increasing energy consumption increases life expectancy, but above this value, the trend saturates. Following Wolfson's [236] presentation, Figs. 2.17 and 2.18 show life expectancy and HDI as a function of per capita energy consumption, respectively. They also present the trend over time, from 1965 to 1991, then to 2018. HDI data for 1965 is not available so only the trend from 1991 to 2021 is shown.

The trends shown in Figs. 2.17 and 2.18 are extremely interesting. At the lower end of the per capita energy consumption-axis, below 30 MWh energy consumption, life

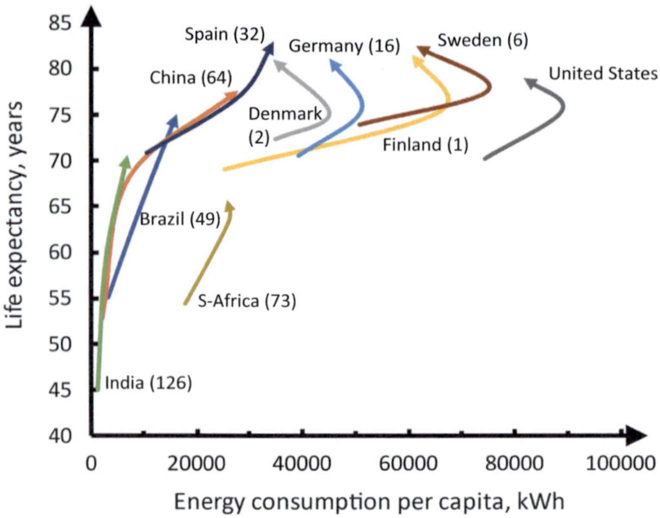

Fig. 2.17 Life expectancy as the function of per capita energy consumption [92]. Start of the arrow corresponds to year 1965, turning point 1991 and arrowhead to 2018, respectively. Numbers in the parentheses indicate countries' position in the World Happiness Report [89]

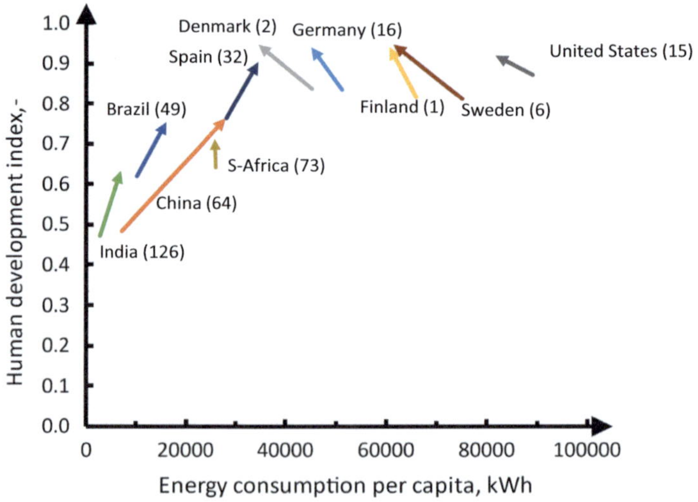

Fig. 2.18 Human development index, HDI, expectancy as the function of per capita energy consumption [89, 92]

expectancy and HDI both increase with energy consumption. For countries that have had energy consumption values above 30–35 MWh per capita, energy consumption had peaked already by 1991 and started to decrease—and yet, life expectancy still increased. Germany and Denmark reached lifespans of 80 + years with 50% less energy compared to the US, while Spain did the same with even less energy. Finland and Sweden, still quite high in their energy consumption, reach 80 + years life expectancy with 25% less energy than the US. For HDI, it also seems clear that countries seem to approach the asymptotic value of energy consumption per capita, and again Denmark reaches the same level with much less energy consumption than Finland, Sweden and the US.

This type of macro measure of social impact is important and useful for comparisons between countries. Macro indicators also give an impressive picture of how much energy consumption has impacted, and continues to impact, country-level welfare and economies. However, several of the social impacts of renewable energy projects can only be evaluated at a local level. Next, we will dive further into the societal benefits and challenges of renewable energy technologies.

2.5.3 Societal Benefits and Challenges of Renewable Energy Technologies

Over 90% of the proposed solutions to address climate change involve renewable energy technologies and energy savings in the form of electrification, energy efficiency, biofuels, and green hydrogen [98]. As we move towards a low-carbon and carbon-negative future, countries, states, regions, corporations, and individuals are adopting renewable energy for a range of different of reasons. In this section, we will discuss some of the benefits and challenges of this transition.

The abundance of renewable resources around the globe makes possible the development of energy systems based on those resources, assuming we properly address the ecological and social sustainability challenges related to their development, construction and operation. This transition is also driven by the increasing cost of obtaining fossil fuels, coupled with the environmental harms linked to their consumption. Using renewable energy resources can decrease pollutants and greenhouse gas emissions, and contribute to a healthier environment. However, different forms of renewable energy also have different types of sustainability impacts, as we have learned in the earlier sections of this chapter—these are further discussed in later chapters of this book. Nonetheless, renewable energy sources are locally available, and these technologies are an important way to decrease dependence on imported fuels. Furthermore, renewable energy sources can help to minimize the geopolitical risks associated with fossil fuels, from trade disputes, to political instability, to pricing wars, all of which are often rooted in access to oil. Renewable energy technologies also present an opportunity to decentralize the electricity system, and reduce the power imbalance between large energy companies and individual energy users ([4]; see also Sects. 2.5.4 and 2.5.5).

The development, implementation, and maintenance of renewable energy technologies (solar, wind, hydro, biofuels) employs 13.7 million people worldwide, of which approximately 60% are in Asia (China alone accounting for 42%), 10% in Europe, 10% Brazil, 7% United states, and 7% India [105, 106]. For comparison, IEA estimates that in 2019, fossil energy industries employed 32 million people [99]. In addition to the jobs directly created in the renewable energy industry, growth in clean energy can create positive economic ripple effects—for example, industries in the renewable energy supply chain will benefit directly, bringing indirect benefits as local businesses increase household and business income. Local governments also can benefit from clean energy, most often in the form of property and income taxes, and other payments from renewable energy project owners. However, these economic incentives are not in use in every state or country. Figure 2.19 shows the number of jobs generated through the green energy transition.

Growing solar and wind energy adoption leads to increased waste from panels and turbine blades (see case study 4.1 for a circular economy concept of the recovery of solar panel materials to batteries). Economic incentives and technology advancements drive panel replacement, but inadequate recycling and rising waste pose serious risks. Challenges persist in responsibly sourcing materials, risking environmental impact and human-rights violations [2]. However, recycling of waste, including wind turbine blades, from renewable energy technologies are being constantly improved and researched.

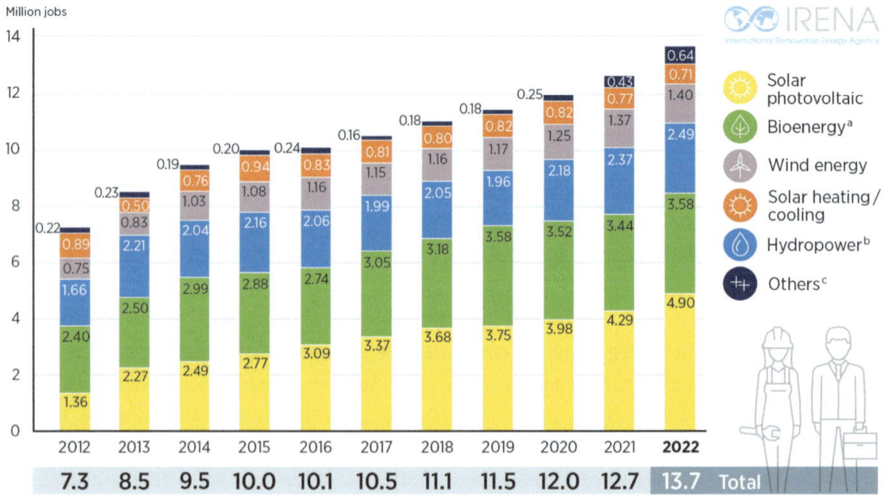

Fig. 2.19 Figure: Global renewable energy employment by technology 2012–22 [105]. [a]Includes liquid biofuels, solid biomass and biogas, [b]Direct jobs only, and [c]Others includes geothermal energy, concentrated solar power, heat pumps (ground based), municipal and industrial waste, and ocean energy

The power density of the technologies used to harness renewable energy sources is considerably lower than for fossil fuels [240]. Consequently, the transition to renewable energy can require significantly larger areas of land, depending on the energy source used [46], which in turn increases competition for land throughout the world. Land-use change in developing countries is often attributed to the spread of bioenergy [21], which has significantly larger land-use impact than other renewable energy technologies. In addition to the variety of negative effects of increasing land use on the environment, including increased water demand, indirect land-use change emissions, and biodiversity loss, it also has several societal impacts. Intense competition for land threatens alternative, less economically-valued but culturally-important use of the land. Land-use changes can have an impact on local livelihoods—that is, on the alternative ways of using the land, whether for local agriculture, forestry or herding—that not only affect community income, but also culture and cultural heritage that is often tied to the land, especially in indigenous cultures. While planning energy projects, the impacts on local livelihoods and culture should be considered, even though it would not produce the same amount of economic benefits as renewable energy production. (See also the previous Sect. 2.4 on land-use impacts of different renewable energy technologies.)

Overall, the transition to renewable energy can be considered a positive phenomenon, and a part of the solutions needed to reduce fossil energy use. The energy system is undergoing radical transformations in the production, distribution and consumption of energy. However, the reconfiguration of climate and energy policies is likely to produce winners and losers in society by distributing the benefits and costs disproportionately. There is evidence that energy transitions can create new injustices and vulnerabilities, while failing to address the existing drivers of injustice in energy markets and in society [205]. How can these negative injustices and vulnerabilities be alleviated? In the next two sections we focus on power disparity in the energy sector and how to increase social participation in energy projects.

2.5.4 Economic and Political Power in Energy Decision-Making

In recent years, there has been a growing recognition among academics, policymakers, and practitioners that the energy transition holds the potential for "sustained economic development, social inclusion, energy security, improved health, job creation, and other societal benefits" [218]. However, to realize these gains, not only must national governments and large industrial companies actively participate in the energy decision-making process, so also must entrepreneurs, municipalities, and wider civil-society actors. Studies show that as the inclusiveness of political processes increases, so does the inclusiveness of their outcomes, which address a wide array of different groups and have a better chance of being more just decisions [182]. It is important for new energy engineers to understand the responsibility and power of privileged groups such as large energy companies, engineers also have an opportunity to influence their organizations from within towards more diverse decision-making processes, engagement with local people affected by new energy projects, and, for example, enhanced safety and user experience in the design processes of specific technologies.

As we observe in countries around the world, political and economic power in the energy sector is often concentrated in the hands of a small group of the most powerful actors, such as large, multinational energy companies and policymakers, and, in some countries, military organizations. National governments control energy policymaking, including the setting of targets and the pace of decarbonization, changes in the energy mix, granting permits for renewable energy investments, reducing subsidies for fossil energy, and licensing electricity distribution. Large energy companies owe their strong position to their amassed financial capital—which they can allocate to lobbying activities—and also to their strong market position and political connections to influential actors [12].

On the one hand, large energy companies are accelerating energy transition and renewable energy technology implementation through divestments to fossil energy, alongside the creation of new business focused on renewable energy technologies. Divestments are the opposite of investments: a strategic decision by a company to let go of assets, values or resources that have previously been important but are not expected to bring benefits in the future—for example, fossil fuels [62]. In this way, representatives of large energy companies can exert a significant impact on the formation of current and future energy pathways in a given region or country.

On the other hand, however, it has been found that countries with close connections between large established firms and governments are less keen on introducing novel technologies [181]. Such countries are more likely to maintain the status quo, for example by keeping fossil-based energy sources, often subsidized by the governments in the form of lower taxes or import support, rather than introducing renewable energy solutions [186, 185].

Thus, while the concentration of political and economic power in the hands of a few actors is to some extent beneficial, enabling an effective coordination of an extremely complex socio-technical energy system, it can nevertheless have unintended consequences that we reflect upon in the remainder of this section. In the following, we outline the most important problems that could arise from such narrow decision making in the context of the energy transition.

First and foremost, the delegation of control to a small group of actors leads to a drastic reduction in the diversity of perspectives from the various social groups that should be considered in decision-making processes. These decisions might concern, for example, the choice of site for the construction of a new large-scale photovoltaic or wind-power plant. This way, energy transitions may contribute to further deepening social inequalities, where privileged actors multiply their wealth while disadvantaged social groups have limited access to the benefits of renewable energy projects. Social groups which already suffer from political and economic marginalization, such as low-income householders, people with disabilities, or ethnic minorities, are at particular risk of being excluded from discussions that may directly affect their well-being

[120]. Notably, about a quarter of the global land area is owned, managed, used, or occupied by indigenous people that are often directly dependent on the land, and are therefore the first ones to suffer from unsustainable land-use practices [102]. The lack of consultation with local communities when planning investments may ultimately lead to widespread public protests and the blocking of renewable energy projects by citizens [170]. This could happen when citizens are convinced that their interests and values are threatened by investments planned over their heads, the profits from which will be taken out of their local communities [135].

Secondly, limiting the debate on the energy transition to the participation of only the most powerful actors could lead to a serious reduction in the transparency of the process and accountability of the decisions taken. In the most extreme cases, decision-makers can be corrupted by investors, which has a profound adverse impact on both the state of the economy and democracy itself, diminishing checks and balances, and public trust in politicians and state institutions. Such events may result in drastic decline of social acceptability of renewable energy projects and corrode the social capital necessary for a just energy transition [71].

Thirdly, limiting the diversity of voices in the debate may have a negative impact on the final decisions regarding both the targets for the energy transition, and the measures needed to achieve those goals. For example, the exclusion of academics or non-governmental organizations specializing in environmental or climate issues could result in a lack of sufficient scientific data on the negative environmental effects of planned investments in renewable energy, such as wind power plants. Moreover, crucially, the current neoliberal capitalist system—bluntly described by Lovering and Turkmen as "bulldozer neoliberalism" [141]—prioritizes short-term financial gains over long-term environmental, social, and economic sustainability [237]. Therefore, public scrutiny is necessary to ensure that renewable energy projects meet the criteria for truly sustainable development.

Finally, as already stated, delegating the responsibility for carrying out the energy transition to a small group of powerful actors may lead to a slower pace of decarbonization and departure from established practices in the energy system. This is because these dominant actors, who are firmly embedded in the current energy system and derive significant benefits from it, may not be willing to take excessive risks or abandon solutions that are beneficial to them [7]. In such cases, international arrangements are crucial—for example, the Paris Agreement, a legally binding international treaty on climate change at the United Nations level, or European Union (EU) measures such as the European Green Deal, which aims to achieve climate neutrality in the EU by 2050. Pressure from civil society, such as the climate movement Fridays for Future initiated by Greta Thunberg, is also crucially important to draw attention to insufficient action, and demand more ambitious measures from world politicians to prevent climate catastrophe [94].

Summing up, the concentration of political and economic power in the energy sector entails wide-ranging consequences for our societies, the environment, and the energy transition process itself. Narrow decision-making processes, dominated by actors in central positions in the energy sectors, limit the participation of diverse stakeholder groups, foster social inequalities, reduce transparency and accountability, may negatively affect the quality of decision-making, and can even jeopardize the state of our democracies or the timely achievement of climate goals, exposing communities globally to the dire consequences of the climate crisis. To counteract these threats, it is essential to conduct a broadly inclusive dialogue, focused on the perspectives and interests of various social groups whose wellbeing depends on the planned projects (see also Case Study 5.3 on the development of a collaborative platform between reindeer farmers and wind power operators in Northern Finland). We must also take into account future generations and planetary boundaries, by adopting an approach geared to achieving the long-term objectives of sustainable development, fairly sharing the benefits and costs of the energy transition among all affected parties.

2.5.5 Social Engagement of Citizens in Energy Systems

The development of distributed renewable energy systems has the revolutionary potential to change power relations within the energy sector, enabling individuals and communities to actively participate in the generation, storage, consumption and exchange of energy. Concepts such as energy prosumers, energy citizens, and energy communities are increasingly appearing in public discussions, and are a part of the ongoing processes of the decentralization and democratization of energy systems. When the flow of energy ceases to run exclusively from producers to consumers, and the number of prosumers (energy self-producer-users) increases dramatically, the role of citizens fundamentally changes from passive consumers, who are only expected to pay energy bills on time, to active and knowledgeable participants in the system [22]. According to a recent Eurostat study, households represent 27% of final energy consumption in the EU [60]. Considering this data, it is clear that citizens must be involved in the transition process by changing their behaviors to become more energy-efficient (for example, buying low-energy light bulbs, reducing air travel, or switching to electric transport), or investing in renewable energy and thermal insulation of their houses.

Importantly, for citizens these new roles also bring new responsibilities and expectations, and it is therefore worth considering what kind of support is needed and what new challenges this increased participation may bring. In that respect, the President of the European Commission, Ursula von der Leyen, declared 2021: "People are at the core of the European Green Deal, our vision to make Europe climate-neutral by 2050. The transformation ahead of us is unprecedented. And it will only work if it is just—and if it works for all. We will support our people and our regions that need to make bigger efforts in this transformation, to make sure that we leave no one behind" (quoted in European [59]). Below, we briefly outline key concepts related to the growing involvement of citizens in the ongoing energy transition processes.

Renewable energy technologies offer citizens new opportunities to engage in the energy transition as 'energy citizens' who play a crucial role in driving the transition towards decentralized and democratically governed energy systems [110]. Citizens can become 'energy prosumers'—that is, individuals or collectives who act simultaneously as both producers and consumers of energy— and this may take various forms, from installing solar panels on a single household rooftop, to community activities such as energy cooperatives aimed at producing renewable energy jointly with other citizens. Moreover, prosumers can contribute to the energy system by storing excess energy or providing demand flexibility—because of this, the term "flexumers" is also gaining popularity as an expanded version of prosumers [64]. According to some estimates, if legislative and technical barriers are removed, every second EU citizen could produce their own energy by 2050, covering a total of 45% of the energy needs of all EU countries [113]. Hence, the EU has introduced legislation aimed at both supporting individual prosumers and enabling the development of so-called energy communities. This kind of decentralized energy system will also create new job opportunities, and roles for energy engineering experts and designers, by widening the scope of how energy production is perceived.

Energy communities refer to a wide range of initiatives owned and managed by local stakeholders engaged in the collective production, consumption, distribution, storage, or sale of renewable energy [229]. According to the International Renewable Energy Agency, such initiatives should aim at generating economic, social, or environmental benefits to the community in addition to financial profits [104]. A range of scholars have observed that energy communities might significantly contribute to reaching European climate and energy targets, forming a cornerstone for renewable energy system development (e.g. [30, 183]).

In practice, the forms that community energy initiatives take, and their main purpose, vary depending on the context in which they operate. For example, energy communities can differ based on their range of stakeholders and level of citizen engagement, technological advancement, rules for decision-making, or profits distribution [93]. They can develop around novel solutions related to, for example, peer-to-peer energy trading [130], raising energy efficiency [139], or co-investment in wind power plants [148]. Furthermore, energy communities can address problematic social issues such as energy poverty [156] or raise public awareness of sustainability transitions [192]. However, energy communities are not a "silver-bullet" solution for all, and have their own challenges, especially in urban contexts due to city space limitations and tensions between different actors [8].

Summing up, energy prosumption could be considered a countermovement to the centralized power of the mainstream, fossil fuel-based systems maintained by large, powerful energy companies, allowing the creation of more democratic and transparent energy systems [235]. Prosumer initiatives may accelerate the transition to renewable sources, foster a sense of citizen empowerment and community, allow citizens to profit financially from renewable energy projects, as well as increase public support for the energy transition. Thus, energy prosumption can bring environmental, social, and economic benefits to individuals and society at large [50]. However, these benefits depend on the specifics of the project and government policy. Prosumer projects can encounter challenges due to high costs, regulatory constraints, lack of volunteers, and the need for diverse expertise. Appropriate policies and support structures can help to overcome these barriers [204]. All in all, to meet climate goals on time, everyone is needed both traditional players in energy sectors, as well as hitherto passive actors, such as households, for whom new opportunities for engagement to renewable energy technologies are only now opening up.

2.5.6 Social Participation as a Key to a Successful Energy Project

As noted in Sects. 2.5.2, participation is one of the best ways to increase social sustainability, and enhance a more just energy system and benefit allocation in renewable energy projects. A good example of early and dynamic development of citizen energy took place in Germany in 2000 with the introduction of the Renewable Energy Sources Act (EEG) to support renewable energy production (primarily through feed-in tariffs) by customers or all sizes, which led to an increase in distributed and bottom-up produced energy [125]. In 2016, energy citizens (individual and collective) owned 42% (42 GW) of total installed renewable energy capacity for power generation in Germany (excluding pumped storage power plants, offshore wind turbines, geothermal or biological waste), including 50% of total installed solar voltaic energy. At the same time, Germany's citizen electricity sales amounted to 79 TWh in 2016, which was more than the electricity sales of the largest Portuguese (EDP) or Norwegian (Statkraft) energy retailers in 2015, at 65 TWh and 56 TWh respectively [22]. Table 2.5 summarizes the benefits of involving the public in the process of planning and implementing sustainable energy solutions.

There are several models for carrying out the participation process when planning renewable energy projects (see also Case Study 5.3 on the establishment of a collaborative platform between reindeer farmers and wind power operators in Northern Finland). Based on an extensive literature review on energy democracy and energy community research, [227] mapped the most common types of citizen participation, and how they distribute across different groups of people and different decision-making levels (Fig. 2.20).

Several studies have shown the benefits of participation in increasing social acceptability [199]. A few studies of the psychological factors of social acceptance of technologies have been carried out. One such approach is the Technology Acceptance Framework, which suggests that people's acceptance is shaped by their beliefs about the technology, perceived personal and social norms, affective responses, and other contextual factors such as social trust [197]. Evidently, acceptance of technology is more complex than mere questions of technological improvement or economic compensation offered to people.

Locations of renewable energy installations, their impacts on the landscape, and their proximity to human settlements, all have an impact on the acceptability of renewable energy. [202] found that in the United States, most citizens support the decarbonization of the energy sector, especially if wind and solar facilities are located at least eight kilometers from their home. There is also a strong preference for

Table 2.5 Economic, political, environmental and health, and social inclusion benefits of citizen engagement in renewable energy production

Economic benefits: Increased energy security; encouraging household investments; pooling information and resources; knowledge exchange; innovation; new business models; jobs; local development; more affordable prices; promoting local ownership; financial benefits from project dividends or community investment funds	Political benefits: Social acceptance for the transition; more accountability; better understanding of bottle-necks and possible solutions; shared decision-making = sense of ownership and responsibility; increasing awareness, knowledge and public acceptance for the transition
Environmental and health benefits: Cleaner air and environment; lower greenhouse gas emissions; increased environmental stewardship; improved health from reduction of emissions; reduced annoyance of noise; improving security and trust	Social inclusion benefits: Addressing energy poverty; building social capital, trust and cooperation; increasing equality and empowering marginalized groups; increasing social cohesion; educational opportunities and raising awareness about clean energy technologies; ensuring the preservation of cultural traditions; supporting energy justice with its three fundamental forms: redistribution, recognition and representation (procedural justice), corresponding to the economic, cultural and political realms of social justice [66]

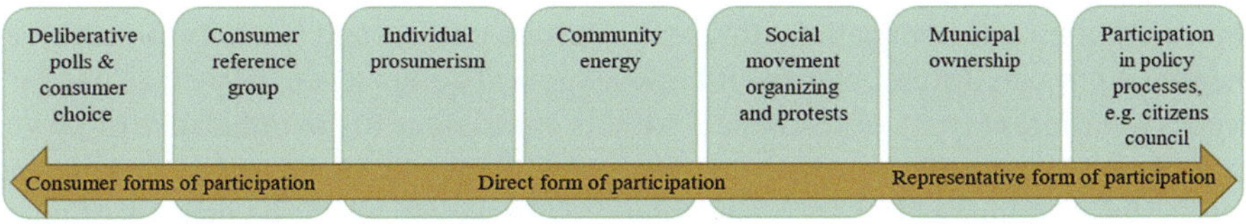

Fig. 2.20 Different forms of social participation based on societal levels [227]

energy production from a mix of renewable sources. Even though the study found significant support for these technologies, some resistance was seen at the local level, and the researchers emphasized the need to analyze the role of socio-demographic characteristics around acceptance of renewable energy [202].

Reasons why people resist technology projects varies, and goes beyond mere technical and economic challenges. Lack of knowledge about the technology and the impacts of the project can also worsen negative attitudes, and discourage stakeholders from participation in discussions and decisions about future energy technologies [202]. The single most important factor for acceptance or resistance is trust. Research shows that, in some cases, a decrease in acceptance when the level of public involvement is high stems from lack of trust. As a consequence, the levels of perceived justice decreases: people can feel that there is a lack of equality in decision-making processes, or that the project's benefits are being amplified by the developers [202].

As we have already demonstrated in the previous sections, although renewable energy brings with it the hope of avoiding the most pessimistic climate catastrophe scenarios, careful planning and design of the transition process is a prerequisite to ensure the development of low-carbon energy is environmentally, economically and socially sustainable. Recognizing that the energy transition goes far beyond purely technological solutions, and that its course and effects may impact the wellbeing of our societies and the environment in which we live, requires collaboration with various different actors and fields of sciences, involving multidisciplinary teams of engineers, natural scientists and social scientists. In the remainder of this section, we took a closer look at the problem of power imbalances and conflicts of interest between different social groups, and possible ways of dealing with them in the planning process for renewable energy projects.

To make the transition process more just, the most important challenge is to create conditions for the involvement of different stakeholder groups (such as NGOs, local communities, regional politicians and administrations, local businesses, etc.) in the early (so-called "upstream") planning stages of renewable energy projects. This is important: only when differing perspectives, interests and values can be expressed from the project planning stage onwards, will the chances of successful implementation increase. Ultimately, solutions designed and implemented on a foundation of a broad social consensus have the best chance of social acceptance and further dissemination. Of course, the involvement of different groups in the decision-making process brings with it new challenges, such as the need to deal with unequal access to information by different actors, different levels of interest and capacity for involvement in a project, different levels of knowledge and competence, as well as unequal access to different types of resources. All this contributes to the imbalance of power in participatory processes, which can lead to conflicts of interest and jeopardize the successful development of adequate solutions. However, when the process of planning and joint reflection on renewable energy projects is carried out in a deliberate manner, many of these problems can be identified and solved early on.

Besides social acceptance and readiness, successful and cost-efficient sustainable energy systems require large-scale investments to enable the transition. Next, we will look at the economics of renewable energy projects, and methods to assess economic feasibility of an investment.

2.6 Economics of Renewable Energy

The economics of renewable energy is important, as the transition to future green energy systems is enabled by new large-scale investments. Generally, the energy sector is very investment-intensive, and the transition of energy infrastructure, including the systems of production, requires a lot of capital. The economics of investments must be well understood in order to analyze systemic aspects, for example how the energy system can be developed, and the profitability of comparative production technology scenarios. Equally, a solid basis in economics is essential for assessing an individual investment from the perspective of an investor or a company, for example when considering the profitability of investing into new production. Both perspectives are important: even if an investment is the cheapest one, it might not be profitable—and thus not implemented by a company. On the other hand, an investment might well be profitable, but investment in some other technology might be even more lucrative.

In this chapter we focus on several different methods to assess economic feasibility of an investment: the net present value (NPV) estimation, the internal rate of return (IRR), and the payback time. We also study estimation of the levelized cost of energy (LCoE), before finally discussing capital investment and learning rates of renewable technologies, and how they are important in realizing the ongoing energy transition. These are the basic methods used by the energy industry and solar and wind power developers in deciding whether the planned project is "bankable" or not.

2.6.1 How Investments Are Assessed—Cashflow Analysis: Concepts of Cost and Revenue

Cashflow analysis is the basis for any financial analysis of an investment or a company. In short, cashflow analysis is listing in detail where money flowing into a company is coming from, and where it is spent to. Costs can be divided into capitals expenditures, CAPEX, and operational expenditures, OPEX.

CAPEX refers to the cost of acquiring or upgrading physical assets such as property, buildings, or equipment. It is a type of cost incurred when a company makes a significant investment in its operations, such as expanding its facilities or purchasing new equipment. In the cashflow analysis CAPEX costs are depreciations of the investment costs divided over the depreciation time of the investment. OPEX refers to the ongoing costs of running a business or a system, including expenses such as wages, fuel use, utilities, rent, and other day-to-day operational costs. It can be further divided to fixed and variable operating costs. Fixed OPEX are costs that are fixed for a time period and are not dependent on amount of electricity or products produced (e.g., staff costs), in contrast to variable OPEX that are directly dependent on the amount of production (e.g., fuel costs).

Revenues or income for the investment can come from the sales of products and services. Naturally, when an energy production investment is considered, the company can get income from selling energy (e.g., electricity, heat or some chemical energy carrier such as hydrogen). Other revenue streams are also available, such as selling services, for example grid balancing and large-scale storage.

The revenue from the sales of the electricity is expressed per unit of electricity delivered to the customers; however, the electricity can be sold in different ways. These categories are discussed further in the technology-specific sections, but can be summarized as follows:

1. **Power exchange**: The produced electricity can be sold directly into a power market exchange, with a consequent variable price defined by the transmission system operator.
2. **Subsidies**: The produced electricity is compensated directly with an applicable subsidy. This might either be a fixed subsidy (feed-in tariff, FIT), a fixed premium above market price (feed-in premium, FIP) or a variable premium above market price (e.g., green certificate).
3. **Power auction**: Governments can create competitive auctions in which the renewable energy system installation bidding for the lowest production price wins the auction. The installation thus receives the (fixed) bid price for the period and volume as defined in the auction.
4. **Power purchase agreements (PPA)**: These agreements are formed directly between power consumers and producers for supply of electricity for a fixed price over a defined period. This is typical approach in many renewable energy projects, and acts as strong evidence for the investors of the project's low risks.

Revenue can involve combinations of the above—for example, a renewable energy system installation may sell the majority of its production through a PPA, but reserve some proportion to be sold in the open power market. This is used when the intent is to maximize returns at relatively low risk. PPA in combination with more volatile open power markets guarantees a fixed return from the PPA, and potentially higher profits from the short-term power market prices.

Calculating the net present value (NPV) is easiest when the price per sold unit of electricity is constant, which from the above categories would only be known for feed-in tariffs, auctions and PPAs. To estimate the price of electricity that can be obtained from the electricity market, some predictions would need to be applied to calculate an assumed NPV. This exposes the renewable energy generation owner to a much higher level of investment risk.

There can also be other forms of "non-energy" revenues for installations, including capital subsidies from governments for the initial plant investment, and capacity payments. Capacity payments are typically annual payments to dispatchable (on-demand) power plants to provide continuously available generation when needed. These are limited to a few national implementations and are not applicable to non-dispatchable renewable energy systems such as wind and solar, but may be available to dispatchable generation such as hydropower and biomass.

2.6.2 Methods for Assessing Profitability of an Investment

The most common methods in assessing profitability of investments are the net present value (NPV), the payback time, and the internal rate of return (IRR) methods. Additionally, levelized cost of production can be used to, for example, compare different investment alternatives. These methods are used by energy producers and the process industry, combined with various kinds of sensitivity assessments.

Time Value of Money and the Net Present Value (NPV) Concept

The value of money decreases over time, and therefore an amount of payment in the future is considered less valuable than the same payment taking place in the present. With NPV, also known as the annuity method, a series of future payments can be made comparable. The payments to be received in the future are multiplied by a discounting factor, to calculate their comparable net present value at the current time. If the net present value of all summed payments related to an investment is positive, the investment is profitable. In addition, if values of several investments are positive, the most valuable is the one with largest net present value.

As an example of the time value of money, a person is told that they can have 10 000 € today, or, if they wait one year, they will get $(1 + r) \times 10\,000$€. What determines the decision? The value of r is an interesting factor, as it determines whether the person chooses to take the money immediately, or to wait. This threshold varies from one person to another. Perhaps a value of $r = 0.05$ (5%) would not be enough for some, but receiving 11 000 € after a year, with $r = 0.1$ (10%), would be sufficient. The value, r, in this context, is referred as the minimum attractive rate of return.

Since the value of money is decreasing with time, this means that the value of money in the future, F, should be larger than the present value, P to make the investment attractive. If the studied time period is N years and the interest rate for one-year increments is r, we can calculate the present value of the money in the future (and vice versa) as follows. Note that the value increases exponentially in geometrical series,

$$F = P(1 + r)(1 + r)(1 + r) \ldots = P(1 + r)^N$$

$$P = \frac{F}{(1 + r)^N}$$

We can also generalize this to several payments, called annuities A_i, paid in different years, to obtain the value of future payments today, the net present value, NPV.

$$P = NPV = \sum_{i=0}^{N} \frac{A_i}{(1 + r)^i}$$

Annuities mean the net amount of money transferred each year, i.e., $A_i = IN_i - OUT_i$, where IN_i and OUT_i refer to incomes and payments (direction defined as positive from the perspective of the project owner) during the year i, respectively. For example, an initial capital cost payment in the beginning is negative, $i = 0$, and all incomes from customers to the bank account of the project's owner are positive.

A positive NPV implies that the investment is profitable, a negative NPV indicates that that the project is a financial loss, and $NPV = 0$ means the project breaks even—that is, the present value of all benefits over the useful lifetime is equal to the present value of all the costs. Note that the selling price can vary during the project years, while the Levelized Cost of Energy, LCoE, is assumed constant over the project lifetime.

In the case of equal annuities, $A_i = A$, the net present value equation becomes simple—from the solution for the geometric series,

$$NPV = \frac{(1+r)^N - 1}{r(1+r)^N} A$$

(Note that this equation cannot be used for variable annuities.)

Next, we consider how the interest rate is defined. Inflation decreases the value of money over time, and therefore the interest rate should be larger than the rate of inflation to avoid losing the value of the invested money. The interest rate in these discounted cash flow calculations is set by the project developer. A typical rate is 7–10%, depending on the investment case. For a high-risk project, a higher value of discount rate should be selected.

As a summary, the Net Present Value calculation gives an indication of whether a studied project is feasible or not—a positive NPV indicates that the project is realizable. For example, when estimating the feasibility of solar PV or wind power plant investment, we should use a price that is as reliable as possible for the electricity to minimize the financial risk of the project. At the time of writing this book, (2021–2024), due to the unstable geopolitical climate, the material costs for wind power plants peaked 40% higher than pre-2021 levels. Time will tell what the new normal will be. This, coupled with high interest rates and increased capacity of new renewable energy generation, resulted in situations where produced power overwhelmed the grid, or where the grid electricity price dropped very low, even becoming negative. In such circumstances, flexible use and consumption are crucial—for example, having large electrolysis plants available to convert and store excess electricity into hydrogen. Otherwise, there is a risk that, to avoid penalties, renewables production must be curtailed. This results in the investment not being paid back as planned—no power generation, no sales, no income. To avoid shutting power plants down, there is a need to develop new kinds of flexible systems and energy storage, such as H_2 energy storage or pumped hydro storage systems, that enable balancing the load requirements of the grid. Energy is stored during the time of overproduction to be used later when there is too little power generation.

Payback Time

The payback time method shows how long it takes for a business to recoup an investment: it is the time at which the net present value becomes zero using the fixed interest rate, calculated by iteration. In other words, payback time is the time taken for the investment to pay for itself and start generating profit (see example 2.1 below).

Internal Rate of Return (IRR)

The internal rate of return (IRR) is the interest rate that gives a net present value of zero at the end of the project lifetime. IRR calculations rely on the same formula as NPV,

$$NPV = \sum_{i=0}^{N} \frac{A_i}{(1 + IRR)^i} - C_0 = 0$$

A_i is the net cashflow during year *i*, and C_0 is the total initial investment cost. Due to the nature of this equation, the IRR cannot be calculated analytically, and instead must be iteratively solved through trial and error, or by using software programmed to calculate IRR (see example 2.1 below).

Levelized Cost of Energy

The main methodology for calculating the costs of energy in general is the Levelized Cost of Energy, LCoE. This method is typically used when we want to compare different technologies for producing the same product. Without knowing the actual unit selling price, it gives a limit above which the selling price gives a positive result, and below which the project is not economically feasible. From here on we denote LCoE as c_L for simplicity.

The LCoE-method allows us to calculate the average net present cost of each unit of electricity generation over a defined period. An alternative definition is "the average revenue per unit of electricity generated, that is needed to recover the cost". The basic concept of the LCoE is defined as the ratio of total cost of production, C_t (€), and the produced energy, E_t (e.g., MWh or kWh),

$$c_L = \frac{C_t}{E_t}$$

Following the discussion above for the discounted cash flow analysis, we can derive the "conservation equation for money". We require that incomes from selling the electricity (or some other product, such as H_2 or heat) across the project lifetime, discounted to the present time, are equal to costs paid to produce it during the project lifetime, also discounted to the present time:

$$\sum_{i=0}^{N} \frac{I_i + M_i + F_i}{(1 + r)^i} = \sum_{i=0}^{N} \frac{E_i c_L}{(1 + r)^i}$$

where, for each year *i*, I_i is the investment, M_i are the operational and maintenance costs, F_i are the operation costs related to fuel or other similar continuously running costs, E_i is the energy sold per year, and c_L is the levelized cost of the sold energy. c_L is assumed constant for the whole project lifetime, and so the equation can be rearranged to give

$$c_L = \frac{\sum_{i=0}^{N} \frac{I_i + M_i + F_i}{(1+r)^i}}{\sum_{i=0}^{N} \frac{E_i}{(1+r)^i}}$$

The interpretation of c_L, the levelized cost of energy, is simply that with this selling price, we would be just able to cover the production cost and reach the required target increase of the value of money based on the interest rate r. If our actual selling price is higher than c_L, as it should be, then we will make profit.

The investment expenditures in a given year, I_i, are in most cases dominated by the initial capital cost of the plant. In simple terms this cost is often described as occurring in "year 0", $i = 0$, of the plant, while the generation begins from "year 1" onwards. In basic calculations, this can be estimated from available sources for the relevant size of the plant, from estimated capacity in €/MW—for example, the cost for installed wind power could be 1.1–1.3 €/W (2023) based on the turbine's power rating (see Fig. 2.21).

In reality, the construction of the power plant takes several years, and the capital costs rise at different times prior to, during, and following the construction. Smaller investments may be made during the plant's lifetime, for example replacement of generators in wind turbines ("repowering"). Finally, there will be decommissioning costs at the end of the power plant lifetime that depend on the location and technology used.

Operations and maintenance expenditures in year M_i cover all operational and maintenance (O&M) costs in terms of, for example, human resources and materials. O&M costs are generally composed of fixed costs (typically in proportion to the generation capacity) and variable costs (per unit of electricity produced). In the

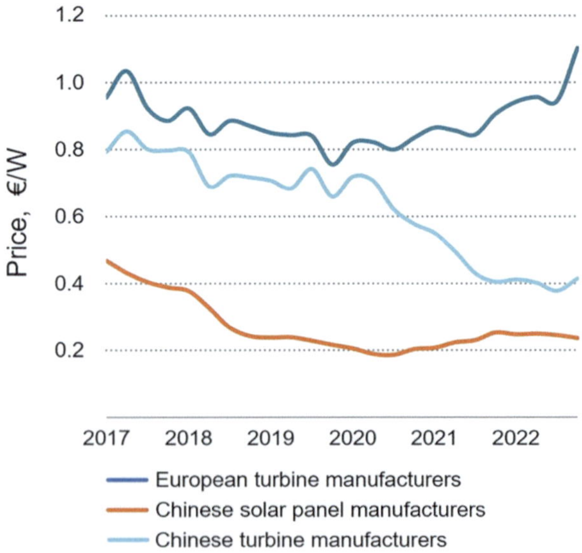

Fig. 2.21 Capital cost estimates (€/W) for European and Chinese manufacturers [100]

worked example below this is simplified to fixed O&M, which tends to be the most important component of O&M for most renewable electricity plants.

Electricity generation in year *i*, E_i, is calculated by multiplying plant capacity, P_{rated}, by the net capacity factor, C_f, and by hours per year,

$$E_i = P_{rated} \times C_f \times 8760\text{ h}$$

C_f is the fraction of full load hours of plant generation across a year (taking account of internal consumption of the plant, and thus only representing electricity supplied to the grid, for which revenue is earned). Fuel expenditures, F_i, in the case of renewables, are only relevant to biomass plant, and will be a function of key factors such as electricity generated, cost of delivered fuel per energy content, and plant efficiency at the load conditions under which the plant operates across the relevant year. Lifetime of the system, N, is the total life of the plant for which costs are incurred. This includes all initial planning and construction years when there are significant expenses, the operational period of the plant, and finally the decommissioning period.

Example 2.1 Economic analysis of a solar PV power plant.

1. Calculate the levelized cost of electricity production for a 10 kWp (p refer to rated power) solar power plant. Capacity factor is assumed 0.14, capacity cost is 1.55 €/Wp, maintenance cost is 7000 €/MW, discount rate is 7%, and project lifetime is 25 years.
2. Carry out *NPV* analysis, using grid price $c_e = 250$ €/MWh.
3. Determine the discounted payback period.

Capital cost is simply 10 000 × 1.55 = 15 500 €. Since 10 kW = 0.01 MW, the annual maintenance cost is 7000 × 0.01 = 70 €, and annual energy production is 10 000 W × 0.14 × 8760 h = 13.14 MWh. Using MS Excel we can construct the following table, also available as electronic supplementary material (years 9–23 are not shown to save space).

The levelized cost is calculated from the totals across the last row: (15,500 + 816)/153.1 = 106.5 €/MWh. NPV can also be calculated (See green columns in Table 2.6) by first multiplying the discounted power production cells by the grid electricity selling price, c_e. The NPV is then calculated at each year as follows:

$$NPV = \sum_{i=0}^{N} \frac{-I_i + M_i + F_i + E_i c_{e,i}}{(1+r)^i}$$

NPV at the end of the 25-year project is 21966 € (blue cell)—a very positive result, indicating that the investment is highly profitable. Discounted payback time can be read from the point when NPV turns positive, in this example slightly longer than six years (yellow cells). Note that if LCoE is larger than grid selling price, the project will never pay itself pack—LCoE has to be always larger than the grid price to make a profit.

Table 2.6 Calculations for Example 2.1. The NPV-specific calculations are highlighted in green, using c_e =250 €/MWh

Year, i	$\frac{1}{(1+r)^i}$ -	$\frac{I_i}{(1+r)^i}$ €	$\frac{M_i}{(1+r)^i}$ €	$\frac{F_i}{(1+r)^i}$ €	$\frac{E_i}{(1+r)^i}$ MW	$\frac{E_i c_e}{(1+r)^i}$ €	NPV €
0	1.000	15500	0.0	0	0.0	0	-15500
1	0.935		65.4	0	12.3	3070	-12495
2	0.873		61.1	0	11.5	2869	-9687
3	0.816		57.1	0	10.7	2682	-7063
4	0.763		53.4	0	10.0	2506	-4610
5	0.713		49.9	0	9.4	2342	-2318
6	0.666		46.6	0	8.8	2189	-176
7	0.623		43.6	0	8.2	2046	1827
8	0.582		40.7	0	7.6	1912	3698
…	…	…	…	…	…	…	…
24	0.197		13.8	0	2.6	648	21374
25	0.184		12.9	0	2.4	605	21966
Total		15500	815.8	0.0	153.1		

Another important analysis tool is the internal rate of return, IRR, the interest rate that makes the $NPV = 0$ at the end of the project lifetime. We can use the same Excel® calculation sheet to find the IRR using a numerical iteration package such as Solver®. Table 2.7 presents the same case as above, but for a solution that searches the value of IRR. This was obtained through iteration, setting the objective $NPV = 0$ and using interest rate as the free parameter to be solved.

To obtain NPV = 0 at the end of the 25-year project, the internal rate of return needs to be as high as 20.5%. Since the interest rate is typically in the order of 10% at most, this investment project can be considered to be risk free; it is very unlikely that the interest rate would be ever so high.

2.6.3 Evaluating the Investment Cost

The estimation of capital investment costs is challenging, especially in the academic world, since detailed, latest price data is available only between business partners in real-world projects. In the early stages of project planning, the size and the capacities of the systems are not yet fixed, and to include them correctly in the engineering and optimization analysis we need to know the cost of the components as a function of their size. For the purpose of a preliminary assessment, the investment and equipment

Table 2.7 Calculations for iterating the internal rate of return. Internal rate of return is 20.5% to reach $NPV = 0$, in 25 years. $c_e = 250$ €/MWh, and all other data are the same as in Example 2.1 and Table 2.1

Year, i	$\frac{1}{(1+r)^i}$ -	$\frac{I_i}{(1+r)^i}$ €	$\frac{M_i}{(1+r)^i}$ €	$\frac{F_i}{(1+r)^i}$ €	$\frac{E_i}{(1+r)^i}$ MW	$\frac{E_i c_e}{(1+r)^i}$ €	NPV €
0	1.000	15500	0.0	0	0.0	0	-15500
1	0.830		58.1	0	10.9	2725	-12833
2	0.688		48.2	0	9.0	2261	-10621
3	0.571		40.0	0	7.5	1875	-8785
4	0.474		33.1	0	6.2	1556	-7263
5	0.393		27.5	0	5.2	1290	-6000
6	0.326		22.8	0	4.3	1070	-4952
7	0.270		18.9	0	3.6	888	-4083
8	0.224		15.7	0	2.9	737	-3362
9	0.186		13.0	0	2.4	611	-2764
...	...	...	...	...	...	...	...
23	0.014		1.0	0	0.2	45	-66
24	0.011		0.8	0	0.1	37	-30
25	0.009		0.7	0	0.1	31	0
Total		15500	337.5	0.0	63.3		

costs can be estimated quickly based on existing knowledge or published data on the investment costs of certain equipment on known size. This requires no detailed design of process equipment, and provides a reasonable investment cost estimate [80]:

$$C_2 = C_1 \left(\frac{S_2}{S_1} \right)^n$$

where C_2 is the capital cost of plant or process unit with the capacity of S_2 of interest in the analyzed system, C_1 and S_1 refer the cost of the known capacity price data, and n is the scale-up exponent which typically varies from 0.6 to 0.9, depending on the type of system and process. This data is available for various systems—for example, Green and Perry [80] and Towler et al. [211] give refence values for different types of raw materials, processes, and plants. Also, simulations software such as Aspen Plus® and Aspen HYSYS® contain large price databases for components. For renewable energy systems, CAPEX values are monitored, for example by IEA [97, 100]. See Fig. 2.21 for an example of price development since 2017.

As Fig. 2.21 clearly shows, at the time of writing of this textbook, the unit price of Chinese turbines is ~ 0.4 €/W, Chinese solar panels are approaching 0.2 €/W, while European manufacture costs are much higher due to the unstable geopolitical situation in Europe.

Learning Rate—Capital Cost Decreases with More Installations

When a new technology is invented, prototyped, and introduced to markets, its production cost is typically high. Over time, when production volumes increase, and all parts of the value chain—materials production, logistics, production, marketing, etc.—develop, the unit cost for production decreases. This behavior can be described by the learning rate, defined as the relative decrease (fraction or %) in unit price (€/kWh, €/ton, etc.) while the installed capacity (MW, tons/day, etc.) doubles. See [193] and McDonald and Schrattenholzer [153] for broader analysis, and literature reviews of the learning rates of several technologies.

We assume that the incremental relative price decrease dc/c is linearly proportional to incremental relative increase in installed capacity, dP/P. This can be presented mathematically as follows

$$\frac{dc}{c} = b\frac{dP}{P}$$

where b is the proportionality constant.

The above equation can be integrated from $c_0 \rightarrow c$ and $P_0 \rightarrow P$ to give

$$\ln\left(\frac{c}{c_0}\right) = b\ln\left(\frac{P}{P_0}\right)$$

Point 0 refers to the initial point of the observation period and can be selected freely based on interest—for example, when there is a change in the slope of the learning curve. If we plot this relationship with the y-axis as $\ln\left(\frac{c}{c_0}\right)$ and the x-axis as $\ln\left(\frac{P}{P_0}\right)$, the model corresponds to a straight line, $y = bx$, that intersects x- and y-axes at zero, and with a slope equal to b. Taking the exponential of both sides gives us

$$\frac{c}{c_0} = \left(\frac{P}{P_0}\right)^b$$

We still need to present the constant b in terms of the learning rate, the fractional decrease in price when capacity is doubled. If the learning rate is X, at the time when capacity has doubled, $P = 2P_0$, the price can be calculated from $c = c_0(1 - X)$. Inserting these into the above equation gives $(1 - X) = 2^b$, from which we can solve $b = \ln(1 - X)/\ln(2)$. Here X is the fractional learning rate (e.g. 0.2 corresponds to 20% learning rate) when the installed capacity was doubled.

Combining all this together gives

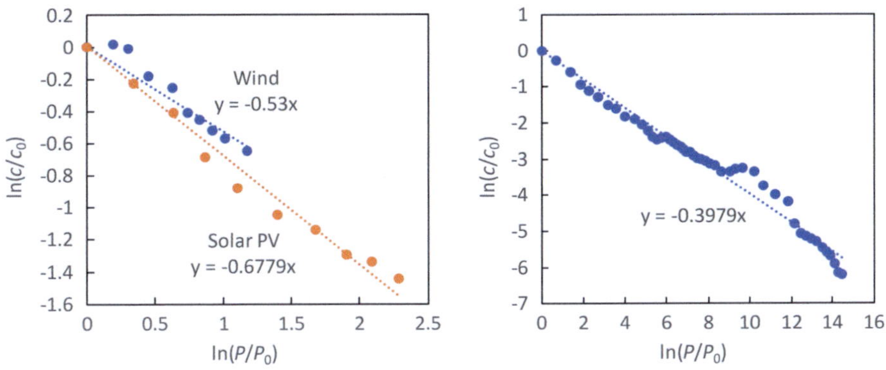

Fig. 2.22 Model fits of solar PV and onshore wind for 2011–2020 (left), [c] = €/kWh, using data from IRENA [106] and [131]. Right figure presents the behaviour for solar PV during period of 1975–2022 with the unit [c] = $/W, data from [173]

$$\frac{c}{c_0} = \left(\frac{P}{P_0}\right)^{\ln(1-X)/\ln(2)}$$

In Fig. 2.22 (left) the data points from global unit prices and installed capacities from the period 2011–2020 are presented for solar PV and onshore wind. Figure 2.22 (right) presents the price development of solar PV from its early market entrance 1975.

The values of b are -0.530 and -0.678 for onshore wind and solar PV, respectively in the left figure, and -0.398 in the right figure for solar PV over a longer period. We can solve for the learning rates from these using $X = 1 - 2^b$, giving $X_{wind} = 1 - 2^{-0.53} = 0.31$ and $X_{PV} = 1 - 2^{-0.6779} = 0.37$ and, for PV over the entire period on the market, $X_{PV} = 1 - 2^{-0.3979} = 0.24$. This means learning rates are 31% and 37% for wind and solar between 2011 and 2020, respectively, and 24% for solar PV panels since their early commercialization. This simple example illustrates the process of estimating learning rates from data.

Normally, learning curves are presented on logarithmic x- and y-scales in dimensional form. Figure 2.23 presents the price development of solar PV since their first commercialization in this way. Each dot presents one-year average value.

As can be seen, there has been a tremendous decrease in the price of solar PV since its market entrance: in 1976 the price was above 100 $/W, and in 2022 it was already well below 1 $/W. Levelized cost for PV-based electricity can be calculated from the unit price, factoring in local sun conditions and other costs. Note also that, in recent years, 2011–2022, the learning rate has still accelerated, as shown in Fig. 2.23 (left).

Example 2.2 Estimation of the learning rate from start and end points.

The task is now to calculate the learning rate from the PV data between 1975 and 2022 using only the start and end points of the overall period, and compare this to the above data, fitted using all the points, that results in a learning rate of 24%.

Fig. 2.23 Price development of solar PV panels ($/W), data from [173]

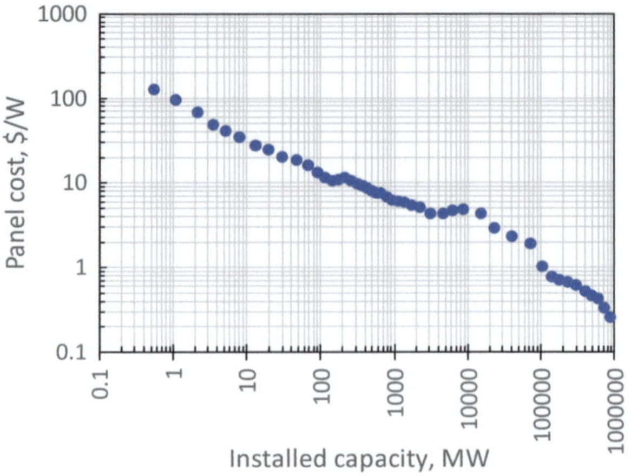

The cost and capacity at start and end points are $c_0 = 125.8$ \$/W, $c = 0.256$ \$/W, $P_0 = 0.54\text{MW}$, $P = 1046614\text{MW}$. The rate constant can then be calculated from

$$b = \frac{\ln\left(\frac{c}{c_0}\right)}{\ln\left(\frac{P}{P_0}\right)} = \frac{\ln\left(\frac{0.256}{125.8}\right)}{\ln\left(\frac{1046614}{0.54}\right)} = -0.428$$

The learning rate is then simply $X_{PV} = 1 - 2^{-0.428} = 0.257$, or 25.7%.

2.6.4 *Internalizing Externalities and Financial Policy Support Mechanisms*

With centuries of experience, fossil-fuel-based thermal power plant technologies are well matured and fine-tuned with respect to technical and economic performance. In 2024 more than 80% of global energy production still came from burning of fossil fuels. The use of fossil fuels has destabilized the climate, resulting in pollution and many other negative environmental and societal impacts, coupled with the imbalance between profit gain and bearing the costs caused by the greenhouse gas emissions. Health issues such as lung disease and other respiratory problems, and environmental issues such as ocean acidification, resulting from the combustion of fossil fuels, presents a significant cost on societies since these costs are not included into the cost / profit opportunity analysis of industry [209]. These costs are called negative externalities. One way to address this imbalance and direct the markets is to implement a penalty tax system these negative externalities, such as the European Emission Trading system, ETS. The ETS has already peaked several times above 100 €/ton of CO_2 emitted (CarbonCredits (2024), decreasing their competitive advantage over renewables.

As discussed in Sect. 2.3, we humans are heterotrophs, and therefore dependent on energy produced by other organisms. These *ecosystem services* include food, drinking water, oxygen, pollination, nutrient recycling, and many other services that make human life possible. We are used to externalizing our impacts on these services, and often treat them as if they were infinite and carry no economic value, which is not the case. Dasgupta [40] published a broad review of the economic value of biodiversity, approaching this question from the economical point of view, and demonstrating that nature is our most valuable asset and we have failed to take care of it. The review showed that between 1992 and 2014 the produced capital per person doubled, and human capital per person increased by about 13% globally. Over the same period, the stock of natural capital per person declined by nearly 40%, risking the welfare of current and future generations.

In the midst of the energy transition, we should actively seek new ways of developing large-scale energy systems and solutions where the value of the nature is an integral part of the economic analysis. Mandatory use of mitigation hierarchy (see Info Box 2.2) in all large-scale projects, including industrial-scale energy projects, would aid in addressing and minimizing negative impacts on nature, wildlife, and secure the functions that ecosystems produce. Policies and regulations are already moving in this direction, but more stringent legislation is needed to redefine the operating environment, for example via subsidies. This is what we discuss next.

Political Intervention in Energy Investments: Subsidies

An increasing amount of new-built renewable power generation technologies are displacing coal and nuclear power without subsidies, since their lifetime costs are often lower than their fossil competitors [209]. However, government-based support mechanisms have also been introduced in the energy sector to make affordable clean energy more widely available. These mechanisms are extremely important in bridging the gap between high production costs and affordable customer prices, to enable the wide use of clean fuels and environmentally friendly energy systems. Likewise, when there is a shortage of large-enough energy storage systems to stabilize the variability of renewable energy production, power producers need to maintain thermal power plants on standby that can cover energy needs in case renewable energy production is not sufficient. In this case, governments can cover the costs of this standby power production to maintain the reliability of the energy supply, while directing production towards renewable energy sources. These types of support or subsidies are called direct or explicit subsidies.

Governments can also reduce taxes for energy producers to secure national energy security. While this is understandable, it also creates a market imbalance and shifts the costs related to pollution, for example, onto society. This type of support is called as indirect or implicit subsidies.

In long-term investments, like energy system technologies, there is an interest in minimizing costs. This requires the price benchmarks used in decision making to be accurate. Looking at the overall benefits and accumulated costs to society, the price benchmark should reflect the total cost of an energy project, including potential health impacts and negative effects on the environment, natural ecosystems, and climate.

Neglecting these can result in a sub-optimal end result with higher long-term impacts and costs than anticipated [209].

The increased implementation of variable renewable energy production complicates the situation further, as the stability of the energy system and the power grid decreases. Periods when the local electrical grid is not able to accept excessive renewable power production, or handle the missing gigawatts when renewable production is low, will become more common. Transmission system operators have to limit overproduction by setting a negative price, which leads power producers to curtail their production. This means economic losses for the power producer, and investments into projects are not paid back as planned, risking projects' profitability if the situation continues. Flexible solutions and large energy storage systems are needed—for example, electrolysis can be used to convert and store energy into hydrogen when there is excess production, and the grid price is low or even negative. The hydrogen or its derivates can then be traded later, when the renewable power production is low and the selling prices are high.

The above-described situation is economically challenging. To phase out fossil energy sources, we need large capacities of wind and solar PV power. But we also need large capacities of electrolyzers to produce hydrogen and hydrogen storage systems (and other storage technologies and flexible systems), to make the system robust and balance the variation in energy production. The problem is, how do we secure the needed investments?

Unless there are no significant customers for the excessive power from variable renewables, investors are hesitant to invest in renewable energy production—if there are production breaks due to curtailment, the payback times become too long. An economically-attractive power plant needs to run at full capacity whenever the weather conditions are favorable. This requires investments on electrolyzers and large-scale hydrogen storage, that in turn requires that there is enough renewable energy production available. This "chicken-and-egg" problem raises the question of who takes the first step in investment: the wind-power industry, or the customer that uses the renewable energy by converting it to hydrogen? In this situation, governments could support the investments, helping to increase installed capacities, and the above-discussed learning rate can then bring the capital cost down into a range that is economically self-sustaining.

The levelized cost of electricity of solar PV and wind power, especially onshore wind, is already in the same range of, and even lower than, electricity produced from fossil fuels. In addition, these costs continue to decrease, as discussed in the previous section on learning rates. To be truly competitive with fossil fuels, the energy from renewables needs to be coupled with storage or a sufficiently large and flexible power grid, which can stabilize the output fed into the load systems. However, this increases the power cost significantly. The key components in this are electrolyzers and hydrogen storage—these are mostly technologically available already, but since they are in their early stages of large-scale implementation, their unit costs are still too high, so that the cost of the hydrogen produced from renewables is much higher (~5 \$/kg) than hydrogen produced from fossil natural gas (1–2 \$/kg).

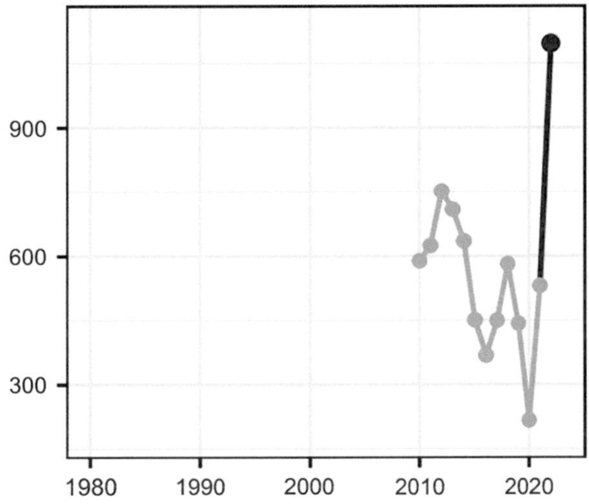

Fig. 2.24 Historical development of direct fossil subsidies, billion $US per year [187]

Subsidies for fossil fuels are monitored systematically, since the aim is to reduce the cash flows that support these kinds of harmful subsidies. Based on IRENA's Energy Subsidies report [209], environmentally-friendly subsidies can help to improve the efficiency of capital allocation across the energy sector. Externalities from fossil fuel use, especially the cost imposed on society from air pollution and climate change, are typically not fully paid. As an example, in 2017, total subsidies for fossil fuels were 3.1 trillion USD, while renewable energy subsidies were only 166 billion USD—just ~ 5% of the fossil subsidies. Direct fossil subsidies were 447 billion USD in 2017, around 2.7 times higher than direct renewable subsidies. At the time of writing this book, the subsidies paid for fossil fuels were still high—in 2023, the direct subsidies were 1.3 trillion $US (increased by a factor of ~ 3 from 2017 to 2023) and the indirect subsidies were more than 6 trillion $USD. As a reference, total global GDP was ~ 100 trillion $US 2022 [174]. See Fig. 2.24 for the historical development of direct subsidies (Ripple et al. [187]).

Subsidy support mechanisms have gained attention in political arenas and there is a lot of scientific discussion about the harmfulness of these subsidies in reaching global emission reduction targets (see for example [56, 108, 109, 154]). One argument for keeping subsidies for fossil fuels is that withdrawing them from natural gas, for example, would result in switching to cheaper but even more harmful fuels, such as coal.

Though the full picture is complex, and this discussion is one example of how we have locked ourselves into a harmful system, these incentives enable continued investment in fossil fuels and the further development of systems running on those fuels. For that reason we need to find a path to a gradual and just transition away from fossil-fuel-based systems, into one built on clean energy sources. For example, the subsidy flow could be directed to support large-scale implementation of electrolyzers and hydrogen storage, to decrease their unit costs and make renewables-based hydrogen production cheaper than fossil-based systems. Based on the discussion of learning rates, implementing a sufficient capacity of electrolyzers will decrease the price sufficiently. After reaching an economically-feasible self-sustaining price level, the support mechanism can be removed. Based on [6], we need to reach 20–30 GW of installed electrolyzer capacity to fall below the fossil hydrogen prices—for reference, the available electrolyzer capacity in 2022 was ~ 0.3 GW.

Already in 2023, the LCoE from solar PV and onshore wind is lower than from fossil-based electricity. And still, the fossil options are supported and subsidized, while being too expensive or non-profitable. Could we speed up the transition to renewable energy by providing more support for renewable energy development? Renewables already have LCoE at the same level as fossil fuels, and their learning curves demonstrate that prices are dropping further. This discussion is well summarized by [209]: "Indeed, given the fact that the negative externalities of fossil fuels remain predominantly unpriced, the subsidies given to fossil fuels today represent a perverse incentive and amplify an already serious market failure with significant socio-economic and environmental costs."

2.6.5 Financing the Energy Investments

Compared to fossil energy sources, renewable energy, such as solar and wind, are very dilute and area-intensive primary energy sources. Fossil sources require heavy mining and drilling activities, logistics and pretreatment, increasing their price. In addition, the better the quality of the fossil fuel, the higher the price. Using bioenergy also requires significant efforts in planting, fertilizing, harvesting and logistics to produce and deliver the fuel to the power plant for combustion, all of which increase its cost—but the actual system that converts these fuels to heat and power is quite compact due to the high temperature and pressure used, and therefore the land use is relatively small. The thermal power plant technologies and materials used in constructing these systems are mature, and this has reduced the unit prices of components to a level where new capacity additions no longer decrease prices, as occurred during the early utilization rates.

For comparison, energy harnessed from wind and sunlight as such is free of charge and available as long as the wind blows and the Sun is shining. However, the energy densities of wind and sunlight are low, requiring construction of large and capital-intensive wind turbines and solar-power plants, covering large areas of land, to collect a meaningful amount of energy to be converted into electricity. The capital cost of a power plant has therefore been higher for a system based on wind and solar power than for a system based on fossil fuels. This is also the reason why wind and solar power plants require large investment capital at the beginning. During operation, the production itself is much cheaper than with fossil fuel plants, which have significantly higher operation costs per unit energy.

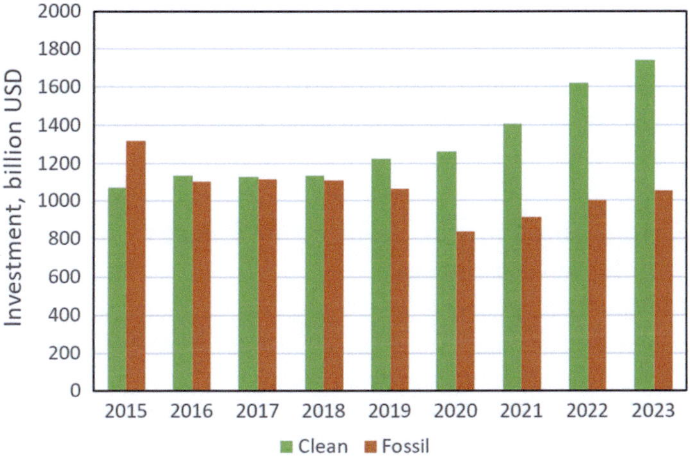

Fig. 2.25 Historical development of clean and fossil energy production infrastructure investments [100]

To give a perspective on the scale of annual energy investments, Fig. 2.25 presents the investments in fossil energy and clean energy systems from the energy producer's side, from 2015 to 2023.

As Fig. 2.25 shows, since 2016 there have been more investment in clean energy technologies (renewables, nuclear, energy efficiency, grids, electric vehicles and low emission fuels, and CCUS systems) than in fossil energy—yet still in 2023, investment in fossil energy was increasing.

To speed up the energy transition, we need to up-scale the materials production and manufacturing value chains of renewable energy systems. According to IEA's Net Zero 2030 scenario [97], we should increase annual investment in production and manufacturing sectors from less than 40 billion USD (2016–2021) to more than 150 billion USD (2021–2030). To have these systems in place and in operation, in order to meet 2030 and 2050 emission targets, the mining sector in particular needs to have their investment decisions ready by 2025, with annual investments in the order of 250–270 billion USD (during 2023–2025) [97].

Renewable energy projects present a fairly clear case for investors. Construction times are relatively short; it takes only a few years after investment decisions are made and permits are granted for even a large unit to be ready and in operation. Wind and solar-power plants can also be deployed in stages, while other parts of the plant are still under construction; similarly, power plants can extend their capacity with an additional permit. The technologies are reliable and robust, and the maintenance of a single turbine or a solar panel has a minor negative impact on the overall production of the plant.

To summarize, to reach our climate targets we should significantly increase investment in clean energy technologies, their materials and value chains. Governments should support the energy transition with financial and legislative tools to enable large-scale implementation of renewable energy systems. Learning rates will then decrease the manufacturing prices of systems to self-sustaining levels. At the same time, harmful subsidies on fossil fuels should be removed gradually, in a just manner, to allow healthy and rapid development of green energy markets.

2.7 A Sustainable Path for Renewable Energy Technologies–Summary

This chapter has touched upon some key topic areas related to sustainable energy transition and sustainable living in general. We examined the three critical planetary boundaries—climate change, biodiversity loss, and land system change—that are linked to the renewable energy transition. We also discussed the social impacts of energy transition in detail. We explained the basic economic tools and metrics for renewable energy project evaluation, and discussed the need to reassess economic success metrics, suggesting, for example, the incorporation of natural capital into accounting systems to make all costs of large projects visible. We underscored the critical need for transformative changes in energy production, resource production, and consumption, to realign human activities within planetary boundaries. Our aim is to equip future energy engineers with the knowledge to design sustainable energy systems. The development of large-scale renewable energy production to replace fossil fuels presents a huge challenge, and an opportunity to strengthen global sustainability efforts.

We first explored the Earth's energy balance, the greenhouse effect, anthropogenic carbon emissions, and carbon cycles impacting atmospheric CO_2 concentrations. Then we examined biodiversity, and discussed methods for quantifying biodiversity loss, and how to mitigate such loss, for example in large renewable energy projects. We briefly outlined the impacts of land-use change driven by different kinds of human activity, and addressed the land-use impacts of renewable energy production, emphasizing the importance of careful land-use planning to mitigate land degradation and destruction of important habitats. Additionally, we addressed the interconnectedness of societal and environmental impacts, emphasizing the need to consider local employment, changes in local economies, and societal consequences in energy transition. With the presented economic analysis tools and examples, it is possible to carry out a basic economic feasibility analysis of renewable energy projects.

In conclusion, with this chapter we want to stress the importance of understanding the various facets and impacts of large-scale energy projects. There is a need to find a fair balance in this socio-ecological-economical-technical system, where renewable energy projects operate in ways that foster development towards sustainability, and that supports intergenerational well-being of both human and non-human species. Such systems are built in cooperation between various actors and across disciplines. Energy engineers have a pivotal role in shaping future energy societies.

Author Contributions Mika Järvinen, Hanna Paulomäki, and Elina Oksanen wrote section 2.1. Mika Järvinen, Dilip Khatiwada, and Hanna Paulomäki wrote the section 2.2. Sami El Geneidy, Maiju Peura and Janne S. Kotiaho wrote the section 2.3. Hanna Paulomäki, Irina Herzon, Elina Oksanen, and Mika Järvinen wrote the section 2.4. Eeva-Lotta Apajalahti, Alicja Dańkowska, Mariana Galvão Lyra, Ayesha Sadiqa, Roosa Talala, and Mika Järvinen wrote the section 2.5. Sam Cross, Mika Järvinen and Antti Arasto wrote the section 2.6. Mika Järvinen and Hanna Paulomäki wrote the section 2.7.

References

1. Afshari H, Agnihotri S, Searcy C, Jaber MY (2022) Social sustainability indicators: a comprehensive review with application in the energy sector. Sustain Prod Consum 31:263–286. https://doi.org/10.1016/j.spc.2022.02.018
2. Agusdinata D, Liu W, Eakin H, Romero H (2018) Socio-environmental impacts of lithium mineral extraction: Towards a research agenda. Environ Res Lett 13(12). https://doi.org/10.1088/1748-9326/aae9b1
3. Alexandratos N, Bruinsma J (2012) World agriculture towards 2030/2050: the 2012 revision. Food and Agricultural Organization of the United Nations. Rome www.fao.org/fileadmin/templates/esa/Global_persepctives/world_ag_2030_50_2012_rev.pdf
4. Algarni S, Tirth V, Alqahtani T, Alshehery S, Kshirsagar P (2023) Contribution of renewable energy sources to the environmental impacts and economic benefits for sustainable development. Sustain Energy Technol Assess 56:103098. https://doi.org/10.1016/j.seta.2023.103098
5. Alkemade R, Van Oorschot M, Miles L, Nellemann C, Bakkenes M, Ten Brink B (2009) GLOBIO3: a framework to investigate options for reducing global terrestrial biodiversity loss. Ecosystems 12(3):374–390
6. Alverà M (2022) The hydrogen revolution. Hodder Studio, Great Britain
7. Apajalahti E-L, Kungl G (2022) Path dependence and path break-out in the electricity sector. Environ Innov Soc Transit 43:220–236. https://doi.org/10.1016/j.eist.2022.03.010
8. Apajalahti E-L, Ruggiero S, Lukkarinen J, Laakso S (2023) Tensions in urban energy communities. In: Sokołowski MM, Visvizi A (eds.) Routledge handbook of energy communities and smart cities. Routledge
9. Archer D (2016) The long thaw: how humans are changing the next 100,000 years of earth's climate, ISBN 9780691169064, Princeton University Press, USA
10. Arlidge WN, Bull JW, Addison PF, Burgass MJ, Gianuca D, Gorham TM, Jacob C, Shumway N, Sinclair SP, Watson JEM, Wilcox C, Milner-Gulland EJ (2018) A global mitigation hierarchy for nature conservation. Bioscience 68(5):336–347
11. Asselin A, Rabaud S, Catalan C, Leveque B, L'Haridon J, Martz P, Neveux G (2020) Product biodiversity footprint—a novel approach to com-pare the impact of products on biodiversity combining life cycle assessment and ecology. J Clean Prod 248:119262. https://doi.org/10.1016/j.jclepro.2019.119262
12. Avelino F, Hielscher S, Strumińska-Kutra M, de Geus T, Widdel L, Wittmayer J, Dańkowska A, Dembek A, Fraaije M, Heidary J, Iskandarova M, Rogge K, Stasik A, Crudi F (2023) Power to, over and with: exploring power dynamics in social innovations in energy transitions across Europe. Environ Innov Soc Transit 48:100758. https://doi.org/10.1016/j.eist.2023.100758
13. BBOP (2012) Business and Biodiversity Offsets Programme Standard on Biodiversity Offsets. Washington, D.C, USA. https://www.forest-trends.org/publications/standard-on-biodiversity-offsets/
14. BP (2022) BP statistical review of world energy. http://www.bp.com/statisticalreview
15. Bailis R, Drigo R, Ghilardi A, Masera O (2015) The carbon footprint of traditional woodfuels. Nature Clim Change 5:266–272. https://doi.org/10.1038/nclimate2491

16. Balvanera P, Pfisterer AB, Buchmann N, He JS, Nakashizuka T, Raffael-li D, Schmid B (2006) Quantifying the evidence for biodiversity effects on ecosystem functioning and services. Ecol Lett 9(10):1146–1156
17. Becker HA (2001) Social impact assessment. Eur J Oper 128(2):311–321
18. Bellard C, Bertelsmeier C, Leadley P, Thuiller W, Courchamp F (2012) Impacts of climate change on the future of biodiversity. Ecol Lett 15(4):365–377
19. Bennun L, van Bochove J, Ng C, Fletcher C, Wilson D, Phair N, Carbone G (2021) Mitigating biodiversity impacts associated with solar and wind energy development. Guidelines for project developers. Gland, Switzerland, IUCN and Cambridge UK: The Biodiversity Consultancy pp 260. https://doi.org/10.2305/IUCN.CH.2021.04.en
20. Berndes G (2011) Bioenergy, land use change and climate change mitigation: background technical report. ExCo:2011:04. IEA Bioenergy. www.ieabioenergy.com/wp-content/uploads/2013/10/Bioenergy-Land-Use-Change-and-Climate-Change-Mitigation-Background-Technical-Report.pdf
21. Berndes G, Ahlgren S, Börjesson P, Cowie AL (2016) Bioenergy and land use change—State of the art. In: Lund PD, Byrne J, Berndes G, Vasalos IA (eds.) Advances in bioenergy: the sustainability challenge. Wiley Interdisciplinary Reviews: Energy and Environment. 2:249–71. https://doi.org/10.1002/wene.41
22. Bertram R, Primova R (eds) (2018) Energy Atlas 2018—Facts and figures about renewables in Europe. Heinrich Böll Foundation, Berlin, Friends of the Earth Europe, Brussels, European Renewable Energies Federation, Brussels, Green European Foundation, Luxembourg
23. Betts RA, McNeall D (2018) How much CO_2 at 1.5 °C and 2 °C? Nat Clim Change 8:546–548
24. Bjelle EL, Kuipers K, Verones F, Wood R (2021) Trends in national biodiversity footprints of land use. Ecol Econ 185:107059. https://doi.org/10.1016/j.ecolecon.2021.107059
25. Boulton CA, Lenton TM, Boers N (2022) Pronounced loss of amazon rainforest resilience since the early 2000s. Nat Clim Change 12:271–278. https://doi.org/10.1038/s41558-022-01287-8
26. Brundtland G (1987) Report of the world commission on environment and development: our common future. United Nations General Assembly document A/42/427
27. Bull JW, Taylor I, Biggs E, Grub HMJ, Yearly T, Waters H, Milner-Gulland EJ (2022) Analysis: the biodiversity footprint of the University of Oxford. Nature 604:420–424. https://doi.org/10.1038/d41586-022-01034-1
28. Bunge M (2003) Emergence and convergence: qualitative novelty and the unity of knowledge. University of Toronto Press, Toronto, p 344
29. Burdge RJ, Vanclay F (1996) Social impact assessment: a contribution to the state of the art series. Impact Assessment 14(1):59–86
30. Burke MJ, Stephens JC (2018) Political power and renewable energy futures: a critical review. Energy Res Soc Sci 35:78–93. https://doi.org/10.1016/j.erss.2017.10.018
31. Butchart SH, Walpole M, Collen B, Van Strien A, Scharlemann JP, Al-mond RE, Watson R (2010) Global biodiversity: indicators of recent declines. Science 328(5982):1164–1168
32. CBD (2022) CBD/COP/DEC/15/5. Decision adopted by the conference of the parties to convention on biological diversity. Monitoring framework for the kunming-montreal global biodiversity framework. https://www.cbd.int/doc/decisions/cop-15/cop-15-dec-05-en.pdf
33. CO_2 Levels (2024) https://www.co2levels.org/
34. CSIRO (2020) State of the climate, ISBN 978-1-4863-1509-3, 2020 Commonwealth of Australia
35. Coleman FC, Williams SL (2002) Overexploiting marine ecosystem engineers: potential consequences for biodiversity. Trends Ecol Evol 17(1):40–44
36. Crawford RH, Bontinck PA, Stephan A, Wiedmann T, Yu M (2018) Hybrid life cycle inventory methods—a review. J Cleaner Prod 172:1273–1288. https://doi.org/10.1016/J.JCLEPRO.2017.10.176
37. Crenna E, Marques A, La Notte A, Sala S (2020) Biodiversity assessment of value chains: state of the art and emerging challenges. Environ Sci Technol 54(16):9715–9728. https://doi.org/10.1021/acs.est.9b05153

38. Crenna E, Sinkko T, Sala S (2019) Biodiversity impacts due to food consumption in Europe. J Clea Prod 227:378–391. https://doi.org/10.1016/J.JCLEPRO.2019.04.054
39. Daewel U, Akhtar N, Christiansen N, Schrum C (2022) Offshore wind farms are projected to impact primary production and bottom water deoxygenation in the North Sea. Commun Earth Environ 3:292. https://doi.org/10.1038/s43247-022-00625-0
40. Dasgupta P (2021) The economics of biodiversity: the dasgupta review. London: HM Treasury pp 610. https://www.gov.uk/government/publications/final-report-the-economics-of-biodiversity-the-dasgupta-review
41. Davis SJ, Liu Z, Deng Z, Zhu B, Ke P, Sun T, Guo R, Hong C, Zheng B, Wang Y, Boucher O, Ciais P (2022) Emissions rebound from the COVID-19 pandemic. Nat Clim Chang 12:412–414
42. Deemer BR, Harrison JA, Li S et al (2016) Greenhouse gas emissions from reservoir water surfaces: a new global synthesis. Bioscience 66(11):949–964. https://doi.org/10.1093/biosci/biw117
43. Denholm P, Hand M, Jackson M, Ong S (2009) Land-use requirements of modern wind power plants in the United States. Technical Report NREL/TP-6A2–4 5834. Golden, Colorado: National Renewable Energy Laboratory. http://www.nrel.gov/docs/fy09osti/45834.pdf
44. Diaz RJ, Rosenberg R (2008) Spreading dead zones and consequences for marine ecosystems. Science 321(5891):926–929
45. Diaz S, Settele J, Brondizio ES et al (2019) Pervasive human-driven decline of life on Earth points to the need for transformative change. Science 366:eaax3100
46. Dijkman TJ, Benders RM (2010) Comparison of renewable fuels based on their land use using energy densities. Renew Sust Energ Rev 14(9):3148–3155. https://doi.org/10.1016/j.rser.2010.07.029
47. Dinesh H, Pearce JM (2016) The potential of agrivoltaic systems. Renew Sustain Energy Rev 54:299–308. https://doi.org/10.1016/j.rser.2015.10.024
48. Dohm R, Drake D (2019) Wind energy effects on birds. In: Moorman CE, Grodsky SM, Rupp SP (eds.) Renewable energy and wildlife conservation. John Hopkins University Press, pp 95–121
49. Díaz S, Pascual U, Stenseke M, Martín-López B, Watson RT, Molnár Z, Shirayama Y (2018) Assessing nature's contributions to people. Science 359(6373):270–272
50. EEA (2022) Energy prosumers in Europe Citizen participation in the energy transition. European Environment Agency, EEA, Report No 01/2022
51. Ehrlich PR, Holdren JP (1971) Impact of population growth: complacency concerning this component of man's predicament is unjustified and counterproductive. Science 171(3977):1212–1217
52. El Geneidy S, Baumeister S, Peura M, Kotiaho JS (2023) Value-transforming financial, carbon and biodiversity footprint accounting. Preprint. arXiv:2309.14186 [econ.GN]. https://doi.org/10.48550/arXiv.2309.14186
53. El Geneidy S, Alvarez Franco D, Baumeister S, Halme P, Helimo U, Kortetmäki T, Latva-Hakuni E, Mäkelä M, Raippalinna L-M, Vainio V, Kotiaho JS (2021) Sustainability for JYU: Jyväskylän yliopiston ilmasto- ja luontohaitat. Wisdom Letters 2/2021. http://urn.fi/URN:NBN:fi:jyu-202104232476
54. Ellis EC (2011) Anthropogenic transformation of the terrestrial biosphere. Phil Trans R Soc A 369(1938):1471–2962. https://doi.org/10.1098/rsta.2010.0331
55. Emilson EJS, Carson MA, Yakimovich KM, Osterholz H, Dittmar T, Gunn JM, Mykytczuk NCS, Basiliko N, Tanentzap AJ (2018) Climate-driven shifts in sediment chemistry enhance methane production in northern lakes. Nat Commun 9:1801. https://doi.org/10.1038/s41467-018-04236-2
56. Erickson P, van Asselt H, Koplow D et al (2020) Why fossil fuel producer subsidies matter. Nature 578:E1–E4. https://doi.org/10.1038/s41586-019-1920-x
57. Van der Esch S, ten Brink B, Stehfest E, Bakkenes M, Sewell A, Bouwman A, Meijer J, Westhoek H, van den Berg M (2017) Exploring future changes in land use and land condition and the impacts on food, water, climate change and biodiversity: Scenarios for the Global Land Outlook. PBL Netherlands Environmental Assessment Agency, The Hague

58. Esteves AM, Franks D, Vanclay F (2012) Social impact assessment: the state of the art. Impact Assess and Proj Apprais 30(1):34–42
59. European commission (2020) Launching the just transition mechanism–for a green transition based on solidarity and fairness. https://ec.europa.eu/info/news/launching-just-transition-mechanism-green-transition-based-solidarity-and-fairness-2020-jan-15_pl
60. Eurostat (2022) Translate energy consumption in households. Available online: https://ec.europa.eu/eurostat/statistics-explained/index.php?title=Energy_consumption_in_households
61. Evelyn J (1961) Fumifumigium: Or the Inconvenience of the Aer and Smoake of London Dissipated, digital archive in https://archive.org/details/fumifugium00eveluoft/mode/2up
62. Ferns G, Lambert A, Günther M (2022) The analogical construction of stigma as a moral dualism: the case of the fossil fuel divestment movement. Acad Manage J 65(4):1383–1415. https://doi.org/10.5465/amj.2018.0615
63. Ferrarini A, Serra P, Almagro M, Trevisan M, Amaducci S (2017) Multiple ecosystem services provision and biomass logistics management in bioenergy buffers: a state-of-the-art review. Renew Sustain Energy Rev 73:277–290. https://doi.org/10.1016/j.rser.2017.01.052
64. Fleschutz M, Bohlayer M, Braun M, Murphy M (2023) From prosumer to flexumer: Case study on the value of flexibility in decarbonizing the multi-energy system of a manufacturing company. Appl Energy 347:121430. https://doi.org/10.1016/j.apenergy.2023.121430
65. Floeter J, Pohlmann T, Harmer A, Möllmann C (2022) Chasing the offshore wind farm wind-wake-induced upwelling/downwelling dipole. Front Marine Sci 9. https://doi.org/10.3389/fmars.2022.884943
66. Fraser N (2009) Scales of justice: reimagining political space in a globalizing world. Columbia University Press, New York, NY, USA
67. Friedl G, Wuest A (2002) Disrupting biogeochemical cycles—consequences of damming. Aquat Sci 64:55–65. https://doi.org/10.1007/s00027-002-8054-0
68. Friedlingstein et al (2022) Global carbon budget 2022. Earth Syst Sci Data 14:4811–4900. https://doi.org/10.5194/essd-14-4811-2022
69. Fritsche U, Berndes G, Cowie A, Johnson F, Dale V, Langeveld H, Sharma N, Watson H, Woods J (2017) Energy and land use. Global land outlook working paper. UNCCD and IRENA. https://doi.org/10.13140/RG.2.2.24905.44648
70. GEF (2016) Sustainable land management and its relationship to global environmental benefits and food security. A synthesis report for the GEF. GEF/STAP/C.50/Inf.03. Washington DC Global Environment Facility
71. Gallop P, Gray E, Nikolovska E, Mustaţă A, Petcu R (2021) PEET: the Political economy of energy transition in Southeast Europe–Barriers and obstacles. Climate change, Energy and Environment, Friedrich-Ebert-Stiftung
72. Garnett T (2009) Livestock-related greenhouse gas emissions: impacts and options for policy makers. Environmen Sci Policy 12(4):491–503. https://doi.org/10.1016/j.envsci.2009.01.006
73. Gasparatos A, Doll CNH, Esteban M, Ahmed A, Olang TA (2017) Renewable energy and biodiversity: implications for transitioning to a Green Economy. Renew Sustain Energy Rev 70:161–184. https://doi.org/10.1016/j.rser.2016.08.030
74. Ghilarov AM (2000) Ecosystem functioning and intrinsic value of biodiversity. Oikos 90(2):408–412
75. Gibbs H, Salmon J (2015) Mapping the world's degraded lands. Appl Geogr 57:12–21
76. Global carbon budget (2023) Global carbon project, with major processing by our world in data
77. Goedkoop M, Spriensma R (1999) The eco-indicator 99. A damage oriented method for life cycle impact assessment. Methodology report and Annex, Pré Consultants, Amersfoort, The Netherlands. http://www.pre.nl/eco-indicator99/
78. Golkowska K, Rugani B, Koster D, van Oers C (2016) Environmental and economic assessment of biomass sourcing from extensively cultivated buffer strips along water bodies. Environ Sci Policy 57:31–39. https://doi.org/10.1016/j.envsci.2015.11.014
79. Gracey E, Verones F (2016) Impacts from hydropower production on biodiversity in an LCA framework–review and recommendations. Int J LCA 21(3):412–428. https://doi.org/10.1007/s11367-016-1039-3

80. Green DW, Perry RH (2008) Chapter 9: Process economics. In: Perry's chemical engineering handbook, 8th edn. McGraw-Hill, New York
81. Grodsky SM, Fritts SR and Hernandez RR (2019) Renewable energy ecology, the next frontier in wildlife science. In: Moorman CE, Grodsky SM, Rupp SP (eds) Renewable energy and wildlife conservation. John Hopkins University Press, pp 247–260
82. Gunderson LH (2000) Ecological resilience, in theory and application. Annu Rev Ecol Sys, pp 425–439
83. Haines-Young R, Potschin M (2010) The links between biodiversity, eco-system services and human well-being. Ecosyst Ecol: New Synth 1:110–139
84. Hakala K, Kontturi M and Pahkala K (2009) Field biomass as global energy source. Agric Food Sci 18:347–365. https://doi.org/10.23986/afsci.5950
85. Hanski I (2015) Habitat fragmentation and species richness. J Biogeogr 42:989–994. https://doi.org/10.1111/jbi.12478
86. Harris NL, Gibbs DA, Baccini A, Birdsey RA, de Bruin S, Farina M, Fatoyinbo L, Hansen MC, Herold M, Houghton RA, Potapov PV, Suarez DR, Roman-Cuesta RM, Saatchi SS, Slay CM, Turubanova SA, Tyukavina A (2021) Global maps of twenty-first century forest carbon fluxes. Nat Clim Chang 11:234–240. https://doi.org/10.1038/s41558-020-00976-6
87. Hartmann H, Grippo M, Heath G, Macknick J, Smith K, Sullivan R, Walston L, Wescott K (2016) Understanding emerging impacts and requirements related to utility-scale solar development. https://doi.org/10.13140/RG.2.2.21140.30087
88. Hein CD, Hale AM (2019) Wind energy effects on bats. In: Moorman CE, Grodsky SM and Rupp SP (eds.) Renewable energy and wildlife conservation. John Hopkins University Press, pp 122–145
89. Helliwell JF, Layard R, Sachs JD, Neve JE, Aknin LB, Wang S (2023) World happiness report 2023. Sustainable Development Solutions Network, New York
90. Hernandez RR, Easter SB, Murphy-Mariscal ML, Maestre FT, Tavassoli M, Allen EB, Barrows CW, Belnap J, Ochoa-Hueso R, Ravi R, Allen MF (2014) Environmental impacts of utility-scale solar energy. Renew Sustain Energy Rev 29:766–779. https://doi.org/10.1016/j.rser.2013.08.041
91. Hernandez RR, Hoffacker MK, Murphy-Mariscal ML, Wu GC, Allen MF (2015) Solar energy development impacts on land cover change and protected areas. PNAS 112(44):13579–13584. https://doi.org/10.1073/pnas.1602975113
92. Herre B, Arriagada P (2023) The human development index and related indices: what they are and what we can learn from them. Published online at OurWorldInData.org
93. Hicks J, Ison N (2018) An exploration of the boundaries of 'community' in community renewable energy projects: navigating between motivations and context. Energy Policy 113:523–534. https://doi.org/10.1016/j.enpol.2017.10.031
94. Hielscher S, Wittmayer J, Dańkowska A (2022) Social movements in energy transitions: the politics of fossil fuel energy pathways in the United Kingdom, the Netherlands and Poland. Extr Ind Soc 10:101073. https://doi.org/10.1016/j.exis.2022.101073
95. Hoffacker M, Allen M, Hernandez R (2017) Land-sparing opportunities for solar energy development in agricultural landscapes: a case study of the Great Central Valley, CA, United States. Environ Sci Technol 51:14472–14482. https://doi.org/10.1021/acs.est.7b05110
96. Huijbregts MAJ, Steinmann ZJN, Elshout PMF, Stam G, Verones F, Vieira M, Zijp M, Hollander A, van Zelm R (2017) ReCiPe2016: a harmonised life cycle impact assessment method at midpoint and endpoint level. Int J Life Cycle Assess 22(2):138–147. https://doi.org/10.1007/s11367-016-1246-y
97. IEA (2023b) Energy technology perspectives 2023. https://www.iea.org/reports/energy-technology-perspectives-2023
98. IEA (2022a) World energy outlook 2022. Paris, France: IEA. https://www.iea.org/reports/world-energy-outlook-2022
99. IEA (2022b) World energy employment. Updated in August 2022. https://www.iea.org/reports/world-energy-employment/overview

100. IEA (2023a) World energy investment 2023. https://www.iea.org/reports/world-energy-investment-2023
101. IPBES (2018) Summary for policymakers of the assessment report on land degradation and restoration of the Intergovernmental Science Policy Platform on Biodiversity and Ecosystem Services, pp 44. https://doi.org/10.5281/zenodo.3237411
102. IPBES (2019) Global assessment report on biodiversity and ecosystem services of the Intergovernmental Science-Policy Platform on Biodiversity and Ecosystem Services. Brondizio ES, Settele J, Díaz S, Ngo HT (eds). IPBES secretariat, Bonn, Germany, pp 1148. https://doi.org/10.5281/zenodo.3831673
103. IPCC (2019) Summary for policymakers. In: Shukla PR, Skea J, Calvo Buendia E, Masson-Delmotte V, Pörtner HO, Roberts DC, Zhai P, Slade R, Connors S, van Diemen R, Ferrat M, Haughey E, Luz S, Neogi S, Pathak M, Petzold J, Portugal Pereira J, Vyas P, Huntley E, Kissick K, Belkacemi M, Malley J (eds) Climate change and land: an IPCC special report on climate change, desertification, land degradation, sustainable land management, food security, and greenhouse gas fluxes in terrestrial ecosystems
104. IRENA (2020) Innovation landscape brief: community-ownership models. International Renewable Energy Agency, Abu Dhabi https://www.irena.org/-/media/Files/IRENA/Agency/Publication/2020/Jul/IRENA_Community_ownership_2020.pdf
105. IRENA (2022) Renewable power generation costs in 2021. International Renewable Energy Agency, Abu Dhabi
106. IRENA and ILO (2023) Renewable energy and jobs: annual review. International Renewable Energy Agency Abu Dhabi and International Labour Organization, Geneva
107. IUCN (2022) The IUCN red list of threatened species. https://www.iucnredlist.org
108. Jewell J, Emmerling J, Vinichenko V et al (2020) Reply to: Why fossil fuel producer subsidies matter. Nature 578:E5–E7. https://doi.org/10.1038/s41586-019-1921-9
109. Jewell J, McCollum D, Emmerling J et al (2018) Limited emission reductions from fuel subsidy removal except in energy-exporting regions. Nature 554:229–233. https://doi.org/10.1038/nature25467
110. Jimenez Iturriza J, Polo L, Beermann M, Lettmayer G, Carrus G, Tiberio L, Biresselioglu ME, Demir MH, Solak B, Dimitrova E, Tasheva-Petrova M, Mutafchiiska I, Burov A, Henriksen IM, Moe Skjølsvold T (2019) Collective energy practices in Europe. ECHOES Project Deliverable 5.4. https://echoes-project.eu/sites/echoes.drupal.pulsartecnalia.com/files/D5.4.pdf
111. Jones IL, Bunnefeld N, Jump AS, Peres CA, Daisy H. Dent DH (2016) Extinction debt on reservoir land-bridge islands. Biol Conserv 199:75–83. https://doi.org/10.1016/j.biocon.2016.04.036
112. Kamermans P, Walles B, Kraan M, van Duren L, Kleissen F, van der Have T, Smaal A, Poelman M (2018) Offshore wind farms as potential locations for flat oyster (*Ostrea edulis*) restoration in the Dutch North Sea. Sustainability 10(11):3942. https://doi.org/10.3390/su10113942
113. Kampman B, Blommerde J, Afman M (2016) The potential of energy citizens in the European Union. Delft, CE Delft
114. Kan D, Patel R, Pillay S, Dumont M, Broer W (2022) ASN Bank Biodiversity Footprint. Biodiversity Footprint For Financial Institutions Impact Assessment 2016–2020. PRé Sustainability. https://www.asnbank.nl/over-asn-bank/duurzaamheid/biodiversiteit/biodiversity-in-2030.html
115. Kaya Y, Yokobori K (1997) Environment, energy and economy: Strategies for sustainability. United Nations University Press, Tokyo
116. Kaza N, Curtis M (2014) The land use energy connection. J Plan Lit 29(4):355–369. https://doi.org/10.1177/0885412214542049
117. Keith DA, Rodríguez JP, Rodríguez-Clark KM, Nicholson E, Aapala K, Alonso A et al (2013) Scientific foundations for an IUCN red list of ecosystems. PLoS ONE 8(5):e62111. https://doi.org/10.1371/journal.pone.0062111

118. Kitzes J (2013) An introduction to environmentally-extended input-output analysis. Resources 2(4):489–503. MDPI AG. https://doi.org/10.3390/resources2040489
119. Kortetmäki T, Puurtinen M, Salo M, Aro R, Baumeister S, Duflot R, Elo M, Halme P, Husu HM, Huttunen S, Hyvönen K, Karkulehto S, Kataja-aho S, Keskinen KE, Kulmunki I, Mäkinen T, Näyhä A, Okkolin MA, Perälä T, Purhonen J, Raatikainen KJ, Raippalinna LM, Salonen K, Savolainen K, Kotiaho JS (2021) Planetary well-being. Humanit. Soc Sci 8:258
120. Kosanic A, Petzold J, Martín-López B, Razanajatovo M (2022) An inclusive future: disabled populations in the context of climate and environmental change. Curr Opin Environ Sustain 55:101159. https://doi.org/10.1016/j.cosust.2022.101159
121. Kotiaho JS, Ten Brink B, Harris J (2016) A global baseline for ecosystem recovery. Nature 532(7597):37–37
122. Kotiaho JS, Kuusela S, Nieminen E, Päivinen J, Moilanen A (2016a) Framework for assessing and reversing ecosystem degradation—report of the Finnish restoration prioritization working group on the options and costs of meeting the Aichi biodiversity target of restoring at least 15 percent of degraded ecosystems in Finland. Reports of the Ministry of the Environment, 15en, 2016. Helsinki, Finland. https://julkaisut.valtioneuvosto.fi/bitstream/handle/10024/74862/YMre_15en_2016.pdf?sequence=1&isAllowed=y
123. Krausmann F, Erb K-H, Gingrich S, Haberl H, Bondeau A, Gaube V, Lauk C, Plutzar C, Searchinger TD (2013) Global human appropriation of net primary production doubled in the 20th century. PNAS 110(25):10324–10329. https://doi.org/10.1073/pnas.1211349110
124. Kreitler J, Schloss CA, Soong O, Hannah L, Davis FW (2015) Conservation planning for offsetting the impacts of development: a case study of biodiversity and renewable energy in the Mojave Desert. PLoS ONE 10(11):e0140226. https://doi.org/10.1371/journal.pone.0140226
125. Kungl G (2015) Stewards or sticklers for change? incumbent energy providers and the politics of the German energy transition. Energy Res Soc Sci 8:13–23. https://doi.org/10.1016/j.erss.2015.04.009
126. Lacey F, Henze DK, Lee CJ, van Donkelaar A, Martin RV (2017) Transient climate and ambient health impacts due to national solid fuel cookstove emissions. PNAS 114(6):1269–1274. https://doi.org/10.1073/pnas.1612430114
127. Lafitte A, Sordello R, de Crespin de Billy V et al (2022) What evidence exists regarding the effects of photovoltaic panels on biodiversity? a critical systematic map protocol. Environ Evid 11(36). https://doi.org/10.1186/s13750-022-00291-x
128. Lambert Q, Gros R, Bischoff A (2022) Ecological restoration of solar park plant communities and the effect of solar panels. Ecol Eng 182:106722. https://doi.org/10.1016/j.ecoleng.2022.106722
129. Lammerant J, Driesen K, Verhelst J, De Ryck J (2022) Assessment of biodiversity measurement approaches for business and financial institutions. EU Business @ Biodiversity Platform. Update Report 4. https://ec.europa.eu/environment/biodiversity/business/assets/pdf/2022/Update%20Report%204_Final.pdf
130. Lavrijssen S, Parra AC (2017) Radical prosumer Innovations in the electricity sector and the impact on prosumer regulation. Sustainability 9(7):1–21. https://doi.org/10.3390/su9071207
131. Lazard (2023) Lazard's levelized cost analysis—Version 16.0
132. Ledec GC, Rapp KW, Aiello RG (2011) Greening the wind: Environmental and social considerations for wind power development. World Bank Studies. World Bank. http://hdl.handle.net/10986/2388
133. Lenzen M, Moran D, Kanemoto K, Foran B, Lobefaro L, Geschke A (2012) International trade drives biodiversity threats in developing nations. Nature 486(7401):109–112. https://doi.org/10.1038/nature11145
134. Leontief W (1970) Environmental repercussions and the economic structure: an input-output approach. Rev Econ Stat 52:262–271. https://doi.org/10.2307/1926294
135. Lis A, Stasik A (2017) Hybrid forums, knowledge deficits and the multiple uncertainties of resource extraction: Negotiating the local governance of shale gas in Poland. Energ Res Soc Sci 28:29–36. https://doi.org/10.1016/j.erss.2017.04.003

136. Lloyd JD, Cameron LA, Taber DA, LeBeau CW, McNew LB, Winder VL (2022) Prairie grouse and wind energy: the state of the science and implications for risk assessment. Wildl Soc Bull 46:e1305. https://doi.org/10.1002/wsb.1305
137. Lo Piano S, Mayumi K (2017) Toward an integrated assessment of the performance of photovoltaic power stations for electricity generation. Appl Energy 186(2):167–174. https://doi.org/10.1016/j.apenergy.2016.05.102
138. Łopucki R, Klich D, Gielarek S (2017) Do terrestrial animals avoid the areas close to turbines in the functioning wind farms in agricultural landscapes? Environ Monit Assess 189:343. https://doi.org/10.1007/s10661-017-6018-z
139. Lorek S, Spangenberg JH (2019) Energy sufficiency through social innovation in housing. Energy Policy 126:287–294. https://doi.org/10.1016/j.enpol.2018.11.026
140. Lovering J, Swain M, Blomqvist L, Hernandez RR (2022) Land-use intensity of electricity production and tomorrow's energy landscape. PLoS ONE 17(7):e0270155. https://doi.org/10.1371/journal.pone.0270155
141. Lovering J, Turkmen H (2011) Bulldozer neoliberalism in Istanbul: the state-led construction of property markets, and the displacement of the urban poor. Int Plan Stud 16(1):73–96. https://doi.org/10.1080/13563475.2011.552477
142. Magurran AE (2021) Measuring biological diversity. Curr Biol 31(19):R1174–R1177. https://doi.org/10.1016/j.cub.2021.07.049
143. Magurran AE, McGill BJ (Eds) (2010) Biological diversity: frontiers in measurement and assessment. OUP Oxford
144. Marquardt SG, Guindon M, Wilting HC, Steinmann ZJN, Sim S, Kulak M, Huijbregts MAJ (2019) Consumption-based biodiversity foot-prints—do different indicators yield different results? Ecol Ind 103:461–470. https://doi.org/10.1016/j.ecolind.2019.04.022
145. Marques A, Verones F, Kok MT, Huijbregts MA, Pereira HM (2017) How to quantify biodiversity footprints of consumption? A review of multiregional input–output analysis and life cycle assessment. Curr Opin Env Sust 29:75–81. https://doi.org/10.1016/j.cosust.2018.01.005
146. Marques AT, Santos CD, Hanssen F, Muñoz A, Onrubia A, Wikelski M, Moreira F, Palmeirim JM, Silva JP (2019) Wind turbines cause functional habitat loss for migratory soaring birds. J Anim Ecol 89(1):93–103. https://doi.org/10.1111/1365-2656.12961
147. Martín-Chivelet N (2016) Photovoltaic potential and land-use estimation methodology. Energy 94:233–242. https://doi.org/10.1016/j.energy.2015.10.108
148. Maruyama Y, Nishikido M, Iida T (2007) The rise of community wind power in Japan: enhanced acceptance through social innovation. Energy Policy 35(5):2761–2769. https://doi.org/10.1016/j.enpol.2006.12.010
149. Masden EA, Haydon DT, Fox AD, Furness RW (2010) Barriers to movement: modelling energetic costs of avoiding marine wind farms amongst breeding seabirds. Mar Pollut Bull 60(7):1085–1091. https://doi.org/10.1016/j.marpolbul.2010.01.016
150. Maxwell SL, Fuller RA, Brooks TM, Watson JEM (2016) Biodiversity: The ravages of guns, nets and bulldozers. Nature 536:143–145. https://doi.org/10.1038/536143a
151. Maxwell AE, Strager MP, Yuill CB, Petty JT (2012) Modeling critical forest habitat in the southern coal fields of West Virginia. Int J Ecol 2012:182683. https://doi.org/10.1155/2012/182683
152. McDonald RI, Fargione J, Kiesecker J, Miller WM, Powell J (2009) Energy sprawl or energy efficiency: climate policy impacts on natural habitat for the United States of America. PLoS ONE 4(8):e6802. https://doi.org/10.1371/journal.pone.0006802
153. McDonald A, Schrattenholzer L (2001) Learning rates for energy technologies. Energy Policy 29:255–261
154. McDowall W (2021) R&D tax credits can be a significant source of taxpayer support for fossil fuel innovation. Environ Res Lett 16:064061. https://doi.org/10.1088/1748-9326/ac0379
155. McKinney ML (2008) Effects of urbanization on species richness: a review of plants and animals. Urban Ecosyst 11(2):161–176
156. Middlemiss L, Ambrosio Albala P, Emmel N, Gillard R, Gilbertson J, Hargreaves T, Mullen C, Tony R, Snell C, Tod A (2019) Energy poverty and social relations: a capabilities approach. Energy Res Soc Sci 55:227–235. https://doi.org/10.1016/j.erss.2019.05.002

157. Millennium ecosystem assessment (2005) Ecosystems and human well-being: Wetlands and water. Synthesis, World Resources Institute, Washington, DC
158. Moilanen A, Kotiaho JS (2018) Fifteen operationally important decisions in the planning of biodiversity offsets. Biol Conserv 227:112–120
159. Moilanen A, Kotiaho JS (2021) Three ways to deliver a net positive impact with biodiversity offsets. Conserv Biol 35(1):197–205
160. Mollot G, Pantel JH, Romanuk TN (2017) The effects of invasive species on the decline in species richness: a global meta-analysis. Adv Ecol Res 56:61–83
161. Moore-O'Leary KA, Hernandez RR, Abella S, Johnston D, Kreitler J, Swanson A, Lovich J (2017) Sustainability of utility-scale solar energy: critical ecological concepts. Front Ecol Environ 15:385–394. https://doi.org/10.1002/fee.1517
162. Muscat A, de Olde EM, de Boer IJM, Ripoll-Bosch R (2020) The battle for biomass: a systematic review of food-feed-fuel competition. Glob Food Sec 25:100330. https://doi.org/10.1016/j.gfs.2019.100330
163. NASA (2011) The carbon cycle. https://earthobservatory.nasa.gov/features/CarbonCycle
164. Nakamura S, Nansai K (2016) Input–output and hybrid LCA. In: Finkbeiner M (Ed) Special Types of Life Cycle Assessment. LCA Compendium—the complete world of life cycle assessment. Springer Dordrecht. https://doi.org/10.1007/978-94-0177610-3_6
165. Natali SM, Holdren JP, Rogers BM, Treharne R, Duffy PB, Pomerance R, MacDonald E (2021) Permafrost carbon feedbacks threaten global climate goals 118(21):e2100163118. https://doi.org/10.1073/pnas.2100163118
166. Niblick B, Landis AE (2016) Assessing renewable energy potential on United States marginal and contaminated sites. Renew Sustain Energy Rev 60:489–497. https://doi.org/10.1016/j.rser.2015.12.045
167. Nicholson E, Watermeyer KE, Rowland JA, Sato CF, Stevenson SL, Andrade A, Watson JE (2021) Scientific foundations for an ecosystem goal, milestones and indicators for the post-2020 global biodiversity framework. Nat Ecol Evol 5(10):1338–1349
168. Norton M, Baldi A, Buda V, Carli B, Cudlin P, Jones MB, Korhola A, Michalski R, Novo F, Oszlányi J, Santos FD, Schink B, Shepherd J, Vet L, Walloe L, Wijkman A (2019) Serious mismatches continue between science and policy in forest bioenergy. Glob change biol. Bioenergy 11(11):1256–1263. https://doi.org/10.1111/gcbb.1264
169. Obermeyer B, Manes R, Kiesecker J, Fargione J, Sochi F (2011) Development by design: mitigating wind development's impacts on wildlife in Kansas. PLoS ONE 6(10):e26698. https://doi.org/10.1371/journal.pone.0026698
170. Ogilvie M, Rootes C (2015) The impact of local campaigns against wind energy developments. Env Polit 24(6):874–893. https://doi.org/10.1080/09644016.2015.1063301
171. Oksanen E, Kontunen-Soppela S (2021) Plants have different strategies to defend against air pollutants. Curr Opin Environ Sci Health 19:100222. https://doi.org/10.1016/j.coesh.2020.10.010
172. Ong S, Campbell C, Denholm P, Margolis R, Heath G (2013) Land-use requirements for solar power plants in the United States. Technical report NREL/TP-6A20–56290. Golden, Colorado: National Renewable Energy Laboratory. www.nrel.gov/docs/fy13osti/56290.pdf
173. Our World in Data (2023) Kaya identity: drivers of CO_2 emissions, World. https://ourworldindata.org/grapher/kaya-identity-co2
174. O'Neill (2024) https://www.statista.com/statistics/268750/global-gross-domestic-product-gdp/
175. Palmeirim AF, Emer C, Benchimol M, Storck-Tonon D, Bueno AS, Peres CA (2022) Emergent properties of species-habitat networks in an insular forest landscape. Sci Adv 8:eabm0397. https://doi.org/10.1126/sciadv.abm0397
176. Parkes D, Newell G, Cheal D (2003) Assessing the quality of native vegetation: the 'habitat hectares' approach. Ecol Manag Restor 4:S29–S38
177. Pistone K, Eisenman I, Ramanathan V (2014) Observational determination of albedo decrease caused by vanishing Arctic sea ice. PNAS 111(9):3322–3326. https://doi.org/10.1073/pnas.1318201111

178. Purvis A, Hector A (2000) Getting the measure of biodiversity. Nature 405(6783):212–219
179. Pörtner HO, Scholes RJ, Agard J, Ngo HT (2021) IPBES-IPCC co-sponsored workshop report on biodiversity and climate change. IPBES and IPCC. https://doi.org/10.5281/zenodo.4782538
180. Raghukumar K, Chartrand C, Chang G, Cheung L, Roberts J (2022) Effect of floating offshore wind turbines on atmospheric circulation in California. Front Energy Res 10:863995. https://doi.org/10.3389/fenrg.2022.863995
181. Ratinen M, Lund P (2014) Growth strategies of incumbent utilities as contextually embedded: examples from Denmark, Germany, Finland and Spain. Technol Soc 38:81–92. https://doi.org/10.1016/j.techsoc.2014.02.005
182. Ratinen M, Lund P (2015) Policy inclusiveness and niche development: examples from wind energy and photovoltaics in Denmark, Germany, Finland, and Spain. Energy Res Soc Sci 6:136–145. https://doi.org/10.1016/j.erss.2015.02.004
183. Reinsberger K, Brudermann T, Hatzl S, Fleiß E, Posch A (2015) Photovoltaic diffusion from the bottom-up: Analytical investigation of critical factors. Appl Energy 159:178–187. https://doi.org/10.1016/j.apenergy.2015.08.117
184. Richardson K, Steffen W, Lucht W, Bendtsen J, Cornell SE, Donges JF, Drüke M, Fetzer I, Bala G, von Bloh W, Feulner G, Fiedler S, Gerten D, Gleeson T, Hofmann M, Huiskamp W, Kummu M, Mohan C, Nogués-Bravo D, Petri S, Porkka M, Rahmstorf S, Schaphoff S, Thonicke K, Tobian A, Virkki V, Wang-Erlandsson L, Weber L, Rockström J (2023) Earth beyond six of nine planetary boundaries. Sci Adv 9:eadh2458. https://doi.org/10.1126/sciadv.adh2458
185. Richter M (2013) German utilities and distributed PV: how to overcome barriers to business model innovation. Renew Energy 55:456–466. https://doi.org/10.1016/j.renene.2012.12.052
186. Richter M (2011) Utilities' business models for renewable energy: evidence from Germany. World Renewable Energy Congress 2011–Sweden 8-13 May 201, Linköping, Sweden
187. Ripple WJ, Wolf C, Gregg JW, Rockström J, Newsome TM, Law BE, Marques L, Lenton TM, Xu C, Huq S, Simons L, King DA Sir (2023) The 2023 state of the climate report: Entering uncharted territory. BioScience 0:1–10 Special report. https://doi.org/10.1093/biosci/biad080
188. Ritchie H (2022) How does the land use of different electricity sources com-pare? Our world in data. https://ourworldindata.org/land-use-per-energy-source
189. Rockström J, Gupta J, Qin D et al (2023) Safe and just earth system boundaries. Nature 619:102–111. https://doi.org/10.1038/s41586-023-06083-8
190. Rockström J, Steffen W, Noone K, Persson Å, Chapin FS, Lambin E, Lenton TM, Scheffer M, Folke C, Schellnhuber H, Nykvist B, De Wit CA, Hughes T, van der Leeuw S, Rodhe H, Sörlin S, Snyder PK, Costanza R, Svedin U, Falkenmark M, Karlberg L, Corell RW, Fabry VJ, Hansen J, Walker B, Liverman D, Richardson K, Crutzen P, Foley J (2009) Planetary boundaries: exploring the safe operating space for humanity. Ecol Soc 14(2): 32. http://www.ecologyandsociety.org/vol14/iss2/art32/
191. Rockström J, Gaffney O (2021) Breaking boundaries: the science of our planet. DK, a division of Penguin Random House LLC. pp 239
192. Rogers JC, Simmons EA, Convery I, Weatherall A (2012) Social impacts of community renewable energy projects: findings from a woodfuel case study. Energy Policy 42:239–247. https://doi.org/10.1016/j.enpol.2011.11.081
193. Rubin ES, Azevedo IML, Jaramillo P, Yeh S (2015) A review of learning rates for electricity supply technologies. Energy Policy 86:198–218. https://doi.org/10.1016/j.enpol.2015.06.011
194. Sadiqa A, Sahrakorpi T, Keppo I (2023) Mapping the gender vulnerabilities in low carbon energy transitions. Environ Res Lett 18:043004. https://doi.org/10.1088/1748-9326/acc819
195. Schaffartzik A, Sachs M, Wiedenhofer D, Eisenmenger N (2014) Environmentally extended Input-Output Analysis. https://www.uni-klu.ac.at/socec/downloads/WP154_webversion_v3.pdf
196. Scherer L, Pfister S (2016) Global water footprint assessment of hydropower. Renewable Energy 99:711–720. https://doi.org/10.1016/j.renene.2016.07.021

197. Scovell MD (2022) Explaining hydrogen energy technology acceptance: a critical review. Int J Hydr Energy 47. https://doi.org/10.1016/j.ijhydene.2022.01.099
198. Searle KR, O'Brien SH, Jones EL, Cook ASCP, Trinder MN, McGregor RM, Donovan C, McCluskie A, Daunt F, Butler A (2023) A framework for improving treatment of uncertainty in offshore wind assessments for protected marine birds. ICES J Mar Sci. https://doi.org/10.1093/icesjms/fsad025
199. Seidl R, von Wirth T, Krütli P (2019) Social acceptance of distributed energy systems in Swiss, German, and Austrian energy transitions. Energy Res Soc Sci 54:117–128. https://doi.org/10.1016/j.erss.2019.04.006
200. Shao S, Huang T, Yang L (2014) Using latent variable approach to estimate China's economy-wide energy rebound effect over 1954–2010. Energy Policy 72:235–248
201. Sharma N, Bohra B, Pragya N, Ciannella R, Dobie P, Lehmann S (2016) Bioenergy from agroforestry can lead to improved food security, climate change, soil quality, and rural development. Food Energy Secur 5(3):165–183. https://doi.org/10.1002/fes3.87
202. Sharpton T, Lawrence T, Hall M (2020) Drivers and barriers to public acceptance of future energy sources and grid expansion in the United States. Renew Sust Energ Rev 126:109826. https://doi.org/10.1016/j.rser.2020.109826
203. Shin Y-J, Midgley GF, Archer ERM, Arneth A, Barnes DKA, Chan L, Hashimoto S, Hoegh-Guldberg O, Insarov G, Leadley P, Levin LA, Ngo HT, Pandit R, Pires APF, Pörtner H-O, Rogers AD, Scholes RJ, Settele J, Smith P (2022) Actions to halt biodiversity loss generally benefit the climate. Glob Chang Biol 28(9):2846–2874. https://doi.org/10.1111/gcb.16109
204. Sovacool BK, Brugger H, Brunzema I, Dańkowska A, Wemyss D, Vernay A-L, Betz R, Avelino F, de Geus T, Dembek A, Duetschke E, Hielscher S, Iskandarova M, Müller L, Musiolik J, Ranville A, Schleich J, Stasik A, Strumińska-Kutra M, Winzer C, Wittmayer J, Rogge KS (2023) Social innovation supports inclusive and accelerated energy transitions with appropriate governance. Commun Earth Environ 4:289. https://doi.org/10.1038/s43247-023-00952-w
205. Sovacool BK, Martiskainen M, Hook A, Baker L (2019) Decarbonization and its discontents: a critical energy justice perspective on four low-carbon transitions. Clim Change 155:581–619. https://doi.org/10.1007/s10584-019-02521-7
206. Statistics Finland (2022) Fuel classification. https://www.stat.fi/tup/khkinv/khkaasut_polttoaineluokitus.html
207. Steffen W, Richardson K, Rockström J, Cornell SE, Fetze I, Bennett EM, Biggs R, Carpenter SR, de Wries W, De Wit CA, Folke C, Gerten D, Heinke J, Mace GM, Persson LM, Ramanathan V, Reyers B, Sörlin S (2015) Planetary boundaries: guiding human development on a changing planet. Science 347:1259855. https://doi.org/10.1126/science.1259855
208. Suh S, Lenzen M, Treloar GJ, Norris G (2004) System boundary selection in life-cycle inventories using hybrid approaches. Environ Sci Technol 38:657–664. https://doi.org/10.1021/ES0263745/SUPPL_FILE/ES0263745SI20021202_090311.PDF
209. Taylor M (2020) Energy subsidies: evolution in the global energy transformation to 2050. Int Ren Energy Agency, Abu Dhabi
210. Tokarska K, Gillett N, Weaver AJ, Arora VK, Eby M (2016) The climate response to five trillion tonnes of carbon. Nature Clim Change 6:851–855. https://doi.org/10.1038/nclimate3036
211. Towler GP, Sinnott RK (2013) Chemical engineering design: principles, practice and economics of plant and process design, 2nd edn. Butterworth-Heinemann.
212. Tsoutsos T, Frantzeskaki N, Gekas V (2005) Environmental impacts from the solar energy technologies. Energy Policy 33:289–296. https://doi.org/10.1016/S0301-4215(03)00241-6
213. Tubiello FN, Salvatore M, Cóndor Golec RD, Ferrara A, Rossi S, Biancalani R, Federici S, Jacobs H, Flammini A (2014) Agriculture, forestry and other land use emissions by sources and removals by sinks, 1990–2011 analysis. FAO Statistics Division Working Paper Series ESS/14–02. Food and Agriculture Organization, Rome
214. UNDP (2022) The 2021/2022 Human development report. United nations development program, eISBN 9789210016407, New York, USA

215. UNECE (2022) Carbon neutrality in the UNECE Region: Integrated life-cycle assessment of electricity sources. https://unece.org/sites/default/files/2022-04/LCA_3_FINAL%20March%202022.pdf
216. United Nations (2015) Transforming our world: the 2030 Agenda for Sustainable Development. Resolution adopted by the General Assembly on 25 September 2015, A/RES/70/1
217. United Nations (2021a) Life cycle assessment of electricity generation options, United Nations economic commission for Europe, UNECE, GENEVA. https://unece.org/sites/default/files/2022-04/LCA_3_FINAL%20March%202022.pdf
218. United Nations (2021b) Theme report on energy transition, Towards the achievement of SDG 7 and net-zero emissions. High level dialogue on energy, United Nations, New York. https://www.un.org/sites/un2.un.org/files/2021-twg_2-062321.pdf
219. United Nations (2022) World population prospects. DESA, Population division. https://population.un.org/wpp/
220. Upham P, Sovacool B, Ghosh B (2022) Just transitions for industrial decarbonisation: a framework for innovation, participation, and justice. Renew Sustain Energy Rev 167:112699. https://doi.org/10.1016/j.rser.2022.112699
221. Vainio V, El Geneidy S (2021) Sustainability for JYU. Jyväskylän yliopiston ilmasto- ja luontohaitat 2020. JYU Reports. https://doi.org/10.17011/jyureports/2021/13
222. Vainio V, El Geneidy S, Peura M, Halme P, Kotiaho J S (forthcoming) Biodiversity impact of the consumption of peat and wood-fired district heating
223. van Zalk J, Behrens P (2018) The spatial extent of renewable and non-renewable power generation: a review and meta-analysis of power densities and their application in the US. Energy Policy 123:83–91. https://doi.org/10.1016/j.enpol.2018.08.023
224. Vanek FM, Albright L, Angenent LT (2016) Energy systems engineering: Evaluation and implementation. Third edition, McGraw Hill, ISBN: 978-1259585098
225. Verones F, Hellweg S, Antón A, Huijbregts MAJ (2020) LC-IMPACT: a regionalized life cycle damage assessment method. J Ind Ecol 24(6):1201–1219. https://doi.org/10.1111/jiec.13018
226. Vivanco DF, Kemp R, Voet E (2016) How to deal with the rebound effect? A policy-oriented approach. Energy Policy 94:114–125
227. Wahlund M, Palm J (2022) The role of energy democracy and energy citizenship for participatory energy transitions: a comprehensive review. Energy Res Soc Sci 87:102482. https://doi.org/10.1016/j.erss.2021.102482
228. Walker BH (1992) Biodiversity and ecological redundancy. Conserv Biol 6(1):18–23
229. Walker G, Devine-Wright P (2008) Community renewable energy: what should it mean? Energy Policy 36(2):497–500. https://doi.org/10.1016/j.enpol.2007.10.019
230. Walston LJ Jr, Rollins KE, Lagory KE, Smith KP, Meyers SA (2016) A preliminary assessment of avian mortality at utility-scale solar energy facilities in the United States. Renew Energy 92:405–414. https://doi.org/10.1016/j.renene.2016.02.041
231. Wang S, Wang S (2015) Impacts of wind energy on environment: a review. Renew Sustain Energy Rev 49:437–443
232. White H (2009) Theory-based impact evaluation: principles and practice. J Dev Effect 1(3):271–284. https://doi.org/10.1080/19439340903114628
233. Willemen L, Barger NN, ten Brink B, Cantele M, Erasmus BFN, Fisher JL, Gardner T, Holland TG, Kohler F, Kotiaho JS, Von Maltitz G, Nangendo G, Pandit R, Parrotta J, Potts MD, Prince SD, Sankaran M, Brainich A, Montanarella L, Scholes R (2020) How to halt the global decline of lands. Nat Sustain 3:164–166
234. Winemiller KO et al (2016) Balancing hydropower and biodiversity in the Amazon, Congo, and Mekong. Science 351:128–129. https://doi.org/10.1126/science.aac7082
235. Wittmayer JM, de Geus T, Pel B, Avelino F, Hielscher S, Hoppe T, Mühlemeier S, Stasik A, Oxenaar S, Rogge KS, Visser V, Marín-González E, Ooms M, Buitelaar S, Foulds C, Petrick K, Klarwein S, Krupnik S, de Vries G, Wagner A, Härtwig A (2020) Beyond instrumentalism: broadening the understanding of social innovation in socio-technical energy systems. Energy Res Soc Sci 70:101689. https://doi.org/10.1016/j.erss.2020.101689

236. Wolfson R (2018) Energy, environment and climate. W.W. Norton & Company, New York, ISBN 9780393622911
237. Wu H et al (2019) Effects of dam construction on biodiversity: a review. J Clean Prod 221:480–489. https://doi.org/10.1016/j.jclepro.2019.03.001
238. Wu G, Torn M, Williams J (2015) Incorporating land-use requirements and environmental constraints in low-carbon electricity planning for California. Environ Sci Technol 49:2013–2021
239. Wuebbles DJ, Fahey DW, Hibbard KA, Dokken DJ, Stewart BC, Maycock TK (eds) (2017) Climate science special report: fourth national climate assessment, Volume I, U.S. Global Change Research Program, Washington, DC, USA
240. Yaka Ö (2018) Rethinking justice: Struggles for environmental commons and the notion of socio-ecological justice. Antipode 51:353–372. https://doi.org/10.1111/anti.12422
241. Zemp D, Schleussner CF, Barbosa H, Hirota M, Montade V, Sampaio G, Staal A, Wang-Erlandsson L, Rammig A (2017) Self-amplified Amazon forest loss due to vegetation-atmosphere feedbacks. Nat Commun 8:14681

Chapter 3
Thermal Power Plant Technologies and Heat Engines for Renewable Sources

Mika Järvinen, Shouzhuang Li, and Hanna Paulomäki

In the heat of the sun, the ocean is the boiler and condenser of a gigantic steam engine, a weather engine that governs crops, floods, droughts, frosts, hurricanes.
—Jacques-Yves Cousteau

The objective of this chapter is to describe the operational principles of the mature, robust and traditional power and heat production processes such as the steam turbine and gas turbine-based power plant concepts. These are mostly still today—in 2025—driven by fossil-fuel energy sources. However, they can be and are already used with several renewable energy sources, such as solar thermal, geothermal and sustainably grown and harvested bioenergy (Fig. 3.1). This chapter will give the tools to optimize design processes and discusses the theoretical limitations of process efficiencies.

The processes explained in this chapter can be used with several energy applications, and to be able to apply them in combination with renewable and sustainable

Supplementary Information The online version contains supplementary material available at https://doi.org/10.1007/978-3-031-69856-9_3.

M. Järvinen (✉) · S. Li · H. Paulomäki
Aalto University, Espoo, Finland
e-mail: mika.jarvinen@aalto.fi

H. Paulomäki
e-mail: hanna.paulomaki@lut.fi

M. Järvinen and H. Paulomäki (eds.), *Designing Renewable Energy Systems within Planetary Boundaries*, Green Energy and Technology,
https://doi.org/10.1007/978-3-031-69856-9_3

Fig. 3.1 Concentrated solar power, Gemasolar Thermosolar 20 MW in Spain, and Hellisheiði geothermal power station in Iceland, 303 MW electricity + 133 MW district heating (Creative Commons Attribution-Share Alike 4.0 International: Kallerna and Sigrg)

sources, engineers need to understand their operation properly. For that reason, a concentrated package of the technologies and required theory are included, though at first glance this might appear very different from the other topics covered by the book. The core of these thermal power plants was developed during the 1800s. The objective of this chapter is to deepen theoretical knowledge, using several numeral examples and give the reader idea of the selection of process variables; for example re-heating and regenerative feed-water heating systems.

Mass and energy balance of open flow systems will be introduced first. After familiarization with these basic tools it is possible to use them in an endless number of applications. Then, we will study the basic Carnot process: a key concept in understanding any kind of heat engine. After this, the basics of the Rankine cycle-based steam power plant is analyzed. Several ways to improve the process efficiency will be introduced as well, including reheating and regenerative feed-water preheating. Some advanced cases can be found in Appendixes A3.1 and A3.2.

Next, the principles of gas turbines and hot air processes, such as the Brayton, Ericsson and Sterling cycles, will be discussed. It will be shown that the thermal efficiency of the gas turbine process is limited largely by the process pressure ratio and the turbine inlet temperature. In addition to this, the exit gas temperature from the basic process is typically very high. We will analyze several ways to improve the thermal efficiency. These process improvements often require efficient heat exchangers that are expensive and challenging to manufacture. It is very interesting to see, what, for example, 3D printing of metals can bring to this field in the future.

One very efficient and already widely used method to increase the efficiency of the gas turbine process is to recover the heat from the turbine exit flue gases and use this energy to run a steam turbine-based bottom process. Using this coupling, the fraction of achievable power from the input heat to the gas turbine can be significantly increased.

Finally, combined heat and power production is discussed. It offers many opportunities for utilizing the waste heat generated during thermal processes. For example, the waste heat in the Rankine cycle with 0.04 bar condenser pressure has very low temperature, close to the environmental value. This temperature level of heat cannot be utilized for power production or for direct heating. By increasing the condenser pressure, the temperature and the value of the waste heat can be increased. Typical applications are district heating and different industrial processes.

Most of the renewable energy systems described in this book are designed to produce continuous 500 MW of electrical power or heat, so they can be easily compared with each other for technical, economical, material and environmental performance. Even though these thermal power plants require a small land area when operating, they can have very large impacts on nature, for example by using significant amounts of biomass that must be collected from a widespread area. Or the in the case of geothermal power plants, the issues related to possible ground water contamination must be carefully analyzed [33].

3.1 Mass and Energy Balances of an Open Flow System

Before thermal power plants and their many required components can be analyzed, the derivation and analysis of the mass and energy balances of an open flow system need to be discussed. We use the control volume approach and derive these balances for time dependent general cases. For traditional power plant systems, relying for example on combustion of biomass or waste, the steady state versions can be used.

Figure 3.2 presents an open flow system with N inlets or exits. We define positive (+) direction as the inflow direction. Summing up all mass flow rates gives us the change rate, kg/s, of mass within the system, defined by the borders of the control volume. Change of mass can take place due to a change in density and/or volume of the observed system.

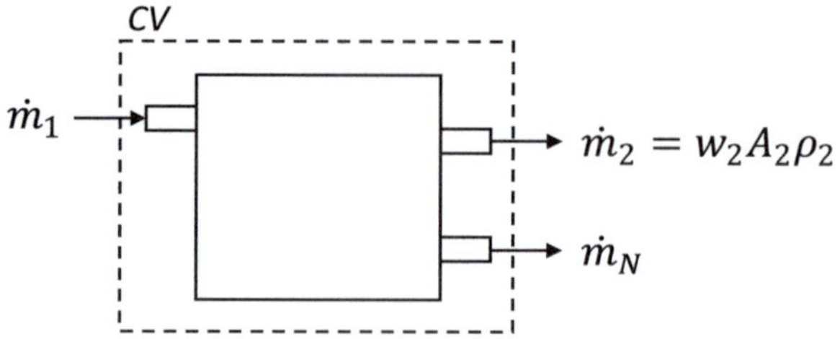

Fig. 3.2 Mass balance of an open flow system

$$\dot{m}_1 - \dot{m}_2 + \cdots - \dot{m}_N = \sum_{i=1}^{N} \dot{m}_i = \frac{dm}{dt} = V\left.\frac{d\rho}{dt}\right|_V + \rho\left.\frac{dV}{dt}\right|_\rho$$

And for a system in steady state, this reduces into.

$$\sum_{i=1}^{N} \dot{m}_i = 0$$

The energy conservation equation for an open flow system is a bit more complex and requires a few additional terms to be defined. One new concept required is the flow work. If the flow is inwards, pressure at an interface pushes incremental fluid elements into the control volume. The work related to this unit process is called the *flow work*. Figure 3.3 presents the system and gives a magnified view of one of its inlets/outlets.

Incremental fluid elements are presented just before they are pushed in and out from the inlet and outlet, respectively. These elements also contain specific total energy (J/kg), e_{tot}, including internal energy as $u(T_i, p_i)$, kinetic energy $w_i^2/2$, and potential energy $z_i \rho(T_i, p_i) g$ and these are transported into and out from the *CV*

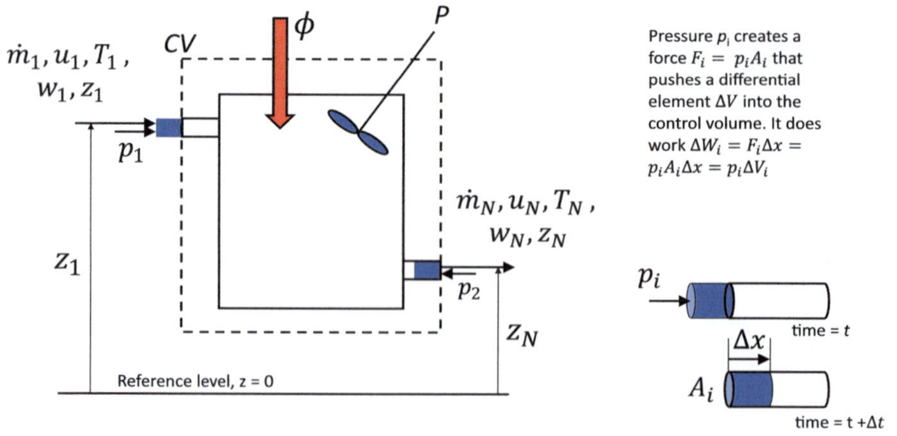

Fig. 3.3 Control volume for the energy balance

(Control Volume) with the moving elements during the time step. In addition to these, during this incremental time step, there is mechanical work done $P\Delta t$ and there is also energy exchange via heat transfer $\phi\Delta t$ taking place. Let us collect all the terms on the left-hand side and mark this as the change of the total energy in the control volume.

$$\sum_{i=1}^{N}\Delta m_i\left(u_i+\frac{w_i^2}{2}+z_i g\right)+\sum_{i=1}^{N}p_i\Delta V_i+P\Delta t+\phi\Delta t=\Delta E$$

The incremental volume corresponds to a mass element $\Delta V_i = v_i\Delta m_i$. We implement this and then collect all the terms containing Δm_i.

$$\sum_{i=1}^{N}\Delta m_i\left(u_i+p_i v_i+\frac{w_i^2}{2}+z_i g\right)+P\Delta t+\phi\Delta t=\Delta E$$

Mass transported via each inlet/outlet is i: $\Delta m_i = \dot{m}_i\Delta t$. We then divide both sides of the equation with Δt and then let $\Delta t \to 0$.

$$\sum_{i=1}^{N}\dot{m}_i\left(u_i+p_i v_i+\frac{w_i^2}{2}+z_i g\right)+P+\phi=\lim_{\Delta t\to 0}\left(\frac{\Delta E}{\Delta t}\right)=\frac{dE}{dt}$$

The sum of specific internal energy and flow work is called as the enthalpy, h. With this definition, we finally have.

$$\sum_{i=1}^{N}\dot{m}_i\left(h_i+\frac{w_i^2}{2}+z_i g\right)+P+\phi=\frac{dE}{dt}$$

We still need to define total energy for the right-hand side. This includes internal energy, potential energy and kinetic energy, as already defined.

$$E=\dot{m}e_{tot}=m\left(u+\frac{w^2}{2}+zg\right)=m\left(h-pv+\frac{w^2}{2}+zg\right)$$

The role of different terms in the accumulated energy within the control volume vary depending on the application; this will be discussed in detail in the coming sections. For steady state power plant calculation, transient terms are absent. If there are several phases present, like in water evaporating, these need to be summed up and multiplied with the masses. We will discuss this in the following examples:

Example 3.1 Flow of water out from a canister.

We will analyze flow measurements from a canister of water, with water flowing freely through a small nozzle (Fig. 3.4). The objective of the example is to show

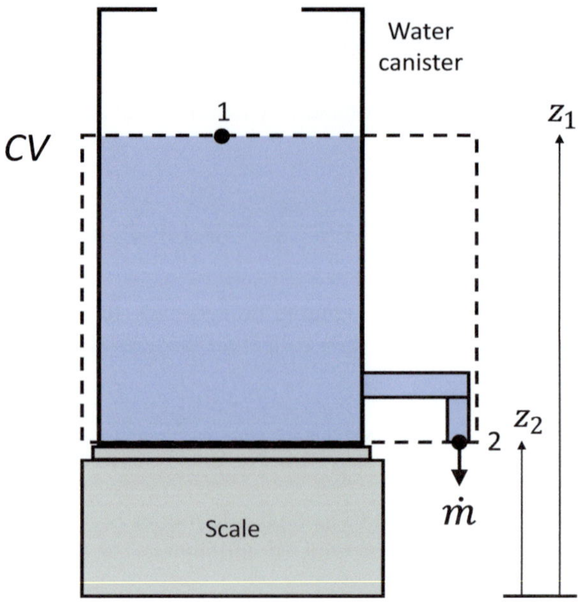

Fig. 3.4 Emptying a water canister

how to use conservation equations in a very simple case. First the mass and energy balances are constructed and then we will analyze the model.

The flowing water has a constant temperature. There is height difference between the upper and lower surfaces, 1 and 2, $\Delta z = h_1 - h_2 = 190\,\text{mm}$. The upper open liquid surface has a diameter of ~ 25 cm and the nozzle has an exit area of 3.5 mm. Water density is assumed 1000 kg/m^3. Imagine that the upper surface is momentarily fixed and the mass flow rate of water at the upper surface is equal to the exit mass flow rate. This assumption is referred to as the quasi-steady state approximation, meaning that when observing the system it seems to stay constant. In this case the upper surface barely decreases during the experiment due to the small mass flow rate, and this assumption is justified. The mass and energy balances are written below.

$$-\dot{m}_2 = \frac{dm}{dt} = -w_2 \rho \frac{\pi d_2^2}{4}$$

$$\dot{m}\left(h_1 + \frac{w_1^2}{2} + z_1 g\right) - \dot{m} e_{loss} = \dot{m}\left(h_2 + \frac{w_2^2}{2} + z_2 g\right)$$

Within the system, the temperature, pressure and density are the same at both points 1 and 2, so enthalpy is the same. The velocity is significantly smaller at the upper surface than at the exit nozzle and it can also be neglected $w_1 \ll w_2$. There are always losses in flow systems. Here, we assume that they are proportional to kinetic energy defined by the largest velocity at the nozzle exit, *i.e.* $e_{loss} = C w_2^2/2$. With these assumptions, the energy equation can be simplified.

$$\Delta zg - \frac{w_2^2}{2} = C\frac{w_2^2}{2}$$

This states that potential energy at the upper surface, minus the flow losses is equal to the kinetic energy at the exit. From this, the loss coefficient can be solved.

$$C = \frac{2\Delta zg}{w_2^2} - 1$$

Based on the stated measurements $\dot{m} = 0.0153$ kg/s, and the velocity at the nozzle is equal to 1.6 m/s. With these values, the coefficient $C = 0.456$.

Example 3.2 Boiling water in a kettle.

Figure 3.5 presents a case where water is boiled on an electric stove. This example is used to explain how to apply mass and energy balance equations in a transient situation, where heat and phase both change and there are two components. Electric power is fed into the control volume and water vapor flows out. The conversion efficiency of electricity to heat is 100%, so we can assume that the power is fully converted into heat. A part of the heat is lost by radiation and convection. Also, a part of the heat goes into heating up the metal and plastic parts of the system, including the kettle and the stove itself. Initially the system is at 25 °C. The model needs to be derived to solve the fraction of heat loss that occurs during the heat-up and boiling phases.

The starting point of our derivation is the mass and energy balance in their full form.

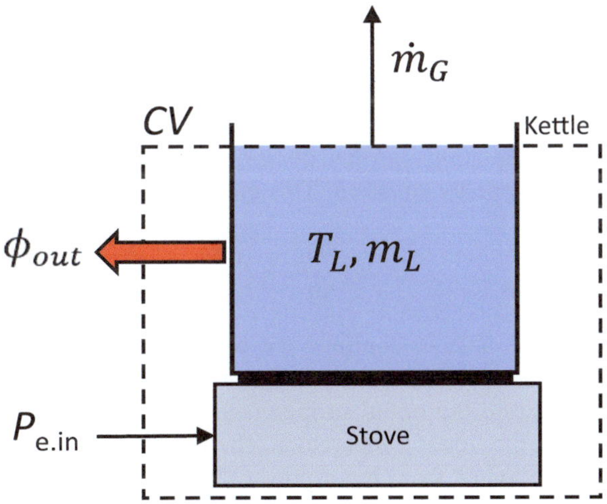

Fig. 3.5 Boiling of water on an electric stove

$$\sum_{i=1}^{N} \dot{m}_i = \frac{dm}{dt}$$

$$\sum_{i=1}^{N} \dot{m}_i \left(h_i + \frac{w_i^2}{2} + z_i \rho g \right) + P - \phi = \frac{dE}{dt}$$

In this example, kinetic and potential energy can be neglected, which greatly simplifies the equations. The accumulation term of the energy equation requires some additional consideration. There is metal, water vapor and water inside the control volume and these all have different thermodynamic properties and masses. Since it is very difficult to formulate an accurate model for the metal and plastic parts of the stove, we will include the heating requirement of these parts into the heat loss term. Let's begin by writing the total energy accumulated inside the control volume. Internal energy is presented here in terms of enthalpy.

$$E = m_L(h_L - p_L v_L) + m_G(h_G - p_G v_G)$$

Assuming that the kettle is mostly filled with water means that the mass of vapor is negligible. In addition, the ratio of liquid and vapor densities is ~ 1000, so this assumption is valid even with only a small amount of water in the kettle. If we compare liquid terms in E, we can see that the enthalpy term clearly dominates, and the $p_L v_L$ term can be omitted ($h_{\mathrm{L}} = 418{,}940$ J/kg, $p_L v_L = 106$ J/kg at 100 °C, 101,325 Pa). Using these assumptions, we have the following simplified equations. Note that $\partial()/\partial t$ refers to a partial derivate, i.e. all other variables are kept constant.

$$-\dot{m}_G = \frac{\partial m_L}{\partial t}$$

$$-\dot{m}_G h_G + P - \phi = \frac{dE}{dt} = \frac{d(m_L h_L)}{dt} = h_L \frac{\partial m_L}{\partial t} + m_L c_{p.L} \frac{\partial T_L}{\partial t}$$

The equations can then be combined by inserting the steam mass flow rate equation into the energy equation. This gives us

$$h_G \frac{\partial m_L}{\partial t} + P - \phi = h_L \frac{\partial m_L}{\partial t} + m_L c_{p.L} \frac{\partial T_L}{\partial t},$$

To be able to continue the temperatures need to be defined. We assume that the water is continuously fully mixed and that the upper control volume surface is attached to the liquid surface (see Fig. 3.4) and as a result the vapor temperature equals the liquid temperature. The following final equation gives us the fraction of energy losses from input power. Note that difference between enthalpies of vapor and liquid is the heat of evaporation at temperature T_L.

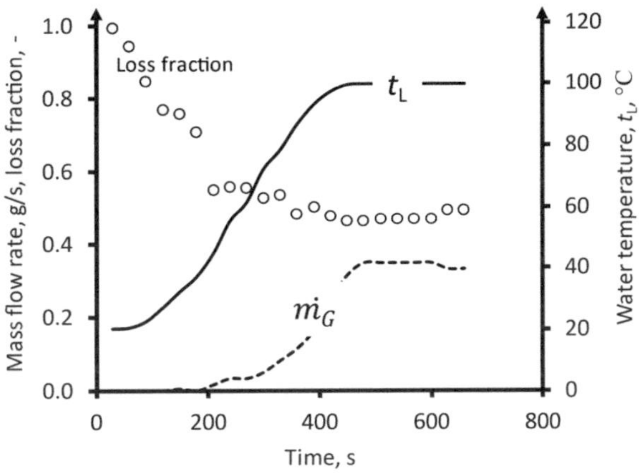

Fig. 3.6 Calculated results from measurements

$$\frac{\partial m_L}{\partial t}(h_G(T_L) - h_L(T_L)) + P - \phi = m_L c_{p.L} \frac{\partial T_L}{\partial t}$$

$$\frac{\phi}{P} = 1 - \frac{\partial m_L}{\partial t} \frac{l(T_L)}{P} - \frac{m_L c_{p.L}}{P} \frac{\partial T_L}{\partial t}$$

The second term on the right is the fraction of power to evaporation and the last term gives the fraction of energy going into heating the water. Figure 3.6 presents measured data and calculates the fraction of energy losses. The initial amount of distilled water was 501 g, and room temperature was 22 °C. The average heating power was 1450 W.

There is no detectable evaporation taking place during the first 170 s (Fig. 3.6). Then, after the temperature increases above 40 °C and approaches boiling point, mass flow increases and finally reaches a constant heat transfer limited value of 0.35 g/s. From these results we can conclude that in the beginning most of the energy is used to heat up the system components and then, when we reach a steady state, 47–50% of the power is lost as heat. If the goal is to minimize heat loss, insulation should be used, or the heater submerged in water.

Example 3.3 Blowing air into a balloon.

In this example we demonstrate a case of using conservation equations for a variable volume and variable density system by blowing air into a balloon with a diameter d (Fig. 3.7). We have measured the pressure and flow rate into the ball accurately with time. The objective of this example is to study a system with varying volume and density, and to discuss the importance of understanding the reference state of measurement data. Pressure can be given in terms of the *absolute pressure* (compared with a vacuum) or as the *gauge pressure* (as a pressure difference to atmospheric pressure). With thermodynamic state equations we always need to use the absolute pressure, as we also use absolute temperature, in K°. On the other hand, when calculating forces or work, care must be taken to use the gauge pressure.

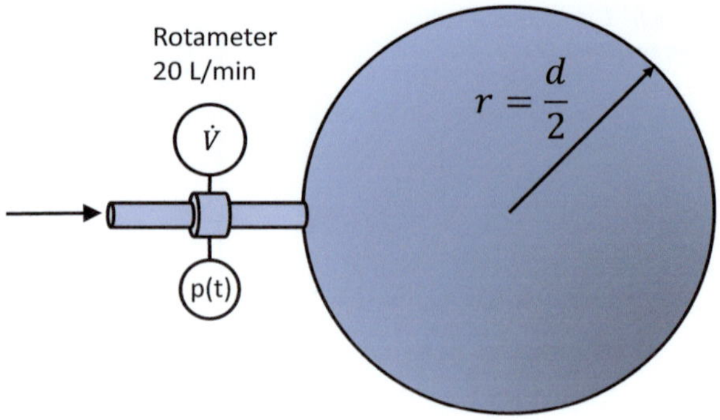

Fig. 3.7 Filling an air balloon

We assume that the heat transfer to the surroundings is efficient and that the temperature in the balloon equalizes quickly with the surrounding temperature. Gauge pressure is used, so in calculating the density, we need to add the value of atmospheric pressure to measured value. The rotameter measuring the volumetric flow rate, is calibrated at 101,325 Pa, so the flow correction for increased pressure is needed, multiplying the measured value with the term $\sqrt{p_0/(p_0 + \Delta p)}$, [52]. The volume of the balloon is $V = \pi d^2/6$. Mass flow rate of air into the system and mass balance are defined as follows.

$$\dot{m} = \dot{V}\rho = \dot{V}_0\sqrt{\frac{p_0}{p_0 + \Delta p}}\frac{(p_0 + \Delta p)M_{air}}{RT} = \frac{dm}{dt}$$

Based on the measured pressure and reading from the rotameter ($\dot{V}_0 = 20$ L/min) the mass and density of the air in the balloon can be calculated as a function of time by numerical integration of the above equation with respect to time. From these, the size of the balloon can be calculated (Fig. 3.8) together with the measured pressure.

$$d(t) = \sqrt[3]{\frac{6m(t)}{\pi\rho(t)}}$$

We can also calculate the work done to the system or stresses on the balloon wall. To calculate this we use the pressure difference, not the absolute pressure. For simplicity we assume a fully symmetrical balloon. The stress in the wall is a function of the linear strain $\lambda = d/d_0$, and d_0 and t_0 is the initial diameter and the wall thickness, respectively. The force balance for the cut section of the balloon is $\sigma 2\pi rt = \Delta p\pi r^2$. If we assume that the rubber is incompressible, the volume of rubber is constant and therefore, during the deformation, the same rubber volume is stretched into a thinner hollow sphere, i.e. $V = 4\pi r^2 t = 4\pi r_0^2 t_0$. Wall thickness can be solved from this. Combining this with the above force balance gives the stress in the wall section, and Fig. 3.9 gives the graphical presentation of the result. We can also integrate the needed inflation work from the measured pdV data. This is presented on the right of Fig. 3.9.

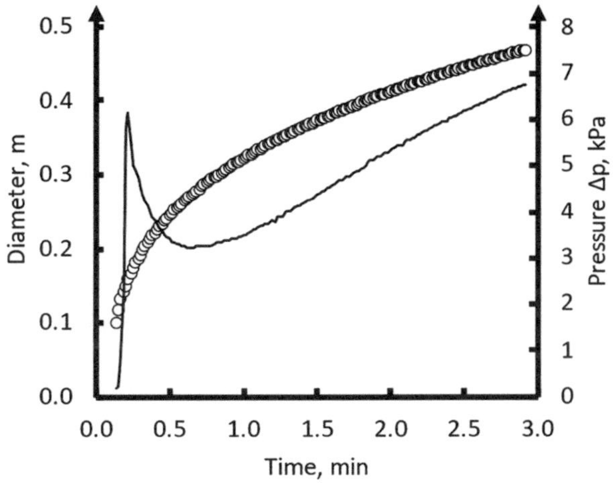

Fig. 3.8 Diameter of the balloon, internal pressure (left) and expansion work (right) as a function of time. Dots present the diameter; the solid line is the internal gauge pressure in kPa

$$\sigma = \frac{\Delta p d}{4t} = \frac{\Delta p d_0}{4t_0}\lambda^3$$

$$W = \int_0^V \Delta p dV$$

The stress at burst is 44 MPa. Based on Nagdi [41], the tensile strength of natural rubber used in the balloons is in the order of 10–25 MPa, indicating that our assumption of a symmetrical balloon is probably over simplified, giving too high strength values, and should be refined for a more detailed analysis. In addition, the thickness of the balloon wall is not constant. The inflation work needed is 250 J. This example describes a system that has both change in size and density, and reminds us to focus on

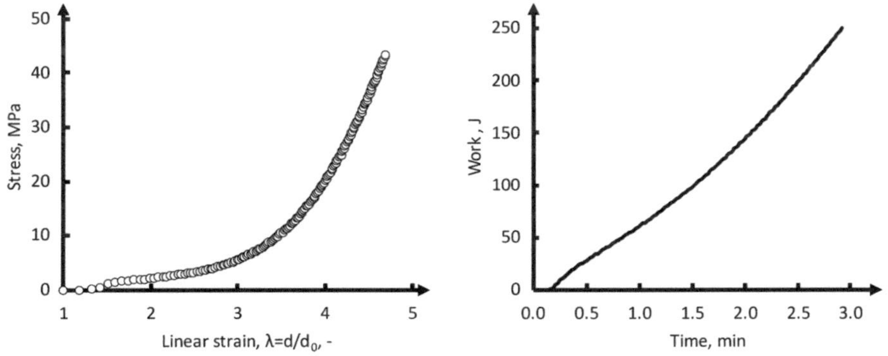

Fig. 3.9 Tension in the rubber balloon during inflation until it bursts, and inflation work required. (Initial size 100 mm and wall thickness 0.4 mm.)

the importance of the pressure reference state definition. Another application would be, for example, an internal combustion engine or a piston compressor.

3.2 Principles of Thermal Power Plants

Many renewable power plants operating today rely on basic thermal power processes, such as the Rankine cycle or gas turbine cycle, or a combination of these, the combined cycle. For example biomass combustion-based power plants use a steam boiler and a turbine to generate heat and power. Geothermal power plants use the Rankine cycle in different ways and use different process fluids, such as water or ammonia as examples, depending on the temperature range of the heat source available. Concentrated solar power plants also utilize steam power plant technologies as their power production process. Before discussing the actual cycles or other processes, we familiarize ourselves with the ideal Carnot's cycle/process that operates between a constant hot and cold reservoir (Fig. 3.10). The reservoirs are assumed to be infinite in size and temperatures remain constant. The cycle is composed of ideal isentropic compression (1-2), isothermal heat addition (2-3), isentropic expansion (3-4) and isothermal heat removal (4-1).

The energy balance of the Carnot process can be written as follows.

$$\phi_{in}(T_H) + P_{in} - P_{out} - \phi_{out}(T_C) = 0$$

From this, the net power produced can be solved, and defines the thermal efficiency of the process as the ratio of power output to input heat flow rate.

$$P = P_{out} - P_{in} = \phi_{in}(T_H) - \phi_{out}(T_C)$$

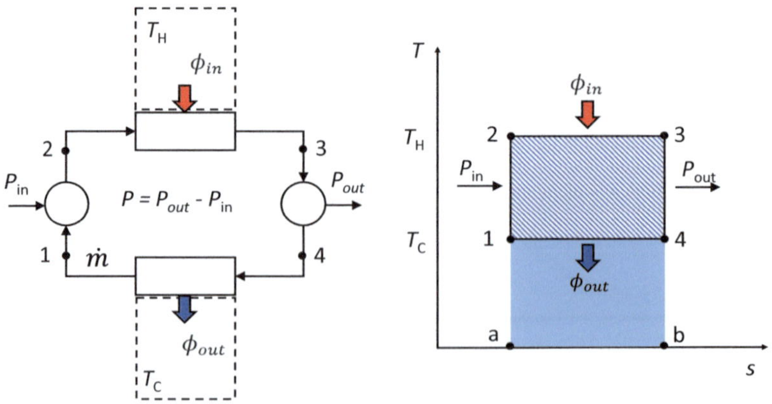

Fig. 3.10 Carnot cycle. 1-2: Isentropic compression, 2-3: isothermal heat input, 3-4: Isentropic expansion and 3-4: Isothermal removal of waste heat

$$\eta = \frac{P}{\phi_{in}(T_H)} = \frac{\phi_{in}(T_H) - \phi_{out}(T_C)}{\phi_{in}(T_H)} = 1 - \frac{\phi_{out}(T_C)}{\phi_{in}(T_H)}$$

From the 2nd law of thermodynamics, we know that the heat transfer rate is, in general form, the integral of the temperature $T(s)$ with respect to entropy s, multiplied by the mass flow rate.

$$\phi = \dot{m} \int_{s_1}^{s_2} T(s)ds = \dot{m}\overline{T}\Delta s$$

If the temperature remains constant during the heat transfer steps, as in the case of the Carnot cycle, this greatly simplifies the calculation.

$$\phi_{in} = \dot{m}T_H \Delta s$$

$$\phi_{out} = \dot{m}T_C \Delta s$$

The last three forms apply for a process where the temperature remains constant during the heat transfer step. From the T-s diagram we can see that the heat transferred into the system is the total area between the path 2-3 and x-axis (a-1-2-3-4-b-a), and the heat removed from the process is the area between the path 4-1 and x-axis (a-1-4-b-a). Noting that the entropy changes during (2-3) and (4-1) are the same due to isentropic work steps, gives us the well-known Carnot equation, defined in terms of the maximum and minimum temperatures of the process.

$$\eta = 1 - \frac{T_C}{T_H}$$

For an ideal Carnot cycle, thermal efficiency is only a function of the temperature of the hot and cold reservoirs. In real processes, there are losses in the compression and expansion steps, and therefore the actual efficiency is much lower. In addition, heat input and removal can take place under non-constant temperatures, as in the case of a real steam power plant with preheating and superheating steps, or a gas turbine process with a combustion chamber.

The efficiency can be increased by increasing the average heat input temperature and decreasing the average heat rejection temperature. Average heat exchange temperature can be defined for a *constant pressure process*, $dp = 0$, as a ratio of change in enthalpy and entropy, since $dh = Tds + vdp = Tds$, and then at an average temperature we have

$$\int_{s}^{s+\Delta s} \overline{T}ds = \overline{T}\Delta s = \int_{h}^{h+\Delta h} dh = \Delta h$$

$$\overline{T} = \frac{\Delta h}{\Delta s}$$

$$\overline{T}_H = \frac{h_3 - h_2}{s_3 - s_2}$$

$$\overline{T}_C = \frac{h_1 - h_4}{s_1 - s_4}$$

These can also be applied for general heating temperature profiles to be able to use the Carnot equation form.

3.3 The Basic Rankine Cycle Steam Power Plant and Its Components

The purpose of this section is to discuss the basic Rankine cycle and its components; feed-water pump, steam boiler/generator, the steam turbine and the condenser. The Rankine cycle was invented by Claus Rankine in 1848 and is probably the most used cycle in large scale heat and power systems. Figure 3.11 presents the basic concept.

The energy balance of the plant is as follows. Note that the input heat flow rate is composed from feed-water preheating in the *economizer*, evaporator, and superheater.

$$\phi_{in} + P_p - P_t - \phi_{out} = 0$$

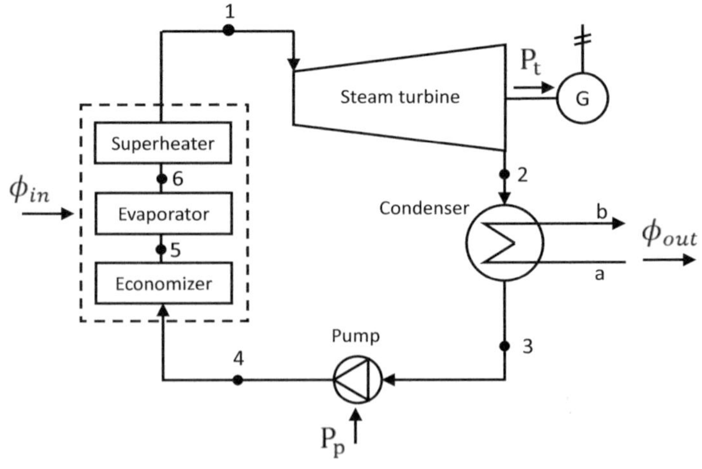

Fig. 3.11 Basic Rankine cycle components

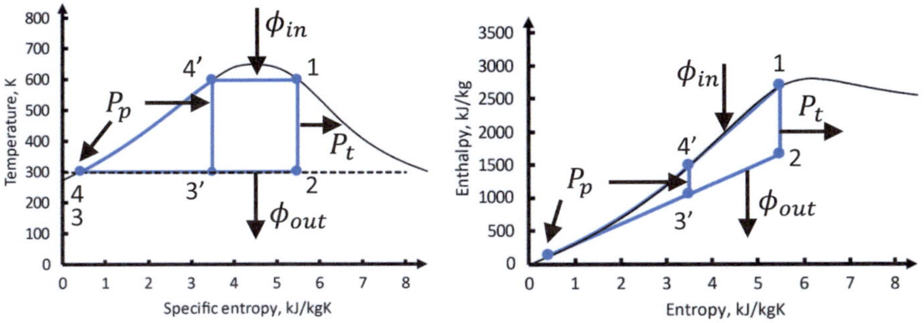

Fig. 3.12 Comparison of the ideal saturated Rankine cycle and Carnot water vapor cycle. Points 3 and 4 are not named in the h–s diagram due to clarity. Points 3′/4′ refers to the pressurization step in the Carnot vapor cycle; the other parts are the same

Which can be rewritten as

$$P_p - P_t = \phi_{in} - \phi_{out} = P$$

The process efficiency is defined as the network P generated, divided by the input heat as follows. This can also be written in terms of heat flow rates.

$$\eta = \frac{P}{\phi_{in}} = \frac{P_t - P_p}{\phi_{in}} = \frac{\phi_{in} - \phi_{out}}{\phi_{in}} = 1 - \frac{\phi_{out}}{\phi_{in}}.$$

Heat comes into the process in the steam generator that is composed of three steps: preheating to boiling temperature, evaporation at a constant temperature and superheating to final steam values. Heat is rejected in the condenser that operates at the pressure p_2. As a starting point, the superheating step is not included, and we focus is on a saturated vapor–liquid system. Figure 3.12 presents the *T-s* and *h–s* diagrams for the ideal water-based Rankine cycle (1-2-3-4-4′-1) and a Carnot vapor cycle 1-2-3′-4′-1 for comparison.

Process values $p_1 = 120$ bar and $p_2 = 0.04$ bar are used for plotting the diagrams. The thermal efficiency of the Carnot cycle is higher (0.495) since the average heat input temperature is higher. In an ideal fully isentropic Rankine cycle, due to required preheating step (4–4′), the average heat input temperature is lower, and this decreases the efficiency (0.398). However, the power requirement for the pressurization step (3-4) is much smaller in the case of ideal Rankine cycle with liquid pumping (see the *h–S* diagram). Pressurization of a two-phase flow fluid in the case of vapor Carnot cycle (3′-4′) would consume a lot of power and it would be very difficult to organize it in practice and avoid mechanical erosion due to water droplets in the flow. The detailed calculation of the process points and thermal efficiency in Table 3.1.

Thermal efficiency of Carnot cycle is simply defined by the ratio of minimum and maximum temperatures of the process; note the K unit.

Table 3.1 Calculation of process values of ideal Rankine and Carnot cycles, values of physical properties are obtained from the Excel-based tool available in electronic supplement

#	*p* (bar)	*t* (°C)	*T* (K)	*h* (kJ/kg)	*s* (kJ/kgK)	Comments
1	120	324.7	597.8	2686	5.494	Saturated vapor at p_1. Read from right-end of p_1 line in vapor dome
2	0.04	29.0	302.1	1654	5.494	Move vertically down from point 1–2 (ds = 0) at p_2
3	0.04	29.0	302.1	121	0.423	Follow p_2 line from point 2 left to saturated water curve
3'	0.04	29.0	302.1	1050	3.497	Find first the point 4′. Move vertically from 4′–3′ ($ds = 0$) to p_2
4	120	30.7	303.9	141	0.443	Add pump enthalpy increase to enthalpy at point 3
4'	120	324.7	597.8	1491	3.497	Follow p_1 line, from point 1 to left to saturated water curve

$$\eta = 1 - \frac{T_3}{T_1} = 1 - \frac{302.1K}{597.8K} = 0.495$$

To be able to calculate the enthalpy at point 4, we need to calculate enthalpy increase in the pump. A very good approximation for the average specific value can be obtained from the saturated value at inlet temperature, this would be 1.004×10^{-3} m^3/kg. For comparison, the true value would be the average between the inlet and outlet specific volumes, $0.5 \times 10^{-3} \times (1.004 + 0.999) \times 10^{-3} = 1.002 \times 10^{-3} \text{m}^3/\text{kg}$. Enthalpy increase of an ideal pump is equal to $\Delta h_p = v_{3-4}(p_4 - p_3) \times 100 = 1.004 \times 10^{-3} \times (120 - 0.04) \times 100 = 12.0$ kJ/kg. This is added to h_3 to get h_4. Efficiency of the ideal Rankine cycle efficiency is as follows, using the indirect method via losses. The resulting process cycles are shown in Fig. 3.12.

$$\eta = 1 - \frac{h_2 - h_3}{h_1 - h_4} = 1 - \frac{1654 - 121}{2686 - 133} = 0.40$$

All modern power plants normally also include a superheating stage that further heats up the saturated steam to higher temperature, to increase the efficiency and to help to keep the steam away from the two-phase region longer during the expansion. Figure 3.13 presents the non-ideal Rankine cycle with a superheater for a 120 bar and 550 °C steam values, all other parameters are kept the same as in the previous example. We consider true processes with losses and use the isentropic efficiencies for pump and turbine. Point symbol 4′ is replaced with 5. From the steam Tables, we can read the point 1 enthalpy, $h_1 = 3480$ kJ/kg. Isentropic expansion (vertically downwards) to point 2 s at pressure 0.04 bar, gives $h_{2s} = 2004$ kJ/kg. But how do we get the true end expansion point and how do we model the true pump? This will be discussed below.

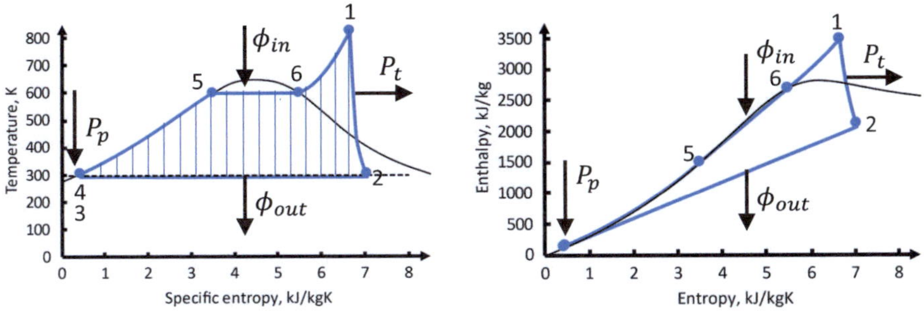

Fig. 3.13 Basic Rankine cycle h–s (right) and T–s (left) diagrams (below). Points 3 and 4 are not named in the h–s diagram due to clarity

Next, we analyze all components separately. We want to design a plant that produces 500 MW of power. We start from the feed-water pump. We often have the pump installed in a constant sized pipeline. Since the liquid compressibility is very small, we assume that the density is constant. Therefore, kinetic energy is also constant over the pump. In addition, if the inlet and the outlet connections are at the same height levels, the potential energy is also the same. The energy equation is then as follows.

$$P_p = \dot{m}(h_{4.s} - h_3) + P_{irr} = \dot{m}\Delta h_s + P_{irr}$$

On the other hand, $P_p = \dot{m}(h_4 - h_3)$. By using this, and the definition of isentropic efficiency of pumping, $\eta_s = (h_{4.s} - h_3)/(h_4 - h_3)$, we can calculate how much mechanical power is converted to internal irreversibility P_{irr} (such as flow friction and turbulence) which then increases the end enthalpy of the fluid in pressurization.

$$P_{irr} = P_p(1 - \eta_s)$$

For an adiabatic, isentropic pump, $\Delta h_s = v\Delta p$. For a true process with an isentropic efficiency, $\Delta h = v\Delta p/\eta_S$. In addition to irreversibility, there are also losses in the motor. The efficiency of the motor is equal to η_m. Therefore, the required true electrical pumping power is equal to.

$$P_p = \frac{\dot{m}\overline{v}(p_4 - p_3)}{\eta_{s.p}\eta_m}$$

At this point, the mass flow rate of water is not yet known, but the enthalpy increase can be calculated. As above, the average specific volume of water is 1.004×10^{-3} m^3/kg. The enthalpy increase of an ideal pump is equal to $\Delta h_p = v_{3-4}(p_4 - p_3) \times 100 = 1.004 \times 10^{-3} \times (120 - 0.04) \times 100 = 12\,\text{kJ/kg}$. The isentropic efficiency of the pump is 0.95 and therefore, the specific mechanical work required is 12/0.95 =

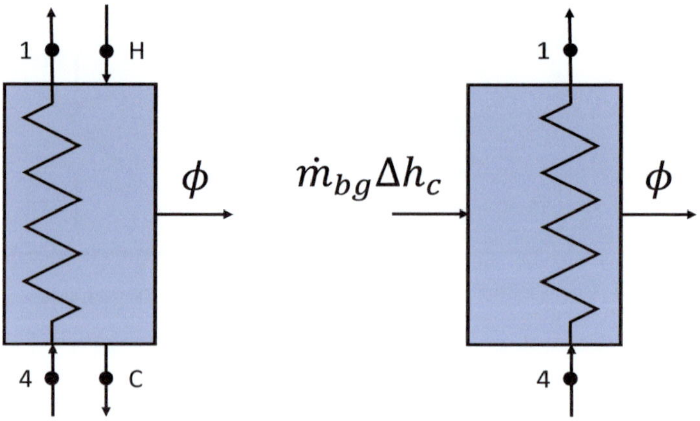

Fig. 3.14 Steam generator energy balance in the CSP power plant (left) and combustion of biogas

12.6 kJ/kg. The power requirement in MW can be calculated when the mass flow rate is known.

The second component in the Rankine cycle is the steam generator or the boiler. It has typically three stages: feed-water preheater (*i.e.* the *economizer*), the evaporator and the superheater. Its objective is to produce reliably high pressure and high temperature steam for the steam turbine. The heat source can be based on combustion of a biogas, thermal energy from the sun, or geothermal heat, just to mention few options. As an example, heat from a molten salt can be used in concentrated solar power plants. Incoming salt is at hot storage temperature T_{H} and it leaves the boiler at cold storage temperature T_{C}, and the mass flow rate is $\dot{m}_s$, (Fig. 3.14).

There are always heat losses, ϕ, and they can be expressed as the fraction $(1 - \eta_b)$ from the input heat, defined by the boiler efficiency η_b. The energy equation for the steam generator efficiency is equal to:

$$\dot{m}_s(h_H - h_C) - \dot{m}(h_1 - h_4) - \phi = 0$$

$$\phi = (1 - \eta_b)\dot{m}_s(h_H - h_C) = (1 - \eta_b)\dot{m}_s c_{p.s}(t_H - t_C)$$

Typical values for these are $t_{\mathrm{H}} = 566$ °C and $t_{\mathrm{C}} = 288$ °C, and the heat capacity of the molten salt is $c_{\mathrm{p}} = 1530$ J/kg K (see Caraballo [10]). So, the specific heat transferred is 1530 × (566–288) = 425.34 kJ/kg of molten salt. Mass flow rate of the molten salt can be solved when the mass flow rate of water and enthalpy change on the water side is known. We already know the enthalpy of superheated steam at 120 bar and 550 °C, is $h_1 = 3482$ kJ/kg. Enthalpy before the evaporator is also known: $h_4 = h_3 + \Delta h_p = 121 + 12.6 = 134$ kJ/kg. Note that electrical efficiency is not used here, just the isentropic efficiency. The ratio of the mass flow rates can be calculated and, for example, if the required steam mass flow rate for power generation is known, the required mass flow rate of molten salt and dimension of the required heliostat heat collection system can be solved.

As another example, if biogas is burnt for heat generation, knowing the combustion heat of the biogas—for example $\Delta h_c = 20$ MJ/kg [26]—the mass flow rate of the fuel can be solved after the steam side is solved. However, also in this case, only a fraction of the chemical energy, defined by the boiler efficiency, can be transferred to water and steam. The steam generator will be analyzed in more detail later together with combustion related applications in the waste and bioenergy combustion applications. In a general case, the energy balance for a boiler is as follows.

$$\dot{m}_{bg}\Delta h_c - \dot{m}(h_1 - h_4) - \phi = 0$$

$$\phi = (1 - \eta_b)\dot{m}_{bg}\Delta h_c$$

From the steam generator, the superheated steam enters the steam turbine, expands, and generates power. It is convenient to express this process in an *h–s* diagram (Fig. 3.15).

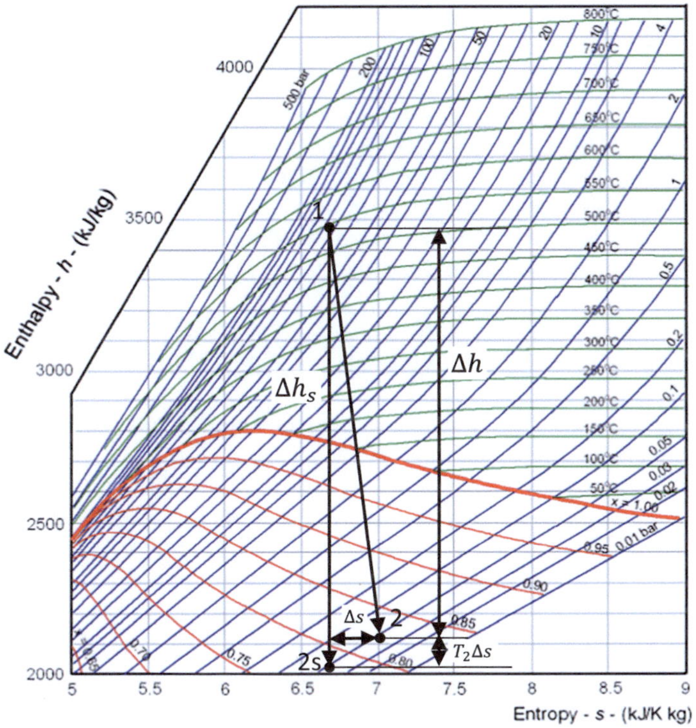

Fig. 3.15 Steam turbine power plotted on an *h–s* diagram, $p_1 = 120$ bar, $p_2 = 0.04$ bar, $h_1 = 3482$ kJ/kg, $h_{2s} = 2004$ kJ/kg, $\eta_s = 0.92$. Background schematic:© EngineeringToolbox

For an ideal case without any irreversible losses, the end point would correspond to 2 s, with isentropic expansion, $h_{2s} = h(p_2, s_1)$. From this, and using the definition of the isentropic turbine efficiency, the actual end point of the expansion can be solved as follows:

$$h_2 = h_1 - \eta_s(h_1 - h_{2s})$$

Using the above values, the specific work done in the turbine is 0.92 × (3482–2004) = 1360 kJ/kg per mass of steam (see the point set in Fig. 3.14). If the end point is in the two-phase region, the vertical enthalpy difference between the isentropic and true end points is equal to:

$$h_2 - h_{2s} = h_1 - \eta_s(h_1 - h_{2s}) - h_{2s} = (h_1 - h_{2s})(1 - \eta_s) = T_2\Delta s$$

The last form comes from the combination of first and second laws of thermodynamics and the definition of enthalpy. This means that as the entropy increases at constant temperature, this increase prevents enthalpy from decreasing to its theoretical maximum value. When moving along the pressure isobar, $dp = 0$, of the end state, we can use the differential form, and at a constant temperature.

$dh = du + pdv + vdp = dq + vdp = dq = T_2ds$, we get by integration.

$$\Delta h_{irr} = T_2\Delta s$$

The end temperature is 29 °C (302.15 K) and $\Delta s = 0.391$ kJ/kg K, and therefore, the $\Delta h_{irr} = 302.15 \times 0.391 = 118$ kJ/kg. From the *h*–*s* diagram $\Delta h_{irr} = 118$ kJ/kg. If we move backwards from the end point of the expansion, the slope of the expansion curve can be solved from the equation below, ***if*** the end point is in the two-phase region, where the curve is linear [29, 53].

$$\frac{\Delta h}{\Delta s} = \frac{\eta_s}{(1 - \eta_s)} T_2$$

In the dry steam region, the non-linear expansion curve can be formed by using the differential from of the isentropic efficiency, $\eta_s = dh/dh_S$.

$$dh = vdp + Tds = dh_s + Tds = dh/\eta_s + Tds$$

$$\frac{dh}{ds} = \frac{-\eta_s}{(1 - \eta_s)} T(h, s)$$

This can be numerically integrated piecewise to create the expansion curve. It is also possible to consider the different efficiencies of the turbine stages in detail. The cycles presented in Fig. 3.13 were generated in this manner, so they are accurate presentations of the cycle, and all cycle curves present the true positions. Note that the differential isentropic efficiency is smaller than the isentropic efficiency—here

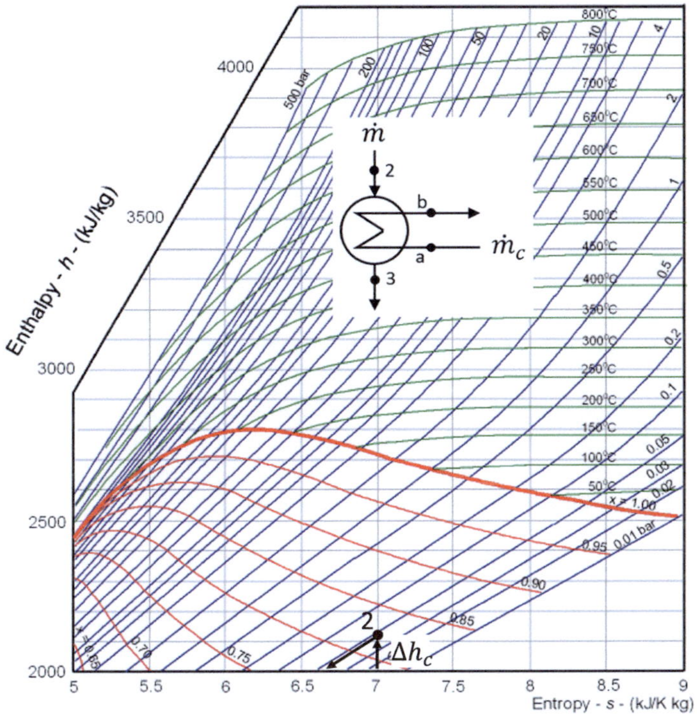

Fig. 3.16 Condenser plotted on an h–s diagram. Background schematic: EngineeringToolbox©

0.882, while $\eta_s = 0.92$. The steam from the turbine exit goes into the condenser. The objective of the condenser is to return the fluid to its liquid form so that it can be efficiently pumped back to the boiler. Pumping of liquid is far more energetically efficient than compression of vapor. Let's discuss the condenser operation using the same *h–s* diagram as for the turbine (Fig. 3.16).

If the end point of expansion is in the wet two-phase zone (here steam mass fraction is ~ 0.82) the enthalpy is a linear function of the steam quality, the ratio of steam and total mass flow rates as follows:

$$h(p_2, x) = h_G(p_2)x + h_L(p_2)(1 - x)$$

This value can be directly obtained from the h–s diagram or calculated by the quality (= vapor mass fraction) and saturated liquid and vapor enthalpies. Energy balance for the condenser is as follows:

$$\phi_{out} = \dot{m}(h_2(p_2) - h_3(p_2, T_3)) = \dot{m}_c c_p (T_a - T_b)$$

Using the values obtained, we can calculate the enthalpy decrease in the condenser as 2123 – 121 = 2001 kJ/kg per combined mass of steam and water flowing into the condenser. Note that at the exit the condensate can also be sub-cooled, requiring the value of the temperature to be able to define the enthalpy. If the temperature of the incoming and outgoing cooling water is fixed, the required mass flow rate of cooling water can be calculated. Normally, we need to have a minimum temperature difference between the inlet streams $\Delta T = T_2(p_2) - T_b = 4 \ldots 5\,\mathrm{K}$, to keep the construction of the condenser economically feasible and surface area not too large. T_b refers here to the exit temperature of the cooling water.

Example 3.4 Let's analyze a power plant of net electrical power production of P_e = 500 MW using a worked example. The steam temperature and pressure values before it enters the turbine are 550 °C and 120 bar, and the condenser pressure is 0.04 bar (absolute pressure). Isentropic efficiency of the turbine and the pump is 0.92 and 0.95, respectively. Inlet and exit temperatures of the cooling water are 20 and 25 °C, respectively. Heat capacity of the cooling water is 4.2 kJ/kg K. Steam generator efficiency is $\eta_b = 1$ and $\eta_b = 0.96$ for concentrated solar power, molten salt and biogas combustion cases, respectively. Calculate:

1. The mass flow rates of steam in (a) molten salt system, (b) biogas combustion system, using the values already analyzed above.
2. Thermal heat input to the steam generator, and heat output from the condenser.
3. Mechanical and electrical power needs for the feed-water pump.
4. Mechanical and electrical power from the steam turbine.
5. The mass flow rate of cooling water.
6. Both the thermal and electrical efficiencies of the process.

1. From the steam turbine and pump energy balances above, we get the steam mass flow rate as follows: (generator and motor efficiencies were both 0.98) Based on the above, $\Delta h_t = 3482 - 2123 = 1359$ kJ/kg and $\Delta h_p = 13$ kJ/kg.

$$\dot{m} = \frac{P_e}{\Delta h_t \eta_g - \Delta h_p / \eta_m} = \frac{500 \times 10^3}{1359 \times 0.98 - 13/0.98} = 379\,\mathrm{kg/s}$$

This same mass flow rate is needed in both (a) and (b). From the evaporator energy balances, the required mass flow rate of the molten salt $\dot{m}_s$ and biogas $\dot{m}_{bg}$ can be solved. The enthalpy decrease of the molten salt is presented in terms of temperature decrease, $\Delta h_s = c_{p.s} \Delta t_s$.

$$\dot{m}_s = \frac{\dot{m}(h_1 - h_4)}{\eta_b(h_H - h_C)} = \frac{379 \times (3482 - 134)}{1.0 \times 1.530 \times (566 - 288)} = 2983\,\mathrm{kg/s}$$

$$\dot{m}_{bg} = \frac{\dot{m}(h_1 - h_4)}{\eta_b \Delta h_C} = \frac{379 \times (3482 - 134)}{0.96 \times 20000} = 66.1\,\mathrm{kg/s}$$

2. Heat input is calculated from the evaporator energy balance, using the steam side balance.

$$\phi_{in} = \dot{m}(h_1 - h_4) = 379 \times (3482 - 134) = 1269\,\text{MW}$$

Heat output is calculated from the condenser energy balance, using steam side.

$$\phi_{out} = \dot{m}_s(h_2 - h_3) = 379 \times (2123 - 121) = 753\,\text{MW}$$

3. Mechanical and electrical power needed for the pump is calculated as follows.

$$P_p = \dot{m}(h_4 - h_3) = 379 \times (134 - 121) = 4.9\,\text{MW}$$

$$P_{p.e} = \frac{P_e}{\eta_m} = \frac{4.8}{0.98} = 5\,\text{MW}$$

4. Mechanical and electrical power from the turbine is calculated as follows.

$$P_t = \dot{m}(h_1 - h_2) = 379 \times (3482 - 2123) = 515.2\,\text{MW}$$

$$P_{t.e} = P_t \eta_g = 515.2 \times 0.98 = 504.9\,\text{MW}$$

5. The required cooling water mass flow rate can be obtained from the energy balance of the condenser.

$$\dot{m}_c = \frac{\dot{m}(h_2 - h_3)}{c_p(t_a - t_b)} = \frac{379 \times (2123 - 121)}{4.2 \times (25 - 20)} = 36131\,\text{kg/s}$$

6. Finally, thermal and electrical efficiencies are calculated as follows.

$$\eta_t = \frac{P_t - P_p}{\phi_{in}} = \frac{515.2 - 4.9}{1269} = 0.402$$

$$\eta_e = \frac{P_{t.e} - P_{p.e}}{\phi_{in}} = \frac{504.9 - 5.0}{1269} = 0.394$$

3.4 Improvement Options for the Basic Rankine Cycle

The efficiency of heat engines can be improved by increasing the average heat input temperature in the boiler and decreasing the average heat removal temperature in the condenser, as discussed above. This is illustrated in Fig. 3.17 and the efficiencies of a basic *saturated* Rankine cycle, and a theoretical Carnot cycle are also shown, for evaporator pressures of 80 and 210 bar and condenser pressures of 1 and 0.04 bar,

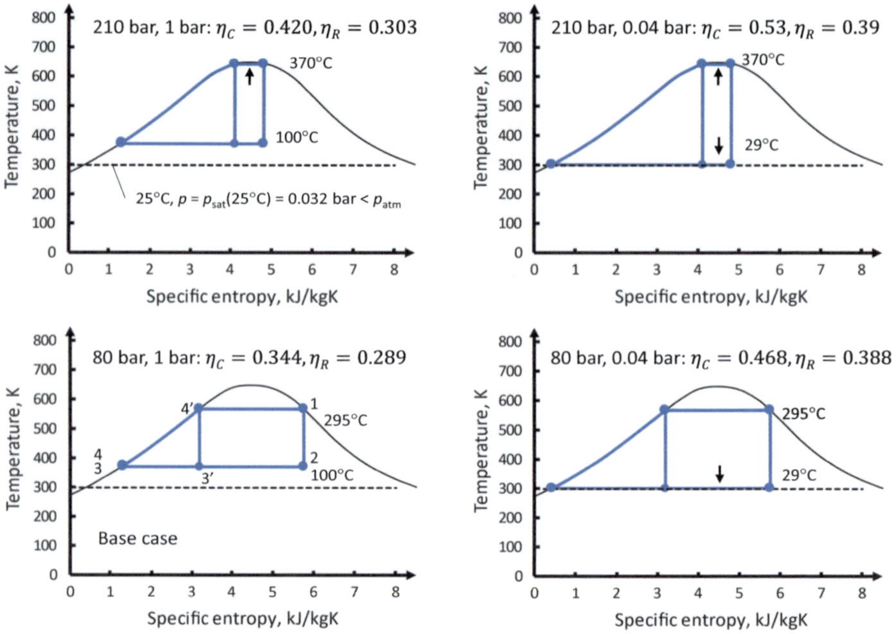

Fig. 3.17 The effect of live steam pressure and condenser pressure on thermal efficiency of an ideal saturated Rankine cycle (1-2-3-4-4′-1) and Carnot cycle (1-2-3′-4′-1). Points 3′ and 4′ refer to the Carnot cycle

respectively. The details of the process points and efficiency calculation are explained above.

Reducing the condenser pressure from the atmospheric value of 1–0.04 bar has a significant impact on the efficiency. The lower we can bring the exit pressure and temperature of the turbine, the larger the work done by the turbine. Therefore, the available temperature of the cooling water plays a major role in a plant's thermal efficiency. Note that bringing the condenser temperature down to 29 °C, means that a vacuum (− 0.96 bar) is formed in the system. A temperature difference of say 4–5 °C must be left in the condenser between the exit temperature of the cooling water and condensation temperature, in order for the condenser not to be too large or expensive.

Reheating

Next the re-heaters are discussed. With high superheated steam pressure the end point of the expansion reaches the high-water content region, since the starting point is on the very left of the *h*–*s* diagram (see, for example Fig. 3.15). The droplets that are formed cause erosion of the turbine blades, and this needs to be avoided. By stopping the expansion at an intermediate pressure—above the vapor dome—and reheating the steam to its live steam value or a bit higher, and then using a low-pressure turbine, we can move the end point of expansion to the right and increase the steam quality, assuming that the condenser pressure is kept constant. Figures 3.18 and 3.19 present an example with a single reheater. External heat is delivered from 6-1, and 2-3. The source of this heat can be for example heat from combustion, heat from the exit gases of a topping gas turbine process, solar thermal energy, or geothermal energy.

Efficiency of this reheated process can be simply calculated as follows, by the direct and indirect method.

$$\eta = \frac{\eta_{s.t}(h_1 - h_{2.s}) + \eta_{s.t}(h_3 - h_{4.s}) - \overline{v}_{5-6}(p_6 - p_5)/\eta_{s.p}}{(h_1 - h_6) + (h_3 - h_2)}$$

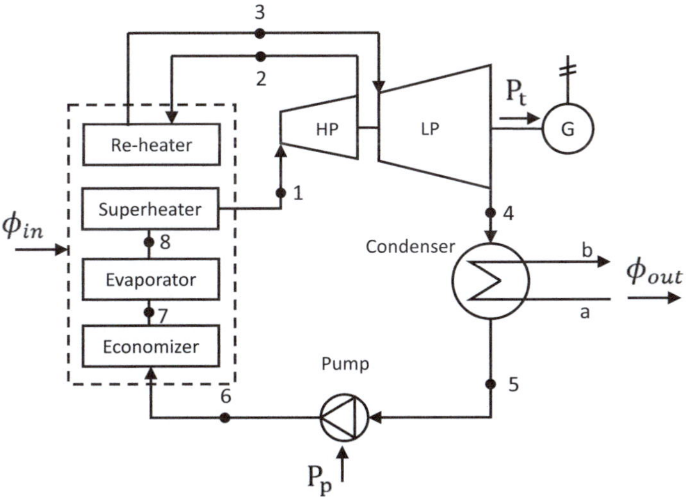

Fig. 3.18 Single reheater system with high- and low-pressure turbines, HP, LP, respectively

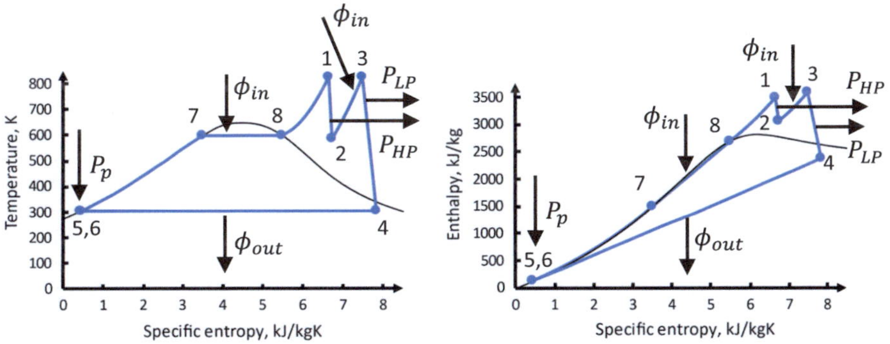

Fig. 3.19 Single reheater system with high- and low-pressure turbines, HP, LP, respectively. External heat input from 6-1 to 2-3. 1:(550°,120 bar), 2:(20 bar), 3:(560°, 20 bar), 4:(0.04 bar), $\eta_{s.t} = 0.92$, $\eta_{s.p} = 1$

$$\eta = 1 - \frac{(h_4 - h_5)}{(h_1 - h_6) + (h_3 - h_2)}$$

End enthalpies of expansion can be calculated as.

$$h_2 = h_1 - \eta_s(h_1 - h_{2s})$$

and

$$h_4 = h_3 - \eta_s(h_3 - h_{4s}),$$

respectively. But how should we determine the optimal re-heating pressure level? If we have developed a mathematical model for the process and thermodynamic properties are automatically calculated, we can easily optimize the efficiency numerically by varying the p_2/p_1 value. Figure 3.20 presents the result for efficiency increase and end-point steam mass fraction *i.e.* the steam *quality*, using the above efficiency equation(s). Process values used are given in the figure caption.

As can be seen from Fig. 3.20, the maximum thermal efficiency increase can be obtained in the range of $p_2/p_1 = 0.15, \ldots, 0.25$, and this could be used as an initial value for process design. Note that the value $p_2/p_1 = 1$ corresponds to the case with no reheating. Very low reheating pressure results in smaller efficiency than with no reheating. In this case also the exit steam of the low- pressure turbine is still superheated, which causes problems for the condenser. A very important benefit of the reheating system is that the end-quality of the steam increases significantly since the end point moves to right (Figs. 3.19 and 3.20). Especially with super-critical and ultra-super-critical processes, having the live steam point on the *h–s* diagram on the very left, the use of the reheating system is mandatory. Real power plants normally use 1-2 reheaters, not more. A larger number of results in a complex system and increases the capital cost of the system.

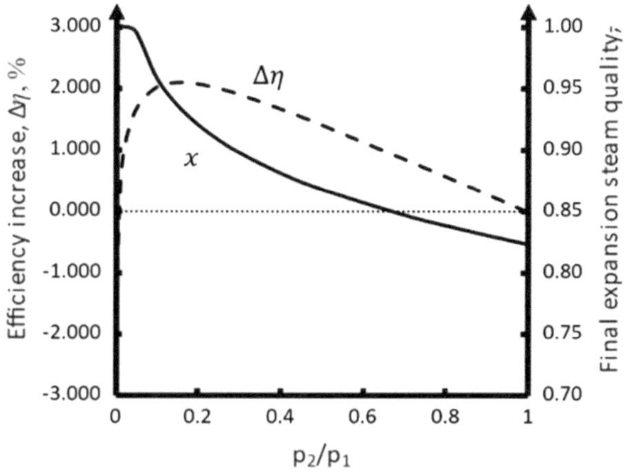

Fig. 3.20 Thermal efficiency increase, $\Delta\eta = 100 \times (\eta - \eta_{noreheat})$ and end-quality of the low-pressure turbine for a single reheater system. 1:(550 °C, 120 bar), 3:(550 °C, varied), 4:(0.04 bar), $\eta_{s.t} = 0.92$, $\eta_{s.p} = 0.95$

Example 3.5 Let's analyze a power plant with net electrical power production of $P_e = 500$ MW, installed with a single re-heat. The values of steam temperature and pressure before the turbine are 550 °C and 120 bar and the condenser pressure is 0.04 bar (absolute pressure). Based on the discussion above, we choose the optimal re-heating pressure at $p_2/p_1 = 0.15$, $p_2 = 18$ bar. The isentropic efficiency of the turbine and the pump is 0.92 and 0.95, respectively.

1. Define all the process points and calculate the mass flow rate of steam
2. Calculate the required thermal heat input to the steam generator, and heat output from the condenser.
3. Calculate mechanical power needed for the feed-water pump.
4. Calculate mechanical power from the high pressure and low-pressure steam turbines.
5. Calculate the thermal efficiency of the process. Compare your results with those from Example 3.4.

1. The process points are best presented in a Table format (see Table 3.2). The main difference to the previous example is the re-heating step and the resulting different end point of expansion.

From point 8, saturated steam passes through the superheater to point 1. We get power from both high-pressure and low-pressure turbines and the same pump power is needed as in Example 3.4. Based on the above, $\Delta h_{HP} = 3482 - 2980 = 502$ kJ/

Table 3.2 Calculation of process values of re-heated Rankine cycle, values of physical properties are obtained from the Excel-based tool available in electronic supplement

#	p (bar)	t (°C)	T (K)	h (kJ/kg)	s (kJ/kg K)	Comments
1	120	550	823	3482	6.655	Read from the cross section of t_1 and p_1
2 s	18	260	533	2936	6.655	Move vertically down from point 1–2s (ds = 0) at p_2
2	18	278	551	2980	6.736	Calculate h_2 from the isentropic expansion equation
3	18	550	823	3581	7.623	Move along the p_2 isobar right to the re-heating temperature
4 s	0.04	29	302	2297	7.623	Move vertically down from point 3–4s (ds = 0) at p_4
4	0.04	29	302	2399	7.963	Calculate h_4 from the isentropic expansion equation
5	0.04	29	302	121	0.423	Move along the isobar to the left side to the saturated water
6	120	30.7	303.7	134	0.466	Add enthalpy increase in the pump
7	120	324.7	597.7	1491	3.497	Saturated water after the economizer, T
8	120	324.7	597.7	2686	5.494	Saturated steam after the evaporator, T

kg, $\Delta h_{LP} = 3581 - 2399 = 1181$ kJ/kg and $\Delta h_p = 134 - 121 = 13$ kJ/kg. With these numbers the mass flow rate of the steam can be calculated.

$$\dot{m} = \frac{P_e}{\Delta h_{HP}\eta_g + \Delta h_{LP}\eta_g - \frac{\Delta h_p}{\eta_m}}$$
$$= \frac{500 \times 10^3}{502 \times 0.98 + 1181 \times 0.98 - 13/0.98} = 306\,\text{kg/s}$$

2. Heat input is calculated from the evaporator energy balance; here using the steam side balance.

$$\phi_{in} = \dot{m}(h_1 - h_6) = 306 \times (3482 - 134) + 306 \times (3581 - 2980) = 1208\,\text{MW}$$

Heat output is calculated from the condenser energy balance.

$$\phi_{out} = \dot{m}(h_4 - h_5) = 306 \times (2399 - 121) = 697\,\text{MW}$$

3. The mechanical power needed for the pump is calculated as follows.

$$P_p = \dot{m}(h_6 - h_5) = 306 \times (134 - 121) = 3.9\,\text{MW}$$

4. Therefore, the mechanical power output from the turbine is calculated as follows:

$$P_t = \dot{m}(h_1 - h_2) + \dot{m}(h_3 - h_4)$$
$$306 \times (3482 - 2980) + 306 \times (3581 - 2399) = 515\,\text{MW}$$

5. The thermal efficiency is as follows.

$$\eta_t = \frac{P_{HP} + P_{LP} - P_p}{\phi_{in}} = \frac{515 - 3.9}{1208} = 0.423$$

The thermal efficiency of the process increases from 40.2 to 42.3%; an increase of 2.1% units, as shown in Fig. 3.19. The re-heated system can generate the same amount of power with a significantly smaller amount of process steam (379 kg/s → 306 kg/s) resulting in smaller condenser losses (759 MW → 696 MW) and a decrease in required heat input (1269 MW → 1206 MW). The same amount of power can be generated with smaller components and smaller flow rates. The downside is that the required additional re-heater creates an additional cost. The final selection of the system configuration is determined by the techno-economic optimum.

Regenerative feed-water preheating

If the Rankine cycle in the T-s plane is divided into vertical slices (Fig. 3.13), and into incremental Carnot cycles, we can see that during the preheating stage, the Carnot's efficiency in these incremental cycles is very low as the input and output temperature are close to each other [29]. This is especially true on the very left-hand corner of the cycle in the *T-s* diagram, where the external heat invested into the heat-up creates only a small increase in thermal efficiency. A method called regenerative feed-water preheating has been developed to increase the feed-water temperature before it enters the economizer. The principal idea of regenerative heating is that we bleed a small fraction of the steam from the turbine and use its condensation heat to preheat the feed-water. Since the latent heat of water is very high, a small amount is sufficient. In addition, as the mass flow rate of water through the condenser and the system heat loss simultaneously decreases, the process efficiency increases. Regenerative feed-water preheating can be done directly, by mixing the bled steam and feed-water flows at the preheating location, or indirectly, using a heat exchanger and then combining the flashed condensate to the main feed-water stream. The direct contact method is analyzed here. See Appendix A3.1 for a more complex system composed from *N* indirect heaters, and Appendix A3.2 for a system that has both feed-water heater types in use.

Direct-contact feed-water preheaters

In the direct method, steam is bled from the turbine for heating the feed-water and the flows are directly mixed, which makes their temperatures at the end equal. As the heat of condensation is large, only a small mass flow rate of bleeding is sufficient. Figure 3.21 presents a single regenerative feed-water heater-based system. A fraction a from the steam is bled into the feed-water heating system. Design criteria is that after the heating process the water should not boil, and therefore, it is assumed that the water is saturated at point 6. How do we solve the parameter a, and how do we choose the optimal bleeding pressure p_2 for maximum thermal efficiency of the process? To make it even more interesting, only point 8 after the economizer is fixed—as saturated water—and point 4 after the condenser. As a result the analysis also defines the thermodynamically optimal size of the economizer (Fig. 3.22).

The efficiency of this regenerative preheating process can be calculated as follows, by both direct and indirect methods, respectively.

$$\eta = \frac{\eta_{s.t}(h_1 - h_{2.s}) - \frac{\overline{v}_{67}(p_7 - p_6)}{\eta_{s.p2}} + (1-a)\left(\eta_{s.t}(h_2 - h_{3.s}) - \frac{\overline{v}_{45}(p_5 - p_4)}{\eta_{s.p1}}\right)}{(h_1 - h_7)}$$

$$\eta = 1 - \frac{(h_3 - h_4)(1-a)}{(h_1 - h_7)}$$

For the following analysis, we rely on the latter, formally simpler, indirect method. Note that in both methods, all process points must still be determined. The fraction of bleeding mass flow rate can be simply solved from the energy balance of the feed-water exchanger.

$$a = \frac{(h_6 - h_5)}{(h_2 - h_5)}$$

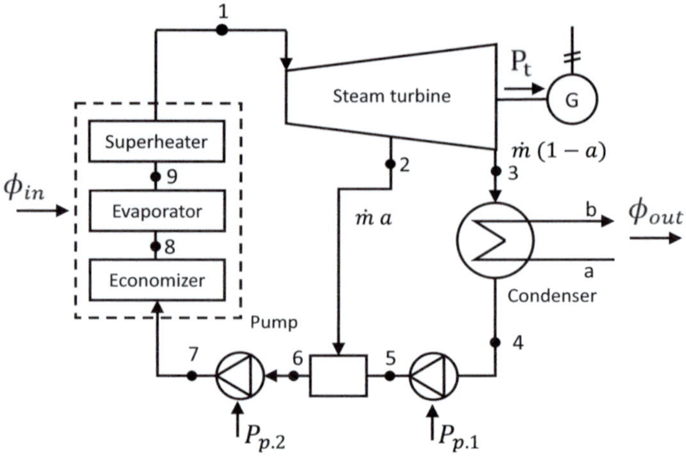

Fig. 3.21 Single direct regenerative feed-water preheating system. External heat input from 7 to 1. 1:(550°,120 bar), 2:(10 bar), 3:(0.04 bar), $\eta_{s.t} = 0.92$, $\eta_{s.p} = 0.95$

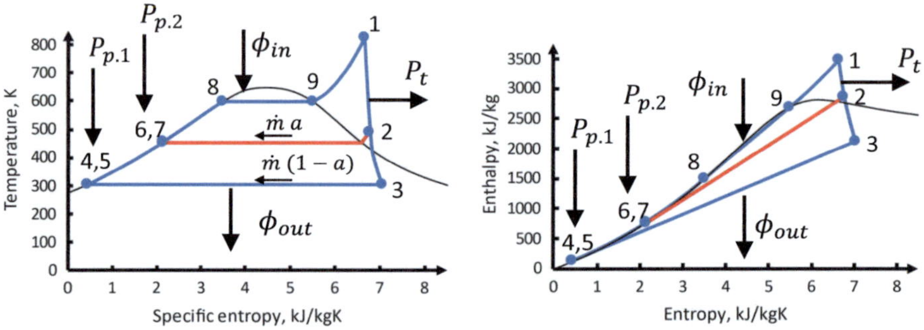

Fig. 3.22 Single direct regenerative feed-water preheating. The optimal bleeding pressure situation is presented, p_2 ~ 10 bar. External heat input 7-1; internal heating from steam condensation between 5 and 6

$$1 - a = \frac{h_2 - h_6}{h_2 - h_5} = \frac{1}{\frac{h_2 - h_5}{h_2 - h_6}} = \frac{1}{1 + \frac{\Delta h_1}{h_2 - h_6}} = \frac{1}{1 + \frac{\Delta h_1}{\Delta h_{b.1}}}$$

Δh_1 is the enthalpy increase of feed-water in the preheater 1, and $\Delta h_{b.1}$ is the enthalpy decrease of the bled steam used in feed-water heater 1. This is then substituted into thermal efficiency equation. Noting that $h_3 - h_4 = \Delta h_c$ is the enthalpy decrease in the condenser and writing the enthalpy h_7 starting from h_4, we get.

$$\eta = 1 - \frac{\Delta h_c}{(h_1 - h_4 - \Delta h_{p.1} - \Delta h_{p.2} - \Delta h_1)\left(1 + \frac{\Delta h_1}{\Delta h_{b.1}}\right)}$$

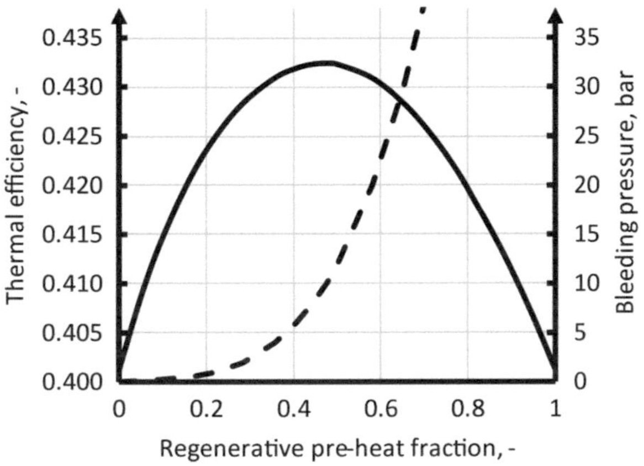

Fig. 3.23 Thermal efficiency and bleeding pressure (solid line) as a fraction of regenerative direct feed-water preheating, $\Delta h_1/(\Delta h_1 + \Delta h_{eco})$

The maximum efficiency is obtained when the denominator of the second term is at a maximum. This extremum can be obtained by simple differentiation of the denominator and setting this equal to 0. Based on this, the optimal enthalpy-increase of the feed-water in the preheater is equal to.

$$\Delta h_1 = \frac{h_1 - \left(h_4 + \Delta h_{p.1} + \Delta h_{p.2}\right) - (h_2 + h_6)}{2}$$

Another way to find the optimal enthalpy increase is to simply carry out calculations for different bleeding pressures and find the optimum graphically or by numerical optimization. Figure 3.23 shows the efficiency as a function of fractional enthalpy increase in the direct feed-water preheater, $\Delta h_1/(\Delta h_1 + \Delta h_{eco})$.

From Fig. 3.23 we can see that the optimal enthalpy increase in the heater is 47% of the total preheating enthalpy increase, including the economizer, and optimal bleeding pressure is 10 bar. It also shows that even by this single regenerative preheater, the efficiency increase is more than 3% units. This type of direct contact preheater is also typically installed with a de-aerator, forming the well-known feed-water-tank concept, the "sub-marine looking component".

Example 3.6 Let's analyze the above-described power plant with a single direct contact feed-water preheater with net electrical power production of $P_e = 500$ MW. As above, values of steam temperature and pressure before the turbine are 550 °C and 120 bar and the condenser pressure is 0.04 bar. Based on the discussion above, we choose the optimal bleeding pressure at $p_2 = 10$ bar. Isentropic efficiency of the turbine and the pump is 0.92 and 0.95, respectively.

1. Define all the process points and calculate the mass flow rate of live steam, the fraction *a*, and the mass flow rate of the bled steam
2. Calculate the required thermal heat input to the steam generator, and heat output from the condenser.

3. Calculate the mechanical power needed for the feed-water pumps.
4. Calculate the mechanical power from the steam turbine.
5. Calculate the thermal efficiency of the process. Compare the results with Example 3.5.

1. The calculation of the process points is shown in detail in Table 3.3. The main difference to the previous example is the direct-contact with the feed-water heater step.

From point 9, the saturated steam passes through the superheater to point 1. First, we need to solve the bled fraction of steam from the mass and energy balances of the preheater.

$$a = \frac{(763 - 122)}{(2863 - 122)} = 0.234$$

It is then possible to solve the required mass flow rate of steam, using the energy equations of the turbine and pumps.

$$\dot{m} = \frac{P_e}{(h_1 - h_2)\eta_g + (h_2 - h_3)(1 - a)\eta_g - \frac{\Delta h_{p1}(1-a)}{\eta_m} - \frac{\Delta h_{p2}}{\eta_m}}$$

Table 3.3 Calculation of process values of Rankine cycle with a single direct regenerative preheater. Values of physical properties are obtained from the Excel-based tool available in electronic supplement

#	p (bar)	t (°C)	T (K)	h (kJ/kg)	s (kJ/kg K)	Comments
1	120	550	823	3482	6.655	Read from the cross section of t_1 and p_1
2 s	10	193	466	2809	6.655	Move vertically down from point 1–2s (ds = 0) at p_2
2	10	215	488	2863	6.768	Calculate h_2 from the isentropic expansion equation
3 s	0.04	29	302	2004	6.768	Move vertically down from point 2–3 s (ds = 0) at p_3
3	0.04	29	302	2123	7.046	Calculate h_3 from the isentropic expansion equation
4	0.04	29	302	121	0.423	Move along the isobar to left side to the saturated water
5	10	29	302	122	0.423	Add enthalpy increase in the pump to get h_5
6	10	180	453	763	2.139	Saturated water at bleeding pressure, no boiling is allowed
7	120	183	456	776	2.150	Add enthalpy increase in the pump to get h_7
8	120	325	598	1491	3.497	Saturated water after the evaporator
9	120	325	598	2686	5.494	Saturated steam after the evaporator

Enthalpy increase in the condenser and feed-water pump is $\Delta h_{p1} = 1$ kJ/kg and $\Delta h_{p2} = 13$ kJ/kg, respectively (see Table 3.3). As above, generator and motor efficiencies are 0.98. Inserting these values, we get $\dot{m} = 435$ kg/s. Bled steam mass flow rate is $\dot{m}_2 = a\dot{m} = 0.234 \times 435 = 102\,\mathrm{kg/s}$.

2. Heat input is calculated from the evaporator energy balance, here using the steam side balance.

$$\phi_{in} = \dot{m}(h_1 - h_7) = 435 \times (3482 - 776) = 1178\,\mathrm{MW}$$

Heat output is calculated from the condenser energy balance.

$$\begin{aligned}\phi_{out} &= \dot{m}(1-a)(h_3 - h_4)\\ &= 435 \times (1 - 0.234) \times (2123 - 121) = 668\,\mathrm{MW}\end{aligned}$$

3. The mechanical power needed for the pumps is calculated as follows:

$$\begin{aligned}P_p &= \dot{m}\Delta h_{p1}(1-a) + \dot{m}\Delta h_{p2}\\ &= 435 \times 1 \times (1 - 0.234) + 435 \times 13 = 6.0\,\mathrm{MW}\end{aligned}$$

4. The mechanical power output from the turbine is calculated as follows:

$$\begin{aligned}P_t &= \dot{m}(h_1 - h_2) + \dot{m}(1-a)(h_2 - h_3)\\ &= 435 \times (3482 - 2863) + 435 \times (1 - 0.234) \times (2863 - 2123)\\ &= 516\,\mathrm{MW}\end{aligned}$$

5. The thermal efficiency is as follows:

$$\eta_t = \frac{P_t - P_p}{\phi_{in}} = \frac{516 - 6}{1178} = 0.433$$

Thermal efficiency of the process increases from 40.2 to 43.3–3.1% units in comparison with the standard Rankine cycle as already shown above. The regenerative feed-water preheating system requires more steam flow to generate the same amount of power—379 kg/s → 435 kg/s—but the condenser loss is smaller (759 MW → 668 MW) and the required heat input also decreases (1269 MW → 1178 MW). The system will be much more complex and requires larger system components. However, it has large potential for increasing efficiency, especially when a larger number of heaters are used. The final selection of the system configuration is determined by the techno-economic combination.

3.5 Principles of Gas Turbine Processes

Figure 3.24 presents a process schematic of a basic closed gas turbine cycle, referred to also as the Brayton cycle. Figures 3.25 and 3.26 present the process in *T-s* and *h–s* diagrams, respectively, for both the ideal and true process with irreversible losses in compression and expansion. The process comprises the following steps: (1) Compressions from pressure p_1 to p_2. Temperature increases. If the compression is ideal and isentropic, the end temperature is T_{2s}. In a real process, where the entropy increases, part of the used power goes to irreversibilities *e.g.* flow friction, and the actual temperature increase is larger. Compression requires power P_c, (2) Heat input, increasing the temperature from T_2 to T_3 at a constant pressure, (3) Expansion in the gas turbine back to pressure p_1. If the process is ideal and isentropic, the end temperature is equal to T_{4s}. and turbine power P_t is produced, (4) A return to starting point 1, where the amount ϕ_{out} of heat must be removed.

The pressure ratio of the process is defined as $\pi = p_2/p_1$. The thermal efficiency of the process is defined as the net power, $P_t - P_c$, divided by the heat input.

$$\eta = \frac{P_t - P_c}{\phi_{in}}$$

Or, in terms of enthalpies or temperatures, we have

$$\eta = \frac{(h_3 - h_{4s})\eta_{st} - (h_{2s} - h_1)\eta_{sc}}{(h_3 - h_2)} = \frac{\dot{m}\overline{c}_{p.t}(T_3 - T_{4s})\eta_{st} - \dot{m}\overline{c}_{p.c}(T_{2s} - T_1)\eta_{sc}}{\dot{m}\overline{c}_{p.h}(T_3 - T_2)}$$

Let us first study the isentropic compression and expansion processes. From the 1st law of thermodynamics definition of enthalpy and entropy we have $dh = dq + vdp = Tds + vdp$. And in the case of isentropic process, $ds = 0$, we have $dh_s = vdp$. Assuming ideal gas behaviour, $dh = c_p dT$ (ideal gas enthalpy is only a function of temperature) and $\Omega = 1$ (compressibility factor), the following equation and its

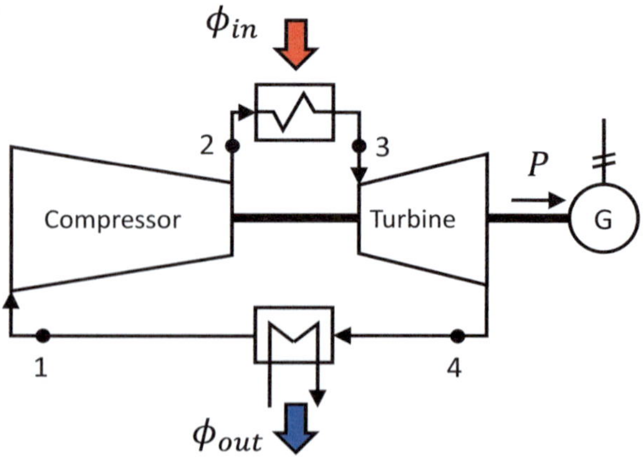

Fig. 3.24 Closed Brayton gas turbine process components

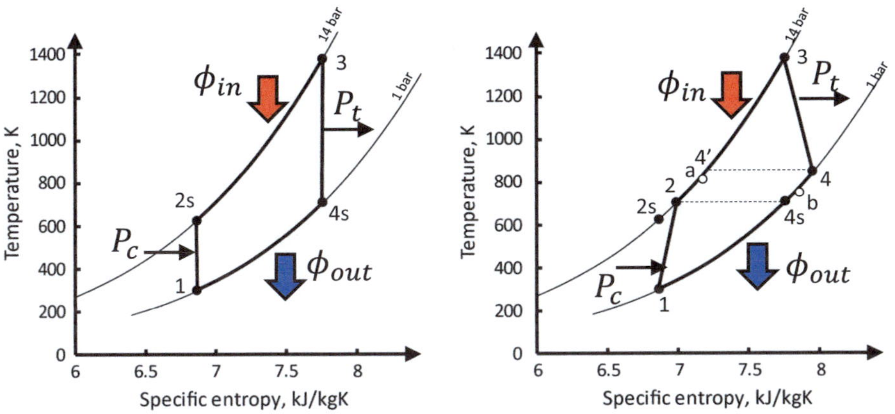

Fig. 3.25 Closed Brayton gas turbine process in *t-s* diagram, $p_1 = 1$ bar, $p_2 = 14$ bar, $t_1 = 25$ °C and turbine inlet temperature is $t_3 = 1100$ °C, compressor and turbine isentropic efficiencies are 0.85 and 0.88, respectively. Thermal efficiencies: left, ideal process $\eta_t = 0.504$ and right, true process $\eta_t = 0.351$. The dashed lines refer to the possibility to use regenerative heating from turbine exit gases; this will be discussed later

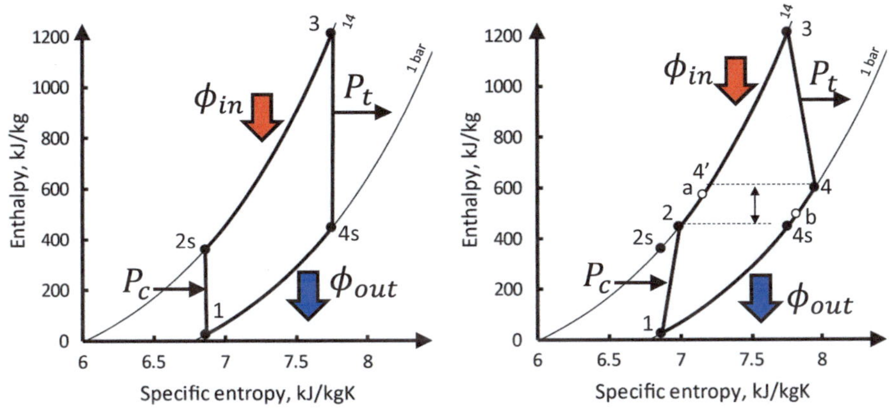

Fig. 3.26 Closed Brayton gas turbine process in h–*s* diagram, $p_1 = 1$ bar, $p_2 = 14$ bar, $t_1 = 25$ °C and turbine inlet temperature is $t_3 = 1100$ °C, compressor and turbine isentropic efficiencies are 0.85 and 0.88, respectively. Thermal efficiencies: left, ideal process $\eta_t = 0.504$ and right, true process $\eta_t = 0.351$. The dashed lines refer to the possibility to use regenerative heating from turbine exit gases; this will be discussed later

integral for solving the isentropic end temperature can be written as follows:

$$\int_{T_1}^{T_{2s}} \frac{c_p(T)dT}{T} = \frac{R}{M} \int_{p_1}^{p_2} \frac{dp}{p}$$

$$ln\left(\frac{T_{2s}}{T_1}\right) = \frac{R}{\overline{c}_p M} ln\left(\frac{p_2}{p_1}\right)$$

$$\overline{c}_p = \frac{c_p(T_{2s}) + c_p(T_1)}{2}$$

As the specific heat capacity is a function of the temperature, some iteration is required to achieve T_{2s}. It is also possible to use a polynomial coefficient equation for the heat capacity such as $c_p(T) = a + bT + cT^2 + dT^3$, resulting in an implicit solution of the temperature.

$$\int_{T_1}^{T_{2s}} (a/T + b + cT + dT^2)dT = \frac{R}{M} \int_{p_1}^{p_2} \frac{dp}{p}$$

$$a\, ln\left(\frac{T_{2s}}{T_1}\right) + b(T_{2s} - T_1) + \frac{c}{2}(T_{2s}^2 - T_1^2) + \frac{d}{3}(T_{2s}^3 - T_1^3) = \frac{R}{M} ln\left(\frac{p_2}{p_1}\right)$$

As can be seen, this form also requires iteration in solving the temperature. When we calculate the thermal efficiency of the process and want to draw the process in *T-s* or in *h-S* diagrams, several integrals need to be calculated and iterated. In this book, we use state functions, described below, and solving the process points using them, as was done earlier with the steam power plant and water vapor and liquid.

First, we however analyze a case using the cold gas approximation as described by [39]. A simplified equation is used to analyze the problem. We can calculate the isentropic end temperature for both compression and expansion (note the reverse pressure ratio). $k = c_p/c_v = C_{pm}/C_{vm}$ is equal to the ratio of constant pressure and volume heat capacities.

$$T_{2s} = T_1 \pi^{\left(\frac{R}{\overline{c}_p M}\right)} = T_1 \pi^{\left(\frac{k-1}{k}\right)}$$

$$T_{4s} = \frac{T_3}{\pi^{\left(\frac{R}{\overline{c}_p M}\right)}} = \frac{T_3}{\pi^{\left(\frac{k-1}{k}\right)}}$$

If using the *h-s* diagram, note that these can also be solved using a similar method to the Rankine cycle. These isentropic relations show that $T_{2s}/T_3 = T_1/T_{4s}$. The mass flow rate is constant through all components, and we assume that heat capacity is constant for all steps, $c_p = \overline{c}_{p.t} = \overline{c}_{p.c} = \overline{c}_{p.h}$. With this so-called cold gas assumption, we can show after some manipulation.

$$\eta = \frac{T_{4s}(T_3/T_{4s} - 1) - T_1(T_{2s}/T_1 - 1)}{T_3(1 - T_2/T_3)}$$

$$= \frac{T_{4s}\left(1 - \frac{T_1}{T_{4s}}\right)\left(\frac{T_{2s}}{T_1} - 1\right)}{T_3\left(1 - \frac{T_1}{T_{4s}}\right)} = \frac{\left(\frac{T_{2s}}{T_1} - 1\right)}{\frac{T_3}{T_{4s}} = \frac{T_{2s}}{T_1}} = 1 - \frac{1}{\pi^{\left(\frac{k-1}{k}\right)}}$$

Thermal efficiency of an ideal Brayton cycle is only a function of the pressure ratio and the heat capacity ratio. Of course, considering the actual heat capacity temperature dependency would give a different result. The higher the pressure after the compressor, the higher the thermal efficiency. As the pressure increases, the temperature also increases. The efficiency can also be presented in terms of temperatures. The last form of the thermal efficiency can be presented in terms of temperature ratios as follows:

$$\eta = 1 - \frac{1}{T_{2s}/T_1} = 1 - \frac{T_1}{T_{2s}} = 1 - \frac{T_3}{T_{4s}}$$

From this equation it can be concluded that the ratio of inlet and outlet temperatures is constant at every vertical line drawn on the *T*-*s* diagram. Note that this is not the Carnot efficiency, even if it has the same form, because the heat is delivered and removed at varying temperatures. Since in the true process compression and expansion are not isentropic, the maximum efficiency is reached as a function of pressure ratio, which is opposite to the idealized case with continuously increasing efficiency.

The cold gas idealization, *i.e.* assuming a constant heat capacity, offers an excellent way to understand the trends in gas turbine processes. But if the absolute values are of interest, then one should work with enthalpies and entropies, and solve other state variables with the help of these variables. When working with air, reasonable accuracy can be obtained by using the ideal gas model, meaning that the enthalpy is not a function of the pressure. First, we need to define the enthalpy as a function of the temperature and then formulate an equation for the entropy in terms of the pressure. With this definition and equation all required process changes can be calculated. From the energy equation and definition of enthalpy, we can show that:

$$ds = \frac{dh}{T} - \frac{R}{M}\frac{dp}{p} = \frac{c_p(T)dT}{T} - \frac{R}{M}\frac{dp}{p}$$

And by integration of this we get entropy and enthalpy, just by using the integral of the heat capacity:

$$\begin{aligned} s(T, p) - s_0 = {} & a \ln\left(\frac{T}{T_0}\right) + b(T - T_0) + \frac{c}{2}\left(T^2 - T_0^2\right) \\ & + \frac{d}{3}\left(T^3 - T_0^3\right) - 0.001 \times \frac{R}{M} \ln\left(\frac{p}{p_0}\right) \end{aligned}$$

$$h(T) - h_0 = a(T - T_0) + \frac{b}{2}\left(T^2 - T_0^2\right) + \frac{c}{3}\left(T^3 - T_0^3\right) + \frac{d}{4}\left(T^4 - T_0^4\right)$$

Here, the reference values are set as $T_0 = 273.15K, p_0 = 1$ bar, $h_0 = 0\,\text{kJ/kg}$ and $s_0 = 6.77634\,\text{kJ/kgK}$. Polynomial coefficients for air are $a = 0.982076$, $b = 1.64395 \times 10^{-5}$, $c = 2.25868 \times 10^{-7}$, $d = -8.81495 \times 10^{-11}$, and fit with the air data from [39]. The above equations can be easily programmed for example with Python or as Visual Basic functions in MS-Excel®. Visual basic versions of these are available as an electronic appendix. Based on the above equations, all variables can be iteratively solved. There are direct functions available for $h(T)$, $T(h)$, $s(T, p)$, $t(p, s)$ and $h(p, s)$. All cases presented below are calculated and plotted by these.

Example 3.7 Analyze a simple Brayton gas turbine power plant of net electrical power production $P_e = 500$ MW. Inlet temperature and pressure conditions to the compressor are 25 °C and 1 bar and the pressure ratio is 14. Isentropic efficiency of the turbine and the compressor is 0.88 and 0.85, respectively, and generator efficiency is 0.98.

1. Define all the process points and calculate the mass flow rate of air.
2. Calculate the required thermal heat input to system, and heat output from the low temperature heat exchanger.
3. Calculate the mechanical power needed for the compressor.
4. Calculate the mechanical power from the turbine.
5. Calculate the thermal efficiency of the process.
6. Calculate the back work ratio.

1. The calculation of the process points is presented in Table 3.4. Use the available Excel© functions in the electronic appendix.

From the point 4, the air passes through the low temperature heat exchanger back to point 1. We get power from the turbine and part of this is used to run the compressor.

Table 3.4 Calculation of process values of basic Brayton process, values of physical properties are obtained from the Excel-based tool available in electronic supplement

#	p (bar)	t (°C)	T (K)	h (kJ/kg)	s (kJ/kg K)	Comments
1	1	25	298	25	6.864	Read from the cross section of t_1 and p_1
2 s	14	350	624	360	6.864	Move vertically up from point 1–2s (ds = 0) to p_2
2	14	406	679	419	6.955	Calculate h_2 from the isentropic compression equation
3	14	1100	1373	1210	7.752	Move along the p_2 isobar right to turbine inlet temperature
4 s	1	431	704	447	7.752	Move vertically down from point 3–4s (ds = 0) at p_4
4	1	516	789	538	7.875	Calculate h_4 from the isentropic expansion equation

Based on the above, $\Delta h_t = 1210 - 538 = 672$ kJ/kg and $\Delta h_c = 419 - 25 = 394$ kJ/kg. With these, the mass flow rate of the air can be calculated. Note that as the turbine and compressor share the same axis, only one electrical efficiency is used.

$$\dot{m} = \frac{P_e/\eta_g}{\Delta h_t - \Delta h_c} = \frac{500 \times 10^3/0.98}{672 - 394} = 1835\,\text{kg/s}$$

2. Heat input is calculated from the high temperature heat exchanger energy balance, here using the air side balance.

$$\phi_{in} = \dot{m}(h_3 - h_2) = 1835 \times (1210 - 419) = 1451\,\text{MW}$$

Heat output is calculated from the low temperature heat exchanger energy balance.

$$\phi_{out} = \dot{m}(h_4 - h_1) = 1835 \times (538 - 25) = 941\,\text{MW}$$

3. Mechanical power needed for the compressor is calculated as follows:

$$P_c = \dot{m}(h_2 - h_1) = 1835 \times (419 - 25) = 723\,\text{MW}$$

4. Mechanical power output from the turbine is calculated as follows:

$$P_t = \dot{m}(h_3 - h_4) = 1835 \times (1210 - 538) = 1233\,\text{MW}$$

5. The thermal efficiency is the ratio of net work to heat input.

$$\eta_t = \frac{P_t - P_c}{\phi_{in}} = \frac{1233 - 723}{1451} = 0.351$$

6. The back work ratio is simply the compressor power divided by the turbine power, 723/1233 = 0.59. This is significantly higher than in the basic steam turbine plant which was only 4.9/515 = 0.0093, where the pressure is raised by pumping the w liquid; see Example 3.4.

3.6 Intercooling and Regenerative Pre-heating

One way to increase the efficiency is to add intercooling to the compressor. By removing the heat created in compression the gas volume decreases, the compression work needed will decrease and the efficiency will increase. Figure 3.27 presents an example of this option. Except for the intercooler at 2-3, the process is identical to the standard gas turbine process discussed above.

Fig. 3.27 Gas turbine process with intercooling

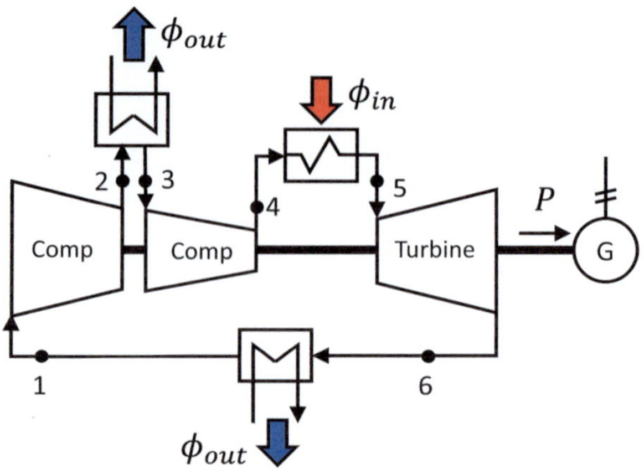

The thermal efficiency of this process can be calculated as follows. Note that point 5 is fixed by the pressure p_4 and turbine inlet temperature and point 3 is fixed by the intercooling pressure and the selected temperature.

$$\eta = \frac{(h_5 - h_6) - (h_2 - h_1) - (h_4 - h_3)}{(h_5 - h_4)} = \frac{(h_5 - h_{6s})\eta_{s.t} - (h_{2s} - h_1)/\eta_{s.c} - (h_{4s} - h_3)/\eta_{s.c}}{h_5 - h_3 - (h_{4s} - h_3)/\eta_{s.c}}$$

Example 3.8 Analyze an inter-cooled gas turbine power plant of net electrical power production $P_e = 500$ MW. The inlet temperature and pressure conditions of the compressor are 25 °C and 1 bar and the pressure ratio is 14. The intercooling pressure is 2.1 bar, and the temperature is 25 °C. The turbine inlet temperature is 1100 °C. Isentropic efficiency of the turbine and the compressor is 0.88 and 0.85, respectively, and generator efficiency is 0.98.

1. Define all process points and calculate the mass flow rate of air.
2. Calculate the required thermal heat input to system, and the heat output from the low temperature heat exchanger.
3. Calculate the mechanical power needed for the compressor.
4. Calculate the mechanical power from the turbine.
5. Calculate the thermal efficiency of the process.
6. Calculate the back flow work ratio.

1. The calculation of the process points is presented in Table 3.5. Use the available Excel© functions.

From point 6, the air passes through the low temperature heat exchanger back to point 1. Based on the above, $\Delta h_t = 1210 - 538 = 672$ kJ/kg and $\Delta h_c = 108 - 25 + 278 - 25 = 336$ kJ/kg. With these, the mass flow rate of the air can be calculated.

Table 3.5 Calculation of process values of inter-cooled gas turbine process, values of physical properties are obtained from the Excel-based tool available in electronic supplement

#	p (bar)	t (°C)	T (K)	h (kJ/kg)	s (kJ/kg K)	Comments
1	1	25	298	25	6.864	Read from the cross section of t_1 and p_1
2 s	2.1	95	368	96	6.864	Move vertically up from point 1–2 s (ds = 0) to p_2
2	2.1	107	380	108	6.898	Calculate h_2 from the isentropic compression equation
3	2.1	25	298	25	6.651	Move along the p_2 isobar left to intercooling temperature
4 s	14	236	509	240	6.651	Move vertically down from point 3–4 s (ds = 0) at p_4
4	14	272	545	278	6.723	Calculate h_4 from the isentropic compression equation
5	14	1100	1373	1210	7.752	Move along the p_4 isobar right to turbine inlet temperature
6 s	1	431	704	447	7.752	Move vertically down from point 5–6s (ds = 0) at p_1
6	1	516	789	538	7.875	Calculate h_6 from the isentropic expansion equation

Note that as the turbine and the compressor share the same axis, only one electrical efficiency is used.

$$\dot{m} = \frac{P_e/\eta_g}{\Delta h_t - \Delta h_c} = \frac{500 \times 10^3/0.98}{672 - 336} = 1518\,\text{kg/s}$$

2. Heat input is calculated from the high temperature heat exchanger energy balance, here using the steam side balance.

$$\phi_{in} = \dot{m}(h_5 - h_4) = 1518 \times (1210 - 278) = 1415\,\text{MW}$$

Heat output is calculated from the low temperature heat exchanger energy balance.

$$\phi_{out} = \dot{m}(h_6 - h_1) = 1518 \times (538 - 25) = 778\,\text{MW}$$

3. Mechanical power needed for the compressor is calculated as follows.

$$P_c = \dot{m}\Delta h_p = 1518 \times 336 = 510\,\text{MW}$$

4. Mechanical power output from the turbine is calculated as follows.

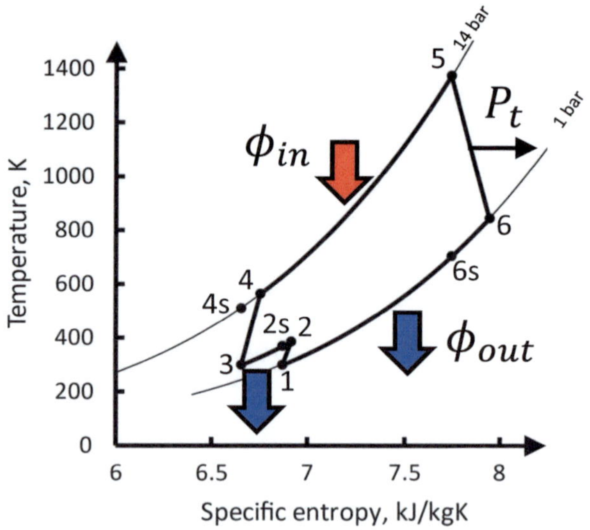

Fig. 3.28 T-s diagram of closed gas turbine process with intercooling. $p_1 = 1$ bar, $p_2 = 2.1$ bar, $p_4 = 14$ bar, $t_1 = 25$ °C and turbine inlet temperature is $t_5 = 1100$ °C. Isentropic efficiency of the turbine and the compressor is 0.88 and 0.85, respectively. The thermal efficiency is $\eta_t = 0.36$

$$P_t = \dot{m}\Delta h_t = 1518 \times 672 = 1020\,\text{MW}$$

5. Thermal efficiency is the ratio of net work to heat input.

$$\eta_t = \frac{P_t - P_c}{\phi_{in}} = \frac{1020 - 510}{1415} = 0.36$$

6. Back flow work ratio is simply the compressor power divided by the turbine power, 510/1020 = 0.5. The *T-s* diagram of this process is presented in Fig. 3.28.

The impact of the intercooling is that the compression work decreases since the gas temperature and therefore the gas volume decreases. In this case, the optimal intercooling pressure is 2.1 bar. In comparison with the standard case (Fig. 3.25) efficiency increased from 0.35 to 0.36.

By using a part of the waste heat—after point 4 of the standard process, or after point 6 in the intercooled process—to preheat the gases after the final compression, the need for external heat can be decreased. Therefore, regenerative preheating is a way to improve processes. Without any heat recovery, the heat from point 4 to 1 is wasted. This could be partially used to heat up the gas before the actual high temperature heat exchanger. The heat recovery is possible when $T_4 > T_2$. This process is presented (Fig. 3.29).

The thermal efficiency of this process can be calculated from the following equation that transfers the heat from 4-b to 2-a (Fig. 3.25).

$$\eta = \frac{(h_3 - h_4) - (h_2 - h_1)}{(h_3 - h_a)}$$

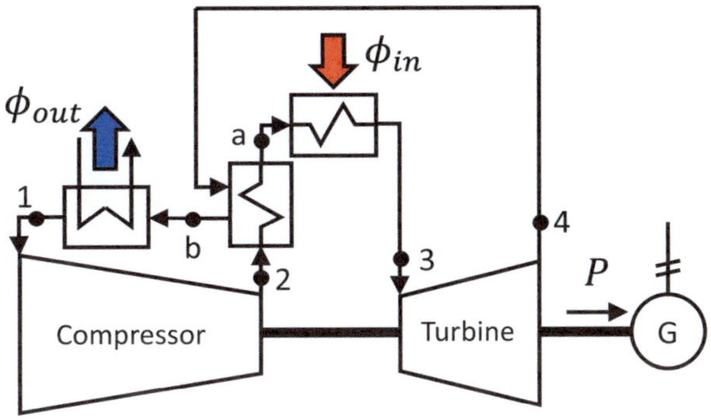

Fig. 3.29 Brayton cycle with re-generative preheating

The temperature/enthalpy value at point a depends on the efficiency of the preheating heat exchanger and the recuperation degree, η_r.

$$\eta_r = \frac{T_a - T_2}{T_4 - T_2}$$

From which the preheating temperature can be solved.

$$T_a = T_2 + \eta_r(T_4 - T_2)$$

In the maximum case of a perfect heat exchanger with infinite surface area $\eta_r = 1$, and $T_a = T_4$ the equation reduces to:

$$\eta = 1 - \frac{(h_2 - h_1)}{(h_3 - h_4)}$$

Using the same process values as in Fig. 3.25, the efficiency of this new improved process would be $\eta = 0.41$. For comparison the efficiency without heat recovery was 0.351, representing an improvement of 5.9%units. Based on the theoretical isentropic process, if we increase the pressure ratio, efficiency increases. However, if we study a process with a recuperator, the lower the pressure ratio, the larger the reheating potential is [39]. When the pressure ratio approaches a value of 1, the efficiency approaches the Carnot efficiency, $\eta = 1 - T_1/T_3$. However, the production of mechanical work from this small pressure difference is not feasible.

Figure 3.30 presents the thermal efficiency of an ideal isentropic process, and three cases for a standard real process (isentropic efficiencies 0.9) with turbine inlet temperatures of 800, 1000 and 1200 °C and the same for processes with recuperator (dashed lines).

The optimal pressure ratio is smaller for a system with a recuperator. The preheated efficiency meets the standard process efficiency curves when $T_4 < T_2$. For an ideal cycle description and discussion, see [51]. Traupel [53] also discusses several options for real processes.

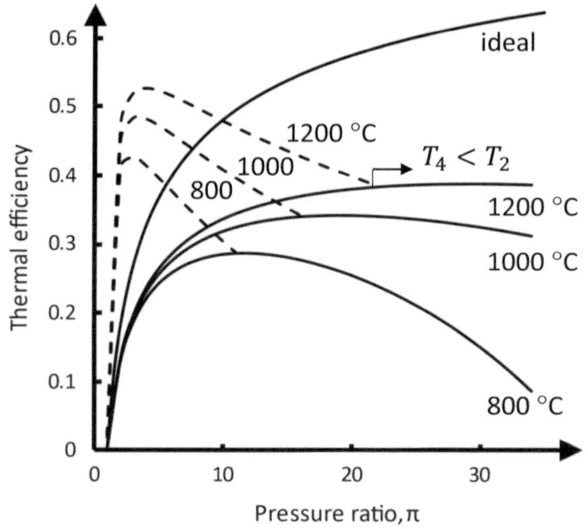

Fig. 3.30 Thermal efficiency of different Brayton processes (no intercooling). Dashed line cases correspond to a system with a recuperator of $\eta_r = 0.9$, that transfers heat from the exit gas to steam before the inlet heat exchanger. Turbine inlet temperatures are T_3 values. The isentropic efficiency of the turbine and the compressor is 0.88 and 0.85, respectively. Ideal refers to the isentropic cold gas approximation of a basic Brayton cycle

3.7 Ericsson and Stirling Cycles

The above examples demonstrated that there are several ways to improve the thermal efficiency of the gas turbine process. As the process complexity increases, the price of the system also increases. Therefore, the final construction of this process component is always a compromise between the technical and economic performance. However, if we forget economic limitations, it is interesting to study two concepts called the Ericson and the Stirling cycles. These processes are composed of a large number of compressors, with each having their own intercooler and many turbine stages, that each have their own re-heaters (Fig. 3.31). The process operates between two approximately constant temperature levels. Therefore, it is possible to transfer heat effectively from the hot exit gas to preheat the gas after the final compression.

Figure 3.32 presents the actual calculated T-s diagrams of this process. The main difference is that Ericsson's recuperator operates at constant pressures, while Stirling's cycle recuperator works at constant volumes. All isentropic efficiencies are assumed to be 0.9 and the recuperator efficiency is $\eta_r = 1$. We have used 10 steps ($N = M$) for both compression and expansion.

Note that heat is delivered into the process from points a-3′-3-4 and removed between points 1 and 2. There is a recuperative heat exchanger between the isobars and isochors that transfers heat effectively from the exhaust gases to the turbine input stream. This means that the efficiency of this process will approach the Carnot efficiency. Since the pressure ratio of each step is kept constant, the calculations are quite easy. Following this, efficiency η_r corresponds to recuperator efficiency. Note that $h_4 = h_{3'}$.

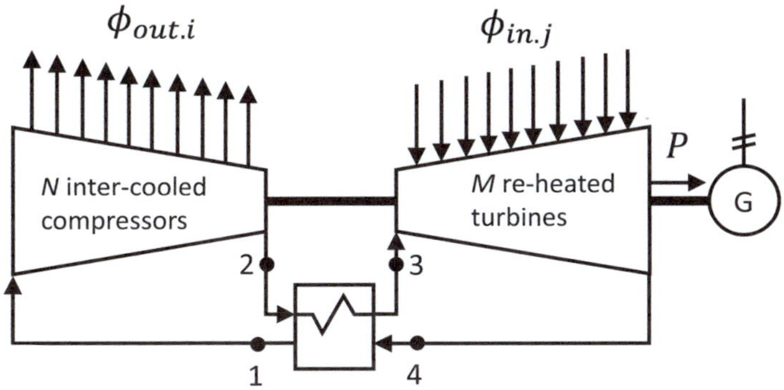

Fig. 3.31 Ericsson cycle with N intercooled compressors and M reheated turbine stages. Each ϕ arrow corresponds to a separate stage heat flow

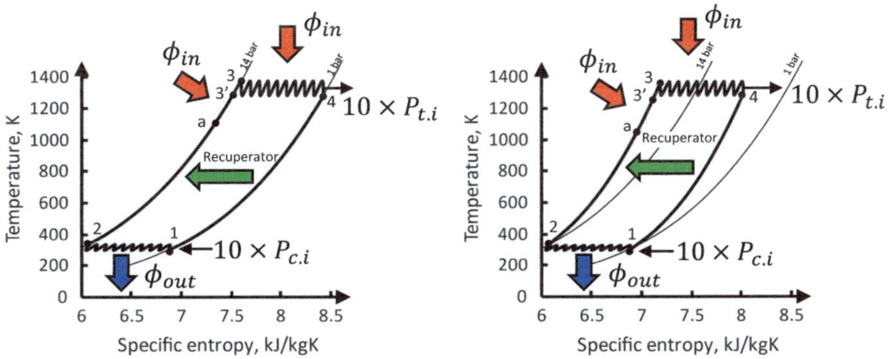

Fig. 3.32 T-s diagram of Ericsson (left) and Stirling cycles with 10 intercooled compressor and 10 re-heated turbine stages. $p_1 = 1$ bar, $p_2 = 14$ bar (Ericson), $p_2 = 100$ bar (Stirling), $t_1 = 25$ °C and the turbine inlet temperature is $t_3 = 1100$ °C, both compressor and turbine isentropic efficiencies are 0.8. The thermal efficiency is $\eta_t = 0.64$ while the Carnot efficiency would be 0.78

$$\eta = \frac{M\,\Delta h_{t.i} - M\,\Delta h_{c.j}}{M\,\Delta h_{t.i} + (h_4 - h_a)} = \frac{\Delta h_t - \Delta h_c}{\Delta h_{t.i} + \frac{h_4 - h(T_2 + \eta_r(T_4 - T_2))}{M}}$$

It is important to notice that all compressors and turbine stages have equal enthalpy increases and decreases $\Delta h_{c.i} = \Delta h_c$ and $\Delta h_{t.j} = \Delta h_t$, respectively. In addition, all intercooling and reheating enthalpy changes correspond to the enthalpy increase of their work phases in compression and expansion steps. This also means that all intercooling final temperatures are equal, and all reheating final temperatures are equal. The above equation can also be presented in terms of pressure ratios as follows. Note that to be accurate this would require iteration, but we assume constant values for heat capacities to simplify the analysis.

$$\eta = \frac{c_p(T_3)T_3\left(1 - \frac{1}{\pi_i^{(k_t-1)/k_t}}\right)\eta_{s.t} - c_p(T_1)T_1\left(\pi_i^{(k_c-1)/k_c} - 1\right)\eta_{s.c}}{c_p(T_3)T_3\left(1 - \frac{1}{\pi_i^{(k_t-1)/k_t}}\right)\eta_{s.t} + \frac{h_4 - h(T_2 + \varepsilon(T_4 - T_2))}{M}}$$

As an extreme, if the recuperator efficiency is assumed to be $\eta_r = 1$, and heat capacity ratios are the same $k_t = k_c = k$, and since $\pi_i = \pi^{1/M}$, we have:

$$\eta = 1 - \frac{1}{\eta_{s.c}\eta_{s.t}}\frac{T_1}{T_3}\pi^{(k-1)/(kM)}$$

If both the compression and the expansion steps approach the isentropic ideal processes, then as the number of compression and expansion steps, M, approaches infinity, the thermal efficiency approaches the Carnot efficiency. This same result applies for the Stirling cycle as well.

The Stirling cycle is applied in the Stirling engine. The construction of a large number of compressors and turbines is replaced by a specially designed piston system, making the concept economically feasible. The Stirling engine is widely used in several renewable heat applications including waste heat and concentrated solar thermal systems. See for example [43] for the details of these systems.

Example 3.9 Analyze a gas turbine power plant of net electrical power production

$P_e = 500$ MW. The inlet temperature and pressure condition of the compressor is 25 °C and 1 bar and the pressure ratio is 24. The isentropic efficiency of the turbine and the compressor is 0.88 and 0.85, respectively, and the generator efficiency is 0.98. The exit temperature of the turbine is 670 °C.

1. Define all process points and calculate the mass flow rate of air. Note that the turbine inlet temperature must be solved first.
2. Calculate the required thermal heat input to system, and heat output from the low temperature heat exchanger.
3. Calculate the mechanical power needed for the compressor.
4. Calculate the mechanical power from the turbine.
5. Calculate the thermal efficiency of the process.
6. Calculate the back work ratio.
7. Discuss the possible role of regenerative heating for this process. Note that we have used the Siemens SGT5-9000HL as a reference, 593 MW, exit temperature 670 °C, exit mass flow rate 1050 kg/s, Siemens [50]. In this example we assume that we use only air. Mass flow rate in the turbine would therefore be larger in a system based on gas combustion.

1. The calculation of the process points is presented in Table 3.6. Use the available Excel© functions.

When the enthalpy at isentropic compression and expansion is known, true end points are easily solved.

Table 3.6 Calculation of process values of industrial 500 MW Brayton process, values of physical properties are obtained from the Excel-based tool available in electronic supplement

#	p (bar)	t (°C)	T (K)	h (kJ/kg)	s (kJ/kg K)	Comments
1	1	25	298	25	6.864	Read from the cross section of t_1 and p_1
2 s	24	448	721	464	6.864	Move vertically up from point 1–2s (ds = 0) to p_2
2	24	519	792	542	6.967	Calculate h_2 from the isentropic compression equation
3	24	1511	1784	1713	7.917	Iterate state 3 from turbine isentropic efficiency equation, see above
4 s	1	547	820	572	7.917	Solved as a part of the iteration
4	1	670	943	709	8.072	Endpoint was fixed beforehand

$$h_2 = h_1 - (h_1 - h_{2s})/\eta_{s.c}$$

$$h_4 = h_3 - \eta_{s.t}(h_3 - h_{4s})$$

Since only the turbine exit temperature is given, we need to solve the turbine inlet temperature iteratively. This can be obtained from isentropic efficiency equation by iteration, solving first s_3 from $f(s_3)$, and then other values from that at point 3.

$$f(s_3) = \eta_{s.t} - \frac{h_3(p_3, s_3) - h_4(p_4, t_4)}{h_3(p_3, s_3) - h_{4s}(p_4, s_3)} = 0$$

To start the iteration, we need to make a guess for s_3; for example we let $s_3 = s_4(p_4, t_4)$. We already know that $t_4 = 670\,°\text{C}, p_4 = 1$ bar, $p_3 = 24$ bar. Using Excel's Solver©, we get $t_3 = 1511\,°\text{C}$, $s_3 = 7.917\,\text{kJ/kgK}$ and $h_3 = 1713\,\text{kJ/kg}$.

From point 4 the air passes through the low temperature heat exchanger back to point 1. Based on the above, $\Delta h_t = 1713 - 709 = 1004$ kJ/kg and $\Delta h_c = 542 - 25 = 517$ kJ/kg. With these, we can calculate the mass flow rate of the air. Note that as the turbine and the compressor share the same axis, only one electrical efficiency is used.

$$\dot{m} = \frac{P_e/\eta_g}{\Delta h_t - \Delta h_c} = \frac{500 \times 10^3/0.98}{1004 - 517} = 1048\,\text{kg/s}$$

2. Heat input is calculated from the hot temperature heat exchanger energy balance, here using the air side balance.

$$\phi_{in} = \dot{m}(h_3 - h_2) = 1048 \times (1713 - 542) = 1227\,\text{MW}$$

Heat output is calculated from the low temperature heat exchanger energy balance.

$$\phi_{out} = \dot{m}(h_4 - h_1) = 1048 \times (709 - 25) = 717\,\text{MW}$$

3. The mechanical power needed for the compressor is calculated as follows:

$$P_c = \dot{m}(h_2 - h_1) = 1048 \times (542 - 25) = 542\,\text{MW}$$

4. The mechanical power output from the turbine is calculated as follows:

$$P_t = \dot{m}(h_3 - h_4) = 1048 \times (1713 - 709) = 1052\,\text{MW}$$

5. The thermal efficiency is the ratio of net work to heat input.

$$\eta_t = \frac{P_t - P_c}{\phi_{in}} = \frac{1052 - 542}{1227} = 0.42$$

6. The back work ratio is simply the compressor power divided by the turbine power; 542/1052 = 0.52
7. For the studied process, the temperature at point 4 and 2 is 670 and 519 °C respectively. This means that the recuperator could be used to utilize the waste heat in heating up the gases before adding external heat to the process. We assume a recuperator efficiency of 0.8, meaning that we use 80% of the potential temperature difference and $t_a = t_2 + \eta_r(t_4 - t_2) = 519 + 0.8 \times (670 - 519) = 640\,°\text{C}$. Enthalpy at this point is $h_a(p_2, t_a) = 675\,\text{kJ/kg}$. And from this, the thermal efficiency is.

$$\eta = \frac{(h_3 - h_4) - (h_2 - h_1)}{(h_3 - h_a)} = \frac{1004 - 517}{1713 - 675} = 0.47$$

This increases the process efficiency by 5% units and decreases the need for external heat from 1227 MW to $1048 \times (1713 - 675) = 1087\,\text{MW}$. The mass flow rate of air was very close to the reference of SGT5 turbine. To reach this power, the turbine inlet temperature should be 1511 °C, using the model presented here. There is no direct reference for this available but based on [33], values as high as 1700 °C are used in commercial turbines, so the value is reasonable.

3.8 Introduction to Combined Power Plants

In the previous sections we have discussed the thermal efficiency of both steam and gas turbine-based power plants. In gas turbines in particular the temperature of the exit gas is always relatively high—650–750 °C—causing considerable heat loss. Recovering this heat in a specially designed heat recovery steam generator and using it to run the bottom steam power plant offers a significant improvement. In general, a combined cycle is the most efficient type of large scale practical thermal power plant, reaching efficiencies higher than 60%. In this book we discuss only gas turbine-Rankine cycle examples. For a more complete analysis on complex cycles including several pressure levels see for instance [33] or [25].

Combined cycles can be designed in numerous ways. The principal idea is that the waste heat of the topping process is used as the input heat for the bottoming process(es). An essential component that connects these processes together is the heat recovery steam generator (in this case assuming that process uses Rankine cycle-based concept). Selection of the process fluid in the Rankine cycle for optimal temperature range is crucial. For example, with gas turbines, water is often used but with internal combustion engines having a low temperature of flue gases, an organic or ammonia-based Rankine cycle can be used.

Figure 3.33 presents a simplified schematic of a gas turbine-driven combined cycle with a Rankine bottoming cycle. Thermal efficiency of the gas turbine cycle, the heat recovery steam generator and the Rankine cycle is η, η_{HRSG}, η_R, respectively.

It is very important to understand the role of the different terms in the schematic. The heat loss terms in particular can be confusing. The heat loss term from the heat recovery steam generator equates to the losses associated with the gases from the gas turbine. The objective is to have gas temperature as low as possible to minimize the loss. If the gas turbine is driven by a combustion process, as in the normal case, then the exit temperature is limited by the dew point of the flue gases, to prevent corrosion [33]. The heat losses from the Rankine cycle refer to heat rejected in the condenser. If the process is maximized with respect to power, then the temperature of the condensate can be as low as ~ 28–29 °C, corresponding to 0.04 bar condenser pressure. If the waste heat is used for district heating, then the temperature of the condenser must be higher—90–100 °C—and the condenser pressure must be 1 bar. The overall thermal efficiency of the process is the sum of all power produced divided

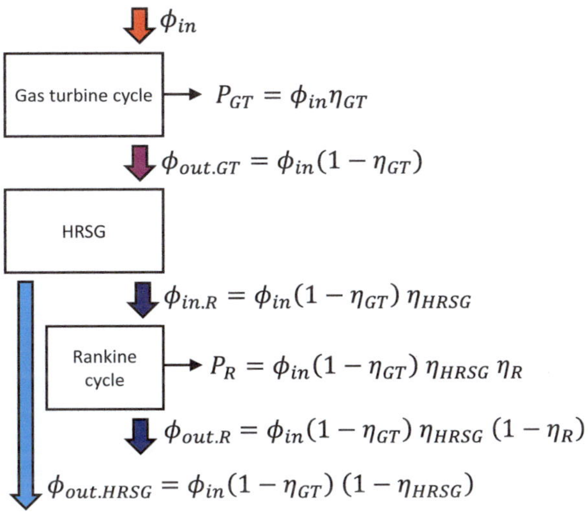

Fig. 3.33 Gas turbine based combined cycle

by the heat input to the gas turbine process [20].

$$\eta = \frac{P_{GT} + P_R}{\phi_{in}} = \frac{\phi_{in}\eta_{GT} + \phi_{in}(1 - \eta_{GT})\eta_{HRST}\eta_R}{\phi_{in}} = \eta_{GT} + (1 - \eta_{GT})\eta_{HRST}\eta_R$$

The following example discusses this equation in more detail and gives a rough understanding of the magnitude of process improvement opportunities when using the combined cycle. After this, we will study the heat recovery steam generator in sufficient detail to be able to understand its main features.

Example 3.10 Analyze a simple combined cycle of net electrical power production of $P_e = 500$ MW. The thermal efficiency of the gas turbine is $\eta_{GT} = 0.43$ and heat recovery steam generator $\eta_{HRSG} = 0.92$. The bottoming Rankine cycle is optimized for power production and has thermal efficiency of $\eta_R = 0.44$. The electrical efficiency of the generators is 0.98 and the pumps of the Rankine cycle are neglected.

1. Calculate the thermal efficiency of the combined plant.
2. Calculate the required thermal heat to run the process.
3. Calculate the mechanical power from the gas turbine and steam power plants.

1. The thermal efficiency can be calculated from the above derived equation.

$$\eta = \eta_{GT} + (1 - \eta_{GT})\eta_{HRST}\eta_R = 0.43 + (1 - 0.43) \times 0.92 \times 0.44 = 0.661$$

2. The heat requirement is calculated simply from the definition of thermal efficiency and considering the total electrical efficiency.

$$\phi_{in} = \frac{P_e}{\eta_g \eta} = \frac{500\,\text{MW}}{0.98 \times 0.661} = 772\,\text{MW}$$

This is of course significantly less than with the other processes studied above as the thermal efficiency is higher.

3. The mechanical power production from the gas turbine process is $P_{GT} = \phi_{GT}\eta_{GT} = 772\,\text{MW} \times 0.43 = 332\,\text{MW}$ and the power generation of the steam cycle is $P_R = \phi_{in}(1 - \eta_{GT})\eta_{HRSG}\eta_R = 772\,\text{MW} \times (1 - 0.43) \times 0.92 \times 0.44 = 178\,\text{MW}$.

Figure 3.34 presents T-S diagrams for both the gas turbine and the Rankine cycle, which will be studied below as an example of a combined cycle. Note that the values of specific entropy of air and steam/water are at different ranges, so these processes cannot be directly drawn as such on the same diagram. There is a way to normalize them and draw to same diagram, the details of this were broadly discussed in [25]. These exact same processes have already been analyzed separately above, and we use the same process values here. The thermal efficiency of the gas turbine and the

Rankine process is 0.42 and 0.402, respectively. To be able to calculate the thermal efficiency of the combined cycle, we need to know the efficiency of the heat recovery steam generator, η_{HRSG}, as discussed above.

The temperatures needed to design the heat recovery steam generator are t_1, t_4, t_5, t_6, t_7 and t_8. The exit gas from the gas turbine cools down to the exit temperature of the heat recovery steam generator, t_a, (not shown in the schematic, since we do not know it at this point) and this heat is used to heat-up and evaporate the water and to superheat the steam, from 5 to 8. Our task is to analyze the capacities (in MW) of the heat transfer surfaces of the heat recovery steam generator, to derive the equation to calculate the mass flow rate of the water in the Rankine cycle and finally, to derive the equation for the η_{HRSG} and overall thermal efficiency of the combined cycle.

Figure 3.35 presents the process schematic of this combined cycle. Note that on the gas side, not all the heat can be recovered and must be dumped into the environment. This is presented here as an additional heat exchanger, but for example with the natural gas fired gas turbine this exhaust gas is just fed into the stack and new combustion air is taken into the compressor. If the gas turbine is to be externally heated using, for instance, solar thermal energy or combustion of solid biomass, and/or if the gas turbine process fluid were to be a special internally recycled gas, we would use the closed coupling i.e., the Brayton cycle. Figure 3.36 presents the temperature-heat flow rate diagram of the heat recovery steam generator.

This temperature—heat flow rate diagram presents temperature changes in different components and the rate of heat transfer. In this book we only calculate the heat transfer rates (in MW) within different components and leave the detailed geometrical design out. See Incropera and DeWitt [27] or VDI Heat Atlas [55] for detailed analysis on heat exchangers.

To make the comparison of different technologies transparent, we set the electrical power production from a combined cycle to the same as all the other technologies in this book: 500 MW. In the combined power plant, power comes from two processes that are coupled together. As discussed above, combined electrical power production for the gas turbine and the Rankine cycle is as follows: (assuming that the role of the

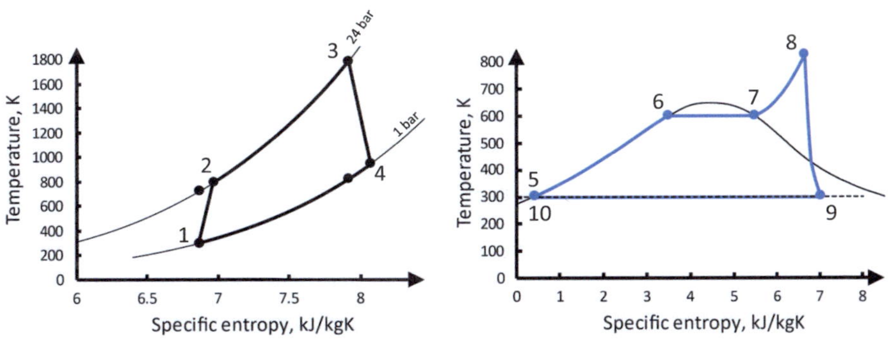

Fig. 3.34 T-S diagrams of the topping gas turbine and the bottoming Rankine cycle. The dashed line represents the temperature of the environment—25 °C

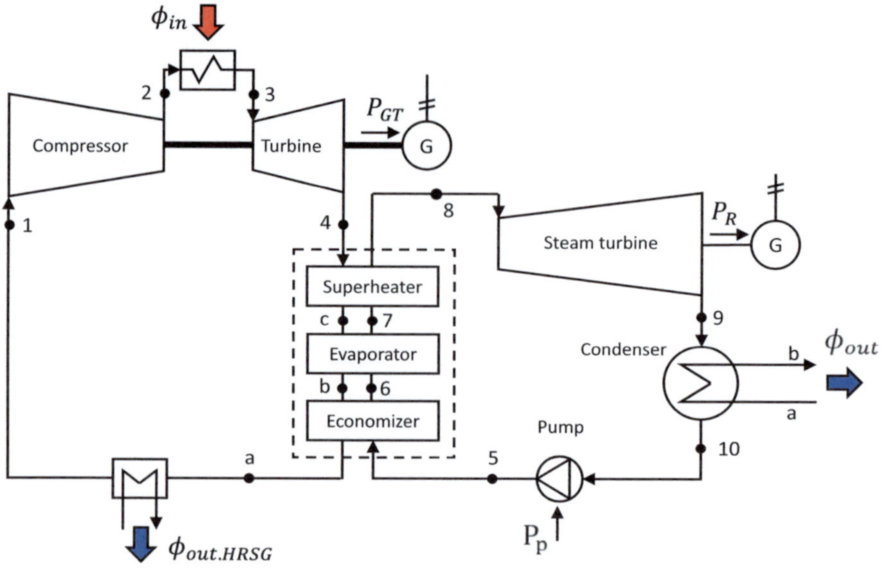

Fig. 3.35 Process schematic of the gas turbine-based combined cycle power plant

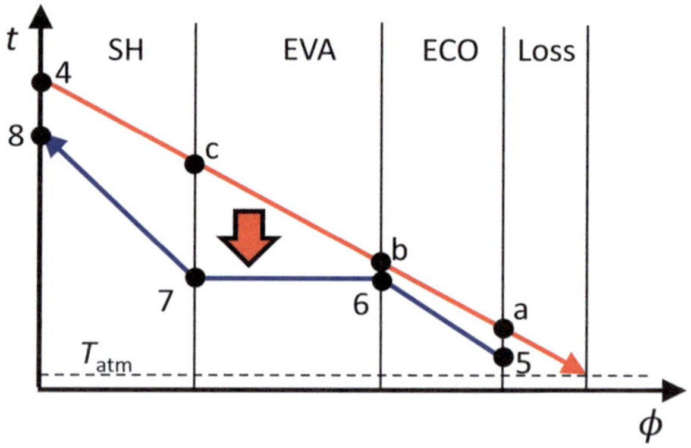

Fig. 3.36 Heat flow rate—temperature diagram of the heat recovery steam generator, SH = superheater, EVA = evaporator, ECO = feed-water preheater *i.e.* Economizer, and Loss = Heat losses from the heat recovery steam generator, not to scale

electrical efficiency of the pump's motor is 1).

$$P_e = \phi_{in}\eta_g(\eta_{GT} + (1 - \eta_{GT})\eta_{HRSG}\eta_R)$$

If we assume that the efficiencies of the gas turbine and Rankine cycles are fixed the only unknowns are the amount of heat required and η_{HRSG}. The equation for the efficiency is derived by considering the mass flow rate of air as a free variable and then we define the water flow rate in relation to this. Then, the calculation of the η_{HRSG} is possible. For the above-described system, the efficiency of the heat recovery steam generator is defined as the ratio of heat transferred to the steam cycle divided by the heat in the exit gases coming out from the gas turbine cycle; here $\dot{m}_s$ and $\dot{m}$ is the mass flow rate of steam and air, respectively.

$$\eta_{HRSG} = \frac{\dot{m}_s(h_8 - h_5)}{\dot{m}(h_4 - h_1)} = \frac{(h_4 - h_a)}{(h_4 - h_1)}$$

Note that neither the ratio of mass flow rates nor the exit enthalpy is known and must be solved. In principle, if the heat exchanger only had a gas–gas configuration it would be simple to make this derivation, by setting the accepted gas exit temperature and then calculating the ratio of mass flow rates. However, since the temperature profile of water changes in steps, and heat capacities of liquid and gas are different, it is important to take care that the hot and cold stream temperatures do not mix in the heat recovery steam generator. To make this safely, we first build an overall energy balance for the evaporator and the superheater, and then set a minimum temperature difference between the streams at the inlet of the evaporator, $\varepsilon = t_b - t_6$. This temperature difference is the pinch temperature, as already discussed above. If this value is set very low, it means that the heat exchanger surface becomes large and the heat recovery steam generator cost increases. To prevent this, we normally set $\varepsilon \sim$ 5–10 K. Therefore, since the evaporator temperature is fixed by the boiler pressure, $t_6 = t'(p)$, $t_b = t_6 + \varepsilon$. The energy balance around the system composed of evaporator and superheater can then be written as follows. The system is assumed to be adiabatic.

$$\dot{m}_s(h_8 - h_6) = \dot{m}(h_4 - h_b)$$

Here, all the enthalpies are fixed, and the ratio of mass flow rates can be solved.

$$\frac{\dot{m}_s}{\dot{m}} = \frac{(h_4 - h_b)}{(h_8 - h_6)}$$

This can be inserted into the efficiency equation to give us.

$$\eta_{HRSG} = \frac{(h_4 - h_b)}{(h_8 - h_6)}\frac{(h_8 - h_5)}{(h_4 - h_1)}$$

And when this is known, the exit gas temperature, t_a, can be solved from the energy balance of the economizer.

$$\dot{m}_s(h_6 - h_5) = \dot{m}(h_b - h_a) = \dot{m}\bar{c}_p(t_b - t_a)$$

$$t_a = t_b - \frac{\dot{m}_s}{\dot{m}}\frac{(h_6 - h_5)}{\bar{c}_p} = t_b - \frac{(h_4 - h_b)}{(h_8 - h_6)}\frac{(h_6 - h_5)}{\bar{c}_p}$$

Or, if programmed functions are used, the enthalpy of the exit gas can be solved first, and then the temperature solved directly, using the function $t(p, h)$ as $t_a =$

$t(1bar, h_a)$.

$$h_a = h_b - \frac{\dot{m}_s}{\dot{m}}(h_6 - h_5) = h_b - \frac{(h_4 - h_b)}{(h_8 - h_6)}(h_6 - h_5)$$

Now that we know all the state variables, we can solve the overall heat needed, from the electrical power requirement and calculated combined cycle thermal efficiency, η.

$$\phi_{in} = \frac{P_e}{\eta_g \eta}$$

From this, the heat flow rate to the heat recovery steam generator can be solved as $\phi_{out.GT} = \phi_{in}(1 - \eta_{GT})$. And the heat flow into and out from the Rankine cycle are, $\phi_{in.R} = \phi_{in}(1 - \eta_{GT})\eta_{HRSG}$, $\phi_{out.R} = \phi_{in}(1 - \eta_{GT})\eta_{HRSG}(1 - \eta_R)$, respectively. The mass flow rate of the air in the gas turbine is calculated from the *mechanical energy balance* of the whole gas turbine process, and the steam mass flow rate can then be calculated from the above derived equation. Combining these equations with the gas turbine cycle energy balance we can get the air mass flow rate as follows:

$$\dot{m} = \frac{\phi_{in}\eta_{GT}}{h_4 - h_3 - (h_2 - h_1)}$$

$$\dot{m}_s = \frac{\dot{m}(h_4 - h_b)}{(h_8 - h_6)}$$

When these terms are known it is easy to calculate the heat loads of different heat transfer surfaces, preheater, evaporator, and superheater, either from the water/steam or gas side.

$$\phi_{ECO} = \dot{m}_s(h_6 - h_5) = \dot{m}(h_b - h_a)$$

$$\phi_{EVA} = \dot{m}_s(h_7 - h_6) = \dot{m}(h_c - h_b)$$

$$\phi_{SH} = \dot{m}_s(h_8 - h_7) = \dot{m}(h_4 - h_c)$$

The exit gas loss is calculated from the enthalpy difference between the exit state and the surroundings: the gas cannot cool further than this.

$$\phi_{Loss} = \dot{m}(h_a - h_{atm})$$

Example 3.11 Analyze a simple combined cycle of net electrical power production of $P_e = 500$ MW. The thermal efficiency of the gas turbine is $\eta_{GT} = 0.416$. The efficiency of the heat recovery steam generator η_{HRSG}, is not known. The bottoming

Rankine cycle has a thermal efficiency of $\eta_R = 0.4$. The electrical efficiency of the generators is 0.98 and the impact of the electrical efficiency of the pump of the Rankine cycle is neglected.

1. Define all process points of the heat recovery steam generator. All the other points are the same as in the basic gas turbine process (Table 3.6 and the Rankine cycle, Example 3.4).
2. Calculate the thermal efficiency of the heat recovery steam generator and the combined plant.
3. Calculate the required thermal heat to run the process.
4. Calculate the mass flow rate of air and steam
5. Calculate the mechanical power from the gas turbine and steam power plants.
6. Calculate the heat transfer loads of the economizer, evaporator, and superheater of the heat recovery steam generator.
7. Draw the temperature-heat flow rate diagram.

1. As above, it is convenient to collect all the points into a Table. The points a, b and c are not initially known. Point b can be calculated based on the boiling temperature of the water, $t_b = t_7 + \varepsilon = 324.7 + 10 = 334.7$ °C and then the enthalpy and entropy can be obtained from the Tables of Excel functions available. From the above derived energy balance for the evaporator and superheater combination, we can calculate the ratio of steam and air mass flow rates.

$$\frac{\dot{m}_s}{\dot{m}} = \frac{(h_4 - h_b)}{(h_8 - h_6)} = \frac{709 - 345}{3482 - 1491} = 0.1833$$

The exit enthalpy of the gas is then the calculated from the energy balance of the economizer.

$$h_a = h_b - \frac{\dot{m}_s}{\dot{m}}(h_6 - h_5) = 345 - 0.1833 \times (1491 - 134) = 95\,\mathrm{kJ/kg}$$

This corresponds to a temperature of $t_a = t(1\,\mathrm{bar}, 95.3\,\mathrm{kJ/kg}) = 94$ °C. The gas temperature before the superheater can be calculated similarly from its energy balance.

$$h_c = h_4 - \frac{\dot{m}_s}{\dot{m}}(h_8 - h_7) = 709 - 0.1833 \times (3482 - 2686) = 563\,\mathrm{kJ/kg}$$

This corresponds to a temperature of $t_c = t(1\,\mathrm{bar},\ 562.9\,\mathrm{kJ/kg}) = 538\,°\mathrm{C}$.

2. Using the enthalpies (see Table 3.7) we can solve the thermal efficiency of the heat recovery steam generator and combined process.

$$\eta_{HRSG} = \frac{(h_4 - h_a)}{(h_4 - h_1)} = \frac{(709 - 95.3)}{(709 - 25)} = 0.897$$

Table 3.7 Process points in the combined cycle and heat recovery steam generator, values of physical properties are obtained from the Excel-based tool available in electronic supplement

#	p (bar)	t (°C)	T (K)	h (kJ/kg)	s (kJ/kg K)	Comments
1	1	25	298	25	6.864	Inlet point to the compressor
2	24	519	792	542	6.967	Exit from the compressor
3	24	1511	1784	1713	7.917	State before the expansion
4	1	670	943	709	8.072	Endpoint of expansion from the gas turbine
5	120	30.7	303.7	134	0.466	After the pump
6	120	324.7	597.7	1491	3.497	Saturated water
7	120	324.7	597.7	2686	5.494	Saturated steam
8	120	550	823	3482	6.655	Superheated steam
9	0.04	29	302	2123	7.046	Exit of turbine
10	0.04	29	302	121	0.423	Exit of condenser
c	1	538	811	653	7.905	Air after the superheater
b	1	334.7	607.7	345	7.595	Boiling temperature + pinch temperature difference $\varepsilon = 10$ °C
a	1	93.9	367.1	95	7.074	Air after the economizer

$$\eta = \eta_{GT} + (1 - \eta_{GT})\eta_{HRST}\eta_R = 0.416 + (1 - 0.416) \times 0.897 \times 0.4 = 0.625$$

3. Heat requirement is calculated simply from the definition of thermal efficiency and considering the total electrical efficiency.

$$\phi_{in} = \frac{P_e}{\eta_g \eta} = \frac{500\,\text{MW}}{0.98 \times 0.626} = 815\,\text{MW}$$

4. Mass flow rates of air can be calculated from the gas turbine energy balance, and steam mass flow rate from the evaporator and superheater energy balance shown above.

$$\dot{m} = \frac{\phi_{in}\eta_{GT}}{h_3 - h_4 - (h_2 - h_1)} = \frac{815000 \times 0.416}{1714 - 709 - (543 - 25)} = 696\,\text{kg/s}$$

$$\dot{m}_s = \dot{m}\frac{(h_4 - h_b)}{(h_8 - h_6)} = 696 \times 0.1833 = 128\,\text{kg/s}$$

5. The mechanical power production from the gas turbine process is $P_{GT} = \phi_{in}\eta_{GT} = 815\,\text{MW} \times 0.416 = 339\,\text{MW}$ and the power generation of the steam cycle is $P_R = \phi_{in}(1 - \eta_{GT})\eta_{HRSG}\eta_R = 815\,\text{MW} \times (1 - 0.416) \times 0.897 \times 0.40 = 171\,\text{MW}$.

6. The heat transfer loads of the economizer, evaporator, and superheater of the heat recovery steam generator can be calculated from the mass flow rate of air and enthalpy changes.

$$\phi_{ECO} = \dot{m}(h_b - h_a) = 696 \times (345 - 95) = 173\,\text{MW}$$

$$\phi_{EVA} = \dot{m}(h_c - h_b) = 696 \times (563 - 345) = 152\,\text{MW}$$

$$\phi_{SH} = \dot{m}(h_4 - h_c) = 696 \times (709 - 563) = 102\,\text{MW}$$

The overall heat transfer rate is 173 + 152 + 102 = 428 MW. Just as a check, 428 × 0.4 = 171 MW power from the Rankine cycle.

7. The temperature-heat load diagram is presented in Fig. 3.37. The x-axis is composed from adding the heat loads of the different components. The y-axis is the corresponding temperatures of the inlets and outlets. The loss term is calculated by cooling the flow down to the inlet temperature of the atmosphere.

As we can see, the selected pinch point location is correct, the temperature curves are not crossing, and it is not possible to have inlet and outlet temperatures closer to each other at this boiling temperature and pressure level.

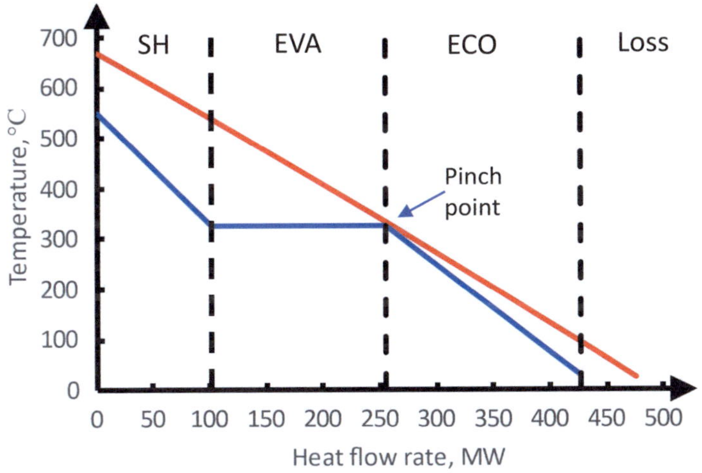

Fig. 3.37 Temperature-heat load diagram of the heat recovery steam generator

3.9 Combined Heat and Power Production

We have learned earlier that thermal power plants always have heat losses. To run a power cycle a part of the heat must be rejected. For example, if a steam power plant is optimized only for power production, the condenser pressure and temperature are then minimized. This means that the condenser pressure is determined by the available cooling temperature. For example, if we have cooling water at 15 °C available, and some minimum temperature difference is defined—say 10 °C—then the condensation temperature is higher than ~ 30 °C, depending on the temperature increase of the cooling water. This corresponds to a condensation pressure of 0.042 bar. At this low temperature, heat has only minimal value unless a heat pump is used to upgrade the temperature higher.

If we want to produce both power and higher temperature heat—say water at 95 °C to be used in district heating—we need to increase the condenser pressure. Normally, the return temperature of district heating water is around 45–55 °C and it is heated up to 95 °C. If the minimum temperature difference in the district heating heat exchanger is 5 °C, then the condensation temperature is 100 °C, corresponding to 1.01325 bar. Or, if we need to heat a chemical process up to 250 °C, the condenser temperature must be already ~ 255 °C, corresponding to 43 bars. Since decreasing the condenser pressure increases the power efficiency, this production of process heat will decrease the power efficiency. On the other hand, since we are generating required heat for a purpose, it can also have significant monetary value—even higher than the price of power. Next, we will analyze the thermal efficiency of a simplified combined heat and power plant and study some theoretical limitations. Here, the *T-s* diagram is the most effective presentation of the process. We use the basic Rankine cycle as shown in Fig. 3.38.

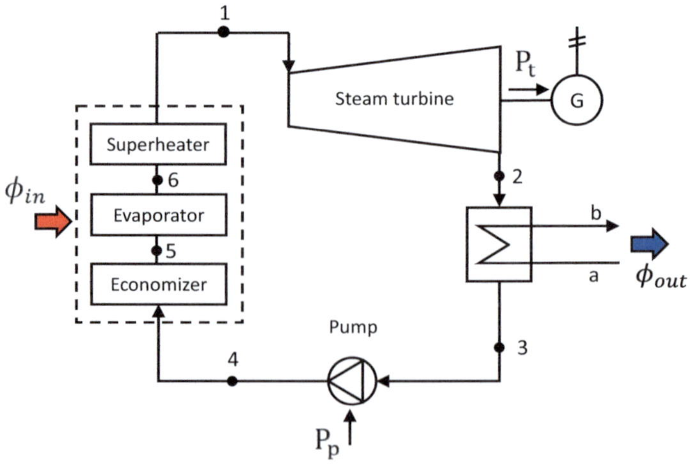

Fig. 3.38 Combined heat and power plant

To be comparable with the previous cases, we use the same live steam values as in the basic Rankine cycle of 120 bar and 550 °C. We let the steam expand to pressures corresponding to the process heat required and then simply pump the water back to boiler temperature. As an example, Table 3.8 presents the process values for process heat requirement temperatures of 95, 200 and 250 °C.

The thermal efficiencies for these processes are calculated using the familiar indirect efficiency equation at 1, 17.2 and 43.2 bars condenser pressure, respectively.

$$\eta = 1 - \frac{h_2 - h_3}{h_1 - h_4} = 1 - \frac{2498 - 417}{3482 - 437} = 0.317$$

$$\eta = 1 - \frac{2790 - 875}{3482 - 891} = 0.261$$

$$\eta = 1 - \frac{3185 - 1110}{3482 - 1122} = 0.121$$

Figure 3.38 shows that when the required process temperature increases, the thermal efficiency for power production decreases. But in this case, we want to produce high temperature heat for process use. Figure 3.39 presents the processes in T-s diagrams, so it is easier to observe the difference in the processes.

If the above cases are compared, we can see that the power output significantly decreases when the condensation temperature increases.

All previous process examples have studied cases where we have produced 500 MW of electricity. In the combined heat and power plants we are producing intentionally both heat and power, so in principle, all heat invested except the losses in motors and generators is utilized. Of course, depending on the source of heat, the

Table 3.8 Process values Rankine cycle for combined heat and power production, values of physical properties are obtained from the Excel-based tool available in electronic supplement

#	p (bar)	t (°C)	T (K)	h (kJ/kg)	s (kJ/kg K)	Comments
1	120	550	823	3482	6.655	Read from the cross section of t_1 and p_1
2 s	1 17.2 43.2	100 255 380	373 528 653	2413 2926 3136	6.655	Move vertically down from point 1–2s (ds = 0) at p_2
2	1 17.2 43.2	100 273 390	373 546 663	2498 2790 3185	6.885 6.738 6.694	Calculate h_2 from the isentropic expansion equation
3	1 17.2 43.2	100 205 255	373 478 528	417 875 1110	1.303 2.377 2.838	Follow the condensation temperature to saturated water
4	120 120 120	102 207 258	375 481 531	437 891 1122	1.320 2.386 2.844	Add enthalpy increase in the pump

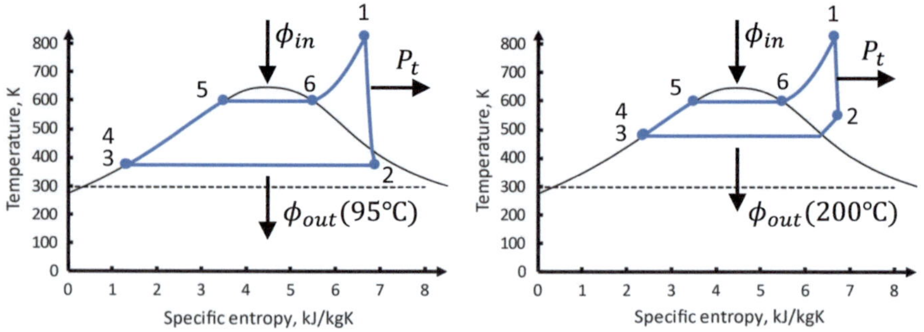

Fig. 3.39 T-s diagrams of combined heat and power plants: the left process is used for district heating at 95 °C and the right used for generating process heat of 200 °C. The dashed line represents the temperature of the environment—25 °C

total efficiency may vary. For example if we are burning biomass to generate heat for steam production, we lose a fraction of the heat as the boiler losses, $(1 - \eta_b)$, plus the losses from electrical power systems.

The above-mentioned system is simple but not very flexible. In practice, we could have had bleeding of steam at various pressure levels for producing heat at required temperatures [15, 51]. If heat is not needed, we can concentrate on full power operation and produce more power, within the limits of the off-design characteristics of the turbine(s). This kind of system is analyzed in Example 3.12.

Example 3.12 As a difference to the above discussed 500 MW electrical power producing power plant concepts, here we analyze a system that produces 500 MW of district heat and produces electrical power (Fig. 3.40). As with earlier examples, steam values of temperature and pressure before the turbine are 550 °C and 120 bar and the condenser pressure is 0.04 bar. The turbine bleeding point is at 1 bar pressure. This is sufficient to produce 95 °C water for a district heating network. The isentropic efficiency of the turbine and the pump is 0.92 and 0.95, respectively.

1. Define all the process points and calculate the mass flow rate of the bled steam to fill the required 500 MW heat with 95 °C water, assuming that the bled stream is cooled to 75 °C. What is the minimum amount of power that must be produced, and what is the bled fraction for this case? Calculate the required thermal heat input to the steam generator. Also calculate the thermal efficiencies for power, and power and heat combined.
2. Calculate the mass flow rate of the live steam, bled fraction a, and the thermal power efficiency for a 500 MW heat + 500 MW electrical power production.

1. The calculation of the process points is shown in detail in Table 3.9.

The mass flow rate of the bled steam can be simply calculated from the district heating heat exchanger energy balance.

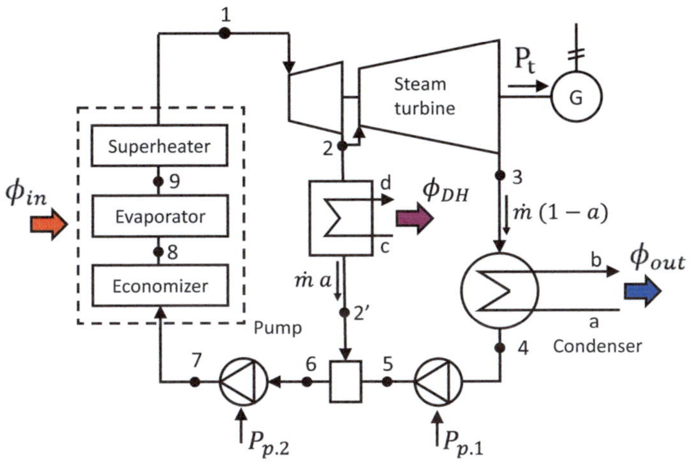

Fig. 3.40 Combined heat and power plant with bleeding for a district heating and condensing tail turbine

Table 3.9 Calculation of process values of combined heat and power steam cycle. *Please note that points 3 s…6 are not needed in task 1

#	p (bar)	t (°C)	T (K)	h (kJ/kg)	s (kJ/kg K)	Comments
1	120	550	823	3482	6.655	Read from the cross section of t_1 and p_1
2 s	1	100	373	2413	6.655	Move vertically down from point 1–2s (ds = 0) at p_2
2	1	100	373	2498	6.885	Calculate h_2 from the isentropic expansion equation
2'	1	75	348	314	1.016	We assume that bled stream condenses and is cooled to 75 °C
3 s	–* 0.04	– 29	– 302	– 2004	– 6.885	Move vertically down from point 2–3s (ds = 0) at p_3
3	– 0.04	– 29	– 302	– 2123	– 7.046	Calculate h_3 from the isentropic expansion equation
4	– 0.04	– 29	– 302	– 121	– 0.423	Move along the isobar to left side to the saturated water
5	– 1	– 29	– 302	– 121	– 0.423	Add enthalpy increase in the pump to get h_5
6	– 1	– 52	– 325	– 218	– 0.727	Calculate a, then calculate h_6 from mixer energy balance
7	120 120	76 53	349 326	327 229	1.017 0.728	Add enthalpy increase in the pump to get h_7
8	120	325	598	1491	3.497	Saturated water before the evaporator
9	120	325	598	2686	5.494	Saturated steam after the evaporator

$$\dot{m}_2 = \frac{\phi_{DH}}{(h_2 - h_{2'})} = \frac{500000}{2498 - 314} = 229\,\text{kg/s}$$

The minimum power is produced when the "condensing tail" is not used ($a = 1$). Then, the system corresponds to the one already calculated in Example 3.11, just that here the bled stream is cooled down more, to 75 °C. The *mechanical* power produced is calculated as the difference of the turbine and pumping power. Note that the enthalpy before the pump is the same as for the bled stream 2. The enthalpy increase in the pump is calculated as is $\Delta h_{P.2} = v_{6-7}(p_7 - p_6) \times 100/\eta_{s.p} = 1.03 \times 10^{-3} \times 100/0.95 = 12.9\,\text{kJ/kg}$.

$$\begin{aligned} P &= \dot{m}_2(h_1 - h_2) - \dot{m}_2(h_7 - h_{2'}) \\ &= 229 \times (3482 - 2498) - 229 \times (327 - 314) \\ &= 222\,\text{MW} \end{aligned}$$

The heat input is calculated simply as the sum of net mechanical power and heat produced.

$$\phi_{in} = P + \phi_{DH} = 222 + 500 = 722\,\text{MW}$$

The thermal efficiency for the power production is 222/722 = 0.307, and for combined heat and power, the efficiency is 1. Note that in real processes there are always losses. For example, if we are using biomass combustion, the boiler losses via flue gases, ash losses and radiation and convection losses are typically 10–15%.

2. Based on the above, we already know the mass flow rate required for the district heating bleeding stream. To be able to calculate what mass flow rate is required to produce 500 MW of electrical power, we need to write the energy balance for the turbines and pumps and solve the mass flow rate from that. Enthalpy increase in the pumps is calculated as $\Delta h_{P.1} = v'(t_4)(p_5 - p_4) \times 100/\eta_{s.p} = 1.004 \times 10^{-3} \times (1 - 0.04) \times 100/0.95 = 0.1\,\text{kJ/kg}$ and $\Delta h_{P.2} = v'(t_6)(p_7 - p_6) \times 100/\eta_{s.p} = 1.01 \times 10^{-3} \times (120 - 1) \times 100/0.95 = 12.7\,\text{kJ/kg}$.

$$P_e = \dot{m}(h_1 - h_2)\eta_g + (\dot{m} - \dot{m}_2)(h_2 - h_3)\eta_g - (\dot{m} - \dot{m}_2)\frac{\Delta h_{p1}}{\eta_m} - \dot{m}\frac{\Delta h_{p2}}{\eta_m}$$

$$\begin{aligned} \dot{m} &= \frac{P_e + \dot{m}_2(h_2 - h_3)\eta_g - \dot{m}_2\frac{\Delta h_{p1}}{\eta_m}}{(h_1 - h_2)\eta_g + (h_2 - h_3)\eta_g - \frac{\Delta h_{p1}}{\eta_m} - \frac{\Delta h_{p2}}{\eta_m}} \\ &= \frac{500000 + 229 \times (2498 - 2123) \times 0.98 - 229\frac{0.1}{0.98}}{(3428 - 2498) \times 0.98 + (2498 - 2123) \times 0.98 - \frac{0.1}{0.98} - \frac{13}{0.98}} = 459\,\text{kg/s} \end{aligned}$$

The fraction of bled steam to the district heating stream is therefore $a = 229/459 = 0.499$. We can now calculate the enthalpy after the mixer: $h_6 = ah_{2'} + (1 - a)h_5 = 0.499 \times 314 + (1 - 0.499) \times 122 = 218\,\text{kJ/kg}$. The process can now be drawn in

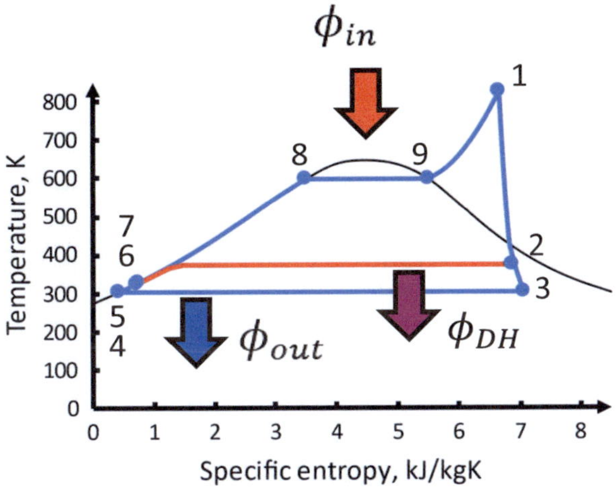

Fig. 3.41 Combined heat and power plant process T-s diagram with bleeding for the district heating and condensing tail turbine (2-3)

T-s diagrams (see Fig. 3.41). The district heating bleeding stream is presented in red, please note the temperature drop from 100 to 75 °C in the sub-cooling part before point 6.

The mechanical power from both turbine stages is as follows:

$$P_{HP} = \dot{m}(h_1 - h_2) = 459 \times (3428 - 2498) = 427\,\text{MW}$$

$$P_{LP} = (\dot{m} - \dot{m}_2)(h_2 - h_3) = (459 - 229) \times (2498 - 2123) = 86\,\text{MW}$$

This example shows that this kind of back-pressure turbine arrangement is a very efficient way to produce process heat together with power production. We will later study different ways of using heat pumps to upgrade low temperature heat to higher temperatures.

3.10 Concentrated Solar Power Plants

The heat source of traditional thermal power plants is usually provided by fossil fuel combustion, which is the major cause for the global warming and climate change. Therefore, the development with renewable energy sources is necessary. Among different renewable energy sources, concentrated solar power (CSP) directly employs the solar radiation to obtain medium or high temperature thermal energy for not only electricity production but also heat source for several chemical processes. CSP technology is safe and reliable, and comparing with wind energy and photovoltaics, thermal energy storage, to later produce electricity is an integral part of the systems by default. CSP technology could be able to contribute 9% of the cumulative emission

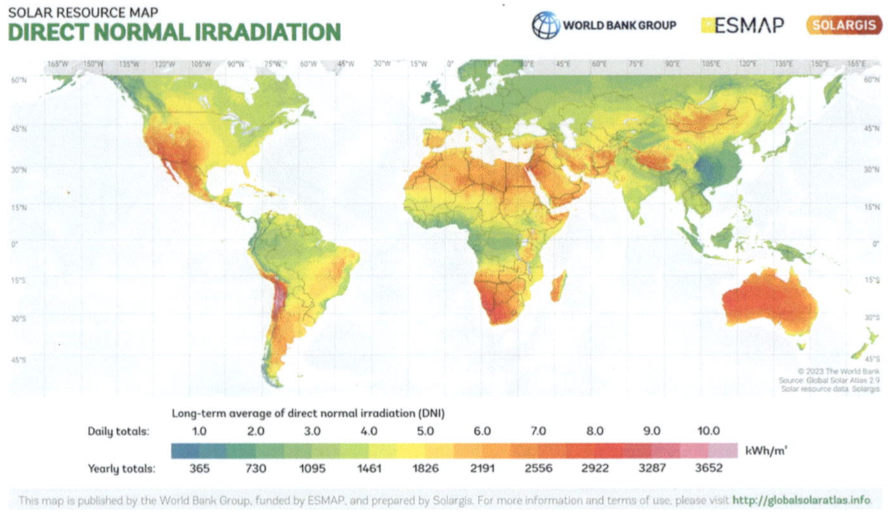

Fig. 3.42 Direct normal irradiation. © 2020 The World Bank, *Source* Global Solar Atlas 2.0, Solar resource data source is Solargis [18]

reduction. Therefore, CSP plants have boomed in recent years. There was around 8 GW of operational and under-construction CSP globally by 2023. [37]

Direct Normal Irradiance (DNI) is a crucial parameter for concentrated solar energy generation. As the global DNI shown in Fig. 3.42, regions between 15° and 40° both north and south latitudes and places at high altitudes are the best places for CSP. Thus, CSP plants are currently built in 22 countries: Algeria, Australia, Canada, Chile, China, Denmark, Egypt, France, Germany, India, Israel, Italy, Kuwait, Mexico, Morocco, Saudi Arabia, South Africa, Spain, Thailand, Turkey, United Arab Emirates and United States. Spain, China, and United States are the leading countries in CSP capacity. The Aalborg CSP-Brønderslev in Denmark is the northernmost CSP plant (Latitude 57.3° N, beyond the optimized DNI region for CSP). It combines CSP with a biomass organic Rankine cycle (ORC) to produce heat and power.

3.10.1 CSP Technologies

Currently, four CSP technologies are widely employed, and they are classified according to the collector and reflector types, as Fig. 3.43 illustrates. These are discussed shortly in below.

Parabolic trough collectors (PTC) are the linear collectors in a parabola shape to focus the sunlight on a tube mounted on the focal line of the parabola. A parabolic trough CSP plant consists of parallel connected collectors. Parabolic trough CSP plants require much less land than other three CSP technologies. The thermal fluid

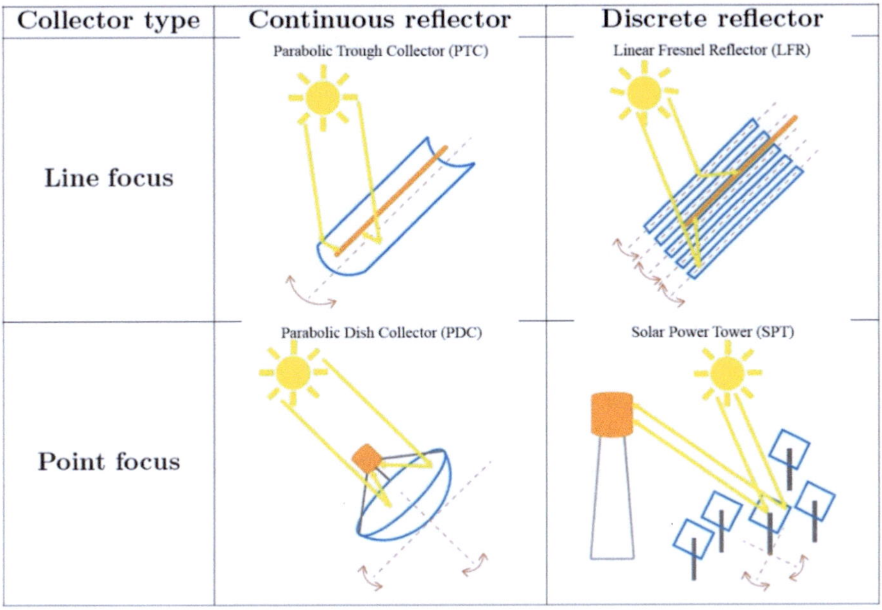

Fig. 3.43 Classification by reflector geometry of the commonly accepted CSP systems. [37]

can be heat transfer fluids (HTF) for further power production and steam directly for power production.

Linear Fresnel reflectors (LFR) are the several long-row flat or slightly curved mirrors arranged in an approximate parabolic shape, and similar as PTC, it is also line focus. It has an easy design with less investment since it usually works for direct steam heating. However, the electrical efficiency of LFR system is lower, and heat storage system is difficult to integrate with LFR system.

Solar power towers (SPT) use a series of reflectors, called heliostats to reflect and concentrate the sunlight onto a central receiver placed on the top of a tower, where the HTF absorbs the heat and further transfers the heat to a Rankine cycle for power production. SPT is flexible with HTF choice and design of heliostats and receivers.

Parabolic dish collectors (PDC) concentrate the sunlight on the focal point of the dish. Although the concentration ratio C (the ratio between the intensity of concentrated beam solar radiation hitting the absorber/receiver and the intensity of beam solar radiation incident on the concentrator aperture, which is approximately equal to the surface area ratio of the collectors and receiver) of PDC is very high (500–1000), each parabolic dish has a low capacity and produces electricity independently. This means that the number of required dishes is very high if a large scale of CSP plant is built.

Figure 3.44 illustrates the share of operational and under-construction CSP plants by technologies, year, and countries. There is no PDC CSP operating or under-construction recently. The reason should be that it is not capable for large-scale CSP

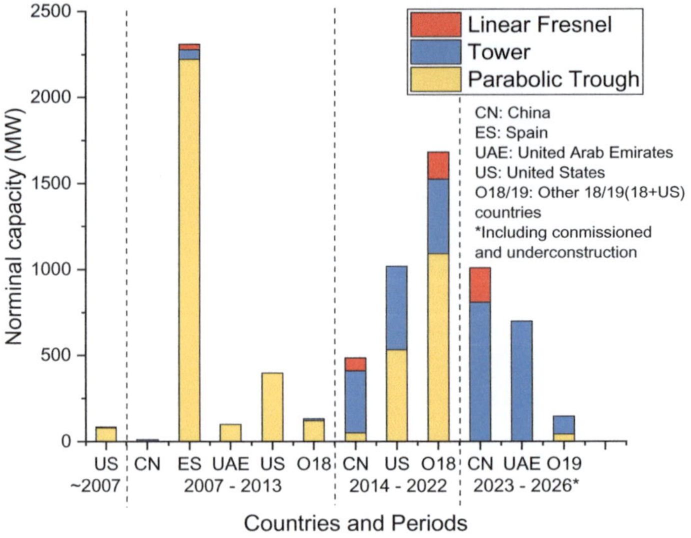

Fig. 3.44 Shares of global CSP plants in operational and under construction by countries, commissioning years and technologies. *Data source* https://solarpaces.nrel.gov/by-country by 2023

plants. Parabolic trough is the most popular CSP technology by 2022. The main contribution is in Spain, who built more than 2000 MW parabolic trough CSP plants between 2007 and 2013. Even though, there is no large new scale CSP plants launched after 2013, Spain still has the largest capacity of CSP worldwide. Between 2014 and 2022, US built the most CSP plants, followed by China. The share of PTC and SPT were almost equal in US, while China built more SPT in these years. After 2022, China has the most capacity of new built or under-constructed CSP, and more than 80% is SPT. Because of the 600 MW under-construction CSP, UAE becomes the second in this period.

In a CSP, heat transfer fluid can be oil, steam, molten salt, gas, or sand particles depending on the target temperature and integrated cycle. In this section, we mainly focus on SPT technology with molten salt as HTF.

3.10.2 Solar Tower Power System with Molten Salt

Figure 3.45 depicts the schematic of a solar tower power plant. As previously mentioned, the sunlight is reflected by the heliostats field and concentrated on a central receiver with a very high temperature. This high temperature heat is transferred to the cold molten salt (inlet temperature of 288 °C in this case), and the temperature of the molten salt is increased to 566 °C. The hot molten salt preheats, evaporates and superheats up the steam in heat exchangers. After exchanging heat, the hot steam goes to Rankine cycle as already discussed above sections and the cold molten salt is pumped to the central receiver for the next cycle. Two molten salt storage tanks are installed for the night-mode operation.

Heliostats

Heliostat field costs typically accounts for 40–50% cost and 40% energy loss of a SPT power plant. Therefore, optimized design of a heliostats and the heliostat field is essential for techno-economic performance improvement of a SPT power plant. A heliostat comprises of the foundation, the driver, the control system, and the reflector. The structure and tasks of optimal design for a heliostat are shown in Fig. 3.46. Larger size will reduce the driver cost, while too large size would result in very high cost of the support structure because of the high weight. Current operational heliostat size is usually between 1.1 and 150 m^2. The 4 mm glass mirror is the standard reflector with a reflectance of 0.935. Thin-glass and film mirrors are two possible improvements for reflectors with reaching reflectances of 0.955 and 0.94, respectively. Novel designs of the driver system and low-weight design of the support structures would be also helpful for heliostat cost reduction [21]. More novel designs of heliostat can be found in the review presented by Pfahl et al. [45].

The layout of heliostat field also determines the overall efficiency of the SPT power plant and the required area of land. The arrangement of heliostats is classified as patterned and non-patterned. Radial staggered and biometric layout are commonly used as patterned layout. For instance, the Gemasolar Thermosolar 20 MW CSP shown in Fig. 3.1 is an example of surrounding radial staggered layout. According to the study of Noone et al. [42], biometric layout can improve the annual efficiency by 0.36% with 15.8% land area reduction. The summary of different heliostat field layouts can be found in the review reported by Rizvi et al. [46]. Many tools such as Campo, MUUEN, RCELL, and Tiesol can be used to design and optimize the heliostat field [21].

Central receivers

The solar central receiver must be capable of resisting the significant fluid temperature difference, for example from 288 to 566 °C in Fig. 3.45, with a high solar heat flux. Conventional central tubular receivers consist of an array of thin-walled tubes for HTF to absorb concentrated solar heat. Figure 3.47 shows two typical tubular receivers: external and cavity receivers. The external receiver efficiencies can reach 84–89% with various heat transfer fluids.

The useful heat rate flow rate of the receiver ϕ_c is equal to the heat loss of the central receiver ϕ_{loss} subtracted from the beam irradiation absorbed by the central receiver ϕ_{abs}.

$$\phi_c = \phi_{abs} - \phi_{loss}$$

The rate of beam irradiation absorbed by the central receiver is ϕ_{abs} as follows:

$$\phi_{abs} = I_b A_{hf} \eta_{opt}$$

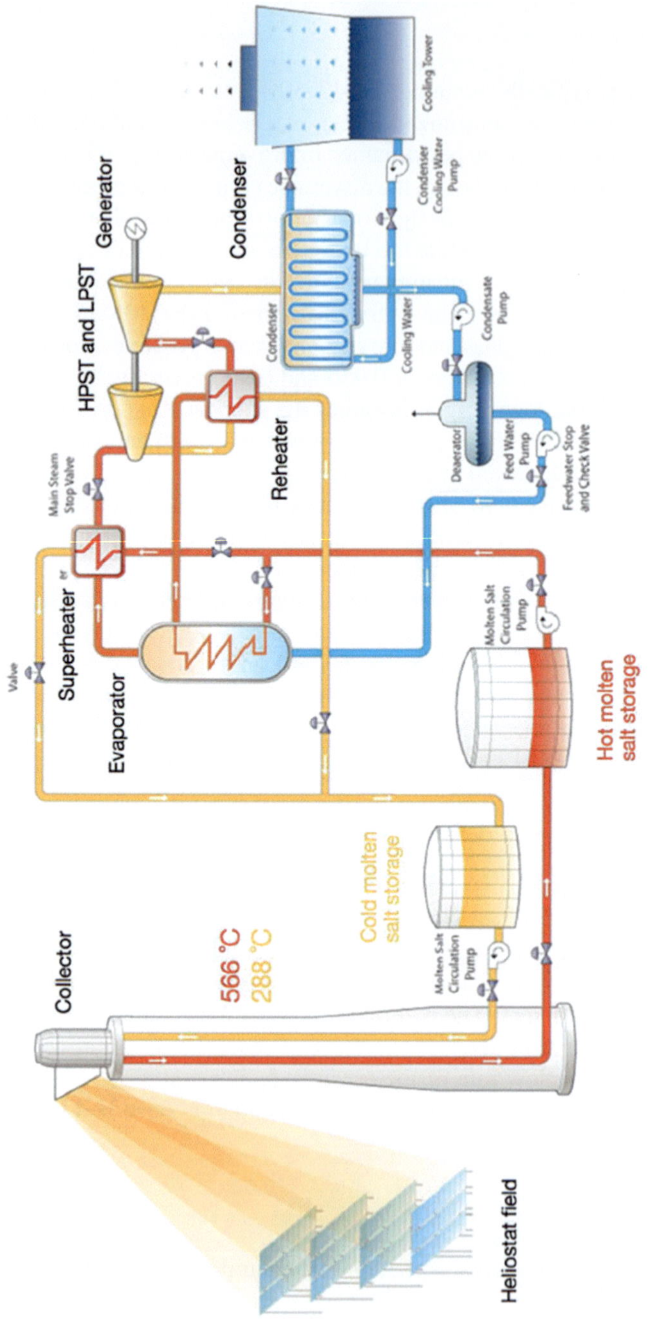

Fig. 3.45 Schematic of a solar tower power plant [1]

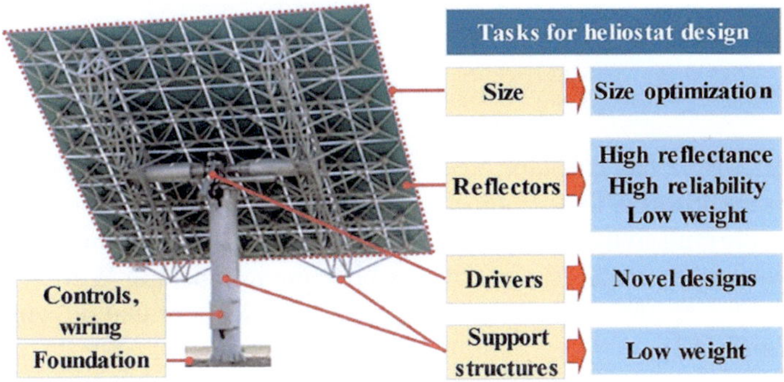

Fig. 3.46 Components of a heliostat and tasks for optimal design [21]

Fig. 3.47 Schematics of tubular (left) external and (right) cavity receivers [24]

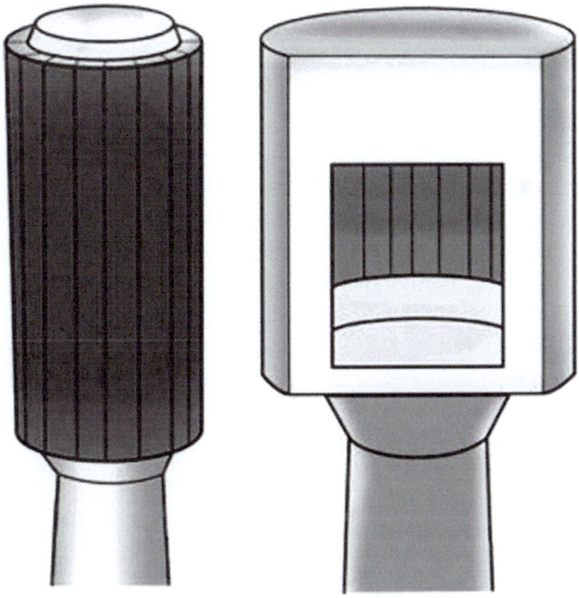

A_{hf} is the area of the reflectors in the heliostat field, I_b is the beam irradiance incident to heliostat field and η_{opt} is the optical collection efficiency of the heliostat-central receiver system. The optical efficiency is defined as

$$\eta_{opt} = \rho_{hs}\gamma\tau\alpha_{abs}$$

where ρ_{hs} is the heliostat reflectivity, γ is the intercept factor of central receiver, τ is the transmissivity of transparent cover (if applied) and α_{abs} is the absorptivity of the central receiver. The heat loss rate of an open type of receiver to the environment is the sum of the heat loss due to the convection ϕ_{conv} and the heat loss by radiation

ϕ_{rad}:

$$\phi_{loss} = \phi_{conv} + \phi_{rad} = A_{abs}\big(h_c(T_{abs} - T_a) + \varepsilon_{abs}\sigma\big(T_{abs}^4 - T_a^4\big)\big)$$

σ is Stefan-Bolzman constant, 5.67×10^{-8} W/m^2 K^4. A_{abs} is the absorber surface area, T_{abs} is the absorber temperature, T_a is the effective temperature of the surroundings, ε_{abs} is the emissivity of the absorber and h_c is the convective heat transfer coefficient between the central receiver and ambient environment in W/m^{2}K. The value of h_c can be for example determined from the empirical correlation with the wind speed V_a (m/s) [12].

$$h_c = 10.45 - V_a + 10\sqrt{V_a}$$

Molten salt and its storage

Molten salt has been widely employed as HTF in STP plants due to its various working temperature, low cost, and thermal stability, and nitrate, carbonate and chloride salts are mainly used. Table 3.10 lists the properties of some typical molten salts used. Melting point (T_m), the maximum temperature of significant mass reduction ($T_{\max}$), latent heat capacity (c_p), density and viscosity are important thermos-physical properties of molten salts. $NaNO_3$-KNO_3 (60:40) is the most used molten salt in STP plant because it is non-hazardous, non-explosive and stable. Its melting point is low while $T_{\max}$ is high, meaning that it can deal with a larger range of temperature with relatively high latent heat capacity. For example, Li-Na-K nitrate has a lower T_m and higher c_p, but its $T_{\max}$ is also much lower than solar salt. The c_{p} of Zn-Na-K chlorides are much lower even their $T_{\max}$ are 850 °C.

The useful heat absorbed by the central receiver transfers to the molten salt, the required mass flow rate can be calculated according to energy balance with known latent heat capacity c_p, inlet and outlet temperatures of the molten salt T_{in} and T_{out}.

$$\phi_c = \dot{m}_{HTF} c_p (T_{out} - T_{in})$$

High-temperature molten salt should be stored for the night mode operation when there is no sun available. The minimum required storage tank volume V can be determined by the night mode operation time (t_n) and the mass flow rate of the molten salt with its density:

$$V = \frac{\dot{m}_{HTF}}{\rho} t_n$$

Example 3.12 A solar tower CSP power plant uses a thermal energy storage system

that operates are follows: Beam solar radiation flux incident on the heliostat $I_{b,N} =$ 800 W/m^2, heliostat-receiver concentration ratio $C = 500$, single heliostat surface area $A_h = 144$ m^2, local wind speed $V_a = 5$ m/s, atmospheric temperature $T_a =$

Table 3.10 Properties of selected molten salts. Summarized from Aljaerani et al. [3]

Salt	Composition (wt%)	T_m(°C)	T_{max} (°C)	c_p (J/gK)	Density (kg/m^3)	Viscosity (cp)
Solar salt	$NaNO_3$-KNO_3 (60:40)	221	588	1.498	1899	3.26
Li-Na-K nitrate	$LiNO_3$-$NaNO_3$-KNO_3 (38:15:47)	100	435	1.6	–	–
Li-K Carbonate	Li_2CO_3/K_2CO_3 (62:38)	488	< 900	1.6	1976 @ 577 °C	5.57 @ 707 °C
Li-Na-K Carbonate	K_2CO_3/Li_2CO_3/Na_2CO_3 (34.52:32.12:33.36)	378	< 800	1.61	2010 @ 600 °C	10.51 @ 600 °C
Zn-Na-K chlorides	$ZnCl_2$-NaCl-KCl (68.6:7.5:23.9)	204	850	0.81 (300…600 °C)	–	–
Mg-Na-K chlorides	$MgCl_2$-NaCl-KCl (68.2:14.0:17.8)	380	> 800	0.85 (500…800 °C)	–	–

35 °C, optical efficiency of the heliostat system $\eta_{opt} = 0.74$, central receiver surface temperature $T_{abs} = 900$ °C, emissivity of the central receiver $\varepsilon_{abs} = 0.8$, power plant electric power output: $P_{el} = 50$ MW, Overall efficiency of the power plant: $\eta_{overall} = 0.2$, TES system enables a 15 h operation of the power plant at a full capacity, the molten salt $NaNO_3$-KNO_3 mentioned in Table 3.10 is employed as the HTF. The inlet and outlet temperatures are the same as Fig. 3.44. Calculate: The number of required heliostats, the mass flow rate of the $NaNO_3$-KNO_3 molten salt and the minimum required volume of the molten salt storage tank.

With the electrical power output and the overall efficiency, the required total solar energy input is:

$$\phi_{total} = \frac{P_{el}}{\eta_{overall}} = \frac{50\,\mathrm{MW}}{0.2} = 250\,\mathrm{MW}$$

The total solar energy input equals to the solar radiation flux multiplying the total area of heliostats, then the number of heliostats can be calculated:

$$n = \frac{\phi_{total}}{A_h I_{b,N}} = \frac{250\,\mathrm{MW}}{144\,\mathrm{m}^2 \times 800\,\mathrm{W/m}^2 \times 10^{-6}} \approx 2171$$

The rate of beam irradiation absorbed by the central receiver ϕ_{abs} is:

$$\phi_{abs} = I_b A_{hf} \eta_{opt} = \phi_{total} \eta_{opt} = 250\,\mathrm{MW} \times 0.74 = 185\,\mathrm{MW}$$

In order to calculate the useful heat rate flow rate of the receiver ϕ_c, the heat loss should be computed, first, the value of convective heat transfer coefficient h_c is determined by the empirical correlation with the wind speed:

$$h_c = 10.45 - V_a + 10\sqrt{V_a} = 10.45 - 5 + 10 \times \sqrt{5} = 27.81\,\mathrm{W/m^2K}$$

The surface area of the central receiver is:

$$A_{abs} = \frac{nA_h}{C} = \frac{2171 \times 144\,\mathrm{m}^2}{500} = 625.25\,\mathrm{m}^2$$

The total heat loss is:

$$\phi_{loss} = A_{abs}\left(h_c(T_{abs} - T_a) + \varepsilon_{abs}\sigma\left(T_{abs}^4 - T_a^4\right)\right)$$

$$= 625.25\mathrm{m}^2 \times (27.81 \times (900 - 35) + 0.8 \times 5.67 \times 10^{-8} \times \left(1173^4 - 308^4\right)) = 68.5\,\mathrm{MW}$$

Then, the useful heat flow of the receiver:

$$\phi_c = \phi_{abs} - \phi_{loss} = 185 - 68.5 = 116.5\,\text{MW}$$

The mass flow rate of the molten salt:

$$\begin{aligned} \dot{m}_{HTF} &= \frac{\phi_c}{c_p(T_{out} - T_{in})} \\ &= \frac{116.5\,\text{MW} \times 10^6\,\text{W/MW} \times 10^{-3}\,\text{kg/g}}{1.5\,\text{J/gK} \times (566 - 288)\,\text{K}} = 279.4\,\text{kg/s} \end{aligned}$$

The required minimum volume of the storage tank is:

$$V = \frac{\dot{m}_{HTF}}{\rho} t_n = \frac{279.4\,\text{kg/s} \times 15\,\text{h} \times 3600\,\text{s}}{1899\,\text{kg/m}^3} = 7944.4\,\text{m}^3$$

3.11 Geothermal Power Plants

Geothermal energy has gained more interest in the energy transition from fossil-fuel dependent to renewable and low-carbon systems. The source of geothermal energy is the heat stored beneath the ground since the Earth was formed 4.5 billion years ago. The heat energy accumulated in the core of the energy over a significant period and formed a thermal gradient (2.5 °C, …, 3 °C/100 m in average) from the core to surface of the earth. This means the temperature rises as the function of the depth from the surface of the earth. Although only a small fraction of geothermal energy is exploitable due to the technical and geological limitations, its availability is still considerable. For instance, the geothermal energy contained from 10,000 m from the surface of the earth equals to 50,000 times of the energy in global oil and gas resources [30]. Moreover, comparing with the intermittent renewable energy sources such as solar and wind, the supply of geothermal energy is stable and continuous, which can meet the requirement for the base load energy demand.

The geothermal energy concentration depends on the geological and geophysical parameters. The geothermal energy is considerable in three areas: the circum-Pacific belt, the Mid-Atlantic Ridge, and the Mediterranean to Himalayan region, as Fig. 3.48 illustrates. Geothermal energy plays a primary role in some countries energy supply system. For example, geothermal energy accounts for 78% of the energy source in Iceland. Geothermal energy can be either directly utilized as a heat source or employed to produce electricity. In direct heat applications, more than half of geothermal energy was used for geothermal heat pumps (55.3%), followed by bath and swimming (20.3%) and space heating (15.0%) [40]. Heat pumps will be explained in Sect. 3.12, and in this section, we focus on geothermal power plants.

Figure 3.49 shows the global electricity production from geothermal energy projection. The electricity produced from geothermal energy in 2023 was 95.6 TWh and expected to increase with an annual rate of 4.1% from 2024 to 2028. The top 5

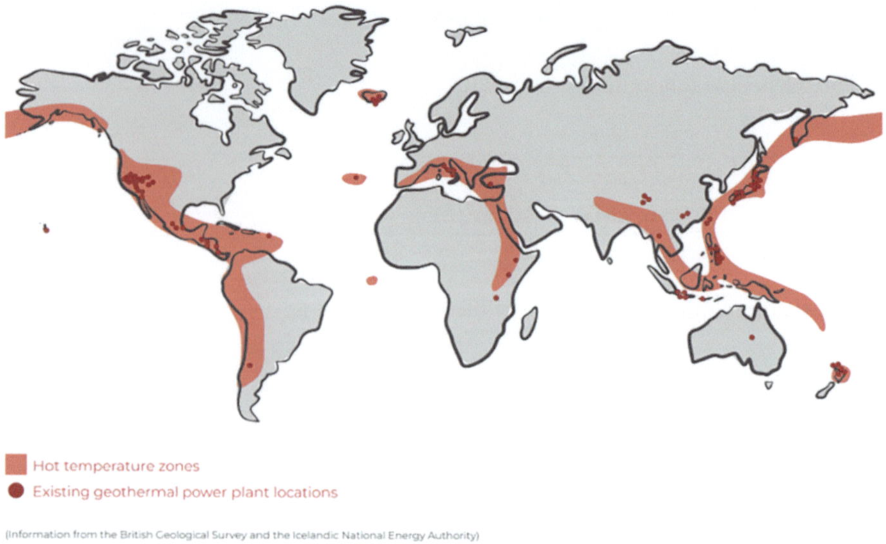

Fig. 3.48 Worldwide geothermal energy potential and existing geothermal power plant distribution. https://www.edengeothermal.com/about/geothermal-energy/worldwide-potential/ [14]

countries with the highest electricity production from geothermal energy are: USA, Indonesia, Turkey, Philippines, and New Zealand, which also meets the geothermal energy potential in Fig. 3.49. The production of top 5 countries represents 65–70% of global electricity production from geothermal energy. According to the projection, USA will be the country with the highest electricity produced from geothermal energy until 2027, and Indonesia becomes the first from 2028. Turkey surpassed Philippines and became the third in 2023. These facts indicate that the growing rate of geothermal energy in Indonesia and Turkey are significant in recent years and future.

Geothermal energy is coupled with the Rankine cycle through back pressure (1% of total electricity production capacity, same for the others), dry steam (23%), flash cycle (62%), and binary system (14%) to produce electricity, in which, flash cycle is classified into single flash (41%), double flash (19%) and triple flash (2%) [40]. Most of them utilize water as a working fluid to absorb heat from the heat reservoirs. The working fluid water is partially evaporated to steam, and the steam is introduced into steam turbines to produce electricity in dry steam and flash system. In a binary system, the heat absorbed by the working fluid water is transferred to a secondary fluid such as ammonia and propane. The process schematics of dry steam, single flash cycle and binary system are depicted in Fig. 3.50, and different systems will be described in detail in following sections.

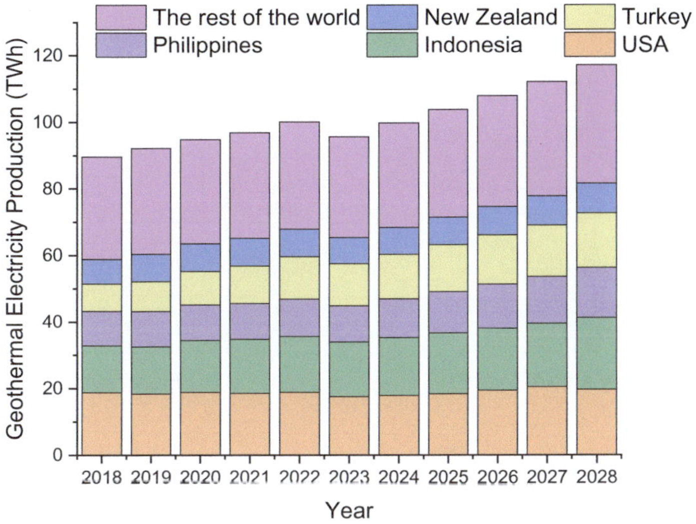

Fig. 3.49 Worldwide and top 5 countries electricity generation from geothermal energy projection from 2018 to 2028. Data source: Statista https://www.statista.com/outlook/io/energy/renewable-energy/geothermal-energy/worldwide

3.11.1 Dry Steam System

Dry steam system uses vapor dominant (saturated or superheated vapor) working fluid at a high temperature from the reservoir at point 1 in Fig. 3.50a. After the well valve, the high-pressure steam is expanded in the steam turbine and produces power for electricity generation. Then the expanded working fluid is condensed in the condenser and injected back to the well to start the next cycle. Due to the high-temperature working fluid, dry steam system is the most efficient among the geothermal power plant concepts. However, achieving such a high temperature requires a deeper drilling or a good location, which is costly. Therefore, the flash system is more popular and as the dominant technology in geothermal power plants.

3.11.2 The Flash Geothermal Powerplants

The schematic of the flash system is shown in Fig. 3.50b. Similarly, the working fluid, which is now pressurized liquid water in the well, absorbs the heat from the reservoir and becomes a high-temperature working fluid at point 1. There is a pressure drop through the well valve to initiate the flash boiling of the liquid mixture at point 2. Then, the vapor phase (point 3) and liquid phase (point 4) are separated in a separator. The vapor phase is expanded in the steam turbine and condensed in the condenser as the dry steam system. The liquid is mixed with the condensed working fluid and

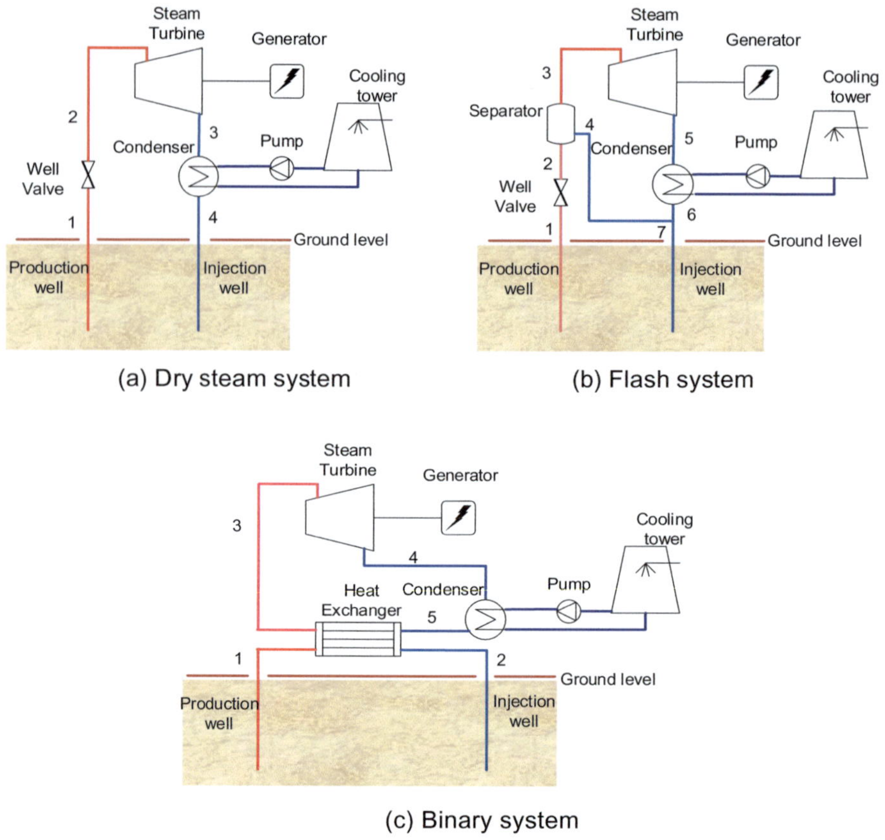

Fig. 3.50 Three main technologies for electricity production from geothermal energy

injected into the ground to start the next cycle. The double flash system has also been developed to improve the efficiency by adding a second pressure drop and an additional separator to obtain more vapor flows. A second steam turbine with low pressure expansion would be also required depending on the design. This section focuses on the single flash system.

Example 3.13 A geothermal power plant in northern Chile is operated with following parameters: Working fluid water mass flow rate 43.66 kg/s, condensing temperature 50 °C, turbine isentropic efficiency 85%, inlet temperature 250 °C and the flashing temperature 180 °C. Calculate the electrical power production.

At point 1, the working fluid is assumed as saturated water at 250 °C, thus, the enthalpy at point 1 is $h_1 = 1085.69$ kJ/kg. Assuming that the well valve expansion is adiabatic, then $h_2 = h_1 = 1085.69$ kJ/kg, which is a mixture of vapor and liquid phases at the flashing temperature 180 °C. Then, the enthalpies of saturated vapor (point 3) and saturated liquid (point 4) are $h_3 = 2777.22$ kJ/kg and $h_4 = 763.19$ kJ/

kg, respectively. Values of physical properties are obtained from the Excel-based tool available in electronic supplement. From the energy balance for the flash unit, we can solve then the steam mass fraction at point 2 as follows.

$$h_1 = xh_3 + (1 - x)h_4$$

$$x = \frac{h_1 - h_4}{h_3 - h_4} = \frac{1085.69 - 763.19}{2777.22 - 763.19} = 0.160$$

Therefore, the mass flow rate of the vapor is:

$$\dot{m}_V = \dot{m}_T x = 43.66\,\text{kg/s} \times 0.160 = 6.99\,\text{kg/s}$$

After the turbine expansion, the enthalpy of point 5 is:

$$h_5 = h_3 - (h_3 - h_{5s})\eta_{is} = 2777.22 - (2777.22 - 2109.54) \times 0.85 = 2209.69\,\text{kJ/kg}$$

where, h_{5s} is obtained with the enthalpy at the same entropy as at the point 3 and condenser pressure of $p_5 = p_{sat}(50\,°\text{C}) = 0.12\,\text{bar}$, giving $h_{5s} = 2109.6\,\text{kJ/kg}$. Then the power production is:

$$P = \dot{m}_V(h_3 - h_5) = (2777.22 - 2209.69)\,\text{kJ/kg} \times 6.99\,\text{kg/s} = 3.97\,\text{MW}$$

The flashing temperature is an important parameter to maximize the power production when other parameters are fixed. A higher temperature would result in a higher enthalpy at point 3. However, with the same enthalpy at point 2, a higher enthalpy at point 3 would reduce the vapor ratio (or the mass flow rate of vapor), and power is the product of mass flow rate and enthalpy difference in the steam turbine. Therefore, there is a compromise between mass flow rate of vapor and enthalpy difference to optimize the power production, which is determined by the flashing temperature. Here below we will derive the solution for the optimized flashing temperature for maximizing the power production in a single flash geothermal power plant. The specific work, w, (kJ/kg) to be maximized is:

$$w = x(h_3 - h_5)$$

The equation is simplified as the work per unit mass flow rate, where x is

$$x = \frac{h_1 - h_4}{h_3 - h_4}$$

Since points 1 and 3 are saturated vapor, we assume that the enthalpy difference can be reformulated into the heat capacity multiplying the temperature difference between point 1 and 3, $\Delta h \sim c_p \Delta T$.

$$x = \frac{c_p(T_1 - T_3)}{h_3 - h_4}$$

Then the work per unit mass flow can be expressed as:

$$w = \frac{c_p(T_1 - T_3)(h_3 - h_5)}{h_3 - h_4}$$

In addition, we assume that $h_3 - h_4 = h_5 - h_6 =$ constant. This assumption is with a negligible error. For example, in Example 3.13, $h_3 - h_4 = 2014.03\,\text{kJ/kg}$, $h_5 - h_6 = 2000.35\,\text{kJ/kg}$, with an 0.68% error. Therefore, we can write.

$$w = \frac{c_{p1}(T_1 - T_3)(h_4 - h_6)}{h_{fg}}$$

Similarly, since point 4 and 6 are both saturated water streams, the enthalpy difference can be expressed as:

$$w = \frac{c_p(T_1 - T_3)(T_3 - T_6)}{h_{fg}} = \frac{c_p}{h_{fg}}\left(T_1 T_3 - T_3^2 + T_6 T_3\right)$$

where c_p is the product of two heating capacities c_{p1} and c_{p2}, and $T_3 = T_4$. To obtain the optimized work, the derivative of w respect to T_3 should be calculated and set equal to zero as follows.

$$\frac{dw}{dT_3} = -\frac{c_p}{h_{fg}}(T_1 - 2T_3 + T_6) = 0$$

From which we obtain the optimal flashing temperature.

$$T_3 = \frac{T_1 + T_6}{2}$$

Therefore, the maximal power production can be obtained when the flashing temperature is the average of injection and condensation temperatures. For the Example 3.13, the optimized flashing temperature is:

$$T_3 = \frac{T_1 + T_6}{2} = \frac{250\,^\circ\text{C} + 50\,^\circ\text{C}}{2} = 150\,^\circ\text{C}$$

In order to verify this conclusion, the power production at flashing temperatures between 100 and 212 °C were calculated, without any simplified assumptions. The power and steam mass fraction are presented in Fig. 3.51.

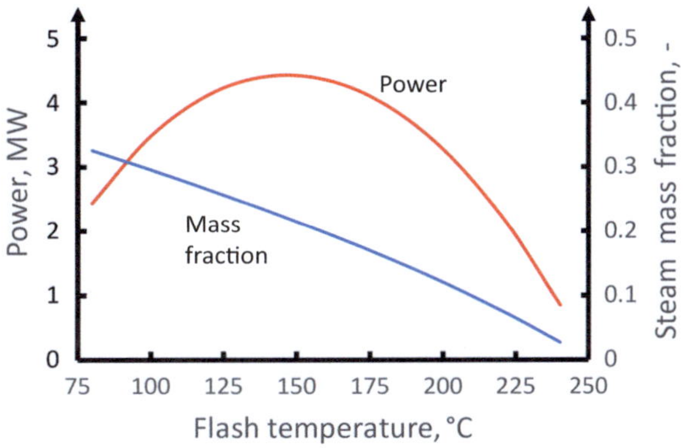

Fig. 3.51 Power production varies with flashing temperature, T_3.

From the Fig. 3.51 we verify the maximum power at ~ 150 °C. As the flash temperature increases, the used potential for flashing decreases, since the temperature drop decreases, and the steam mass fraction also drops almost linearly.

3.11.3 Binary Geothermal Power Plants

If the heat in the reservoir is not enough to produce steam from water as the process fluid, the hot water (95...180 °C in most binary power plants) can still be used to heat and evaporate the secondary fluid to generate electrical power through Rankine cycle in a separate loop [49]. This is called the binary system, as Fig. 3.50c shows. The heated secondary fluid from the heat exchanger follows the Rankine cycle described for steam process to produce power. The heat exchange between the working fluid water and the secondary fluid includes preheating, evaporation and/or superheating stages. As mentioned previously, the secondary fluid such as ammonia or propane can utilize the low temperature heat effectively. The binary system is the reverse process of heat pumps which will be introduced in Sect. 3.12.

3.11.4 Hybrid Configurations

It can be observed that only a small fraction of working fluid is used for power generation in a single flash system. For example, the vapor fraction in Example 3.13 is 0.16, which means 84% of working fluid is not utilized and a lot of heat is wasted. Hybrid systems are designed to solve this problem and improve the performance of geothermal power plant. This section shortly discusses flash-binary and solar-flash systems as examples, presented below, respectively.

As Fig. 3.52 shows, the flash-binary system exploits the saturated water after separator to provide heat for a secondary working fluid instead of discharging it

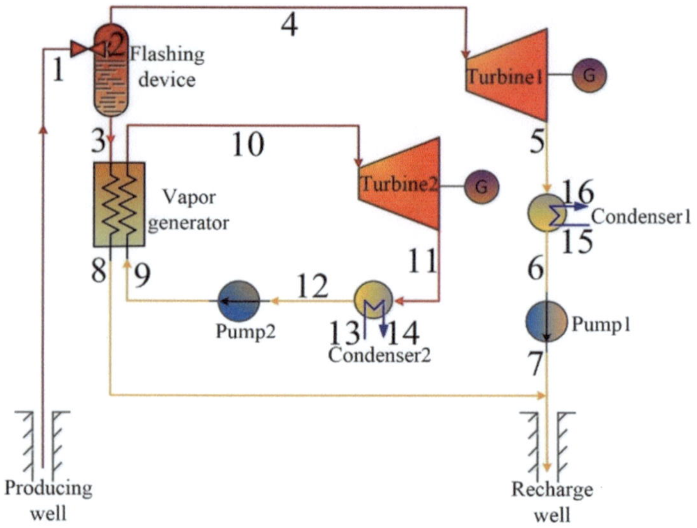

Fig. 3.52 Schematic of a flash-binary geothermal power plant [60]

to the injection well directly. This recovers the heat in the saturated water, and a second turbine is required to produce electricity. According to [60], the integration of a R245fa binary system with a flash system would produce additional 22…36% of power.

Cardemil et al. [11] simulated a single flash geothermal power plant integrated with a parabolic trough solar heat collectors, as Fig. 3.53 illustrates. The heat transfer fluid absorbs the solar heat in the parabolic through collectors and heats both vapor and liquid phases. Part of the liquid phase is evaporated. After the phase separation, the new vapor phase is mixed with the initial vapor phase. The new vapor phase temperature is further increased. The flow rate and inlet temperature of steam are improved by integrating the solar energy. Cardemil et al. reported that the power output was improved by 44% with integrating the solar heating system.

However, the integration of a binary system or a concentrated solar heat system would increase the cost of the plant. Therefore, the economic performance should be also evaluated to optimize the operating conditions.

3.12 Heat Pumps

Heat pump is a device that absorbs heat from a low-temperature or waste heat source and upgrades this energy to a high temperature for heating. The development of heat pump technology started in 1930s in Switzerland when Germany ceased the coal supply during the Second World War [56]. Comparing with a coal combustion furnace, indoor pollutants can be avoided by heat pumps, especially the poisonous

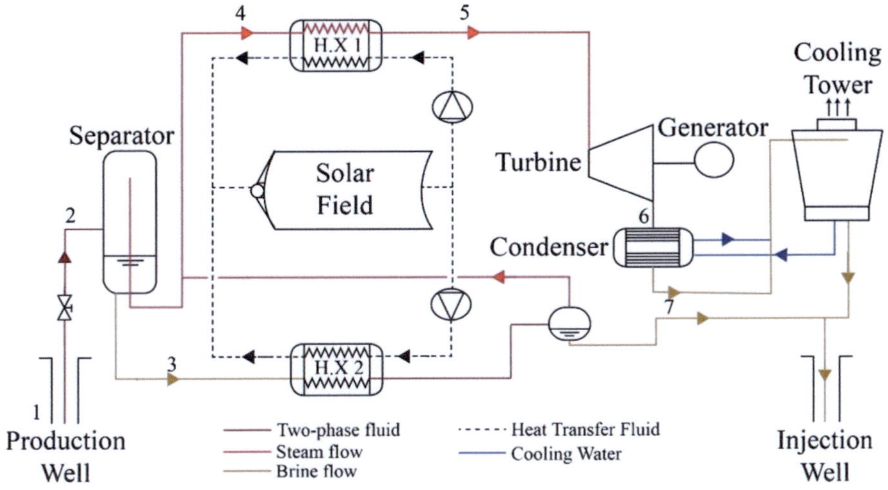

Fig. 3.53 Schematic of a hybrid single-flash geothermal power plant [11]

carbon monoxide, which used to be problem when using combustion for indoor heating in poor quality furnace.

Nowadays, global warming caused by the combustion fossil fuels is a main concern of humans. Energy transition to a system based on renewable energies propels the development of renewable heating sources, since heat pump is a reliable, cost competitive, easy to construct and retrofit with existing systems. The utilization of heat pumps has been boosted in recent decades. The share of heat pumps in global space heating was only 10% in 2017, but the installation increased significantly comparing with last decade, and it will continue to accelerate [17].

3.12.1 Heat Pump Types and Applications

Technically, heat pumps can be divided into the vapor compression-based cycles and heat pumps based on absorption, adsorption and chemical heat pumps. In the book, we only focus on the vapor compression cycle. In addition to this technical operational principle division, heat pump technologies can also be classified by heat sources, applications, and products. For heat sources and sinks we can utilize air, water, geothermal, solar energy, wastewater, etc. For products, heat pumps can be used for generating heating, cooling or combined heating and cooling, and there are several applications scales such as residential, industrial, commercial, district heating/cooling systems. Some details are summarized in Table 3.11.

The heat pump is operated in the heating mode when it provides heat, and in the cooling mode when it removes heat. Therefore, heating and cooling as "service" can be produced with the same cycle at different seasons. Being precise, when a heat pump

Table 3.11 Typical heat pump applications. Summarized from [13]

	Residential	Industrial	Commercial	District heating/ cooling
Capacity (kW)	1–400	100–30,000	20–1000	400–10,000
Heat source	Air, water, geothermal	Warm water, air, or geothermal	Air, water or geothermal heat	Water, wastewater
Heat sink	Air or water	Process heat, waste heat	Air or water	Air or water
Utilization	Space heating/ cooling	Heating or cooling for processes such as drying, evaporation, and distillation processes; space heating/ cooling	Greenhouse farming space heating and drying of agricultural products	Space heating/ cooling

is used, always the cold reservoir cools down and hot reservoir heats up. It therefor acts as a reverse heat engine. Residential application is the largest utilization purpose of heat pumps, and air is the most common medium as the heat source and sink [13]. With the development of geothermal energy technology as Sect. 3.11 introduced, the use of geothermal heat pumps is also growing. The industrial utilization of heat pumps aims in avoiding rejecting low-grade waste heat and the thermal efficiency of the plant can be significantly improved. The low-temperature heat source can be upgraded by heat pumps and used, e.g., for steam production, drying and distillation processes. Commercial applications in agriculture includes greenhouse farming and drying of different products. For example, in South Korea, local special plants such as fruits and flowers are planted in the greenhouse farming, and the space is heated by air-sourced heat pumps during winter [35]. Heat pumps for district heating and cooling are suitable for the base load in systems without CHP generation, low load with low-temperature summer operation for domestic hot water production, the places where supply/return temperatures are low, and water is available as a heat source.

3.12.2 Working Fluids

Working fluids, or refrigerants, are the fluids that can absorb low-temperature heat during evaporation and provide the cooling effect, and then condensing at the higher pressure, releasing heat. Primary working fluids in heat pumps are halocarbons, hydrocarbons, inorganic compounds, azeotropic mixtures and non-azeotropic mixtures.

Inorganic compounds and hydrocarbons were initially employed as refrigerants, including ammonia, CO_2, SO_2, methane and so on. Some of them such as ammonia (R-717) and CO_2 (R-744) are still used. The halocarbons contain one or more of halogens: fluorine, chlorine, and bromine. They are the most commonly used working fluids, and the most famous one is Freon (R22). Due to their greenhouse and ozone layer depletion effect, the use of halogens as working fluids in heat pump decreased significantly, for example, Freon has been phased out in 2020 [48].

Azeotropic mixtures are the two substances with different properties but similar boiling points that cannot be separated by distillation. Non-azeotropic mixtures are the mixtures of different volatiles and the composition can be varied during evaporation or condensation. The aim of both mixtures is to combine the advantages of different refrigerates and improve the performance of heat pump. For example, the most common azeotropic mixture R-502 is the mixture of 48.8% R22 and 51.2% R-115, providing less toxicity and better performance [13].

Some new substitutes such as R-134a have been introduced since it is nontoxic and nonflammable. R-134a has a good potential to replace R22 but R-134a requires 50% more displacement than R22 in the compressor at the same cooling capacity. The properties of R134a and other refrigerants are listed in Table 3.12. As the data shows, although R134a can avoid the ozone layer depletion problem, its global warming effect is still remarkable if released in the atmosphere.

When analyzing and dimensioning heat pumps, the most important tool is the log(p)-enthalpy diagram. Figure 3.54 presents this for the R134a refrigerant (schematics for other refrigerants are available from SWEP).

In this diagram, the vapor dome (thicker black curve) separates the saturated two-phase region in the middle. The left border of the vapor dome is the saturated water and right border is the saturated vapor, respectively. The vertically tilted black curves inside the vapor dome present constant vapor fractions, x, increasing from left to right from 0 to 1. Constant temperature isotherms are presented by the red color, decreasing on the right and left sides and being constant in the vapor dome at the saturation pressure that can be read from the left axis. Green lines present the specific volume and finally, enthalpy and entropy can be read from the vertical grey and tilted blue curves, respectively. We will discuss the use of the diagram in the

Table 3.12 Selected refrigerants and their properties [48]

Refrigerants	Boiling point (°C)	Critical point (°C)	Critical pressure (MPa)	ODP R11 = 1	GWP $CO_2 = 1$
R134a	– 26.07	101.06	4.06	0	1430
R32	– 51.65	78.11	5.78	0	675
R600a	– 11.75	134.7	3.65	0	3
R1233yf	– 29.45	94.7	3.38	0	4
R290	– 42.11	96.75	4.25	0	20
R451A	– 29.6	95.4	3.50	0	149

ODP: Ozone depletion potential, it is a number that reflects the amount of stratospheric ozone depletion caused by a substance and the ODP of R11 is defined as 1 as the reference

GWP: Global warming potential, it is a number that indicates the global warming caused by a substance and the GWP of CO_2 is defined as 1 as the reference

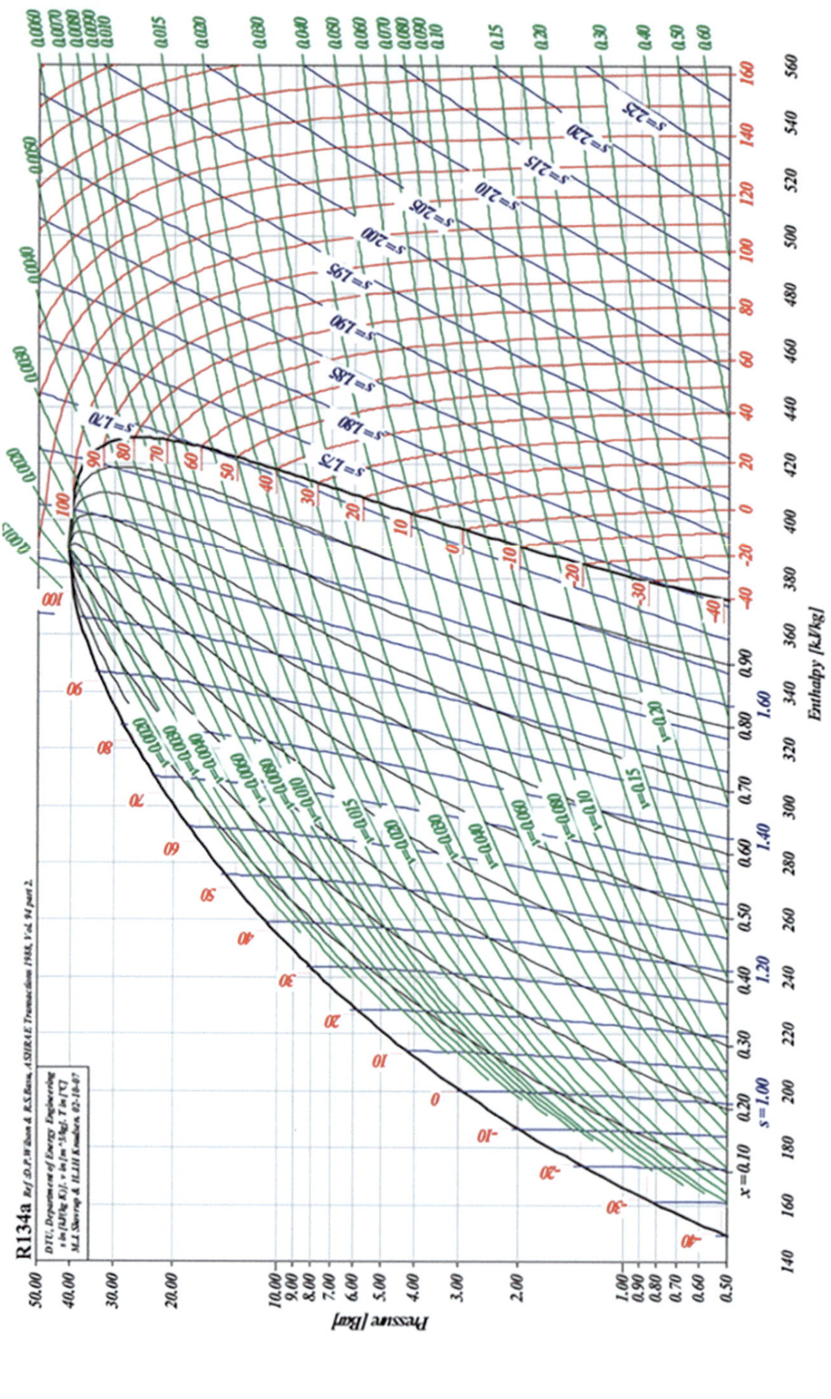

Fig. 3.54 Example of the log(p)-enthalpy diagram of R134a refrigerant © SWEP International. www.swep.net

examples below. Please also note that the highest point of the vapor dome presents the critical point, compare the value for R134a given in the Table 3.12 and the diagram ($T_c = 101.06\,°\mathrm{C}$).

3.12.3 Carnot Refrigeration and Heat Pump Cycle

The Carnot cycle is a theoretical model, where work processes have no losses and entropy is constant during compression and expansion. This is useful in understanding the heat pump operation. In thermodynamics, Carnot power cycle is a heat engine where the work is produced by the heat energy. The Carnot heat pump cycle is operated with the same energy transfer but in an opposite direction, as Fig. 3.55 illustrates.

Even the heat flows from the cold to hot direction in the heat pump on the overall process level, locally, inside the evaporator and condenser the heat is transferred from the cold reservoir which temperature is higher than the local temperature of the process fluid in evaporator. Similarly, in the condenser, the process fluid must be hotter than the hot reservoir temperature for heat to flow. We assume that the temperatures of the cold and hot reservoirs remain constant, their entropy change rates therefore are

$$\Delta \dot{S}_C = -\frac{\phi_C}{T_C}$$

$$\Delta \dot{S}_H = \frac{\phi_H}{T_H}$$

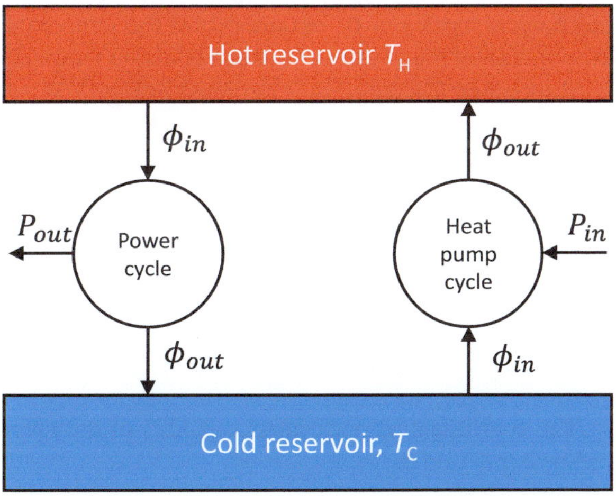

Fig. 3.55 Comparison of Carnot power and heat pump cycles, H and C refer to hot and cold temperature reservoirs, respectively

respectively. Since the heat pump system itself operates at the steady state, the entropy of the fluid in it is constant with time, and therefore, the total entropy change of the overall system can be presented as follows.

$$\Delta \dot{S} = \dot{m}\Delta s = \Delta \dot{S}_C + \Delta \dot{S}_H = \frac{\phi_H}{T_H} - \frac{\phi_C}{T_C}$$

For the same reason, internal energy of the fluid in the process is also constant, $dU/dt = 0$.

$$\frac{dU}{dt} = \phi_C - \phi_H + P = 0 \rightarrow P = \phi_H - \phi_C$$

Heat from the cold reservoir can be solved from entropy balance and substituted into energy balance to give:

$$\phi_C = \frac{T_C}{T_H}\phi_H - \Delta \dot{S} T_C$$

$$P = \Delta \dot{S} T_C + \phi_H \left(1 - \frac{T_C}{T_H}\right)$$

In reversible processes, $\Delta \dot{S} = 0$, and in a heat pump $T_H > T_C$. Since $\phi_H > 0$, the power P has to be positive. In a real, irreversible process, $\Delta \dot{S} > 0$, further increasing the amount of work is needed. Figure 3.56 shows the schematic of the Carnot refrigeration and heat pump cycle. Four stages are included: 1-2 is an isentropic compression with work input, and temperature increases from a lower temperature T_C to a higher temperature T_H; 2-3 is the heat rejection of the working fluid condensation at a constant temperature T_H; 3-4 is the isentropic expansion through a turbine, mechanical work can be obtained, and the temperature decreases to T_C; 4-1 is the evaporation of the working fluid at a constant temperature T_C by absorbing heat. The corresponding T-s diagram is depicted in Fig. 3.56.

In a reversible cycle, the rejected heat is calculated as the area under the process line 2-3:

$$\phi_H = T_H(s_2 - s_3)\dot{m} = T_H(s_1 - s_4)\dot{m}$$

The required power is the area within the cycle line 1-2-3-4-1:

$$P = (T_H - T_C)(s_1 - s_4)\dot{m}$$

The coefficient of performance (COP) of the reversible Carnot heat pump cycle can be expressed as:

$$\varepsilon_h = \frac{\phi_H}{P} = \frac{T_H(s_1 - s_4)}{(T_H - T_C)(s_1 - s_4)} = \frac{T_H}{T_H - T_C}$$

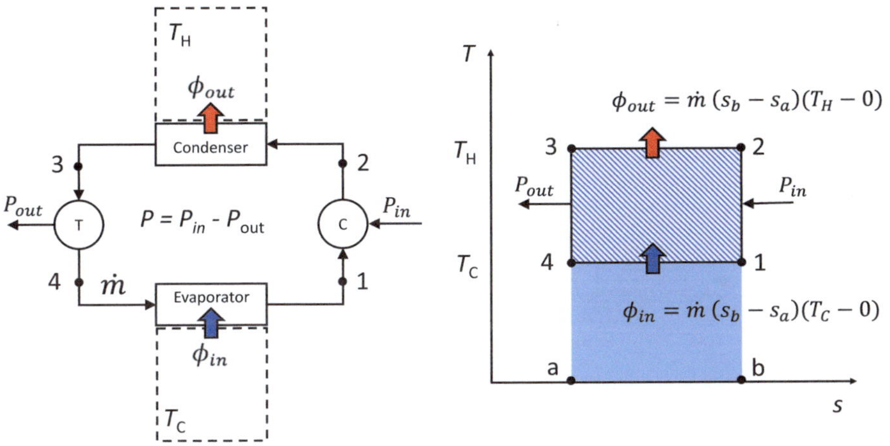

Fig. 3.56 Schematic of the Carnot heat pump cycle and its T-s diagram, C and T refer to compressor and turbine, respectively (temperatures in K)

Similarly, the COP of the reversible Carnot refrigeration cycle can be calculated as:

$$\varepsilon_c = \frac{\phi_C}{P} = \frac{T_C(s_1 - s_4)}{(T_H - T_C)(s_1 - s_4)} = \frac{T_C}{T_H - T_C}$$

Thus, the relation between ε_h and ε_c is:

$$\varepsilon_c = \varepsilon_h - 1$$

ε_h and ε_c are the maximal COPs for the Carnot heat pump and refrigeration cycle, respectively. The ratio between the real COPs and the reversible COPs shows how much usable energy has been recovered. It is worthy to notice that all temperatures for COP calculations should be in Kelvin.

3.12.4 Single-Stage Heat Pump

The above discussed reversed Rankine cycle-based heat pump has some practical issues that need improvement. The actual heat pump process is therefore a modification of the Carnot heat pump cycle. In the above cycle, fluid might be partly liquid at the compressor inlet. This is not acceptable, since any liquid would be very harmful for the compressor, causing erosion and washing of oil from the walls of cylinder. Some superheat before the compression is required, but not too much, since the density decreases which increases the compressor power. Superheating is typically limited to 5–10 K, depending on the process fluid. Sub-cooling of the fluid starts at

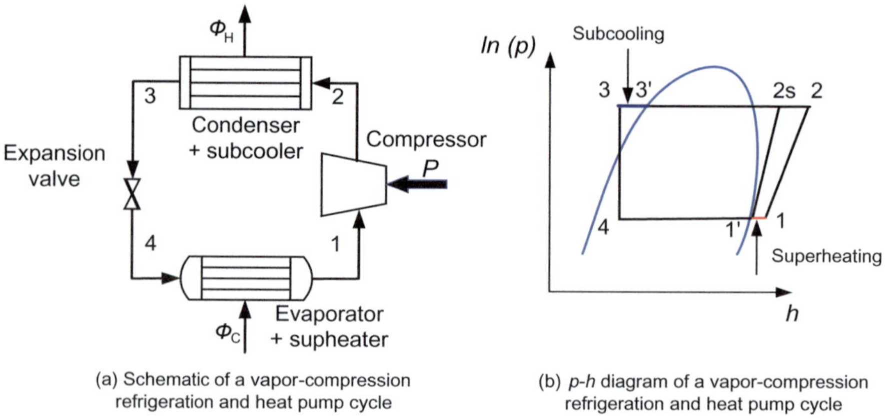

Fig. 3.57 Schematic of a vapor-compression refrigeration and heat pump cycle and its ln(p)-h diagram

point the 3' and continues to point 3. These low temperatures are not available in the surroundings. As much as possible of sub-cooling is required to maximize the amount of condenser heat flow rate. The isentropic power from the turbine using compressed liquid is very small (note small energy consumption in the liquid pump in the Rankine power cycle) and therefore, it is not required. The pressure can be simply dropped by a throttling valve or even a long capillary tube with a small diameter. For an isenthalpic expansion, $h_3 = h_4$. In reality, expansion is not adiabatic, and its entropy increases. Denoting all the modifications, our new process looks like the one presented in Fig. 3.57.

Example 3.14. A heat pump system is employed to provide district heating. The oil field wastewater is used as heat source in the evaporator of the heat pump, and it is assumed to heat the district heating water from 55 to 80 °C in the condenser. R134a is selected as the working fluid, and its flow rate is 100 kg/s. The evaporation and condensation temperatures of the process fluid are 20 °C and 80 °C, respectively. The working fluid is subcooled in the condenser by 5 °C. The isentropic efficiency of the compressor is assumed as 0.85. Calculate: (a) the heat produced in condenser; (b) the mass flow rate of the district heating water; (c) ΔT_{min} in the condenser; (d) the COP of the heat pump.

According to p–h diagram of R134a, the enthalpy after evaporation at point 1 in Fig. 3.55b is $h_1 = 410$ kJ/kg. If the compression is isentropic, the ideal enthalpy of point 2 (to condensation pressure 25 bar, temperature 80 °C) should be $h_{2s} = 445$ kJ/kg. Therefore, the actual enthalpy at point 2 with considering isentropic efficiency is:

$$h_2 = \frac{h_{2,s} - h_1}{\eta_c} + h_1 = \frac{445 - 410}{0.85} + 410 = 451 \text{ kJ/kg}$$

The working fluid is subcooled by 5–75 °C at the same condensation pressure, so $h_3 = h_4 = 317$ kJ/kg. With knowing these enthalpies, we can calculate:

(a) the heat produced in condenser:

$$\phi_H = \dot{m}_R(h_2 - h_3) = 100\ \text{kg/s}\ \times (451 - 317)\ \text{kJ/kg}\ \times 10^{-3} = 13.4\ \text{MW}$$

(b) the mass flow rate of the district heating water:

$$\dot{m}_W = \frac{\phi_H}{c_p(T_{out} - T_{in})} = \frac{13400\ \text{kW}}{4.2\text{kJ/kg}\cdot\text{K} \times (80 - 55)\text{K}} = 128\text{kg/s}$$

(c) $\Delta T_{\min}$ in the condenser. According to Fig. 3.55b, there are 3 stages in the condenser: de-superheating, condensation and subcooling. The saturated vapor and liquid enthalpies at 80 °C can be read from the *p–h* diagram as $h_{SV} = 430$ kJ/kg and $h_{SL} = 322$ kJ/kg, respectively. The district heating water temperatures after each stage can be calculated, taking the subcooling stage as an example. The energy balance at this stage should be:

$$\dot{m}_W c_p \left(T_{out,1} - T_{in}\right) = \dot{m}_R(h_{SL} - h_3)$$

Then, the temperature $T_{out,1}$ can be calculated:

$$\begin{aligned} T_{out.1} &= \frac{\dot{m}_R(h_{SL} - h_3)}{\dot{m}_W c_p} + T_{in} \\ &= \frac{100\text{kg/s} \times (322 - 317)\ \text{kJ/kg}}{128\text{kg/s} \times 4.2\text{kJ/kg}\cdot\text{K}} + 55\,^{\circ}\text{C} = 56\,^{\circ}\text{C} \end{aligned}$$

Similarly, the temperature $T_{out.2}$ can be calculated:

$$\begin{aligned} T_{out,2} &= \frac{\dot{m}_R(h_{SV} - h_{SL})}{\dot{m}_W c_p} + T_{in} \\ &= \frac{100\ \text{kg/s} \times (430 - 322)\ \text{kJ/kg}}{128\ \text{kg/s} \times 4.2\ \text{kJ/kgK}} + 56\,^{\circ}\text{C} = 76\,^{\circ}\text{C} \end{aligned}$$

The $T - \phi$ diagram can be plotted, as shown in Fig. 3.58. It can be seen that the minimum temperature difference appears in the de-superheating stage, and it is:

$$\Delta T_{min} = 80\,^{\circ}\text{C} - 76\,^{\circ}\text{C} = 4\,^{\circ}\text{C}$$

(d) In order to calculate the COP, the required compressor work should be computed:

$$P = \dot{m}_R(h_2 - h_1) = 100\ \text{kg/s} \times (451 - 410)\ \text{kJ/kg} \times 10^{-3} = 4.1\ \text{MW}$$

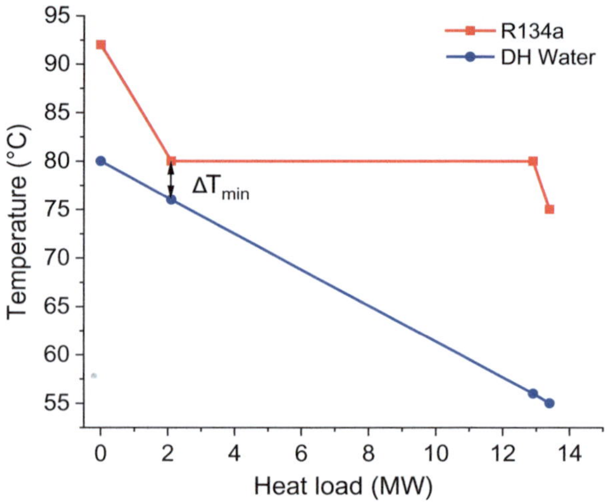

Fig. 3.58 Temperature-heat load ($T - \phi$) diagram of the condenser heat exchanger in the heat pump

Then, the COP is:

$$\varepsilon_H = \frac{\phi_H}{P} = \frac{13.4\text{MW}}{4.1\text{MW}} = 3.27$$

In Example 3.15, we will compare the CO_2 emissions of a heat pump and other heat resources.

Example 3.15 The heat pump in Example 3.14 is operated in an equivalent full load 1500 h for the district heating. Compare the CO_2 emissions if the required electrical work is provided by a coal power plant with a 0.35 efficiency and the district heating is directly provided by a coal boiler with a 0.9 thermal efficiency. The CO_2 emission factor of coal is 0.34 t/MWh.

The CO_2 emission from the coal boiler is:

$$\text{CO}_{2,boiler} = \frac{1500\,\text{h} \times 13.4\,\text{MW}}{0.9} \times 0.34\,\text{t/MWh} = 7.56\,\text{kt}$$

The CO_2 emission from the coal power plant is:

$$CO_{2,powerplant} = \frac{1500\text{h} \times 4.1\text{MW}}{0.35} \times 0.34\,\text{t/MWh} = 5.97\text{kt}$$

The CO_2 emission difference is:

$$\frac{7.56 - 5.97}{7.56} \times 100\% = 21\%$$

which means introducing heat pump can reduce 21% CO_2 emission comparing with the same amount of heat provided by coal furnace. This result indicates that implementing heat pumps would motivate CO_2 emission reduction with the same energy sources.

3.12.5 Two-Stage Heat Pump

Compared with a single-stage heat pump, two-stage heat pumps can achieve a higher COP because of the less compression ratio in each stage with fewer losses [9]. A typical two-stage heat pump is designed as Fig. 3.59a depicts. Instead of expanding the working fluid to the evaporation pressure as the single-stage heat pump, it is expanded to an intermediate pressure where the mixture of liquid and vapor phase is formed. The two phases are separated in a flash tank. The liquid phase continues to expand to the evaporation pressure. It is evaporated, compressed to the same intermediate pressure, and mixed with the vapor phase from the flash tank. The mixed working fluid is compressed to the condensation pressure and provides heat through condensation (and subcooling). Figure 3.59b illustrates the p–h diagram of the two-stage heat pump. This design has been employed in many Nordic countries for district heating. For example, the Katri Vala heat pump in Helsinki, Finland produces both district heating (126 MW) and cooling (80 MW) with the same technology for Helsinki metropolitan area (www.helen.fi).

Example 3.16 As the same conditions in Example 3.14, but operating it in a two-

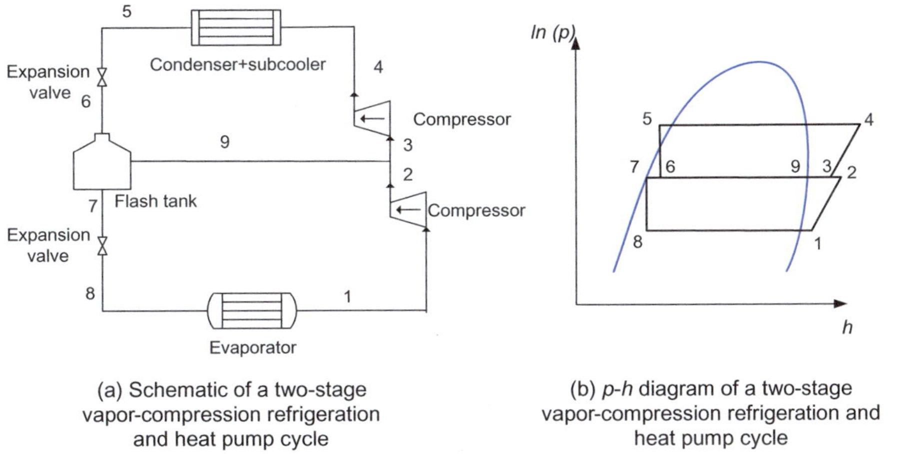

Fig. 3.59 Schematic of a two-stage vapor-compression refrigeration and heat pump cycle and its *p–h* diagram

stage heat pump system with an intermediate pressure for expansion or compression at 12 bar, recalculate questions (a)(b)(d) in Example 3.15.

h_1 is the same as Example 3.14, so it is $h_1 = 410$ kJ/kg, with similar idea in Example 3.15, after first stage compression can be calculated with reading $h_{2s} = 428$ kJ/kg:

$$h_2 = \frac{h_{2,s} - h_1}{\eta_c} + h_1 = \frac{428 - 410}{0.85} + 410 = 431\ \text{kJ/kg}$$

The flow at point 3 is the mixture of 2 and 9, so the enthalpy should be calculated based on energy balance. Therefore, it is essential to determine the mass ratio of vapor and liquid phases. The enthalpy of point 5 is the same as point 3 in Example 3.14, which is $h_5 = h_6 = 317$ kJ/kg. After expansion, point 6 is the mixture of saturated liquid and vapor, the vapor mass ratio $x = 0.3$ can be read from the *p–h* diagram directly. Thus, the saturate vapor after intermediate pressure expansion (point 9) is 30% of the total flow rate of the working fluid. The remained 70% goes through the second expansion, evaporation and first compression. The enthalpy of saturated vapor phase at point 9 can be read from the diagram $h_9 = 422$ kJ/kg. The enthalpy of liquid phase h_7 can be calculated based on the energy balance:

$$xh_9 + (1 - x)h_7 = h_6$$

Substituting the values, $h_7 = 264$ kJ/kg, which is the same as h_8. With knowing the mass ratio of streams 2 and 9, the enthalpy of point 3 can be calculated with the same mass balance:

$$xh_9 + (1 - x)h_2 = h_3$$

and we can obtain $h_3 = 428$ kJ/kg. Similarly, h_4 can be calculated as h_2:

$$h_4 = \frac{h_{4,s} - h_3}{\eta_c} + h_3 = \frac{448 - 428}{0.85} + 428 = 452\ \text{kJ/kg}$$

All enthalpies have been obtained, then questions (a), (b), and (d) can be solved:

(a) the heat produced in condenser:

$$\phi_H = \dot{m}_R(h_4 - h_5) = 100\,\text{kg/s} \times (452 - 317)\,\text{kJ/kg} \times 10^{-3} = 13.5\,\text{MW}$$

(b) the mass flow rate of the district heating water:

$$\dot{m}_W = \frac{\phi_H}{c_p(T_{out} - T_{in})} = \frac{13500\,\text{kW}}{4.2\,\text{kJ/kg K} \times (80 - 55)\,\text{K}} = 129\,\text{kg/s}$$

(c) the total required power in two compressors is:

$$P = \dot{m}_R(1 - x)(h_2 - h_1) + \dot{m}_R(h_4 - h_3) = 3.9\,\text{MW}$$

The COP of the two-stage heat pump is:

$$\varepsilon_h = \frac{\phi_H}{P} = \frac{13.5\,\text{MW}}{3.9\,\text{MW}} = 3.46$$

Comparing with the single stage, the COP is improved by:

$$\frac{3.46 - 3.27}{3.27} \times 100\% = 6\%$$

The improvement would be more significant when the difference between condensation and evaporation temperatures is remarkable.

3.13 Environmental Impacts of Thermal Power Plants and Their Mitigation

The objective of this section is to shortly discuss the most typical impacts of the thermal power plants on nature and environment, and to discuss how they can be mitigated. Since the CO_2 emissions were already discussed in Sect. 2.2, we focus here only on the concentrated solar and geothermal power plants technologies.

3.13.1 Concentrated Solar Power Plants

As discussed above, concentrated Solar Power (CSP) plants are large-scale power plants with generating capacities of 10–500 MW. A CSP system requires a relatively large areas of land, up to 10 m^2/MWh of electricity. The majority of the power plant consists of the mirrors in the heliostat field (see Fig. 3.1). In addition it also contains the thermal storage system, the Rankine cycle power plant components, electrical infrastructure, such as the transmission power lines, transformers, substation and connection to the grid, mounting structures and a security fence around the power facility. These constructions take about 80% of the total land use of the CSP system [54]. During the construction, if relevant, the vegetation is cleared and land modification is needed for the foundations, access roads and other facilities like the substation. Large-scale solar power development can cover large areas, and there can several similar projects in the same region, creating cumulative impacts across the landscape. Previously continuous habitats can get fragmented, get degraded or are completely lost. The habitat modification, loss and degradation, and fences around the power plants can create barriers for animal movement. Birds and bats are also known to collide with CSP structures and powerlines, or get burned and killed by the thermal radiation heat flux [57]. Mortalities caused by heat flux have been documented from several CSP farms in Israel, Spain and the USA [23, 32]. Though the collisions are less frequent than in some other human activities, with increase and concentration of solar power production in certain areas, cumulative collision incidents can become locally or regionally significant. Soil modification can result in erosion and loss of organic carbon. Studies conducted in semiarid regions have shown that the loss of carbon after vegetation removal can be considerable in the soil organic crust and remain low for even after 15 years compared to undisturbed sites [6, 34]. Such a loss of organic carbon is detrimental to health of the soil: in the undisturbed sites also the microbial biomass and activity were found to be higher than on disturbed area [6]. Land degradation and erosion results in the loss of biodiversity, worsening the degradation further. Loss of vegetation results in increased dust flux [22]. Disturbance, new roads and increased traffic can also facilitate introductions of invasive alien species that can outcompete local wildlife [7]. Moreover, chemicals used in operation, such as dust suppressant, anticorrosive chemicals and herbicides can have unwanted negative impacts on nature, accumulate in food chain, and pollute waterbodies and ground water [19, 36].

Depending on the cooling system technology (wet cooling, dry cooling or a hybrid system), CSP plants can consume considerable amounts of water. Wet cooling can consume water over 3.07 m^3/MWh, which is more than required by coal and natural gas power plants [22]. The use of dry cooling technologies can reduce water consumption by up to 95% [22]. Water is also needed to wash the dust from the mirrors. This is essential to keep the optical efficiency as high as possible. In arid and water-stressed regions excessive use of water can deplete the surface and groundwater sources and influence local habitats and threaten species survival, and compete with other water use needs, such as agriculture [22, 58]. Water withdrawal can also destroy intermittent or perennial aquatic habitats by reducing frequency and duration of wet periods. This can increase the habitat fragmentation for species dependent on these seasonal water bodies [19].

Wastewater emissions can create thermal and chemical pollution that degrade habitats. Release of cooling water into natural waterbodies can heat the environment that can induce algal blooms and cause mortality among fish and other animals [7]. Wastewaters can also contain hazardous substances that contaminate water bodies and pose a risk to wildlife. Wastewater streams can also be collected and stored into evaporation ponds where chemicals are concentrated into contents that are appropriate for collection and recycling. If properly handled this method mitigates chemical pollution in the environment. But there is a risk related to the evaporation ponds as well. These ponds can attract animals and create a death trap for them. As an example, in South Africa 37 carcasses, including birds, reptiles and mammals that were drowned or poisoned, were found from an evaporation pond during a four-month study on a 50 MW CSP plant [28].

Mitigation Measures

The most efficient way to mitigate and avoid negative impacts on nature from CSP construction is to direct construction on areas already altered by human uses. With a careful planning and proper siting it is possible to develop more renewable energy production while safe-guarding the value of important areas for nature and nature conservation areas [16, 22]. Protected areas and ecologically important areas like breeding, wintering and feeding areas and migration routes should be left outside CSP and any other energy development. Methods have been developed to assess, plan and implement CSP construction on areas that do not overlap with valuable nature areas. Alone in the South Africa it has been estimated that a nominal capacity of 548 GW of CSP could be constructed without a need to compromise areas supporting endangered or vulnerable vegetation [16]. Siting of the CSP power plants outside migratory routes, stopover sites, and important habitats is also the single most efficient measure to mitigate bird and bat collisions [57]. Mortality caused by the thermal radiation heat flux can be mitigated by altering the standby positioning of heliostats to reduce the amount of heat flux around the receiver tower [57]. Also, use of parabolic (curved) mirrors instead of flat heliostats minimize potential bird collisions [7].

After the construction the area can be re-vegetated with locally occurring species to mitigate harmful impacts. This type of restoration measure is important in order to restore and maintain the soil productivity, local biodiversity, reduce dust (and the need for cleaning) and prevent spread and introduction of invasive alien species.

Construction on sensitive areas and during sensitive seasons should be avoided. We should also avoid soil sealing by favoring foundations that do not damage natural soil functioning and maintains species inhabiting the soil, and above ground [7]. We should limit the clearance of nature to minimum, use restoration and offset residual damage. We need to implement measures to minimize soil erosion (drainage for access roads, avoid flooding land and/or damaging nearby water bodies). Finally, it is important that we modify security fences so that animals can pass through and won't get trapped in or between the fencing.

There are several measures that can help to reduce and mitigate the negative impacts resulting from the water use. First, dry and hybrid cooling systems and cleaning methods should be favored over technologies needing vast amounts of water to safeguard water reserves and mitigate negative impacts on nature, including aquatic biodiversity. Water use can also be reduced by modifying operations for example by storing steam in pressurized tanks or to molten salt storages to generate energy at night when the there is less need for cooling [7, 8]. Wastewater needs to be managed to mitigate and avoid thermal and chemical pollution of the terrestrial and aquatic environment and use measures that prevent animals from entering the evaporation ponds.

3.13.2 Geothermal Power Plants

Theoretically, the geothermal energy production potential is significant. Based on Aghahosseini and Brayer [2] the annual potential for geothermal power generation could be as high as 52,600 TWh and 946,000 TWh for 2030 and 2050, respectively. If economic and water stress constraints are excluded, the global geothermal potential during the next 30 years could be as much as 1,752,000 TWh, exceeding the global primary energy consumption of ~ 180 000 TWh significantly. Aghahosseini and Brayer [2] also estimate the sustainable limit of geothermal energy as 256 GW being equal to 0.256 TW × 8760 h = 2240 TWh. Sustainable here refers to the amount of heat that is continuously available in the Earth's crust in the layer on 10 km. The impacts on the nature climate, ground water and societies were not included in the analysis, which still significantly decreases the estimate. As for comparison, based on the updated study of Perez and Perez [44], the finite total resources of combined coal, oil and natural gas would be 13,750,000 TWh.

Development of geothermal energy has proceeded slowly due to scarcity of geologically suitable sites, technological limitations, long construction times, high price, and no guarantee of obtaining enough steam resources for power generation [30]. At current geothermal energy contribution is incremental, less than 1% of the global electricity production [31] but as demand for reliably produced, stable, clean and affordable energy is increasing it is expected that the use of geothermal energy will grow as well. This has increased the need to understand and address better the potential negative impacts construction and operation of geothermal energy production can have on nature and wildlife, and improve development processes to avoid and mitigate unwanted ecological and social impacts.

Negative impacts on nature can happen through-out the life-cycle of the geothermal power plant starting from exploratory drilling activities and well-testing [59]. The exact impacts are site specific but in general terms it can be said that areas in natural or semi-natural state experience more negative impacts than areas already altered by human uses. The construction can result in habitat loss, fragmentation and disruption of ecosystems, their function and wildlife. Especially during the construction can disturb wildlife and impact their behavior as animals avoid the construction site. Exploratory drilling, construction and operation can result also in soil erosion, dust, emissions and effluents from drilling and diesel-based drilling power generation, noise, steam venting and air pollution from non-condensable gas in steam, brine contamination of soil and water resources, and visual impacts among others [30]. The land use and land use change created by geothermal energy production is relatively modest when compared to other energy production technologies, also including wind and solar power (see Fig. 2.15). Land is required for drilling the wells, constructing the infrastructure, roads, powerlines and installing power plant systems. However, as the construction of geothermal energy increases, also its total land use and potential cumulative impacts will grow as well. At the local scale siting decisions and timing of the construction are important aspects, not the create locally significant negative impacts on nature and wildlife.

Geothermal power plants require water for various purposes: drilling, fluid circulation, and cooling. High water consumption can strain local water resources, especially in the areas already facing water scarcity or drought conditions. Additionally, the extraction of geothermal fluids can alter local hydrology and impact groundwater resources. Geothermal power plants can also result in thermal pollution in nearby water bodies if the excess heat is released into the environment during operation. Elevated water temperatures can degrade habitats by disrupting aquatic ecosystems, induce algal blooms, affect species composition or increase mortality [7]. Drilling and fluid circulation processes can also release naturally occurring chemicals and heavy metals from deep underground reservoirs in concentrations that lead into chemical pollution of the environment. The most common non-condensable gases in steam include carbon dioxide, methane, hydrogen sulfide, ammonia, hydrogen, nitrogen, argon, and radon. If not properly managed, these chemicals can contaminate surface- and groundwater or soil, leading to environmental pollution, habitat degradation and harm the aquatic and terrestrial organisms [19, 36].

In certain cases, geothermal operations, especially those involving injection and extraction of fluids, can induce seismic activity or earthquakes as the land above sinks over time [30]. While most induced seismicity is minor and goes unnoticed, larger events can cause damage to infrastructure and pose risks to nearby communities. This potential risk for surface instability is one of the highest concerns of geothermal power plants. The risk can be reduced by reinjection of geothermal fluid into the earth via a reinjection well. Reinjection well is one of the major measures that can be taken to improve the sustainability of an operating geothermal power plant [30].

Kabeyi and Olanrewaju [30] raise the concern that accessible geothermal energy resources are often located on ecologically valuable and sensitive areas, or areas having socio-economic importance for local communities, such as productive agricultural lands. For example in Japan most of the geothermal resources are inside protected areas [38]. The large network of hot pipelines needed for geothermal central power plants could create barriers for movement for wild animals' migratory paths and degrade habitats. Also new road networks and development of new high voltage transmission and distribution powerlines fragment the environment and can open up previously intact areas for other human uses.

Local communities can also suffer from negative impacts created by the geothermal energy production. In Kenya at Eburru wellhead power plant the local community has been negatively impacted by emissions of hydrogen sulfide, high ambient noise levels and high boron concentrations [4]. Moreover venting of geothermal fluid from production well have released silica and brine that ended up on agricultural lands of neighboring farms and resulted in protests and requests for compensation [5]. These deposits can pose a threat also on wild flora and fauna. The venting of wellhead was due to high water concentration in the steam and low voltage network connection that resulted in tripping because of a frequent overload. To ensure successful siting of geothermal power plants it is therefore important to plan carefully how to control and manage potential negative ecological and social impacts of geothermal energy development [31].

Mitigation Measures

To mitigate these negative impacts, careful site selection, thorough environmental impact assessments, sustainable water management practices, proper waste disposal, and community engagement are essential. Implementing advanced technologies and best practices can help minimize the environmental footprint of geothermal energy projects and ensure sustainable energy production without compromising the nature. To avoid and mitigate the negative impacts of geothermal energy production, it is important to time the construction of the power plants outside important seasons, such as the breeding and migration periods, and avoid construction on areas that have high ecological value or which are hosting rare, sensitive or threatened species or ecosystems, or are nature conservation areas. Emissions of harmful substances from both central and wellhead powerplants can be reduced with technical applications, like reinjection of geothermal fluids or using organic Rankine cycle as a binary cycle, with no direct contact with the water underground. Extracting and reinjecting the geothermal fluid in right quantities help to avoid negative environmental impacts like land subsidence, drying up of surface and ground water, or destruction of geysers [38, 59]. Releases of hydrogen sulfide, that exists naturally in geothermal reservoirs can be reduced by using gas scrubbers [30].

Careful planning and application of the most effective methods and best technologies in geothermal power plant projects is needed to protect the people and nature from harmful impacts including pollution and seismic activities. Important areas for nature and human settlements should be protected with appropriate buffer zones to keep a safe distance to the power plant. Reducing negative impacts is essential for increasing the acceptability of this energy source since properly implanted geothermal powerplants have several benefits compared to fossil fuel sources and therefore can form a central piece in the future low-carbon energy pallet.

3.14 Summary

The objective of this chapter was to describe the theory and operational principles of "traditional" power and heat production processes including steam and gas turbine-based power plant concepts. We also briefly discussed the Ericsson and the Stirling cycles. We introduced a simple combined cycle; the combined process of the topping gas turbine plant and then the bottoming Rankine steam power plant, which utilizes waste heat from a gas turbine. Finally, we discussed the benefits of technical aspects of combined heat and power production.

As a technical summary, state of the art steam power plants can reach a thermal efficiency of around 50% and this already seems to be on the limits of what is achievable. The best large scale gas turbines can reach efficiencies in the order of 45%, being limited by the temperature resistance materials available for the first turbine blades. Ericsson and Stirling cycles could offer higher efficiencies—as much as 60–70%—but there are still challenges in the construction and material choices. In addition, they offer feasible solutions in small-scale applications. In practice the highest thermal efficiencies for power production can be obtained using the combined cycle, reaching efficiencies above 60%. In addition, if we are also utilizing waste heat and having a high value for this, the overall efficiency can be very high—over 90%—even in combustion-based systems.

Today more than 80% of the heat and power produced globally is from burning fossil fuels. Whilst there is an urgent need to move away from fossil fuels, the traditional thermal cycles discussed in this chapter will not become outdated. These systems can be effectively applied to several renewable energies, such as concentrated solar energy, thermal energy in the oceans and geothermal energy systems. For example, gas turbines can be run with green hydrogen, biogas, or using several e-fuels that have been produced from green hydrogen, and captured CO_2, such as e-methane. It will be very interesting to see the future of these processes as a part of our energy systems, and how they manage to compete with the continuously decreasing production cost of solar PV and wind power.

As we have learned, all these process cycles require that a fraction of heat is rejected. In addition, the lower the cooling or heat rejection temperature, the higher the process efficiency. In the case of a condensing power plant optimized for power production only, the environmental impacts (including impacts on nature) of the waste heat needs to be understood. Waste heat can create locally significant impacts by disturbing the local ecosystems in rivers, lakes, and sea bays. In the absence of natural water for cooling we need to use cooling towers or other heat removal systems. In this case, proper water treatment and careful use of the chemicals required in the process are needed.

The renewable energy-specific cases will be explored in the following chapters. We will introduce and discuss the technical applications of each type of renewable energy and detail the potential impacts on society and nature that need to be considered when designing a renewable energy system that also meets social and ecological sustainability and is not simply narrowly optimized from a techno-economic perspective. The benefit of these traditional processes is that they are very durable, and their components are made from common raw materials—mainly steel—so they can be fully recycled. This makes them functional and easily combined with existing and emerging renewable energy technologies as well.

Author contributions 3.1-3.9 Mika Järvinen and Shouzhuang Li, 3.10-3.12 Shouzhuang Li and Mika Järvinen, 3.13 Hanna Paulomäki, and 3.14 Mika Järvinen and Hanna Paulomäki. Subsections can be referenced accordingly.

Appendix A3.1 A system with Number of *N* Indirect Feed-Water Preheaters

In the indirect method, steam is bled from the turbine and this in fed into a shell-tube type heat exchanger. Steam flows on the shell-side and condenses to liquid. The heat released is transferred to the feed-water flowing on the tube side. There are several ways to deal with the condensate flow. For example, the condensate can be cascaded through the lower temperature feed-water preheaters, in a counter-current way and finally—after the last preheater before the condenser—the combined condensate can be fed into the condenser. Unfortunately, the heat in the condensate stream is lost to the environment within the cooling water. The discharge of cooling water can significantly impact the environment. Apart from the emission of heat, cooling waters can contain chemicals and contaminate nature. A much better option is to combine

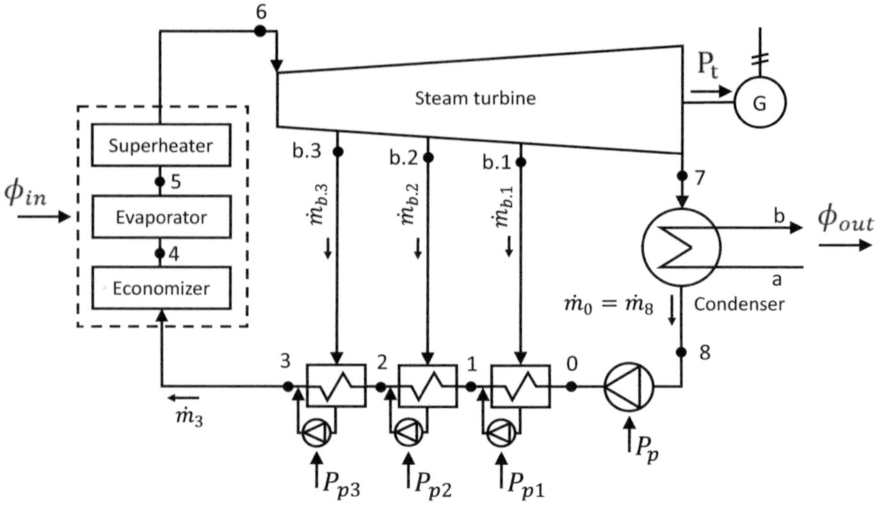

Fig. 3.60 General regenerative feed-water heating system. Subscript *b* refers to bleeding

the final combined condensate into the main feed-water flow with a pump, keeping all the energy inside the process. It is also possible to combine the exit condensate stream after each heater with the main feed-water flow locally. This requires a small booster pump at each heater to pressure the flow to the main feed-water stream, but energy losses are then avoided. In the following we study a case of three regenerative feed-water preheaters. The system has an indirect heat exchanger first, followed by a condensate booster pump (that upgrades the pressure to feed-water pressure before combining it with the main flow upstream). In Fig. 3.60 we follow the approach discussed by Jotuni et al. [29].

The efficiency of this regenerative preheating process can be calculated from the indirect method as follows.

$$\eta = 1 - \frac{(h_7 - h_8)\,\dot{m}_0}{(h_6 - h_3)\,\dot{m}_3} = 1 - \frac{\Delta h_c}{\Delta h_b}\frac{1}{\frac{\dot{m}_3}{\dot{m}_2}\frac{\dot{m}_2}{\dot{m}_1}\frac{\dot{m}_1}{\dot{m}_0}}$$

The mass flow ratio terms can be solved from the energy balances of individual feed-water preheaters. Let us study heater 1 separately first and write the energy and mass balances.

$$\dot{m}_0 + \dot{m}_{b.1} = \dot{m}_1$$

$$\dot{m}_0 h_0 + \dot{m}_{b.1} h_{b.1} = \dot{m}_1 h_1$$

These can be combined simply to solve the ratio of $\dot{m}_1/\dot{m}_0$.

$$\dot{m}_0 h_0 + \dot{m}_{b.1} h_{b.1} = \dot{m}_0 h_1 + \dot{m}_{b.1} h_1$$

$$\frac{\dot{m}_{b.1}}{\dot{m}_0} = \frac{(h_1 - h_0)}{(h_{b.1} - h_1)} = \frac{\dot{m}_1 - \dot{m}_0}{\dot{m}_0} = \frac{\dot{m}_1}{\dot{m}_0} - 1$$

$$\frac{\dot{m}_1}{\dot{m}_0} = 1 + \frac{(h_1 - h_0)}{(h_{b.1} - h_1)} = 1 + \frac{\Delta h_1}{\Delta h_{b.1}}$$

This can also be written for a general heater i as follows.

$$\frac{\dot{m}_i}{\dot{m}_{i-1}} = 1 + \frac{(h_i - h_{i-1})}{(h_{b.i} - h_i)} = 1 + \frac{\Delta h_i}{\Delta h_{b.i}}$$

Here, Δh_i and $\Delta h_{b.i}$ refer to the feed-water enthalpy increase in the preheater i and enthalpy decrease of the bled steam when the flow is combined with the feed-water. This can now be substituted into the equation, and we then get the final equation for the efficiency:

$$\eta = 1 - \frac{\Delta h_c}{\Delta h_b} \frac{1}{\left(1 + \frac{\Delta h_1}{\Delta h_{b.1}}\right)\left(1 + \frac{\Delta h_2}{\Delta h_{b.2}}\right)\left(1 + \frac{\Delta h_3}{\Delta h_{b.3}}\right)}$$

Point 3 before the economizer is now assumed to be fixed for a moment, but we will study later how the solution changes if we make this a free variable and search for the thermodynamically optimum inlet point of the economizer. But first, our task is to find values for the preheater enthalpy increases that maximize the thermal efficiency. In this case, to maximize the efficiency, we need to maximize the function:

$$f(\Delta h_1, \Delta h_2, \Delta h_3) = \left(1 + \frac{\Delta h_1}{\Delta h_{b.1}}\right)\left(1 + \frac{\Delta h_2}{\Delta h_{b.2}}\right)\left(1 + \frac{\Delta h_3}{\Delta h_{b.3}}\right)$$

with the following constraint equation:

$$g(\Delta h_1, \Delta h_2, \Delta h_3) = \Delta h_1 + \Delta h_2 + \Delta h_3 - \Delta h_h$$

To find the optimum enthalpy increases, we introduce the Lagrange function, Råde and Westergren [47]:

$$\begin{aligned} L(\Delta h_1, \Delta h_2, \Delta h_3, \lambda) &= f(\Delta h_1, \Delta h_2, \Delta h_3) + \lambda g(\Delta h_1, \Delta h_2, \Delta h_3) \\ &= \left(1 + \frac{\Delta h_1}{\Delta h_{b.1}}\right)\left(1 + \frac{\Delta h_2}{\Delta h_{b.2}}\right)\left(1 + \frac{\Delta h_3}{\Delta h_{b.3}}\right) + \lambda(\Delta h_1 + \Delta h_2 + \Delta h_3 - \Delta h_h) \end{aligned}$$

For finding the optimum, partial differentiates of L with respect to all variables $(\Delta h_1, \Delta h_2, \Delta h_3, \lambda)$ need to be calculated, and all set equal to 0. The main assumption here is that $\Delta h_{b.i}$ values are not functions of Δh_i. Of course this is not correct if we are being accurate, but serves as a fair approximation, that will be analyzed later.

$$\frac{\partial L}{\partial \Delta h_1} = \frac{1}{\Delta h_{b.1}}\left(1 + \frac{\Delta h_2}{\Delta h_{b.2}}\right)\left(1 + \frac{\Delta h_3}{\Delta h_{b.3}}\right) + \lambda = 0$$

$$\frac{\partial L}{\partial \Delta h_2} = \frac{1}{\Delta h_{b.2}}\left(1 + \frac{\Delta h_1}{\Delta h_{b.1}}\right)\left(1 + \frac{\Delta h_3}{\Delta h_{b.3}}\right) + \lambda = 0$$

$$\frac{\partial L}{\partial \Delta h_3} = \frac{1}{\Delta h_{b.3}}\left(1 + \frac{\Delta h_1}{\Delta h_{b.1}}\right)\left(1 + \frac{\Delta h_2}{\Delta h_{b.2}}\right) + \lambda = 0$$

$$\frac{\partial L}{\partial \lambda} = \Delta h_1 + \Delta h_2 + \Delta h_3 - \Delta h_h = 0$$

These form a linear system of equations, that can be directly solved. We get the following values for the optimal enthalpy increases:

$$\Delta h_1 = \frac{\Delta h_h - 2\Delta h_{b.1} + \Delta h_{b.2} + \Delta h_{b.3}}{3} \approx \frac{\Delta h_h}{3}$$

$$\Delta h_2 = \frac{\Delta h_h + \Delta h_{b.1} - 2\Delta h_{b.2} + \Delta h_{b.3}}{3} \approx \frac{\Delta h_h}{3}$$

$$\Delta h_3 = \frac{\Delta h_h + \Delta h_{b.1} + \Delta h_{b.2} - 2\Delta h_{b.3}}{3} \approx \frac{\Delta h_h}{3}$$

If we assume in the above equations that all values of $\Delta h_{b.i}$ are the same these simplify into $\Delta h_1 = \Delta h_2 = \Delta h_3 = \Delta h = \Delta h_h/3$. This will be compared with numeral values below. But note that when calculating the efficiency, true values of $\Delta h_{b.i}$ are still needed.

But how do we define the inlet condition of the Economizer? For this, we need to let the h_3 value move freely. To simplify our analysis, we assume that all heater enthalpy increases are equal, so $h_3 = h_0 + 3\Delta h$. Substituting these into our previous equation of efficiency gives us:

$$\eta = 1 - \frac{\Delta h_c}{(h_6 - h_0 - 3\Delta h)} \frac{1}{\left(1 + \frac{\Delta h}{\Delta h_{b.1}}\right)\left(1 + \frac{\Delta h}{\Delta h_{b.2}}\right)\left(1 + \frac{\Delta h}{\Delta h_{b.3}}\right)}$$

The maximum of this is achieved when the denominator of the second term gets its maximum. This can be easily found by differentiation of the denominator and setting this to 0 gives us:

$$f(\Delta h) = (h_6 - h_0 - 3\Delta h)\left(1 + \frac{\Delta h}{\Delta h_{b.1}}\right)\left(1 + \frac{\Delta h}{\Delta h_{b.2}}\right)\left(1 + \frac{\Delta h}{\Delta h_{b.3}}\right)$$

$$\frac{df}{d\Delta h} = -3\left(1+\frac{\Delta h}{\Delta h_{b.1}}\right)\left(1+\frac{\Delta h}{\Delta h_{b.2}}\right)\left(1+\frac{\Delta h}{\Delta h_{b.3}}\right) + \frac{(h_6 - h_0 - 3\Delta h)}{\Delta h_{b.1}} \times \left(1+\frac{\Delta h}{\Delta h_{b.2}}\right)\left(1+\frac{\Delta h}{\Delta h_{b.3}}\right) + \frac{(h_6 - h_0 - 3\Delta h)}{\Delta h_{b.2}}\left(1+\frac{\Delta h}{\Delta h_{b.1}}\right)\left(1+\frac{\Delta h}{\Delta h_{b.3}}\right) + \frac{(h_6 - h_0 - 3\Delta h)}{\Delta h_{b.3}}\left(1+\frac{\Delta h}{\Delta h_{b.1}}\right)\left(1+\frac{\Delta h}{\Delta h_{b.2}}\right) = 0$$

$$\frac{\Delta h_{b.1} + \Delta h_{b.2} + \Delta h_{b.3} + 3\Delta h}{(\Delta h_{b.1} + \Delta h)(\Delta h_{b.2} + \Delta h)(\Delta h_{b.3} + \Delta h)} - \frac{3}{(h_6 - h_0 - 3\Delta h)} = 0$$

From which we can solve Δh numerically by iteration. After each iteration cycle values of $\Delta h_{b.i}$ need to be updated. After the solution is found, the optimal enthalpy increase for the economizer can easily be calculated:

$$\Delta h_{ECO} = h_4 - h_0 - 3\Delta h$$

If greater accuracy is required then the complete form, including actual $\Delta h_{b.i}$ values, needs to be used and a numerical solution is required. This general case is briefly discussed below. For a complete solution with true $\Delta h_{b.i}$ values, the efficiency in terms of the Δh_i values can be written as follows:

$$\eta(\Delta h_h) = 1 - \frac{\Delta h_c}{(h_6 - h_0 - \Delta h_h)}\frac{1}{\left(1+\frac{\Delta h_1}{\Delta h_{b.1}}\right)}\frac{1}{\left(1+\frac{\Delta h_2}{\Delta h_{b.2}}\right)}\frac{1}{\left(1+\frac{\Delta h_3}{\Delta h_{b.3}}\right)}$$

This must be optimized in terms of the Δh_1, Δh_2 and Δh_3. As earlier, the maximum of this efficiency is obtained when the denominator is maximized. Our objective function is then:

$$f = (h_6 - h_0 - \Delta h_1 - \Delta h_2 - \Delta h_3)\left(1+\frac{\Delta h_1}{\Delta h_{b.1}}\right)\left(1+\frac{\Delta h_2}{\Delta h_{b.2}}\right)\left(1+\frac{\Delta h_3}{\Delta h_{b.3}}\right)$$

During the iteration the values of all $\Delta h_{b.i}$ are constantly updated with new bleeding pressure values obtained from Δh_i values solved in the previous iteration. To have economically feasible heat exchangers a small minimum temperature difference between the streams, $\varepsilon_i = 4$–5 °C, is needed as discussed above. All $\Delta h_{b.i}$ are calculated as follows. The enthalpy of the bled steam is defined by the turbine expansion curve at bleeding pressure. The exit enthalpies of the feed-water after the preheaters are defined by the saturation temperatures at the bleeding pressures minus the pinch temperatures:

$$\Delta h_{b,1} = h_6 - \eta_s(h_6 - h_s(s_6, p_{b.1})) - h(T(p_{b.1}) - \varepsilon_1, p_0)$$

$$\Delta h_{b,2} = h_6 - \eta_s(h_6 - h_s(s_6, p_{b.2})) - h(T(p_{b.2}) - \varepsilon_2, p_0)$$

$$\Delta h_{b,3} = h_6 - \eta_s(h_6 - h_s(s_6, p_{b.3})) - h(T(p_{b.3}) - \varepsilon_3, p_0)$$

All three of these functions are only variables of three bleeding pressures. Similarly, we could also write Δh_i values in terms of bleeding pressures as follows:

$$\Delta h_1 = h(T(p_{b.1}) - \varepsilon_1, p_0) - h_0$$

$$\Delta h_2 = h(T(p_{b.2}) - \varepsilon_2, p_0) - \Delta h_1 - h_0$$

$$\Delta h_3 = h(T(p_{b.3}) - \varepsilon_3, p_0) - \Delta h_2 - \Delta h_1 - h_0$$

This means that our objective function is only a function of bleeding pressures $f(p_{b.1}, p_{b.2}, p_{b.3})$. If the economizer size is fixed, it also fixes the $p_{b.3}$. But if the economizer size is a free variable, additional constrained equations $p_{b.1} < p_{b.2} < p_{b.3} < p(T(p_0) + \varepsilon_3)$ need to be defined. It is assumed that the bleeding steam is not used to boil any feed-water.

Figure 3.61 presents, as a summary, the calculated efficiency increases for different numbers of preheaters as preheating temperature increases. Enthalpy increases in each preheater are assumed to be the same. The size of the economizer is not fixed. Steam pressure and temperature values are 120 bar and 550 °C, and the condenser pressure is 0.04 bar. All variables are also the same as in the previous examples. The maximum temperature increase is 296 °C since the water should not boil before the evaporator.

With a single heater, the optimal temperature increase in the heater is ~ 50% of the whole range 29…325 °C. When the number of heaters increase, the optimal size of the economizer gets smaller. For example, for a system with 3 and 13 heaters, the optimal preheater enthalpy increase is 316 and 104 kJ/kg, respectively, while the optimal enthalpy increase in the economizer is 409 and 146 kJ/kg, respectively. In practice, the maximum economically feasible number of heaters is 9-11 [29]

Example A3.1 Analyze the above-described power plant with three indirect contact feed-water preheaters of net electrical power of $P_e = 500$ MW. As above, steam values of temperature and pressure before the turbine are 550 °C and 120 bar and the condenser pressure is 0.04 bar. Based on the discussion above, use the optimal size of the economizer—409 kJ/kg—and assume that the total enthalpy increase in the preheating systems is equally divided between the three preheaters: 316 + 316 + 316 kJ/kg. The isentropic efficiency of the turbine and the pump is 0.92 and 0.95, respectively. Note the numbering; the point before the turbine is 6, see Fig. 3.60.

1. Define all process points and calculate the mass flow rate of live steam, fractions a_i and mass flow rates of the bled steam.
2. Calculate the required thermal heat input to the steam generator, and heat output from the condenser.
3. Calculate the mechanical power needed for the feed-water pumps.

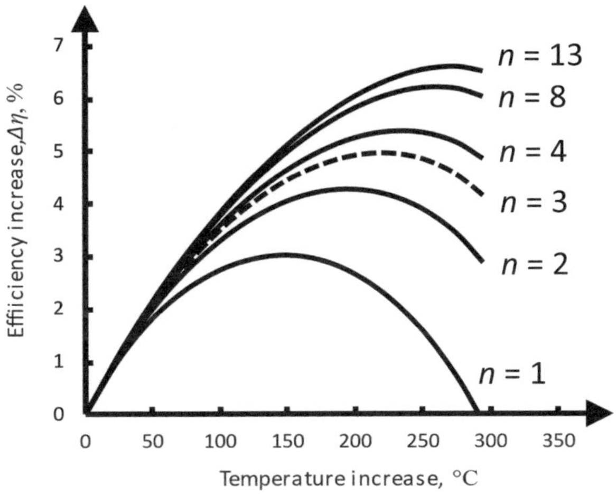

Fig. 3.61 Increase of the thermal efficiency as a function of the number of the preheaters, *n*, and temperature increase in the preheating system. Live steam values are 120 bar 550 °C, condenser pressure 0.04 bar. The boiling temperature at 120 bar is 325 °C and as temperature is 29 °C after the main feed-water pump, maximum temperature increase is 325–29 = 296 °C. The dashed line shows the result for the three-heater system analyzed below

4. Calculate the mechanical power from the steam turbine.
5. Calculate the thermal efficiency of the process. Compare and discuss results with Example 3.6.

The calculation of the process points is shown in detail in Table 3.13. Use the Excel-based tool available in electronic supplement to get the values of physical properties. The enthalpy increase in the main feed-water pump is $\Delta h_P = v_{8-0}(p_0 - p_8) \times 100/\eta_{s.p} = 1.004 \times 10^{-3} \times (120 - 0.04) \times (120 - 0.04) \times 100/0.95 = 12.7\,\text{kJ/kg}$. From point 5 the saturated steam passes through the superheater to point 6. Table 3.14 presents the calculation of three bleeding points. Condensation takes place at the bleeding pressures.

The enthalpy of the bled steam can be directly calculated for the expansion curve using the calculated bleeding pressures:

$$\begin{aligned} h_{b,1} &= h_6 - \eta_s(h_6 - h(s_6, p_{b.1})) = 3482 - 0.92 \times (3482 - 2469) \\ &= 2550\,\text{kJ/kg}. \\ h_{b,2} &= h_6 - \eta_s(h_6 - h(s_6, p_{b.2})) = 3482 - 0.92 \times (3482 - 2831) \\ &= 2883\,\text{kJ/kg}. \\ h_{b,3} &= h_6 - \eta_s(h_6 - h(s_6, p_{b.3})) = 3482 - 0.92 \times (3482 - 3157) \\ &= 3183\,\text{kJ/kg}. \end{aligned}$$

From these, the enthalpy decrease of each bled steam stream can be calculated.

Table 3.13 Calculation of process values of regenerative preheated Rankine cycle, Fig. 3.22

#	p (bar)	t (°C)	T (K)	h (kJ/kg)	s (kJ/kg K)	Comments
6	120	550	823	3482	6.655	Read from the cross section of t_6 and p_6
7 s	0.04	29	302	2004	6.655	Move vertically down from point 6–7s (ds = 0) at p_7
7	0.04	29	302	2123	7.046	Calculate h_7 from the isentropic expansion equation
8	0.04	29	302	121	0.423	Move along the p_7 isobar left to the saturated water border
0	120	29	302	134	0.424	Add enthalpy increase in the pump to get h_0
1	120	105	378	450	1.356	Add $\Delta h_1 = 316\,\text{kJ/kg}$ to h_0, read temperature at (p, h)
2	120	179	522	767	2.120	Add $\Delta h_2 = 316\,\text{kJ/kg}$ to h_1, read temperature at (p, h)
3	120	249	453	1083	2.769	Add $\Delta h_3 = 316\,\text{kJ/kg}$ to h_2, read temperature at (p, h)
4	120	325	456	1491	2.150	Saturated water after the economizer
5	120	325	598	2686	3.497	Saturated steam after the evaporator

Table 3.14 Calculation of the bleeding steam process values; pinch temperature is $\varepsilon = 5$ °C

#	p (bar)	t (°C)	T (K)	$h_{b.i}$ (kJ/kg)	s (kJ/kg K)	Comments
b.1	1.4	110	383	2550	6.867	Condensation pressure is $p\prime(t_1 + \varepsilon)$, $h(p_{b.1})$ from expansion curve
b.2	11.1	225	498	2883	6.762	Condensation pressure is $p\prime(t_2 + \varepsilon)$, $h(p_{b.2})$ as above
b.3	42.8	389	662	3183	6.695	Condensation pressure is $p\prime(t_3 + \varepsilon)$, $h(p_{b.3})$ as above

$$\Delta h_{b.1} = h_{b.1} - h_1 = 2550 - 450 = 2100\,\text{kJ/kg}$$

$$\Delta h_{b.2} = h_{b.2} - h_2 = 2883 - 767 = 2116\,\text{kJ/kg}$$

$$\Delta h_{b.2} = h_{b.3} - h_3 = 3183 - 1083 = 2100\,\text{kJ/kg}$$

As discussed above, these are all close to each other so the assumption of equal $\Delta h_{b.i}$ is justified. The fraction of the bled steam can be calculated as follows:

$$a_i = \frac{\frac{\Delta h_i}{\Delta h_{b.i}}}{1 + \frac{\Delta h_i}{\Delta h_{b.i}}}$$

The following results are obtained: $a_1 = 0.131$, $a_2 = 0.130$, $a_3 = 0.131$. From these, and energy balances of the turbine and pumps, we can solve the live steam mass flow rate and all bled mass flow rates as follows. The enthalpy increase in the booster pumps is calculated as: $\Delta h_{P.1} = v'(p_{b.1})(p_0 - p_{p.1}) \times 100/\eta_{s.p} = 1.001 \times 10^{-3} \times (120 - 1.4) \times 100/0.95 = 12.5\,\text{kJ/kg}$, $\Delta h_{P.2} = v'(p_{b.2})(p_0 - p_{p.2}) \times 100/\eta_{s.p} = 1.001 \times 10^{-3} \times (120 - 11.1) \times 100/0.95 = 11.5\,\text{kJ/kg}$ and $\Delta h_{P.3} = v'(p_{b.1})(p_0 - p_{p.1}) \times 100/\eta_{s.p} = 1.001 \times 10^{-3} \times (120 - 42.8) \times 100/0.95 = 8.1\,\text{kJ/kg}$.

$$\dot{m}_3 = \frac{P_e}{\begin{array}{l}(h_6 - h_{b.3})\eta_g + (h_{b.3} - h_{b.2})(1 - a_3)\eta_g \\ \quad +(h_{b.2} - h_{b.1})(1 - a_3)(1 - a_2)\eta_g \\ \quad +(h_{b.1} - h_7)(1 - a_3)(1 - a_2)(1 - a_1)\eta_g \\ \quad \frac{-\Delta h_p(1-a_3)(1-a_2)(1-a_1)}{\eta_m} \\ \quad \frac{-\Delta h_{p3}a_3}{\eta_m} - \frac{\Delta h_{p2}(1-a_3)a_2}{\eta_m} - \frac{\Delta h_{p1}(1-a_3)(1-a_2)a_1}{\eta_m}\end{array}} = 473\,\text{kg/s}$$

$$\dot{m}_{b.3} = \dot{m}_3 a_3 = 473 \times 0.131 = 61.9\,\text{kg/s}$$

$$\dot{m}_{b.2} = \dot{m}_3(1 - a_3)a_2 = 473 \times (1 - 0.131) \times 0.130 = 53.4\,\text{kg/s}$$

$$\dot{m}_{b.1} = \dot{m}_3(1 - a_3)(1 - a_2)a_1 = 46.8\,\text{kg/s}$$

1. The heat input is calculated from the evaporator energy balance, using the steam side balance.

$$\phi_{in} = \dot{m}_3(h_6 - h_3) = 473 \times (3482 - 1083) = 1134\,\text{MW}$$

The heat output is calculated from the condenser energy balance. Mass flow rate to the condenser is:

$$\dot{m}_0 = (\dot{m}_3 - \dot{m}_{b.3} - \dot{m}_{b.2} - \dot{m}_{b.1}) = (473 - 61.9 - 53.4 - 46.8) = 310.9\,\text{kg/s}.$$

$$\phi_{out} = \dot{m}_0(h_7 - h_8) = 310.9 \times (2123 - 121) = 622\,\text{MW}$$

2. The mechanical power needed for the pumps is calculated as follows:

$$\begin{aligned} P_p &= \dot{m}_0 \Delta h_p + \dot{m}_{b.1}\Delta h_{p.1} + \dot{m}_{b.2}\Delta h_{p.2} + \dot{m}_{b.3}\Delta h_{p.3} \\ &= 310.9 \times 12.7 + 61.9 \times 12.5 + 53.4 \times 11.5 + 46.8 \times 8.1 \\ &= 5.7\,\text{MW} \end{aligned}$$

3. The mechanical power output from the turbine is calculated as follows:

$$P_t = \dot{m}_3(h_6 - h_{b.3}) + \dot{m}_3(1 - a_3)(h_{b.3} - h_{b.2}) + \dot{m}_3(1 - a_3)(1 - a_2)(h_{b.2} - h_{b.1})$$
$$+ \dot{m}_3(1 - a_3)(1 - a_2)(1 - a_1)(h_{b.1} - h_6)473 \times (3482 - 3183)$$
$$+ 473 \times (1 - 0.131) \times (3183 - 2883) + 473 \times (1 - 0.131) \times (1 - 0.130)$$
$$\times (2883 - 2550) + 473 \times (1 - 0.131) \times (1 - 0.130)$$
$$\times (1 - 0.131) \times (2550 - 2123) = 517\,\text{MW}$$

The thermal efficiency is as follows:

$$\eta_t = \frac{P_t - P_p}{\phi_{in}} = \frac{517 - 5.7}{1134} = 0.451$$

The thermal efficiency of the process increases from 40.2 to 45.1%, *i.e.* 4.9% units in comparison with the standard Rankine cycle as already shown above. The regenerative feed-water preheating system requires more steam flow to generate same amount of power, 381 kg/s → 473 kg/s, but still the condenser loss is smaller, 763 MW → 622 MW and the required heat input also decreases (from 1273 to 1135 MW). The system will be much more complex and requires larger system components. However, it has a large potential for efficiency increase, especially when a larger number of heaters are used.

Appendix A3.2 Supercritical and Ultra-supercritical Processes

State of the art modern steam power plants utilize high temperature resisting high strength steels that make it possible to use high live steam values, above the critical point of water; 220 bar and 374 °C. This poses new challenges for the whole system. Firstly, there is no longer a separate evaporation step. As the temperature increases above the critical temperature and the feed-water already has super-critical pressure, there is just a transition to supercritical state. There is no surface tension, no heat of evaporation nor boiling phenomena. Therefore, the whole boiler concept must be different, and the so-called Benson once-through boilers are used.

Next, we will discuss a simplified system that utilizes a steam turbine with inlet pressure and temperature values of 300 bar and 700 °C, similar to the cases discussed by [20] as an example of state-of-the-art technology. Steam expands in a high-pressure turbine (4–5) and is then reheated back to 700 °C (6–7). A fraction of this steam (extracted from a bleeding point in the high-pressure turbine) is bled to the last feed-water heater stage to raise the water temperature to 310 °C. This temperature also defines the pressure of the last bleeding point, $p_{b.3} = p\prime(310\,°\text{C} + \varepsilon_3)$, where ε_3 is the pinch temperature of the heat-exchanger (temperature difference between the saturation temperature and feed-water temperature after the last heater). Re-heated steam then expands in the low-pressure turbine (6–7) and has two bleeding points:

first to the de-aerating open feed-water heater (feed-water tank b_2) and second to the indirect feed-water preheater, b_1. Note that $\varepsilon_2 = 0\,°C$. The condensate after each indirect preheater is pumped to the main flow by booster pumps P_1 and P_2. The main condensate is pumped from the condenser to the first heater by condensate pump P_C and the main feed-water pump is P_M. Index b_i refer to bleeding points. A real power plant would contain many more heaters in the feed-water line, but for simplicity we have here only 1 + 1 + 1 (low pressure, de-aerator and high pressure). As an example, the Lünen Power plant in Germany, which was commissioned 2013 [20], has three high pressure preheaters, five low pressure heaters and one de-aerating direct contact preheater, Gülen [20]. See the description of our simplified system in Fig. 3.62.

We assume that the pinch temperature of the heaters 1 and 3 are 5.6 °C, [20]. The pinch temperature of the feed-water tank is 0 °C. The isentropic efficiencies of all turbines are assumed to be 0.92, and 0.95 for all pumps. Generator and motor efficiencies are assumed to be 0.98. The condenser pressure is set to 0.04 bar. The process is presented in *T-s* and *h–s* diagrams in Fig. 3.63. Note that these also present the case analyzed below and are at true scale.

The efficiency of this reheated and regenerative feed-water preheating process can be expressed using the indirect method as follows. Heat comes into the process via the boiler and re-heater and is only lost in the condenser. As above Δh_c, Δh_b and Δh_r present the enthalpy change in the condenser, the boiler and the re-heater, respectively.

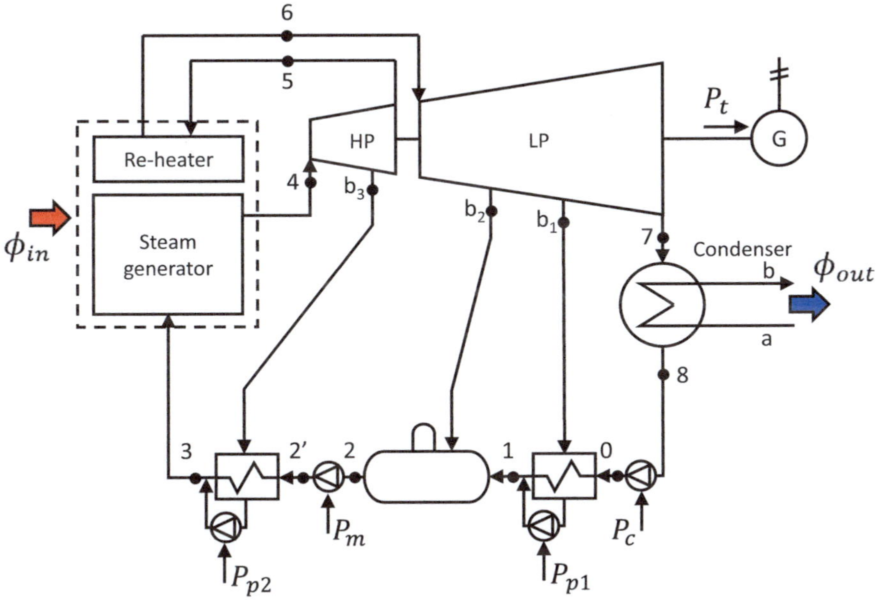

Fig. 3.62 Ultra-super-critical, reheated system with regenerative feed-water preheating

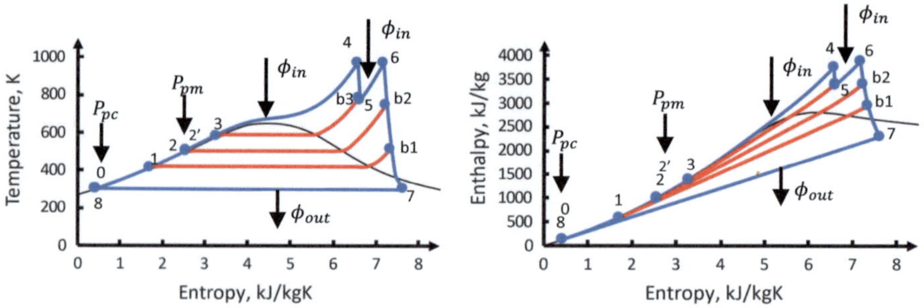

Fig. 3.63 Super-critical, reheated system with regenerative feed-water preheating

$$\eta = 1 - \frac{\Delta h_c \dot{m}_0}{\Delta h_b \dot{m}_3 + \Delta h_r(\dot{m}_3 - \dot{m}_{b.3})} = 1 - \frac{\Delta h_c}{\Delta h_b \frac{\dot{m}_3}{\dot{m}_0} + \Delta h_r \frac{\dot{m}_2}{\dot{m}_0}}$$

Note that $\dot{m}_3 - \dot{m}_{b.3} = \dot{m}_2$. The efficiency can be further rewritten in a similar format as discussed above, in terms of preheater output/input mass flow ratios.

$$\eta = 1 - \frac{\frac{\Delta h_c}{\Delta h_b}}{\frac{\dot{m}_3}{\dot{m}_2}\frac{\dot{m}_2}{\dot{m}_1}\frac{\dot{m}_1}{\dot{m}_0} + \frac{\Delta h_r}{\Delta h_b}\frac{\dot{m}_2}{\dot{m}_1}\frac{\dot{m}_1}{\dot{m}_0}}$$

By inserting here the above derived equation for $\dot{m}_i/\dot{m}_{i-1}$ we have:

$$\eta = 1 - \frac{\frac{\Delta h_c}{\Delta h_b}}{\left(1 + \frac{\Delta h_1}{\Delta h_{b.1}}\right)\left(1 + \frac{\Delta h_2}{\Delta h_{b.2}}\right)\left(1 + \frac{\Delta h_3}{\Delta h_{b.3}}\right) + \frac{\Delta h_r}{\Delta h_b}\left(1 + \frac{\Delta h_1}{\Delta h_{b.1}}\right)\left(1 + \frac{\Delta h_2}{\Delta h_{b.2}}\right)}$$

The maximum efficiency is obtained by setting the denominator as the objective function to be maximized and using the fixed total preheating enthalpy increase Δh_h as the constraint equation. The final preheating temperature is set to 310 °C.

$$f = \left(1 + \frac{\Delta h_1}{\Delta h_{b.1}}\right)\left(1 + \frac{\Delta h_2}{\Delta h_{b.2}}\right)\left(1 + \frac{\Delta h_3}{\Delta h_{b.3}}\right) + \frac{\Delta h_r}{\Delta h_b}\left(1 + \frac{\Delta h_1}{\Delta h_{b.1}}\right)\left(1 + \frac{\Delta h_2}{\Delta h_{b.2}}\right)$$

$$g = \Delta h_1 + \Delta h_2 + \Delta h_3 - \Delta h_h$$

The optimal preheater enthalpy increases can be obtained using the Lagrange's method. To simplify, we mark $\Delta h_{b.1}$, $\Delta h_{b.2}$, $\Delta h_{b.3}$ as *a*, *b* and *c*, respectively. To find the optimum enthalpy increases, we introduce the Lagrange function as follows:

$$L(\Delta h_1, \Delta h_2, \Delta h_3, \lambda) = f(\Delta h_1, \Delta h_2, \Delta h_3) + \lambda g(\Delta h_1, \Delta h_2, \Delta h_3)$$

We need to calculate partial differentiates of L with respect to all variables, here $(\Delta h_1, \Delta h_2, \Delta h_3, \lambda)$ and set these all equal to 0.

$$\frac{\partial L}{\partial \Delta h_1} = \frac{1}{a}\left(1 + \frac{\Delta h_2}{b}\right)\left(1 + \frac{\Delta h_3}{c}\right) + \frac{\Delta h_r}{\Delta h_b}\frac{1}{a}\left(1 + \frac{\Delta h_2}{b}\right) + \lambda = 0$$

$$\frac{\partial L}{\partial \Delta h_2} = \frac{1}{b}\left(1 + \frac{\Delta h_1}{a}\right)\left(1 + \frac{\Delta h_3}{c}\right) + \frac{\Delta h_r}{\Delta h_b}\frac{1}{b}\left(1 + \frac{\Delta h_1}{a}\right) + \lambda = 0$$

$$\frac{\partial L}{\partial \Delta h_3} = \frac{1}{c}\left(1 + \frac{\Delta h_1}{a}\right)\left(1 + \frac{\Delta h_2}{b}\right) + \lambda = 0$$

$$\frac{\partial L}{\partial \lambda} = \Delta h_1 + \Delta h_2 + \Delta h_3 - \Delta h_h = 0$$

These form a linear system of equations, that can be directly solved. The simple algebra is left is for the enthusiastic reader. Values for the optimal enthalpy increases are as follows:

$$\Delta h_1 = \frac{\Delta h_h - 2\Delta h_{b.1} + \Delta h_{b.2} + \Delta h_{b.3}\left(1 + \frac{\Delta h_r}{\Delta h_b}\right)}{3}$$

$$\Delta h_2 = \frac{\Delta h_h + \Delta h_{b.1} - 2\Delta h_{b.2} + \Delta h_{b.3}\left(1 + \frac{\Delta h_r}{\Delta h_b}\right)}{3}$$

$$\Delta h_3 = \frac{\Delta h_h + \Delta h_{b.1} + \Delta h_{b.2} - 2\Delta h_{b.3}\left(1 + \frac{\Delta h_r}{\Delta h_b}\right)}{3}$$

These are similar to a system with no re-heating, except the term $(1 + \Delta h_r/\Delta h_b)$. When using these equations, remember that iteration is required in solving the $\Delta h_i - \Delta h_{b.i} - p_{b.i}$ connection, since there are complex linkages that are not considered in the differentiation. At the end, the optimization is about finding the optimal bleeding pressures that maximize the thermal efficiency.

Example A3.2 Analyze the above-described ultra-supercritical power plant of net electrical power $P_e = 500$ MW. The system has two indirect contact feed-water preheaters and one direct contact preheater/feed-water tank. Steam temperature and pressure values before the turbine are 700 °C and 300 bar and the condenser pressure is 0.04 bar. The power plant is utilizing one preheater above the re-heating pressure, which makes it possible to raise the feed-water temperature up to 310 °C. The steam for this is taken from the high-pressure turbine, before reheating at 100 bar. The isentropic efficiency of the turbine and the pump is 0.92 and 0.95, respectively. Based on derived optimal enthalpy increases we can show that $\Delta h_1 = 459\,\text{kJ/kg}$ and $\Delta h_2 = 386\,\text{kJ/kg}$.

1. Define all process points and calculate the mass flow rate of live steam, fractions a_i and mass flow rates of the bled steam.
2. Calculate the required thermal heat input to the steam generator, and heat output from the condenser.

3. Calculate the mechanical power needed for the feed-water pumps.
4. Calculate the mechanical power from the steam turbine.
5. Calculate the thermal efficiency of the process. Compare and discuss results with Example A3.1.

The calculation of the process points is shown in detail in Table 3.15. Use the Excel-based tool available in electronic supplement to get the properties.

The last preheater enthalpy increase can be calculated by difference, $\Delta h_3 = h_3 - h_{2'} = 378\,\text{kJ/kg}$. Note that enthalpy increase in the first two heaters is given here. If these are unknown, these must be solved first by optimization or set otherwise. From point 3 steam passes through the Benson type of boiler and transition takes place to super-critical state point 4. A major difference with previous cases is that there is no evaporation or superheating steps, just the transition to super-critical state. Table 3.16 presents the calculation of three bleeding points. Condensation takes place at the bleeding pressures.

The enthalpy of the bled steam can be calculated directly from the expansion curve using the calculated bleeding pressures. Note different expansion starting point for b.3.

$$h_{b,1} = h_6 - \eta_s(h_6 - h(s_6, p_{b.1})) = 3870 - 0.92 \times (3870 - 2865)$$

Table 3.15 Calculation of process values of regenerative preheated Rankine cycle, Fig. 3.62

#	p (bar)	t (°C)	T (K)	h (kJ/kg)	s (kJ/kg K)	Comments
0	100	29	302	132	0.424	Add enthalpy increase in the pump to get h_0
1	100	139	412	591	1.718	Add $\Delta h_1 = 459 kJ/kg$ to h_0, read temperature at (p, h)
2	100	227	500	977	2.567	Add $\Delta h_2 = 386 kJ/kg$ to h_1, read temperature at (p, h)
2’	300	231	504	1002	2.570	Add enthalpy increase in the pump to get $h_{2'}$
3	300	310	583	1380	3.265	Pre-defined preheating temperature, $\Delta h_3 = 378\,\text{kJ/kg}$
4	300	700	923	3744	6.560	Read from the cross section of t_4 and p_4
5 s	100	488	761	3345	6.560	Move vertically down from point 4–5 s (ds = 0) at p_5
5	100	501	774	3377	6.602	Calculate h_5 from the isentropic expansion equation
6	100	700	973	3870	7.170	Move along the p_5 isobar to re-heating temperature
7 s	0.04	29	302	2160	7.170	Move vertically down from point 6–7 s (ds = 0) at p_7
7	0.04	29	302	2297	7.623	Calculate h_7 from the isentropic expansion equation
8	0.04	29	302	121	0.423	Follow p_7 isobar left to saturated water

Table 3.16 Calculation of the bleeding steam process values, pinch temperatures are $\varepsilon_1 = 5$ °C, $\varepsilon_2 = 0$ °C and $\varepsilon_3 = 5$ °C

#	p (bar)	t (°C)	T (K)	$h_{b.i}$ (kJ/kg)	s (kJ/kg K)	Comments
b.1	4.1	241	514	2945	7.332	Condensation pressure is $p\prime(t_1 + \varepsilon_1)$, $h_{b.1}$ from expansion curve
b.2	26.4	475	748	3404	7.224	Feed-water tank pressure is $p\prime(t_2)$, $h_{b.2}$ from expansion curve
b.3	106.4	511	784	3395	6.599	Condensation pressure is $p\prime(t_3 + \varepsilon_3)$, $h_{b.3}$ from expansion curve

$$
\begin{aligned}
&= 2945\,\text{kJ/kg}.\\
h_{b,2} &= h_6 - \eta_s(h_6 - h(s_6, p_{b.2})) = 3870 - 0.92 \times (3870 - 3364)\\
&= 3404\,\text{kJ/kg}.\\
h_{b,3} &= h_4 - \eta_s(h_4 - h(s_4, p_{b.3})) = 3744 - 0.92 \times (3744 - 3365)\\
&= 3395\,\text{kJ/kg}.
\end{aligned}
$$

From these, the enthalpy decreases of each bled steam stream can be calculated:

$$
\begin{aligned}
\Delta h_{b.1} &= h_{b.1} - h_1 = 2945 - 591 = 2354\,\text{kJ/kg}\\
\Delta h_{b.2} &= h_{b.2} - h_2 = 3404 - 977 = 2427\,\text{kJ/kg}\\
\Delta h_{b.2} &= h_{b.3} - h_3 = 3395 - 1380 = 2015\,\text{kJ/kg}
\end{aligned}
$$

Note that if all $\Delta h_{b.i}$ are assumed equal this would result in a significant error. Therefore, a solution for the complete system is required. The fraction of bled steam can be calculated as follows:

$$
a_i = \frac{\frac{\Delta h_i}{\Delta h_{b.i}}}{1 + \frac{\Delta h_i}{\Delta h_{b.i}}}
$$

We have $a_1 = 0.1631$, $a_2 = 0.1373$, $a_3 = 0.1579$. From these and energy balances of the turbine and pumps, the live steam mass flow rate and all bled mass flow rates can be solved:

$$\dot{m}_3 = \frac{P_e}{\begin{array}{l}(h_4 - h_{b.3})\eta_g + (1-a_3)(h_{b.3} - h_5)\eta_g \\ +(1-a_3)(h_6 - h_{b.2})\eta_g \\ +(1-a_3)(1-a_2)(h_{b.2} - h_{b.1})\eta_g \\ +(1-a_3)(1-a_2)(1-a_1)(h_{b.1} - h_7)\eta_g \\ \dfrac{-(1-a_3)(1-a_2)(1-a_1)\Delta h_{pc}}{\eta_m} \\ \dfrac{-(1-a_3)(1-a_2)a_1\Delta h_{p1}}{\eta_m} - \dfrac{(1-a_3)\Delta h_{pm}}{\eta_m} - \dfrac{a_3\Delta h_{p2}}{\eta_m}\end{array}} = 352\,\text{kg/s}$$

$$\dot{m}_{b.3} = \dot{m}_3 a_3 = 352 \times 0.1579 = 55.6\,\text{kg/s}$$

$$\dot{m}_{b.2} = \dot{m}_3(1-a_3)a_2 = 352 \times (1 - 0.1579) \times 0.1373 = 40.7\,\text{kg/s}$$

$$\dot{m}_{b.1} = \dot{m}_3(1-a_3)(1-a_2)a_1 = 41.7\,\text{kg/s}$$

1. The heat input is calculated from the steam generator energy balance, using the steam side energy balance.

$$\phi = \dot{m}_3(h_4 - h_3) + (\dot{m}_3 - \dot{m}_{b.3})(h_6 - h_5) = 352 \times (3744 - 1380) \\ + (352 - 55.6)(3870 - 3377) = 978\,\text{MW}$$

The heat output is calculated from the condenser energy balance. The mass flow rate to the condenser is:

$$\dot{m}_0 = (\dot{m}_3 - \dot{m}_{b.3} - \dot{m}_{b.2} - \dot{m}_{b.1}) = (352 - 55.6 - 40.7 - 41.7) = 214\,\text{kg/s}.$$

$$\phi_{out} = \dot{m}_0(h_7 - h_8) = 214 \times (2123 - 121) = 428\,\text{MW}$$

2. The mechanical power needed for the pumps is calculated next. Enthalpy increases for the condensate pump and the main feed-water pumps are simply calculated by enthalpy differences, $\Delta h_{pc} = h_0 - h_8$ and, $\Delta h_{pm} = h_{2'} - h_2$. The calculation of booster pumps, P1 and P2, requires some additional discussion. If we are to be accurate, we must remember that these were neglected from the energy balances of the heaters, and it was assumed that enthalpy increase in the pump was negligible in comparison with the latent heat released. Here, we want to know how much power they consume. The enthalpy increases are calculated as follows. The saturated value for specific volumes can be used at inlet temperatures.

$$\Delta h_{p1} = \frac{v'(T_{b.1})(p_1 - p_{b.1})}{\eta_p} = \frac{1.08 \times 10^{-3} \times 100 \times (100 - 4.1)}{0.95} = 10.9\,\text{kJ/kg}$$

$$\Delta h_{p2} = \frac{v'(T_{b.2})(p_3 - p_{b.3})}{\eta_p} = \frac{1.48 \times 10^{-3} \times 100 \times (300 - 106.4)}{0.95} = 30.1\,\text{kJ/kg}$$

the total power needed for the pumps is therefore as follows:

$$\begin{aligned} P_p &= \dot{m}_0 \Delta h_{pc} + \dot{m}_{b.1} \Delta h_{p.1} + \dot{m}_2 \Delta h_{pm} + \dot{m}_{b.3} \Delta h_{p2} \\ &= 214 \times 11 + 41.7 \times 10.9 + 296 \times 25 + 55.6 \times 30.1 \\ &= 11.9\,\text{MW} \end{aligned}$$

3. The mechanical power output from the turbine is calculated as follows:

$$\begin{aligned} P_t &= \dot{m}_3(h_4 - h_{b.3}) + \dot{m}_3(1 - a_3)(h_{b.3} - h_5) \\ &\quad + \dot{m}_3(1 - a_3)(h_6 - h_{b.2}) \\ &\quad + \dot{m}_3(1 - a_3)(1 - a_2)(h_{b.2} - h_{b.1}) \\ &\quad + \dot{m}_3(1 - a_3)(1 - a_2)(1 - a_1)(h_{b.1} - h_7)352 \\ &\quad \times (3744 - 3395) + 352 \times (1 - 0.1579) \\ &\quad \times (3395 - 3377) + 352 \times (1 - 0.1579) \\ &\quad \times (3870 - 3404) + 352 \times (1 - 0.1579) \\ &\quad \times (1 - 0.1373) \times (3404 - 2945) + 352 \times (1 - 0.1579) \\ &\quad \times (1 - 0.1373) \times (1 - 0.1631) \times (2945 - 2297) \\ &= 523\,\text{MW} \end{aligned}$$

4. The thermal efficiency is calculated as follows:

$$\eta_t = \frac{P_t - P_p}{\phi_{in}} = \frac{523 - 11.9}{978} = 0.524$$

The thermal efficiency of the process increases from 40.2 to 52.4%; 12.2% units in comparison with the standard Rankine cycle as already shown above. Therefore, ultra-supercritical process can generate the same amount of power with significantly smaller amount of steam and heat input.

"Author contributions: 3.1-3.9 Mika Järvinen and Shouzhuang Li, 3.10-3.12 Shouzhuang Li and Mika Järvinen, 3.13 Hanna Paulomäki, and 3.14 Mika Järvinen and Hanna Paulomäki. Subsections can be referenced accordingly."

References

1. Abbas M, Merzouk NK, Belgroun Z, Tigrine Z, Aburidah H (2013) Parametric Study of the installation of a Solar Power Tower plant under Saharan Climate of Algeria: case study

of Tamanrasset. In: Proceedings of the first international conference on nanoelectronics, communications and renewable energy. ISBN: 978-81-925233-8-5
2. Aghahosseini A, Breyer C (2020) From hot rock to useful energy: a global estimate of enhanced geothermal systems potential. Appl Energy 279:115769. https://doi.org/10.1016/j.apenergy.2020.115769
3. Aljaerani HA, Samykano M, Pandey AK, Kadirgama K, Saidur R (2020) Thermo-physical properties and corrosivity improvement of molten salts by use of nanoparticles for concentrated solar power applications: a critical review. J Mol Liq 314:113807. https://doi.org/10.1016/j.molliq.2020.113807
4. Barasa PJ, Magut PKS (2018) Environmental risk assessment: case study of Eburru geothermal wellhead power plant. Int J Dev Sustain 7(10):2570–2584. https://isdsnet.com/ijds-v7n10-20.pdf
5. Barasa PJ, Mathenge RW (2015) Stakeholder engagement through participatory research: a case study of Eburru geothermal wellhead generator in Nakuru County, Kenya. GRC Trans 39:233–238. http://pubs.geothermal-library.org/lib/grc/1032155.pdf
6. Bastida F, Moreno L, Hernández T, García C (2006) Microbiological activity in a soil 15 years after its de-vegetation. Soil Biol Biochem 38(8):2503–2507
7. Bennun L, van Bochove J, Ng C, Fletcher C, Wilson D, Phair N, Carbone G (2021) Mitigating biodiversity impacts associated with solar and wind energy development. Guidelines for project developers. Gland, Switzerland, IUCN and Cambridge UK: The Biodiversity Consultancy 260 pp. https://doi.org/10.2305/IUCN.CH.2021.04.en
8. Bielecki A, Ernst S, Skrodzka W, Wojnicki I (2019) Concentrated solar power plants with molten salt storage: economic aspects and perspectives in the European Union. Int J Photoenergy 1–10. https://doi.org/10.1155/2019/8796814
9. Cao XQ, Yang WW, Zhou F, He YL (2014) Performance analysis of different high-temperature heat pump systems for low-grade waste heat recovery. Appl Therm Eng 71(1):291–300. https://doi.org/10.1016/j.applthermaleng.2014.06.049
10. Caraballo A, Galán-Casado S, Caballero Á, Serena S (2021) Molten salts for sensible thermal energy storage: a review and an energy performance analysis. Energies 14:1197. https://doi.org/10.3390/en14041197
11. Cardemil JM, Cortés F, Díaz A, Escobar R (2016) Thermodynamic evaluation of solar-geothermal hybrid power plants in northern Chile. Energy Convers Manag 123:348–361. https://doi.org/10.1016/j.enconman.2016.06.032
12. Demir ME, Dincer I (2018) Development and analysis of a new integrated solar energy system with thermal storage for fresh water and power production. Int J Energy Res 42(9):2864–2874. https://doi.org/10.1002/er.3846
13. Dincer I (2017) Refrigeration systems and applications. Wiley
14. Eden Geothermal (2024) https://www.edengeothermal.com/about/geothermal-energy/worldwide-potential/
15. El-Wakil MM (1984) Powerplant technology. McGraw-Hill, Singapore. ISBN-0-07-066274-6
16. Fluri TP (2009) The potential of concentrating solar power in South Africa. Energy Policy 37(12):5075–5080. https://doi.org/10.1016/j.enpol.2009.07.017
17. Gaur AS, Fitiwi DZ, Curtis J (2021) Heat pumps and our low-carbon future: a comprehensive review. Energy Res Soc Sci 71:101764
18. Global Solar Atlas (2024)
19. Grippo M, Hayse JW, O'Connor BL (2015) Solar energy development and aquatic ecosystems in the Southwestern United States: potential impacts, mitigation, and research needs. Environ Assess 55:244–256. https://doi.org/10.1007/s00267-014-0384-x
20. Gülen SC (2021) Steam turbine-quo Vadis? Front Energy Res 8
21. He YL, Qiu Y, Wang K, Yuan F, Wang WQ, Li MJ, Guo JQ (2020) Perspective of concentrating solar power. Energy 198:117373. https://doi.org/10.1016/j.energy.2020.117373
22. Hernandez RR, Easter SB, Murphy-Mariscal ML, Maestre FT, Tavassoli M, Allen EB, Barrows CW, Belnap J, Ochoa-Hueso R, Ravian S, Allen MF (2014) Environmental impacts of utility-scale solar energy. Renew Sust Energ Rev 29:766–779. https://doi.org/10.1016/j.rser.2013.08.041

23. Ho CK (2016) Review of avian mortality studies at concentrating solar power plants. AIP Conf Proc 1734:070017. https://doi.org/10.1063/1.4949164
24. Ho CK, Iverson BD (2014) Review of high-temperature central receiver designs for concentrating solar power. Renew Sustain Energy Rev 29:835–846. https://doi.org/10.1016/j.rser.2013.08.099
25. Horlock JH (1992) Combined power plants: including combined cycle gas turbine plants. Pergamon Press, Great Britain. ISBN 0-08-040502-9
26. IEA (2020) World energy outlook special report, outlook for biogas and biomethane—prospects for organic growth. IEA Publications, International Energy Agency, Paris
27. Incropera FP, DeWitt DP (1990) Fundamentals of heat and mass transfer. Wiley, Singapore
28. Jeal C, Perold V, Ralston-Paton S, Ryan PG (2019) Impacts of a concentrated solar power trough facility on birds and other wildlife in South Africa. Ostrich 90(2):129–137. https://doi.org/10.2989/00306525.2019.1581296
29. Jotuni P, Ryti H and Pöyhönen O (1966) Handbook of technology 2, E. thermodynamics (In Finnish: Tekniikan käsikirja 2, E. Termodynamiikka), Gummerus, Jyväskylä, Finland
30. Kabeyi MJB, Olanrewaju OA (2022) Geothermal wellhead technology power plants in grid electricity generation: a review. Energ Strat Rev 39:100735. https://doi.org/10.1016/j.esr.2021.100735
31. Kabeyi MJB (2019) Geothermal electricity generation, challenges, opportunities and recommendations. IJASRE 5(8):53–95. https://doi.org/10.31695/IJASRE.2019.33408
32. Kagan RA, Viner TC, Trail PW, Espinoza EO (2014) Avian mortality at solar energy facilities in Southern California: a preliminary analysis. National Fish and Wildlife Forensics. https://www.ourenergypolicy.org/wp-content/uploads/2014/04/avian-mortality.pdf
33. Khartchenko NV, Khartchenko VM (2014) Advanced energy systems. CRC Press. ISBN-13:978-1-4398-8658-8 (Taylor & Francis, Boka Raton, USA)
34. Li J, Okin GS, Alvarez L, Epstein H (2007) Quantitative effects of vegetation cover on wind erosion and soil nutrient loss in a desert grassland of southern New Mexico, USA. Biogeochemistry 85(3):317–332. https://doi.org/10.1007/s10533-007-9142-y
35. Lim T, Baik YK, Kim DD (2020) Heating performance analysis of an air-to-water heat pump using underground air for greenhouse farming. Energies 13(15):3863. https://doi.org/10.3390/en13153863
36. Lovich JE, Ennen JR (2011) Wildlife conservation and solar energy development in the Desert Southwest, United States. Bioscience 61:982–992. https://doi.org/10.1525/bio.2011.61.12.8
37. Merchán RP, Santos MJ, Medina A, Hernández AC (2022) High temperature central tower plants for concentrated solar power: 2021 overview. Renew Sustain Energy Rev 155:111828. https://doi.org/10.1016/j.rser.2021.111828
38. Mizuno E (2013) Geothermal power development in New Zealand—lessons for Japan. Japan Renewable Energy Foundation. https://www.renewable-ei.org/en/images/pdf/20130220/20120912_lessonfromNewZealand.pdf
39. Moran MJ, Shapiro HN (1993) Fundamental of engineering thermodynamics. Wiley, USA. ISBN 0-471-59275-7
40. Moya D, Aldás C, Kaparaju P (2018) Geothermal energy: power plant technology and direct heat applications. Renew Sustain Energy Rev 94:889–901. https://doi.org/10.1016/j.rser.2018.06.047
41. Nagdi K (1993) Rubber as an engineering material: guideline for users. Hanser Publishers, Munich
42. Noone CJ, Torrilhon M, Mitsos A (2012) Heliostat field optimization: a new computationally efficient model and biomimetic layout. Sol Energy 86(2):792–803. https://doi.org/10.1016/j.solener.2011.12.007
43. Organ JA (2013) Stirling cycle engines. Wiley, ISBN, p 9781118818404
44. Perez M, Perez R (2022) Update 2022—a fundamental look at supply side energy reserves for the planet. Solar Energy Adv 2:1–7. https://doi.org/10.1016/j.seja.2022.100014
45. Pfahl A, Coventry J, Röger M, Wolfertstetter F, Vásquez-Arango JF, Gross F, Liedke P (2017) Progress in heliostat development. Sol Energy 152:3–37. https://doi.org/10.1016/j.solener.2017.03.029

46. Rizvi AA, Danish SN, El-Leathy A, Al-Ansary H, Yang D (2021) A review and classification of layouts and optimization techniques used in design of heliostat fields in solar central receiver systems. Sol Energy 218:296–311. https://doi.org/10.1016/j.solener.2021.02.011
47. Råde L, Westergren B (1990) Beta mathematics handbook. 2nd edn. Chartwell-Bratt, Lund. ISBN 0-86238-140-1
48. Savitha DC, Ranjith PK, Talawar B, Rana Pratap Reddy N (2022) Refrigerants for sustainable environment–a literature review. Int J Sustain Energy 41(3):235–256. https://doi.org/10.1080/14786451.2021.1928129
49. Sharmin T, Khan NR, Akram MS, Ehsan MM (2023) A State-of-the-art review on for geothermal energy extraction, utilization, and improvement strategies: conventional, hybridized, and enhanced geothermal systems. Int J Thermofluids 100323. https://doi.org/10.1016/j.ijft.2023.100323
50. Siemens Energy (2021) Siemens Energy gas turbine portfolio. https://www.siemens-energy.com/gasturbines
51. Strauss K (1998) Kraftwerkstechnik: zur Nutzung fossiler, regenerativer and nuklearer Energiequellen, 4th edn. Springer, Berlin
52. Streeter VL, Wylie EB (1983) Fluid mechanics. McGraw-Hill, Singapore. ISBN 0-07-548015-8
53. Traupel W (1077) Thermische Turbomachinen, 3rd edn. Springer, Berlin. ISBN 3-540-07939-4
54. United Nations (2021) Life cycle assessment of electricity generation options, United Nations economic commission for Europe, UNECE, GENEVA. https://unece.org/sites/default/files/2022-04/LCA_3_FINAL%20March%202022.pdf
55. VDI Heat Atlas (2010) VDI-Buch, vol 2010, Springer. ISBN: 978-3-540-77876-9
56. Von Cube HL, Steimle F (2013) Heat pump technology. Elsevier. ISBN: 9781483102474
57. Walston LJ, Rollins KE, LaGory KE, Smith KP, Meyers SA (2016) A preliminary assessment of avian mortality at utility-scale solar energy facilities in the United States. Renew Energy 92:405–414. https://doi.org/10.1016/j.renene.2016.02.041
58. Webb RH, Leake SA (2006) Ground-water surface-water interactions and long-term change in riverine riparian vegetation in the southwestern United States. J Hydrol 320(3–4):302–323. https://doi.org/10.1016/j.jhydrol.2005.07.022
59. Windrem PF, Marr GL (1982) Environmental problems and geothermal permitting. Nat Resour Lawyer 14(4):675–685. https://www.jstor.org/stable/40922700
60. Zhao Y, Wang J (2016) Exergo-economic analysis and optimization of a flash-binary geothermal power system. Appl Energy 179:159–170. https://doi.org/10.1016/j.apenergy.2016.06.108

Chapter 4
Solar Energy

Mika Järvinen, Hanna Paulomäki, Eetu Laitila, Georgi Popov, Saeed Rahimpour Golroudbary, Mari Lundström, Benjamin P. Wilson, Roosa Talala, Ayesha Sadiqa, Eeva-Lotta Apajalahti, and Alicja Dańkowska

I'd put my money on the sun and solar energy, what a source of power. I hope we don't have to wait until oil and coal run out before we tackle that.
—Thomas Edison, 1931

Supplementary Information The online version contains supplementary material available at https://doi.org/10.1007/978-3-031-69856-9_4.

M. Järvinen (✉) · H. Paulomäki · S. R. Golroudbary · M. Lundström · B. P. Wilson
Aalto University, Espoo, Finland
e-mail: mika.jarvinen@aalto.fi

H. Paulomäki
e-mail: hanna.paulomaki@lut.fi

E. Laitila
Finland, Helsinki, Finland

G. Popov
Helsinki University, Helsinki, Finland

R. Talala · A. Sadiqa · E.-L. Apajalahti · A. Dańkowska
LUT University, Lappeenranta, Finland

M. Järvinen and H. Paulomäki (eds.), *Designing Renewable Energy Systems within Planetary Boundaries*, Green Energy and Technology,
https://doi.org/10.1007/978-3-031-69856-9_4

4.1 Introduction

Energy from the Sun drives most processes on Earth, such as photosynthesis, climate and weather systems, the water cycle and the global circulation of winds, and forms the basis of life on Earth. The continuous flow of solar energy, at a rate of more than 174,000 TW, is spread across the hemisphere facing the Sun, while the hemisphere facing away from the Sun cools by radiating thermal energy outwards. The highest energy density is received at the equator where the sun shines from the *zenith*, directly overhead. As we move further away from the equator towards the poles, the path of the Sun's rays through the atmosphere gets longer, and the irradiation spreads over a larger area of the Earth's surface. This means the most optimal conditions for solar energy generation are towards the equator—even so, solar PV can be economically feasible even at latitudes higher than 60°.

Based on the energy resource analysis of Perez and Perez [152], solar energy is the most abundant energy source in the world. In their fully electrified scenario, solar PV power plants could theoretically generate on average ~ 2, 400,000 TWh annually over the next 30 years, around 13 times the total global energy consumption in 2022 (~180,000 TWh). Other calculations, for example by de Vries [205, 143], make similar predictions. Solar PV power plants already exist at the industrial-level GW scale—in 2023, the largest installation was the 2.25 GW peak power Bhadla Power plant in Rajasthan, India. Its grand scale is shown from above in Fig. 4.1.

Fig. 4.1 The world's largest solar PV power plant in Bhadla village in the Jodhpur district of Rajasthan, India, 2023. Installed peak power is 2.25 GW_p and the power plant spreads over an area of 56.7 km^2. © Business Wire

This chapter focuses on solar photovoltaic (PV) systems. However, sections discussing solar resources apply to thermal technologies as well. Thermal concentrated power plants (CSP) based on traditional heat engines were discussed in Chap. 3.

As long ago as 1839, Ernest Becquerel discovered that illuminating one of a pair of metal plates immersed in a dilute acid solution affected the electromotive force (emf) between the plates. In 1876 Adams and Day also observed that emf was produced by illuminating a junction of selenium and platinum. In modern photovoltaic cells, emf is produced when solar irradiation is absorbed by a semiconductor material containing a positive–negative (p-n) junction. As will be discussed in Sect. 4.6, several semiconductor materials can be successively used, such as silicon and germanium.

The first real silicon solar cell was developed at Bell Laboratories in 1953–1954, by Gerald Pearson, Daryl Chapin and Calvin Fuller. At that time, the production costs were astronomical and, as a result, there were no reasonable or economically attractive applications available. 1955, the US Government announced its intention to launch a satellite using solar energy as its power source, and the first real-world application of silicon solar cells was in 1958, as a power source for the radio transmitter of the USA's first satellite, Vanguard 1. By 1962, the Mariner 2 carried 10,700 solar cells to operate its electrical systems. By that stage, solar cell efficiency was around 10%, already approaching one-third of their theoretical maximum [28, 38].

Since the 1950s, solar cells and their manufacturing technologies have developed extremely quickly. Based on Ember [51], 4.6% of world's electricity was produced by solar panels 2022. Today China is by far the largest producer of solar panels, with annual available production capacity of more than 500 GW of panels per year—about 80% of global production capacity. However, only 40% of this manufacturing capacity was exploited in 2022, showing that there is huge potential to increase panel production if installations increase accordingly. In their Net Zero Emission scenario, IEA [87] predicts that power production from PV panels could increase from 1300 TWh in 2022 to around 30,000 TWh (corresponding to ~ 19 GW installed capacity) by 2050.

Solar energy production has many environmental and socio-economic benefits over the use of fossil fuels, namely those related to reduction of greenhouse gas emissions (CO_2, NO_x) and other pollution (SO_2, particulate emissions). Solar PV is also suitable for installation into human infrastructure, such as rooftops, noise barriers, and airports, or they can be installed on former industrial land or other polluted areas, and coupled with the rehabilitation of degraded land. For societies, solar power improves energy independence, diversifies energy supply and security, and creates new jobs, among other benefits [194].

However, as with any human activity, production of solar energy can also create adverse and unwanted impacts, both on nature and wildlife, and on communities. The main onsite impacts of solar PV development results from land use and land-use change. A solar PV power plant requires relatively large amounts of space, resulting in changes in the landscape and the local natural environment. The impacts of construction are dependent on local conditions, the nature and wildlife present in the area, and distance from nature conservation areas and sensitive ecosystems. Impacts are often less severe in landscapes previously modified for human use. The more valuable the construction area is from a natural perspective, the greater is the need to avoid, mitigate or offset negative impacts, to achieve a goal of "no net loss of biodiversity".

Clean energy production is growing rapidly, and clean energy solutions are needed to tackle the climate change; at the same time, it is vital that the on-going energy transition does not compromise efforts to reach nature-conservation and social-sustainability goals. As we discussed in Chaps. 1 and 2, climate, nature and societies are closely interconnected. Healthy and resilient ecosystems remove and store carbon, while degraded lands can become a significant source of land-based greenhouse gas emissions. Therefore, when planning large-scale solar PV power plants, the development team should consist of experts from different fields, including biologists familiar with the local environment. On agricultural lands, solar power development can displace food production [78]; combined agricultural/solar energy production sites, known as agrivoltaics, have been developed and should be considered [46, 57].

This chapter discusses the principles of harnessing solar energy, the impacts, both positive and negative, of large-scale solar power plants on nature and society, as well as some ways to alleviate the negative impacts. First, we will discuss the Earth's solar resources, irradiance and atmospheric attenuation, and how the energy is captured by fixed or moving PV panels, or thermal collectors. We then discuss the operational principle of a photovoltaic panel and how it can be mathematically modelled using an equivalent circuit. We also analyze the effect of operating conditions, such as atmospheric and panel temperatures, the level of irradiance, and possible shading effects. We discuss broadly the design of industrial-scale solar power plants, their impacts on nature and society, and how to avoid and minimize these negative impacts. Finally, we discuss research developments and future directions in PV panel technologies.

4.2 Solar Resource Analysis

We start our analysis by discussing the irradiance reaching the outer parts of Earth's atmosphere, and then analyze the fraction that reaches the ground. As discussed already in Chap. 2, the Sun's surface corresponds to a black-body emitter at ~ 5800 K. Energy from the Sun spreads spherically in all directions, and when it reaches the Earth, its intensity has significantly decreased due to this dilution effect. Planck's law gives the spectral distribution of the emissive power *at the surface* of the black-body emitter, or the spectral emissivity, as

$$E_{b.\lambda}(T) = \frac{C_1}{\exp(C_2/\lambda T) - 1}$$

where $C_1 = 2\pi hc^2$ and $C_2 = hc/k$, $h = 6.62607 \times 10^{-34}$Js is Planck's constant, $c = 2.9979 \times 10^9$m/s is the speed of light in vacuum, and $k = 1.38065 \times 10^{-23}$ J/K is Boltzmann's constant. The constants are then $C_1 = 3.742 \times 10^8$ W μm^4/m^2 and $C_2 = 14388$ μm/K, respectively (taking care with conversions between different units). To obtain the spectral irradiance reaching Earth's atmosphere, G_λ, we multiply the emissive power at Sun's surface, $E_{b.\lambda}(T)$, , by the ratio of the Sun's surface area (radius $r_S = 695.7 \times 10^6$m) to the area of a sphere of radius equal to the Earth's distance from Sun (radius $s = 149.747 \times 10^9$m):

$$G_\lambda = E_{b.\lambda}(T)\left(\frac{r_S}{s + r_S}\right)^2$$

The total irradiance below wavelength λ can be obtained by integrating this equation from 0 to λ,

$$G_{0-\lambda} = \int_0^\lambda E_{b.\lambda}(T)\left(\frac{r_S}{s + r_S}\right)^2 d\lambda$$

Total irradiation over the whole solar spectrum is obtained by integrating up to $\lambda = \infty$, and using the Stefan-Boltzmann law, $E_b = \sigma T^4$, where $\sigma = 5.67 \times 10^{-8}$ W/m^2 K^4 is the Stefan-Boltzmann constant. Both the spectral irradiance and its integral are presented in Fig. 4.2 for both extraterrestrial irradiance (AM0, air mass zero) and irradiance reaching ground level on a clear day at a reference angle of 41.8° declination (AM1.5, air mass 1.5). The energy spectrum is divided into ultraviolet (0–0.38 μm), visible (0.38–0.78 μm) and infrared (0.78–3 μm) regions.

Today we know that the behavior of light is characterized by the properties of both waves and particles, known as wave-particle duality. Therefore, sunlight can be treated as electromagnetic waves composed of photons with different energies traveling at constant velocity, the speed of light, c. The energy of a photon is inversely proportional to its wavelength λ, $E = hc/\lambda$. Inside a mono-crystalline silicon solar cell, for an absorbed photon to excite an electron to jump across the P–N junction, they must have energy larger than 1.11 eV or 1.778×10^{-19} J (1 eV = 1.6×10^{-19} J), corresponding to wavelengths smaller than 1.12 μm [66]. This limit is presented as a red thick vertical line in Fig. 4.2.

As Fig. 4.2 shows, Planck's law approximates the integral of the emissive power very well. Sunlight is significantly attenuated by different mechanisms during its passage through the atmosphere. Scattering by molecules attenuates sunlight across the entire spectrum, but mostly at short wavelengths. Sunlight is also attenuated by scattering from aerosols and dust particles. Molecular absorption by H_2O, CO_2 and O_3 reduces the irradiance at certain wavelength ranges, the absorption bands, also shown in the Fig. 4.2. As we discuss later, solar panels convert only a fraction of this solar energy to electrical power, and the rest is converted to heat.

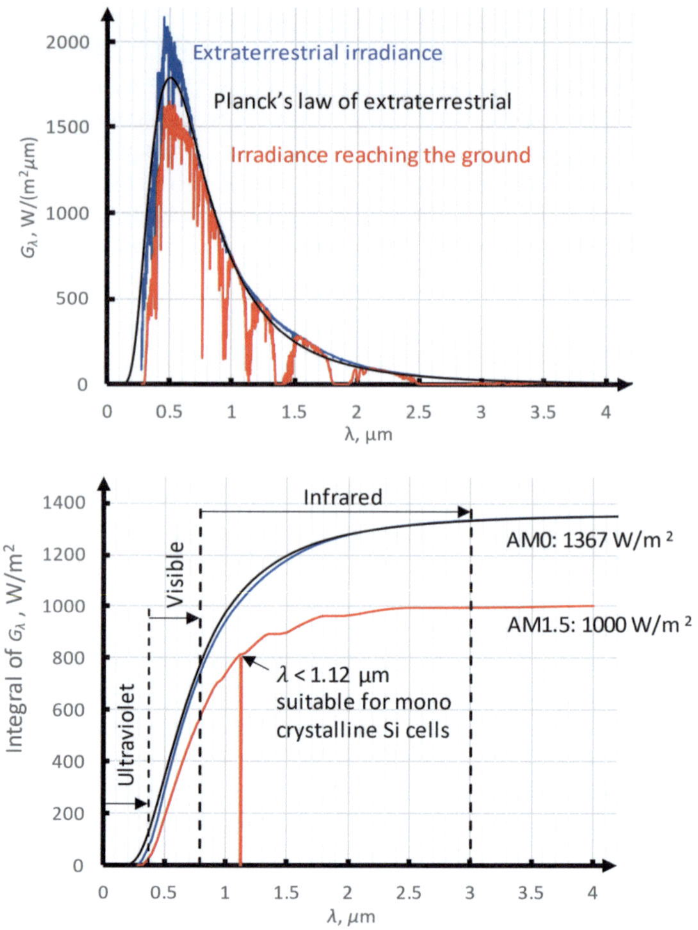

Fig. 4.2 Solar spectral irradiance (upper figure), and integrated irradiance up to wavelength λ (lower figure). Blue curves are the measured value outside the atmosphere (AM0) and black curve is the Planck's law approximation. Red curves correspond to power reaching the Earth's surface on a clear day at a declination angle of 41.8° (AM1.5)

4.2.1 Sun-Earth Geometry

To better understand the solar resource and optimal installation direction of solar panels, it is important to know the geometry of the Sun-Earth system. Figures 4.3 and 4.4 presents the orientation of the Earth around the Sun at the summer solstice, the autumn equinox, the winter solstice and the spring equinox.

The angle between the planes of the Ecliptica (Earth's rotational plane around the Sun) and the Equator is approximately constant, $\delta' = 23.5°$ (green triangle in Fig. 4.3) To calculate solar resource, we need to know the angle between the plane of the Equator and a line from the Sun's location to Earth's location, also called the

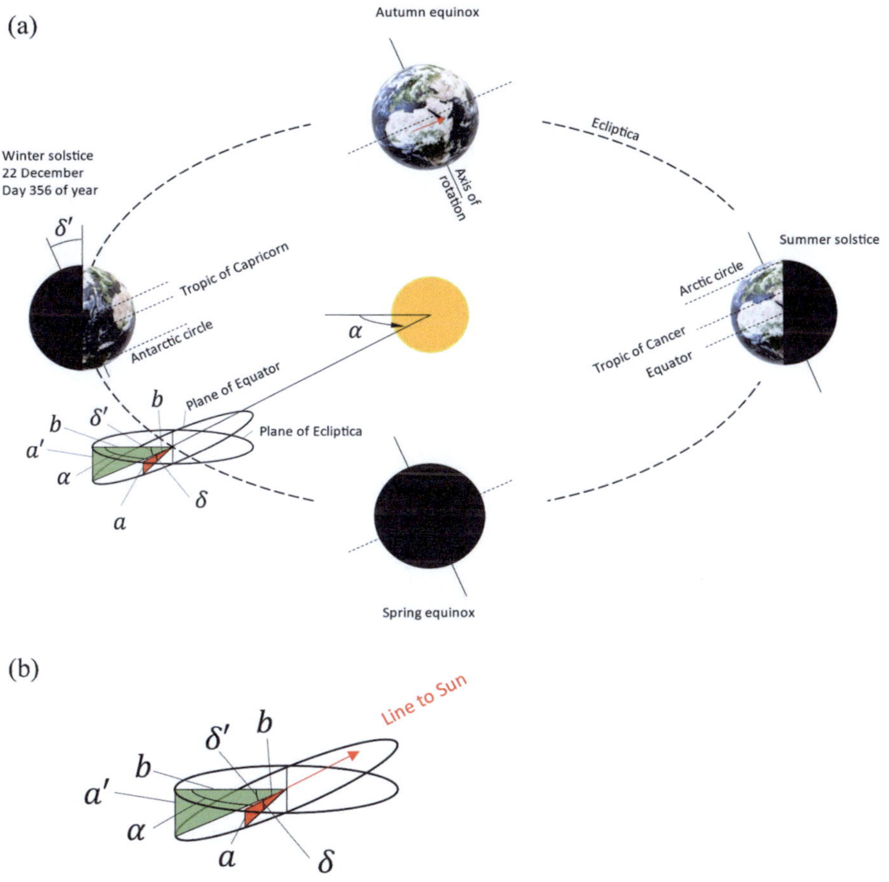

Fig. 4.3 **a** Orientation of Earth around the Sun. Winter solstice date and time varies from year to year—in 2023 the winter solstice was 22 December, 3:27 AM GMT. **b** Orientation of Earth around the Sun. Magnification of the angle descriptions

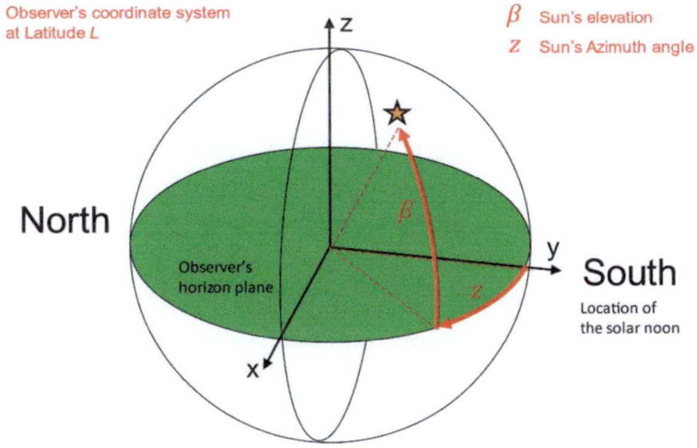

Fig. 4.4 The location of the Sun in the observer's coordinate system

Sun's declination, δ (red tringle in Fig. 4.3). The value of the declination is − 23.5° at the winter solstice, 0° at equinoxes and 23.5° at the summer solstice, and it varies as a function of the angle $\alpha = (n + 9.85)2\pi/365$, where n is the number of the calendar day of the year, 1st of January being $n = 1$. (A value of 9.85 days is added to account for the difference between the winter solstice and the end of the calendar year: the winter solstice is $31 - 21 = 10$ full days, minus 3.5 h or $3.5/24 = 0.146$ days, earlier than 1 Jan, and $10 - 0.145 = 9.85$.)

The radius b of the imaginary circular ecliptical plane is constant but the value of the vertical component a vary. We can write the following relations:

$$\frac{a}{b} = \tan(\delta)$$

$$\frac{a'}{b} = \frac{a}{x} = \tan(\delta') = \frac{a}{b\cos(\alpha)}$$

From these we can find the solar declination,

$$\tan(\delta) = \tan(\delta')\cos\left(\frac{n + 9.86}{365}2\pi\right)$$

Since $\cos(\alpha) = \sin(\alpha - \pi/2)$, and $\pi/2$ corresponds to $365/4 = 91.25$ days, we can write,

$$\begin{aligned}\tan(\delta) &= \tan(\delta')\cos\left(2\pi\left(\frac{n}{365} + \frac{9.85}{365}\right)\right)\\ &= \tan(\delta')\sin\left(2\pi\left(\frac{n}{365} + \frac{9.85}{365} - \frac{91.25}{365}\right)\right)\\ &= \tan(\delta')\sin\left(2\pi\left(\frac{n + 284}{365}\right)\right)\end{aligned}$$

Note that n has a maximum value of 365. We should also consider the definitions for the Arctic and Antarctic circles and the Tropics of Cancer and Capricorn. The Arctic circle defines the maximum area in which the Sun does not set during the summer solstice, and does not rise during the winter solstice; the opposite condition holds for the Antarctic circle. The Tropics of Cancer and Capricorn are defined as the latitudes where the Sun shines from the zenith (directly overhead) at solar noon on the summer solstice (Cancer) and winter solstice (Capricorn). In addition, at the Equator, the Sun shines from the zenith at solar noon on both of the Equinoxes.

Figure 4.4 presents the location of the Sun in local observer's coordinates, with elevation and Azimuth angles, β and z, respectively. Figure 4.5 gives the Sun's location in Equatorial coordinates, as declination, δ and hour angle h. Figure 4.6 combines these coordinate systems together.

The simplified model for the Sun's declination was already derived previously. The hour angle is $h = 2\pi t/24$, where t is time in hours from solar noon (here defined

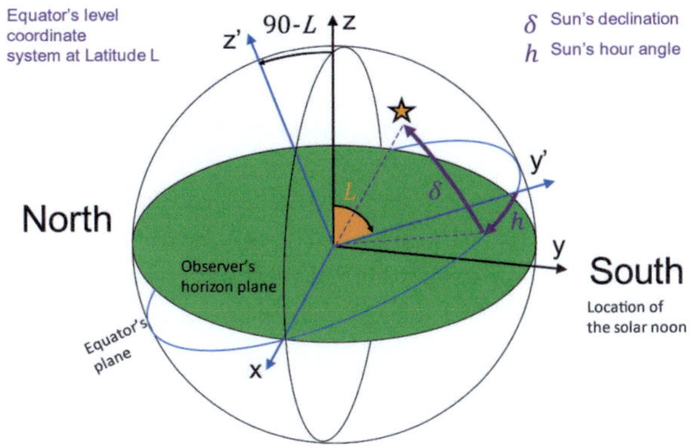

Fig. 4.5 The location of the Sun in the Equator's coordinate system

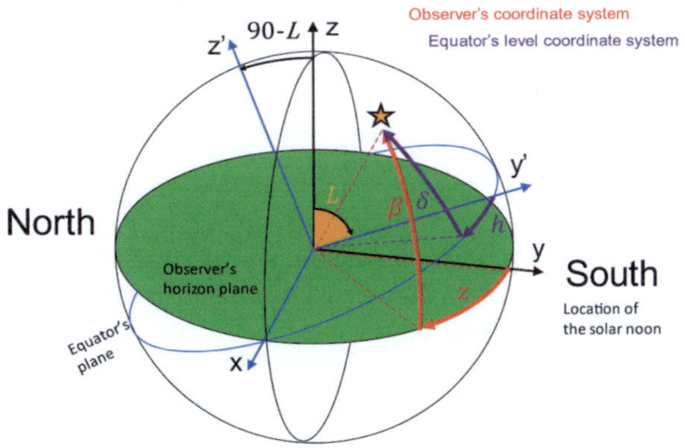

Fig. 4.6 The location of the Sun in the combined coordinate systems of Figs. 4.4 and 4.5

as positive forwards from the solar noon). These Equatorial coordinates (δ, h) are not useful for locating the Sun in actual situations; instead we need to use the local coordinate system (β, z). The coordinate transformation (Fig. 4.6) is quite complex, and a detailed derivation can be found, for example, in [102]. However, with the use of Equatorial coordinates, the calculation is as follows,

$$\sin(\beta) = \cos(\delta)\cos(h)\cos(L) + \sin(\delta)\sin(L)$$

$$\sin(z) = \frac{\cos(\delta)\sin(h)}{\sin(\beta)}$$

where L is the latitude of the location. Note that all angles should be consistently calculated either in radians or in degrees.

We can simply calculate β from the first equation, and the azimuth, z, from the second equation. Based on Fig. 4.6, the Sun's elevation at solar noon ($h = 0$) and optimal vertical tilt angle (perpendicular to the Sun) of a fixed solar panel, α_{panel}, can be calculated as follows:

$$\beta_{noon} = 90^\circ - L + \delta$$

$$\alpha_{panel} = 90^\circ - \beta_{noon} = L - \delta$$

Horizontally, the panel should be installed towards the south, $z = 0$. The tilt angle is usually selected so that the panel faces the Sun at the equinoxes at solar noon, $\delta = 0$, and so the optimal tilt angle is equal to $\alpha_{panel} = L$. We will discuss this in more detail below for fixed and tracking solar PV systems.

Example 4.1 Approximate the length of the day in Helsinki (latitude $L = 60.192°$ $= 1.05055$ rad), and in New York ($L = 40.731° = 0.71089$ rad) on 15 March 2023. Tilt angle of the axis of rotation is 23.5° (0.41015 rad).

March 15 is day 74 of the year during a normal year. From this we can calculate the Sun's declination,

$$\begin{aligned}\delta &= \mathrm{atan}\left(\tan(0.41015)\sin\left(2\pi\left(\frac{74+284}{365}\right)\right)\right) \\ &= -0.05522 \text{ rad}\end{aligned}$$

This is same for all locations in the Earth. At sunrise and sunset, its elevation $\beta = \beta_0 = 0°$. By inserting this into the $\sin(\beta)$ equation above, we can solve the corresponding hour angle, h_0.

$$\sin(\beta) = 0 = \cos(\delta)\cos(h_0)\cos(L) + \sin(\delta)\sin(L)$$

from which we can solve for h_0 for Helsinki,

$$\begin{aligned}h_0 &= \mathrm{acos}\left(-\frac{\sin(\delta)\sin(L)}{\cos(\delta)\cos(L)}\right) = \mathrm{acos}(-\tan(\delta)\tan(L)) \\ &= \mathrm{acos}(-\tan(-0.05522)\tan(1.05055)) = 1.47943 \text{ rad}\end{aligned}$$

and for New York,

$$\begin{aligned}h_0 &= \mathrm{acos}(-\tan(-0.05522)\tan(0.71089)) \\ &= 1.52577 \text{ rad}\end{aligned}$$

These present only half of the day's length, from noon to sunset, so the value for the whole day is twice as long. These need to be converted into hours, and so we multiply the values above by $2 \times 24/(2\pi)$, giving 11.3 h (~11 h 18 min) for Helsinki, and 11.656 h (~11 h 39 min) for New York. For comparison, the actual values for day length on 15 March 2023 were 11 h 42 min in Helsinki, and 11 h 54 min in New York.

Better accuracy can be obtained by considering the equation of time, which includes the non-ideal trajectory of the Earth due to interactions between the planets, and with the Moon. From the equation of time, the following relationship can be obtained:

$$\begin{aligned}\delta = 0.006918 - 0.399912 \times \cos(\gamma) + 0.070257 \\ \times \sin(\gamma) - 0.006758 \times \cos(2\gamma) + 0.000907 \times \sin(2\gamma) \\ - 0.002697 \times \cos(3\gamma) + 0.00148 \times \sin(3\gamma)\end{aligned}$$

$$\gamma = \frac{2\pi}{365}\left(n - 1 + \frac{t - 12}{24}\right)$$

where γ is the fractional year, n is the day of the year, and t is the hour of the day. Let's repeat the above calculations with this model. We neglect the variation of declination during a day (assuming that the Earth does not significantly move during the day). Fractional year is $\gamma = 1.27385$ radians and declination $\delta = -0.03564$ radians. We then can calculate half day lengths.

$$\begin{aligned}h_0 &= \text{acos}(-\tan(-0.03564)\tan(1.05055)) \\ &= 1.50852 \text{ rad}\end{aligned}$$

$$\begin{aligned}h_0 &= \text{acos}(-\tan(-0.03564)\tan(0.71089)) \\ &= 1.54009 \text{ rad}\end{aligned}$$

for Helsinki and New York, respectively. These correspond to day-lengths of 11.524 h (11 h 31 min) and 11.765 h (11 h 46 min), which more closely agree with the true values (11 h 42 min, and 11 h 54 min).

4.2.2 *Attenuation in the Atmosphere*

Where the Sun's radiation reaches the outer surface of the Earth's atmosphere, its intensity is ~ 1367 W/m^2, as discussed in Sect. 4.2.1 above. The irradiation attenuates in the atmosphere due to reflection, scattering and absorption from clouds, dust particles and various gases, decreasing the intensity to a maximum of ~ 1000 W/m^2 when it reaches the Earth's surface, on a sunny day with no clouds. As particles and gases such as H_2O, CO_2 and O_3 absorb thermal radiation, the spectrum (spectral energy distribution) changes accordingly.

Solar energy reaching the ground has two modes: direct beam irradiation, and diffuse irradiation. In the beam irradiation mode all rays of the sun travel in parallel, and can be effectively concentrated with mirrors or lenses into a small area to significantly increase the energy density and temperature of the irradiation. This mode of energy is mostly used in concentrated solar power plants. Diffuse irradiation is irradiation that is scattered, reflected or emitted from clouds and particles, and can

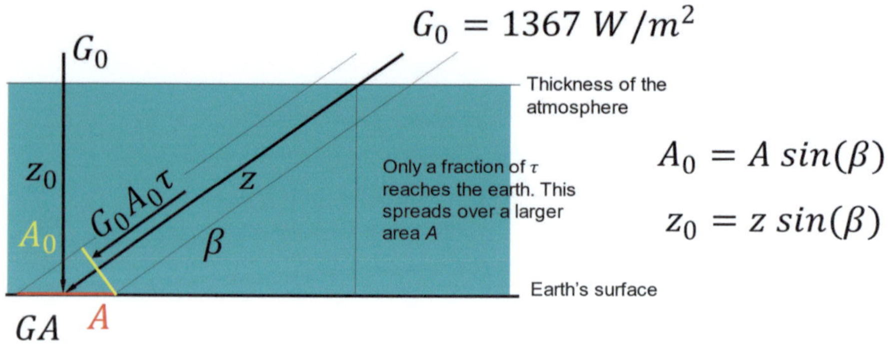

Fig. 4.7 Attenuation of solar irradiance through a cross section of the Earth's atmosphere

only be utilized by thermal non-concentrating heat collectors and photovoltaic PV panels. Concentrating solar systems are less common than flat plate collectors, and are not effective at higher latitudes where diffuse-irradiation-mode heat transfer is dominant.

To compare efficiencies of different solar panels, reference values (known as Standard Test Conditions, STC) have been defined. These are: irradiance 1000 W/m^2, temperature of the environment 25 °C, wind speed 1 m/s, and air mass of 1.5 (AM1.5). An air mass of AM1.5 means a configuration where the actual path length to the Earth's surface is 1.5 times the longer than the thickness of the atmosphere, z_0.

Figure 4.7 shows a cross section of the atmosphere. Solar irradiance outside the atmosphere, G_0, is 1367 W/m^2, and the value reaching ground level is G.

Assuming that a fraction τ of the extraterrestrial radiative energy (W) reaches the ground, the energy balance of the 1D plate of the atmosphere can be written as follows:

$$\phi = GA = G_0 A_0 \tau = G_0 A \sin(\beta)\tau$$

from which we can solve for the irradiance G,

$$G = G_0 \sin(\beta)\tau$$

τ needs to be expressed as a function of the Sun's elevation angle, since the longer the radiative path length, the smaller the fraction of energy passing through. We have $z_0 = z\sin(\beta)$, and AM1.5 is defined in the Standard Test Conditions such that an irradiance of $G = 1000$ W/m^2 reaches the ground—this means that $\tau = G/G_0 = 1000/1367 = 0.73153$. This corresponds to irradiation coming from an angle β, and so we can write $\sin(\beta) = 1/1.5$, from which the angle can be solved as $\beta = 41.8°$.

Next, we discuss the derivation of the so-called exponential decay model for irradiation passing through a gaseous, semi-transparent liquid, or solid medium [95]. If we assume that attenuation at any point in the medium is proportional to the irradiance at that point, then

$$\frac{dG(x)}{dx} = -aG(x)$$

for attenuation coefficient a (in units of 1/m, assumed independent of wavelength). This can be rearranged and integrated as

$$\int_{G_0}^{G} \frac{dG(x)}{G(x)} = -a \int_{0}^{z} dx$$

Performing the integral gives the exponential decay model, also referred to as the Beer's law,

$$G = G_0 \exp(-az) = G_0 \tau$$

where τ is the fraction of irradiation that passes through the atmosphere. We can write the above equation in terms of the irradiation from the zenith (subscript 0) to solve for the attenuation coefficient a.

$$\tau_0 = \exp(-az_0)$$

$$a = -\frac{\ln(\tau_0)}{z_0}$$

This definition of a can be then substituted into the general equation for τ to get

$$\tau = \exp(-az) = \exp\left(\frac{\ln(\tau_0)}{\sin(\beta)} \frac{z_0}{z_0}\right)$$
$$= \exp\left(\frac{\ln(\tau_0)}{\sin(\beta)}\right) = \tau_0^{1/\sin(\beta)}$$

The fraction of irradiation coming directly from the zenith on a clear day reaching the ground (AM1) can be calculated as $\tau_0 = 0.73153^{1/1.5} = 0.812$, and the corresponding irradiance reaching ground level would be 1110 W/m^2. And so we define AM1.5 such that a fraction of 0.732 of extraterrestrial irradiance coming from elevation angle $\beta = 41.8°$ reaches the ground, resulting in the irradiance of 1000 W/m^2.

Transmissivity τ_0 varies significantly for example with cloudiness, cloud optical thickness, atmospheric absorption characteristics, scattering, sunshine duration, and a location-specific geographic features [7, 17, 135, 178, 196]. For example, Srivastava et al. [183] studied the latitudinal distribution of atmospheric transmissivity, and found no clear trend—Fig. 4.8 presents a sample of their results.

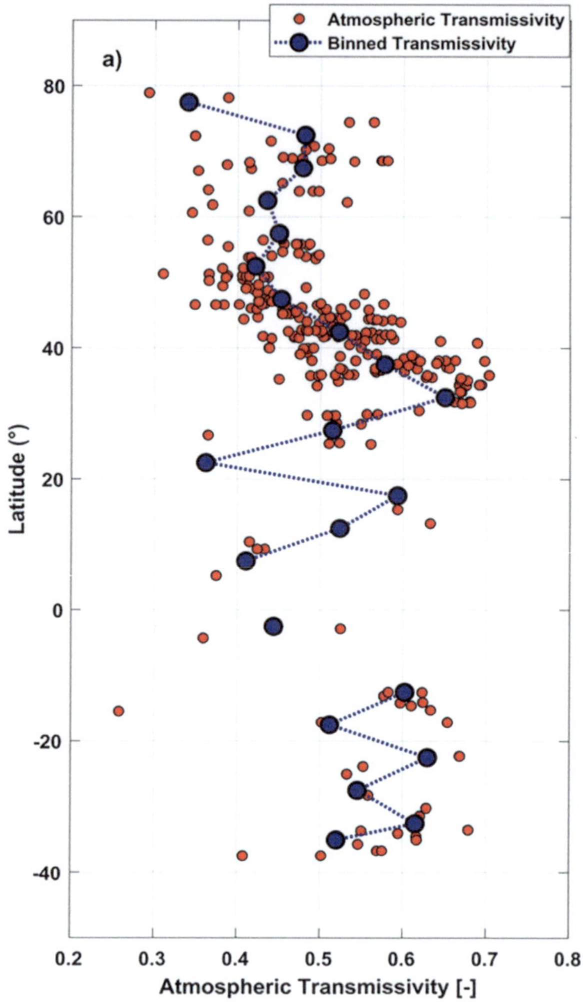

Fig. 4.8 Latitudinal variation of atmospheric transmissivity [183]

However, this data is not accurate enough to be used in actual solar resource estimation. In a real project, accurate, location-specific, long-term measurements over several years are needed to produce a representative reference year. If τ_0 is available as a function of time and location, it can be directly implemented and used in a solar resource simulator.

4.2.3 Impact of Collector Orientation

To reach the maximum energy capture efficiency, a solar panel should directly face the Sun, so that the incident energy flow (W) is spread over the minimum surface area, increasing the panel's power production per area, W/m^2. Figure 4.9 presents an example where a panel of area A_p is facing in direction (β_p, z_p) while the Sun is shining from the direction (β_s, z_s).

The geometrical capture efficiency can be defined as the ratio of the energy flowing through the panel's projection perpendicular to sun's rays, and the energy flow in the optimal energy capture configuration (i.e., panel directly facing the Sun),

$$\eta_g = \frac{GA}{GA_p} = \frac{A}{A_p} = \frac{A_p\cos(\theta)}{A_p} = \cos(\theta) = \overline{a}\cdot\overline{b}$$

where θ is the angle between unit vectors $\overline{a}$ and $\overline{b}$ as described in Fig. 4.9. These are given by the relations.

$$\overline{a} = [\cos(\beta_s)\sin(z_s),\ \cos(\beta_s)\cos(z_s), \sin(\beta_s)]$$

$$\overline{b} = \left[\cos(\beta_p)\sin(z_p),\ \cos(\beta_p)\cos(z_p), \sin(\beta_p)\right]$$

If the angle between the panel's normal vector and the Sun's direction vector is known, the efficiency can be calculated from the above equation containing $\cos(\theta)$. However, for modeling purposes, usually the local elevation and azimuth coordinates

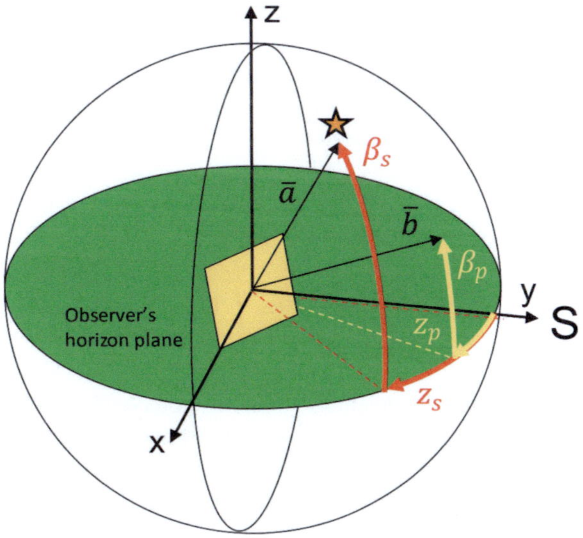

Fig. 4.9 Orientation of the solar panel. The panel front's normal vector is in the direction (β_p, z_p) while the Sun is shining from the direction (β_s, z_s)

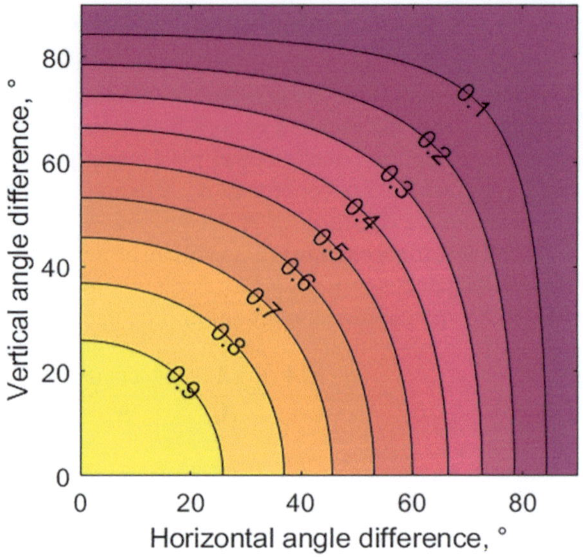

Fig. 4.10 Geometrical efficiency of a single panel, η_g

are used. Figure 4.10 shows η_g for different values of the angle between the Sun and the panel. The x-axis is $(z_{Sun} - z_{Panel})$ and the y-axis is $(\beta_{Sun} - \beta_{Panel})$.

From Fig. 4.10 it is clear that optimal alignment is important: highest efficiencies are only obtained only with a properly aligned panel.

4.2.4 Modeling the Transient Solar Resource

The amount of irradiance at time t, $G(t)$, and the integrated daily insolation, W, reaching a collecting surface at the Earth's surface across a day from sunrise to sunset, $[-t_0, t_0]$, are equal to:

$$G(t) = G_0\tau(t)\eta_g(t)$$

$$W = \int_{-t_0}^{t_0} G(t)dt = \int_{-t_0}^{t_0} G_0\tau(t)\eta_g(t)dt = \int_{-t_0}^{t_0} G_0\tau_0^{1/\sin(\beta)}\eta_g(t)dt$$

where τ is the atmospheric transmission coefficient along the actual path through the atmosphere, τ_0 is the atmospheric transmission coefficient for the thickness of the atmosphere (z_0), and η_g is the geometric capture efficiency of the receiving surface. Note that in this example we assume that the weather conditions and the sky are clear; in reality, in cloudy weather the value of τ_0 decreases significantly.

There is no analytical solution for the insolation integral above, so we need to use numerical methods—but first, we need to present the integral in terms of known parameters. The Sun's elevation can be solved from the coordinate transformation equation,

$$\sin(\beta) = \cos(\delta)\cos(h_0)\cos(L) + \sin(\delta)\sin(L)$$

As discussed in Example 4.1, the hour angles, $\pm h_0$, of sunset and sunrise can be solved from the above equation by setting $\beta = 0$, and then solved from $\cos(h_0) = -\tan(\delta)\tan(L)$. For the cosine function only values between $[-1, 1]$ can be accepted—if $\cos(h_0) < -1$ the Sun is circumpolar and does not set below the horizon, while if $\cos(h_0) > 1$ the Sun does not rise above the horizon [102].

For integration, insolation must be written in terms of the hour angle h. This can be achieved by changing variables, parameterizing time t in terms of the hour angle, $t = h \times 24/2\pi = 12 \times h/\pi$, and so $dt = 12/\pi\, dh$. Note that time is in hours, so the energy is Wh. After inserting these into the equation of insolation we obtain

$$W = \frac{12 \times G_0}{\pi} \int_{-h_0}^{h_0} \tau_0^{\frac{1}{\sin(\beta(h))}} \eta_g(t)dh$$

This equation must be numerically integrated. For the sake of stability, if the hour angle at sunset or sunrise is smaller than $\varepsilon = 10^{-5}$, we assume that the Sun is below the horizon. Matlab® model is available for free download at the book www-page. This can be used to calculate annual insolation to a solar panel, at different locations and panel configurations. In all these examples we assume that the sky is clear. As examples of the models results, Fig. 4.11a–c show annual variation of daily insolation at different locations for (1) a panel installed flat on the ground, $\beta = 0$; (2) a panel directed towards $\beta = 90 - L$; and (3) a fully Sun-tracking panel, respectively.

Several interesting observations can be made from these examples. A flat-installed panel in the polar regions receives more solar energy in the summertime than one at the Equator (Fig. 4.11a). Comparing Fig. 4.11a, c we see that a panel directly facing the Sun at the Equinoxes at solar noon has stable energy generation in the Equator region, but for other locations, even though the production peak decreases, the total energy generated increases, (that is, the total area under the curve increases) due to the longer optimal alignment to the Sun at both equinoxes. When a fully tracking panel is used, the energy generation ineases significantly, since the tracking panel is able to capture solar energy efficiently from sunrise to sunset (Fig. 4.11c).

The main idea of this exercise was to illustrate how to examine the changing geometry of the Sun and Earth in resource estimation. Throughout we assumed that the weather is clear and there are no shading losses, which means that the energy generation of a real solar power plant may be much smaller. However, this basic model can be further developed with the use of spatially-referenced local weather data to produce reliable solar resource estimations. There are several tools to perform

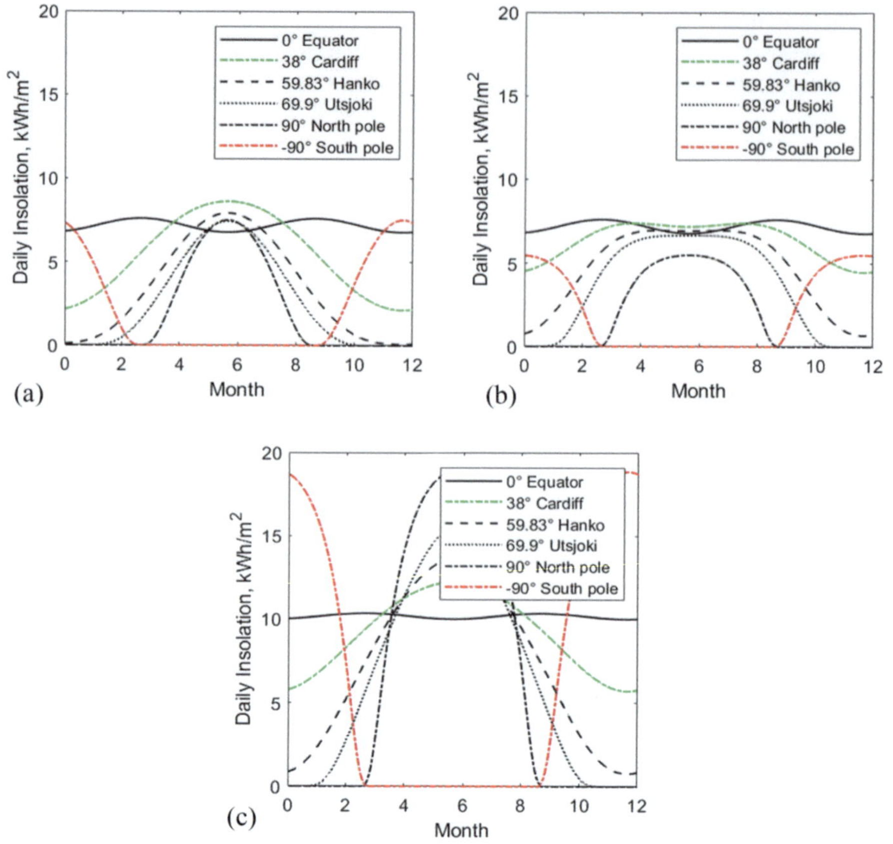

Fig. 4.11 **a** Annual variation of daily insolation for a clear sky, panel flat on the ground, $\alpha_{panel} = 0°$. **b** Annual variation of daily insolation for a clear sky, panel installed at angle $\alpha_{panel} = L$. **c**. Annual variation of daily insolation for a clear sky, Sun-tracking panel

more accurate analysis, such as the above-mentioned Global Solar Atlas (https://globalsolaratlas.info) or the Photovoltaic Geographical information system (https://re.jrc.ec.europa.eu/pvg_tools/en/).

4.2.5 *Panel Geometrical Efficiency and Shading in Grid Arrangement*

Until now we have discussed a single-panel setup that receives irradiation from the Sun without any shading. However, in solar power plants the panels are assembled in a grid arrangement with a horizontal spacing L_s. The side length of the panel row is L_p—see Fig. 4.12 for an actual geometry in a real industrial-scale PV power plant.

Fig. 4.12 PV panel grid geometry between the rows, x and y-axes are presented with a thick line

To consider the geometry of this setup, we set our coordinate system's origin at the upper corner of the first, left hand panel as shown. As shown in the figure, the lower part of the panel has some shading, of length a, shown on Fig. 4.12 as the red dot. The location of this point, where Sun's lowest rays just graze the first panel and hit the next panel over, needs to be calculated to solve the efficiency of each of the shaded panels to the right.

The horizonal location of the red dot, x, can be found by setting the line of Sun's lowest rays and surface level of the next panel meet each other, as shown. From this vertical location definition, we find

$$-x\tan(\beta) = -L_p\sin(\alpha_p) - L_s\tan(\alpha_p) + x\tan(\alpha_p)$$

$$x = \frac{L_p\sin(\alpha_p) + L_s\tan(\alpha_p)}{\tan(\beta) + \tan(\alpha_p)}$$

The shaded length a is then found from

$$a = \frac{x - L_s}{\cos(\alpha_p)} = \max\left(\frac{L_p\sin(\alpha_p) + L_s\tan(\alpha_p)}{\cos(\alpha_p)(\tan(\beta) + \tan(\alpha_p))} - \frac{L_s}{\cos(\alpha_p)}, 0\right)$$

where the max function is used to avoid non-physical solutions. The geometric efficiency of all rows of panels behind the first open row, η_r, is defined then

$$\eta_r = \frac{L_p - a}{L_p} = 1 - \frac{a}{L_p}$$

$$= 1 - \max\left(\frac{\sin(\alpha_p) + L_s/L_p \tan(\alpha_p)}{\cos(\alpha_p)(\tan(\beta) + \tan(\alpha_p))} - \frac{L_s/L_p}{\cos(\alpha_p)}, 0\right)$$

This equation makes the solution independent on the actual panel array dimensions, since we only need to specify the ratio of the spacing and panel length, $L_s/L_p = \xi$.

Since the efficiency of the first row is 1, and all other rows have equal efficiency if we assume that panels are installed on the flat ground with equal spacing, we can calculate the total grid efficiency for N rows of panels,

$$\eta_{grid} = \frac{1 + (N - 1)\eta_r}{N}$$

Figure 4.13 shows the η_{grid} for different panel tilt angles and Sun elevations, and for different spacings $\xi = L_s/L_p = 1, 1.5, 2, 2.5$.

How do we interpret these charts? First, select the panel tilt to be the same as the latitude, $\alpha_p = L$. Then follow horizontally from the vertical axis to the right to see how grid efficiency increases as Sun elevation increases. For example, if the panel tilt is 60° and spacing is $\xi = 2.5$, we reach full efficiency for Sun elevation of ~ 20°, while for panels packed more densely at $\xi = 1.0$, the Sun has to rise above to 41° to reach full grid efficiency. In Helsinki, Finland, during the winter solstice at solar noon, the Sun only rises up to $\beta = 90° - 60° - 23.4° = 6.6°$, significantly decreasing the efficiency. Better efficiency could be obtained by optimally decreasing the tilt angle, considering the geometrical panel efficiency discussed previously.

More generally, shading effects can be included by introducing a vector or a function that gives the elevation angle of the horizon from the lower end of the panel's edge, $\beta_h(h)$ (note that the actual angle is defined in the observer's coordinate, z, instead of the hour angle, h). If the Sun's elevation is smaller than this angle, irradiation to the panel is equal to zero. Mathematically, we find

$$W = \frac{12 \times G_0}{\pi} \int_{-h_0}^{h_0} \tau_0^{\frac{1}{\sin(\beta(h))}} \eta_g(h)\eta_h(h)dh$$

$$\eta_h(h) = \begin{cases} 0, \beta(h) \leq \beta_h(h) \\ 1, \beta(h) > \beta_h(h) \end{cases}$$

The elevation of the horizon could be defined, for example, with image analysis of calibrated fish-eye optics, as illustrated in Fig. 4.14.

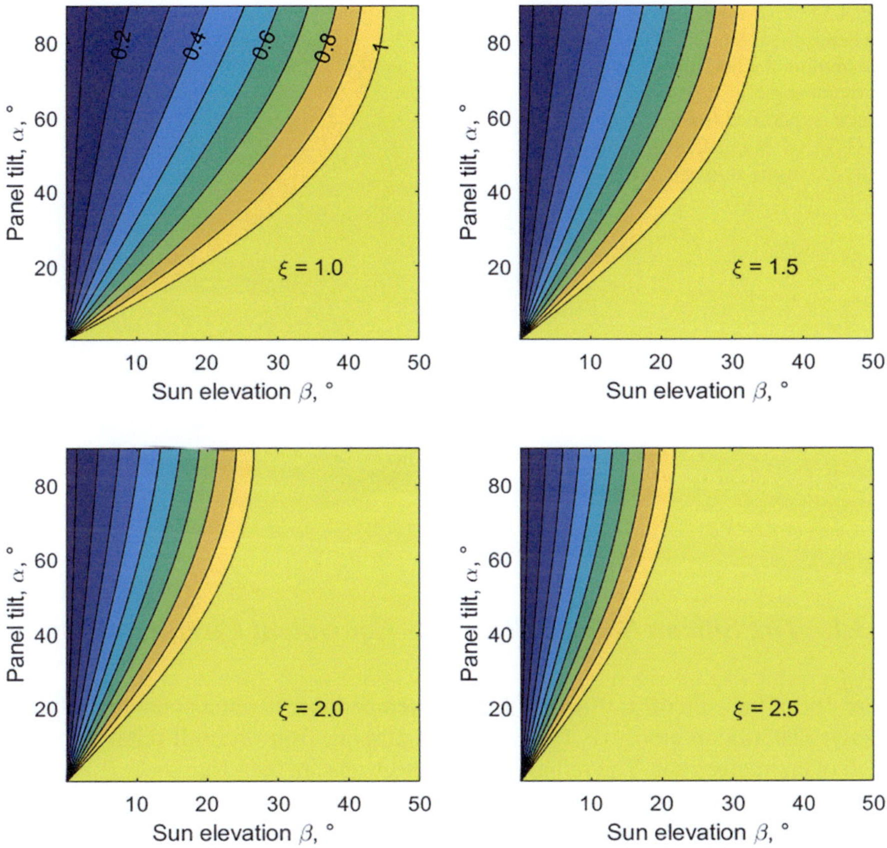

Fig. 4.13 Grid efficiency, η_{grid}, for $N = 20$ panel rows. The color scale is same in all figures, with yellow corresponding to efficiency of 100%, and dark blue to efficiency range of 0–10%

4.3 Principles of Solar Photovoltaic Panels

This section focuses on the basic principles of solar PV panels. We first focus on the physical principles of the dark diode and the p–n junction, then we extend our analysis to the crystalline silicon solar cell: a solar cell is simply an illuminated diode with a large, flat cross-sectional surface area fixed along the p–n junction. We then discuss modeling the electrical characteristics of solar panels and introduce the equivalent circuit concept. Finally, we analyze the effects of operational conditions such as irradiance, shading, and temperature.

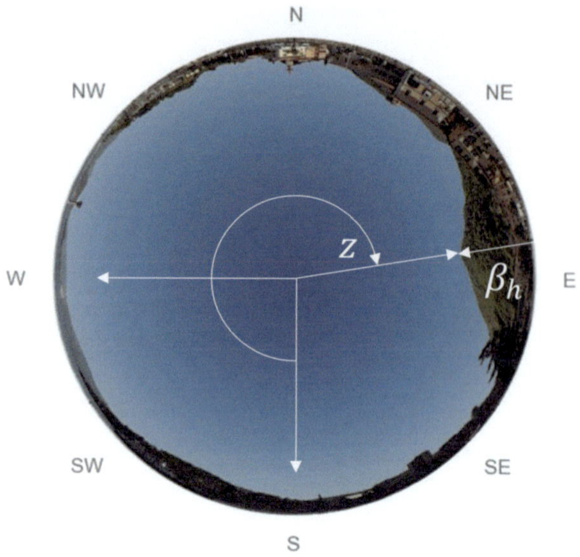

Fig. 4.14 Fish-eye image of the hemisphere for determining the horizon elevation angle, $\beta_h(z)$. Figure is modified from PVGIS User Manual, https://re.jrc.ec.europa.eu/pvg_tools/en/tools.html

4.3.1 The Silicon Solar Cell and Its Equivalent Circuit

Pure crystalline silicon at 0 K (–273.15 °C) temperature is not a conductor of electricity. The silicon atom has four electrons missing from the full outer (valence) shell of eight electrons. The remaining four electrons are therefore seeking partners, and pair willingly with another four Si atoms, forming strong electrical bonds. As a result, since all valence electrons are accounted for and locked into their bonds, there are no free electrons to allow electricity to flow.

If the material is heated, the vibrational energy is enough the loosen electrons into the conduction band, while at the same time forming electron deficits (or 'holes'). When a voltage difference is applied to pure silicon at this exited state, electrons can move through the lattice, and an electrical current is established. As such, pure silicon is an intrinsic semiconductor with an equal number of holes and free electrons. Figure 4.15 illustrates the electron configurations for pure silicon and other types of semiconductors: exited silicon, boron-doped silicon and phosphorus doped silicon. Note that the actual system is a 3-dimensional lattice of atoms instead of the planar presentation used here.

To produce an extrinsic semiconductor material that has naturally positive or negative charge, the base Si-material must be doped with atoms of a suitable species that creates either free electrons or electron holes. In a solar cell, the thicker base silicon sheet is made from Boron- or Gallium-doped Silicon. These elements both have three outer layer electrons, and when mixed into the Si lattice, they create one electron deficit per Boron atom, giving the material an overall positive charge, and therefore the tendency to attract and bind electrons. The electron deficits are able to move through the lattice: an electron in a covalent bond jumps into the hole, filling

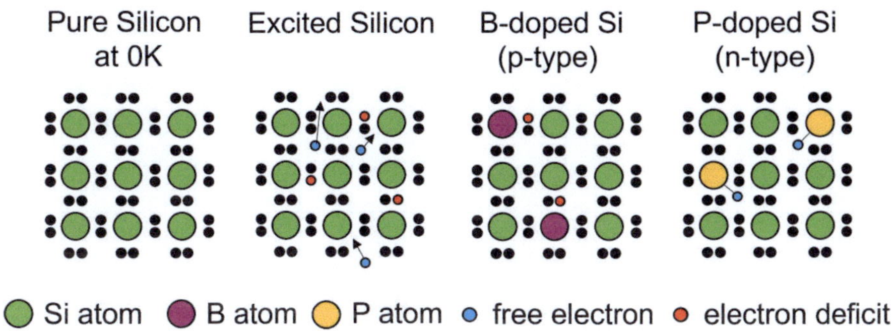

Fig. 4.15 Silicon in different states and doping compositions

that deficit but leaving another hole behind; this is then filled with another electron, and so on—and in this way, the hole moves from atom to atom like dominos falling in a line, transporting the positive charge carrier (Fig. 4.15, second figure from the right). These forms of doped silicon are known as p-type semiconductor.

To produce a negatively-charged (n-type) semiconductor material, the silicon is doped with atoms of a species with more than four outer-layer electrons, typically phosphorus (P) with five valence electrons (others, such as arsenic (As), can also be used). When phosphorus is mixed with silicon, this leaves one extra electron per phosphorus atom unable to form a covalent bond in the crystal. This electron is actually not completely free, since it is lightly bound to the phosphorus atom (shown by the thin line in the right-hand image of Fig. 4.15). However, this relatively weakly bound electron can be more easily exited and become mobile, for example by hitting it with the energy of an incoming photon.

When the p- and n-type materials are combined together, a p–n junction is formed. At the boundary, the closest excess electrons from the n-type material migrate into the holes in the p-type material side This creates a reverse potential difference of ~ 0.6 V—this is illustrated in Fig. 4.16a, b. For simplicity, in this figure we simply show the charge carriers (+ for holes, – for electrons) rather than the atomic lattice as a whole.

If we attach the positive terminal of a voltage source to the p-type side (anode) and negative terminal to the n-type side (cathode), and gradually increase the applied voltage, the depletion region thickness decreases—but current is not able to flow until the applied potential difference is larger than the reverse potential of 0.6 V (Fig. 4.16c). When the voltage source is reverse-biased, connecting the source's negative terminal to the diode's anode and positive terminal to cathode, the holes are drawn towards the negative connection and the electrons towards the positive connection (Fig. 4.16d)—this increases the width of the depletion region, and current is not able to flow. We call this a "dark diode" as we are not yet illuminating the p–n junction with photons.

In Fig. 4.16e the diode is connected with a load resistor, and we shine light onto the surface—it is now acting as a solar cell. Due to the depletion zone an electrical field

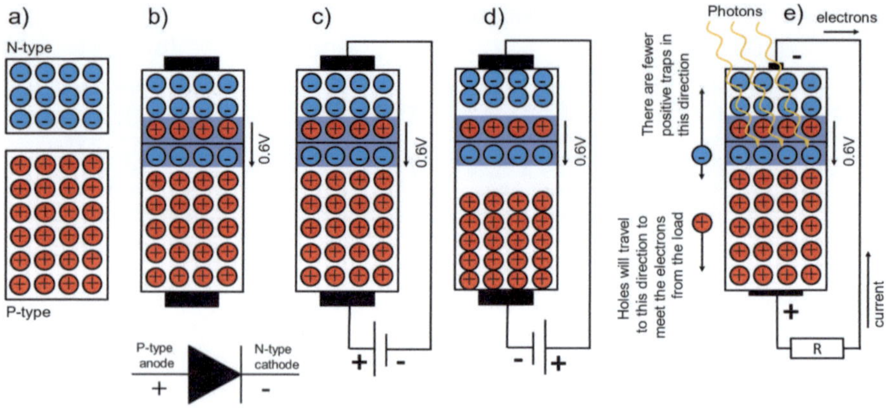

Fig. 4.16 Formation of the p–n junction in a dark diode (**a**) and (**b**), forward biased dark diode (**c**), reverse biased dark diode (**d**), and solar cell (**e**). Note that the electrons in the depletion zone are fixed to a hole, making a strong covalent bond, but the electrons in the n-type material are in excess, having only a mild bond to the "parent" phosphorus atoms

is created in the opposite direction towards p- and n-type materials. When photons hit the electrons throughout the cell and kick them free, electrons that move towards the p-type material will be rapidly recombined with the positive holes and become fixed there. The junction acts as a barrier for electrons in the n-p direction. Electrons that are kicked upwards can migrate through the thinner n-type material towards the negative electrical connection, and then travel easily through the wires and load to the positive terminal, where they re-combine with the electron holes that have traveled there through the p-type material. The reversed electrical field of the depletion zone pushes electrons in the n-direction (opposite to the conventional positive-to-negative direction of the potential). Consequently, when illuminated with photons, current flows from the positive terminal to the negative terminal through the load.

Before going into the equivalent-circuit analysis, let's look at the voltage-current characteristic of an illuminated diode (Fig. 4.17). For a p–n junction acting as a (dark) diode, the resistor is replaced with a voltage source connected in the forward-biased direction (Fig. 4.16c) to produce the characteristic curve shown as a dashed line in the graph in Fig. 4.17.

Electrons excited by the energy of the incident photons are "harvested" from across the entire cell, but mostly from the depletion zone and the n-type material. As discussed above, due to the asymmetric and rectifying nature of the p–n junction (that is, the reversed electrical field in the depletion zone), electrons are only able to travel in the p-n direction, and holes can only travel in the n-p direction. The holes move to the positive terminal of the cell to recombine with electrons that have passed through the lode resistor, ensuring that charge is balanced. (Note that in the dark diode, current flows in the opposite (n-p) direction (current on the + side) due to the external potential difference.)

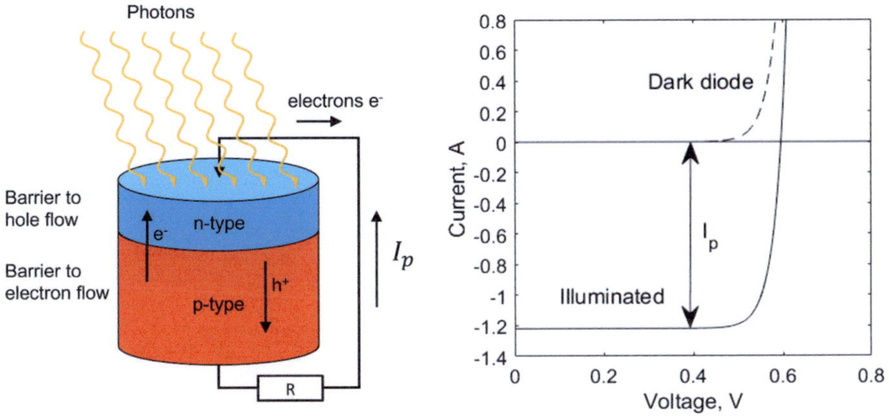

Fig. 4.17 Voltage-current characteristic of a dark and illuminated diode. I_p is the photoelectric current from illumination

The above voltage-current graph is often turned upside down, with the current presented as positive upwards. X-axis is the potential difference across the cell terminals. Figure 4.18 presents the U-I characteristic of a solar panel and the most important parameters: the maximum power point voltage and current (MP, green dot); the short circuit current, I_{SC}; and the open circuit voltage, U_{OC}. $I_R = U/R_L$ curves present characteristics of the load resistor. The operational point of the system is at the intersection of the curves.

The power obtained from the panel is simply the product of the voltage and current, $P = UI$. The maximum value is indicated by the green dot in Fig. 4.18; red dots represent operational scenarios that are not optimal. When no load is connected between the terminals of the panel, clearly no current will flow, and the voltage at the terminals is at its maximum, U_{OC}. If the terminals are short-circuited, the potential difference is zero and this point defines the maximum current that the panel can develop. This is called the short-circuit current, I_{SC}, and it is equal to the photovoltaic current of an ideal panel with no resistive losses. The performance of the panel is influenced by the ratio of the areas defined by these two extreme conditions ($U_{MP}I_{MP}$ and $U_{OC}I_{SC}$), and the closer the actual U-I curve is to the area defined the product of $U_{OS}I_{SC}$, the better is the performance of the panel. This ratio is known as the fill factor, f:

$$f = \frac{U_{MP}I_{MP}}{U_{OC}I_{SC}}$$

The operation of a solar panel described above can also be modeled by a series of equivalent circuits (Fig. 4.19): normal operation with a finite load (a), short-circuited system (b), open-circuit system (c), and ideal cell with no resistive losses (d). Note that the role of the diode in these circuit diagrams is different from the illuminated diode operation described in the Fig. 4.17—in the equivalent circuit, the diode functions

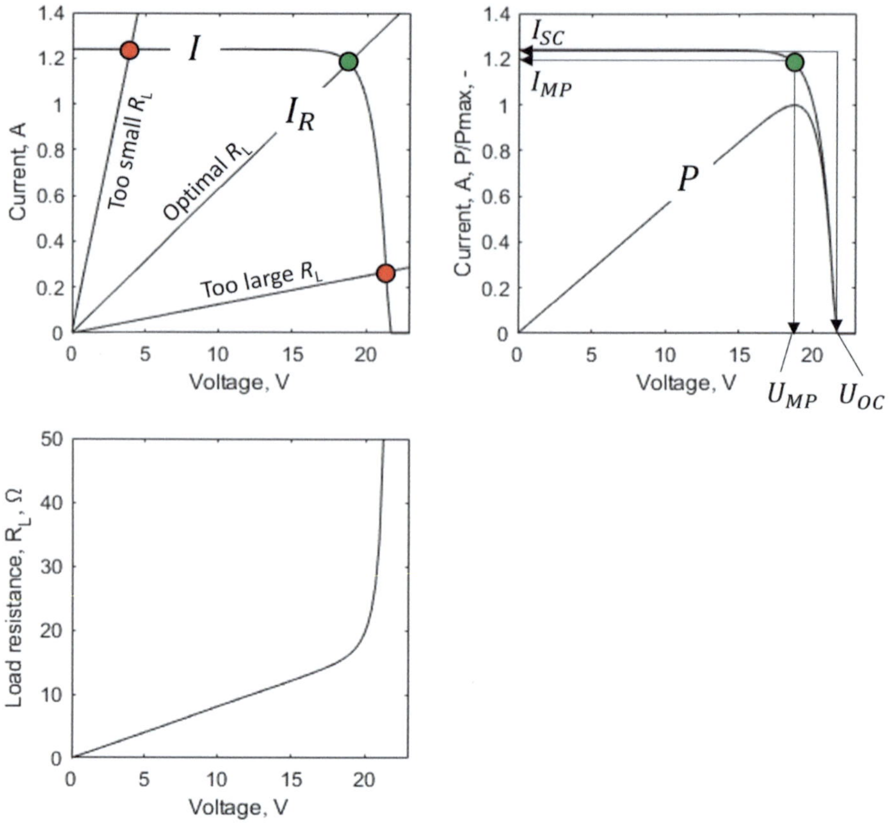

Fig. 4.18 Characteristics of a solar panel, and the most important parameters. MP indicates the maximum power point

as a normal dark diode, and an additional photoelectric current source is required as shown (the arrow inside the circle, far left of each circuit).

The equivalent circuit is composed of a photovoltaic current source, i_P, a diode, and series and shunt resistors, R_s and R_{sh} respectively, that describe the different resistive losses in the system. Current and voltage values i and u refer to a single cell, while I and U are values for the whole panel.

In this model presented in Fig. 4.19, the photoelectric current source is directly proportional to irradiance, G—that is, to the flow rate of incoming photons,

$$i_p = aG(1 - \rho)\tau$$

where ρ is the reflectance, τ is the transmissivity of the anti-reflective coating, and a is a panel-specific constant. The photovoltaic current is divided between the currents through the diode, the shunt, the series and load resistors. The equation describing the diode's voltage-current behavior (as shown graphically in Fig. 4.17) is

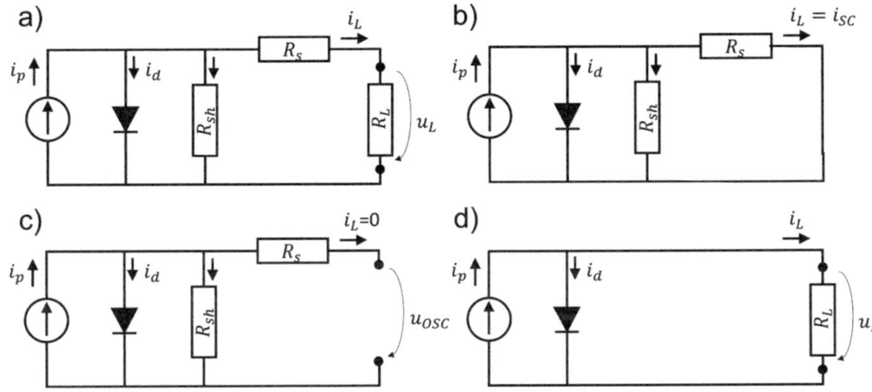

Fig. 4.19 Equivalent circuit of a solar cell for normal operation (**a**), short-circuited system (**b**), open circuit system (**c**) and ideal cell with no resistive losses (**d**)

$$\frac{i_d}{i_0} = \exp\left(\frac{u_d}{kT/q}\right) - 1 = \exp\left(\frac{u_d}{u_T}\right) - 1$$

where i_d is the current (A) through a dark diode, q is electron charge (1.602×10^{-19} C), k is the Stefan Boltzmann constant (1.38×10^{-23} J/K), u_d is the voltage drop across the diode (V), T is the absolute temperature (K), and i_0 is the reverse saturation current, which is system dependent. For example, at 25 °C the thermal voltage is $u_T = 0.025683$ V, while at 50 °C, $u_T = 0.02784$ V. The voltage drop across the diode is the same as the voltage across the shunt resistor, and the combined voltage across the shunt and load resistors:

$$R_{sh} i_{sh} = R_{sh}\left(i_p - i_d - i_L\right) = u_d = (R_s + R_L) i_L$$

From this we derive equations for i_d and u_d:

$$i_d = i_p - i_L\left(1 + \frac{R_s + R_L}{R_{sh}}\right)$$

$$u_d = i_L (R_s + R_L)$$

Inserting these into the diode equation, the current through the load resistance is

$$i_L = i_p - i_0\left(\exp\left(\frac{u_L + i_L R_s}{u_T}\right) - 1\right) - \frac{u_L + i_L R_s}{R_{sh}}$$

From this equation, the load current i_L can be iteratively solved when i_p is known (along with the constant a and the irradiance G). The power can then be calculated from $P_L = R_L i_L^2$. If the shunt resistance is infinite (if there is no connection) and the series resistance is 0 (wire connection)—corresponding to an ideal cell as shown in

Fig. 4.19d—the following explicit form for the produced power can be derived:

$$P_L = u_L i_L = u_L\left(i_p - i_0\left(\exp\left(\frac{u_L}{u_T}\right) - 1\right)\right)$$

To use the above model efficiently, we need to find the value of a to relate irradiance G to photoelectric current i_p. In addition, the values of the resistances have to be determined. In an ideal short-circuited panel, we know that the load current is equal to short-circuit current, $i_L = i_{SC}$. Inserting this into the previous equation, we can solve for the photoelectric current:

$$i_p = i_0\left(\exp\left(\frac{i_{SC}R_s}{u_T}\right) - 1\right) + i_{SC}\left(1 + \frac{R_s}{R_{sh}}\right)$$

For an ideal cell where $R_s = 0$ and $R_{sh} = \infty$, we get the approximation $i_p = i_{SC}$. For an open-circuited ideal cell, we have:

$$\frac{i_p}{i_0} = \frac{i_{SC}}{i_0} = \exp\left(\frac{u_{OC}}{u_T}\right) - 1$$

from which the reverse saturation current can be solved. Note that u_{OC} is for a single cell—the open circuit voltage for a panel, U_{OC}, divided by the number cells in series, N_S. Similarly, i_{SC} is the panel's short circuit current, I_{SC}, divided by the number cells in parallel, N_P. Hence we have

$$i_0 = \frac{i_{SC}}{\exp\left(\frac{u_{OC}}{u_T}\right) - 1} = \frac{\frac{I_{SC}}{N_P}}{\exp\left(\frac{U_{OC}}{u_T N_S}\right) - 1}$$

Example 4.2 Calculate the reverse saturation current and determine the irradiance-photoelectric current conversion factor, $a = I_p/G$, for a 20 W PV panel with the following data at STC: $I_{SC} = 1.22\text{A}$, $U_{OC} = 21.6\text{V}$, cell temperature 25 °C, and the panel has 36 cells in series and none in parallel. We assume reflectance $\rho = 0$, and transmissivity $\tau = 1$. Single-cell dimension is 22 mm × 157 mm. Air mass is 1.5, meaning that the angle of the irradiance is 41.8° from the normal, affecting the fraction of irradiance reflecting away from the panel.

Solution Based on above, we can calculate the thermal voltage.

$$u_T = \frac{kT}{q} = \frac{1.38 \times 10^{-23}}{1.602 \times 10^{-19}} \times (25 + 273.15) = 0.02568\text{V}$$

$$i_0 = \frac{\frac{1.22}{1}}{\exp\left(\frac{21.6}{0.02568 \times 36}\right) - 1} = 8.695 \times 10^{-11}\text{A}$$

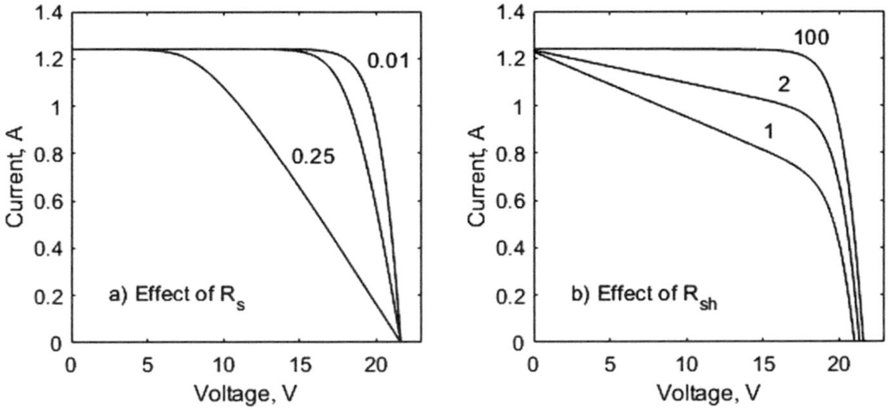

Fig. 4.20 Effect of parasitic resistances, **a** $R_s = 0.01, 0.05, 0.25\ \Omega$ for $R_{sh} = 100\ \Omega$, and **b** $R_{sh} = 100, 2, 1\ \Omega$ for $R_s = 0.01\ \Omega$

Under the standard test conditions irradiance is 1000 W/m^2. The conversion factor a can be calculated using

$$a = \frac{I_p}{G} = \frac{I_{SC}}{G} = \frac{1.22\text{A}}{1000\text{W/m}^2} = 0.00122\ \frac{\text{A}}{\frac{\text{W}}{\text{m}^2}}$$

The same current flows through all cells in this case, since they are in series. In the case of a panel with parallel cells, the generated photoelectric current must be divided across the number of parallel rows.

The effects of series and shunt resistances are presented in Fig. 4.20.

The series resistance, R_s, can be obtained from the measured slope of the U-I curve at U_{OC}, and R_{sh} from the slope at the $U = 0$, or they can be obtained by fitting our model to the data by keeping them as free variables.

Example 4.3 Analyze the following measurement data from a 20 W panel. As above, we have at STC: $I_{SC} = 1.22$A, $U_{OC} = 21.6$V, cell temperature is assumed to be the same as the surrounding temperature, 19 °C, and there are 36 cells in series, none in parallel. Fit the following equivalent circuit model of a single cell to the data given in Table 4.1:

$$i_L = i_p - i_0\left(\exp\left(\frac{u_L + i_L R_s}{U_T}\right) - 1\right) - \frac{u_L + i_L R_s}{R_{sh}}$$

Solution Examining the data, we see that there is no slope in the current at small voltages around the point of short-circuit current. We can therefore assume that the shunt resistance is infinite, $R_{sh} \to \infty$, and feeding this into the model provided eliminates the third term. In addition, as there are 36 cells in series, the cell voltage

Table 4.1 Measurement data, at atmosphere temperature of 19 °C, clear-sky conditions

i	U	I	P
–	V	A	W
1	0.0	1.24	0.00
2	5.6	1.24	6.94
3	8.1	1.24	10.04
4	9.9	1.24	12.28
5	11.7	1.24	14.51
6	17.0	1.20	20.40
7	19.5	0.96	18.72
8	20.8	0.50	10.40
9	21.1	0.25	5.28
10	21.4	0.08	1.71

can be obtained by dividing the panel voltage by the number of cells. We therefore have,

$$I = i_p - i_0\left(\exp\left(\frac{U/N_s + IR_s}{u_T}\right) - 1\right)$$

where capital letters refer to panel values. If we assume that the panel temperature is equal to atmospheric temperature, we can calculate the reverse saturation current,

$$u_T = \frac{kT}{q} = \frac{1.38 \times 10^{-23}}{1.602 \times 10^{-19}} \times (19 + 273.15) = 0.02517\text{V}$$

$$i_0 = \frac{\frac{1.24}{1}}{\exp\left(\frac{21.4}{0.02517 \times 36}\right) - 1} = 6.865 \times 10^{-11}\text{A}$$

If we assume that photoelectric current is equal to the short circuit current, $i_p = I_{SC} = 1.24\text{A}$ then the only remaining unknown variable is the series resistance, R_s. This can be obtained by fitting the model to the measurement data in Table 4.1 using series resistance as a free parameter—more specifically, we want to minimize the sum of the errors between the model and the data, for $N = 10$ measurements. This can be written mathematically as

$$\min(F(R_s)), R_s > 0$$

$$F(R_s) = \sum_{i=1}^{10} \left(I_i(U_i) - I_{exp.i}(U_i)\right)^2$$

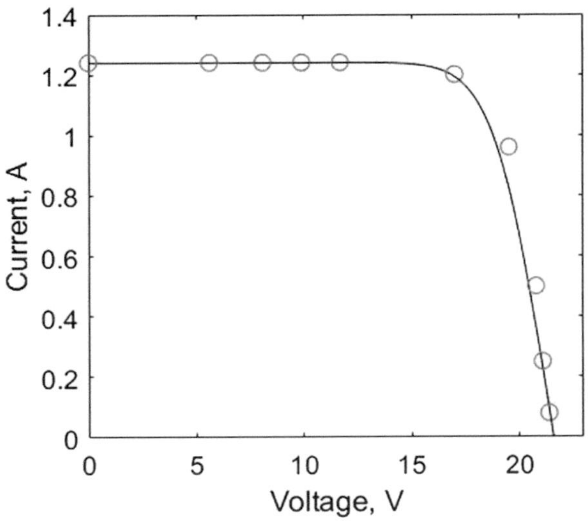

Fig. 4.21 Fitted model versus measured data for Example 4.3, $R_s = 0.0368\ \Omega$ and $R_{sh} \rightarrow \infty$

This calculation can be performed using mathematical software, such as Matlab® or Excel®, giving $R_s = 0.0368\ \Omega$. Figure 4.21 presents the results.

As Fig. 4.21 shows, the correlation is very good and there is no visible error resulting from the assumption of infinite shunt resistance in this case.

4.3.2 Limits for the PV Panel Efficiency

The efficiency of a photovoltaic cell and a panel, η_p, is defined as the ratio of output power at the maximum point of power, to incoming irradiance. The efficiency can be defined by three main variables—short circuit current I_{SC}, open circuit voltage U_{OC}, and fill factor f—so that

$$\eta_p = \frac{P_{MP}}{G} = \frac{U_{MP} I_{MP}}{G} = \frac{f U_{OC} I_{SC}}{G}$$

Under standard test conditions, input irradiance is equal to 1000 W/m^2.

Based on Green [66], the maximum value of the fill factor is mainly a function of the open circuit voltage, and can be calculated from an experimentally-determined empirical correlation, as a function of the normalized voltage $v_{OC} = U_{OC}/u_T$, such that

$$f = \frac{v_{OC} - \ln(v_{OC} + 0.72)}{v_{OC} + 1}$$

What are the physical mechanisms that limit the efficiency? We will first discuss the main theoretical limitations for U_{OC} and I_{SC}, and then discuss their temperature dependency.

In an ideal case, every photon having more energy than the band gap value, $E_{bc} = hc/\lambda_{bc}$—that is, with wavelength shorter than λ_{bc}—is able to create a free electron–hole pair. For silicon, the band gap is 1.1 eV, which corresponds to a wavelength of 1.13 μm. Boosted electrons travel to the negative terminal and, from there, through the load to the positive cell terminal, where they meet the holes that have travelled through the p-type semiconductor material. This results in electrical current flowing from positive to negative terminal through the load.

Electrical current is defined as the number flux of electrons, $\dot{N}_e$ (1/m^{2}s), multiplied by electron charge, q, and cell area, $i = \dot{N}_e q A_{cell}$. The number flux of electrons at different wavelengths can be calculated by dividing the spectral radiance distribution by the energy of a photon at each wavelength,

$$\dot{N}_{e.\lambda} = \frac{G_\lambda}{hc/\lambda}$$

where $h = 6.62607 \times 10^{-34}$Js is Planck's constant, and $c = 2.9979 \times 10^9$m/s is the speed of light in vacuum. By integrating this with respect to wavelength from 0 to λ_{bc}, the photocurrent density per cell surface area can be obtained,

$$\frac{i_p}{A_{cell}} = \dot{N}_e q = \int_0^{\lambda_{bc}} \frac{G_\lambda q}{hc/\lambda} d\lambda$$

The photoelectric current essentially defines the theoretical maximum of the short-circuit current. The result of this is integration is shown in Fig. 4.22 as a function of band gap. The values for both AM0 and AM1.5 are calculated using their respective irradiative power densities from Fig. 4.2.

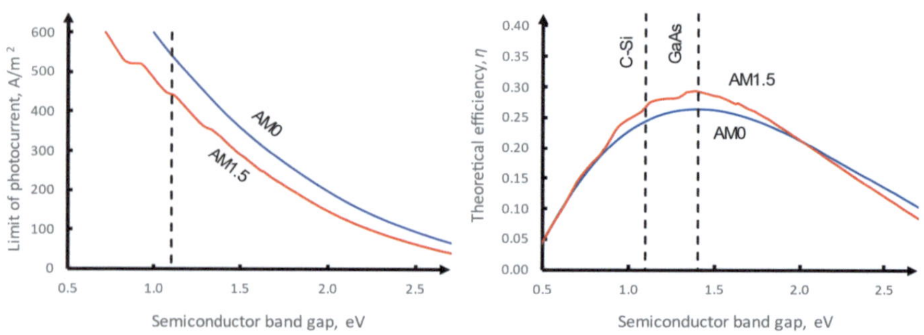

Fig. 4.22 Left: Theoretical maximum for photocurrent density, vertical dashed line indicates the 1.1 eV band gap for silicon cells. Right: theoretical efficiency. Blue lines show extraterrestrial (AM0) and red lines show terrestrial (AM1.5) conditions

Based on the data in Fig. 4.22, the theoretical maximum current for silicon is around 450 A/m^2. Modern mono-crystalline silicon cells are capable of generating 300–400 A/m^2 *per single cell surface area* under standard test conditions. Note that for cells connected in series, the overall current is the same value as that for a single cell. As Fig. 4.22 shows, the smaller the band gap, the larger is the amount of photons able to participate in generating electrical energy.

The definition of the theoretical maximum of the open-circuit voltage is more complicated. From the open-circuit model equation (no current to the load, see Fig. 4.19c) of an ideal panel with no parasitic resistive losses, we can solve for the value of the open circuit voltage,

$$u_{OC} = u_T \ln\left(\frac{i_p}{i_0} + 1\right) = \frac{kT}{q} \ln\left(\frac{i_{SC}}{i_0} + 1\right)$$

The reverse saturation current is the only unknown parameter that impacts this result: the smaller the value of i_0, the larger the open circuit voltage. Reverse saturation current can be calculated in principle from the physical properties of the panel materials [66]. A simplified equation for the theoretical minimum value for i_0 (A) can be obtained from the following model, where the relevant parameters have been combined into a single constant:

$$i_0 = 1.5 \times 10^9 \times \exp\left(-\frac{E_{bc}}{kT}\right) A_{cell}$$

$$u_{OC} = \frac{kT}{q} \ln\left(\frac{i_{SC}}{1.5 \times 10^9 \times \exp\left(-\frac{hc/\lambda_{bc}}{kT}\right) A_{cell}} + 1\right)$$

Based on this equation, smaller band gaps give a larger reverse saturation current, and a smaller open-circuit voltage.

The results discussed above can be combined into a model for panel efficiency as follows:

$$\begin{aligned}\eta_p &= \frac{N_s u_{MP} N_p i_{MP}}{G N_p N_s A_{cell}} = \frac{f(u_{oc}) u_{oc} i_{sc}}{G A_{cell}} \\ &= \frac{U_{MP} I_{MP}}{G N_p N_s A_{cell}} = \frac{f(u_{oc}) U_{oc} I_{sc}}{G N_p N_s A_{cell}}\end{aligned}$$

where $N_p N_s$ is the total number of the (parallel and series) cells in a panel.

At this point we assume that there are no losses or protection diodes connected between the cells, so series-connected voltages and parallel-connected currents are additive—in other words, the number of cells used does not affect the panel efficiency. Using this and the above-mentioned models for maximum current, voltage, and fill factor, we can draw the theoretical maximum efficiency curve for a cell (Fig. 4.22, right-hand graph). As can be seen, crystalline silicon cell efficiency is slightly off from the peak of the curve, while the GaAs cell has an almost optimal band gap of 1.4 eV.

Next, we briefly discuss the effect of temperature on cell performance. Since cell temperature varies as a function of ambient temperature and irradiance, it is important to understand the effect of cell temperature on the produced power and efficiency.

Open circuit voltage is primarily impacted by temperature [66]. Without going into a detailed derivation, the temperature dependency of the open circuit voltage and short circuit current *of a single cell* can be derived from the equations for i_{sc} and u_{os}, and the following result is obtained (temperatures here in °C):

$$\frac{du_{oc}}{dT} = -0.0023V/°\text{C} \approx \frac{u_{oc}(t) - u_{oc}^{STC}}{t - 25}$$

This means that for example if the temperature increases 10 °C, u_{oc} decreases 0.023 V. Standard test conditions are used here as the reference, and any other value $u_{oc}(t)$ can be solved from this linear equation. Temperature also affects the fill factor as a function of open-circuit voltage, $f(u_{oc}(t))$. For a panel with N_s series cells, this temperature coefficient must be multiplied by N_s. If better accuracy is required, the temperature dependence of the short circuit current can be calculated from $di_{sc}/dT = 0.0006V/°\text{C}$, which is significantly smaller than the effect of temperature on voltage.

To use these models, the temperature of the panel must be known. Panel manufacturers often report the panel's normal operating temperature, t_p^*, in relation to the reference surrounding air temperature t_∞^*, reference irradiance G^*, and wind speed. The panel's energy balance for these normal operating conditions can be written in the steady state as

$$G^* = \alpha_e\left(t_p^* - t_\infty^*\right)$$

This equation in interpreted as the incoming irradiance heating the panel, while the panel cools by convection and thermal radiation to the surroundings—the effective heat transfer coefficient, α_e, combines both of these heat transfer mechanisms.

We then write the energy balance for any other conditions as

$$G = \alpha_e\left(t_p - t_\infty\right)$$

If we assume that the effective heat transfer coefficient is constant, solving for α_e from these two equations allows us to obtain the following model for the panel temperature:

$$\frac{G^*}{\left(t_p^* - t_\infty^*\right)} = \alpha_e = \frac{G}{\left(t_p - t_\infty\right)}$$

$$t_p = t_\infty + \frac{G}{G^*}\left(t_p^* - t_\infty^*\right)$$

As an example, Green [66] estimates that under an irradiance of 1000 W/m^2, panel manufacturers typically allow the panel temperature to increase to 30 °C above atmospheric temperature—as a result, $\left(t_p^* - t_\infty^*\right)/G^* = 0.030$ and we find $t_p = t_\infty + 0.03 \times G$. Unless more accurate data is available, this estimate can be used as a starting point. A more detailed analysis could be carried out using computational fluid dynamics simulations on the conduction and convection heat transfer with the surroundings, and conduction in the panel structure, or by carrying out detailed temperature measurements.

4.3.3 Additional Loss Effects

The theoretical efficiencies discussed above give a guide to the characteristics of cell behavior when there are no losses. As already shown, parasitic resistive losses should be minimized, and they are often indications of imperfections in the manufacturing process. When cells are assembled into a PV module, panel, or multiple panels connected into a larger grid in a solar PV power plant, additional requirements must be met for optimal system operation.

One of the most significant losses comes from the mismatch of cell and panel voltage-current characteristics when connected in series. If one of the cells in a series-connected group of cells is broken or shaded, the total voltage of the group is still the sum of all cells—but since the same current must flow through all cells connected in series, the current of a single broken or shaded cell sets the limit for the current through the series-connected string of cells as a whole. Power generated in the operating cells will be dissipated in the problematic cell as heat. This creates a hot spot in the panel which can, in the worst case, break the cell or the entire panel.

Let us study a panel made from four cells connected in series, of which one cell is shaded so that its photovoltaic current is reduced by 50% (value 0.5 in the second step equation below) compared to the other three fully-illuminated cells. Based on the above, the current through the shaded cell must be equal to the current flowing through all other cells, so

$$\begin{aligned} I(U) &= 0.5 \times i_p - i_0\left(\exp\left(\frac{\Delta u}{u_T}\right) - 1\right) \\ &= i_p - i_0\left(\exp\left(\frac{U - \Delta u}{3 \times u_T}\right) - 1\right) \end{aligned}$$

Note the number 3 in the exponential, as that is the number of properly operating cells. To examine this situation, first we can solve for the voltage drop over the shaded cell, Δu, from right-hand equality; then we can find the corresponding current, $I(U)$, using either of the equalities in the equation. Power dissipated as heat is then calculated as $P_{heat} = I(U)\Delta u(U)$, noting that Δu is a function of U.

Example 4.3 Analyze a panel with four series-connected cells, in which one cell is first 30% shaded, and then when it is 90% shaded. As in Example 4.3, $I_p = 1.22$ A, $u_T = 0.02517$ V, $i_0 = 6.865 \times 10^{-11}$A. Draw the U-I curves for a fully illuminated panel, for the single shaded cell, and for the panel overall including the shaded cell.

Solution To draw the curves as a function of the panel voltage, we first need to solve for the voltage drop of the shaded cell, Δu, as a function of the panel voltage, $U = [0, 2.5\text{V}]$,

$$f(\Delta u) = 0.7 \times i_p - i_0\left(\exp\left(\frac{\Delta u}{u_T}\right) - 1\right) \\ - i_p + i_0\left(\exp\left(\frac{U - \Delta u}{3 \times u_T}\right) - 1\right) = 0$$

Solving this requires numerical iteration, and with the obtained Δu values, the current produced by the panel can be easily calculated,

$$I(U) = 0.7 \times i_p - i_0\left(\exp\left(\frac{\Delta u(U)}{u_T}\right) - 1\right)$$

Finally, the produced power and wasted power are calculated as $P = UI$ and $P_{heat} = I(U)\Delta u(U)$, respectively. Note that value 0.7 corresponds to 30% shaded cell, and 0.1 should be used for 90% shaded cell.

For the 30% shaded cell, for example, with the panel in the short-circuit condition, $U = 0$, and the solution of the above equation $f(\Delta u) = 0$ gives $\Delta u = -1.73$V. This means that, for a short-circuited panel, the shaded cell is reverse biased. Power wasted to heat is equal to $P_{heat} = -1.73 \times 0.868 = -1.482$W, and again the negative sign indicates that the shaded cell is reverse biased. Figure 4.23 presents these results.

To summarize, a single shaded or poorly operating cell within a panel spoils the production of the panel as a whole—a significant share of the production of properly operating cells is wasted as heat in the shaded cell. In the short-circuited system, $U = 0$, the shaded cell is reverse-biased, and the other cells feed power to this heat dissipation. We can also see from the bottom row of Fig. 4.23 that when the panel voltage increases above ~ 1.7 V, the shaded cell begins to produce power.

4.4 Design of a PV Power Plant

In this section we discuss the design of an actual PV power plant. We start with small off-grid DC systems, and a brief analysis of connecting a PV panel to a battery and more advanced maximum power point controllers. We then extend the analysis to large-scale industrial plants with AC output to a grid, sharing some practical design ideas. Finally, we discuss issues of sustainability in the siting of PV power plants:

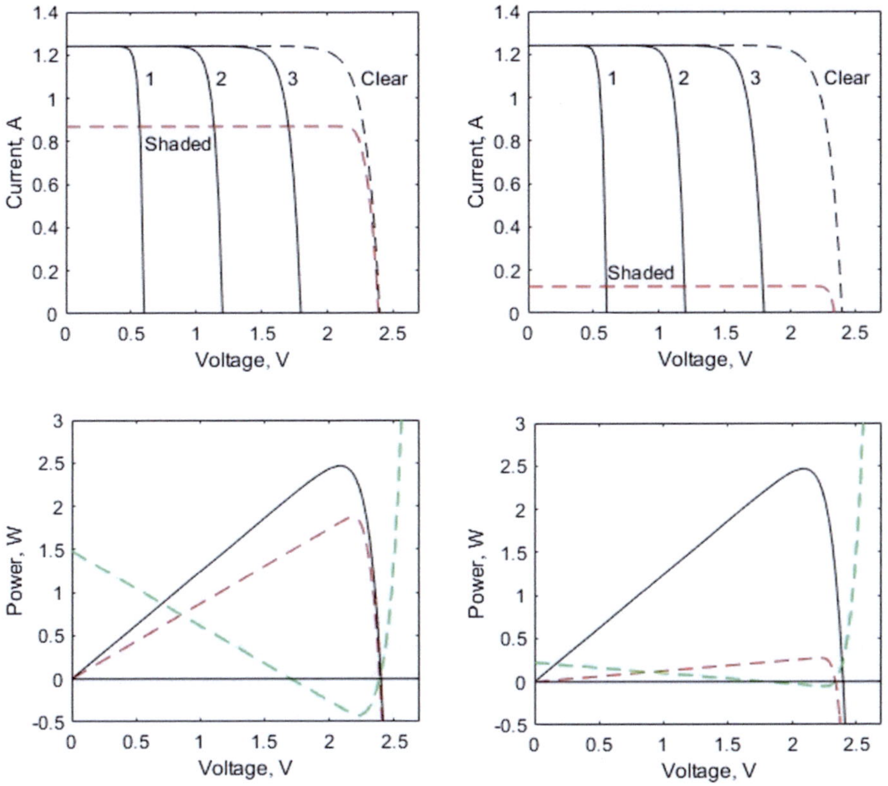

Fig. 4.23 (Top row) Voltage-current curves of 1 to 4 fully illuminated cells in series—the dashed black lines indicate operation for a fully unshaded panel. The red dashed lines show operation with one shaded cell, with shading at 30% (left) and 90% (right). (Bottom row) Power curves: black solid line is the fully illuminated panel, red dashed line is the shaded panel, and green dashed line is the dissipated heat (sign changed to positive) in the shaded cell

what are the negative impacts on nature and communities, and how they can be minimized.

4.4.1 Stand-Alone, Battery-Coupled PV Systems

Remote locations with no grid connection, and many electric-powered mobile applications, often use batteries as energy storage. Different kinds of batteries will be discussed in detail in Chap. 7; here we focus only on the electric characteristics of a traditional lead-acid battery as part of a solar energy system.

We discussed in Sect. 4.3 the voltage-current characteristic of solar photovoltaic (PV) cells and panels. A PV panel is a *current* source, and even when the terminals

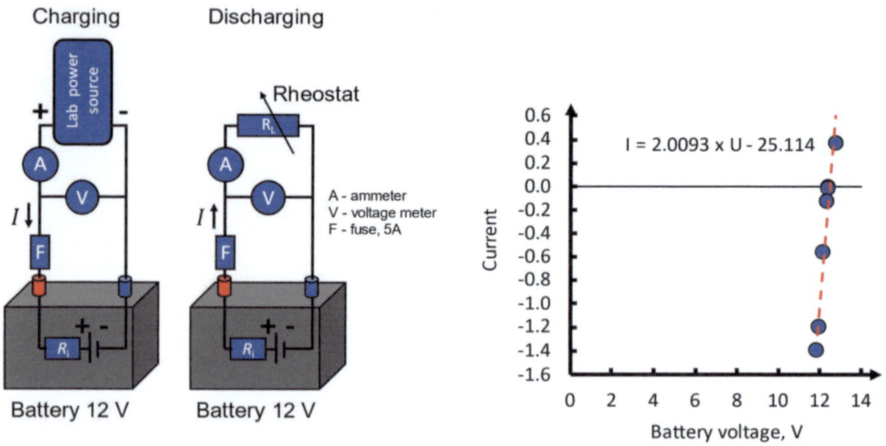

Fig. 4.24 Battery connected with a variable load resistor (rheostat), laboratory power source, voltage meter, and ammeter. The graph shows a small set of experimental results (positive direction of the current is into the battery)

are short-circuited, the current can only reach a finite value—the short-circuit current. In contrast, a battery is a *voltage* source, maintaining a constant voltage as the load increases. This means that with infinite load (that is, short-circuiting the battery) forces the current to approach infinity, which in the worst case can result in the battery exploding. For this reason, systems with large batteries must always be fused for safety. If the current grows higher than a predefined value (say, 16 A) the fuse will burn, saving the wires and battery from damage.

To understand how a battery works when connected to a solar panel, let us consider a set of measurements for a series-connected 12 V battery, an adjustable load resistor, a laboratory power source, and a fuse. Fig. 4.24 presents measured data for a U-I curve for this system.

The more negative the discharging current, the smaller the battery terminal voltage, U. There is no current through the voltmeter branch, and so (noting that sign of the current from the positive battery terminal is negative outwards) we can write the equation for the potential drop over the circuit as

$$U_0 - (-R_i I) - U = 0$$

From this we can solve the current equation and set it equal to the fitted trendline in Fig. 4.24,

$$I = \frac{U}{R_i} - \frac{U_0}{R_i} = 2.0093 \times U - 25.114$$

From this we can solve for the internal battery resistance, getting $R_i = 1/2.0093 = 0.5\ \Omega$, and the open circuit voltage of the unloaded battery from $U_0 = 0.5 \times 25.114 =$

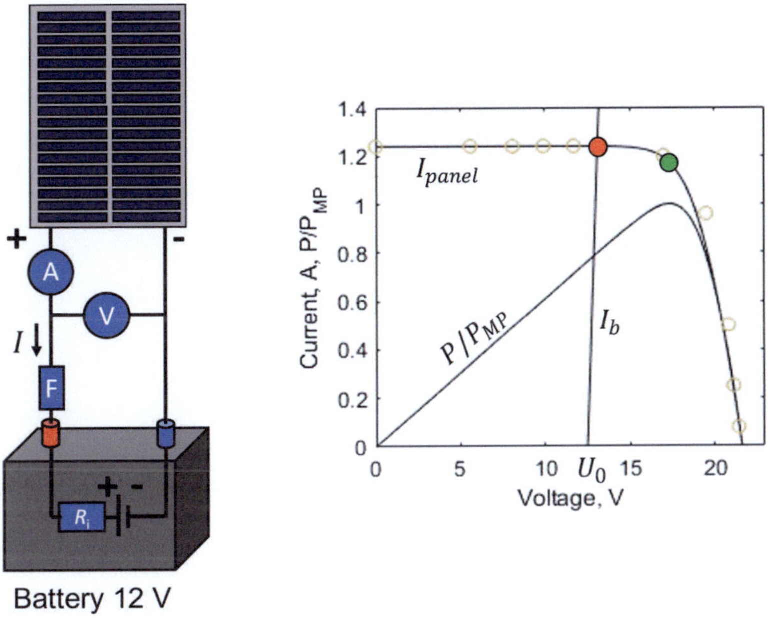

Fig. 4.25 Battery connected with a PV panel, and the U-I characteristics of both components. Red dot indicates the actual operation point, green dot is the maximum operational point

12.5V. We can now use this model to determine the operational point of a PV panel and battery in combination.

The current from the panel is same as the charging current—mathematically, this means that the U-I characteristics of the panel and battery intersect. This is illustrated in Fig. 4.25.

As Fig. 4.25 shows, the operational point of this panel-battery system is not at the maximum available power point, indicated by the green dot—being on the left-side of the maximum, however, guarantees that even if the irradiance drops, then operational point is still is on the correct side. If the open circuit voltage of the battery is too close to the open circuit voltage of the panel, there is a risk that when the panel heats up, moving U_{OC} left, the curves no longer meet on the charging mode, but the battery starts to feed power to the panel. As an extreme example, in Nordic countries the temperature of a panel across a year can vary from 60 to − 25 °C—as a result, the open circuit voltage may drop as much as 0.2 V per cell, so for a panel with 36 cells, the voltage drop is already more than 7 V.

The benefit of the direct panel-battery connection discussed above is its simplicity, but there is no mechanism to limit charging when the battery is full, and the system is not operating at the maximum power point. In addition, if the panel is shaded and produces no power, current starts to flow into the panel, heating it up. Actual solar PV plants have a special maximum power point controller, charge regulator, and low voltage disconnector, which together control the system, taking it to the maximum

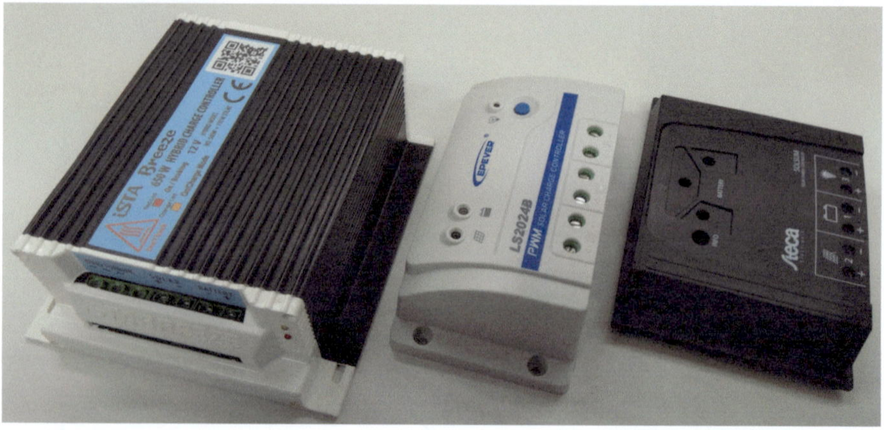

Fig. 4.26 Maximum power point controllers. The leftmost model is for a hybrid system with solar PV of 150 W and a maximum 500 W wind turbine

power point, and disconnecting the panel when there is no production—see some examples in Fig. 4.26. These controllers are robust and very easy to use. Larger than recommended loads by the controller manufacturer should be directly connected to battery terminals.

The regulator and the low voltage disconnect are systems that disconnect the battery when its voltage level gets too high or too low, respectively. Since using diodes or transistors always results in a voltage drop of 0.6–0.9 V per component, a mechanical relay is often used to minimize these losses in the regulator.

These controllers generate DC (direct current) voltage and current, which can be used to charge a battery, and to supply mobile chargers and any other systems that operate with DC power. However, to power normal AC (alternating current) electrical devices, a DC/AC inverter is required. The inverter is simply connected with the load connection—we will discuss the operation of inverters in more detail in Sect. 4.4.2 on industrial- and utility-scale grid connected systems.

In addition to the components discussed above, the PV system also needs a blocking diode to prevent current from flowing from the battery to the panels in the case of no production. Bypass diodes are also used in all panels to prevent them from overheating if reverse-biased when shaded or malfunctioning. An example of a simple 24 V system is shown in Fig. 4.27.

4.4.2 Industrial Large-Scale PV Systems

In 2022, 4.6% of world's electricity was produced by solar panels [51] with the construction of increasing numbers of large, up-to-GW-scale, grid-connected photovoltaic power plants. The ongoing, continual decrease in the costs of solar panels

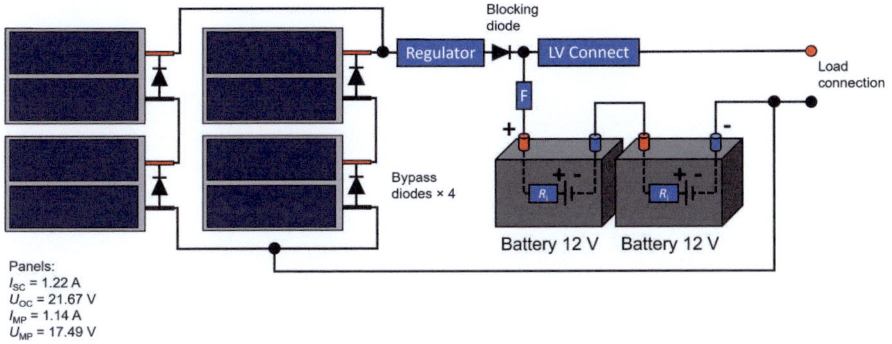

Fig. 4.27 An example 24 V off-grid system. Bypass diodes are usually built into the panels, but are shown here as separate components for clarity

has hastened this development—in fact, in 2023, solar PV and onshore wind were the cheapest renewable energy generation technologies, with a lower limit of 24 $/MWh [115]. This price does not include the cost of storage, however, and so the actual price of continuously-available power is higher.

The grid connection is a large energy sink and provides a stable reference frequency to the inverter that converts the DC power to AC. Grid-connected PV power plants continuously monitor their grid condition, its voltage level and frequency. Similar to converter-connected wind power, solar PV power plants cannot operate without the grid, and they are shut down if the grid loses its power supply, so as not to cause damage to other systems connected to the grid. And unlike coal-fired power stations, for example, in PV systems there is no rotating masses to provide physical inertia or grid-forming capacity.

To connect the power plant to a national grid, it must satisfy the grid code requirements—country-specific standards set by the power transmission system operator (TSO) that define allowed voltage and frequency variations and required recovery rates. These are discussed in more detail in the Chap. 8 on Electrical power generation, conversion and transmission systems.

In PV power plants the panels are installed in long rows, called arrays, and then a sufficient number of panels (typically a single row) are connected to a maximum power point controller and a DC/AC inverter. This generates grid-synchronized AC power and feeds it to the powerhouse and switch, and finally to the main transformer and grid connection. Figure 4.28 illustrates an example of a panel arrangement in a relatively small 5 MW solar PV power plant, with the inverter unit shown at the end of the row.

For the Vertex bifacial panels used in the plant shown in Fig. 4.28, the maximum power is 665 W, efficiency is 21.4%, open circuit voltage is 46.1 V and short circuit current is 18.5 A (all values at STC). Panel dimensions are 2384 × 1303 mm, and each contains 132 cells. There are four horizontal rows of horizontal panels installed, and so the vertical dimension of the rows is $4 \times 1.3 \sim 5.2$m. Panels are connected electrically so that, in the morning and evening when lower rows of panels are shaded

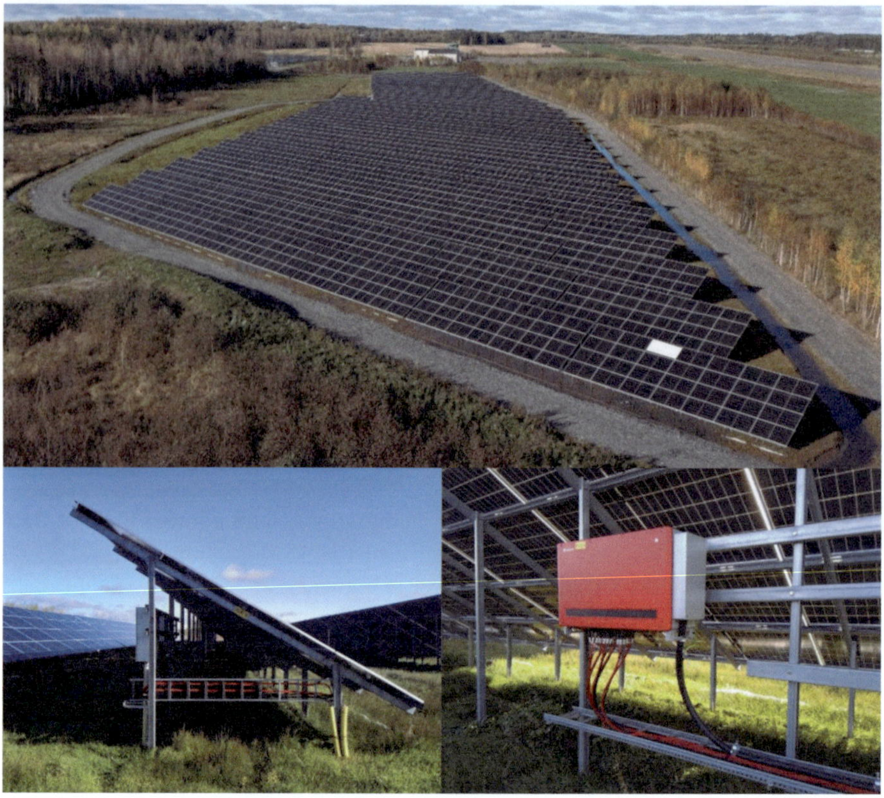

Fig. 4.28 Industrial 5 MW PV power plant with bifacial solar panels. A single panel is highlighted with a white rectangle. Four panel rows are vertically installed in each row. The bottom-left figure shows the steel support structure and panel wiring (red and black for positive and negative terminals, respectively) entering the red inverter box shown at bottom-right. The thick black cable on the right is the AC power connection, which is fed to a local switch, and from there to the switch boxes using underground cable. (Pictures by Mika Järvinen and Kalle Kataila)

by the upper rows, the illuminated upper rows can still continue full production. In practice this means that panels across each of the four rows are connected in series, while the four rows are connected in parallel.

As can be seen from the top image in Fig. 4.28, the ground area around the panels is cleared to avoid any shading effects. Panels are sufficiently raised from the ground (~1 m) to avoid shading from vegetation in summer, and from snow in winter. In this location the space under the panels has not been deliberately revegetated, but the land has been taken over by herbaceous annuals and perennials, such as grasses and other flowering plants. Small animals can also pass through the power plant under the arrays, though deep ditches along the arrays might form a barrier for movement for some animals. The location is near to an old airport, so there has been no need for additional forest clearance since the operation of the airport requires that the vegetation is controlled. There are service and safety exit roads around the area that allow access to all panels. The connections are collected at the end of the rows for easy access and maintenance. Ground mounting is made very stable by using steel foundations, with a horizontal plate at the base of the vertical, corrosion-free steel profiles. To prevent ground frost from twisting the panels in winter, the lowest point of the foundations is installed below the frost depth (e.g., 1–1.5 m in central Finland, Syke [188]).

4.4.3 Identification of Sites for PV Production

Identifying possible sites for PV plants, also known as the screening, is a multi-step process. Unlike rooftop solar, ground-mounted PV plants occupy vast land areas, and so it is imperative to consider the kind of land used, and the impacts, including ecological and social, of occupying that land. PV-based electricity production also has different kinds of very significant upstream impacts related to material extraction and manufacturing of the PV modules, inverters, cables and other components, as discussed in Sect. 4.4.2—however, these impacts are not included in the screening process of a solar PV plant, which considers only local conditions. The negative ecological and social impacts of mineral extraction is, however, a growing concern, and though these impacts cannot be mitigated by the choices made in the screening process, it is nonetheless important to consider the sustainability of the entire value chain of the power plant, and to avoid or minimize negative impacts at every step. In the screening process, the focus is on the site of the PV power plant itself: effects related to construction, land occupation, and changes in the land use or landscape that can impact nature and local communities (see Sects. 4.7 and 4.8).

Potential negative impacts can be mitigated by taking them into account in the screening process from the beginning. There are two basic screening approaches: a Boolean logic method, which applies different constraints to exclude unfavorable areas; and a suitability-score or index-based approach, which evaluates the potential of different areas by calculating a score based on how well they fulfill certain criteria. Boolean logic was used in studies to identify sites for PV plants in Italy [140], Bangladesh [176] and Sweden [126], for example. Examples of the score-based approach include an implementation of a "Compatibility Index" to identify sites with low biological impact [186], "suitability classes" that consider operational constraints and local populations [222], and a "least conflict area" classification that includes the evaluation of environmental and agricultural constraints [151].

To better understand the design process, let us consider a case study from Finland, which introduces a geospatial screening analysis carried out with QGIS, a free open-source geographic information system (GIS) software tool [114]. The study was based on Boolean logic with the purpose of identifying potential locations in Finland for large-scale PV plants with minimal adverse impacts. The analysis began with the definition of boundary conditions, i.e., the minimum solar irradiation and maximum distance to the point of possible grid connection. Next, the so-called "NO-area" was defined, comprising different areas where PV plants could not or should not be built. The NO-area was formed through technical, legislative, environmental, social, cultural, and practical constraints, which may vary between countries or even between regions and municipalities. The first steps of the process are explained in Fig. 4.29.

The study sought to mitigate environmental impacts by including in the NO-area regions with high environmental, ecological, cultural, or social value (see also Sect. 4.7 Impacts on nature and mitigation of negative effects, and 4.8 Social impacts of solar energy). Conservation areas and other locations with identified ecological or environmental value, such as important bird areas, were included as NO-areas with a 500 m buffer zone. The inclusion of a buffer zone around nature conservation areas is important, as the clearance of the power plant site and related power lines, construction of new roads, and increased traffic during the construction period, all create disturbance that can conflict with the nature conservation goals. After all, the man-made borders of these areas are not visible for species living within the ecosystem. To ensure areas in a natural or semi-natural state were not considered for PV, productive forest land and undrained wetlands were included in the NO-area. Forests, even when intensively managed, provide habitats for some forest species, and maintain connectivity between forests with higher natural value, such as old growth and protected forests. Moreover, clearing forests for solar power development results in forest loss, and may also create land-use change elsewhere as new, replacement forestry areas are established.

These same principles can also be applied to agricultural land. Using arable land for PV plants might contribute to clearing new land for agriculture elsewhere, unless the site incorporates an agrivoltaic system [57, 78]. In any case, it is not ideal to compete with food production—however, due to a lack of data, analyses carried out for the Finland case study did not include agricultural lands as NO-areas.

The NO-area also included urban regions, houses, and culturally or socially valuable areas such as valuable landscapes. Proximity to homes and holiday houses is important for social acceptability. In addition, infrastructure such as roads, gas pipelines and powerlines, may require safety distances. In the case study analysis, safety distances were considered for roads, but not for gas pipelines and powerlines, as the Finnish gas grid is not extensive and suitable data was not available. Figure 4.29 presents all the criteria, boundary conditions and limitations for the Finland example.

After defining the criteria and collecting relevant data, the spatial analysis can be carried out. In GIS software, the spatial data can be processed with built-in tools. In the case study, the analysis was based on Boolean logic, such that the NO-areas were deducted from the area formed by the boundary conditions. Areas below 20 hectares were also removed, as the study focused on large-scale PV plants. The results were then manually checked and cleaned, as not all the data were available or up to date.

With the Boolean approach, the analysis does not classify or rank the results based on their suitability, and so the suitability of each resulting area must be independently considered. The Finland study, for example, resulted in areas that were too small, and located too far from the grid, to be profitable. This is a weakness of the Boolean approach compared to the suitability-score approach, which could give fewer points

	Define parameters	Look for available data	Define screening criteria
		• What kind of data are available? • Can the format be used in spatial analysis? • Is the data up to date?	• Is a buffer needed? • How heavy the analysis is, can the data be simplified?
• What are parameters that directly affect the profitability? • Which of them can be included in the screening criteria?	Solar irradiation	Maps of Global Horizontal Irradiation can be georeferenced	GHI > 850 kWh/m2
	Distance to the grid	Substation locations can be manually collected from TSO map service	Distance to substation < 15 km
	Size of the project	Available after performing the analysis	Size of the area > 20 ha
• What kind of areas should not be used for PV? • What kind of areas have high environmental or ecological value? • Where the project could have significant impacts on the environment?	Conservation areas	Nature protected areas, Natura 2000 areas, National Conservation Programme areas available in shapefile	500-metre buffer
	Important habitats	Important Bird Areas, important habitats defined in the Forest Act, high biodiversity value forests defined in other studies available in shapefile	500-metre buffer
	Natural or semi-natural habitats	Spatial data with forest land peatland classification available. Peatland data old so much be checked after analysis	No overlapping
• What kind of areas should not be used for PV? • What kind of areas are culturally or socially valuable? • Where the project could have significant impacts on local people?	Landscape areas	Nationally Valuable Landscapes and Special Landscape Management Areas available in shapefile	500-metre buffer
	Cultural environments and historical sites	Nationally significant built cultural environments and ancient monuments available in shapefile	500-metre buffer
	Urban areas and houses	Corine Land Cover data for areas related to urban actions and data for addresses available in shapefile	500-metre buffer to everything but roads, overlapping tackled with another criteria
• What kind of areas cannot be used for PV? • What kind of restrictions come from legislation? • What kind of technical constraints are applied?	Safety distances to roads, powerlines, gas pipelines	Spatial data available for roads but not for powerlines or gas pipelines but they can be manually checked from the results	Respect safety distances to roads Powerlines, gas pipelines can be manually checked after the analysis
	Slope of the terrain	Elevation model from which the slope can be calculated available	Below 15 %
	Water bodies	Spatial data available for the Baltic sea, lakes, ponds, rivers and ditches.	500-metre buffer to rivers and lakes over 5 ha to simplify analysis

Fig. 4.29 The beginning of the screening process and formation of the criteria for the Finland case study. The process is greatly dependent on available data and should be adapted and modified accordingly. Sometimes the required data is not available—criteria may then need to be finetuned or, if it is not too laborious, to be manually defined

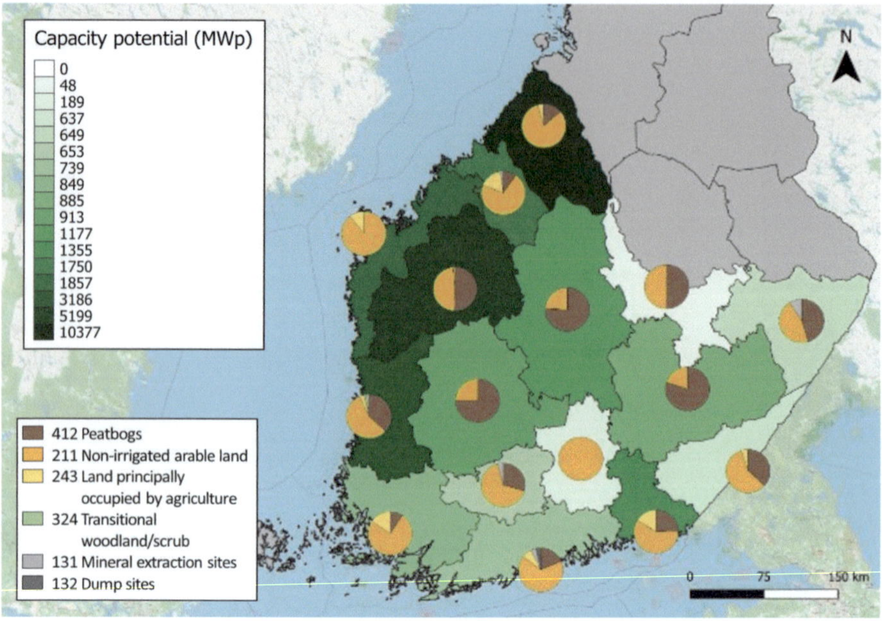

Fig. 4.30 Results of the screening analysis by region. Capacity potential of a region is shown in shades of green. Pie charts show the distribution of land-use types for each area, categorized according to Corine Land Cover 2018. Åland (excluded in the analysis) and regions or parts of regions that did not fulfill the boundary conditions are shown in grey. Background map: © OpenStreetMap contributors, licensed under ODbL

to small areas far from the grid, and therefore present them as less favorable areas. Figure 4.30 presents a map of the Finland results.

The results of the analysis show good potential for PV in Finland—in total, the predicted potential was ~ 30 GW. As can be seen in Fig. 4.30, the results are concentrated on the western coast, likely due to the fact that the grid is denser there, as wind power has been developed in that region for many years.

A weakness of the study is also clearly shown, as the results include a lot of agricultural land that should be avoided but could not be categorized due to the lack of suitable data. The second-most-common type of land use, however, is peat production areas—and these are exactly what should be considered to minimize adverse impacts. They have already been in industrial use and the land can be considered degraded, which means that using them for PV plants would not cause significant land-use changes. These so-called brownfield areas should be prioritized to prevent larger land-use changes and habitat losses. Old industrial sites, landfills, and mines have also been suggested for PV plants to mitigate environmental impacts [79]. However, these sites can be difficult to identify in nationwide analyses, as the information is scattered across local sources. In the study, a few landfills and old mines were identified.

Spatial analyses always have data-related limitations and inaccuracies. Finding the best-possible data might be too time consuming, and sometimes the data can be overly detailed, consisting of large files that make the analysis too demanding. Performing the spatial analysis, however, is only the first step in site identification. As the analyses are never perfect, the next step is to identify the best sites from the results and examine those more closely. The limitations of the analysis should always be kept in mind; here, for example, most of the agricultural land in the results should probably not be used for PV plants, and only unproductive land should be targeted. As powerlines were not included in the analysis, there is also a possibility that some areas are closer to the grid than the results suggest, if a direct connection to a powerline is possible. On the other hand, the analysis most likely does not identify all the suitable areas near powerlines, as they might be too far away from the substations.

4.5 Brief Overview of Solar Cell Technologies

Emerging solar cell technologies aim to drive wider implementation of solar energy. The motivation behind developing alternative solar cell technologies is to address the challenges related to the so-called first generation of solar cells, which rely on silicon and III-V compounds such as gallium arsenide (GaAs).

To function efficiently, first-generation solar cells require high-purity, defect-free materials that are expensive to produce; nonetheless, at the time of writing this textbook, first-generation solar cells are still the most widespread solar cells, and account for the majority of generated solar electricity. Silicon dominates the photovoltaic production market with over 90% of annual share [153], whereas GaAs, other III-V and their tandems occupy niche areas such as space and aerospace applications, where best-available performance takes precedence over cost factors.

The dominance of first-generation photovoltaics originates from the semiconductor industry, which produces devices that are tightly intertwined with our daily lives, like smartphones and computers, with much larger markets compared to photovoltaics [55, 157]. The semiconductor industry relies on the same materials, with similar purity requirements, as first-generation photovoltaics, and so the development of PVs benefitted from mature production technologies and resource streams already developed for the semiconductor industry.

The connection of first-generation solar cells to the semiconductor industry turned out to be a double-edged sword, and gave additional motivation to the pursuit of alternative technologies. The access that the early solar industry had to the knowledge and resource streams of the semiconductor industry made competition with first-generation PVs difficult. However, this connection has also exposed the first-generation solar cell industry to the crises of the semiconductor industry, and silicon supply shortages and price fluctuations have had negative effects on the silicon photovoltaics industry [42, 45, 75, 132, 231].

To be competitive with silicon photovoltaics, emerging PV technologies need to surpass silicon qualities, or to be applicable in circumstances where silicon photovoltaics are not. Silicon solar cells are rigid and non-transparent; silicon itself has suboptimal properties for light absorption (it has an indirect band gap, which means that a thicker cell is required for proper absorption) that place a limit on how thin the silicon layer can be made, and hence how much the use of costly high-purity silicon can be reduced. These features of silicon photovoltaics influenced the development of alternative photovoltaic technologies.

Cadmium telluride (CdTe) photovoltaics are currently (in 2022) the biggest commercial competitor of silicon cells: CdTe photovoltaics have held the largest market share after silicon PV since 2007, and accounted for 6% of produced capacity in 2020 [153]. Commercial CdTe solar cells are rigid and non-transparent, and so they compete with silicon for the same application areas. Unlike silicon, CdTe has a direct band gap, which allows for efficient light absorption by films just several micrometers thick, compared to wafers several hundred micrometers thick in the case of silicon (Fig. 4.31a, b).

Key factors that allowed CdTe solar cells to claim their market share are the high efficiency, low cost, and low environmental impact of production and recycling [25, 131, 145, 167]. CdTe efficiency is on par with silicon (Fig. 4.32), the cells are more cost-efficient (shorter energy payback time), and their environmental impact is smaller. The main disadvantage of CdTe cells is the toxicity of CdTe and supplementary materials (CdS). First Solar, the largest CdTe photovoltaics manufacturer, played a significant role in the establishment of the technology. First Solar constantly dedicated efforts to the improvement of the efficiency of CdTe solar cells, and developed a recycling program to address the toxicity issue.

Other well-known alternatives to silicon photovoltaics are copper indium gallium selenide (CIGS), dye-synthesized (DSSC), and organic (OPV) photovoltaics (Fig. 4.31c–e). CIGS solar cells are commercially available, and accounted for 2% of produced capacity in 2020; DSSCs and OPVs are yet to make their entry to the market. These technologies can provide compatibility with flexibility and transparency, but suffer from poorer efficiency and stability compared to commercially available technologies.

4.5.1 Halide Perovskite Solar Cells

Halide perovskite PV is the most recent (in 2022) alternative technology to silicon. Halide perovskites are not newly discovered materials, research on them dates back to the fifties (MØLLER 1958); in the nineties they were studied as potential materials for optoelectronic applications [138]. The scientific community expressed tremendous interest and dedicated a lot of research effort to perovskite photovoltaics after pioneering work on them was published in 2009–11 [37, 108]. In just a few years the annual number of publications on halide perovskites exceeded one thousand, and

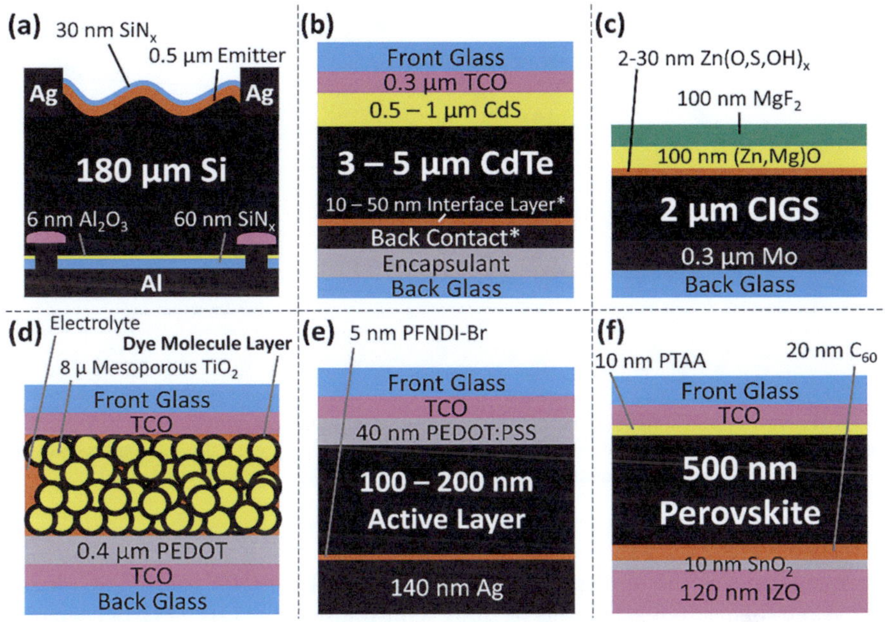

Fig. 4.31 Example structures of **a** silicon solar cell [230], **b** CdTe solar cell [112, 117], **c** copper indium gallium selenide (CIGS) solar cell [103], **d** dye-sensitized solar cell (DSSC) solar cell [31], [229, 228], **e** organic solar cell (OPV) [229, 228], and **f** perovskite solar cell [229, 228]. In (**b**) the asterisk denotes proprietary layers whose exact composition is unknown, for example, the interface layer could be ZnTe or CuTe, and the back contact could be Mo [41, 83]). In (**b**, **d**–**f**), TCO is transparent conductive oxide, for example SnO_2:F or In_2O_3:Sn. In (**d**) PEDOT denotes Poly(3,4-ethylenedioxythiophene), and in (**e**) PEDOT:PSS denotes poly(3,4-ethylenedioxythiophene) polystyrene sulfonate. In (**e**) active layer denotes a blend of organic electron donor and acceptor materials, and PFNDI-Br denotes poly[(9,9-bis(3′-((N,N-dimethyl)-N-ethylammonium)propyl)-2,7-fluorene)-alt-5,5′-bis(2,2′-thiophene)-2,6-naphthalene-1,4,5,8-tetracaboxylic-N,N′-di(2-ethylhexyl)imide]dibromide. In (**f**) perovskites denotes $Cs_{0.05}[(CH(NH_2)_2)_{0.92}(CH_3NH_3)_{0.08}]_{0.95}Pb(I_{0.92}Br_{0.08})_3$, PTAA denotes poly[bis(4-phenyl)(2,4,6-trimethylphenyl)amine, C_{60} denotes fullerene-C_{60}, and IZO denotes indium zinc oxide

currently is over 3000. The efficiency of perovskite solar cells exhibited the fastest improvement in the history of solar cell development (Fig. 4.32).

Halide perovskites combine advantageous features of established solar cell materials, and have some unique features of their own. Halide perovskites consist of available elements with annual production at the same level or higher than silicon [197], whereas their energy payback time is faster and environmental impact lower than those of silicon [64, 202].

Like III-V semiconductors, halide perovskites are a family of compounds with a similar crystal structure that form solid solutions; this in turn allows their properties to be tuned over a wide range [65, 208, 225]. Like CdTe, halide perovskites have a direct band-gap and large absorption coefficient, which enables the use of thin

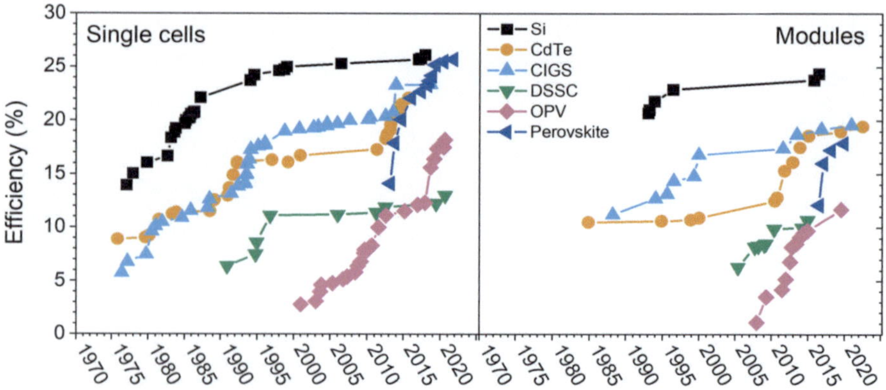

Fig. 4.32 History of record efficiencies of single cells and modules for different solar cell technologies. CIGS denotes copper indium gallium selenide, DSSC denotes dye-synthesized solar cells, and OPV denotes organic photovoltaics. Data courtesy of the National Renewable Energy Laboratory, Golden, CO

films and reduces the amount of material required. Like OPV and DSSC, halide perovskite solar cells can be made flexible, semi-transparent [116, 121], or even fully transparent [127]. Finally, the chemical and physical properties of halide perovskites make them solution-processable, defect-tolerant semiconductors, this in turn enables wide accessibility and rapid prototyping.

Halide perovskites are ternary compounds with crystal structures similar to the mineral perovskite (Fig. 4.33). The name perovskite has its roots in the nineteenth century and refers to a specific oxide mineral (calcium titanate, $CaTiO_3$); later, the term was adopted for other compounds whose structure was similar to calcium titanate [104]. Currently the term is used quite loosely in the halide perovskite literature [4]. Like $CaTiO_3$, halide perovskites have three constituents and similar general chemical formula, ABX_3, where: X is a halide, typically iodine, bromine or chlorine; B is tin or lead; and A can be an element like cesium or a small organic amine, like methyl amide ($CH_3NH_3^+$) or formamidinium ($[CH(NH_2)_2]^+$).

Halide perovskites are a large group of compounds whose properties are tunable within the properties of the group—that is, when mixed they form solid solutions just like III-V compounds from first-generation photovoltaics. The previous paragraph listed the most common constituents of halide perovskites, which in fact includes 18 different compounds with different properties. Mixing two or more halide perovskites in different ratios allows tuning of the properties of the resulting compound. For instance, halide content has the greatest effect on optical properties; and introducing bromide or chloride into an iodine perovskite shifts absorption to lower wavelengths enabling semi-transparent or even fully transparent (UV only) solar cells [116, 125, 127, 163]. The motivation behind developing such cells is to integrate them with

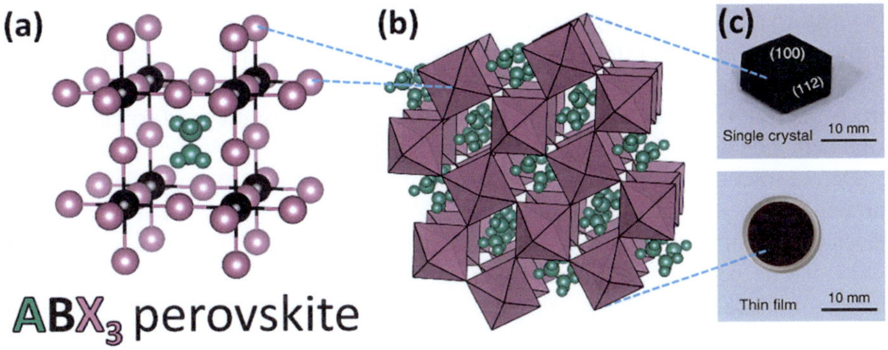

Fig. 4.33 **a** Unit cell of an ideal perovskite. **b** Arrangement of unit cells depicted as polyhedra in tetragonal (distorted) $CH_3NH_3PbI_3$ halide perovskite. **c** Digital photographs of a single crystal and a thin film of $CH_3NH_3PbI_3$ halide perovskite. Crystal structures in (**a**) and (**b**) were generated with VESTA software (ref) using crystallographic data from ICSD entry 255644 and 428898 [181]. Hydrogen, carbon and nitrogen are colored with the same bluish-green color for clarity. The photographs in (**c**) reproduced with permission from Xia et al. [215], Copyright © 2021 The Authors, distributed under the Creative Commons Attribution 4.0 International (CC BY 4.0) License

windows in buildings, and thus decrease the amount of land needed for photovoltaics deployment. B- and A- site constituents have minor effects on optical properties, but strong effects on the performance and stability of the resulting compound [99, 130, 200, 226].

The high absorption coefficient and direct band-gap of halide perovskites, as in the case of CdTe, enables the use of thin films in solar cells (Fig. 4.31f); this in turn minimizes the amount of material used. Iodide perovskites, in particular, have high absorption coefficients in the visible range, which allows them to absorb light efficiently with films thinner than those of most other materials [216, 221, 227].

In addition to efficient light absorption, halide perovskites excel in its conversion to usable power. The electronic properties and band structure of halide perovskites are such that charge carriers generated by light absorption can be extracted with relatively minor losses to competing parasitic processes. Similar properties in other solar cell semiconductors, such as silicon or CdTe, require very high material purities and expensive processing methods to minimize the number of defects. In halide perovskite compounds these properties come 'out of the box' due to their defect tolerance. Not all halide perovskites are defect tolerant, but studies show that relatively effortless post processing can passivate defects in the absence of inherent defect tolerance [123, 217].

Different types of defects exist, and they are detrimental to solar cell performance. The smallest are point defects: crystal structure constituents (Fig. 4.6a) in wrong positions, or the lack of the constituent in its correct position. A larger agglomeration of point defects turns into larger-scale defects, such as dislocations and grain boundaries. Such defects are extremely difficult to avoid because thermodynamics favor their formation. The defects modify the electronic structure of the host material by creating additional energy states. In defect-tolerant semiconductors the energies of these states are such that the properties of the host material remain unchanged, whereas in semiconductors that are not defect tolerant, defect states can enhance processes that compete with charge extraction and decrease the performance of the solar cell.

Defect tolerance is an inherent property of the compound that is independent of the deposition method, and is one of the key factors that contributed to the rapid development of halide perovskite photovoltaics. Another is their chemical properties: halide perovskites dissolve in a variety of organic solvents. Solubility and defect tolerance allow the use of inexpensive and accessible liquid-phase deposition methods to make high-quality, solar-cell suitable halide perovskites. In the simplest case halide perovskite thin films can be quickly made with a few inorganic salts, organic solvent, and a beaker. The ability to use liquid-phase methods made halide perovskites readily accessible to researchers around the world, and gave access to rapid prototyping and optimization of the solar cells. Groups working on DSSCs and OPV had a low entry-barrier into halide perovskites, where they could apply their accumulated expertise on solar cells and knowledge on supplementary materials.

Environmental and economic impact studies agree that halide perovskite photovoltaics have excellent potential for becoming the sustainable and environmentally friendly solar cell technology of choice. Life-cycle assessment analyses show that halide perovskite solar cells have the shortest energy payback time (EPBT) of the established solar cell technologies [64, 94], and that they have strong economic potential [30, 179]. The environmental impact of perovskite photovoltaics is comparable to established technologies with relatively higher CO_2 emissions, originating from shorter lifetimes and supplementary production materials [32, 64]. Cell lifetimes will very likely increase with future developments, this stems from the poor stability of perovskites, which will be discussed later. More recent research emphasizes that producing sustainable perovskite solar cells with low environmental impact efforts requires pushing the efficiency closer to the theoretical maximum limit, as well as the choice and development of deposition methods for supplementary materials in the perovskite solar cell [63, 94].

Scalability studies have become one of the major directions for perovskite solar cell development. The goal is to produce perovskite solar cells with a large area on an industrial scale, while maintaining the high efficiency demonstrated by small laboratory prototypes, as well as keeping production costs reasonable. While defect tolerance takes care of crystal structure scale defects, micro- and macroscopic defects such as pinholes and thickness nonuniformity can still exist. Liquid-phase methods struggle with these types of defects when they are scaled from small laboratory prototypes to larger modules for real-life photovoltaic systems.

The efficiency of the (2021) record-holding perovskite solar cell is 25%, measured from a 0.0804 cm^2 area, equivalent to a 2.8×2.8 mm^2 [98]. All layers in the cell, except for one compact TiO_2 layer, are made with spin-coating, in which a substrate is rotated several thousand revolutions per minute and a precursor solution is added drop-wise onto the rotating substrate. The centrifugal force spreads the precursor solution, and the solvent of the solution evaporates to leave a thin solid film of material on the substrate. Spin-coating is a sol–gel method that offers quick and inexpensive deposition of thin films in a laboratory, but it struggles to deliver high-quality, macroscopic-defect-free films on larger areas—hence the typical small size of the most efficient halide perovskite solar cells to date. NREL's Champion Photovoltaic Module Efficiency Charts include a small sized (802 cm^2) perovskite solar cell module with an 18% efficiency—however, that record is held by a commercial company, who did not disclose the manufacturing process and solar-cell details. Published research papers show lower efficiency (14%) mini-modules (140 cm^2) made with scalable sol–gel methods, such as doctor blading or ink-jet printing, the closest relatives of spin coating [61, 166]. Alternatively, mini-modules (51 cm^2) with higher efficiency (17%) can be made by resorting to vacuum-deposition methods that provide higher film quality at an increased cost [166]. The upscaling of perovskite solar cell fabrication methods still needs further improvement.

The Achilles' heel of halide perovskites is their stability [34, 47]. Halide perovskites degrade in ambient air, and the solar cells stop working without proper protection. A lot of studies focus on addressing this problem, literature demonstrates that solutions based on surface passivation of the perovskite and/or its encapsulation will likely address this stability issue.

Of the range of encapsula-tion techniques used to protect halide perovskites, atomic layer deposition (ALD) deserves a special mention. ALD is an established method for making thin protective coatings for a wide range of products, from jewelry to organic light emitting diode (OLED) displays [100, 149, 165]. The commercialization of ALD was pioneered in Finland, the key ALD inventor, Tuomo Suntola, received the Millennium Technology Prize in 2018. Lately ALD has been used to deposit perovskites' functional layer, or as encapsulation solution for the stability issue [12, 27, 44, 173, 209, 219, 224].

The poor stability of halide perovskites has inspired studies on a myriad of non-perovskite materials that perform supplementary functions in the solar cell structure. Encapsulation materials are one group of such compounds; another is junction materials, which enable extraction of light-generated charge carriers from the absorber layer. Depositing a layer of such a material on top of the halide perovskite is challenging, because halide perovskites have poor thermal stability: they degrade at temperatures above 100 °C. Since most established junction materials need higher temperatures for deposition, this has stimulated research into the use of alternatives with lower processing temperatures, such as conductive polymers.

The low processing temperature limits material selection, but enables the development of flexible solar cells. Halide perovskites are soft, flexible and ductile [72, 187]—these properties, along with defect tolerance, make halide perovskites suitable for flexible solar cells. The thermal budget of the solar cell fabrication process is already limited by the poor thermal stability of the halide perovskite, and so replacing the rigid glass substrate with one based on flexible polymer only adds minor additional restrictions on the process. Indeed, flexible halide perovskite solar cells have demonstrated impressive efficiencies over 24%, which are retained after 10,000 bending cycles [121]. Flexible solar cells have better portability and easier deployment than then rigid ones. They allow solar power generation on complex-shaped surfaces like roof tiles, vehicles and wearables.

Some researchers believe that, instead of competing with established solar cell technologies, halide perovskites can be combined with them to surpass the fundamental limits of solar cells employing a single light-absorbing material. The maximum efficiency of a one-absorber solar cell is limited to a little over 30%. Tandem solar cells with multiple absorbers can be made, with III-V compounds for example, but the solar cells become too expensive and are only worth using in a few niche application areas. Halide perovskites, however, are potential low-cost tandem components. For instance, impressive efficiencies of over 29% were demonstrated by perovskite-silicon tandems [5], and recently the Swiss Center for Electronics and Microtechnology (CSEM) and the École polytechnique fédérale de Lausanne (EPFL) claim to have breached 30% [20].

There is some concern in the field over the possible toxicity of halide perovskite solar cells, and their impact on the environment in event of solar cell damage and material leaching. Most effective halide perovskites contain toxic lead, and the lead content of products is met with caution and suspicion after the use of tetraethyllead as a fuel additive in twentieth century, which led to an epidemic occurrence of lead poisoning and ultimately the phasing out and banning of tetraethyllead [80], Seyferth 2003a, 2003b). There is still no consensus on the impact of lead in halide perovskite solar cells [15, 164] but clearly the toxicity problem must be addressed.

Lead is likely to remain in the composition of halide perovskite solar cells, and so this potential toxicity needs to be addressed by other means. Lead-free perovskites exist but currently have lower performance and even poorer stability [105]. Reports on the toxicity of iodine [86] adds a further challenge for these lead-free solutions. CdTe, the most successful thin-film solar cell technology, also contains toxic cadmium-containing materials (Fig. 4.31b). CdTe solar cell manufacturers, like First Solar, were able to address toxicity concerns by investing in encapsulation technologies and recycling programs [192]. Cadmium compounds in CdTe solar cells are relatively stable, and thus immobile, compared to the lead compounds in halide perovskite solar cells. Safe use of halide perovskite solar cells would thus require an even more advanced encapsulation than CdTe solar cells.

Halide perovskites possess a combination of unique properties, but they still require effort from the research community in multiple areas of their development before they can contribute to our sustainable future (Fig. 4.34). For halide perovskites the main development directions are scalability, stability, addressing toxicity, and integration with other solar cell technologies. First and foremost, halide perovskites are semiconductors with an outstanding set of properties and issue to match those. The use of halide perovskites is not limited to solar cells, and studies on their application in light emitting diodes (LEDs), microelectronic components, sensors, and even X-ray imaging are gaining traction. In these applications, halide perovskites can achieve higher energy efficiencies, and contribute to a sustainable future from different directions (Fig. 4.34).

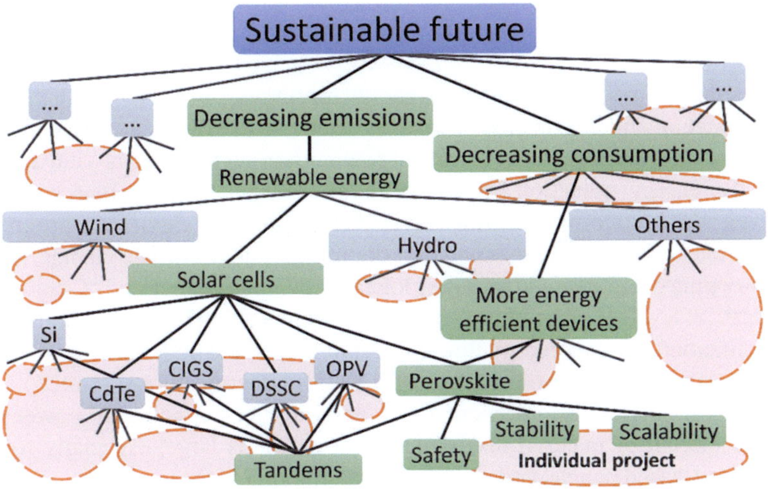

Fig. 4.34 Hierarchical view on how halide perovskites can become part of the sustainable future, and the role of individual projects focused on solving challenges or improving performance

4.5.2 Summary and Outlook

Material science is the key field that advances solar cell technologies. Technologies would be stuck at the fundamental limits of established materials without new materials, and engineering solutions on how to apply them. The research follows a limitation – attempted solution cycles, not all attempted solutions become established immediately or at all. Some, however, bear fruit decades after their development, frequently in fields outside of their initial purpose.

It is difficult to predict which research direction will become successful and prominent in future; rather, it is more relevant to identify potential directions and explore them. For instance, organic solar cells (OPV) and dye-sensitized solar cells (DSSC), despite decades of research, are still not on the market. Yet processing technologies, materials and research infrastructure developed for OPVs and DSSCs were used for the development of the halide perovskite solar cell, contributing significantly to their prominence. Halide perovskites themselves are not new materials, and were originally studied as potential semiconductors for microelectronics. Two decades later they are still not employed in microelectronics, but instead are shaping the landscape of the solar cell research. Next, we will take a brief look at the availability and recycling potential of critical materials needed in green energy applications.

4.6 Material Aspects of Solar PV Panels

The transition from fossil to renewable energy resources—in parallel with the growth of electrification and digitization—requires extensive use of often critical raw materials. Ensuring the availability of robust low-carbon technology is vital for accelerating the energy transition towards a more ecologically conscious society and advancing socioeconomic progress. Over the last decade, significant progress has been made towards decarbonization and the mitigation of climate change, with the goal of keeping the increase in average global temperatures to below 2 °C compared to pre-industrial levels. In part, this progress has been enabled by a significant increase in the utilization of solar energy through technologies like photovoltaic (PV) panels in conjunction with lithium-ion batteries (LiBs) for energy storage [160].

Mining of materials is a particularly important topic for solar PV power plants, since the amount of material needed for solar panels is large, especially for traditional silicon-based panels, and processing and refining of these materials has a large impact on nature. Mining causes direct damage to nature, and mining and processing of minerals can contaminate soil and water, and produce hazardous solid and liquid waste, and atmospheric pollutants via particle emissions and gaseous carbon, sulfur and nitrogen oxide emissions [48]. The environmental impacts depend on the materials needed, and where they are being extracted. Some of the materials needed in solar PV panels are more abundant than others, and their recyclability varies considerably. Nonetheless, impacts of the acquisition of materials must be made transparent when assessing the environmental impacts of a solar PV power plant. Likewise, panel manufacturing and the disposal of used panels can both generate pollution and greenhouse gas emissions, and the disposal of old or damaged solar panels can pose environmental risks. Case Study 4.1 later in this chapter presents a sample study of seeking circular solutions for materials recycling. The impacts on nature of solar power plants will be discussed in Sect. 4.7.

According to the International Renewable Energy Agency (IRENA), global installed solar PV capacity reached approximately 850 gigawatts (GW) in 2021 (IRENA 2022a). Nevertheless, to achieve the goals of a green economy transition, a projected growth rate of over 25% in the annual deployment of PVs is expected by 2030 [201]. Additionally, the demand for LiBs needed for electric vehicles (EVs) is estimated to increase from 215 gigawatt-hours (GWh) in 2020 to 1525 GWh by 2030 [113].

Consequently, the growth of PV technologies—more specifically, monocrystalline and polycrystalline silicon solar panels, which today account for around 90% of the worldwide PV market—has led to a notable increase in the consumption of key minerals such as silicon (Si). Over the last twenty to thirty years, there has been a steady increase in the extraction, manufacturing, and use of silicon metal grades, such as metallurgical, electronic, and solar. Since 2000, the annual global production of silicon metal has expanded by a factor of 2.4. The EU Commission and the US Geological Survey (USGS) have identified high-purity silica (> 99.95% purity) as a critical material since it is both relatively scarce and economically significant [113],

EU [49] and [198]). Each crystalline silicon (c-Si) PV module requires approximately 2–4 kg of solar-grade silicon (SoG-Si). Moreover, when the 20-to-30-year lifespan of PVs is taken into consideration, a significant amount of waste will be generated—an estimated 8 million tonnes (Mt) by 2030, and 78 Mt by 2050—which will require proper treatment [36]. On the other hand, about 1.3 Mt of silicon will be required by 2030 to meet the demands of PVs, LiBs, and other energy storage technologies for the global energy transition [218]. Furthermore, it should be noted that as the world's resources are becoming exhausted, there is less and less high-quality raw quartz sand available, which is necessary for the manufacture of silicon. Given these challenges, the adoption of circular practices for silicon materials presents a significant opportunity to ensure a sustainable supply for the energy sector [50].

4.6.1 Circularity of Material and Their Environmental Impact

The circular economy is an economic system that revolves around three fundamental principles: life cycle thinking, sustainability, and systems thinking. At its core, the circular economy aims to maintain the value of materials by keeping them in use for as long as possible, recycling them, and utilizing secondary raw materials to minimize waste generation and prevent the loss of valuable elements, including critical raw materials. For example, the European Union has taken significant action towards promoting these principles through the Circular Economy Action Plan, as outlined by the European Commission. This plan encompasses the entire life cycle of a product, from production and consumption to repair, remanufacturing, waste management, and the utilization of secondary raw materials. By considering the entire life cycle, this action plan ensures that resources are used efficiently, and waste is minimized through various strategies and initiatives.

Circular economy strategies provide a framework to support and design circularity, allowing materials to circulate in the economy for as long as possible. Research is also ongoing to create techniques for substituting crucial raw materials and using alternative materials to reduce raw resource requirements. This will help alleviate the shortage of resources and lessen the overall impact on the environment.

Here, we concentrate on recycling practices that adhere to the circularity principle and augment silicon supply through recovery from photovoltaic waste. As the core principle of the circular economy is to minimize environmental impacts, it is crucial to grasp its implications across various system levels. This encompasses understanding the effects of circular products and services, as well as potential macro impacts, such as rebound effects, where increased efficiencies at the product level result in increased resource consumption or environmental impact at the macro-level. While there are multiple environmental assessment methods for different system levels, it's uncertain which ones are optimal and how they can be effectively combined.

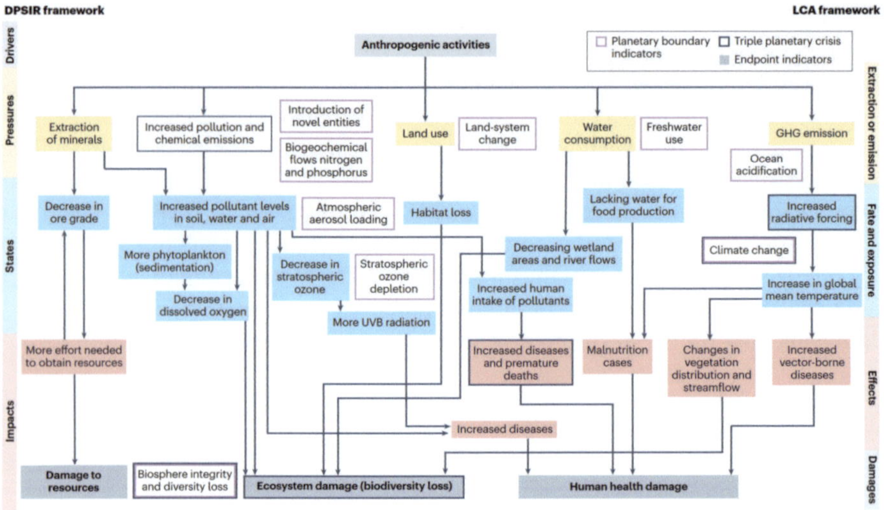

Fig. 4.35 Simplified life-cycle assessment (LCA) impact pathways and planetary boundaries situated in the drivers–pressure–state–impact–response (DPSIR) framework. The triple planetary crisis (solid black boxes) aspects of climate change, biodiversity loss, and human health impacts from pollution are situated in terms of placement along the DPSIR framework and in relation to LCA impact pathways. Endpoint indicators (dark grey boxes) include damage to human health or marine, terrestrial, or freshwater ecosystems. Responses from the DPSIR framework are not shown, because the outcomes of LCAs support the development of responses (such as policies), but responses as such are not included in the LCA framework. Midpoint indicators are not shown explicitly. GHG = greenhouse gas; UVB = ultraviolet B radiation. [77]

For example, Helweg et al. (2023) proposed a framework for how emerging life-cycle assessment (LCA)-based approaches would be beneficial for use of the planetary boundaries' framework and other environmental assessment tools to support decisions—see Fig. 4.35.

Case Study 4.1 Photovoltaic solar panels and lithium-ion batteries.

The emergence of the energy transition gives rise to three significant impacts. Firstly, there is a growing requirement for materials in larger volumes. Secondly, the quantity of elements that are in demand is rising. Finally, the demand for novel materials is urgent. Here we present a holistic analysis that considers the three primary grades of silicon: metallurgical-grade (MG-Si), solar-grade (SoG-Si), and electric-grade (EG-Si). This involves the recovery of silicon from PV waste and the subsequent use of recycled silicon for the creation of new PVs, as well as innovative LiBs (Rahman et al. 2021)—that is, Si-based anode LiBs such as LMR-NMC-Gr-Si (cathode ($0.5Li_2MnO_3{\cdot}0.5LiNi_{0.44}Co_{0.25}Mn_{0.31}O_2$ [LMR-NMC]) and a graphite-silicon blend anode) [6, 169] and [124]).

We evaluated the indirect link between the proliferation of technologies and damage to the environment by focusing on flows of silicon used for green energy technologies (Fig. 4.36). Subsequently, the contribution from the sustainable management of material circularity has been assessed, with an emphasis on the recovery of various grades of silicon by optimizing the mass flows of photovoltaic panels and battery technologies. Using system dynamics modeling [54], as a first step we quantified the global mass flows of silicon through all stages of its supply chain. The silicon requirements for PV and LiBs technologies from mining to recycling, including exports and imports, were simulated. The geographical distribution of flows was used across GDP per capita, along with the silicon utilization factor for green energy technologies. The life cycle assessment and environmental assessment of each material flow through all stages of the supply chain was carried out, addressing energy consumption, water use, greenhouse gas (GHG) emissions and related pollution such as particulate organic carbon (POC), black carbon (BC), nitrous oxide (N_2O), methane (CH_4), sulfur oxides (SO_x), particulate matter with sizes smaller than 2.5 μm (PM2.5), airborne particulate matter with sizes smaller than 10 μm (PM10), nitrogen oxides (NO_x), carbon monoxide (CO), volatile organic compounds (VOC), and carbon dioxide (CO_2) emissions.

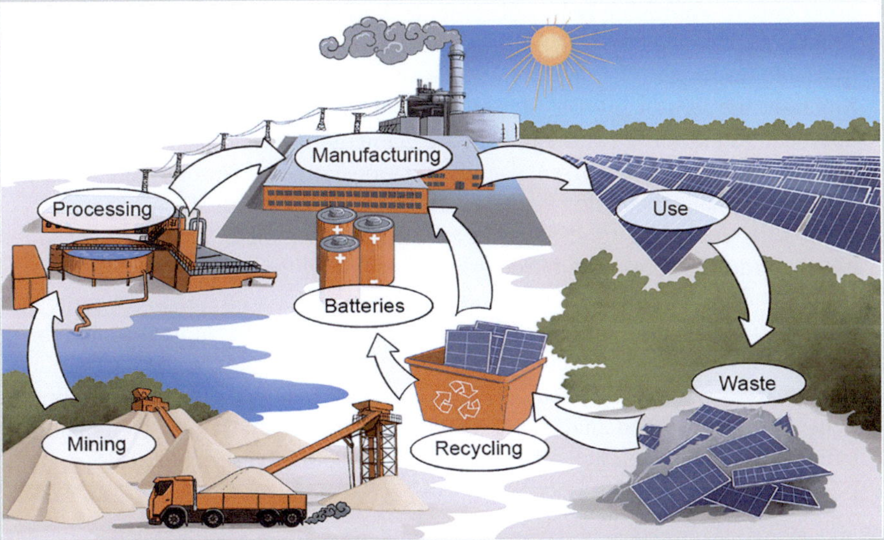

Fig. 4.36 Conceptual model of supply chain of silicon used in solar photovoltaic panels and batteries. © Maija Karala

Our study [160] shows that increasing silicon recycling can play a significant role in achieving environmental sustainability and promotion of a more circular economy. We demonstrated that the energy and water consumed, and emissions generated, in primary silicon production remain substantial. The results show that an effective global circularity of silicon (use in PV and LiB production) would contribute to a 54% reduction in energy consumption (~ 3.5 billion GJ) and 69% of water use (equivalent to ~ 3.1 million m^3) by 2030. Secondary production of silicon mitigates around 45% of total GHG emissions (~ 65 Mt CO_2 eq). Furthermore, possible synergizes between the flows from the recycling of end-of-life PVs and production of new LiB technologies, using an optimum scenario for the recovered volumes of different silicon grades (50% metallurgical, 25% solar, 25% electronic), could potentially save around 38 M GJ of energy, over 0.01 million m^3 of water, and mitigate 4 Mt CO_2 eq of GHG emissions by 2030 that result from green technologies waste management.

The findings of our study indicate how the circularity of critical materials like silica can be managed to minimize their environmental impacts through the propagation of green technologies. In particular, the data obtained address the feasibility and environmental sustainability challenges of PV recycling and related processing of silicon as a key material for future battery and energy storage technologies. These results aim to improve technology designs for energy systems and formalize theoretical frameworks for material use and recycling, and these benefits contribute to more effective utilization of materials and technologies. Also, the environmental assessment results contribute to improving policy development, and identifying risks and opportunities to balance the primary and secondary sources of silicon, based on available resources with minimum negative environmental impact. The results highlight the significant role played by a systematic sustainable management approach to the waste resulting from green energy technologies, when carried out in conjunction with current decarbonization activities. Furthermore, policy support to enable technology synergies and scale-up of critical materials recycling will reduce the total amounts of primary raw materials required, and alleviate the environmental burdens associated with their production.

The model developed in our study can be used as a reference for policymakers to understand the role of silicon in green energy technologies, and the possible development in circularity of other critical materials by the integration of primary and secondary sectors. This approach is applicable for a comprehensive environmental analysis of technology innovation trends in a dynamic competitive energy market, and for assessing the effectiveness of development policies.

4.7 Impacts on Nature and Mitigation of Negative Effects

Solar power is rapidly evolving and, together with wind, is the most promising technology to answer the need to decarbonize energy production. Globally, solar is the most abundant source of energy [152], and has the potential to become a significant source of electricity, even at higher latitudes, like northern Europe [143, 205]. Photovoltaic power plants vary greatly in size, from small-scale, residential projects typically installed on rooftops, to ground-mounted, utility-scale projects covering large areas of land. In addition, during recent years, floating photovoltaic systems—solar panels mounted on a structure that floats on a body of water—have also been developed. This section summarizes the impacts of solar PV systems on nature and wildlife, and introduces ways to mitigate negative impacts (concentrated solar power (CSP) was discussed in Chap. 3).

Solar power is generally considered to have many benefits over fossil fuel-based energy production. Solar-based energy production helps to reduce CO_2 emissions and other air pollutants, but as this form of energy production expands at an accelerating rate, there is a growing need to understand the potential negative impacts of large-scale solar power plants on nature and wildlife. To avoid the most negative impacts on nature, the construction of utility-scale solar power plants can be restricted to previously-altered landscapes, and solar PV development on brownfields or degraded lands can provide an opportunity for land rehabilitation and the creation of biodiversity benefits. Solar PV systems can also be integrated on the roofs and walls of buildings and other urban structures, such as highway noise barriers, or on parking spaces. In these cases, the direct land-use impact is close to zero if no new powerlines are needed, and therefore most of the land-use impacts are indirect [57, 199]. With ground-mounted solar power plants, environmental impacts happen both on-site where the power plant is constructed (direct land use), and outside the area due to material extraction and manufacturing of the components. See Fig. 4.37 on the different types of impact areas. The full impacts of large-scale renewable energy development, including solar energy, are not fully understood; knowledge of cumulative impacts is still poor, for example. We need to increase our knowledge base to develop best practices for sustainable renewable energy development, and to make maximum use of the potential for positive impacts of this sector, to reach targets to halt climate change, and to create nature-positive effects. The main focus of this section is on nature impacts created on-site, mitigation measures for those impacts, and opportunities of co-locating solar power plants with other human uses, such as agriculture. Availability of materials and circular economy solutions were discussed in Sect. 4.6; and will be further discussed in Sect. 9.2.1.

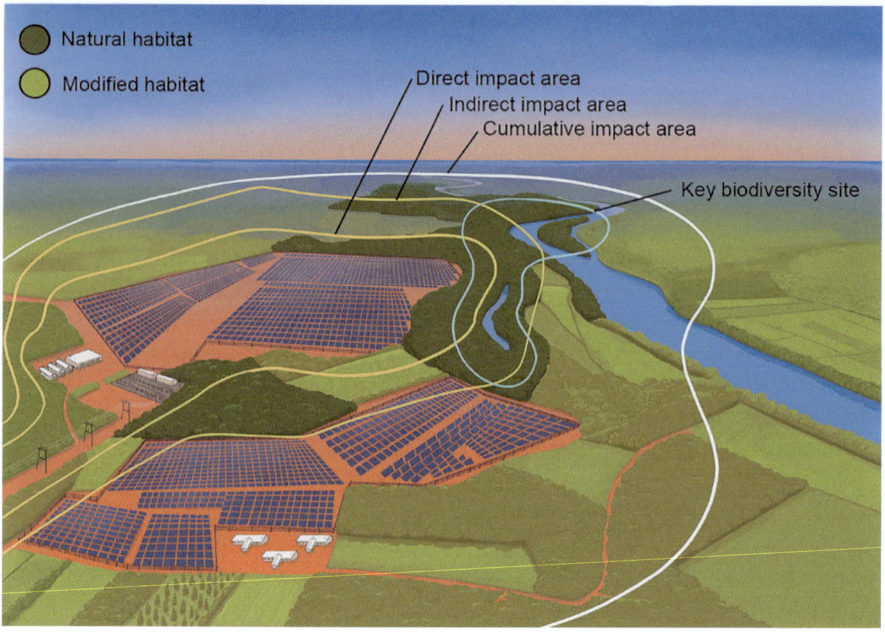

Fig. 4.37 The relationship between direct, indirect, and cumulative impacts on biodiversity—an illustrative example of a solar energy development project. © Maija Karala

4.7.1 Impacts of Ground-Mounted Solar PV Power Plants on Nature

Compared to wind power plants, the impacts of solar PV power plants on nature are not well understood. As the construction of ground-mounted solar power plants increases, their land-use impacts will become more significant, and consequently there is a growing need to better understand their potential impacts on nature [26]. The affected land area of a solar energy system depends on what kind of solar system is used, and the local insolation. A system requiring up to 1.5 m^2/MWh of annual electricity production at high latitudes might only occupy 0.5 m^2/MWh close to the equator [57, 128, 134].

The land-use footprint of ground-mounted panels is due to the power plant itself: the arrays of panels, access roads, substation and powerlines. The construction area is heavily modified, and a large-scale solar power plant can lead to habitat loss and fragmentation (including soil disturbance and degradation of aquatic ecosystems), cause changes in species composition or increased mortality in local populations, create barriers for animal movement, cause changes in microclimate, light and noise, and enable introductions of invasive alien species [60, 68, 79, 142, 194, 207].

The exact impacts on nature are site specific, and depend on the total area affected and the status of nature in the site prior to construction. Next, we take a closer look at some of the main ways solar PV power plants impact environment and nature during construction and operation. See also summary Table 4.2 at the end of this Sect. at 4.7.4.

The construction phase is the largest on-site disturbance to nature during the life cycle of a ground-mounted PV power plant. Often the original environment is lost completely during construction, and habitats can become fragmented [79]. The impacts are most severe when large areas in natural or semi-natural condition, such as forests or shrublands, are cleared for solar power development [156, 204]. Construction can also affect aquatic environments through pollution, erosion, run-off, and physical destruction of waterbodies or wetlands [67]. The least impacts are experienced in areas that are already modified by human activities, such as brownfield sites, former peat production areas, agricultural lands, and built environments. Co-location of PV power plants with other human activities, and combining solar power production with the rehabilitation of nature in degraded lands, are discussed below.

Besides the power plant itself, the construction of access roads, power station, and grid connection create disturbance. Cabling requires digging new trenches, creating disturbance during construction [79, 129] and increasing the risk of sediment load, soil erosion, changes in groundwater quality and availability, and flooding [195]. Maintenance roads are also needed inside the power plant, and their construction and the increased traffic create noise, dust, and other kinds of pollution, such as greenhouse gas emissions and potential leakages of lubricants or oil [69, 155].

The foundations for the solar arrays require modification of the land, the extent of which depends on the site characteristics and the type of foundation used. Vegetation is cleared and sometimes the ground is covered with gravel or sand to even the surface around the foundations. Removal of vegetation and land modification usually disturbs the wildlife and affects local ecosystems.

In particular, land removal and construction often destroy the habitats of species living in the construction site. For burrowing animals, adapted to live primarily underground, the loss can be severe if their burrows—and therefore their entire habitat—are destroyed. These include badgers (Mustelidae), rodents such as naked mole-rats (*Heterocephalus glaber*), many mongoose (Herpestidae), and desert tortoises (*Gopherus agassizii* and *G. morafkai*), as well as many snakes, lizards, insects and spiders, and even one bird: the burrowing owl (*Athene canicular*). [3] have described how the burrows of one animal, the desert tortoise, also provide a habitat for a number of other species. Within their territory, desert tortoises act as ecosystem engineers, and the destruction of tortoise burrows have an impact on number of other species and the impacts can cascade across the local ecosystem.

Soil disturbance, clearance of vegetation, land tillage, and transfer of soil due to construction may also facilitate the introduction of invasive alien species that can outcompete locally occurring species [21, 89]. Invasive species are often generalists that can thrive in many kinds of environments, and are able to establish themselves quickly on disturbed landscapes. Seeds are often transported into the construction site via soil used in construction and landscaping [89].

Besides habitat loss, utility-scale solar PV power plants can also fragment formerly continuous environments, or form barriers for animal movement—this effect is larger on a natural or semi-natural landscape than on an already-modified environment. A solar PV power plant is typically fenced, and this effectively prevents animal movement [79, 195]. If double fences are used, those can also form a trap for animals such as birds and lizards [204]. Barriers for animal movement can also be formed due to changes in the landscape: fragmentation and physical loss of vegetation create edges and open spaces that can prevent some animals from using the newly created environment. The clearance of forested areas for a power plant, for example, fragments continuous habitats, and creates an edge that disrupts for some forest species [76, 136, 212]. The barrier effect caused by solar power development is not well studied, but it is known that large scale construction, including energy developments and their related infrastructure such as fences and roads, impact species movement and can reduce their range. For example, large mammal migrations can be severely affected [210, 214].

During the operation of the solar PV power plant, impacts on nature and wildlife are site-dependent, and the effects vary from one species to another. After construction, human disturbance in the area is relatively low, as maintenance activities are not major and require little traffic and equipment. This creates an environment that is favorable to some species—for example, in the UK, plant diversity and abundance of butterflies, bumblebees and other pollinators have been found to be greater on solar PV power plants than on control sites [141]. Impacts on birds is mixed: while some species thrive in human-built environments like solar arrays, other species become displaced. Studies in California, for example, showed that solar PV development displaced most raptors (predatory birds), but benefitted species that found the structures suitable for nesting and perching, provided shelter, food and refuge from pray—species that are in general adapted to live in human environments [26, 73]. In South Africa, where a savanna covered in shrubs was cleared for solar PV development, and the vegetation replaced with grassland, the bird species community was changed. There was a loss of shrub and woodland species, while other species became more abundant [204]. In total, both the species richness (number of different species) and density were found to be lower at the solar PV site compared to the surrounding area.

Power transmission lines create a risk for birds, in the form of collisions and electrocutions [21]. Collisions with power transmission lines can be mitigated by burying cables underground—however, burying might create further habitat loss and fragmentation, and therefore each case must be carefully considered to decide which measures are the best in which area, especially if sensitive or threatened species are present on the site.

Solar PV development can increase bird mortality because of collisions with the panels, especially in dry (and often open) landscapes [203, 204, 207]. This may be because some insects are attracted to solar panels, which in turn attracts birds to forage in the PV facility area, making them vulnerable to colliding with the panels and other structures [58, 97]. It has also been suspected that birds and insects mistake the panel arrays as water bodies [67, 101, 129]—this so called "lake effect" was first described in the southwestern United States where stranded, injured, and dead water birds such as herons (Pelecaniformes), loons (Gaviiformes), and grebes (Podicipediformes) were observed at a utility-scale solar PV power plant [101]. However, there is little evidence on how widespread and severe this impact on water birds is, and the conditions that lead to this problem [111, 207], and different solar technologies, siting decisions, and local conditions may play a part [207]. Therefore, it is important to carefully monitor and assess the impacts of utility-scale solar power plants to increase our understanding on the causal factors that contribute to bird fatalities.

The impact on bats of land-use changes related to solar power development have not been closely studied, but recent studies suggest a negative effect [189, 193]. Due to fragmentation and loss of foraging habitats, bats start to avoid solar PV power plants. In England, of eight bat species or species groups studied, negative effects due to solar PV panels were observed in the activity of six species/groups [193]. A comparison study in Hungary revealed that some bat species avoid solar power plants, while other species that had adapted to human-altered environments solar power plant sites as well [189]. Moreover, there is some suggestion that solar panels might act as sensory traps for bats and insects, similar to other smooth surfaces, and therefore increase the risk of collisions [189]. Though knowledge gaps exist, this research demonstrates that solar power development can impact local ecosystems in complex ways. As the construction of large-scale renewable energy production expands, there is a growing need to understand these interactions to inform siting decisions and plan mitigation measures accordingly.

Operational solar PV power plants are also known to alter the local micro-climate and soil temperature, and these affects are greater when forest is cleared for a large solar PV power plant than when the project is constructed on an open environment [60, 213]. The shading of panels (the so-called shadow effect) together with the changes in the microclimate can reduce species diversity and alter composition [11]. On the other hand, on agricultural lands, this shading has helped some crops to survive during heatwaves or drought, as shading mitigates temperature extremes and reduces water evaporation [18]. Large-scale solar power plants can also form a “heat island” that increases the local temperature, and impacts the ecosystem and species living there as a result [223]. In general, vegetated areas tend to heat up less than areas without vegetation and covered with gravel [223]. The glare caused by optical reflection of sunlight on the surface of the panels may also be a source of discomfort to fauna or residents nearby [168].

In the operation phase, chemicals are used to keep modules free from dust, sand, and pollen, and to control vegetation. Many of the chemicals used are highly toxic and can negatively affect local water resources and terrestrial and aquatic ecosystems [79, 129]. Especially in areas experiencing water shortages, using water for cleaning panels can result in changes in hydrology, water availability, and water quality, causing habitat degradation [172]. For example, at a solar facility in Jodhpur, India, panels needed to be cleaned four times per month in summer, and two times per month during the winter. Each cleaning required 20,000 l of water per 0.5 MW [172]. In areas with more rainfall and higher panel tilt angle, the need for cleaning is reduced [88]. Development of dry-cleaning methods and non-fouling materials also helps to save water.

Cumulative impacts of solar PV production, or the combined impacts of many solar power plants and other human development in the same area, are not well-studied or understood [204]. Cumulative impacts in the environment are caused by the combination of several human activities, where no individual activity would have a large effect, but together they can result in significant impacts on nature and the function of the ecosystem. Cumulative impacts are often difficult to predict or manage, because of the diversity of human activity, the lack of environmental data, and the complexity of ecological processes. As utility-scale solar PV power production is developing rapidly, we need to understand better the effects of building larger numbers of power plants, and producing energy together with other human activities [142]. This is especially relevant in regions where conditions for renewable energy production are favorable, and many solar and wind power developments can concentrate in one area. For example, in the Mojave Desert, California, the expansion of solar and wind energy production between 2009 and 2017 was found to be responsible for over 60% of the observed downgrade in conservation value [150].

The rapid expansion of solar power also raises questions about cumulative impacts on different species—for example, more information is needed to predict cumulative effects on birds [111]. While solar power contributes less to bird mortality than many other human activities, the combined mortality from solar development might become significant [204, 207]. In South Africa, the rapid expansion of solar power in the Northern Cape Provence has raised concerns on the cumulative impacts on the highly specialized bird species found from the region, some of which are endemic and of conservation concern [204]. Similarly, in the Southwestern United States the growth of solar power development together with other human activity may threaten aquatic ecosystems, in particular for small and/or temporary water bodies and streams. These cumulative effects include fragmentation and loss of aquatic habitats, alterations in water flow, and sedimentation, which also threatens surface–groundwater connectivity [67]. Proper planning, including the collection of solid environmental and ecological data, forms the basis for informing siting decisions, and is key to avoiding negative cumulative impacts when developing solar power [68, 79].

Changes in the environment due to renewable energy expansion—the resulting displacement, reduction, disappearance or addition of species—can also cause changes to the entire ecosystem; a process called trophic cascades. A classic example of a top-down trophic cascade is the removal or reduction of a top predator that results in changes in the prey population through the entire ecosystem structure. Trophic cascades caused by solar power are not well understood. For example, Grippo et al. [67] argues that, although cumulative impacts of solar power on birds and insects can be large, we do not know if the mortality is great enough to produce population-level, or cascading, effects. Examples of trophic cascades related to wind power development are better understood, yet still remain under-studied (see Sect. 5.8).

Solar energy production can also compete with food production and nature conservation through land displacement [78]. Construction of solar power plants on agricultural land can result in land-use impacts away from the construction site, since new agricultural lands are subsequently cleared elsewhere. Construction of solar power on remote areas can make these places accessible for other human uses and lead to harvesting of natural resources, such as collecting or hunting of species, and increased pollution [21]. Incorporating broad expertise and local knowledge in siting decisions becomes important when balancing trade-offs between different goals, such as energy, food, and conservation.

Replacing carbon-intensive energy sources, like coal, with solar and other renewable energy sources is pivotal in addressing climate change by directly decreasing greenhouse gas emissions. However, clearing vegetation and disturbing the soil during construction and decommissioning stages can reduce the site's capacity to sequester carbon, and so increase the release of CO_2 to the atmosphere [43, 133]. In semiarid regions, clearance of vegetation can result in significant loss of organic carbon compared to undisturbed areas, and this can also affect microbial biomass and activity. Bastida et al. [19] and Li et al. [122] have shown that total organic carbon can decrease by as much as 25–30% in disturbed areas, and this reduction remains even 15 years after the original clearance. In forested locations and wetlands, the installation of solar power plants can result in the emission of 2–4 times more CO_2 to the atmosphere, ranging from 16 to 86 gCO_2/kWh, due to deforestation and removal of vegetation compared to solar power installations in desert areas [195]. The negative impacts of deforestation have directed solar PV power plant development towards desert areas in the United States and Australia [58, 71, 194]. In Finland, solar power plants are being planned on decommissioned peat-production sites, where lowered water levels can cause CO_2 emissions through aerobic decomposition of the peat [107]. These negative impacts can be reduced via revegetation and rewetting, which slows down the aerobic decomposition and reduces greenhouse emissions. Also, constructing renewable energy on previously altered or contaminated landscapes can aid in reducing land-based carbon emissions when compared to development on undisturbed lands [43, 79].

As we have shown, while the demand for clean energy sources is urgent in tackling climate change, substantial knowledge gaps remain regarding the impacts of large-scale solar power development on species, ecosystems, their functions, and the services they provide to us [133, 191]. It is crucial to acknowledge these gaps in the initial planning stages of solar power projects, and to establish monitoring programs that allow re-evaluation of practices as new information becomes available. Next, we will examine how to avoid and mitigate some of the negative impacts on nature.

4.7.2 *Avoiding and Mitigating Negative Impacts on Nature in Ground-Mounted Solar PV Power Plants*

Development of solar PV power plants can affect nature in many ways, as we learned in the previous section. Some impacts, such as the destruction of a habitat, can be severe, and disrupt the entire ecosystem and its function. Any large-scale energy development project needs to comply with national legal requirements on environmental impact assessments and meet required social criteria. These legal requirements may or may not request the use of ecological compensation to avoid, minimize and offset the potential negative effects on nature and wildlife. Whether ecological compensation is required or not, it is still a good practice to include measures aimed at reducing the negative impacts of large-scale development projects. This section focuses on measures that can be taken during the development of a utility-scale energy project to mitigate negative impacts on nature.

As shown in the mitigation hierarchy in Fig. 4.38, the most important step is to avoid negative impacts in the first place. Any negative impacts that cannot be avoided should be minimized and damage restored. Finally, any residual impacts that cannot be avoided or restored are offset.

Note that offsetting is always the last step of the mitigation hierarchy, and should only be used after all other steps are carried out. Ideally, the target is that the project causes 'No Net Loss' (NNL) of biodiversity, or preferably, creates a 'Net Gain' (NG). (See more on the mitigation hierarchy Sect. 2.3).

The higher the natural value of the construction area is, the more mitigation and offsetting are required. Mitigation of negative impacts on nature needs to be included in the early stages of project planning: alongside technical perspectives, insolation, possible shading, the landscape, and the assessment of the impacts on nature and

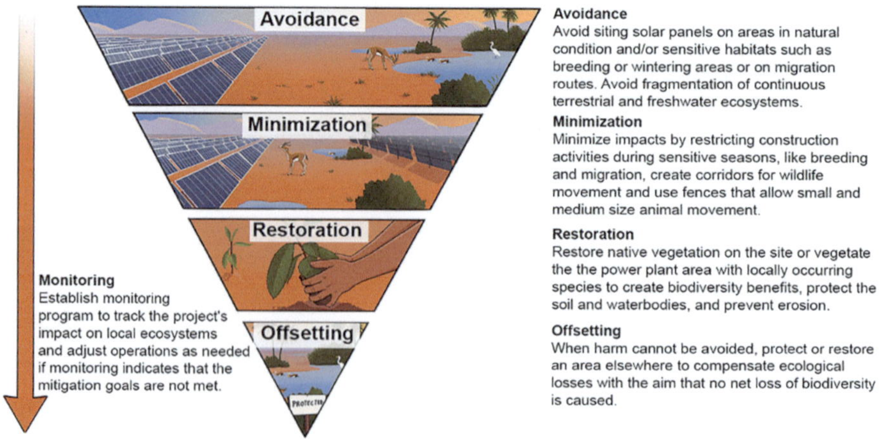

Fig. 4.38 Mitigation hierarchy for negative impacts of large-scale solar projects. © Maija Karala

wildlife must be included into project plans. Assessments should also include cumulative impacts stemming from other energy development projects or human activities in the area. Prior to construction, to avoid and mitigate negative impacts on nature, it is vital to understand important habitats and species, and key areas for different species, such as migration corridors, breeding, feeding, and wintering areas [70]. Generally, areas with natural or seminatural status should be left outside solar power development [78]. This kind of broader planning requires taking a multidisciplinary perspective (see also the team T-profile, Chap. 1), and should guide any large-scale land-use development to allocate the most suitable areas for different human activities, and to prioritize sites that create the least conflicts with nature [79, 151].

Careful siting is the single most efficient measure to avoid and mitigate negative impacts of solar PV power plant development. Nature conservation areas, areas in a natural state, and ecologically sensitive areas should always be left outside project development. Protected nature conservation areas should be considered with particular care to avoid causing degradation of nature values inside these areas; buffer zones can be used to avoid locating a power plant too close. Other critical habitats or sensitive environments, for example small freshwater bodies, wetlands, old growth forests, and burrowing species habitats, should also be protected from destruction and fragmentation due to their importance for many species, and therefore for the entire ecosystem. Disturbance of such habitats should always be avoided and minimized, and damage restored where possible [26, 40, 67]. During construction, negative impacts can be mitigated by timing construction work to be outside important seasons, such as breeding or migration.

Impacts of habitat fragmentation can be mitigated with well-designed wildlife-corridors to allow animal movement. If the power plant is being constructed in an area with important migration routes, the area can be divided into smaller units, leaving corridors for movement in-between. The movement of smaller animals can be assisted by designing fences with large-enough mesh size, and a gap at the lower edge [195]. In the Topaz Solar Farm in southern California, fences were altered to protect the endangered San Joaquin Kit Fox (*Vulpes macrotis mutica*)—the fence was modified to let the kit foxes pass through the fence, but keep out predator coyotes (*Canis latrans*) [2, 177].

Collisions can be mitigated by not routing new power lines through important bird areas, like wetlands or migration corridors [22]. Bernardino et al. [23] concluded in their meta-analyses that wire-marking power lines reduced bird collisions globally by 50% on average. Burying power lines is also an effective measure to reduce collisions, but burying can be costly, and excavation can destroy or fragment important habitats for different species, like amphibians or reptiles [21]. The most effective measure, or combination of measures, for mitigating bird collisions with power lines is case dependent and should be based on an on-site risk assessment.

From research on window collisions, it has been suggested that collisions with solar panels could be reduced with the use of contrasting bands or gaps between the panels [67, 82]. This measure can reduce energy production, however, and therefore can increase land-use impact. Visser et al. [204] recommend more research on the effects of bird collisions with solar panels to identify the best mitigation measures.

Erosion and pollution control are also important factors in the mitigation of harmful nature impacts and land-based carbon emissions. Sediment control measures, for example refraining from unnecessary vegetation removal, are important in areas prone to erosion. Pollution from chemical spills during maintenance activities can be controlled and avoided with established protocols and appropriate infrastructure. Technological choices and design planning also help in managing negative impacts. Bifacial solar panels and the use of tracking systems can reduce land use, and therefore minimize the impacted area per produced energy unit.

Restoration measures after construction or decommissioning depend on the state of nature prior to construction, and restoration plans should be included in the project decommissioning plan. The higher the natural status of the area prior to development, the harder and more expensive it is to restore. Also, some environments can be restored easier than others—for example, a previously forested landscape cannot be restored to its natural state after construction, and even after decommissioning the restoration will take decades. In effect, construction of solar power plants in forested areas will necessarily result in increased deforestation. The best opportunity for restoration after the power plant's decommissioning exists in areas that are already altered by human activity, such as eroded landscapes, old mines, landfills, and so on. Revegetation with locally occurring species can bring biodiversity benefits for degraded lands (see Sect. 4.7.3 below).

Successful restoration planning sets specific restoration goals—that is, defining whether the aim is to enhance the status of previously degraded land, or a complete restoration of the area to its natural state. A monitoring program is needed to track progress and guide follow-up actions to maintain the restored habitat on a long-term basis [26]. The restoration plan should also list actions to minimize soil erosion, protect water resources and aquatic habitats, and manage risk of wildfire in areas where this is relevant [26].

As the last step of the mitigation hierarchy, after negative impacts have been avoided, minimized, and restored, any residual damage should be offset. There are different methods available for calculating the amount of offset so that the project will result in no net loss of biodiversity (NNL), or better, creates net positive impacts on nature, known as net gain (NG) (BBOP [13], see Sect. 2.3). Offsets can be realized via ecological restoration, establishment of new protected areas, or some form of habitat management that balances ecological losses caused by the construction of the power plant with other ecological gains [139].

All mitigation measures, avoidance, minimization, restoration, and offsetting should always be coupled with an environmental monitoring program to track the project's impact on local ecosystems, and to adjust operations if monitoring indicates that the ecological goals are not met [26].

Though there are no industry standards yet available for using mitigation hierarchies in solar energy development projects, some general guidelines can be summarized from the existing literature:

- Mitigation and compensatory measures, including monitoring, need to be included in planning early on, at latest in the environmental impacts assessment.

- Avoid constructing solar power in areas in a natural or near-natural state. The best way to mitigate negative impacts on nature is to prioritize sites that are already altered by human uses and have lower ecological value. Note that areas that have naturally lower species diversity are still valuable in their natural state.
- Apply buffer zones around important areas for nature, e.g., nature conservation areas, areas in their natural state, or important areas for nature, such as breeding or wintering grounds, and migration routes.
- Minimize barrier effects by creating corridors inside large solar power plants, and modify fences in accordance with the local wildlife to allow animal movement.
- Unnecessary land work and soil sealing should be avoided.
- Involve local experts to develop ecologically-sound mitigation plans and measures.
- Use only locally occurring species in revegetation, and monitor and manage potential spread of invasive alien species.
- Establish a monitoring program that guides management decisions as new information becomes available.

All human activities incur a cost of some sort to wildlife, and there is no single solution to protect all species from all impacts stemming from renewable energy production, except for total avoidance. Mitigation techniques are also typically species-specific and site-dependent, and it is not possible to address the entire affected wildlife community [142]. Nonetheless, applying the mitigation hierarchy early in the planning phase can effectively reduce the adverse effects of solar PV development on nature, and foster a more sustainable approach to renewable energy generation. Next, we consider the potential of co-locating solar PV systems with other human activity, and incorporating solar energy power plants in the rehabilitation of degraded or polluted lands.

4.7.3 Co-location and Land Rehabilitation Options of Solar PV Power Plants

Land use can be optimized, and negative impacts on nature minimized, by selecting locations that are already in human use, such as urban settlements, grass fields around airports, and agricultural lands, or areas that already have a reduced importance for wildlife or are contaminated, such as different kinds of brownfields. Unlike many other energy systems, solar PV power is well suited for installation on buildings and other human settlements (parking spaces, noise barriers, airports, etc.), thereby decreasing direct land-use pressure and impacts on nature. Solar panels can also be installed into agricultural infrastructure, and arrays can be sited on spaces that are not prime agricultural lands, such as field and road edges [81]. Development of these kinds of decentralized energy systems have gained attention and interest as a response to the transition away from fossil fuel-based energy systems [1, 56].

Although installing solar PV systems on urban settlements is not entirely problem-free—for example, due to the heat-island effect [9] or issues from glare [182]—it nonetheless is an important way to alleviate pressure and competition over land, and thus also impacts on nature.

The physical foundations of ground-mounted solar arrays directly take up less than 1% of the plant's land area, leaving space for other uses. The panels are raised around 2 m above the ground, with tilt approximately the same as the latitude—as a result, the space between the panels varies with the latitude, and approximately 20–40% of the ground is shaded by the panels [26, 14]. One evolving co-location option of ground-mounted, utility-scale solar PV systems is to combine power plants with agriculture [46, 57]. These are known as agrivoltaic systems, and the idea is similar to agroforestry (see Chap. 6), which combines trees and shrubs with cropping or animal farming to create environmental, economic, and social benefits [46]. Agrivoltaics combine agricultural practices with solar PV energy production to create more benefits than either of these systems could generate alone.

The benefits of agrivoltaics are twofold. First, it helps to increase land-use efficiency and alleviate competition over land by combining two land-use intensive activities together. Second, agrivoltaics can provide farmers with increased security: in the case of crop failure, they will still generate income from electricity production [172]. The spacing between the rows of solar arrays usually needs to be a bit wider than the minimum spacing required for electricity generation, but overall, this combined system increases land-use efficiency [223]. Several crops have been identified, including vegetables, fruits and herbs, that can tolerate shade and are not too tall to overshadow panels, and are therefore suitable for cultivation under solar panels [172].

Agrivoltaics is a relatively new concept, and it represents a promising synergy between renewable energy production and sustainable agriculture. However, as with any other large-scale renewable energy project, agrivoltaic systems can also cause habitat fragmentation and barriers for movement of wildlife, or if not well planned can displace food production [81]. Careful planning and implementation of agrivoltaics can support the development of clean energy and help minimize negative impacts on nature (see Fig. 4.39).

Another way to mitigate negative effects of land occupation by solar PV power plants is to install the system on pastures [141]. The area between the panels can be vegetated with locally-occurring plants, providing benefits for biodiversity, and the vegetation is then managed by grazing animals.

Other co-location options include hybrid wind-solar plants both on land and at sea, which also provide an opportunity to increase land- and sea-use efficiency. This kind of hybrid power plant, coupled with storage solutions such as hydrogen, are likely to become increasingly common in the future (see also Chap. 7 on energy storage). Floating photovoltaics are also suggested as a method to decrease land-use pressure and halt deforestation [35, 118]. However, the impacts of floating photovoltaics on natural ponds, lakes, coastal waters, and other water bodies, and the effects on aquatic life, are not well studied or understood [67]. Shading created by the panels can have different impacts on aquatic plants and animals, anchoring might disturb sediments,

Fig. 4.39 Example of an agrivoltaic system. © Maija Karala

cause changes in water quality and disrupt bottom-dwelling species; and accidental oil or chemical spills can pollute water [155]. To avoid these negative impacts, floating photovoltaics are better suited to water bodies created by human activities, such as artificial lakes and ponds, water bodies created for aquaculture, abandoned mining lakes, and so on [180].

Solar power development sites can be used to enhance soil recovery and maintenance. Solar PV plants have been found to be bring positive biodiversity benefits in areas where human uses are intensive and natural habitats are degraded or fragmented. In California, for example, land and water use is intensive because of urban settlements and agriculture, resulting in salt accumulation in the soil and loss of species and habitats. In these areas, solar PV development can bring significant positive change for nature if the soil is restored during the construction, habitats for different species are provided, and the connectivity between habitat areas is increased [40, 81, 154].

Solar PV power plants provide an opportunity to establish meadows for pollinators, and threatened species such as monarch butterflies (*Danaus plexippus*) [26]. In UK, for example, solar power plants have been shown to foster greater diversity of plants, invertebrates, and birds than surrounding areas used for agriculture and other human activities [141]. Similarly, in Chesapeake Bay, US, Evans et al. [53] observed that the rapid expansion of solar power in the bay area during the past decade had mostly not replaced natural areas, but had instead been constructed on agricultural or degraded lands, and provided an opportunity to create biodiversity benefits and restore ecosystem services, such as carbon sequestration, in the watershed. However, areas in their natural state should not be converted into combined solar power production sites and meadows, this kind of revegetation is best suited for areas already disturbed by previous human activity. Only locally occurring species should be used for revegetation, and the spread of invasive alien species should be controlled and managed.

Degraded lands, such as landfills, abandoned open mines, or eroded landscapes that can be rehabilitated and used for solar energy production also provide an opportunity for combined renewable energy production and the creation of biodiversity benefits. In the United States alone there is nearly 20 000 km^2 of degraded land that could be used for renewable energy production, potentially generating 114 TWh (3% of US electricity supply US in 2016) with no new direct land occupation [147]. Revegetation helps to restore and conserve the soil, which also helps to protect water resources [79, 81]. At a global scale, combining electricity production and rehabilitation of degraded lands could be an important and cost-efficient way to manage degraded or contaminated lands, create biodiversity benefits and restore ecosystem services.

4.7.4 Summary Table

Table 4.2 summarizes the different impacts of ground-mounted solar PV power plants on nature and wildlife. Examples and ways to mitigate negative impacts are drawn from scientific literature, and discuss habitat loss, collisions, avoidance and barrier effects, cumulative impacts, trophic cascades, and introduction of invasive alien species. The main purpose of the table is to summarize existing scientific literature of the impacts of solar power development on nature, covering a wide range of species, habitats and geographical locations. The list is not exhaustive, so when commencing development of a wind power project, always consult local information sources.

Table 4.2 Summary of the potential impacts of utility-scale ground-mounted photovoltaic (PV) solar power plants on species and their populations, and ways to mitigate these impacts. The impacts discussed include habitat loss, collisions, avoidance and barrier effects, cumulative impacts, trophic cascades, and introduction of invasive alien species

Impact	Explanation	Impacts with PV panels	Mitigation measures
Habitat loss: destruction, fragmentation, degradation	Habitat is lost when it no longer can support its original species. Habitat can be lost via physical destruction, fragmentation, or degradation. The loss results in reduced biodiversity and changes in species abundance. The species can also be replaced by other, sometimes more generalist species, as a result of changes in land use	Vegetation is often totally cleared or altered significantly under the solar arrays to facilitate operation. Clearance of vegetation (bushes, trees) changes species composition. For example, the diversity and density of bird species have been found to be lower at the power plant than in the surrounding untransformed landscape[a] Solar PV development has been shown to fragment bat foraging and commuting habitats[b] Solar power development can also degrade aquatic ecosystems. Construction can impact flow alteration and sedimentation, displace aquatic biota, and cause contaminant runoff[c] Cleaning of dust from panels can degrade the availability and quality of surface and groundwater sources, and affect ecosystems dependent on them[d] Solar panels can change the air and soil microclimate, and this may alter the original habitat and species composition or diversity beneath the PV panels[e]	Best way to mitigate the most severe negative impacts on nature is to avoid all construction in or close to sensitive areas, and during sensitive seasons In construction, avoid unnecessary sealing of soil and favor foundations that don't damage natural soil functioning, to maintain the species diversity both below and above ground[f]. Limit clearance of nature to a minimum, and use restoration and offsets to compensate residual damage, for example by planting vegetation that maintains foraging and commuting habitats for bats[g] Implement measures to minimize soil erosion (construct drainage for access roads, avoid flooding land and damaging nearby water bodies) To protect aquatic environments, identify all stream crossings and make a plan that allows uninterrupted movement and safe passage of fish during construction, operation and decommissioning[h] To preserve water resources, use manual dry brushing for cleaning Limit the use of chemicals in managing vegetation and dust abatement After construction, restore biodiversity features and/or revegetate damaged temporary-use areas

(continued)

Table 4.2 (continued)

Impact	Explanation	Impacts with PV panels	Mitigation measures
Collisions	The PV panels can pose a collision risk for birds and bats. To date there are few studies on this issue, so the significance of this threat is largely unknown. Collisions with power distribution lines can cause death via electrocution	A summary study of bird mortality at PV power plants in the United States estimated that the average annual fatality is 2.49 birds/MW/year[i]. Bird species using pylons (or poles) from low- or medium-voltage power lines as perches or for nesting can be significantly affected by electrocutions[j]. Collisions and electrocutions can cause population decline, in particular for long-lived species, such as Egyptian vultures, the chaco eagle[k], or great bustards[l] Solar power plants can also attract birds and aquatic insects that mistake the solar panels for water bodies. Evidence on this "lake-effect" hypothesis is still limited and requires more research[m] Solar panels can act as a sensory trap for bats and insects, increasing collisions with the panels[n]	Avoid constructing power lines on major migration routes. Depending on location, re-routing, marking, or burying powerlines helps to avoid collisions[o]. Burying of transmission lines must be planned carefully to avoid destroying terrestrial and aquatic habitats. Usually a mix of measures (burying and marking) adapted to local conditions works best[p] Electrocutions can be reduced by insulation of low- and medium-voltage poles and wires[q] To mitigate bird collisions with panels, use of contrasting bands has been proposed to increase panel visibility. This can reduce energy production, and therefore increase land intake, and so more research is needed to understand the severity and cause of collisions with panels[r] [204] Use of non-polarizing white tape on panel borders, and/or in a grid-like pattern across panels, may help to reduce aquatic insect collisions with the panels[s] [82]

(continued)

Table 4.2 (continued)

Impact	Explanation	Impacts with PV panels	Mitigation measures
Avoidance and barrier effect	Utility-scale solar PV power plants can disrupt wildlife movement and/or migrations, and therefore act as a barrier. A barrier effect is formed when the construction physically prevents an individual's movement from one area to another (for example to feeding, wintering, and breeding areas). Some species may also avoid the construction site due to noise or changes in the environment, and therefore the power plant forms a barrier for movement even when there is no physical barrier. Avoidance can be temporary and may be restricted only to construction period	Noise created by construction can disturb animals, causing them to avoid the area The barrier effect caused by solar power development is not well studied, but it is known that large scale construction, including energy developments and related infrastructure such as fences and roads, impact species movement and can reduce their range. For example, large mammal migrations can be impacted severely[t]. Solar power plants also disturb the movement of smaller animals: mammals, birds and lizards can get trapped in the fence, or between a double-fence[u] Many bats avoid anthropogenic structures, and avoidance behavior has also been found in solar PV power plant sites[v]	Avoidance of construction noise can be mitigated by timing the operations outside sensitive seasons, like migration or breeding With fences that form a barrier for movement, consider whether the fences are in fact necessary. Especially for large solar power plants, leave passageways for larger animals to facilitate movement. Movement of small and medium size animals can be enabled my making modifications to fencing, such as using a larger mesh-size and leaving a gap between the base of the fence and the ground [w] Limit clearance of nature to a minimum, and use restoration and offsets to compensate residual damage, for example by planting vegetation that maintains foraging and commuting habitats for bats[x]

(continued)

Table 4.2 (continued)

Impact	Explanation	Impacts with PV panels	Mitigation measures
Cumulative impacts	Cumulative impacts are changes in the environment caused by the combined effect of multiple human activities and natural processes. They are often difficult to predict and manage because of a lack of sufficient environmental data, complexity of ecological processes, and the large scale of human activities	Cumulative impacts caused by solar power development are not well-studied. But areas with suitable conditions for solar energy production might experience intensive development that together create significant pressure via land-use change. For example, solar panels can cumulatively kill substantial numbers of birds and insects, but it is not yet known if this is severe enough to have population-level effect[y]. It has also been suggested that migratory birds can lose important stopover sites due to cumulative impacts from several large PV plants along their flyway[z]. For example, the Egyptian vulture, a long-distance migrant, is exposed to human-induced mortality along its migration routes. Any additional mortality (such as increased collisions on power transmission lines) can be significant[aa] Cumulative loss of aquatic ecosystems caused by many solar power plants and related infrastructure can be significant[ab]	Solar power production should be considered in relation to other energy projects and human activities in the area. It is important to have data on the species in the area, particularly migrants and wide-ranging species. Impacts can be mitigated by including corridors for species movement in planing[ac], and depending on the species, creating habitats inside or outside the solar power plant[ad] To avoid creating cumulative loss of local aquatic environments, prior to construction local surface water networks should be mapped and their vulnerability evaluated, including stream reaches where mitigation and monitoring measures are most needed[ae]. Implement stormwater management plans and protect streams with buffer zones

(continued)

Table 4.2 (continued)

Impact	Explanation	Impacts with PV panels	Mitigation measures
Trophic cascades	Indirect, strong interactions that can affect the entire ecosystem are called trophic cascades. Trophic cascades affect at least three levels of a food web via predator–prey interaction. In its simplest form, predators limit the density of their prey, and this in turn enhances the survival of the next lower trophic level. Human disturbance can disturb these interactions, resulting in changes to the entire ecosystem	Trophic cascades caused by solar power are not well studied. Though cumulative impacts of solar power on birds and insects can be large, it is not known if the mortality is great enough to produce population-level, or cascading, effects[af]. Due to rapid development of large-scale solar energy production, further research is needed to understand ecosystem level changes related to solar power development	Any areas unused after construction should be restored to allow the original nature to return. Particularly important areas include grasslands, forests, wetlands, aquatic environments and coastal ecosystems, such as dunes. Take preventive measures to protect water bodies from erosion. Gather and develop sufficient knowledge of the local nature and biodiversity to intervene appropriately to lessen or eliminate negative outcomes of the project
Invasive alien species, AIS	Invasive alien species are plants, animals, pathogens, or other organisms that are non-native to an ecosystem. They can have an adverse impact on biodiversity, ecosystem functions, the economy, and human health, by causing a decline or loss of species in local ecosystem	Creation of new open spaces and habitats, land disturbance and new access roads can facilitate the dispersal of invasive alien species. IAS are transported via people, machinery, or soil[ag]	Prevent introduction and movement of IAS by decontaminating vehicles and equipment before entering new areas. In landscaping, use soil from local or otherwise known sources to prevent spread of invasive species. In revegetation, use locally occurring species. Establish monitoring programs for invasive species, and eradicate these species when found[ah]

[a]Pocewicz et al. [156], Visser et al. [204], Bennun et al. [21], [b]Tinsley et al. [189, 193]. [c]Grippo et al. [67]. [d]Bennun et al. [21]. [e]Armstrong et al. [11]. [f]Bennun et al. [21]. [g]Tinsley et al. [193]. [h]Grippo et al. [67], [i]Kosciuch et al. [110]. [j]Bennun et al. [21]. [k]Angelov et al. [8]. [l]Bernardino et al. [22]. [m] Kagan et al. [101], Huso et al. [85], Visser et al [203], Taylor et al. [190], Kosciuch et al. [111, 110]. [n]Szabadi et al. [189]. [o]Raab et al. [159], Bennun et al. [21]. [p]Bernardino et al. [22]. [q]Bennun et al. [21], [r]Visser et al. [203]. [s]Horváth et al. [82]. [t]Wingard et al. [210], Wyckoff, [214]. [u]Visser et al. 2019. [v]Tinsley et al. [193]. [w]Bennun et al. [21]. [x]Tinsley et al. [193]. [y]Grippo et al. [67], [z]Bennun et al. [21]. [aa]Angelov et al. [8]. [ab]Grippo et al. [67], [ac]Guiller et al. 2017, [22]. [ad]Montag et al. 2016, [21, 193]. [ae]Grippo et al. [67], [af]Grippo et al. [67], [ag]Bennun et al. [21]. [ah]Grippo et al. [67]

4.8 Social Impacts of Solar Energy

Solar energy is a relatively new and growing industry, and so the social impacts of solar energy production, especially large installations, are not as well-known as those for wind energy. Compared to other renewable energy sources, solar energy (particularly rooftop installations) has generally enjoyed a high level of socio-political acceptance. However, increasing solar capacity and size of solar installations has an effect on social acceptance, and the impacts of solar energy need to be considered when developing the technology [39]. The solar energy industry stimulates local and global economies, and increases employment opportunities for local communities and global industry, creating jobs in many sectors such as manufacturing, installation, maintenance, and research [106]. Additionally, increasing solar power capacity promotes energy independence by decreasing reliance on imported energy and fossil fuels, which have different negative ecological and geopolitical impacts (Çetin and Egrican 2011).

In Sect. 2.5 we discussed power dynamics in energy systems, demonstrating how more resourceful actors, such as large energy corporations, can use their power to fulfill their interests, while local communities in more vulnerable positions might lack the opportunity to participate in decision-making processes. Therefore, solar energy is also expected to create democratizing benefits, as households, municipals, and medium-sized companies can become solar self-producer-users (energy prosumers) and thus participate energy decision-making. This is very much context-dependent, and applies in strong market economies and the Global North, but might vary in different national contexts and energy systems.

The negative social impacts of the solar energy industry are often tied to global social justice issues and the exploitation of minority groups and vulnerable communities [170, 220]. These can arise around mining of critical minerals for solar panels, manufacture of components, or dislocation of local social groups due to large installations. In this section we look further into both the positive and negative social impacts of solar energy, and provide some suggestions on how to alleviate the downsides.

4.8.1 Employment and Community Benefits of Solar Energy Field

Climate change mitigation and improving energy security by reducing the reliance on energy imports and fossil fuels are well known positive impacts of solar energy (Çetin and Egrican 2011). The rapidly growing solar energy industry is creating new direct and indirect employment opportunities, which are essential for livelihoods and social welfare in a market economy [106]. Throughout the 2010s the solar energy industry has provided employment opportunities for people with a wide range of skills, from experts in design, manufacturing, and research, to vocational jobs in installation, maintenance, and production of solar energy technologies (Çetin and Egrican 2011).

The International Renewable Energy Agency (IRENA) estimated that in 2021 the global solar PV industry employed 4.3 million people, and this number is growing year by year [93]. The employment opportunities in the solar energy industry vary by geographical regions; developing nations have greater potential for employment in the solar industry than developed nations, because of reduced labor and manufacturing costs [33]. Most of solar product manufacture originates in China and

Southeast Asia; China is the dominant country in PV manufacturing, producing 96% of wafers, 84% of cells, and 81% of modules [93].

Solar energy, especially when implemented by communities, wields many positive social impacts. Citizens and local communities can engage in numerous ways with the energy system, for example, as investors, innovators, prosumers, flexumers (who consume and store energy, and distribute surplus energy to the grid), participants in local energy initiatives, or supporters of climate movements. Active participation of different actors, including traditional energy players and previously passive actors such as households, is essential for reaching climate neutrality and more just and equal energy systems. Households have a large potential for contributing to the energy transition—for example in Europe, household share of final energy consumption is 27% [52]. EU legislation is supporting citizen-led initiatives and requiring member states to facilitate the establishment of, for example, energy communities, in which, energy prosumers can jointly produce, store, exchange, and trade energy [206].

Energy communities can be small initiatives, involving only a few households, but they can also be larger endeavors, bringing together various private and public organizations. A good example in Europe is Denmark, where energy communities have been set up by different coalitions or associations who own wind power plants and distribute electricity at least partially amongst their members. Among the most important positive social impacts of community energy initiatives is the distribution of energy, income and other benefits in a more socially just manner. Such endeavors are also expected to democratize energy systems [91, 211] and make them more transparent, flexible and resilient.

There are several compelling examples of energy communities that demonstrate the positive impacts such initiatives can have on local communities. One notable example comes from a housing cooperative in a residential block in Zwoleń in Poland. By 2022, the cooperative had installed photovoltaic panels on the building's roof, carried out thermal insulation of the building, and installed ground-source heat pumps coupled with a borehole heat exchanger. These investments enabled the building to disconnect from the district fossil-fuel based heating network. As a result, residents experienced significantly reduced energy bills, with heating costs and loan installments totaling 35–40% less than the heating costs of a neighboring building without such improvements [16].

Another noteworthy example is in Budapest, Hungary, where the "Kazán" Community House installed a rooftop 36 kW solar PV system in 2022. Located in an old factory building, Kazán hosts various grassroots initiatives, including a nursery, ceramic workshop, workout room, and radio studio, collectively consuming a substantial amount of electricity. Kazán tenants partially crowdfunded the PV installation, but rather than using the electricity savings directly, the community decided to allocate them to a common fund dedicated to renovating the energy-inefficient building. These investments significantly reduce energy bills for social organizations, enabling them to allocate funds towards activities benefiting the local community [29].

Community solar energy examples are widespread, and the origins of the community can vary—existing communities such as housing cooperatives, islands or villages can form energy communities, motivated by legislation, crisis, orientation towards sustainable energy, desire for energy saving and income, utilizing a local skilled workforce, or by external intervention where some actor provides the energy solutions for the community. One example of the accumulation of a skilled workforce stimulated by a crisis is the "Solar Town", a renewable energy community established in Slavutych, Ukraine, near Chernobyl. The city was founded in 1986 for power plant workers evacuated after the nuclear disaster. In 2016, an energy cooperative was established there by city officials and residents, launching a successful crowdfunding campaign. Leveraging a skilled local workforce comprising energy experts, lawyers, and economists, the cooperative installed a 200 kW rooftop solar system connected to the grid, benefiting from favorable feed-in tariffs. Each cooperative member earned an annual dividend, and the cooperative supported local development by allocating 5% of its income to fund local projects [62]. As a further example, WEnCoop stands out as the first all-female energy community in Greece. It was founded in 2021 by the Greek Association of Women Entrepreneurs to promote inclusivity in the energy transition by engaging more women. The cooperative encourages women without expertise in energy matters to participate in its activities, such as the recent completion of an investment in a 1 MWp solar park [158].

It is important to note that, while the above examples occurred in Europe, similar examples can be found globally. Although global north examples highlight energy democracy and distribution of power, local solar energy production and communities exist that highlight welfare impacts rather than democratic impacts. An example from the Tanzanian villages east of Arusha shows that self-production of solar, whether at the individual, family or community level, offers a means for good life with access, affordability, and social equity, especially in countries with low development of nation-wide electricity grids, and where grid access is expensive [74]. The 9177 installed solar and battery systems for 646 villages were part of a state welfare program where capital was collected from large international corporations, equity and consultant firms, energy utility EDF, and oil company TOTAL [74]. In their research paper, [74] also described a Moroccan example where the government provided solar modules for nomads, adapted to their mobile lifestyles to increase electrification rates. Masdar Clean Energy, an Emirati company, provided the solar modules.

The solar energy communities and energy prosumers described above can reduce power inequalities in energy systems, and thus give voice to previously neglected social groups. Nevertheless, renewable energy investments, even community-based, might also bear some negative impacts. These are discussed next, from the impacts of land use and problems in global solar value chains, to unequal distribution of employment benefits and unjust opportunities for engaging in community solar initiatives.

4.8.2 Negative Social Impacts Linked to Land-Use Changes and Global Value Chains

Although the solar energy industry provides positive socio-economic impacts that have the potential to improve the welfare of the society, the fairness and social impacts of the full global supply chain of solar products cause uncertainty. The supply chain risks linked to raw materials used in solar PV production encompass human rights violations, conflict financing, transparency issues, child labor, safety standards, and labor health problems; many energy companies struggle to identify the origin of the raw material used for their PV cells, complicating efforts to trace suppliers and assess compliance with ethical practices and human rights [84]. The equity impacts of the entire supply chain can be significant depending on the required minerals, metals and other natural resources [24], and the geographical location of mines, extraction and refining.

Negative attitudes towards solar projects are often related to the traditional and emotional attachment of local communities to the land, and aesthetic change to the landscape [96], and also to impacts of large-scale energy projects on local livelihoods, value of the land for local communities, and even to forceful direct dislocations of communities. In addition to evaluation of environmental impacts, it is essential to evaluate the socio-economic and cultural impacts of land acquisition for solar park development, this is because social justice concerns and the welfare of minority groups and local communities may be overlooked in solar energy projects for the purpose of national development [220], or due to a too-narrow focus on profit-seeking behavior of powerful large energy corporations.

Land-use changes can affect the rights and livelihoods of vulnerable people, including rural communities, women, indigenous people, and other minority groups whose livelihoods and welfare are often very dependent on the resources provided by the land [170, 220]. Case studies conducted in India show that land acquisition by the government for renewable energy projects and land-use changes ultimately affect livelihoods of local communities and amplify pre-existing inequalities, because land is often acquired from "low-value users" and redistributed to economic classes and industries more capable of developing it [119, 120, 220]. Solar power plants can be deployed unjustly without adequate consent and fair planning with the local communities [144]. Another study of solar deployment in India found that demographic, dispossession, interspecies, and temporal injustices stemmed from the loss of local livelihood, job insecurity, water pollution, deforestation, air pollution, habitat loss, and gendered resource dislocation [184].

It is important to note that local land-use impacts of solar energy are often gendered. For example, [185] found in their research on female disempowerment that solar parks disproportionately impacted the women from lower cast and class in India. Moreover, land enclosures for solar power plants have excluded local resource-dependent women's access to prime fire-wood foraging lands. The physical components of the solar panels are manufactured in factories that produce emissions and use large quantities of water compared to other renewable energy sources [137], which in turn mostly have an impact on local communities. Although land-use rights and energy justice issues are often tied to national energy development projects in developing countries, these issues have resonance globally and apply to energy transitions of both developing and developed countries [220].

Most of the unjust local social impacts of solar power projects link to power differences in the energy system, as extensively discussed in Sect. 2.4. These power imbalances—where politically and economically powerful actors control planning and profits while local communities bear the environmental and land-use impacts—limit the diversity of perspectives, potentially excluding marginalized groups and people in a vulnerable positions, including for example low-income households, people with disabilities, or ethnic minorities, from decision-making processes [109]. This type of societal exclusion risks burdening ordinary citizens with the financial, environmental, and health costs, while the privileged actors benefit from renewable energy investments. Furthermore, limiting the diversity of voices in the energy transition may result in inadequate consideration of environmental and social aspects. For example, excluding academics or environmental organizations could lead to insufficient scrutiny to ensure that renewable energy projects meet sustainability criteria. This may furthermore decrease transparency and accountability of energy decision-making, which can further decrease social acceptability of renewable energy projects and increase social inequalities [59], potentially leading to widespread public protests and opposition in some global contexts [148].

As discussed earlier, the promise of energy communities and prosumerism has been offered as a solution to the imbalance of power, and a means to more equal distribution of the benefits. This is very much the case when assuming that the country or geographical location is otherwise already acting democratically, and many studies on energy communities have a Global North context. The challenges of energy communities vary in different cultural contexts. In Nordic countries, where people are expected to act independently, these challenges include the equal involvement of different societal groups, communities or individuals, which might significantly differ in their investment capacities, technical knowledge and skills, or the resources available to become individual or collective energy prosumers. Conditions vary within European countries, and this already creates tensions in forming energy communities [10]—not to mention other continents and cultural contexts.

4.8.3 Solutions to Mitigate the Negative Impacts

To address these challenges, a broadly inclusive dialogue that considers the perspectives and interests of all affected parties is essential. Different stakeholder groups, including NGOs, local communities, regional politicians, and local businesses, should be involved in the early planning stages of renewable energy projects. This inclusive approach increases the chances of successful implementation and social acceptance of renewable energy. In Chap. 5 the Sect. 5.9 on the social impacts of wind energy provides step-by-step suggestions for this type of a dialogue model, and recommendations on how to design and plan more inclusive energy projects.

To have a socially sustainable solar energy industry, the entire lifecycle of photovoltaic panels must be sustainable, from material extraction through to e-waste management of the panels. Raw material extraction and PV e-waste management pose sustainability challenges for the solar energy industry, but also create economic opportunity to develop these industries [90]. Given the growing solar PV market and the average panel lifetime of 30 years, focus must be directed towards end-of-life management of PV panels [90]. Countries with inadequate regulations, waste management infrastructure, and e-waste recycling systems are at risk from various negative social and environmental impacts from e-waste created by the solar energy industry [171].

4.9 Summary

Global utility-scale solar power development is growing rapidly as a response to the urgent need to move away from fossil fuels. IEA [87] estimates that wind and solar PV will together account for 95% of global renewable expansion by 2028. Technically this is possible, and as we have discussed in this chapter, solar power development also provides positive prospects for local communities in the form of prosumerism and agrivoltaic systems, and offers opportunities for biodiversity benefit creation through restoration and rehabilitation of degraded lands while producing energy.

In this chapter, we started with solar resource analysis, and discussed optimal panel orientation. Since solar power plants are area-intensive, and require a lot of materials, careful site selection and optimal design are needed to minimize the use of land and materials. Several free simulation programs are available for the initial design of solar power plants, such as the PVGIS (EU Science Hub) and the Global Solar Atlas, which provide useful insights for siting and in technical performance. However, before using these tools, it is important to understand the theory at work behind the colorful graphical user interface—and so we then discussed the physical phenomena on which these programs are based, to gain a better understanding of how they work, and not just to use them as "black boxes". To make studying these topics more interesting, several MS Excel© and Matlab© tools are available for download which can be freely used and developed by readers and students in their own projects. We then discussed the operational principles of solar panels and their maximum theoretical efficiencies and limitations. We examined the connection of a solar PV panel with a battery in off-grid systems, and briefly considered the design of larger MW scale solar power plants.

Finally, we broadly discussed the environmental aspects and impacts on nature of solar power plants, as well as the societal impacts and their mitigation. While the technical solutions often already exist, the impacts on nature and surrounding societies are often the key to successful implementation of solar energy projects, and should be well understood by the energy engineers and designers of these systems. The rapid expansion of solar power construction to meet the objectives of decarbonizing the energy sector and increasing the share of renewable energy sources also increases the risk of improper siting of solar power plants. There is a need to increase our understanding, inclusion, and mitigation of negative ecological and social impacts caused by solar power development. Solar energy production can impact terrestrial and aquatic ecosystems and local communities in many ways. Overlooking the impacts on nature, or local societies' rights, can result in costly and long-lasting legal battles and prevent construction of energy projects.

It is difficult to predict which research directions will become successful and prominent in the future; rather, it is more relevant to identify potential directions and explore them. For instance, organic photovoltaic panels (OPV) and dye-synthesized solar cells (DSSC), despite decades of research, are still not on the market. Yet processing technologies, materials, and research infrastructure developed for OPVs and DSSCs have been instrumental in the development of the halide perovskite solar cells, significantly contributing to their prominence. Halide perovskites themselves are not new materials, and were originally studied as potential semiconductors for microelectronics. Two decades later they are still not employed in microelectronics, but instead are shaping the landscape of solar cell research.

To conclude, the potential of solar energy is great, and new technological solutions will increase efficiency and generate cost reductions, making solar energy increasingly accessible. Properly sited solar PV power plants will be one of the major solutions to combat climate change and reduce our reliance on fossil fuels. Furthermore, the creation of new jobs and increased energy democratization underscore the role of solar power in fostering sustainable development.

Author Contributions 4.1–4.3 Mika Järvinen, 4.4 Eetu Laitela and Mika Järvinen, 4.5 Georgi Popov, 4.6. Saeed Rahimpour Golroudbary, Benjamin P. Wilson and Mari Lundström, 4.7 Hanna Paulomäki and Eetu Laitela, 4.8 Roosa Talala, Ayesha Sadiqa, Eeva-Lotta Apajalahti and Alicja Dankowska, 4.9 Mika Järvinen and Hanna Paulomäki. Subsections can be referenced accordingly.

References

1. Adil AM, Yekang Ko Y (2016) Socio-technical evolution of decentralized energy systems: a critical review and implications for urban planning and policy. Renew Sust Energ Rev 57:1025–1037. https://doi.org/10.1016/j.rser.2015.12.079
2. Agha M, Lovich JE, Ennen JR, Todd BD (2020) Wind, sun, and wildlife: do wind and solar energy development 'short-circuit' conservation in the western United States? Environ Res Lett 15(7):075004. https://doi.org/10.1088/1748-9326/ab8846
3. Agha M, Smith AL, Lovich JE, Delaney D, Ennen JR, Briggs J, Fleckenstein LJ, Tennant LA, Puffer SR, Walde A, Arundel TR, Price SJ, Todd BD (2017) Mammalian mesocarnivore visitation at tortoise burrows in a wind farm. J Wildl Manage 81:1117–1124. https://doi.org/10.1002/jwmg.21262
4. Akkerman QA, Manna L (2020) What defines a halide perovskite? ACS Energy Lett 5(2):604–610. https://doi.org/10.1021/acsenergylett.0c00039
5. Al-Ashouri A et al (2020) Monolithic perovskite/silicon tandem solar cell with >29% efficiency by enhanced hole extraction. Science 370(6522):1300–1309. https://doi.org/10.1126/science.abd4016

6. Almeida A, Sousa N, Coutinho-Rodrigues J (2019) Quest for sustainability: Life-cycle emissions assessment of electric vehicles considering newer Li-ion batteries. Sustainability 11. https://doi.org/10.3390/su11082366
7. Almorox J, Hontoria C, Benito M (2011) Models for obtaining daily global solar radiation with measured air temperature data in Madrid (Spain). Appl Energy 88:1703–1709
8. Angelov I, Hashim I, Oppel S (2013) Persistent electrocution mortality of Egyptian Vultures *Neophron percnopterus* over 28 years in East Africa. Bird Conserv Int 23(1):1–6. https://doi.org/10.1017/S0959270912000123
9. Ansar K, Mattheos S (2023) On the local warming potential of urban rooftop photovoltaic solar panels in cities. Sci Rep 13:15623. https://doi.org/10.1038/s41598-023-40280-9
10. Apajalahti E-L, Ruggiero S, Lukkarinen J, Laakso S (2023) Exploring tensions in urban energy communities. In: Sokołowski MM, Visvizi A (eds) Routledge handbook of energy communities and smart cities. Routledge
11. Armstrong A, Ostle NJ, Whitaker J (2016) Solar park microclimate and vegetation management effects on grassland carbon cycling. Environ Res Lett 11(7):074016. https://doi.org/10.1088/1748-9326/11/7/074016
12. Asundi AS, Raiford JA, Bent SF (2019) Opportunities for atomic layer deposition in emerging energy technologies. ACS Energy Lett 4(4):908–925. https://doi.org/10.1021/acsenergylett.9b00249
13. BBOP, Business and Biodiversity Offsets Programme (2012) Standard on biodiversity offsets. Washington. https://www.forest-trends.org/publications/standard-on-biodiversity-offsets/
14. BRE, Building Research Establishment (2014) Biodiversity guidance for solar developments. Parker GE, Greene L (eds). https://app.croneri.co.uk/feature-articles/biodiversity-guidance-solar-developments
15. Babayigit A et al (2016) Toxicity of organometal halide perovskite solar cells. Nat Mater 15(3):247–251. https://doi.org/10.1038/nmat4572
16. Bachowski M (2022) Ten blok odłączył się od ciepłowni. Rachunki spadły o 90 proc., Noizz.pl. https://noizz.pl/ekologia/ten-blok-odlaczyl-sie-od-cieplowni-rachunki-spadly-o-90-proc/xjzb3p7
17. Baigorria GA, Villegas EB, Trebejo I, Carlos JF, Quiroz R (2004) Atmospheric transmissivity: distribution and empirical estimation around the central Andes. Int J Climatol J R Meteorol Soc 24:1121–1136
18. Barron-Gafford GA, Pavao-Zuckerman MA, Minor RL, Sutter LF, Barnett-Moreno I, Blackett DT, Thompson M, Dimond K, Gerlak AK, Nabhan GP, Macknick JE (2019) Agrivoltaics provide mutual benefits across the food–energy–water nexus in drylands. Nat Sustain 2(9):848–855. https://doi.org/10.1038/s41893-019-0364-5
19. Bastida F, Moreno L, Hernández T, García C (2006) Microbiological activity in a soil 15 years after its devegetation. Soil Biol Biochem 38(8):2503–2507. https://doi.org/10.1016/j.soilbio.2006.02.022
20. Bellini E (2022) CSEM, EPFL achieve 31.25% efficiency for tandem perovskite-silicon solar cell. pv magazine. https://www.pv-magazine.com/2022/07/07/csem-epfl-achieve-31-25-efficiency-for-tandem-perovskite-silicon-solar-cell/
21. Bennun L, van Bochove J, Ng C, Fletcher C, Wilson D, Phair N, Carbone G (2021) Mitigating biodiversity impacts associated with solar and wind energy development. Guidelines for project developers. The Biodiversity Consultancy, Gland, Switzerland, IUCN and Cambridge UK, 260 pp. https://doi.org/10.2305/IUCN.CH.2021.04.en
22. Bernardino J, Bevanger K, Barrientos R, Marques R, Martins R, Shaw J, Silva J, Moreira F (2018) Bird collisions with power lines: state of the art and priority areas for research. Biol Conserv 222:1–13. https://doi.org/10.1016/j.biocon.2018.02.029
23. Bernardino J, Martins R, Bispo R, Moriera F (2019) Re-assessing the effectiveness of wire-marking to mitigate bird collisions with power lines: A meta-analysis and guidelines for field studies. J Environ Manage 252. https://doi.org/10.1016/j.jenvman.2019.109651
24. Boer L, Pescatori MA, Stuermer M (2021) Energy transition metals. International Monetary Fund

25. Bomgardner M (2018) How first solar recycles its CdTe panels. C&EN Global Enterprise 96(15):40–40. https://doi.org/10.1021/cen-09615-cover2
26. Boroski BB (2019) Solar energy—A technology with multi-scale opportunities to integrate wildlife conservation. In: Moorman CE, Grodsky SM, Rupp SP (eds) Renewable energy and wildlife conservation. Johns Hopkins University Press, pp 177–197
27. Brinkmann KO, Gahlmann T, Riedl T (2020) Atomic layer deposition of functional layers in planar perovskite solar cells. Solar RRL 4(1):1900332. https://doi.org/10.1002/solr.201900332
28. Britannica (1956) Encyclopedia Britannica. 1956 edn., Jan 1
29. CAN Europe (2022) Communities for RES—Energy communities in Central Eastern Europe. https://caneurope.org/cee-energy-communities
30. Cai M et al (2017) Cost-performance analysis of perovskite solar modules. Adv Sci 4(1):1600269. https://doi.org/10.1002/advs.201600269
31. Cao Y et al (2018) Direct contact of selective charge extraction layers enables high-efficiency molecular photovoltaics. Joule 2(6):1108–1117. https://doi.org/10.1016/j.joule.2018.03.017
32. Celik I et al (2016) Life Cycle Assessment (LCA) of perovskite PV cells projected from lab to fab. Sol Energy Material Sol Cells 156:157–169. https://doi.org/10.1016/j.solmat.2016.04.037
33. Çetin M, Eĝrican N (2011) Employment impacts of solar energy in Turkey. Energy Policy 39(11):7184–7190. https://doi.org/10.1016/j.enpol.2011.08.039
34. Chen B et al (2022) A critical review on the moisture stability of halide perovskite films and solar cells. J Chem Eng 430:132701. https://doi.org/10.1016/j.cej.2021.132701
35. Choi Y (2014) A study on power generation analysis of floating PV system considering environmental impact. Int J Softw Eng its Appl 8:75–84. https://doi.org/10.14257/ijseia.2014.8.1.07
36. Chowdhury MS, Rahman KS, Chowdhury T, Nuthammachot N, Techato K, Akhtaruzzaman M et al (2020) An overview of solar photovoltaic panels' end-of-life material recycling. Energy Strat Rev 27. https://doi.org/10.1016/j.esr.2019.100431
37. Chung I et al (2012) All-solid-state dye-sensitized solar cells with high efficiency. Nature 485(7399):486–489. https://doi.org/10.1038/nature11067
38. Cleveland CJ (2019) Concise Encyclopedia of the history of energy. Academic Press Inc. ISBN 9780123751171
39. Cousse J (2021) Still in love with solar energy? Installation size, affect, and the social acceptance of renewable energy technologies. Renew Sustain Energy Rev 145. https://doi.org/10.1016/j.rser.2021.111107
40. Cypher BL, Phillips SE, Kelly PA (2013) Quantity and distribution of suitable habitat for endangered San Joaquin kit foxes: conservation implications. Canid Biol Conserv 16:25–31
41. Damjanovic D et al (2016) Photovoltaic devices and method of manufacturing. United States. US10529883B2
42. Daniel W (2022) The skyrocketing price of one critical metal could put a stop to new solar projects worldwide. Fortune. https://fortune.com/2022/07/14/solar-projects-could-stop-polysilicon-prices/
43. DeFries R, Field C, Fung I, Collatz G (1999) Combining satellite data and biogeochemical models to estimate global effects of human-induced land cover change on carbon emissions and primary productivity. Global Biogeochem Cycles 13(3):803–815. https://doi.org/10.1029/1999GB900037
44. Deng K, Li L (2016) Advances in the application of atomic layer deposition for organometal halide perovskite solar cells. Adv Mat Interfaces 3(21):1600505. https://doi.org/10.1002/admi.201600505
45. Desai P (2021) Solar industry demand raises temperature in silicon market. Reuters. https://www.reuters.com/business/energy/solar-industry-demand-raises-temperature-silicon-market-2021-09-21/
46. Dinesh H, Pearce JM (2016) The potential of agrivoltaic systems. Renew Sust Energ Rev 54:299–308. https://doi.org/10.1016/j.rser.2015.10.024

47. Dipta SS, Uddin A (2021) Stability issues of perovskite solar cells: a critical review. Energy Technol 9(11):2100560. https://doi.org/10.1002/ente.202100560
48. Dudka S, Adriano DC (1997) Environmental impacts of metal ore mining and processing: a review. J Environ Qual 26(3):590–602. https://doi.org/10.2134/jeq1997.00472425002600030003x
49. EU Commission (2020) CRM list 2020. https://rmis.jrc.ec.europa.eu/?page=crm-list-2020-e294f6
50. El-Khawad L, Bartkowiak D, Kümmerer K (2022) Improving the end-of-life management of solar panels in Germany. Renew Sust Energy Rev 168. https://doi.org/10.1016/j.rser.2022.112678
51. Ember (2022) Yearly electricity data. European Electricity Review, Ember
52. Eurostat (2022) Translate energy consumption in households. https://ec.europa.eu/eurostat/statistics-explained/index.php?title=Energy_consumption_in_households
53. Evans MJ, Mainali K, Soobitsky R, Mills E, Minnemeyer S (2023) Predicting patterns of solar energy buildout to identify opportunities for biodiversity conservation. Biol Conserv 283:110074. https://doi.org/10.1016/j.biocon.2023.110074
54. Forrester JW (1997) Industrial dynamics. J Oper Res Soc 48:1037–1041
55. Fortune Business Insights (2022) Semiconductor market size, share & COVID-19 impact analysis, by component (memory devices, logic devices, analog IC, MPU, discrete power devices, MCU, sensors, and others), by application (networking & communications, data processing, industrial, cons). https://www.fortunebusinessinsights.com/semiconductor-market-102365
56. Freitas S, Catita C, Redweik P, Brito MC (2015) Modelling solar potential in the urban environment: State-of-the-art review. Renew Sust Energ Rev 41:915–931. https://doi.org/10.1016/j.rser.2014.08.060
57. Fritsche U, Berndes G, Cowie A, Johnson F, Dale V, Langeveld H, Sharma N, Watson H, Woods J (2017) Energy and land use. Global land outlook working paper. UNCCD and IRENA. https://doi.org/10.13140/RG.2.2.24905.44648
58. Fthenakis V, Blunden J, Green T, Krueger L, Turney D (2011) Large photovoltaic power plants: wildlife impacts and benefits. In: IEEE photovoltaic specialists conference Seatle (WA, USA): IEEE, p. 2011–2016
59. Gallop P, Gray E, Nikolovska E, Mustață A, Petcu R (2021) The political economy of energy transition in Southeast Europe—Barriers and obstacles. Friedrich Ebert Stifung and CEE Bankwatch Network
60. Gasparatos A, Doll CNH, Esteban M, Ahmed A, Olang TA (2017) Renewable energy and biodiversity: implications for transitioning to a Green Economy. Renew Sustain Energy Rev 70:161–184. https://doi.org/10.1016/j.rser.2016.08.030
61. Giacomo F Di et al (2018) Large area >140 cm^2 perovskite solar modules made by sheet to sheet and roll to roll fabrication with 14.5 % efficiency. In: 2018 IEEE 7th World conference on photovoltaic energy conversion (WCPEC) (A Joint Conference of 45th IEEE PVSC, 28th PVSEC & 34th EU PVSEC). IEEE, pp. 2795–2798. https://doi.org/10.1109/PVSC.2018.8548157
62. Giovannini S (2019) Slavutych, Ukraine—The community energy story you need to hear!, Energy cities. energy-cities.eu/slavutych-ukraine-the-community-energy-story-you-need-to-hear
63. Goetz KP et al (2021) Sustainability in perovskite solar cells. ACS Appl Mater Interfaces 13(1):1–17. https://doi.org/10.1021/acsami.0c17269
64. Gong J, Darling SB, You F (2015) Perovskite photovoltaics: life-cycle assessment of energy and environmental impacts. Energy Environ Sci 8(7):1953–1968. https://doi.org/ https://doi.org/10.1039/C5EE00615E
65. Grancini G, Nazeeruddin MK (2019) Dimensional tailoring of hybrid perovskites for photovoltaics. Nat Rev Mater. Springer US 4(1):4–22. https://doi.org/10.1038/s41578-018-0065-0
66. Green MA (1982) Solar cells. Prentice-Hall, USA. ISBN 0-13-822270-8

67. Grippo M, Hayse JW, O'Connor BL (2015) Solar energy development and aquatic ecosystems in the Southwestern United States: potential impacts, mitigation, and research needs. Environ Assess 55:244–256. https://doi.org/10.1007/s00267-014-0384-x
68. Grodsky SM, Moore-O'Leary KA, Hernandez RR (2017) From butterflies to bighorns: multi-dimensional species-species and species-process interactions may inform sustainable solar energy development in desert ecosystems. In: Reynolds RL (ed) Proceedings of the 31st annual desert symposium
69. Guerin T (2017) A case study identifying and mitigating the environmental and community impacts from construction of a utility-scale solar photovoltaic power plant in eastern Australia. Sol Energy 146:94–104. https://doi.org/10.1016/j.solener.2017.02.020
70. Guiller C, Affre L, Deschamps-Cottin M, Geslin B, Kaldonski N, Tatonia T (2017) Impacts of solar energy on butterfly communities in Mediterranean agroecosystems. Environ Prog Sustain Energy 36(6):17–1823. https://doi.org/10.1002/ep.12626
71. Gunerhan H, Hepbasli A, Giresunlu U (2009) Environmental impacts from the solar energy systems. Energy Sources 31:131–138. https://doi.org/10.1080/15567030701512733
72. Guo L, Tang G, Hong J (2019) Mechanical properties of formamidinium halide perovskites $FABX_3$ ($FA=CH(NH_2)_2$; B=Pb, Sn; X=Br, I) by first-principles calculations. Chin Phys Lett 36(5):056201. https://doi.org/10.1088/0256-307X/36/5/056201
73. HT Harvey & Associates (2015) California Valley Solar Ranch, San Luis Obispo County, California: Avian activity surveys final report, Oct 2011–Oct 2014: Project 3326-03. Prepared for High Plains Ranch II, LLC, Carlsbad, CA
74. Haines MB, Moore S, Adornetto T (2023) Suspending democratic (dis)belief: Nonliberal energy polities of solar power in Morocco and Tanzania. Energy Res Soc Sci 96:102942. https://doi.org/10.1016/j.erss.2023.102942
75. Hamberg J (2011) Falling silicon prices shakes up solar manufacturing industry. Down to Earth. https://www.downtoearth.org.in/news/falling-silicon-prices-shakes-up-solar-manufacturing-industry-34045
76. Hanski I (2015) Habitat fragmentation and species richness. J Biogeogr 42:989–994. https://doi.org/10.1111/jbi.12478
77. Hellweg S, Benetto E, Huijbregts MAJ, Verones F, Wood R (2023) Life-cycle assessment to guide solutions for the triple planetary crisis. Nat Rev Earth Environ 4:471–486. https://doi.org/10.1038/s43017-023-00449-2
78. Hernandez RR, Hoffacker MK, Murphy-Mariscal ML, Wu GC, Allen MF (2015) Solar energy development impacts on land cover change and protected areas. PNAS 112(44):13579–13584. https://doi.org/10.1073/pnas.1602975113
79. Hernandez RR, Easter SB, Murphy-Mariscal ML, Maestre FTe, Tavassoli M, Allen EB, Barrows CW, Belnap J, Ochoa-Hueso R, Ravian S, Allen MF (2014) Environmental impacts of utility-scale solar energy. Renew Sust Energ Rev 29:766–779. https://doi.org/10.1016/j.rser.2013.08.041
80. Hernberg S (2000) Lead poisoning in a historical perspective. Am J Ind Med 38(3):244–254. https://doi.org/10.1002/1097-0274(200009)38:3%3c244::AID-AJIM3%3e3.0.CO;2-F
81. Hoffacker MK, Allen MF, Hernandez RR (2017) Land-sparing opportunities for solar energy development in agricultural landscapes: a case study of the great central valley, CA. US. Environ Sci Technol 51(24):14472–14482. https://doi.org/10.1021/acs.est.7b05110
82. Horváth G, Blahó M, Egri Á, Kriska G, Seres I, Robertson B (2010) Reducing the maladaptive attractiveness of solar panels to polarotactic insects. Conserv Biol 24(6):1644–1653. https://doi.org/10.1111/j.1523-1739.2010.01518.x
83. Hotz C et al (2009) Improved back contact in thin solar cells. WIPO (PCT): WO2009117072A1
84. Huber ST, Steininger KW (2022) Critical sustainability issues in the production of wind and solar electricity generation as well as storage facilities and possible solutions. J Clean Prod 339:30720
85. Huso M, Dietsch T, Nicolai C (2016) Mortality monitoring design for utility-scale solar power facilities. U.S. Geological Survey Open-File Report 2016-1087. https://doi.org/10.3133/ofr20161087

86. Hutter EM et al (2022) Metal halide perovskite toxicity effects on Arabidopsis thaliana plants are caused by iodide ions. iScience 25(1):103583. https://doi.org/10.1016/j.isci.2021.103583
87. IEA (2023) Renewables 2023, Analysis and forecasts to 2028. https://iea.blob.core.windows.net/assets/96d66a8b-d502-476b-ba94-54ffda84cf72/Renewables_2023.pdf
88. IFC (2015) Utility-scale solar photovoltaic power plants—A project developer's guide. Guidebook, International Finance Corporation, Washington
89. IPBES (2023) Summary for policymakers of the thematic assessment report on invasive alien species and their control of the intergovernmental science-policy platform on biodiversity and ecosystem services. IPBES Secretariat, Bonn, Germany. https://doi.org/10.5281/zenodo.7430692
90. IRENA, IEA-PVPS (2016) End-of-life management: solar photovoltaic panels. International Renewable Energy Agency and International Energy Agency Photovoltaic Power Systems
91. IRENA (2020) Innovation landscape brief: community-ownership models. International Renewable Energy Agency, Abu Dhabi https://www.irena.org/-/media/Files/IRENA/Agency/Publication/2020/Jul/IRENA_Community_ownership_2020.pdf
92. IRENA (2022a) Renewable energy statistics. https://www.irena.org/publications/2022/Jul/Renewable-Energy-Statistics-2022
93. IRENA (2022b) Renewable energy and jobs—Annual review 2022. https://www.irena.org/publications/2022/Sep/Renewable-Energy-and-Jobs-Annual-Review-2022
94. Ibn-Mohammed T et al (2017) Perovskite solar cells: an integrated hybrid lifecycle assessment and review in comparison with other photovoltaic technologies. Renew Sust Energy Rev 80:1321–1344. https://doi.org/10.1016/j.rser.2017.05.095
95. Incropera FP, DeWitt DP, Bergman TL, Lavine AS (2017) Fundamentals of heat and mass transfer. John Wiley & Sons, Singapore. ISBN 9781119382911
96. Ioannidis R, Koutsoyiannis D (2020) A review of land use, visibility and public perception of renewable energy in the context of landscape impact. Appl Energy 276. https://doi.org/10.1016/j.apenergy.2020.115367
97. Jenkins AR, Ralston S, Smit-Robinson HA (2015) Birds and solar energy best practice guidelines, South Africa. https://www.birdlife.org.za/wp-content/uploads/2020/03/BLSA-Guidelines-Solar-and-Energy.pdf
98. Jeong J et al (2021) Pseudo-halide anion engineering for α-FAPbI3 perovskite solar cells. Nature 592(7854):381–385. https://doi.org/10.1038/s41586-021-03406-5
99. Ji L et al (2019) Band alignment of Pb–Sn mixed triple cation perovskites for inverted solar cells with negligible hysteresis. J Mater Chem A 7(15):9154–9162. https://doi.org/10.1039/C8TA11891D
100. Justin Kunene T et al (2022) Review of atomic layer deposition process, application and modeling tools. Mater Today: Proc 62:S95–S109. https://doi.org/10.1016/j.matpr.2022.02.094
101. Kagan RA, Viner TC, Trail PW, Espinoza EO (2014) Avian mortality at solar energy facilities in southern California: a preliminary analysis. National Fish and Wildlife Forensic Laboratory 28
102. Karttunen H, Kröger P, Oja H, Poutanen M, Donner KJ (2016) Fundamental astronomy. ISBN 978-3-662-53044-3
103. Kato T et al (2017) Enhanced efficiency of Cd-Free Cu(In, Ga)(Se, S)2 minimodule ia (Zn, Mg)O second buffer layer and alkali metal post-treatment. IEEE J-PV 7(6):1773–1780. https://doi.org/10.1109/JPHOTOV.2017.2745710
104. Katz EA (2020) Perovskite: name puzzle and German-Russian Odyssey of discovery. Helv Chim Acta 103(6). https://doi.org/10.1002/hlca.202000061
105. Ke W, Kanatzidis MG (2019) Prospects for low-toxicity lead-free perovskite solar cells. Nat Commun 10(1):965. https://doi.org/10.1038/s41467-019-08918-3
106. Khalid A, Hawila D, Feng J, García-Baños C, Pollitt H, Stenning J, Alexandri E, Chewpreecha U, Smith A (2019) Analysis built on Renewable energy benefits: Measuring the economics
107. Kirkinen J, Minkkinen K, Penttilä T, Kojola S, Sievänen R, Alm J, Saarnio S, Silvan N, Laine J, Savolainen I (2007) Greenhouse impact due to different peat fuel utilisation chains in Finland—A life-cycle approach. Boreal Environ Res 12(2):211–223

108. Kojima A et al (2009) Organometal halide perovskites as visible-light sensitizers for photovoltaic cells. J Am Chem Soc 131(17):6050–6051. https://doi.org/10.1021/ja809598r
109. Kosanic A, Petzold J, Martín-López B, Razanajatovo M (2022) An inclusive future: disabled populations in the context of climate and environmental change. Curr Opin Environ Sust 55:101159. https://doi.org/10.1016/j.cosust.2022.101159
110. Kosciuch K, Riser-Espinoza D, Moqtaderi C, Erickson W (2021) Aquatic habitat bird occurrences at photovoltaic solar energy development in Southern California, USA. Diversity 13(11):524. https://doi.org/10.3390/d13110524
111. Kosciuch K, Riser-Espinoza D, Gerringer M, Erickson W (2020) A summary of bird mortality at photovoltaic utility scale solar facilities in the Southwestern U.S. PLoS ONE 15(4):e0232034. https://doi.org/10.1371/journal.pone.0232034
112. Kranz L et al (2013) Doping of polycrystalline CdTe for high-efficiency solar cells on flexible metal foil. Nat Commun 4(1):2306. https://doi.org/10.1038/ncomms3306
113. Lai X, Chen Q, Tang X, Zhou Y, Gao F, Guo Y et al (2022) Critical review of life cycle assessment of lithium-ion batteries for electric vehicles: a lifespan perspective. E Transportation 2022:12. https://doi.org/10.1016/j.etran.2022.100169
114. Laitila E (2023) Screening of large-scale industrial solar PV plants in Finland. Master's thesis. School of Engineering, Master's Programme in Advanced Energy Solutions. Aalto University. Espoo, Finland. http://urn.fi/URN:NBN:fi:aalto-202305283431
115. Lazard (2023) Lazard's levelized cost of energy analysis. Version 16.0
116. Lee JH et al (2020) Reproducible dry stamping transfer of PEDOT:PSS transparent top electrode for flexible semitransparent metal halide perovskite solar cells. ACS Appl Mater Interfaces 12(9):10527–10534. https://doi.org/10.1021/acsami.9b22657
117. Lee TD, Ebong AU (2017) A review of thin film solar cell technologies and challenges. Renew Sustain Energy Rev 70:1286–1297. https://doi.org/10.1016/j.rser.2016.12.028
118. Lee Y, Joo H, Yoon S (2014) Design and installation of floating type photovoltaic energy generation system using FRP members. Sol Energy 108:13–27. https://doi.org/10.1016/j.solener.2014.06.033
119. Levien M (2011) Special economic zones and accumulation by dispossession in India. J Agrar Change 11(4):454–483
120. Levien M (2015) Social capital as obstacle to development: brokering land, norms, and trust in rural India. World Dev 74:77–92
121. Li L et al (2022) Flexible all-perovskite tandem solar cells approaching 25% efficiency with molecule-bridged hole-selective contact. Nat Energy. https://doi.org/10.1038/s41560-022-01045-2
122. Li J, Okin GS, Alvarez L, Epstein H (2007) Quantitative effects of vegetation cover on wind erosion and soil nutrient loss in a desert grassland of southern New Mexico, USA. Biogeochem 85(3):317–332. https://doi.org/10.1007/s10533-007-9142-y
123. Li B et al (2021) Efficient passivation strategy on Sn related defects for high performance all-inorganic CsSnI3 perovskite solar cells. Adv Funct Mater 31(11). https://doi.org/10.1002/adfm.202007447
124. Liao C (2021) Fundamentals of rechargeable lithium ion and beyond lithium-ion batteries. Batteries. IOP Publishing. https://doi.org/10.1088/978-0-7503-2682-7ch1
125. Lim S-H et al (2021) Semi-transparent perovskite solar cells with bidirectional transparent electrodes. Nano Energy 82:105703. https://doi.org/10.1016/j.nanoen.2020.105703
126. Lindberg O, Birging A, Widén J, Lingfors D (2021) PV park site selection for utility-scale solar guides combining GIS and power flow analysis: a case study on a Swedish municipality. Appl Energy 282(A):11. https://doi.org/10.1016/j.apenergy.2020.116086
127. Liu D, Yang C, Lunt RR (2018) Halide perovskites for selective ultraviolet-harvesting transparent photovoltaics. Joule 2(9):1827–1837. https://doi.org/10.1016/j.joule.2018.06.004
128. Lo Piano S, Mayumi K (2017) Toward an integrated assessment of the performance of photovoltaic power stations for electricity generation. Appl Energy 186(2):167–174. https://doi.org/10.1016/j.apenergy.2016.05.102

129. Lovich JE, Ennen JR (2011) Wildlife conservation and solar energy development in the Desert Southwest, United States. Bioscience 61:982–992. https://doi.org/10.1525/bio.2011.61.12.8
130. Lv S et al (2022) Stability of Sn-Pb mixed organic–inorganic halide perovskite solar cells: progress, challenges, and perspectives. J Energy Chem 65:371–404. https://doi.org/10.1016/j.jechem.2021.06.011
131. Maani T et al (2020) Environmental impacts of recycling crystalline silicon (c-SI) and cadmium telluride (CDTE) solar panels. Sci Total Environ 735:138827. https://doi.org/10.1016/j.scitotenv.2020.138827
132. Maisch M (2020) Polysilicon price spike puts pressure on solar developers in Australia. pv magazine. https://www.pv-magazine-australia.com/2020/08/28/polysilicon-price-spike-puts-pressure-on-solar-developers-in-australia/
133. De Marco A, Petrosillo I, Semeraro T, Pasimeni MR, Aretano R, Zurlini G (2014) The contribution of utility-scale solar energy to the global climate regulation and its effects on local ecosystem services. Glob Ecol Conserv 2:324–337. https://doi.org/10.1016/j.gecco.2014.10.010
134. Martín-Chivelet N (2016) Photovoltaic potential and land-use estimation methodology. Energy 94:233–242. https://doi.org/10.1016/j.energy.2015.10.108
135. Matsuda Y, Fujita K, Ageta Y, Sakai A (2006) Estimation of atmospheric transmissivity of solar radiation from precipitation in the Himalaya and the Tibetan Plateau. Ann Glaciol 43:344–350
136. Maxwell AE, Strager MP, Yuill CB, Petty JT (2012) Modeling critical forest habitat in the southern coal fields of West Virginia. Int J Ecol 2012:182683. https://doi.org/10.1155/2012/182683
137. Mekonnen MM, Gerbens-Leenes PW, Hoekstra AY (2015) The consumptive water footprint of electricity and heat: a global assessment. Environ Sci-Wat Res 1:285–297. https://doi.org/10.1039/c5ew00026b
138. Mitzi DB, Chondroudis K, Kagan CR (2001) Organic-inorganic electronics. IBM J Res Dev 45(1):29–45. https://doi.org/10.1147/rd.451.0029
139. Moilanen A, Kotiaho JS (2021) Three ways to deliver a net positive impact with biodiversity offsets. Conserv Biol 35(1):197–205. https://doi.org/10.1111/cobi.13533
140. Mondino EB, Rabrizio E, Chiabrando R (2015) Site selection of large ground-mounted photovoltaic plants: a GIS decision support system and an application to Italy. Int J Green Energy 12(5):515–525. https://doi.org/10.1080/15435075.2013.858047
141. Montag H, Parker DG, Clarkson T (2016) The effects of solar farms on local biodiversity: a comparative study. Clarkson and Woods and Wychwood biodiversity. https://www.solar-trade.org.uk/wp-content/uploads/2016/04/The-effects-of-solar-farms-on-local-biodiversity-study.pdf
142. Moore-O'Leary KA, Hernandez RR, Abella S, Johnston D, Kreitler J, Swanson A, Lovich J (2017) Sustainability of utility-scale solar energy: critical ecological concepts. Front Ecol Environ 15:385–394. https://doi.org/10.1002/fee.1517
143. Moriarty P, Honnery D (2012) What is the global potential for renewable energy? Renew Sust Energ Rev 16:244–252. https://doi.org/10.1016/j.rser.2011.07.151
144. Mulvaney D (2019) Solar power: innovation, sustainability, and environmental justice. University of California Press
145. Munshi AH et al (2018) Thin-film CdTe photovoltaics—The technology for utility scale sustainable energy generation. Sol Energy 173:511–516. https://doi.org/10.1016/j.solener.2018.07.090
146. Møller CK (1958) Crystal structure and photoconductivity of cæsium plumbohalides. Nature 182(4647):1436–1436. https://doi.org/10.1038/1821436a0
147. Niblick B, Landis AE (2016) Assessing renewable energy potential on United States marginal and contaminated sites. Renew Sustain Energy Rev 60:489–497. https://doi.org/10.1016/j.rser.2015.12.045
148. Ogilvie M, Rootes C (2015) The impact of local campaigns against wind energy developments. Environ Polit 24(6). https://doi.org/10.1080/09644016.2015.1063301

149. Ommen JR, Goulas A, Puurunen RL (2021) Atomic layer deposition. In Kirk-Othmer Encyclopedia of Chemical Technology Wiley, pp 1–42. https://doi.org/10.1002/0471238961.koe00059
150. Parker SS, Cohen BS, Moore J (2018) Impact of solar and wind development on conservation values in the Mojave Desert. PLoS ONE 13(12):e0207678. https://doi.org/10.1371/journal.pone.0207678
151. Pearce D, Strittholt J, Watt T, Elkind EN (2016) A path forward: Identifying least-conflict solar PV development in California's San Joaquin Valley. Prepared by Conservation Biology Institute and Center for Law, Energy and the Environment. Berkeley, CA: UC Berkely School of Law. https://escholarship.org/uc/item/543174qd
152. Perez M, Perez R (2022) Update 2022—A fundamental look at supply side energy reserves for the planet. Solar Energy Adv 2:1–7. https://doi.org/10.1016/j.seja.2022.100014
153. Philipps S, Warmuth W (2022) Photovoltaics report. Fraunhofer ISE and PSE Projects GmbH, Freiburg. https://www.ise.fraunhofer.de/content/dam/ise/de/documents/publications/studies/Photovoltaics-Report.pdf
154. Phillips SE, Cypher BL (2015) Solar energy development and endangered upland species of the San Joaquin Valley: Identification of conflict zones. Prepared for the California Department of Fish and Wildlife. California State University, Stanislaus Endangered Species Recovery Program
155. Pimentel Da Silva GD, Branco DA (2018) Is floating photovoltaic better than conventional photovoltaic? Assessing environmental impacts. Impact Assess Project Appraisal 36(5):390–400. https://doi.org/10.1080/14615517.2018.1477498
156. Pocewicz A, Copeland H, Kiesecker J (2011) Potential impacts of energy development on shrublands in western North America. Nat Resour Environ Issues 17(14). http://digitalcommons.usu.edu/nrei/vol17/iss1/14
157. Precedence Research (2022) Solar Power Market. https://www.precedenceresearch.com/solar-power-market
158. Proka A (2022) European citizen energy academy best practice guide for Southeast Europe—Inspiring community energy initiatives. https://www.rescoop.eu/uploads/rescoop/downloads/2023_EUCENA-Balkan-Best-Practice-Guide-EN.pdf
159. Raab R, Schütz C, Spakovszky P, Julius E, Schulze C (2012) Underground cabling and marking of power lines: conservation measures rapidly reduced mortality of West-Pannonian Great Bustards *Otis tarda*. Bird Conserv Int 22(3):1–8. https://doi.org/10.1017/s0959270911000463
160. Rahimpour Golroudbary S, Lundström M, Wilson BP (2024) Synergy of green energy technologies through critical materials circularity. Renew Sustain Energy Rev 191:114180. https://doi.org/10.1016/j.rser.2023.114180
161. Rahman MM, Mateti S, Sultana I, Hou C, Falin A, Cizek P et al (2021) End-of-life photovoltaic recycled silicon: a sustainable circular materials source for electronic industries. Adv Energy Sustainability Res 2:2100081. https://doi.org/10.1002/aesr.202100081
162. Rahman MM, Sultana I, Fan Y, Yu B, Tao T, Hou C et al (2021) Strategies, design and synthesis of advanced nanostructured electrodes for rechargeable batteries. Mater Chem Front 5:5897–5931. https://doi.org/10.1039/d1qm00274k
163. Rahmany S, Etgar L (2020) Semitransparent perovskite solar cells. ACS Energy Lett 5(5):1519–1531. https://doi.org/10.1021/acsenergylett.0c00417
164. Ravi VK et al (2020) Don't let the lead out: new material chemistry approaches for sustainable lead halide perovskite solar cells. ACS Omega 5(46):29631–29641. https://doi.org/10.1021/acsomega.0c04599
165. Ritala M, Niinistö J (2008) Chapter 4. Atomic layer deposition. In: Chemical vapour deposition. Royal Society of Chemistry, Cambridge, pp 158–206. https://doi.org/10.1039/9781847558794-00158
166. Ritzer DB et al (2022) Upscaling of perovskite solar modules: the synergy of fully evaporated layer fabrication and all-laser-scribed interconnections. Prog Photovolt: Res Appl 30(4):360–373. https://doi.org/10.1002/pip.3489

167. Rix A et al (2015) First solar's CdTe module technology—Performance, life cycle, health and safety impact assessment. https://www.firstsolar.com/-/media/First-Solar/Sustainability-Documents/Sustainability-Peer-Reviews/CRSES2015_06_First-Solar-CdTe-Module-Technology-Review-FINAL.ashx
168. Rose T, Wollert A (2015) The dark side of photovoltaic—3D simulation of glare assessing risk and discomfort. Environ Impact Assess Rev 52:24–30. https://doi.org/10.1016/j.eiar.2014.08.005
169. Ruther RE, Hays KA, An SJ, Li J, Wood DL, Nanda J (2018) Chemical evolution in silicon-graphite composite anodes investigated by vibrational spectroscopy. ACS Appl Mater Interfaces 10:18641–18649. https://doi.org/10.1021/acsami.8b02197
170. Sadiqa A, Sahrakorpi T, Keppo I (2023) Gender vulnerabilities in low carbon energy transitions: a conceptual review. Environ Res Lett 18(4). https://doi.org/10.1088/1748-9326/acc819
171. Samarakoon S, Munro P, Zalengera C, Kearnes M (2022) The afterlives of off-grid solar: The dynamics of repair and e-waste in Malawi. Environ Innov Soc Tr 42:317–330
172. Santra P, Pande PC, Kumar S, Mishra D, Sing RK (2017) Agrivoltaics or solar farming: the concept of integrating solar PV based elecricity generation and crop production in a single land use system. Int J Renew Energy Res 7(2):694–699. https://doi.org/10.20508/ijrer.v7i2.5582.g7049
173. Seo S et al (2019) Atomic layer deposition for efficient and stable perovskite solar cells. Chem Comm 55(17):2403–2416. https://doi.org/10.1039/C8CC09578G
174. Seyferth D (2003a) The rise and fall of tetraethyllead. 1. Discovery and slow development in European universities 1853−1920. Organometallics 22(12):2346–2357. https://doi.org/10.1021/om030245v
175. Seyferth D (2003b) The rise and fall of tetraethyllead. 2. Organometallics 22(25):5154–5178. https://doi.org/10.1021/om030621b
176. Shiraishi K, Shirley RG, Kammen DM (2019) Geospatial multi-criteria analysis for identifying high priority clean energy investment opportunities: a case study on land-use conflict in Bangladesh. Appl Energy 235:1457–1467. https://doi.org/10.1016/j.apenergy.2018.10.123
177. Sinha P, Hoffman B, Sakers J, Althouse L (2018) Best practices in responsible land use for improving biodiversity at a utility-scale solar facility. Case Stud Environ 2:1–12. https://doi.org/10.1525/cse.2018.001123
178. Sjoberg RW, Horn BK (1983) Atmospheric effects in satellite imaging of mountainous terrain. Appl Opt 22:1702–1716
179. Song Z et al (2017) A technoeconomic analysis of perovskite solar module manufacturing with low-cost materials and techniques. Energy Environ Sci 10(6):1297–1305. https://doi.org/10.1039/C7EE00757D
180. Song J, Choi Y (2016) Analysis of the potential for use of floating photovoltaic systems on mine pit lakes: case study at the ssangyong open-pit limestone mine in Korea. Energies 9:1–13. https://doi.org/10.3390/en9020102
181. Spanopoulos I et al (2018) Unraveling the chemical nature of the 3D "Hollow" hybrid halide perovskites. J Am Chem Soc 140(17):5728–5742. https://doi.org/10.1021/jacs.8b01034
182. Sreenath S, Sudhakar K, Yusop AF, Cuce E, Solomin E (2020) Analysis of solar PV glare in airport environment: potential solutions. Results Eng 5:100079. https://doi.org/10.1016/j.rineng.2019.100079
183. Srivastava A, Rodriguez JF, Saco PM, Kumari N, Yetemen O (2021) Global analysis of atmospheric transmissivity using cloud cover, aridity and flux network datasets. Remote Sensing 13(9):1716. https://doi.org/10.3390/rs13091716
184. Stock R, Sovacool BK (2024) Blinded by sunspots: revealing the multidimensional and intersectional inequities of solar energy in India. Glob Environ Change 84:02796
185. Stock R, Birkenholtz T (2020) Photons versus firewood: Female (dis) empowerment by solar power in India. Gend Place Cult 27(11):1628–1651
186. Stoms DM, Dashiell SL, Davis FW (2013) Siting solar energy development to minimize biological impacts. Renew Energy 57:289–298. https://doi.org/10.1016/j.renene.2013.01.055

187. Sun S et al (2015) Mechanical properties of organic–inorganic halide perovskites, $CH_3NH_3PbX_3$ (X = I, Br and Cl), by nanoindentation. J Mat Chem A 3(36):18450–18455. https://doi.org/10.1039/C5TA03331D
188. Syke (2023) Depth of ground frost. Finish Environmental Institute
189. Szabadi KL, Kurali A, Abdul Rahman NA, Froidevaux JSP, Tinsley E, Jones G, Görföl T, Estók P, Zsebők S (2023) The use of solar farms by bats in mosaic landscapes: implications for conservation. Glob Ecol Conserv 44:e02481. https://doi.org/10.1016/j.gecco.2023.e02481
190. Taylor R, Conway J, Gabb O, Gillespie J (2019) Potential ecological impacts of ground-mounted photovoltaic solar panels. UK: BSG ecology, pp 1–12. BSG ecology. https://infrastructure.planninginspectorate.gov.uk/wp-content/ipc/uploads/projects/EN010085/EN010085-000610-Appendix%204%20-%20Potential%20Ecological%20Impacts%20of%20Ground-Mounted%20Solar%20Panels.pdf
191. Terrapon-Pfaff J, Fink T, Viebahn P, Jamea EM (2019) Social impacts of large-scale solar thermal power plants: assessment results for the NOORO I power plant in Morocco. Renew Sust Energ Rev 113:109259. https://doi.org/10.1016/j.rser.2019.109259
192. The Virginia Center for Coal and Energy Research (2019) Assessment of the risks associated with thin film solar panel technology. Blacksburg, Virginia, USA. https://www.firstsolar.com/-/media/First-Solar/Sustainability-Documents/Sustainability-Peer-Reviews/Virgina-Tech-Peer-Review.ashx
193. Tinsley E, Froidevaux JSP, Zsebők S, Szabadi KL, Jones G (2023) Renewable energies and biodiversity: impact of ground-mounted solar photovoltaic sites on bat activity. J Appl Ecol 60(9):1752–2176. https://doi.org/10.1111/1365-2664.14474
194. Tsoutsos T, Frantzeskaki N, Gekas V (2005) Environmental impacts from the solar energy technologies. Energy Policy 33:289–296. https://doi.org/10.1016/S0301-4215(03)00241-6
195. Turney D, Fthenakis V (2011) Environmental impacts from the installation and operation of large-scale solar power plants. Renew Sustain Energy Rev 15:3261–3270. https://doi.org/10.1016/j.rser.2011.04.023
196. Tyson P, Gasse F, Bergonzini L, D'Abreton P (1997) Aerosols, atmospheric transmissivity and hydrological modelling of climatic change over Africa south of the equator. Int J Climatol J R Meteorol Soc 17:1651–1665
197. U.S.G.S (2022) Mineral commodity summaries 2022. U.S. Geological Survey. https://doi.org/10.3133/mcs2022
198. U.S.G.S (2023) Mineral commodity summaries 2023. U.S. Geological Survey 2023. https://doi.org/10.3133/mcs2023
199. United Nations (2021) Life cycle assessment of electricity generation options, United nations economic commission for Europe. UNECE, GENEVA. https://unece.org/sites/default/files/2022-04/LCA_3_FINAL%20March%202022.pdf
200. Ünlü F et al (2020) Understanding the interplay of stability and efficiency in A-site engineered lead halide perovskites. APL Mater 8(7):070901. https://doi.org/10.1063/5.0011851
201. Verlinden PJ (2020) Future challenges for photovoltaic manufacturing at the terawatt level. J Renew Sustain Energy 12. https://doi.org/10.1063/5.0020380
202. Vidal R et al (2021) Comparison of perovskite solar cells with other photovoltaics technologies from the point of view of life cycle assessment. Adv Energy Sustainability Res 2(5):2000088. https://doi.org/10.1002/aesr.202000088
203. Visser E (2016) The impact of South Africa's largest photovoltaic solar energy facility on birds in the Northern Cape. University of Cape Town, Rondebosch, South Africa
204. Visser E, Perold V, Ralston-Paton S, Cardenal AC, Ryan PG (2019) Assessing the impacts of a utility-scale photovoltaic solar energy facility on birds in the Northern Cape, South Africa. Renew Energ 33(C):1285–1294. https://doi.org/10.1016/j.renene.2018.08.106
205. de Vries BJ, van Vuuren DP, Hoogwijk MM (2007) Renewable energy sources: their global potential for the first-half of the 21st century at a global level: an integrated approach. Energy Policy 35:2590–2610. https://doi.org/10.1016/j.enpol.2006.09.002
206. Walker G, Devine-Wright P (2008) Community renewable energy: what should it mean? Energy Policy 36(2):497–500

207. Walston LJ Jr, Rollins KE, Lagory KE, Smith KP, Meyers SA (2016) A preliminary assessment of avian mortality at utility-scale solar energy facilities in the United States. Renew Energy 92:405–414. https://doi.org/10.1016/j.renene.2016.02.041
208. Wang J, Yin WJ (2022) Effect of alloying on the carrier dynamics in high-performance perovskite solar cells. J Energy Chem 68:267–274. https://doi.org/10.1016/j.jechem.2021.11.016
209. Weiß A et al (2022) Atomic layer deposition of CsI and $CsPbI_3$. Chem Mater. https://doi.org/10.1021/acs.chemmater.2c01202
210. Wingard J, Zahler P, Victurine R, Bayasgalan O, Bayarbaatar B (2014) Guidelines for addressing the impact of linear infrastructure on large migratory mammals in Central Asia. UNEP Convention on Migratory Species (CMS), Bonn, Germany. https://www.cms.int/sites/default/files/publication/cms-cami_pub_linear-infrastructure_wcs_e.pdf
211. Wittmayer JM, de Geus T, Pel B, Avelino F, Hielscher S, Hoppe T, Mühlemeier S, Stasik A, Oxenaar S, Rogge KS, Visser V, Marín-González E, Ooms M, Buitelaar S, Foulds C, Petrick K, Klarwein S, Krupnik S, de Vries G, Wagner A, Härtwig A (2020) Beyond instrumentalism: broadening the understanding of social innovation in socio-technical energy systems. Energy Res Soc Sci 70:101689
212. Wood PB, Bosworth SB, Dettmers R (2006) Cerulean warbler abundance and occurrence relative to large-scale edge and habitat characteristics. The Condor 108(1):154–165. https://doi.org/10.1093/condor/108.1.154
213. Wu Z, Hou A, Chang C, Huang X, Shi D, Wang Z (2014) Environmental impacts of large-scale CSP plants in northwestern China. Environ Sci Process Impacts Internet 16:2432–2441. http://xlink.rsc.org/?DOI=C4EM00235K
214. Wyckoff TB, Sawyer H, Albeke SE, Garman SL, Kauffman MJ (2018) Evaluating the influence of energy and residential development on the migratory behavior of mule deer. Ecosphere 9(2):e02113. https://doi.org/10.1002/ecs2.2113
215. Xia CQ et al (2021) Limits to electrical mobility in lead-halide perovskite semiconductors. J Phys Chem Lett 12(14):3607–3617. https://doi.org/10.1021/acs.jpclett.1c00619
216. Xie Z et al (2017) Refractive index and extinction coefficient of $NH_2CH = NH_2PbI_3$ perovskite photovoltaic material. J Physics: Condens Matter 29(24):245702. https://doi.org/10.1088/1361-648X/aa6e6c
217. Xu P et al (2014) Influence of defects and synthesis conditions on the photovoltaic performance of perovskite semiconductor $CsSnI_3$. Chem Mater 26(20):6068–6072. https://doi.org/10.1021/cm503122j
218. Xu C, Dai Q, Gaines L, Hu M, Tukker A, Steubing B (2020) Future material demand for automotive lithium-based batteries. Commun Mater 1. https://doi.org/10.1038/s43246-020-00095-x
219. Yang Y et al (2022) Research progress of atomic layer deposition technology to improve the long-term stability of perovskite solar cells. J Mater Chem C 10(3):819–839. https://doi.org/10.1039/D1TC05035D
220. Yenneti K, Day R, Golubchikov O (2016) Spatial justice and the land politics of renewables: dispossessing vulnerable communities through solar energy mega-projects. Geoforum 76:90–99. https://doi.org/10.1016/j.geoforum.2016.09.004
221. Yin W-J, Shi T, Yan Y (2014) Unique properties of halide perovskites as possible origins of the superior solar cell performance. Adv Mater 26(27):4653–4658. https://doi.org/10.1002/adma.201306281
222. Yushchenko A, de Bobo A, Chatenoux B, Patel MK, Ray N (2018) GIS-based assessment of photovoltaic (PV) and concentrated solar power (CSP) generation potential in West Africa. Renew Sustain Energy Rev 81(2):2088–2103. https://doi.org/10.1016/j.rser.2017.06.021
223. Zainol Abidin MA, Mahyuddin MN, Mohd Zainuri MA (2021) Solar photovoltaic architecture and agronomic management in agrivoltaic system: a review. Sustainability 13(14):27. https://doi.org/10.3390/su13147846
224. Zardetto V et al (2017) Atomic layer deposition for perovskite solar cells: research status, opportunities and challenges. Sustain Energy Fuels 1(1):30–55. https://doi.org/10.1039/C6SE00076B

225. Zarick HF et al (2018) Mixed halide hybrid perovskites: a paradigm shift in photovoltaics. J Mater Chem A 6(14):5507–5537. https://doi.org/10.1039/C7TA09122B
226. Zhang M et al (2017) High-efficiency rubidium-incorporated perovskite solar cells by gas quenching. ACS Energy Lett 2(2):438–444. https://doi.org/10.1021/acsenergylett.6b00697
227. Zhang F et al (2020) Comparative studies of optoelectrical properties of prominent PV materials: Halide perovskite, CdTe, and GaAs. Mater Today 36:18–29. https://doi.org/10.1016/j.mattod.2020.01.001
228. Zhang M et al (2021) Single-layered organic photovoltaics with double cascading charge transport pathways: 18% efficiencies. Nat Comm 12(1):309. https://doi.org/10.1038/s41467-020-20580-8
229. Zhang D et al (2021) A molecular photosensitizer achieves a Voc of 1.24 V enabling highly efficient and stable dye-sensitized solar cells with copper(II/I)-based electrolyte. Nat Comm 12(1):1777. https://doi.org/10.1038/s41467-021-21945-3
230. Zhu H et al (2022) Design investigation on 100 μm-thickness thin silicon PERC solar cells with assistance of machine learning. Mater Sci Semicond Process 137:106198. https://doi.org/10.1016/j.mssp.2021.106198
231. Zoco E (2020) 65% surge in polysilicon prices triggers unexpected increase in module production costs and affects solar installation outlook in 2020. IHS Markit. https://ihsmarkit.com/research-analysis/surge-in-polysilicon-prices-triggers-unexpected-increase-in.html

Chapter 5
Wind Energy

Mika Järvinen, Hanna Paulomäki, Rogier Ralph Floors, Christian Pansch, Paul Thompson, Eeva-Lotta Apajalahti, Mariana Galvão Lyra, Emma Luoma, and Jonna Kangasoja

First, there is the power of the wind, constantly exerted over the globe. Here is an almost incalculable power at our disposal, yet how trifling the use we make of it.
—Henry D. Thoreau

Supplementary Information The online version contains supplementary material available at https://doi.org/10.1007/978-3-031-69856-9_5.

M. Järvinen (✉) · H. Paulomäki
Aalto University, Espoo, Finland
e-mail: mika.jarvinen@aalto.fi

R. R. Floors
Technical University of Denmark, Wind & Energy Systems, Kongens Lyngby, Denmark

C. Pansch
Åbo Akademi University, Turku, Finland

P. Thompson
University of Aberdeen, Aberdeen, Scotland

E.-L. Apajalahti · M. Galvão Lyra
LUT University, Lappeenranta, Finland

E. Luoma · J. Kangasoja
Akordi Oy, Helsinki, Finland

M. Järvinen and H. Paulomäki (eds.), *Designing Renewable Energy Systems within Planetary Boundaries*, Green Energy and Technology,
https://doi.org/10.1007/978-3-031-69856-9_5

5.1 Introduction

Wind power is, compared to many other emerging energy sources, already a mature technology that has been harnessed by humans for millennia. The Sun is the enabler of this power source. The Sun first heats the Earth's surface, which then heats the air above—creating density differences—which finally produces wind in the atmosphere and makes wind energy systems viable, over land and sea. Wind power has many environmental advantages over the use of fossil fuels. It is seen as one of the key solutions to mitigate climate change. During the past decade the use of wind power has expanded rapidly. Relatively short construction times, reducing installation and operating costs and increased generation per MW installed decreasing costs per kilowatt hour, have made wind power a thriving technology on the road to carbon neutral future. It is estimated that wind and solar together will produce about half of world's electricity by 2030 [1]. In the previous chapter we discussed solar power technologies that form, together with wind, a promising combination of renewable energy that can also be installed in the same plant and used to complement each other. Wind is a variable energy source, and therefore its increasing installed capacities increases flexibility needs in the power system, that can be provided by demand, generation (including wind) and storage technologies that are discussed in Chap. 7. However, wind power, whilst being a promising vehicle to bring greenhouse gas emission back within safe boundaries, can have also negative environmental and social impacts that need to be addressed to create truly sustainable energy solutions.

The kinetic energy of the wind has been systematically used for at least 3000 years to produce mechanical energy, transport vessels on the seas, and help humans to carry out different tasks. As an example, vertical axis drag-type windmills were built by the Persians in 900 A.D. Windmills have been used in Europe since the Middle Ages, typically with horizontal axes. They were used to run different mechanical tasks such as grinding of grain [2]. European examples of these historic windmills are shown in Fig. 5.1.

Before large-scale industrialization, or the so-called "industrial revolution", (approximately 1760–1840) wind power and hydropower were the main energy source for mechanical work production. As the technology to utilize the combustion of coal in power production developed in steam engines and turbines, the share of wind power dropped. The spread of the use of wind power was limited by the lack of robust energy storage systems to manage the variability of the wind and the power transportation technology was not sufficiently developed or didn't even exist. The mechanical power produced by the wind had to be used in situ [2, 3].

When electrical generators and electrical power transmission systems were developed, by the end of the nineteenth century, they were also installed into wind turbines, and this opened new opportunities for wind power technology to spread and develop further. During the mid-1900's turbine size started to grow, and the features seen

Fig. 5.1 Examples of horizontal axis turbines used in 1800–1900's. From the left: Lumijoki, Finland (1830), Humppila, Finland (1929) and close-up illustrating the pitch angle of ~ 30° of the flat blades of the Humppila mill. In the Lumijoki windmill, the whole structure was turned from above the stand against the wind. In comparison to the Humppila mill, only the top part is turned making it easier to use, since the heavy grain packages no longer need to be lifted to the upper floor. As a detail, note the pitch angle of the Humppila mill rotor on the right. After studying the chapter, as a little sidekick, you should be able to estimate the power from these types of historical windmills

in today's turbines, such as the three-bladed horizontal axis technology, were developed. It's remarkable that the power generated by wind was used to produce hydrogen via electrolysis already back then, with the hydrogen used in lightning [2, 4]. Today we walk in the same footsteps as those visionaries. One of the most studied Power2X (electrical power converted to for example hydrogen by electrolysis) energy storage systems are based on coupling wind power with different types of electrolyzers to make this connection the backbone of the hydrogen economy. Denmark has been at the forefront of wind energy development and one landmark was the 200 kW Gedser turbine, built in the late 1950s, after World War II [5]. It had an aerodynamical stall control system implemented and used an induction generator. These developments made it possible for Denmark to be one of the leaders of wind power in the 1980's and it retains this lead still today [4].

Wind power was developed in other countries as well. For example, in the 1930s in Vermont, US, the largest wind turbine was the famous Smith-Putnam turbine that had a rotor diameter of 53 m and produced rated power—$P_r = 1.25$ MW. Unfortunately, this turbine was too large for the materials used at that time and its rotor broke down in 1945. Wind turbines were 're-invented' in the US in mid-1970's during the energy crises. As one example, the $P_r = 3.2$ MW Boeing research turbine, with a 98 m diameter rotor, was the largest system built in the 1980's [4].

Following several decades of intensive material and technology development, industrially manufactured turbines were at the 2 MW range by 2002 and 5 MW units were under development [4]. Today, in 2023, several turbine manufacturers are developing test versions of $P_r = 14–16$ MW turbines, with commercial operation anticipated to start in 2023–24. These large systems are designed for offshore use and have rotors close to 240 m in diameter and tip heights of up to 280 m. Land based turbines of 6–7 MW are commercially available and installed especially in the Northern Europe.

Wind energy has many health, environment and economic benefits compared with fossil fuel-based energy systems. Wind as a resource can be harnessed in many kinds of environments and it helps to reduce emissions of air pollutants like sulfur oxides, nitrogen oxides and particulate matter. And, as there is no need for cooling water, the water footprint of wind energy production is low, making it a competitive source of energy in water stressed parts of the world. For comparison, coal power plants can use millions of liters of water per day [6].

As with any energy production technology, wind energy also has adverse environmental impacts though the research to understand these effects has not developed as fast as the production of wind energy [7]. Bird collisions are the best described and most discussed impact resulting from wind power. However, like industrial scale solar power, wind power plants can also have large-scale impacts on nature. The impacts can be manifested via land use change, or other direct or indirect effects on species or ecosystems and the processes they maintain. The direct land or seascape impacted by wind power is relatively small, around 1 m^2/MWh of electricity (depending on wind conditions and topography) when compared to many other energy production systems, even when the impacts of extraction of critical materials are included [8]. Over 90% of the area occupied by the wind power plant is not directly used or changed by the turbines, roads, power lines, power station and other structures leaving this space for other uses like grazing or cropping [9, 10]. However, the indirectly impacted area can be much larger, and the disturbance on wildlife can sometimes stretch for kilometers away from the turbines. This depends on the location of the wind power plant and wildlife that is present in the area. For example, many species, such as prairie chickens [11], black kite [12], European roe deer and hare [13], wolf [14], some sea birds [15] and migrating raptors [16] have been shown to change their behavior and avoid locations near wind turbines, or die in collisions with the turbine blades. Avoidance behavior decreases the living spaces of different species. In forested areas, clearance of forest around turbines fragments the earlier continuous habitat and can create an "edge effect", exceeding even hundreds of meters in all directions from the turbine and creating disruption for some forest species [17, 18, 19]. In offshore areas the wind wake effect can result in impacts that span over tens of kilometers away from the turbines creating a potential threat for marine ecosystems' function [20–22]. With globally growing demand of land—and seascape—for wind power development, there is without careful siting and mitigation of negative impacts, a risk of worsening the loss of biodiversity in our efforts to solve the problems contributing to human induced climate change. Environmental concerns, including impacts on wildlife, have also been identified as a growing source of opposition and barrier to renewable energy development [23]. The impacts of wind power development on nature will be discussed further under Sect. 5.8.

In this chapter, we will first briefly study the operating principles of modern wind turbines and their main components. We focus only on the horizontal axis technology. We will then discuss the principles of power production from wind, introduce the concept of power coefficient, and derive the theoretical limits for the fraction of how much power can be obtained from the wind (the Betz-limits for the flow with and without tangential rotation in the wake). We will introduce the basic airfoil characteristics and introduce the blade element theory to be used in geometrical rotor design. We will present simple blade design methods, initially without the drag force, but later including drag force in the analysis. We will discuss the wind resource and statistics, discuss the use of Wind Atlases, and combine these with the predicted power coefficient to be able to estimate the capacity factor, that is, how much of the theoretically rated annual production potential of the turbine can be captured in practice? Finally, we will discuss the basics of turbine site selection in wind power plant design, wake effects analysis, impacts on nature and society and have a look at critical materials and land use.

5.2 Components of a Modern Wind Turbine, Materials Requirements, and the Power Equation

All modern horizontal axis industrial wind turbines are based on the variable rotational speed drive technology. These rotors have blades that can be pitched for maximum performance. A schematic of this kind of a pitch-controlled system is presented in Fig. 5.2. Figure 5.2 illustrates wind blowing from the left. To be able to capture maximum fraction of the specific kinetic energy of the wind, $U_\infty^2/2$, the rotor fixed to the nacelle must be accurately turned towards the wind using the yaw system. To give a reference for the size, the length of a modern 4–5 MW turbine nacelle is ~ 10–15 m. The yaw system obtains the required wind direction information from the controller which also measures the approximate wind speed.

The pitch system installed in the hub optimizes the rotation of the blades for the rotational speed of the rotor and the wind speed. The aerodynamic principles of control systems will be discussed later. Typically, the maximum tip speed of a blade is 80 m/s, due to sound generation reasons. This speed value was obtained from [24] as an average value of 14 turbine sizes, ranging from 800 kW to 7.6 MW. From this it is easy to calculate the maximum rounds per minute (rpm) for different rotor sizes d as $60 \times U_{tip}/(\pi d)$. For example, for a 15 MW turbine with a 236 m diameter rotor, the maximum rotational speed would be $60 \times 80/(\pi \times 236) = 6.5$ rpm. However, to be able to operate the generator at its maximum efficiency, a much higher rpm is required. For this purpose, a gearbox that increases the low rpm of the

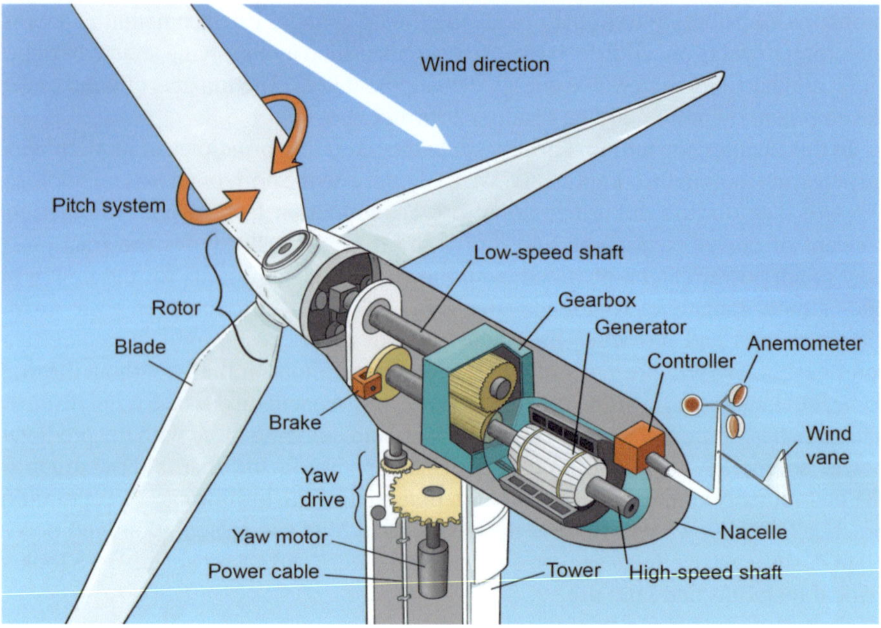

Fig. 5.2 Components of a pitch-controlled turbine © Maija Karala

primary axis ("low-speed shaft in Fig. 5.2 attached to the rotor) to the high rpm of the secondary axis ("high-speed shaft") is required. There are also direct drive permanent magnet generators available, but these must have much larger generator rotor and stator diameters, in order to have space for a sufficiently large number of permanents magnet and poles fitted into the perimeter to reach the required frequency.

As the name variable speed implies, the rpm of modern wind turbines is allowed to vary to be optimized to the wind speed. On the other hand, the frequency of the electrical grid is fixed at 50 Hz or 60 Hz. To make the grid connection possible an AC/DC/AC converter is needed to adjust the variable frequency from the generator to a fixed value in the grid with minimal losses. Different generator and converter types will be discussed in Chap. 8. The voltage level of a typical modern generator is 690 V. The internal voltage level of the wind power plant is normally 20 or 33 kV, so each turbine also needs a separate transformer to raise the voltage level to sufficiently minimize losses in the power plant's internal electrical grid. All electrical systems have losses and there are also losses in the generator. These losses are dissipated into heat and therefore the nacelle always has a cooling system to keep the temperature of the components low enough so as to prolong the lifetime of the system. For emergency situations there must be a process to stop the rotor safely. Therefore, all turbines have a brake system, which is also required during maintenance breaks.

To ensure that the turbine can resist the wind loads it must be rigidly fixed on the ground. There are several types of foundation structures in use. Figure 5.3 presents the most common ones used for onshore and offshore wind turbines.

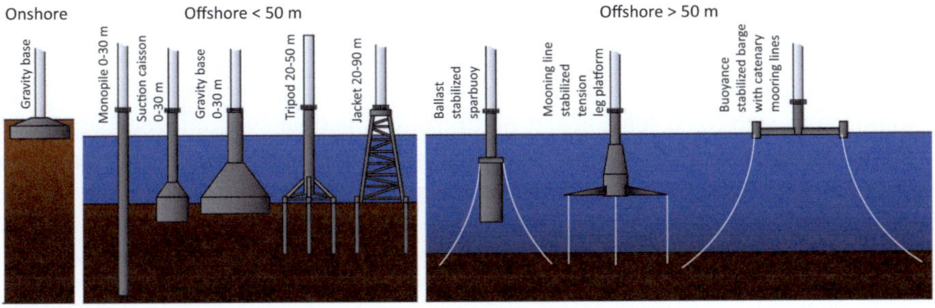

Fig. 5.3 Wind turbine foundations

The most common onshore foundation type is the gravity base, a large steel reinforced concrete slab that has sufficient weight and diameter to resist the turning moment coming from the thrust force multiplied by the height of the nacelle. If the ground is not sufficiently rigid, the foundation needs to be reinforced with piles. If the turbine is located on bedrock, it can be directly fixed to the bedrock and save concrete.

Offshore, there are various options available depending on the water depth and seafloor type. These can be divided into fixed bottom (<50 m water depth) and floating foundations (>50 m water depth). For example, in the North Sea the mono-pile type is the most common foundation type used typically for depths up to 25 m [25], though these have also been used in much deeper locations, even 50–60 m [26]. Mono-piles can be used in sandy and clay seafloor bottoms, but they are not suitable for shallow bedrocks or strata with boulders or coarse gravel [26].

The gravity base with ballast weights can be used on offshore depths up to 25–30 m. As the name implies, it uses its own weight to anchor to seabed. The larger the volume of the ballast, the larger the buoyancy force. This type is applicable for seafloor composed of sand, medium to stiff clays, bedrock, and strata with cobbles, boulders, or coarse gravel, excluding very soft soils or weak clays [26].

A less massive type of foundation is the suction caisson, where the hollow bottom structure is emptied of water by pumping, thus creating a suction anchoring force. This type of foundation is also applicable at depths up to 25–30 m [26].

Both the jacket and the tripod foundations provide very good stiffness and stability and are a relatively light weight option for depths of 20–50 m, and can be used even up to 90 m [25]. They both utilize piles or suction buckets for the bottom fixing. Pile-based fixing is applicable in stiff clays and medium to dense sand. It can also be used in softer silts and clays, and in very soft sediments overlying stiffer soils or bedrock, but is less well suited for locations containing many boulders. Suction caisson fixing is suitable for medium stiff clays and fine to medium sand, but it is not suitable for strata with cobbles, boulders, or coarse gravel layers, or in very soft soils [26].

Normally the wind conditions are the best at open sea. Often this also means deeper waters, and for locations deeper than 50 m, floating foundations are used. Offshore shallow banks and reefs are attractive for wind power development, but these areas are often also ecologically important areas. Figure 5.3 presents the most common types of floating offshore wind turbine foundations: ballast weight stabilized buoy, mooning line stabilized tension leg platform and buoyancy stabilized barge with catenary mooring lines. The method of fixing the cables to seabed vary based on the bottom type. More technical details and especially environmental impacts can be found in [26].

Wind power plants require various materials in large quantities. Table 5.1 presents the approximate amount of materials required for a 500 MW wind turbine plant. The number of turbines is 250, 112 and 34 for 2 MW 4.5 MW and 15 MW turbines respectively. No foundations and cabling are included Table 5.1. Note that for example in the case of massive gravitational foundation, the CO_2 footprint of the system significantly increases. For a location that has a capacity factor of 0.5 (just as an example to explain the concept) the amounts in Table 5.1 should be multiplied by a factor of 2, to have continuous power production of 500 MW.

As can be seen, the need for all except copper, alloys, glass and carbon fibers are highest for the smallest turbines. So, from a material efficiency perspective, larger sizes should be favored. The estimation of materials required for foundations is challenging since their construction varies case by case. According to Carrara [28], 70–75% of the total weight of an onshore wind power unit is in the concrete foundation (gravity-based). The total mass of turbines in a 500 MW power plant is ~ 50,000 t, this would correspond to117,000 t–150,000 t of concrete. An offshore turbine, for example, an 8 MW Siemens-Gamesa (model SG8.0–167 DD) turbine, requires a steel mono-pile foundation of 925 t, corresponding to ~ 58,000 t of steel per 500 MW. Note, though, that these values are site specific and as above, the final figures must be divided by the capacity factor of the location. It is important to note that 100% of these metallic materials in the turbine can be recycled.

Wind power units also need comparatively high amounts of certain critical materials. Table 5.2 presents a comparison in terms of rare earth metals of two different

Table 5.1 Materials required (tons) for manufacturing 500 MW (rated power) of wind turbines, no foundations and cabling included [27]

	V120-2 MW	V150-4.5 MW	V236-15 MW
Turbine power	2 MW	4.5 MW	15 MW
Total mass	62,000	49,556	51,000
Steel and iron	52,700	41,379	42,024
Aluminum and alloys	930	545	663
Copper and alloys	310	347	459
Polymer materials	3224	1784	255
Glass and carbon fiber	3782	4956	4845
Electronics and electrics	558	396	459
Lubricants	186	149	204

Table 5.2 Rare earth materials required (tons) for a 500 MW (rated power) wind power plant for PMSG and DFIG generators [29]

	DD-PMSG	GB-DFIG
Dysprosium	8.5	1
Neodymium	90	6
Boron	3	0
Praseodymium	18	0
Molybdenum	55	49
Terbium	3.5	0.5

generator system types: the direct drive permanent magnet synchronous generator (DD-PMSG, common in offshore turbines) and the double fed induction generator with gearbox (DFIG-GB, the most commonly used technology in onshore turbines in 2022).

As can be seen from Table 5.2, the type of generator has a huge impact on material consumption. Rare earth metals are needed in permanent magnets, including Dysprosium, Neodymium and Praseodymium.

Each component along the power line creates losses from the net kinetic energy of wind itself, to the final power fed into the grid (see Fig. 5.4).

Figure 5.4 shows that there are energy losses at many points along the power-line that must be minimized to achieve maximum power production. At first, we need to be able to turn the maximum fraction of the wind's kinetic energy to mechanical energy in the rotor. It will be shown later that the maximum power coefficient of a fast-rotating rotor is $C_P = 0.593$. A fraction of this mechanical energy is lost as heat by friction in the gearbox and all bearings, and the mechanical energy reaching the generator is $P_{kin}C_P\eta_m$. Another fraction of this is lost as heat by electromagnetic and resistive losses, and power output from the poles of the generator is $P_{kin}C_P\eta_m\eta_g$. Often the wind power plant is spread over a large area and there are additional losses $(1 - \eta_c)$ in the cables and transformers. The final electrical power from a single turbine at the border of the wind power plant is:

$$P = P_{kin}C_P\eta_m\eta_g\eta_c$$

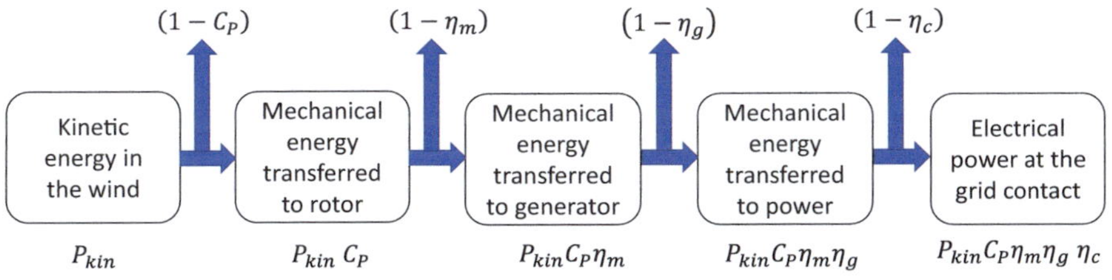

Fig. 5.4 'Power-line' from the kinetic energy of the wind to the grid connection point

On the other hand, this output power must be equal to the load on the grid side. This means that the mechanical load created by the generator must be equal to the generation of mechanical power by the rotor, which is equal to the torque times the angular velocity, $P = M\Omega$. This balance also defines the rotational speed of the rotor.

When the wind blows freely at a speed of U_∞ it carries specific kinetic energy of amount $e_{kin} = 0.5 \times U_\infty^2$, having the unit of J/kg $= \text{m}^2/\text{s}^2$. We assume that this wind passes through a wind turbine rotor disk with a diameter of d, and area $A = \pi/4d^2 = \pi R^2$. The mass flow rate passing the swept area A is then $\dot{m} = U_\infty A\rho$. If the velocity remains constant, then the total kinetic energy passing through the swept area of the rotor is:

$$P_{kin} = \dot{m}\frac{1}{2}U_\infty^2 = \frac{1}{2}A\rho U_\infty^3$$

This simple equation shows that the available kinetic energy is proportional to the 3rd power of the free wind velocity, and indicates the huge impact that wind condition has on potential power resource. For example, if the average wind speed is 5 m/s and then it increases to the typical rated wind speed of large turbines—say 15 m/s—the available power increases by a factor of $3^3 = 27$.

If we were to try to capture all the kinetic energy in free wind it would require that the wind would be completely stopped. This would mean that at the rotor, the wind would no longer have any velocity nor ability to do work. Therefore, only a fraction, defined by the power coefficient C_p, of the total kinetic energy can be utilized. The final power equation is written as follows:

$$P = C_p\frac{1}{2}A\rho U_\infty^3$$

Modern large industrial turbines can have maximum power coefficients of ~ 0.45…0.5. A typical example of a power coefficient curve as a function of the tip speed ratio is shown in Fig. 5.5.

The tip speed ratio is one of the most important parameters of rotors. It defines the ratio of the blade tip speed to the speed of the free wind. Ω is the angular velocity of the rotor (rad/s) and n is the rotational speed (1/s).

$$\lambda = \frac{U_{tip}}{U_\infty} = \frac{\Omega R}{U_\infty} = \frac{2\pi nR}{U_\infty}$$

We are operating between zero and storm values of wind speeds. There are three important wind speeds defined: cut-in, rated and cut-out wind speeds. Below the cut-in speed there is very little power available in the wind and a typical value is somewhere around 3 m/s. Above this, the maximum production potential is utilized between the cut-in and the rated wind speed. A typical rated speed is around 10–15 m/s. Above this wind speed, we need to limit the power production so as not to exceed the performance of the turbine's components, both mechanical and electrical.

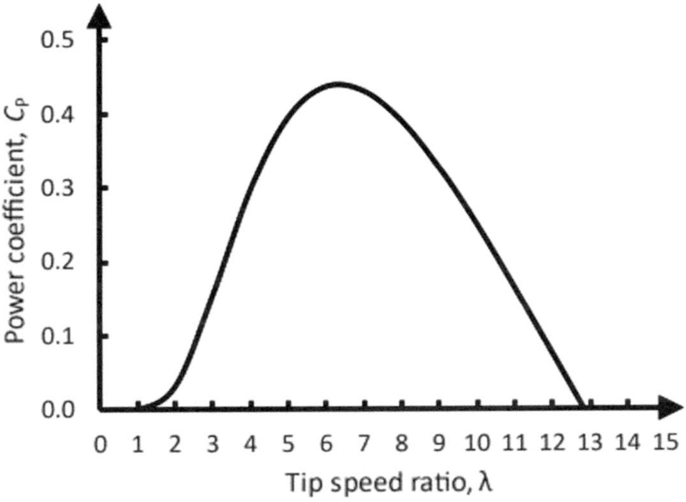

Fig. 5.5 Power coefficient curve of a modern wind turbine

Finally, manufacturers define a cut-out wind speed when the turbine must be shut down to prevent any damage. For example, for Vestas V236-15.0 MW® turbine this is as high as 30 m/s.

There are two ways to limit the power: (1) ensuring that the rotor stalls above the rated wind speed, or (2) turning the blades so that the power production remains constant even when the wind speed increases. Both of these methods will be discussed later. Figure 5.6 presents the power curve for a modern 15 MW wind turbine with rotor diameters of 168, 236 and 312 m. The unlimited power from the general power equation is presented using dashed lines.

Looking at the power equation, the only parameter that can be adjusted to keep the power below the rated value is the power coefficient. For example, for a pitch-controlled system this simply means that the blades are turned to decrease the torque to keep the power production at the constant rated value. Figure 5.6 also shows that the rotor diameter has a huge impact on the rated wind speed; the speed at which the rated power is achieved. The example commercial turbine mentioned above with a 234 m rotor is assumed produce rated power at 11 m/s wind but if, for example, we decrease the rotor size by 30% to 168 m, the rated power is not reached until wind speed reaches ~ 14 m/s. The lower rated wind speed would increase the annual energy production significantly, since the constant power production time increases, making the energy system more stable, and the need for energy storage also decreases.

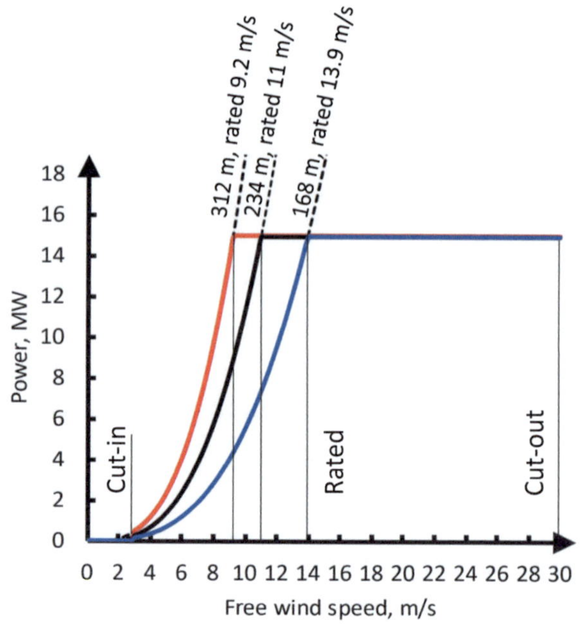

Fig. 5.6 Power curve for a modern 15 MW offshore wind turbine (C_p is assumed constant 0.4 between cut-in and rated power) with rotor diameters 168, 234 and 312 m. The cut-in speed is 3 m/s

5.3 Wind Resource Analysis

The speed and the direction of the wind varies constantly with time. In addition, the average wind speed increases with height. It is important to understand how these factors affect the performance of the wind turbines and for example, how to predict the annual power production and analyze the system efficiency. As an example Fig. 5.7 presents the modelled hourly variation of the wind speed outside of Hanko peninsula in Finland, in January and July 2020 (source for data: [30] and [31], www.renewables.ninja).

As shown in Fig. 5.7, there is a significant wind speed variation per hour and throughout the month. To calculate the amount of energy produced accurately, the wind history together with the power equation must be numerically integrated for the period of interest, τ (one month or a year for example).

$$E = \int_0^\tau P(t)dt = \int_0^\tau C_P(t)\frac{1}{2}\rho(t)\pi R^2 U_\infty^3(t)dt$$

$$\approx \sum_{j=1}^{N} C_P(t_j)\frac{1}{2}\rho(t_j)\pi R^2 U_\infty^3(t_j)\Delta t$$

In this example Δt is one hour and τ presents the total hours during the observation period. The power coefficient, C_P, varies with tip speed ratio and blade pitch angle.

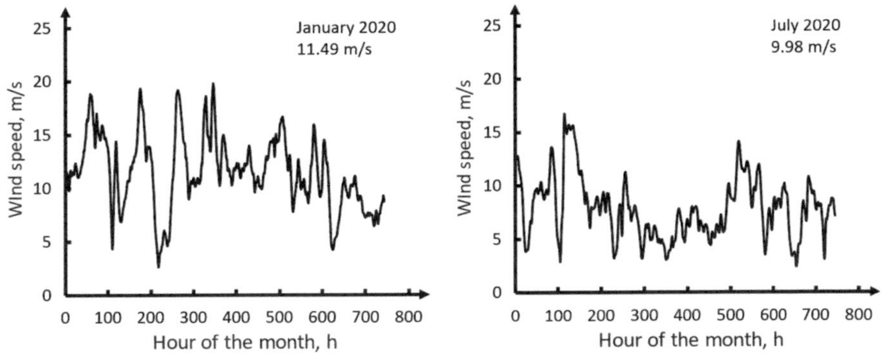

Fig. 5.7 Modelled wind speed variation outside of Hanko peninsula, Finland, in January and July 2020, at a height of 150 m. The average wind speeds are also presented

Note that the density of air also can vary as the atmospheric temperature varies throughout the year.

We are often interested in the annual energy production—to be able to carry out economic analysis for example. In this case the average capacity factor, C_f, needs to be defined. This is the ratio of actual energy produced divided by the energy that would have been produced if the turbine had run the whole observation period on rated power, P_r.

$$C_f = \frac{1}{\tau P_r}\sum_1^N P(t_j)\Delta t = \frac{1}{\tau P_r}\sum_1^N C_P(t_j)\frac{1}{2}\rho(t_j)\pi R^2 U_\infty^3(t_j)\Delta t$$

A simplified method to analyze the capacity factor is the 'Method of bins' [4]. Here, the wind history is arranged into N 'baskets' from 0 to cut-out speed, typically every 1 m/s, and then the duration of the speed in each 1 m/s bin is summed together to get the duration of the wind speeds of the entire basket, Δt_i. Next the summation is carried out for each of the N bins. It is assumed that the density is constant for each bin. $C_{P.i}$ and P_i are defined at the average wind speed of the bin, $\overline{U} = 0.5 \times (U_l + U_u)$. For example, for 0–1 m/s, $\overline{U}$ is 0.5 m/s. U_l and U_u refer to low and upper wind bin speed values, respectively. Note that Δt_i is not constant as it was in the previous hourly based integration.

$$C_f = \frac{1}{\tau P_r}\sum_{i=1}^N P_i\Delta t_i = \frac{1}{\tau P_r}\sum_{i=1}^N C_{P.i}\frac{1}{2}\rho\pi R^2\overline{U}_i^3\Delta t_i$$

Example 5.1 Monthly Production Estimation Calculate the capacity factor for the above-mentioned Hanko case, both in January and June 2020 for a 15 MW, 236 m rotor offshore turbine. Assume a power coefficient of 0.4. U_l U_u and Δt_i are given in

Table 5.3. The average bin speed, power, and energy production are calculated from these.

In Table 5.3 the energy from each wind bin is calculated as.

$$E_i = P(\hat{U}_i)\Delta t_i = \min\left(\frac{C_{p.i}}{2 \times 10^6}\rho\pi R^2\overline{U}_i^3,\ 15\ \text{MW}\right)\Delta t_i(\hat{U}_i)$$

Note that min-function must be used to limit the power to the rated value and 10^6 is used to convert from W to MW. Based on the sum of all 20 bins of E_i shown on the last row the theoretical monthly production potential 744 h $\times$ 15 MW = 11,160 MWh, and the capacity factors in January and July are 8768/11160 = 0.786 and 4582/11160 = 0.411, respectively. This difference is very large, considering that the average wind speed is only 1.5 m/s greater in January.

Table 5.3 Production of monthly energy at Hanko wind power plant; comparison of January and July 2020, for a 15 MW, 236 m rotor with power coefficient 0.4, over 31 days with $\tau = 744$, $N = 20$ and Δt_i values refer to [Data]/[Weibull model fitted to data]

U_l, m/s	U_u, m/s	Δt_i, h Jan	Δt_i, h July	$\overline{U}$, m/s	P_i, MW	E_i, MWh Jan	E_i, MWh July
0	1	0/0	0/1	0.5	0.00	0	0
1	2	0/1	0/8	1.5	0.04	0	0
2	3	2/2	5/21	2.5	0.17	0	1
3	4	5/6	50/39	3.5	0.47	2	23
4	5	22/12	78/60	4.5	1.00	22	78
5	6	24/20	74/80	5.5	1.82	44	135
6	7	19/31	94/95	6.5	3.00	57	282
7	8	49/44	103/101	7.5	4.61	226	475
8	9	38/58	119/96	8.5	6.72	255	799
9	10	63/71	79/83	9.5	9.38	591	741
10	11	111/81	35/64	10.5	12.66	1405	443
11	12	95/85	41/44	11.5	15.00	1425	615
12	13	82/83	21/27	12.5	15.00	1230	315
13	14	56/75	14/14	13.5	15.00	840	210
14	15	71/62	7/7	14.5	15.00	1065	105
15	16	33/46	21/3	15.5	15.00	495	315
16	17	27/31	3/1	16.5	15.00	405	45
17	18	17/18	0/0	17.5	15.00	255	0
18	19	23/10	0/0	18.5	15.00	345	0
19	20	7/4	0/0	19.5	15.00	105	0
20	21	0/3	0/0	20.5	15.00	0	0
		744	744	Total E		8768	4582

Often the hourly based data is not available, and maybe just the annual average wind speed is given. In this case, it is possible to use correlations for calculating the duration when the wind is blowing. The most often used model is the continuous cumulative Weibull's correlation/distribution, which gives the fraction of time that the wind speed is below the speed U. c and k are the Weibull scale and shape parameters.

$$F(U) = 1 - exp\left(-\left(\frac{U}{c}\right)^k\right)$$

If this were to be fitted into the data presented above, the parameters c and k would be 12.69 m/s and 3.83 for January and 8.66 m/s and 3.01 for June. This data is shown in Table 5.3.

In addition to the temporal variation average wind speed increases with height. The higher the rotor installation, the greater the power available. The third power velocity dependency has a huge impact on wind power production. There are several ways to approximate the wind speed profile in the vertical direction. In the lowest part—the surface layer where $z = 0$–100 m—the logarithmic law is often recommended to predict the horizontal wind speed because it is based on physical arguments [4, 32, 33]

$$U(z) = \frac{u^*}{\kappa} ln\left(\frac{z}{z_0}\right)$$

where u^* is the friction velocity, κ is the von Karman constant of ~ 0.4, and z_0 is the surface roughness length that depends on the structure of the terrain. See Table 5.4 for roughness values. The wind speed is often measured at the reference height h—typically 10 m—and we can then re-write as follows:

$$\frac{U(z)}{U(h)} = \frac{ln\left(\frac{z}{z_0}\right)}{ln\left(\frac{h}{z_0}\right)}$$

Or, if we need to estimate the friction velocity, this can be solved by using the logarithmic law as follows, when two points have been measured:

Table 5.4 Surface roughness length for different types of terrains [4, 34]

Terrain type	Roughness length z_0, m
Rough sea, desert	0.001–0.005
Grass land	0.008–0.01
Crops field	0.03–0.05
Forests	0.5–0.7
Cities	0.7–3

$$u^* = \kappa \frac{(U(z_2) - U(z_1))}{ln\left(\frac{z_2}{z_1}\right)}$$

If there is a more refined vertical velocity distribution measured the least squares method can be used or the data can be plotted as a straight line in the following format, where the horizontal axis is $U(z)$ and the vertical axis is $ln(z)$, respectively:

$$ln(z) = \frac{\kappa}{u^*} U(z) + ln(z_0)$$

The slope is equal to κ/u^* and the intersection with the vertical axis is equal to $ln(z_0)$. This is studied in the example below. Another simpler way of describing the wind profile is the power law, where the exponent n varies as a function of several parameters. A typical value of $n = 7$.

$$\frac{U(z)}{U(h)} = \left(\frac{z}{h}\right)^{1/n}$$

Example 5.2 Vertical Profile Estimation from the Logarithmic Model There is data available on the vertical wind profile from an offshore location near the Hanko peninsula, coordinates (59.54175N, 22.36095E). The data is presented in Table 5.5 below. Determine the parameters in the logarithmic model, friction velocity u^* and roughness length z_0, by drawing the data in a $U(z) - ln(z)$ diagram and then plot the data and model in the same figure.

Based on the data, we first calculate the $ln(z)$—see Table 5.5. When this is plotted, in Fig. 5.8, the parameters can be calculated from the constants of the line, $y = ax+b$, where $a = \kappa/u^*$ and $b = ln(z_0)$.

The model constants, the friction velocity $u^* = 0.4/0.7034 = 0.569$ m/s and the roughness length $z_0 = exp(-4.1149) = 0.0163$ m can be obtained from Fig. 5.8. Using these, the original data and the model are plotted on the right-hand side of Fig. 5.8. For example, if the hub height increases from 100 to 250 m, the wind speed

Table 5.5 Average vertical velocity wind profile in December, Hanko peninsula, (59.54175, 22.36095) (Data from [35])

z, m	$U(z)$, m/s	$ln(z)$, –
50	11.5	3.9120
75	12.0	4.3175
100	12.4	4.6052
125	12.6	4.8283
150	12.9	5.0106
200	13.4	5.2983
300	14.0	5.7038
400	14.4	5.9915

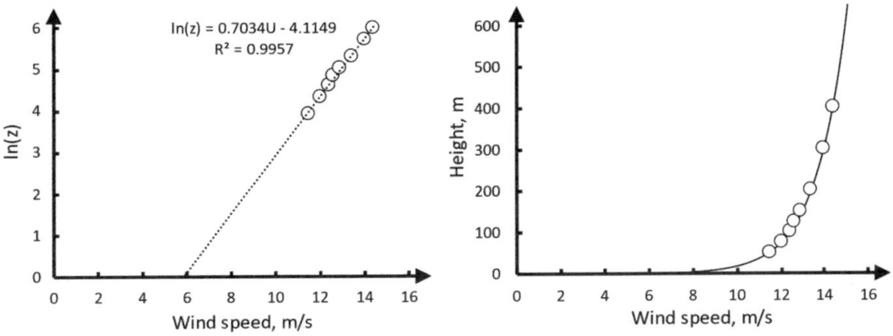

Fig. 5.8 Fitting the model parameters (left) and vertical velocity data and fitted model (right)

increases from 12.4 m/s to 13.7 m/s and therefore, the kinetic energy in the wind increases ~ 35%.

5.4 Theoretical Limitations for Harnessing Power from the Wind

In this section the theoretical limitations for the power coefficient are discussed. As explained above, only a fraction of the kinetic energy in the free wind can be captured by the rotor. The flow of the rotor can't be stopped because that would stop the work. Figure 5.9 presents a schematic of a flow through the rotor disk that captures energy from the wind. When the free wind approaches and hits the disk, its velocity decreases by the factor of a. The velocity in the wake has already decreased by a factor of b. It will be shown below that $b = 2a$, meaning that 50% of the velocity drop takes place before and 50% after the wind has passed the rotor. The value of a at this point is not known and will be determined. We follow here the idea presented almost 100 years ago by [36].

In Fig. 5.9 the stream tube borders are presented with a dashed line. The stream tube is an imaginary expanding tube, that shows how the cross section of the flow increases as the velocity decreases in the axial direction. There is no mass flow rate through the stream tube side walls, so the mass flow rate at each axial location is constant.

$$\dot{m} = U_\infty A_\infty \rho_\infty = U_D A_D \rho_D = U_w A_w \rho_w$$

Note that the density can be assumed constant, ρ. The axial pressure profile is presented in the lower section of Fig. 5.9. As the disk restricts the flow, the pressure in front of the disk increases. The pressure then sharply drops after the disk and increases back to the pressure of the environment in the wake. Work can be obtained from the pressure difference acting across the rotor disk. Note that eventually the

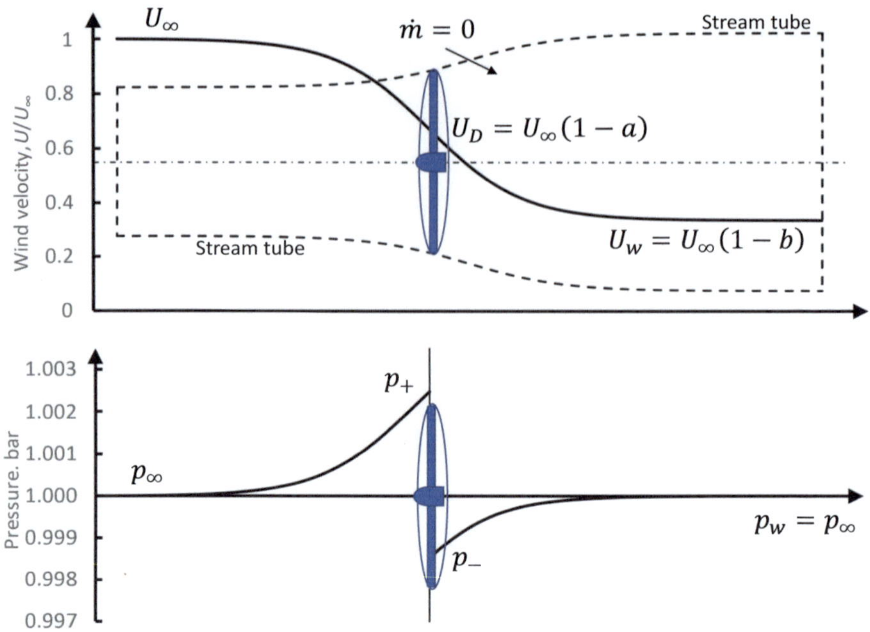

Fig. 5.9 Schematic of wind flow through the disk or rotor (note that the vertical axis of the figure is to scale)

velocity increases back to free wind value, after it mixes with the flow outside of the stream tube in the wake.

Let us draw a control volume (dashed line) around the stream tube shown in Fig. 5.9 and then write the *axial momentum equation* for the system. There will also be another equation in the angular direction: the *angular momentum equation.* The pressure around the control volume is constant since pressure is returned to its environmental value in the wake. Therefore, the only force acting on our system is the force created by the pressure difference across the rotor disk. In the momentum balance this force must be balanced by the change in the axial momentum.

$$\dot{m}(U_\infty - U_w) = U_\infty(1 - a)A_D\rho(U_\infty - U_w) = (p_+ - p_-)A_D$$

Pressure before and after the disk can be simply calculated from the Bernoulli equation, written separately from the free wind to the disk and after the disk to wake.

$$p_\infty + \frac{1}{2}\rho U_\infty^2 = p_+ + \frac{1}{2}\rho U_D^2$$

$$p_- + \frac{1}{2}\rho U_D^2 = p_\infty + \frac{1}{2}\rho U_w^2$$

The pressure difference across the disk can be calculated from these and then this is substituted back into the axial momentum equation.

$$(p_{+}-p_{-})=\frac{1}{2}\rho\left(U_{\infty}^{2}-U_{\mathrm{w}}^{2}\right)$$

$$U_{\infty}(1-a)(U_{\infty}-U_{w})=\frac{1}{2}\left(U_{\infty}^{2}-U_{\mathrm{w}}^{2}\right)=\frac{1}{2}(U_{\infty}-U_{w})(U_{\infty}+U_{w})$$

The last form was obtained using the rule $\left(x^{2}-y^{2}\right)=(x+y)(x-y)$. From this, the wake velocity $U_{w}=U_{\infty}(1-2a)$ can be solved. So, as mentioned previously, 50% of the velocity drop takes place upstream and 50% downstream. But what is the value of a that gives maximum power from the rotor? This can be obtained by writing the equation for power production in terms of a and then maximizing this. The power extracted from the wind is equal to the thrust force times the velocity of the wind at the disk. We could think of this as how much power is required to move the disk with a thrust force at the speed of the wind. The axial thrust force on the rotor disk can be simply calculated from the pressure difference.

$$\begin{aligned}T=(p_{+}-p_{-})A_{D}&=U_{\infty}(1-a)A_{D}\rho(U_{\infty}-U_{\infty}(1-2a))\\&=2A_{D}\rho U_{\infty}^{2}a(1-a)=4a(1-a)\frac{1}{2}A_{D}\rho U_{\infty}^{2}\end{aligned}$$

wherein $4a(1-a)=C_{T}$ is the thrust coefficient. This can be written:

$$P=2A_{D}\rho U_{\infty}^{3}a(1-a)^{2}=\left(4a(1-a)^{2}\right)\frac{1}{2}A_{D}\rho U_{\infty}^{3}=C_{P}\frac{1}{2}A_{D}\rho U_{\infty}^{3}$$

where the power coefficient is equal to $C_{p}=4a(1-a)^{2}$. The maximum power can be obtained by differentiating the power coefficient, C_{p}, with respect to a and setting this to zero. The roots are $a=1/3$ and 1. A value of $a=1$ would mean that the flow has stopped—see also Fig. 5.10. The maximum power coefficient value is therefore:

$$C_{p}=\frac{4}{3}\left(1-\frac{1}{3}\right)^{2}=\frac{16}{27}$$

This is the well- known Betz constant, $C_{p}=0.593$. Both the power and thrust coefficient are presented in Fig. 5.10.

The maximum point and value for the power coefficient is $a=1/3$ and 0.593, respectively. Note that the thrust coefficient has a maximum at $a=0.5$, and at this point velocity at the wake is equal to zero. If the axial induction factor, a, further increases from this, velocity will turn and become negative, and the theory is no longer valid. There are several approaches to handle this situation occurring in highly loaded turbines. For example, [37] presented the following model applicable for thrust coefficients higher than 0.96, corresponding to $a>0.4$. This is graphically

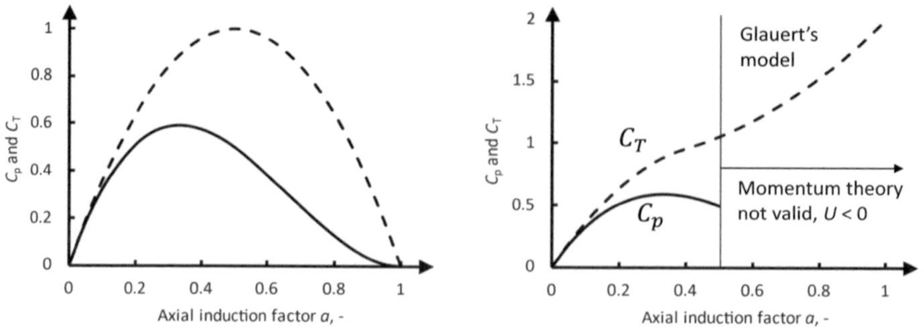

Fig. 5.10 Power and thrust coefficients

presented in Fig. 5.10.

$$a = 0.143 + \sqrt{0.0203 - 0.6427(0.998 - C_T)}$$

When $a < 0.4$, the turbine is working in the windmill state and when $a > 0.4$, the turbine operates at the turbulent wake state. In practice under the turbulent wake state the flow structures are far too complex to be described with the simple analytic models discussed here. Even though this experimental correlation was derived for the overall rotor, it can be also used for the radial segments. In this traditional [36] limit derivation, we assume that the wake has no rotation and there is an infinite number of blades. It is important to realize that the structure of the disk is not limited to a horizontal axis turbine, but could be of any design.

In practice all large scale industrial sized turbines are based on the horizontal axis concept and therefore, to be able to better understand the maximum power that can be obtained and the impact on wake rotation, we need to discuss the implications of a rotating wake. The angular momentum balance must be studied in detail to understand this properly. Since all the parameters vary as a function of radial position, we need to write the angular momentum balance for an incremental radial annulus ring (see Fig. 5.11). We follow here the derivation by [37] and [4], opening the ideas in a bit more detail compared to the original publications.

The situation on the right-hand side of Fig. 5.11 presents the velocity triangles for a system fixed to the moving blade coordinates. The radial increment of the blade is moving downwards in the figure with tangential velocity Ωr. At the inlet edge of the blade, the blade experiences air velocity of equal magnitude Ωr but in the opposite direction. From the axial left the wind blows at velocity $U_\infty(1 - a) = U$. It is assumed that the axial velocity remains constant over the rotor disk (see Fig. 5.11). After the rotor, the wake is rotating tangentially at a speed of ωr. As the direction of the wake rotation is opposite to rotor movement, this velocity must be added to the velocity of the rotor. From these, the relative velocity magnitudes at the inlet and outlet can be solved.

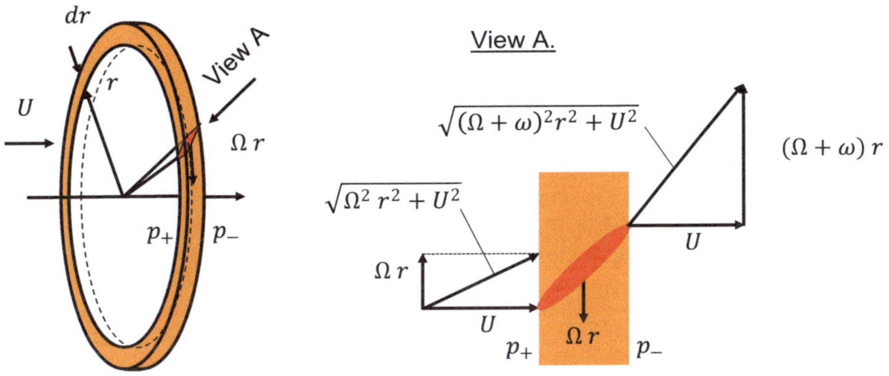

Fig. 5.11 Radial annulus ring of the rotor (left) and view from the point A towards the axis (right) of rotation along the blade element with the velocity triangles before and after the blade

$$U_1 = \sqrt{U^2 + \Omega^2 r^2} \quad \text{and} \quad U_2 = \sqrt{U^2 + (\Omega + \omega)^2 r^2}$$

The pressure difference across the annulus can be solved from the energy balance along a streamline, i.e. using the Bernoulli equation.

$$p_+ + \frac{1}{2}\rho\left(U^2 + \Omega^2 r^2\right) = p_- + \frac{1}{2}\rho\left(U^2 + (\Omega + \omega)^2 r^2\right)$$

$$p_+ - p_- = \rho\left(\Omega + \frac{1}{2}\omega\right)\omega r^2$$

This pressure difference is required to accelerate the flow in a tangential direction from 0 to ωr. In presenting the pressure difference using the energy equation, the related velocities can be used as scalars with no direction. In the momentum balance, vectors are required. By multiplying the above pressure difference with the cross-sectional annulus flow area, $2\pi\, rdr$, the incremental axial force required to accelerate the flow can be obtained. We also define the tangential induction factor as $a' = \omega/2\Omega$, $\omega = 2a'\Omega$.

$$\begin{aligned} dF &= (p_+ - p_-)dA = \rho\left(\Omega + \frac{1}{2}\omega\right)\omega r^2 2\pi\, rdr \\ &= 4a'(1 + \text{a}')\frac{1}{2}\rho\Omega^2 r^2 2\pi\, rdr \end{aligned}$$

On the other hand, this force must be equal to the force that was obtained from the axial momentum balance, already derived above.

$$dF = 4a(1-a)\frac{1}{2}\rho U_\infty^2 2\pi r dr = 4a'\left(1+\mathrm{a}'\right)\frac{1}{2}\rho\Omega^2 r^2 2\pi r dr$$

This can be re-written as

$$\frac{\Omega^2 r^2}{U_\infty^2} = \frac{a(1-a)}{a'(1+a')} = \lambda_r^2$$

The right-hade side parameter λ_r is referred as the local speed ratio.

$$\lambda_r = \frac{\Omega r}{U_\infty} = \lambda\frac{r}{R}$$

To be able to calculate the power obtained from the rotor we need to derive the angular momentum equation. It is assumed here that the tangential velocity of the incoming air flow is $U_{t.1} = \omega r$ before the rotor is zero. The second form is obtained by using the definition of the angular induction factor.

$$dM = \dot{m}U_t(r)r = U_\infty(1-a)\rho 2\pi r(\omega r)rdr$$

$$dM = U_\infty(1-a)\rho 2\pi r\left(a' 2\Omega r\right)rdr = 4a'(1-a)\frac{1}{2}\rho U_\infty \Omega r^2 2\pi r dr$$

The power generated form the differential annular ring is simply the moment multiplied by the angular velocity Ω.

$$dP = 8a'(1-a)\frac{1}{2}\rho U_\infty \Omega^2 r^2 \pi r dr$$

In this we have a', a, U_∞, Ω and, to be able to combine this with the previous definition of the local speed ratio, this needs to be written in terms of λ_r, and to present the integration with respect to λ_r. We therefore need to present dr in terms of $d\lambda_r$ as $dr = R/\lambda d\lambda_r$. Inserting this gives us:

$$\begin{aligned} dP &= 8a'(1-a)\frac{1}{2}\rho U_\infty \Omega^2 r^2 \pi r \frac{R}{\lambda} d\lambda_r \\ &= \frac{1}{2}\rho U_\infty^3 \pi R^2 \left(8a'(1-a)\frac{\lambda_r^3}{\lambda^2} d\lambda_r\right) \end{aligned}$$

In this form we can see that there is the total kinetic energy passing through the rotor disk in front of the parenthesis, so the remaining part in the parenthesis must be the contribution of the differential annular ring to the power coefficient. Note that the local speed ratio at the tip is equal to the overall tip speed ratio. To get the power coefficient of the whole rotor this must be integrated from $\lambda_r = 0$ to $\lambda_r = \lambda$.

$$C_p = \frac{8}{\lambda^2} \int_0^{\lambda} a'(1-a)\lambda_r^3 d\lambda_r$$

The maximum value of this can be obtained when the term $a'(1-a) = f(a, a')$ achieves its maximum value. To be able to proceed and finally carry out the integration all variables need to be presented in terms of a single variable [4]. From the above solved equation for the local speed ratio, we can re-write it as a second-order equation for a' and solve it as follows (only a positive solution is acceptable):

$$a'^2 + a' - \frac{a(1-a)}{\lambda_r^2} = 0$$

$$a' = \frac{1}{2}\sqrt{1 + \frac{4a(1-a)}{\lambda_r^2}} - \frac{1}{2}$$

This is then substituted in $f(a, a') = a'(1-a)$. The equation can be divided by 1/2 since it does not affect the optimal a value.

$$\left(\sqrt{1 + \frac{4a(1-a)}{\lambda_r^2}} - 1\right)(1-a) = f(a)$$

The maximum of this can be obtained by differentiating $f(a)$ with respect to a and setting this to zero.

$$(1-a)\frac{\frac{2-4a}{\lambda_r^2}}{\left(\sqrt{1+\frac{4a(1-a)}{\lambda_r^2}}\right)} - \sqrt{1 + \frac{4a(1-a)}{\lambda_r^2}} + 1 = 0$$

Then, to make it even more interesting (and challenging), this needs to be solved for λ_r^2. An enthusiastic student can use some paper to do this, but we used Mathcad© (version 15.0) to do it, and the result is:

$$\lambda_r^2 = \frac{16a^3 - 24a^2 + 9a - 1}{3a - 1} = \frac{(1-a)(4a-1)^2}{3a-1}$$

This simple equation relates the axial induction factor, a, to the maximum power as the local speed ratio for each annular ring at the distance of r from the axis of rotation. Looking above we see that there are now two equations that define the local speed ratio. Setting these equal gives us.

$$\lambda_r^2 = \frac{(1-a)(4a-1)^2}{1-3a} = \frac{a(1-a)}{a'(1+a')}$$

From which we can build the relationship between a' and a.

$$\frac{(4a-1)^2}{1-3a} = \frac{a}{a'(1+a')}$$

This is again a second order equation in terms of a' and has only a single physically relevant solution.

$$a' = \frac{1-3a}{4a-1}$$

This angular induction factor gives the maximum power at each location. We now have all the terms required inside the integral, but the integration is carried out with respect to the local speed ratio. The required change of variable from λ_r to a can be done by using the above derived equation for the local speed ratio.

$$2\lambda_r d\lambda_r = \frac{6(2a-1)^2(4a-1)}{(3a-1)^2} da$$

$$d\lambda_r = \frac{3(2a-1)^2(4a-1)}{(3a-1)^2} \frac{1}{\lambda_r} da$$

Now we have all the components of the integration.

$$C_p = \frac{8}{\lambda^2} \int_0^\lambda a'(1-a)\lambda_r^3 d\lambda_r = \frac{8}{\lambda^2} \int_0^\lambda \frac{1-3a}{4a-1}(1-a)\lambda_r^2 d\lambda_r$$

$$\frac{8}{\lambda^2} \int_0^\lambda \frac{1-3a}{4a-1}(1-a)\frac{(1-a)(4a-1)^2}{3a-1} \frac{3(2a-1)^2(4a-1)}{(3a-1)^2} da$$

$$C_p = \frac{24}{\lambda^2} \int_{a_1}^{a_2} \frac{(1-a)^2(4a-1)^2(2a-1)^2}{(1-3a)^2} da$$

The limits of integration can be obtained from the local speed ratio equation at the axis of rotation and at the tip as follows:

$$\lambda_{r=0}^2 = 0 = \frac{(1-a_1)(4a_1-1)^2}{1-3a_1}$$

$$\lambda^2 = \frac{(1-a_2)(4a_2-1)^2}{1-3a_2}$$

It can easily be seen that the lower limit of integration is at the rotation axis $r = 0$, when $a_1 = 0.25$. Meanwhile, the infinite value of the local speed ratio is obtained when the axial induction factor approaches $a_2 = 1/3$. Or vice versa, the asymptotic value of axial induction factor is $a_2 = 1/3$ when the local speed ratio approaches infinity. These values can then be used to draw the power coefficients between $a_2 = 1/4 \ldots 1/3$.The actual integration of the power coefficient equation can be done again analytically, though we used Mathcad© for this. See also the solution provided by [38] for comparison. And note that this is only defined between the range $0.25 < a_2 < 1/3$. Exact values of 0.25 and 1/3 are not defined. Use a small increment at the borders to stay away from these limits.

$$C_p = \frac{24}{\frac{(1-a_2)(4a_2-1)^2}{1-3a_2}} \left(\begin{array}{l} \frac{737(a_2-a_1)}{729} - \frac{970(a_2^2-a_1^2)}{243} + \frac{20(a_2^3-a_1^3)}{3} \\ -\frac{136(a_2^4-a_1^4)}{27} + \frac{64(a_2^5-a_1^5)}{45} \\ +\frac{4^{\circ}}{729} \ln\left(\frac{1-3a_2}{1-3a_1}\right) \\ -\frac{4}{6561}\left(\frac{1}{(a_2-\frac{1}{3})} - \frac{1}{(a_1-\frac{1}{3})}\right) \end{array} \right)$$

The result is shown in Fig. 5.12, where on the *x*-axis for this, the tip speed ratio is defined as

$$\lambda(a_2) = \sqrt{\frac{(1-a_2)(4a_2-1)^2}{1-3a_2}}$$

As can be seen from Fig. 5.12, the higher the tip speed ratio, the higher the theoretical maximum power coefficient. At higher tip speed ratios, the power coefficient approaches the Betz limit of 0.593. So, the higher the speed of rotation, the higher the theoretical maximum. What does this mean for different radial locations on the blades? In Fig. 5.13 both the axial and the tangential induction factors are presented as a function of the radial position on the blade, *r*/*R*, for tip speed ratios of 0.25, 1, 4 and 16.

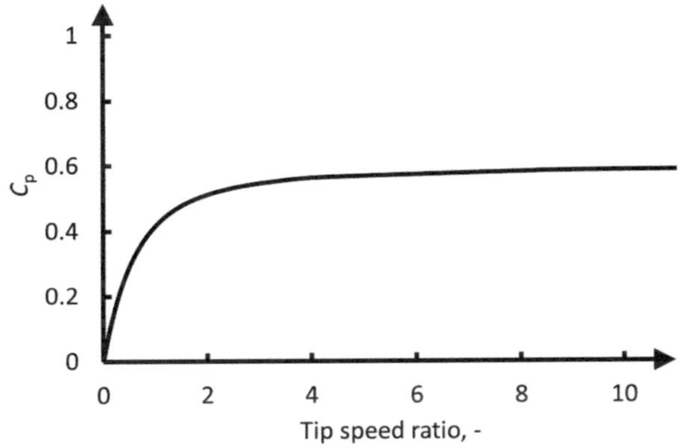

Fig. 5.12 Power coefficient for an actuator disk with wake rotation

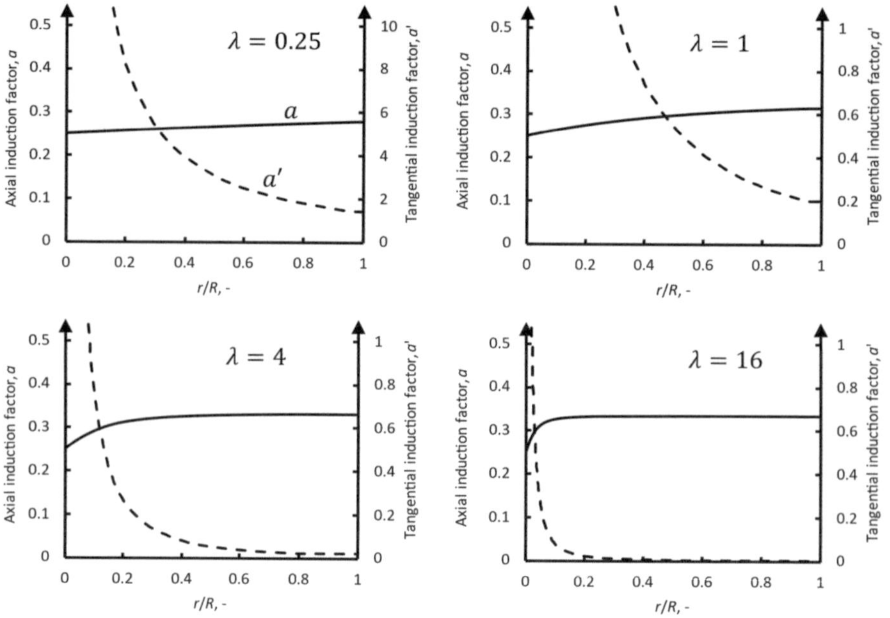

Fig. 5.13 Axial and tangential induction factors for tip speed ratios of 0.25, 1, 4 and 16. Note the 10 × scale for the tangential induction factor at $\lambda = 0.25$

As can be seen in Fig. 5.13, when $\lambda = 0.25$ the axial induction factor is far from the optimum value of 1/3 and there is also significant tangential rotation at all blade positions. If the wind speed remains constant and the rotational speed (i.e. tip speed ratio) increases, the achievable maximum power increases. This can also be seen as a significant decrease in the tangential induction factor and almost constant axial induction factor of 1/3 throughout the blade length. Only the root section deviates from the optimal. Perhaps trivial, but because of the position of the hub it is not possible to have the blades reaching all the way to axis of rotation.

The above results already provide important guidelines for blade design. Without any knowledge of the detailed rotor geometry, we can already see that higher rotational speeds favor higher efficiencies, and at the root section there seems to always be some wake rotation created. If we compare modern fast rotating 3-blade turbines with slow rotating traditional multiblade American water-pumping turbines, it can be seen that in addition to higher frictional losses, multi-blades also have lower efficiency due the higher wake rotation created. However, these are already designed to create high pumping torque at low rotational speeds and in low winds, so this is the sacrifice required.

In the following, the airfoils and blade design principles are discussed in sufficient detail to understand their operational and dimensioning principles.

5.5 Principles of Blade Design

5.5.1 Airfoil Characteristics

This section discusses airfoil parameters, lift, drag coefficients, and blade element theory. First a simple blade design without drag force is presented, then with drag force and finally some practical examples for blade design are presented. Figure 5.14 presents the parameters of an airfoil. Note that lift and drag forces are always defined as parallel and perpendicular to the relative velocity of the wind. The relative wind vector is not the same as the free wind vector in a rotating rotor blade.

The camber line divides the airfoil into two equal parts and the chord line is the direct line connecting the leading and trailing edges. Area is defined as $A = cL$ where L presents the length of the blade section. The overall lift and the drag forces are defined as follows:

$$F_L = C_L(\alpha, Re)A\frac{1}{2}\rho U^2$$

$$F_D = C_D(\alpha, Re)A\frac{1}{2}\rho U^2$$

And for an incremental radial segment of area, $dA = cdr$, dF_L and dF_D are as follows:

$$dF_L = C_L(\alpha, Re)\frac{1}{2}\rho U^2 cdr$$

$$dF_D = C_D(\alpha, Re)\frac{1}{2}\rho U^2 cdr$$

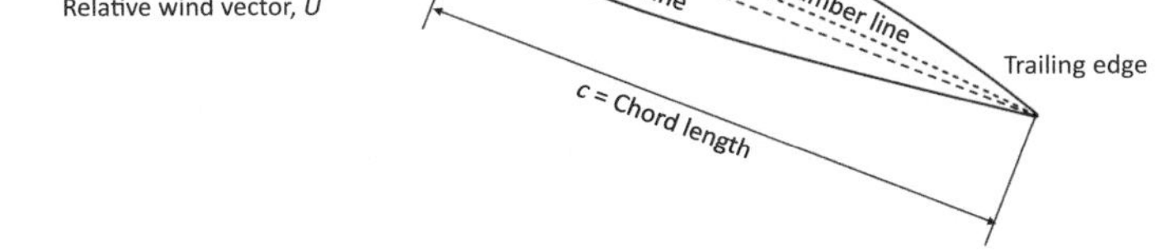

Fig. 5.14 Airfoil characteristics

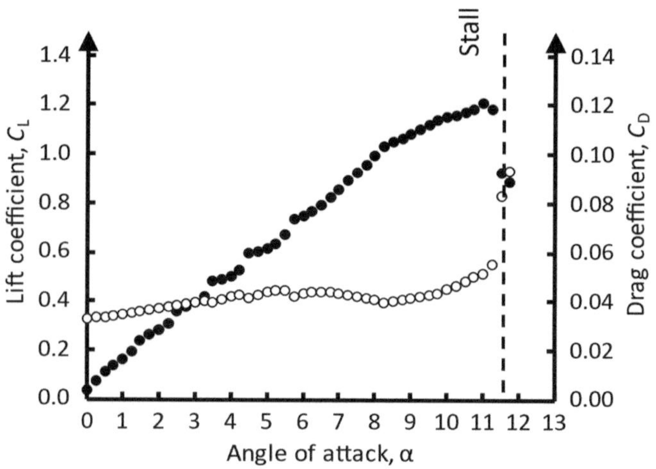

Fig. 5.15 Lift and drag coefficients of NACA 2415 airfoil at $Re = 50{,}000$. The point of stall is shown as a dashed vertical line. Filled circles present the lift coefficient. Data from http://airfoiltools.com/airfoil/index

The Reynolds number is calculated at each location separately as $Re = U(r)c(r)/\nu$ where ν is the kinematic viscosity of the air. To be able to carry out accurate calculations, $C_L(\alpha, Re)$ and $C_D(\alpha, Re)$ must be defined in detail using measurements or state-of-the-art CFD modelling. Figure 5.15 present one example of the lift and the drag coefficients of the airfoil NACA 2415 at Reynolds number 50,000.

As can be seen in Fig. 5.15, the lift coefficient increases almost linearly up to 1.2 and the point of stall and then suddenly drops at ~ 11.5°. Below this critical angle of attack the drag coefficient is almost constant at 0.04 but then also rapidly increases at the stall. When the blade is aligned with the flow, values of α are small, and the streamlines are tightly attached to the airfoil. When the angle of attack is increased the flow starts to separate from the upper airfoil surface, starting at the trailing edge, and as the critical angle is approached the flow on the complete upper side of the airfoil is separated, generating high vortices and turbulence.

5.5.2 *Blade Element Theory*

The basic idea of the blade element theory is that the turbine blades are divided into N radial increments of Δr, each having their own characteristics such as the pitch angle, θ, the chord length, c, and even the airfoil profile, which can be changed to different blade positions for optimal performance [37]. Figure 5.16 presents a schematic of the concept.

Figure 5.16 showcases blades with 18 radial increments. Note that the drag force is parallel, and the lift force is perpendicular to the relative velocity with *respect to*

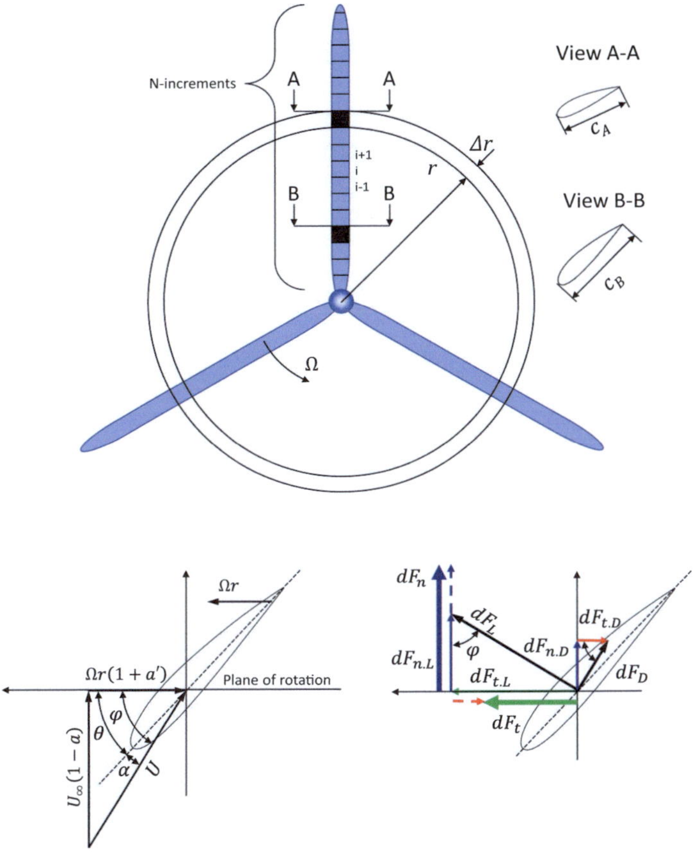

Fig. 5.16 **a** Principles of blade element theory **b** Principles of blade element theory, velocity triangle at the blade and force vectors (right). Note that the tangential velocity component of air experienced by the blade is larger than the velocity coming from the blade movement only and also has the opposite direction. Thick arrows show the total forces

the blade. Sections A–A (11th segment) and B–B (4th segment) illustrate the twist of the blade to optimize the lift coefficient and prevent stall conditions closer to the root section of the blade. This will be discussed in more detail in the section on control methods.

One important observation from Fig. 5.16 is that the tangential velocity component of the air experienced by the blade is larger and has opposite direction to the movement of the blade. If you imagine yourself being small enough to be able to sit on the leading edge of the blade and facing towards the moving direction of the blade, the wind will blow at a greater velocity than that coming from the rotation, Ωr, due to the rotating wake. This velocity is $\Omega r(1 + a')$ in the tangential direction. The velocity from the free wind in the axial direction has decreased by a factor of a when reaching the rotor. The total wind would then blow from your left, from ~ 10 o'clock (Fig. 5.16a and b, left lower schematic). Close your eyes and enjoy the ride! We will next derive the expressions for these forces and build the model to calculate the torque and power.

From the force triangle the forces acting on tangential and axial directions of the blade, dF_t and dF_n, can be easily derived. In the axial direction, components from the drag and the lift forces are simply added, but in the tangential direction, the contribution of the drag force to tangential working force is negative. U is the relative air velocity.

$$\begin{aligned} dF_t &= dF_L \sin(\varphi) - dF_D \cos(\varphi) \\ &= (C_L \sin(\varphi) - C_D \cos(\varphi))\frac{1}{2}\rho U^2 c dr \end{aligned}$$

$$\begin{aligned} dF_n &= dF_L \cos(\varphi) + dF_D \sin(\varphi) \\ &= (C_L \cos(\varphi) + C_D \sin(\varphi))\frac{1}{2}\rho U^2 c dr \end{aligned}$$

The local twist angle of the blade, θ_T, is defined as the difference between the local pitch angle and the pitch angle of the tip of the blade, θ_0. The angle of the relative wind is simply the sum of the local pitch angle and the angle of attack (Fig. 5.15).

$$\theta_T = \theta - \theta_0 \quad \text{and} \quad \varphi = \theta + \alpha$$

Note that although the blades are pitched during the operation, the twist angle remains constant. The angle of the relative wind can be calculated from the axial and tangential velocity components (Fig. 5.16).

$$\tan(\varphi) = \frac{U_\infty(1-a)}{\Omega r(1+a')} = \frac{(1-a)}{\lambda_r(1+a')}$$

If the angle of the relative wind and axial induction factor is known, the relative wind velocity can be calculated simply as follows.

$$U = \frac{U_\infty(1-a)}{\sin(\varphi)}$$

To get the total force acting on the radial increment at r from all blades, the force resulting from a single blade must be simply multiplied by the number of blades, B. This requires an assumption that the blades do not interact aerodynamically. We insert the above defined relative velocity and move the lift coefficient to the front.

$$dF_t = BC_L\left(\frac{1}{\sin(\varphi)} - \frac{C_D}{C_L}\frac{1}{\sin(\varphi)\tan(\varphi)}\right)\frac{1}{2}\rho U_\infty^2(1-a)^2 c dr$$

$$dF_n = BC_L\left(\frac{1}{\sin(\varphi)\tan(\varphi)} + \frac{C_D}{C_L}\frac{1}{\sin(\varphi)}\right)\frac{1}{2}\rho U_\infty^2(1-a)^2 cdr$$

The final BEM (Blade Element Method) equation for torque and power production can be obtained when the incremental tangential force, dF_t, is multiplied with radial distance to the axis of rotation and then this torque by angular velocity of the rotor.

$$dM(r) = BC_L\left(\frac{1}{\sin(\varphi)} - \frac{C_D}{C_L}\frac{1}{\sin(\varphi)\tan(\varphi)}\right)\frac{1}{2}\rho U_\infty^2(1-a)^2 crdr$$

$$dP(r) = BC_L\left(\frac{1}{\sin(\varphi)} - \frac{C_D}{C_L}\frac{1}{\sin(\varphi)\tan(\varphi)}\right)\frac{1}{2}\rho U_\infty^2(1-a)^2 c\Omega rdr$$

If the blade geometry is available and the rotational speed and wind speed are known, then in principle we have all the information we need to solve the torque and power of the rotor when above equations are numerically integrated over the blade length. If the blade design is not known an optimal design can be explored, where the blade shape is allowed to vary.

As a summary of the above discussions, we have two equations for axial force: one for the axial momentum balance, dT, and the other one from the forces calculated from the lift and drag, dF_n.

$$dT = 4a(1-a)\frac{1}{2}\rho U_\infty^2 2\pi rdr$$

$$dF_n = BC_L\left(\frac{1}{\sin(\varphi)\tan(\varphi)} + \frac{C_D}{C_L}\frac{1}{\sin(\varphi)}\right)\frac{1}{2}\rho U_\infty^2(1-a)^2 cdr$$

In addition, we also have two equations for the generated torque.

$$dM = 4a'(1-a)\frac{1}{2}\rho U_\infty \Omega r^2 2\pi rdr$$

$$dM(r) = BC_L\left(\frac{1}{\sin(\varphi)} - \frac{C_D}{C_L}\frac{1}{\sin(\varphi)\tan(\varphi)}\right)\frac{1}{2}\rho U_\infty^2(1-a)^2 crdr$$

In these equations the induction factors a and a' are not known. By setting the thrust forces equal in the first two and the torques equal in the last two, this results in two equations that are sufficient to solve both induction factors. It is also common practice to assume that the ratio of C_D and C_L approaches zero in the blade shape design phase [39]. This is shown in detail below. We first set the thrust force from force balance equation equal to the thrust solved from axial momentum equation, and after some cleaning we have:

$$\frac{BC_L c}{\sin(\varphi)\tan(\varphi)8\pi r}(1-a) = a$$

Then, the torques are set equal from both equations, clean a bit and we have the equation for a'.

$$\frac{BC_L c U_\infty}{\sin(\varphi) 8\Omega r^2 \pi}(1-a) = a'$$

a can be easily solved from the first equation and then a' from the second equation is as follows:

$$a = \frac{1}{1 + \frac{\sin(\varphi)\tan(\varphi) 8\pi r}{BC_L c}}$$

$$a' = \frac{U_\infty}{\Omega r}\left(\frac{\tan(\varphi)}{1 + \frac{\sin(\varphi)\tan(\varphi) 8\pi r}{BC_L c}}\right) = \frac{1}{\lambda_r}\left(\frac{\tan(\varphi)}{a}\right)$$

But how do we use these in blade design effectively? This will be discussed after we have learned how to calculate the power coefficient and understood the prerequisites for this.

5.5.3 *Power Coefficient of the Rotor*

The total power of the rotor can be obtained by integrating/summing the incremental powers from each radial annulus. r_h is the radius of the hub.

$$P = \int_{r_h}^{R} dP = \int_{r_h}^{R} \Omega dM$$

Based on the definition of the power coefficient we have:

$$C_P = \frac{\int_{r_h}^{R} \Omega dM}{\frac{1}{2}\rho\pi R^2 U_\infty^3}$$

Inserting the integral term for the torque from above, we get:

$$C_P = \frac{\Omega \int_{r_h}^{R} BC_L c \frac{1}{\sin(\varphi)}\left(1 - \frac{C_D}{C_L}\cot(\varphi)\right)\frac{1}{2}\rho U_\infty^2 (1-a)^2 r dr}{\frac{1}{2}\rho\pi R^2 U_\infty^3}$$

And after some cleaning, the following form is obtained:

$$C_P = \frac{B\Omega}{\pi R^2 U_\infty} \int_{r_h}^{R} \frac{C_L c}{\sin(\varphi)} \left(1 - \frac{C_D}{C_L} \cot(\varphi)\right)(1-a)^2 r dr$$

To calculate C_P, the profiles of chord length, $c(r)$, relative wind angle, $\varphi(r)$, axial induction factor, $a(r)$, and the drag and the lift coefficients need to be known. In principle, we first need to determine the blade geometry. If this is not already available it must be determined in order to achieve the design objectives, for example to maximize the power coefficient or to maximize the torque at low wind speed. This will be discussed below with three examples explored in Sect. 5.5.4. Once the profiles of chord length and pitch angles are available the performance can be estimated from the above presented equation. For this, the profiles of axial and tangential induction factors and relative wind speeds need to be iterated. The iteration also gives us the values for lift and drag coefficients.

This is only a part of the solution. As already discussed in the introduction, the mechanical power generated by the rotor must match the load generated by the generator. The combination of the rotor and the generator will find a balance and rotational speed that eventually sets the system in mechanical equilibrium.

5.5.4 *Three Blade Design Examples*

This section discusses three blade designs: (1) A theoretical case with no drag and no wake rotation i.e. the so-called Betz-rotor, (2) No drag in the shape optimization but including the wake rotation, and considering the drag in the power production, (3) A practical rough blade design of a 15 MW offshore wind turbine. All of these utilize the combined linear and angular momentum equations and the equations from the blade element theory. Momentum equations give us the flow field boundary around the blade and the BET (Blade Element Theory) equation relates these to the forces, torque and produced power.

Case 5.1: Ideal Betz Rotor with No Drag and Wake Rotation

In this first case, we aim to design a blade that fully satisfies the Betz limit. It was already shown above that the maximum power is obtained when the axial induction factor $a = 1/3$. In addition, this theoretical maximum is obtained when there is no angular wake rotation, $a' = 0$ and no drag force, $C_d = 0$. We also assume that the number of blades, B, is sufficient to capture the required kinetic energy from the passing wind. As a starting point, we need to select the tip speed ratio λ, the number of blades and the radius of the rotor R. From the linear momentum equation and BET (Blade Element Theory) equations we have two equations for the normal thrust force. By setting these equal and using the above definitions $a' = 0$ and $C_d = 0$, a model for the blade shape can be built. Collecting the required force equations from above, inserting the above assumed values and setting the forces equal gives us the following:

$$dF_n = BC_L\left(\frac{1}{\sin(\varphi)\tan(\varphi)} + \frac{C_D}{C_L}\frac{1}{\sin(\varphi)}\right)\frac{1}{2}\rho U_\infty^2(1-a)^2 cdr$$
$$= BC_L\left(\frac{1}{\sin(\varphi)\tan(\varphi)}\right)\frac{2}{9}\rho U_\infty^2 cdr$$

$$dT = 4a(1-a)\frac{1}{2}\rho U_\infty^2 2\pi rdr = \frac{8}{9}\rho U_\infty^2 \pi rdr$$

By equating these, $dT = dF_n$, the chord length can be solved.

$$BC_L\left(\frac{1}{\sin(\varphi)\tan(\varphi)}\right)\frac{2}{9}\rho U_\infty^2 cdr = \frac{8}{9}\rho U_\infty^2 \pi rdr$$

$$c = \frac{4\pi r}{BC_L}\sin(\varphi)\tan(\varphi)$$

The relative wind angle can be solved from, $a' = 0$ and $a = 1/3$.

$$\tan(\varphi) = \frac{(1-a)}{\lambda_r(1+a')} = \frac{2}{3\lambda_r}$$

Resulting in the following model for the blade shape:

$$c(r) = \frac{8\pi r}{3\lambda_r BC_L}\sin(\varphi(r)) = \frac{8\pi R}{3\lambda BC_L}\sin(\varphi(r))$$

$$\varphi(r) = \tan^{-1}\left(\frac{2}{3\lambda_r}\right) = \tan^{-1}\left(\frac{2}{3\lambda}\frac{R}{r}\right)$$

This model is discussed in Example 5.3.

Example 5.3 Betz Rotor Design Design a Betz rotor of 1 m in diameter with 3 blades and for a wind speed of 10 m/s. The hub diameter is 10 cm. The rotational speed of the turbine is 1000 rpm. The air density is 1.225 kg/m^3 (US standard atmosphere) and the kinematic viscosity is 1.506×10^{-5} m^2/s. The airfoil profile is NACA 2415 (National Advisory Committee for Aeronautics) and its C_L and C_D/C_L data is given below for Re values 50,000 and 100,000 in Fig. 5.17.

1. Calculate the tip speed ratio. Assume that the Reynolds number is constant for all segments and define the optimum angle of attack.
2. Using a tabular format, calculate local tip speed ratios, relative angles of wind, chord lengths, relative air velocities, Reynold's numbers and section pitches for 10 segments.
3. Calculate the power production from the rotor and compare this with the value from the Betz turbine. Also discuss the manufacturing aspects of this design.

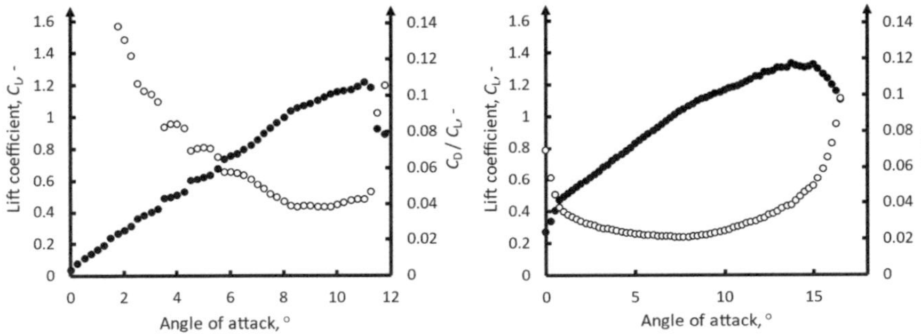

Fig. 5.17 Properties of the NACA 2415 airfoil at $Re = 50{,}000$ and 100,000. Open circles present the ratio of drag and lift coefficients. Data from http://airfoiltools.com/airfoil/index

1. The tip speed ratio is simply calculated as the ratio of blade tip and wind speeds. The corresponding angular velocity is equal to $\Omega = 1000/60\ \text{s} \times 2\pi = 104.7\ \text{rad/s}$ and $\lambda = (1000 \times 2\pi \times 0.5\ \text{m})/(60\ \text{s} \times 10\ \text{m/s}) = 5.24$. We assume that $Re = 50{,}000$. The maximum performance of the rotor is obtained when the C_D/C_L is at a minimum *i.e.*, the C_L/C_D at maximum. In this case the optimal angle of attack is 9.5° and $C_L = 1.123$ (read from Fig. 5.17).
2. Local tip speed ratios, relative angles of wind, relative wind speed, chord lengths and section pitches are calculated using the following equations: Please note that the values are defined at the centre of each blade element in a radial direction.

$$\lambda_r = \lambda \frac{r}{R}$$

$$\varphi(r) = \tan^{-1}\left(\frac{2}{3\lambda_r}\right)$$

$$U = \frac{U_\infty(1-a)}{\sin(\varphi)}$$

$$c(r) = \frac{8\pi r}{3\lambda_r B C_L}\sin(\varphi(r))$$

$$\theta(r) = \varphi(r) - \alpha$$

As can be seen the blade shape is very dramatic; the chord is 16 cm at the blade root, while the blade tip radius is 50 cm. The blades are also quite twisted with a total twist angle of $31.8° + 1.9° = 33.7°$. Due to rotation the relative wind speed is significantly larger at the tip compared with the root of the blades. One very interesting observation is that the Reynolds number is constant at all blade positions. So, our assumption of constant C_L at all positions was correct. However, the assumed Re value was too low. To get correct answers we should use the new $Re = 1.1 \times 10^5$ value obtained and repeat all steps. New values are given in Table 5.7. We read the new values for C_L and angle of attack at the minimum drag/lift ratio. In this case the optimal angle of attack is 7.5° and $C_L = 1.031$ (Table 5.7).

The total twist angle is 33.8° − 0.1° = 33.7°, and the chord length at the hub end is 17.1 cm. The power produced by each segment was calculated as follows (see Table 5.7 for calculated values):

$$\Delta P(r) = BC_L \frac{1}{\sin(\varphi_i)} \frac{2}{9} \rho U_\infty^2 c_i \Omega r_i \Delta r$$

3. The total power is simply the sum of all segments, and is equal to 288 W. As a comparison, using the Betz limit for the power coefficient we get:

Table 5.6 Profiles of blade geometry parameters, $\lambda = 5.24, B = 3, \alpha = 9.5°$. The rightmost column is the cumulative sum power element powers from the hub end to different locations. Reynold's number is the value calculated from new results

r m	r/R –	λ_r –	φ °	c m	θ °	U m/s	Re –	P_i W	$\sum P_i$ W
0.073	0.145	0.759	41.3	0.157	31.8	10.1	1.1×10^5	7.6	7.6
0.118	0.235	1.230	28.4	0.113	18.9	14.0	1.1×10^5	12.3	19.9
0.163	0.325	1.702	21.4	0.087	11.9	18.3	1.1×10^5	17.0	36.9
0.208	0.415	2.173	17.1	0.070	7.6	22.7	1.1×10^5	21.7	58.6
0.253	0.505	2.644	14.2	0.058	4.7	27.3	1.1×10^5	26.4	85.1
0.298	0.595	3.115	12.1	0.050	2.6	31.9	1.1×10^5	31.1	116.2
0.343	0.685	3.587	10.5	0.043	1.0	36.5	1.1×10^5	35.9	152.1
0.388	0.775	4.058	9.3	0.038	−0.2	41.1	1.1×10^5	40.6	192.6
0.433	0.865	4.529	8.4	0.035	−1.1	45.8	1.1×10^5	45.3	237.9
0.478	0.955	5.000	7.6	0.031	−1.9	50.4	1.1×10^5	50.0	287.9

Table 5.7 Profiles of converged blade geometry parameters, $\lambda = 5.24$, $B = 3$, $\alpha = 7.5°$

r m	r/R –	λ_r –	φ °	c m	θ °	U m/s	Re –	P_i W	$\sum P_i$ W
0.073	0.145	0.759	41.3	0.171	33.8	10.1	1.1×10^5	7.6	7.6
0.118	0.235	1.230	28.4	0.123	20.9	14.0	1.1×10^5	12.3	19.9
0.163	0.325	1.702	21.4	0.094	13.9	18.3	1.1×10^5	17.0	36.9
0.208	0.415	2.173	17.1	0.076	9.6	22.7	1.1×10^5	21.7	58.6
0.253	0.505	2.644	14.2	0.063	6.7	27.3	1.1×10^5	26.4	85.1
0.298	0.595	3.115	12.1	0.054	4.6	31.9	1.1×10^5	31.1	116.2
0.343	0.685	3.587	10.5	0.047	3.0	36.5	1.1×10^5	35.9	152.1
0.388	0.775	4.058	9.3	0.042	1.8	41.1	1.1×10^5	40.6	192.6
0.433	0.865	4.529	8.4	0.038	0.9	45.8	1.1×10^5	45.3	237.9
0.478	0.955	5.000	7.6	0.034	0.1	50.4	1.1×10^5	50.0	287.9

$$P_{Betz} = C_p \frac{1}{2} \rho A U_\infty^3 = 0.593 \times \frac{1}{2} \times 1.225 \times \pi \times 0.5^2 \times 10^3 = 291 \text{ W}.$$

The above design has a 10 cm hub that takes away part of the area, which decreases the power from the theoretical Betz value. Only a small number of elements were used; this also causes error in the summation/integration. Also, using more accurate numerical integration method, such as the trapezoidal rule, would improve the accuracy.

Most of the power is produced at the outer sections of the blade, and only 30% of the power is produced within the inner half of the blade length. The chord and mass of the blade at the root can therefore be reduced, saving materials and cost. Even though the above design in massive and difficult to manufacture, it gives a nice idea of what an ideal design would look like. In the following we add complexity to the model by also considering the angular wake rotation and drag force in the power production estimation.

Case 5.2: Design of the Rotor with Drag Force and Wake Rotation

In this second case, we consider the impact of drag force and wake rotation on the blade design. Typically, the drag force is not considered in the shape optimization, but it is included in the calculation of the power coefficient using the optimal shape obtained. As a starting point, we need to select the tip speed ratio λ, the number of blades and the radius of the rotor, R. The basis for the shape optimization is the maximization of the power coefficient. The following equation was derived above, but here we also show the derivation steps in a form that gives the power coefficient in terms of the local tip speed ratio.

$$\begin{aligned} C_P &= \frac{B\Omega}{\pi R^2 U_\infty} \int\limits_{r_h}^{R} \frac{C_L c}{\sin(\varphi)} \left(1 - \frac{C_D}{C_L} \cot(\varphi)\right)(1-a)^2 r dr \\ &= \frac{8}{\lambda^2} \int\limits_{\lambda_h}^{\lambda} \frac{\Omega r B C_L c}{U_\infty 8\pi r \sin(\varphi)} \left(1 - \frac{C_D}{C_L} \cot(\varphi)\right)(1-a)^2 \lambda_r d\lambda_r \\ &= \frac{8}{\lambda^2} \int\limits_{\lambda_h}^{\lambda} \frac{B C_L c}{8\pi r \sin(\varphi)} \left(1 - \frac{C_D}{C_L} \cot(\varphi)\right)(1-a)^2 \lambda_r^2 d\lambda_r \end{aligned}$$

This final form is the one we use to calculate the power coefficient once the optimal or desired shape is determined. As already mentioned above, the drag force is not normally included in the shape optimization and therefore, the above model is greatly simplified.

$$C_P = \frac{8}{\lambda^2} \int\limits_{\lambda_h}^{\lambda} \frac{B C_L c}{8\pi r \sin(\varphi)} (1-a)^2 \lambda_r^2 d\lambda_r$$

The following equally valid form was derived from the momentum equations in Sect. 5.4, for a case with no drag.

$$C_p = \frac{8}{\lambda^2} \int_{\lambda_h}^{\lambda} a'(1-a)\lambda_r^3 d\lambda_r$$

De Vries [40] showed that these drag free equations can also be presented only as a function of relative wind angle and local tip speed ratio.

$$C_p = \frac{8}{\lambda^2} \int_{\lambda_h}^{\lambda} \{\sin^2(\varphi)(\cos(\varphi) - \lambda_r \sin(\varphi)) \\ (\sin(\varphi) + \lambda_r \cos(\varphi))\}\lambda_r^2 d\lambda_r$$

This form is extremely useful for our purposes, since the only free variable is the relative wind angle, φ, because the local speed ratio is pre-determined when the tip speed ratio is locked. The basic assumption of blade element theory is that there is no flow interaction between the elements in the radial direction. The maximum value of the power coefficient can therefore be obtained by maximizing only the term in the brackets { } for each radial position/element separately. These maximum values can be obtained by setting the differentiate of the { }-term to zero and then solving the corresponding optimal value of φ.

$$\frac{\partial}{\partial \varphi}\left\{\sin^2(\varphi)(\cos(\varphi) - \lambda_r \sin(\varphi))(\sin(\varphi) + \lambda_r \cos(\varphi))\right\} = 0$$

We obtain the following simple result, from which the optimal relative wind angle at each location can be calculated.

$$\lambda_r = \frac{\sin(\varphi)(2\cos(\varphi) - 1)}{(1 - \cos(\varphi))(2\cos(\varphi) + 1)} = \frac{\cos(3\varphi) + 1}{\sin(3\varphi)}$$

$$\varphi = \frac{2}{3}\tan^{-1}\left(\frac{1}{\lambda_r}\right)$$

As above for the ideal Betz rotor, the pitch angle of the blade can be obtained by subtracting the angle of attack from the relative wind angle, $\theta = \varphi - \alpha$, for each radial position. To obtain the solution for the optimal chord length the above result needs to be used and combined with the equations that relate local speed ratio and induction factors to the optimal chord. To begin with, we have the following four equations:

$$\tan(\varphi) = \frac{(1-a)}{\lambda_r(1+a')}$$

$$\lambda_r = \frac{\sin(\varphi)(2\cos(\varphi) - 1)}{(1 - \cos(\varphi))(2\cos(\varphi) + 1)}$$

$$\frac{a}{1-a} = \frac{BC_L c \cos(\varphi)}{\sin^2(\varphi) 8\pi r}$$

$$\frac{a'}{(1+a')} = \frac{\frac{BC_L c}{8\pi r}}{\cos(\varphi)}$$

Inserting these last two equations into the first one we get new equation for the local speed ratio.

$$\begin{aligned}\tan(\varphi) &= \frac{(1-a)}{\lambda_r(1+a')} = \frac{1}{\lambda_r}\frac{a}{a'}\frac{\frac{BC_L c}{8\pi r}}{\cos(\varphi)}\frac{1}{\frac{BC_L c}{8\pi r}\frac{\cos(\varphi)}{\sin^2(\varphi)}} \\ &= \frac{1}{\lambda_r}\frac{a}{a'}\tan^2(\varphi)\end{aligned}$$

$$\lambda_r = \frac{a}{a'}\tan(\varphi)$$

Then, inserting a and a' (solved from two last equations) into this new local speed ratio equation we have:

$$\lambda_r = \frac{\frac{1}{1+\frac{\sin^2(\varphi)}{\frac{BC_L c}{8\pi r}\cos(\varphi)}}}{\frac{1}{\frac{\cos(\varphi)}{\frac{BC_L c}{8\pi r}} - 1}}\tan(\varphi) = \frac{\sin(\varphi)\left(\cos(\varphi) - \frac{BC_L c}{8\pi r}\right)}{\frac{BC_L c}{8\pi r}\cos(\varphi) + \sin^2(\varphi)}$$

From which we can solve the optimal chord length.

$$c = \frac{8\pi r}{BC_L}(1 - \cos(\varphi))$$

Please compare this and optimal relative wind angle with the equations from the theoretical Betz model. Now that the optimal blade shape $(\theta(r), c(r))$ is determined, the power coefficient can be calculated. This can be numerically integrated using Euler's first approximation and equal radial increment size, $\Delta r = (R_t - R_h)/N$. Note that the first element starts from the hub, R_h, and all parameters are calculated at the middle of the elements.

$$C_P = \frac{B\Omega}{\pi R^2 U_\infty}\sum_{i=1}^{N}\frac{C_{L.i}c_i}{\sin(\varphi_i)}(1 - a_i)^2\left(1 - \frac{C_{D.i}}{C_{L.i}}\cot(\varphi_i)\right)r_i\Delta r$$

For greater accuracy, there are also other more accurate numerical integration schemes available. The model presented here is discussed in Example 5.4.

Example 5.4 Rotor Design with Wake Rotation and Drag Force Design a rotor of 1 m in diameter with 3 blades and for a wind speed of 10 m/s. Consider the wake rotation and the drag force, following the theory presented above. The hub diameter is 10 cm, the rotational speed of the turbine is 1000 rpm, the air density is 1.25 kg/ m^3 and the kinematic viscosity is 1.506×10^{-5} m^2/s. The air foil profile is NACA 2415. To simplify the calculation process approximate functions for C_L, C_D and α at the point of minimum C_D/C_L as a function of the Reynold's number are given by the following power law correlations, $\alpha = 53.47 \times Re^{-0.166}$, $C_L = 2.2543 \times Re^{-0.067}$ and $C_D = 8.4475 \times Re^{-0.506}$, respectively.

1. Use a tabular format and calculate optimal relative angles of wind, chord lengths, relative air velocities, induction factors a, a', Reynold's numbers and section pitch angles for 10 segments.
2. Calculate the power coefficient profile, total power coefficient and power production from the rotor and compare this with the value from the Betz turbine. Also discuss the manufacturing aspects of this more realistic design.

   1. Local tip speed ratios are the same as in Example 5.1. Relative angles of wind, relative wind speed, chord lengths and section pitches are calculated using the following equations (note that the values are defined at the centre of each blade element in the radial direction):

$$\varphi = \frac{2}{3}\tan^{-1}\left(\frac{1}{\lambda_r}\right)$$

$$c = \frac{8\pi r}{BC_L}(1 - \cos(\varphi))$$

$$a = \frac{1}{1 + \frac{\sin(\varphi)\tan(\varphi)8\pi r}{BC_L c}} = \frac{1}{1 + \frac{\sin^2(\varphi)8\pi r}{\cos(\varphi)BC_L c}}$$

$$a' = \frac{1 - 3a}{4a - 1}, \quad U = \frac{U_\infty(1 - a)}{\sin(\varphi)}$$

$$\theta = \varphi - \alpha$$

The equation for a' gives the maximum value for the local power coefficient and was derived in Sect. 5.4. Table 5.8 presents the first iteration results using $Re =$ 110,000, $\alpha = 7.5°$, $C_L = 1.031$ and $C_D = 0$ as initial guess values, obtained from the Betz rotor design.

Note that optimal angle of attack is a function of the Reynolds number and therefore varies with location. The total $C_P$is 0.501 and power production is 246 W. The following iteration steps are calculated by using the recently solved Reynolds number profile and updating α, C_L and $C_D$etc. Table 5.9 presents the converged solution after five iterations.

Table 5.8 Profiles of blade geometry parameters, $\lambda = 5.24$, $B = 3$, $\alpha = 7.5°$.

r, m	r/R, –	λ_r, –	φ, °	c, m	θ, °	a	a'	U, m/s	Re	α °	C_L, –	C_D, –	C_P, –
0.073	0.145	0.759	35.2	0.108	27.7	0.310	0.288	12.0	85,585	8.1	1.053	0.027	0.012
0.118	0.235	1.230	26.1	0.097	18.6	0.321	0.128	15.4	99,618	7.9	1.043	0.025	0.021
0.163	0.325	1.702	20.3	0.082	12.8	0.326	0.071	19.4	105,741	7.8	1.038	0.024	0.031
0.208	0.415	2.173	16.5	0.069	9.0	0.329	0.045	23.7	108,813	7.8	1.036	0.024	0.039
0.253	0.505	2.644	13.8	0.059	6.3	0.330	0.031	28.1	110,537	7.8	1.035	0.024	0.048
0.298	0.595	3.115	11.9	0.052	4.4	0.331	0.022	32.5	111,590	7.8	1.035	0.024	0.056
0.343	0.685	3.587	10.4	0.046	2.9	0.331	0.017	37.1	112,277	7.8	1.034	0.024	0.063
0.388	0.775	4.058	9.2	0.041	1.7	0.332	0.013	41.7	112,748	7.8	1.034	0.023	0.070
0.433	0.865	4.529	8.3	0.037	0.8	0.332	0.011	46.3	113,085	7.8	1.034	0.023	0.077
0.478	0.955	5.000	7.5	0.034	0.0	0.332	0.009	50.9	113,333	7.7	1.034	0.023	0.084

Table 5.9 Converged profiles of blade geometry parameters

r, m	r/R, –	λ_r, –	φ, °	c, m	θ, °	a	a'	U, m/s	Re	α °	C_L, –	C_D, –	C_P, –
0.073	0.145	0.759	35.2	0.105	27.0	0.310	0.288	12.0	83,647	8.1	1.055	0.027	0.012
0.118	0.235	1.230	26.1	0.096	18.1	0.321	0.128	15.4	98,429	7.9	1.043	0.025	0.021
0.163	0.325	1.702	20.3	0.081	12.4	0.326	0.071	19.4	104,927	7.8	1.039	0.024	0.030
0.208	0.415	2.173	16.5	0.069	8.7	0.329	0.045	23.7	108,198	7.8	1.037	0.024	0.039
0.253	0.505	2.644	13.8	0.059	6.0	0.330	0.031	28.1	110,037	7.8	1.036	0.024	0.047
0.298	0.595	3.115	11.9	0.051	4.1	0.331	0.022	32.5	111,161	7.8	1.035	0.024	0.055
0.343	0.685	3.587	10.4	0.045	2.6	0.331	0.017	37.1	111,894	7.8	1.035	0.024	0.063
0.388	0.775	4.058	9.2	0.041	1.5	0.332	0.013	41.7	112,398	7.8	1.034	0.023	0.070
0.433	0.865	4.529	8.3	0.037	0.5	0.332	0.011	46.3	112,758	7.8	1.034	0.023	0.077
0.478	0.955	5.000	7.5	0.033	-0.2	0.332	0.009	50.9	113,023	7.7	1.034	0.023	0.084

The total converged C_P is 0.499 and the power production is 245 W. The chord is 11 cm at the root and 3.3 cm at the tip. The total twist angle is $27° + 0.2° = 27.2°$. As a difference to Betz's rotor, the Reynolds number varies as a function of radial position. In addition, especially at the root of the blade, the angular induction factor is quite high (~0.3) but decreases at the tip of the blade. In manufacturing this design would require fewer raw materials and would be much easier to build than the Betz rotor. As a comparison Fig. 5.18 presents the blade profiles. The blades are shown as looking from the front and the airfoils are shown as they look when observed from the tip towards the hub, with the wind coming from the left. The direction of rotation is anticlockwise.

However, the designs and geometries presented in Fig. 5.18 are not possible in large industrial turbines. The mass and cost of the blades would just become large. And due to the large mass, the forces coming from the large rotating mass would exceed the strength of the material. Therefore, a practical design must be a

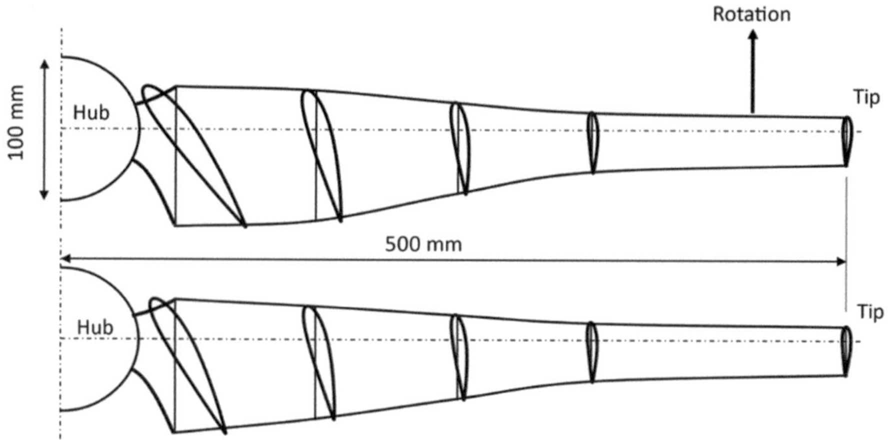

Fig. 5.18 Blade profiles from examples 5.2 and 5.3. Top: Betz rotor, bottom: optimized rotor

compromise, where the chord length is smaller than predicted from the above theory. Below, a 15 MW turbine is analyzed. This is the maximum size range available in 2022 for offshore use. We use public data from current commercial turbines and make our own simplified design proposal. Note that this is a very rough design, and much more refined tools such as CFD are required to optimize the design.

Example 5.5 Design of Industrial 15 MW Turbine Sketch a rough design of an industrial size 15 MW 3-blade turbine rotor. Use the VESTAS V236 −15 MW model as an inspiration. It has a rotor diameter of 236 m, blade length of 115.5 m and a rated power of 15 MW. The hub diameter is ~ 5 m. At the design stage we assume that the wind speed is 10 m/s and the rotational speed is 6.5 rpm. The air density is 1.25 kg/m^3 and the kinematic viscosity is 1.506×10^{-5} m^2/s. The airfoil profile used is the high camber NACA 6409. The functions for C_L, and C_D are as follows (we assume that these are not functions of the Reynolds number for simplicity and that there is no high Re data available):

$$C_L = 0.00007175 \times \alpha^3 - 0.006199 \times \alpha^2 + 0.1372 \times \alpha + 0.6503$$

$$C_D = 0.00002264 \times \alpha^3 - 0.0001299 \times \alpha^2 + 0.0003314 \times \alpha + 0.007$$

1. Determine first the optimal blade shape design $(\theta(r), c(r))$. Use a tabular format and determine *optimal* relative angles of wind, chord lengths and section pitch angles. Use 20 elements,

2. Using the optimal design obtained above, propose a more realistic construction. You can then use a linear chord profile that meets the optimal design at locations $r/R = 0.7$ and 0.9.
3. Using the linear chord profile calculate the power coefficient at the design point and for tip speed ratios between 2 and 12.

1. We first need to calculate the design tip speed ratio; it is expressed simply as. The local speed ratios can be obtained by multiplying this with r/R. The optimal relative wind angle and chord are calculated from the same equations, as with Example 5.2.

$$\varphi(r) = \frac{2}{3}\tan^{-1}\left(\frac{1}{\lambda_r}\right)$$

$$c_o(r) = \frac{8\pi r}{BC_L}(1 - \cos(\varphi))$$

The lift coefficient needs to be selected from a point where the C_D/C_L has its minimum. Based on the above given correlations, the minimum occurs at $\alpha = 4.6°$, $C_L = 1.16$ and $C_D = 0.008$. Please note that if C_D and C_L were functions of the Reynolds number, then an iteration would be required. From the solved φ, the blade pitch can be calculated, $\theta(r) = \varphi(r) - \alpha(r)$. We now have the optimal blade profile available as $(\theta(r/R), c(r/R)$. The results are given in Table 5.10.

2. As shown in the previous example, this optimal design has too large a chord at the root. We therefore need to use the practical design; a linear chord profile that meets the optimal profile at $r/R = 0.7$ and 0.9 (please note that the closest values to these were used below). The optimal pitch profile is used as such. The equation for the chord profile is therefore defined as follows (the results are given in Table 5.10): Please note that the blade starts at 2.5 m and the radial increment is 5.775 m.

$$c\left(\frac{r}{R}\right) = c\left(\frac{r}{R} = 0.682\right) + \frac{c_o(0.927) - c_o(0.682)}{0.927 - 0.682}\left(\frac{r}{R} - 0.682\right)$$

3. Finally, we determine the performance of the designed rotor. At first, we study the performance at the design point, $\lambda = 8$. We need to first set the initial guess values for induction factors, $a = 0.333$ and $a' = 0$. From these, we can start the iteration. This procedure is explained in Table 5.11 in detail. The definition of the axial thrust coefficient can be obtained from the above BEM equation for the normal force.

Table 5.12 presents the converged profiles. See Fig. 5.19 for the graphical presentation of the blade shape.

Table 5.10 Optimal profile of blade geometry and practical design, $\lambda = 8$, $B = 3$, $\alpha = 4.6°$. The points where the optimal profile and practical shape profiles cross are given in bold

r, m	r/R, –	λ_r, –	φ_o, °	θ, °	c_o, m	c, m
5.39	0.046	0.4	46.6	42.0	12.162	7.010
11.16	0.095	0.8	35.2	30.6	14.726	6.794
16.94	0.144	1.2	27.3	22.7	13.617	6.579
22.71	0.192	1.5	21.9	17.3	11.870	6.363
28.49	0.241	1.9	18.2	13.6	10.278	6.147
34.26	0.290	2.3	15.5	10.9	8.968	5.932
40.04	0.339	2.7	13.4	8.8	7.911	5.716
45.81	0.388	3.1	11.9	7.3	7.055	5.500
51.59	0.437	3.5	10.6	6.0	6.355	5.285
57.36	0.486	3.9	9.6	5.0	5.774	5.069
63.14	0.535	4.3	8.7	4.1	5.286	4.854
68.91	0.584	4.7	8.0	3.4	4.871	4.638
74.69	0.633	5.1	7.4	2.8	4.515	4.422
80.46	0.682	5.5	6.9	2.3	4.207	4.207
86.24	0.731	5.9	6.4	1.8	3.937	3.991
92.01	0.780	6.3	6.0	1.4	3.698	3.775
97.79	0.829	6.7	5.7	1.1	3.487	3.560
103.56	0.878	7.0	5.4	0.8	3.298	3.344
109.34	0.927	7.4	5.1	0.5	3.128	3.128
115.11	0.976	7.8	4.8	0.2	2.975	2.913

It took ~ 20 iterations to reach convergence of the induction factors. The total power coefficient, C_P, is the sum of last column terms and equal to 0.501. The power is equal to:

$$P = C_P \frac{\pi D^2}{8} \rho U_\infty^3 = 0.501 \times \frac{\pi \times 118^2}{8} \times 1.25 \times 10^3 = 14 \text{ MW}.$$

To achieve the designed power of 15 MW, the rated wind speed must increase slightly from 10 m/s up to 10.3 m/s. We can see that the whole rotor operates at windmill state since all $C_T < 0.96$. Using the above model, the power coefficient curve can be produced using the same method and letting the tip speed ratio vary from 2 to 12. This was done by keeping the wind speed at a constant 10 m/s and varying the rpm from 1.75 to 13.5. It should be noticed that already at $\lambda = 8$, the tip speed is 80 m/s. At rpm 13.5, the tip speed is 167 m/s. This would probably be far too much for the mechanical strength of the blades. Figure 5.19 shows the calculated power coefficient curve.

Table 5.11 Iteration of the rotor operation (the geometry is available)

0	Initialize guess values, $a = 0.333$ and $a' = 0$, calculate local speed ratios	
1	Set $\hat{a} = a$ and $\hat{a}' = a'$	
2	Calculate relative wind angle, $\tan(\varphi) = \frac{(1-a)}{\lambda_r(1+a')}$	
3	Calculate angle of attack, $\alpha(r) = \varphi(r) - \theta(r)$, $\theta(r)$ as already determined and locally constant during iteration	
4	Determine $C_L(\alpha, Re)$ and $C_D(\alpha, Re)$. Please note that these are most probably strong functions of the Reynolds number	
5	Calculate local thrust coefficients to see whether the turbine operates at the windmill state or at the turbulent wake state. σ is the solidity of the rotor $C_T = \frac{\sigma(1-a)^2(C_L\cos(\varphi)+C_D\sin(\varphi))}{\sin^2(\varphi)}$ $\sigma = \frac{Bc}{2\pi r}$	
6	$C_T < 0.96$	$C_T > 0.96$, turbulent wake state
7	Calculate axial induction factor: $a = \frac{1}{1+\frac{4\sin^2(\varphi)}{\cos(\varphi)\sigma C_L}}$	$a = \left(0.143 + \sqrt{0.0203 - 0.6427(0.889 - C_T)}\right)$
8	Calculate tangential induction factor: $a' = \frac{1}{\frac{4\cos(\varphi)}{\sigma C_L} - 1} = \frac{a}{\lambda_r}\tan(\varphi)$	
9	To make the convergence more stable, use under-relaxation, and calculate new guess values for induction factors from the new values (7/8. and 1.) $a' = \left((a' + \hat{a}')\right)/2$	
10	If induction factors are sufficiently close to previously obtained values, continue to final step 11. If not, return to point 1 and repeat until convergence is reached	
11	Calculate rotor performance, C_p and P $C_P = \frac{B\Omega}{\pi R^2 U_\infty}\int_{\lambda_h}^{\lambda}\frac{BC_L c}{8\pi r\sin(\varphi)}\left(1 - \frac{C_D}{C_L}\cot(\varphi)\right)(1-a)^2\lambda_r^2 d\lambda_r$ $P = C_P\frac{\pi D^2}{8}\rho U_\infty^3$	

As Fig. 5.19 shows, the maximum power coefficient is obtained at the design point. This is logical since the pitch was optimized for this condition. Below the optimum tip speed ratio of 8.6, the turbine operates fully under windmill state. When the tip speed ratio increases from the optimal value—here ~ 8.6—the fraction of the rotor disk operating under the turbulent wake state increases. On the top right and below in Fig. 5.19 the blade shape is presented. The optimal blade shape is significantly larger than the practical design and it would not be possible to manufacture this in practice. For a more detailed blade design CFD (Computational Fluid Dynamics) and more refined tools should be used. The design shown in Fig. 5.19 (bottom) would be mechanically too thin and weak at the hub end and should be made more rigid using a thicker airfoil. This design also looks quite clumsy and massive in comparison with the blades in the latest turbines.

Table 5.12 Converged profiles parameters, $\lambda = 8$ $B = 3$

r/R, –	a	a'	φ, °	α, °	C_L, –	C_D, –	C_T, –	C_P, –
0.046	0.178	0.657	53.5	11.5	1.518	0.028	0.601	0.001
0.095	0.146	0.183	43.5	12.9	1.543	0.039	0.512	0.003
0.144	0.158	0.092	33.8	11.1	1.507	0.026	0.539	0.006
0.192	0.176	0.057	26.7	9.4	1.452	0.017	0.585	0.008
0.241	0.196	0.040	21.7	8.1	1.396	0.013	0.632	0.011
0.290	0.215	0.030	18.1	7.2	1.345	0.011	0.677	0.014
0.339	0.234	0.024	15.4	6.5	1.302	0.010	0.718	0.017
0.388	0.251	0.019	13.3	6.0	1.266	0.009	0.754	0.020
0.437	0.268	0.016	11.6	5.6	1.237	0.009	0.786	0.023
0.486	0.284	0.013	10.3	5.3	1.212	0.008	0.815	0.026
0.535	0.299	0.011	9.2	5.0	1.193	0.008	0.839	0.029
0.584	0.312	0.010	8.3	4.8	1.178	0.008	0.859	0.032
0.633	0.323	0.008	7.5	4.7	1.166	0.008	0.875	0.034
0.682	0.331	0.007	6.9	4.6	1.158	0.008	0.887	0.037
0.731	0.338	0.006	6.4	4.6	1.153	0.008	0.895	0.039
0.780	0.341	0.006	6.0	4.5	1.151	0.008	0.900	0.042
0.829	0.342	0.005	5.6	4.5	1.151	0.008	0.900	0.043
0.878	0.338	0.004	5.3	4.6	1.153	0.008	0.896	0.044
0.927	0.332	0.004	5.1	4.6	1.158	0.008	0.887	0.041
0.976	0.322	0.004	4.9	4.7	1.164	0.008	0.873	0.029

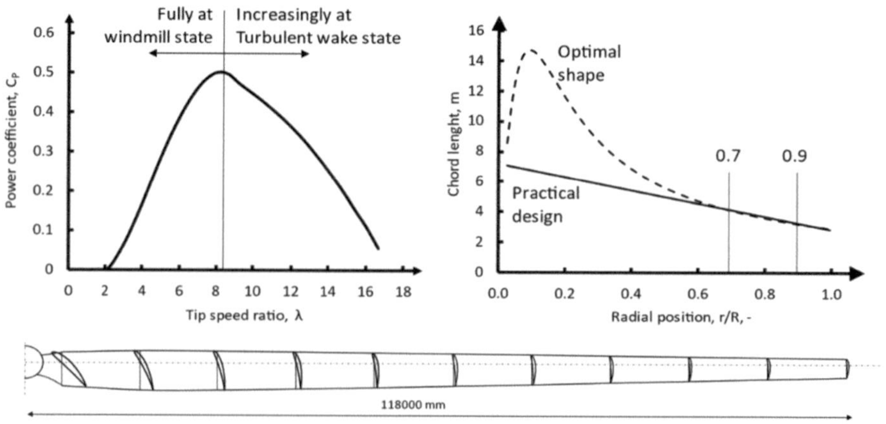

Fig. 5.19 Power coefficient of the 15 MW turbine (top left) and the blade shape (top right). Below the blade profile, front view and airfoils are drawn, as seen from the tip towards the hub

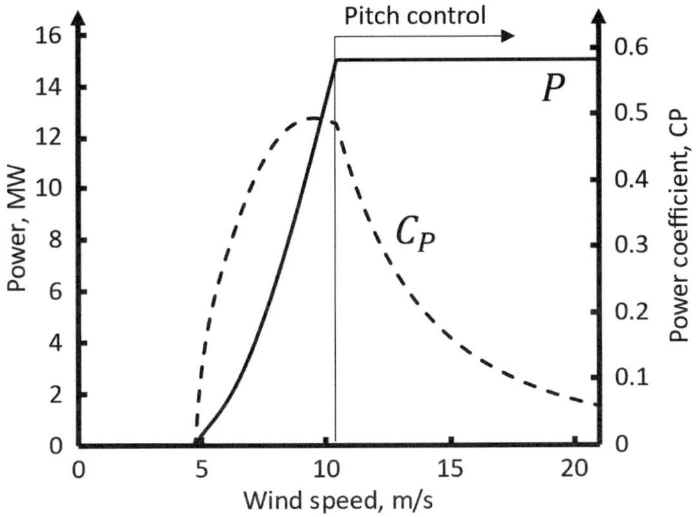

Fig. 5.20 Power curve of the fixed speed variable pitch turbine, 15 MW, with constant 6.5 rpm

It is important to understand that the maximum power is not typically reached at the maximum power coefficient. The maximum power coefficient is reached earlier, to leave some room for limiting the maximum power to the rated constant power as the wind speed increases. It is also interesting to plot the power curve of this turbine. For example, if we use a fixed speed turbine rotating at 6.5 rpm, we can plot the power curve as a function of the wind speed, as shown in Fig. 5.20. The point whereupon the power production is limited, by pitching the blades to keep the power constant, is shown. Notice that all modern turbines utilize the variable speed drive system, so the example presented here is only used as a simplified case. Different power control methods are discussed in Sect. 5.6.

5.5.5 Number of Blades

We have now derived and discussed the models for power generation from the rotor, with B blades rotating with an angular velocity of Ω. The models assume that the blades do not interact; that they act as individual blades in the flow. However, as the number of blades increase, the flow gets "squeezed" between the blades (Fig. 5.21).

The number of blades is one design variable that is affected by the application for which the turbine is to be used. Since American multiblade turbines are used for water pumping, the main design requirement is that they should start rotating and operate at low wind speeds with high torque, to be able to pump water for irrigation. By contrast, turbines used for energy production need to have maximal power and efficiency. Therefore, the rotational speed must be maximized, and the number of

Fig. 5.21 Examples of a one-bladed, a modern 3-bladed and an *American multi-blade* wind turbine (Permissions from left: photo by Paul Gipe, Riva Calzoni MS30 AS1 in the vicinity of Collarmele east of Avezzano in the central Appenines, 250 kW, 30 m rotor, 1998. Middle: CC Attribution-Share Alike 3.0: Andrés Franchi Ugart. Right: Creative Commons Attribution-Share Alike 2.5 Generic)

blades kept to a minimum. Based on [4] for high torque turbines a tip speed ratio 1…3 is recommended with 8…24 ($\lambda = 1$), 6…12 ($\lambda = 2$) and 3…6 ($\lambda = 3$) blades. For high-speed power production turbines 1–3 blades and tip speed ratios larger than 4 can be used.

We assume that the torque generated by a single blade is constant, M_b, with every blade acting without affecting the other blades, and that the torque itself is not a function of the rotational speed. Then, as the number of blades is increased, the rotor generates more torque, and the equation for the power generation is as follows (0 refers to the reference case):

$$P_0 = B_0 M_b \Omega_0 = B M_b \Omega = P$$

If the number of identical blades is increased the same amount of power with a smaller speed of rotation can be generated. Conversely, the smaller the number of blades we have, the higher the speed of rotation must be to generate the same amount of power. As an example, a one-blade rotor must rotate 3 times faster compared with a 3-blade rotor. Of course, this analysis is very rough, but it gives an idea of the relationship between the blades and rotational speed and their impact on power production.

The rotor can be designed with any number of blades but as mentioned, a rotor with fewer blades must rotate faster. The noise generation is proportional to the 5th power of the tip speed [4]. This limits the tip speed of onshore turbines to 60–80 m/s. Higher speeds and louder noise levels are usually allowed at sea, although the impact of noise on wildlife is not well-studied. Higher rotational speeds increase centrifugal forces and mechanical strength limits the maximum allowed speed.

5.6 Power Control of Wind Turbines

The aim of wind turbine control systems is to maximize the production of energy in a safe way while minimizing the excessive wear of turbine components, thereby assuring maximal system lifetime. In addition, the aim should also be to minimize harm to people, nature, and wildlife. Bird and bat collisions with blades and power lines are common. There are advanced camera and radar -based systems available that can stop the turbine if there is a risk of collision with birds [41]. Unfortunately, these systems are still not in wide use.

There are several ways to control the power production of a wind turbine. These include the aerodynamic control, the generator torque control, the brake torque control, and the yaw orientation control [4]. In the aerodynamic control approach the variables that can be changed are the speed of rotation and the rotor power coefficient (which is defined by the blade design and the pitch). The principle of generator torque control is that the load/torque created by the generator is electrically adjusted. Grid connected synchronous generators do not allow sudden load variations. However, there is constant variation in the wind speed, gusts etc. that can create significant torque and power peaks. This must be handled somehow. One solution is to use an induction generator that allows speed variations of 1–2% resulting in a damped and softer response and lower torque and power peaks. Today most turbine systems rely on the variable speed drive system. This means that even as the rotational speed of the rotor varies, the frequency of the output matches the grid frequency. The benefit is that the speed of rotation can be adjusted, and the system can easily follow the maximum performance point of the system. Finally, the brake torque system is required to allow safe stopping of the rotor during shutdown, maintenance and in the case of an emergency.

5.6.1 Fixed Speed, Fixed Pitch (FSFP)

The first industrial sized wind turbines were rotating at a constant speed. The rotor had rigidly fixed blades that would stall at the rated wind speed and at the rated power, dropping the power production non-linearly after this. The rotational speed of the rotor on the primary axis was increased by a gearbox to the secondary axis of the generator rotating at 50 or 60 1/s, to match the constant grid frequency of 50 or 60 Hz. An induction generator was usually used with this turbine type.

This system was very simple and easy to control, there were practically no moving parts in the control system. The nacelle had a yaw system to turn the rotor towards the wind and a system for wind speed measurement that indicated the cut-in wind speed at around 5 m/s and the cut-out wind speed—for example at 25 m/s to stop the rotor and prevent breakage of the components.

These turbines were only able to operate when attached to the grid with a fixed frequency, and they also required power from the grid to run the auxiliary systems.

Since the rotational speed was locked to the grid frequency, the power was increased by increasing the aerodynamic torque from the rotor at increasing wind speeds. When the wind speed got too strong, the rotor stalled, and the torque and power production dropped. As already shown in Fig. 5.14 when the angle of attack increases, the lift force of the airfoil increases linearly until the flow separates from the upper surface and a stall takes place, significantly reducing the lift force and increasing the drag force. This phenomenon, which is fatal for aviation applications, is very useful for controlling wind turbines.

Let's assume that there is a *negligible wake rotation* after the rotor, so $a' = 0$. In this case the tangential velocity at a selected point of the blade, at a distance of r from the axis of rotation, is equal to $U_t = \Omega r = 2\pi nr$ (where n is the rotational speed (1/s)). When the velocity of the axial wind at the rotor, U, increases, the angle of the relative wind, φ, increases (see Fig. 5.22). And if the pitch angle, θ, is locally constant, as it is for fixed speed turbines, this then means that the angle of attack, α, will increase and that this location will reach the point of stall eventually.

$$\varphi = \tan^{-1}\left(\frac{U}{2\pi nr}\right)$$

$$\alpha = \tan^{-1}\left(\frac{U}{2\pi nr}\right) - \theta$$

As an example, if the local pitch angle (turned away from the plane of rotation) of the blade is $\theta = 2°$, the rotational speed is $n = 0.3$ 1/s, the local position is $r =$ 25 m, and U increases from 5 to 15 m/s, the angle of attack increases as follows:

$$\alpha = \tan^{-1}\left(\frac{5\frac{m}{s}}{2\pi \times 0.3\frac{1}{s} \times 25m}\right) - 2° = 4.1°$$

$$\alpha' = \tan^{-1}\left(\frac{15\frac{\mathrm{m}}{\mathrm{s}}}{2\pi \times 0.3\frac{1}{\mathrm{s}} \times 25\ \mathrm{m}}\right) - 2° = 15.7°$$

For the airfoil NACA 2415, this would already be well above the stall point of 11.5°. Figure 5.22 presents this same situation in terms of the velocity triangles.

In Fig. 5.22 the axial wind speed increases by a factor of 3 from left to right. The rotational speed and tangential velocity component are constant. Note that for simplicity no wake rotation is assumed. This means that the magnitude of the wind speed experienced by the leading edge in a tangential direction is the same as the velocity of the blade, but in the opposite direction.

In the above system, the power is simply calculated from the familiar power equation discussed above.

$$P = C_p(\lambda, \beta)\frac{1}{2}A\rho U_\infty^3$$

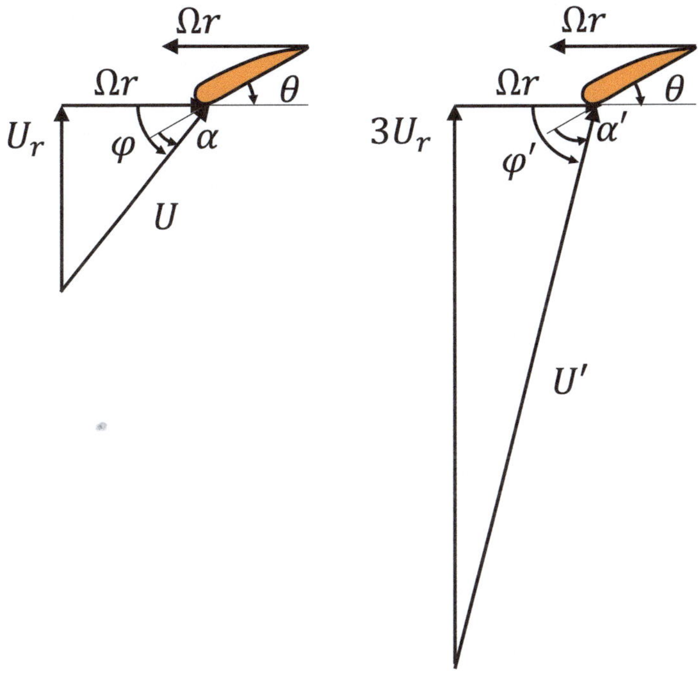

Fig. 5.22 Velocity triangles for a blade cross section of a constant speed turbine, at a distance r from the axis of rotation

A general form of rotor power coefficient can be described by a correlation based on measurements. One example of this is shown below [42].

$$C_p(\lambda, \beta) = a_1\left(\frac{a_2}{a} - a_3\beta - a_4\right)exp\left(\frac{a_5}{a}\right)$$

$$\frac{1}{a} = \frac{1}{\lambda + a_6\beta} + \frac{a_7}{\beta^3 + 1}$$

Coefficients are turbine specific and here values are $a_1 = 0.22$, $a_2 = 116$, $a_3 = 0.4$, $a_4 = 5$, $a_5 = -12.5$, $a_6 = 0.08$ and $a_7 = 0.035$. β is the blade angle from the maximum power coefficient C_P point and optimal tip speed ratio condition, where $\beta = 0$. When β is changed, local pitch angles change equally; $\theta = \theta_0 + \beta$, θ_0 refers to pitch at the tip of the blade. Table 5.13 summarizes the ranges of operation variables for a fixed speed, fixed pitch turbine.

Table 5.13 Control parameters and ranges for a FSFP turbine, n_0 is the fixed speed of rotation

Operations range	Rotational speed	Power coefficient
$U_\infty < U_{cut-in}$	$n = 0$	$C_P = 0$
$U_{cut-in} < U_\infty < U_{cut-out}$	$n = n_0$	$C_P = C_P(\lambda, \beta = 0)$
$U_\infty > U_{cut-out}$	$n = 0$	$C_P = 0$

5.6.2 *Fixed Speed, Variable Pitch (FSVP)*

The next step in developing constant speed turbines is the addition of a blade pitch system to the constant speed rotor, to turn the blades and to have a more sophisticated way by which the power can be limited. Instead of letting the blades stall at the rated power (which is smaller than the peak power of FSFP system), the blades were turned to limit power production and to keep it constant all the way to the cut-out wind speed. In this case a part of the power potential is left unused, since the power is limited to a constant value, instead of letting it peak and then drop due to stall. As you can imagine, this step significantly improves the controllability of the system.

The objective of the FSVP turbine is to keep the power at rated value, even when the wind speed increases above that value. While the wind speed further increases from rated value, the power is limited to the rated power by pitching the blades so that they produce less torque. Or if presented as an objective function to be iterated to zero, $f(\beta) = 0$, we need to find the value of β that satisfies the below equation.

$$C_p\left(\frac{\Omega R}{U_\infty}, \beta\right)\frac{1}{2}\rho A^2 U_\infty^3 - P_r = f(\beta)$$

This must be numerically solved. Care must the taken with the solution since at some tip speed ratio and pitch angle combinations there are several roots, but only the positive ones and the ones that give a continuous pitching profile are allowed. In addition, there is a jump in the value of β at lower tip speed ratios, when there is only a single solution available (see Fig. 5.25) (Table 5.14).

Table 5.14 Control parameters and ranges for a FSVP turbine. n_0 (1/s) is the fixed speed of rotation

Range and conditions	Rotational speed	Power coefficient
$U_\infty < U_{cut-in}$	$n = 0$	$C_P = 0$
$U_{cut-in} \le U_\infty \le U_r$	$n = n_0$	$C_P = C_P(\lambda, \beta = 0)$
$U_\infty > U_r$	$n = n_0$	$C_P = \frac{P_r}{\frac{1}{2}\rho A U_\infty^3} = C_P(\lambda, \beta)$ Solve required blade pitch β numerically
$U_\infty > U_{cut-out}$	$n = 0$	$C_P = 0$

5.6.3 Variable Speed, Variable Pitch (VSVP)

Most modern wind turbines today use a system that has variable rotational speed and pitch control. The main operational basis for this control type is that there is only a single value of tip speed ratio, with a certain pitch setting of the blades, that provides the maximum power coefficient. We mark these as λ^* and C_p^*, respectively. By using these, and by accurately measuring the wind speed, and assuming that the rotational speed can be maintained to match the optimal tip-speed ratio λ^* we can re-write the power equation as follows:

$$P = \left(\frac{C_p^*}{2\lambda^{*3}} A\rho R^3\right)\Omega^3 = a\Omega^3$$

This is the power that can be obtained by keeping the tip speed ratio constant and letting the rotational speed of the turbine increase directly in proportion with the wind speed. Please note that the value of a is constant for a fixed blade pitch setting. We can see that by following this 3rd order curve, the maximum power can always be obtained for a certain wind speed. The load power is electrically controlled to match the production curve. For a certain value of rotational speed, the load is adjusted so that the above equation is satisfied.

Variable speed systems were made possible by the development of power converter technology, and they allow the coupling of a varying frequency generator into a fixed frequency electrical grid. This means that the optimal tip speed ratio can be always followed when the wind speed changes, so as to have a maximum power coefficient available under all conditions. In this system, after the wind speed gets higher than the cut-in speed, the rotor is released, and an electrically controlled load adjusts the rpm to reach the optimal power coefficient. When the wind speed increases, the rpm is increases respectively to follow the constant power coefficient and optimal tip speed ratio, $C_P(\lambda^*)$. At some point, the maximum rotational speed is reached and then it must be electrically limited so as not to cause damage to systems. The rotational speed is constant above this point. The blade pitch angle is kept constant until the rated wind speed is reached. Then, the blades are turned to decrease the torque generated at the maximum rotational speed. When the wind speed increases above the cut-out speed, the rotor is stopped, and blades turned to minimum torque position. Table 5.15 summarizes the ranges and conditions for VSVP turbines. Figs. 5.23, 5.24 and 5.25 present examples of profiles of operational parameters in all control approaches.

Figure 5.25 Note the sudden jump in the blade pitch angle of FSVP turbine. The variable speed turbine does not have this sudden jump because it operates closer to the optimal tip speed ratio.

Table 5.15 Control parameters and ranges for a VSVP turbine

Range and conditions	Rotational speed	Power coefficient
$U_\infty < U_{cut-in}$	$n = 0$	$C_P = 0$
$U_{cut-in} < U_\infty < U_r, n < n_{\max}$	$n = \frac{\lambda^* U_\infty}{2\pi R}$	$C_P = C_P^*(\lambda^*, \beta = 0)$
$U_{cut-in} < U_\infty < U_r, n > n_{\max}$	$n = n_{\max}$	$C_P = C_P(\lambda, \beta = 0)$
$U_r < U < U_{cut-out}$	$n = n_{\max}$	$C_P = \frac{P_r}{\frac{1}{2}\rho A U_\infty^3} = C_P(\lambda, \beta)$ Solve the required blade pitch β numerically
$U_\infty > U_{cut-out}$	$n = 0$	$C_P = 0$

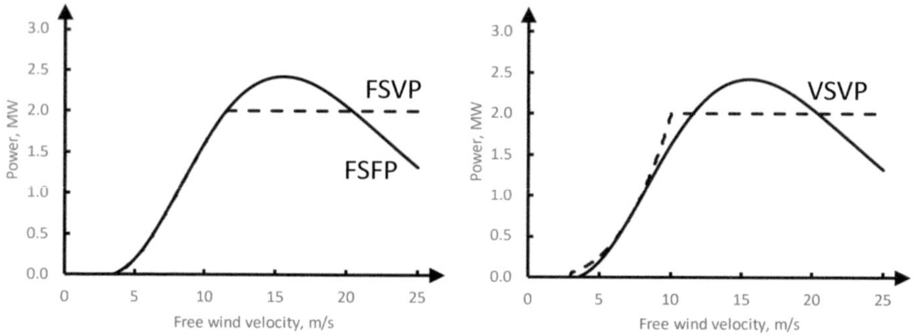

Fig. 5.23 Power production for fixed speed, fixed pitch (FSFP, solid line), fixed speed, variable pitch (FSVP, dashed line) and variable speed, variable pitch (VSVP, dashed line) turbines

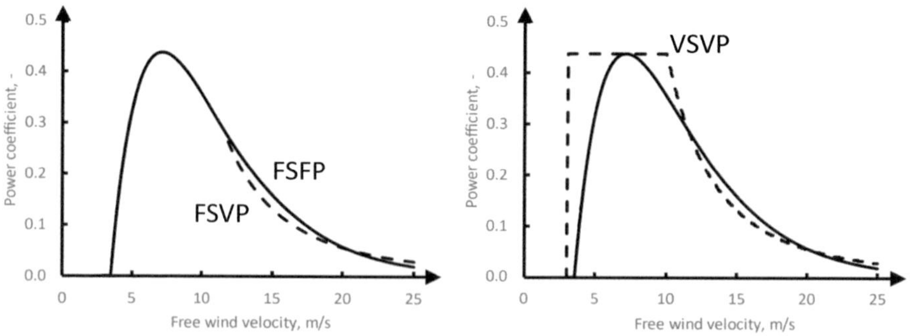

Fig. 5.24 Power coefficients for fixed speed, fixed pitch (FSFP, solid line), fixed speed, variable pitch (FSVP, dashed line) and variable speed, variable pitch (VSVP, dashed line) turbines

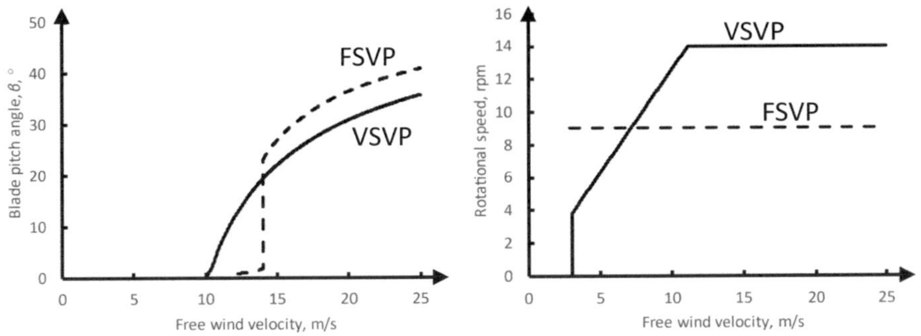

Fig. 5.25 Blade pitch angle and rotational speed for fixed speed, fixed pitch (FSFP, solid line), fixed speed, variable pitch (FSVP, dashed line) and variable speed, variable pitch (VSVP, solid line) turbines

5.7 Wind Power Plant Design and Development

Planning, constructing, and decommissioning a wind power plant requires broad and multi-disciplinary understanding. The objective of this section is to give general guidelines on designing wind power plants. Given the complexity of the topic we can only scratch at the surface here; a whole book could be dedicated to this topic. With the information provided here we want to highlight the importance of forming multidisciplinary teams (see the T-profile, Chap. 1, Fig. 1.12) for developing wind power plant projects. Local knowledge of the area, understanding of the natural environment and collaboration with local communities are prerequisites for any large-scale wind energy project. Wind power should be seen as embedded within the ecosystem and human society, and designed so that minimal harm occurs during the construction and operation.

This section discusses what it means in practice if 500 MW wind power plants are built in different locations, both onshore and offshore. Use of land- and seascape is discussed. The focus is on site selection, wind resource evaluation, impacts on nature (and ways to avoid, minimize, and compensate these impacts), social acceptability (such as sound and flickering) and local socio-economies including compatibility with other forms of land use and livelihood. Finally, we briefly summarize the steps in wind power plant development in practice.

5.7.1 Site Selection, Turbine Siting and Land or Sea Area Requirements

There are several wind atlases, which are used in wind power plant design and site selection. A wind atlas is a database of long-term wind climates for given locations. They are typically created by running numerical weather prediction models using

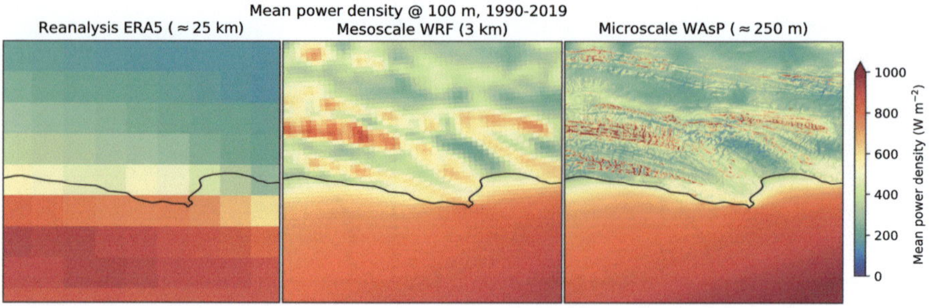

Fig. 5.26 Power density over the same area in a coastal region in South Africa from a global model (ERA5, left), a limited area model (WRF, middle) and a microscale model (WAsP, right)

weather data from the past. These models solve prognostic equations for momentum, temperature, and moisture on a specified grid. Using past observations from near-surface weather stations, radiosondes and satellite data, 'reanalysis' products can be generated. Examples include ERA5 [43] and CFSR [44]. Due to their coarse model grids, these products miss important details that are relevant for estimating the wind power available in a specific area. Figure 5.26 (left) shows an example of such a global model. 'Mesoscale models' solve the same equations as the global models used for reanalysis, but are instead run over a limited area to obtain a wind climate with higher model resolution. Using mesoscale models typically gives a better (higher) wind resource, because they can better represent local mesoscale speed-up effects (see Fig. 5.26 middle). The higher red areas onshore (that are not visible in the reanalysis data) are generated by greater resolution of terrain undulations and stand out clearly.

Even mesoscale models cannot fully resolve the effect of terrain on the flow and, for onshore wind power plant design in particular, a microscale model is often needed (see Fig. 5.26 right). An example of a linearized microscale flow model that has frequently been applied is the Wind Atlas Analysis and Application Software (WAsP), which combines submodules for microscale flow effects; for example, the speed-up effects associated with orographic and roughness length change. With such models individual hills and their associated speed-up effects become visible, helping to identify the locations which are ideal for siting wind turbines.

To illustrate the effects of using time-series from different type of models to calculate the wind power resource we sort the pixels in Fig. 5.26 from the windiest to the least windy locations. It can be clearly (Fig. 5.27) seen that the ERA5 model illustrates lower wind resource in this area for both the least windy and most windy areas. When comparing a microscale model with a mesoscale model the difference between the not-so-windy sites is quite small, but using a high-resolution model we are able to find the best locations for siting wind turbines within our area of interest.

So, to give us useful information about the annual energy capacity factor, C_f, in different wind climates, different surface conditions, and variable bathymetry for example, we need a model setup that includes all the steps from global models down

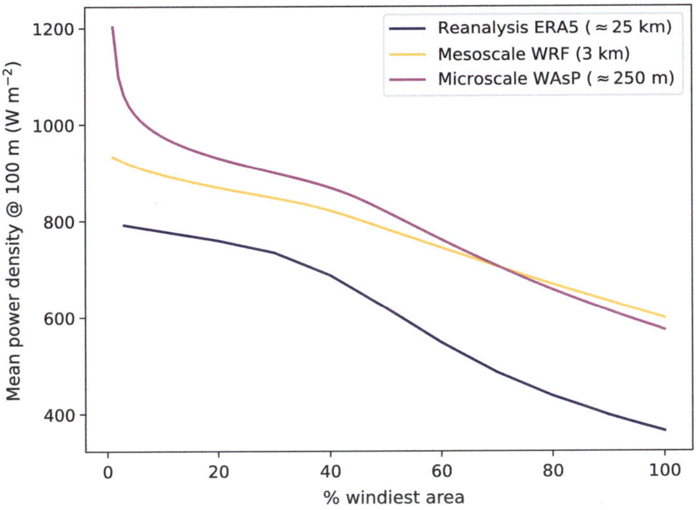

Fig. 5.27 Power densities over the same area in a coastal region in South Africa from a global model (ERA5), a limited area model (WRF) and a microscale model (WAsP)

to microscale models. Figure 5.28 gives an example of the output of such a model chain from the Global Wind Atlas (https://globalwindatlas.info/).

Wind atlases provide a detailed site analysis that can be used in wind power plant planning. Placing the cursor on the map brings up local values of the studied parameter. For example, Fig. 5.28 shows the energy capacity factor according to the International Electrotechnical Commission (IEC) Class III. These IEC wind classes indicate the limits of safe operation for turbines at a particular site. The average and 50-year wind gust speeds for Classes I, II and III are 10/70 m/s, 8.5/59.5 m/s and 7.5/52.4 m/s, respectively. Class I turbines are designed with a strong structure for very rough conditions, while class III are maximized for power production over the year [46].

Even though numerical wind atlases give indications for possible wind power plant locations, the uncertainty in the modelling chain is typically too large to use them for accurate planning of a wind power plant. For a wind resource assessment that will convince investors measurements are still required. Wind measurements can be done with light detection and ranging (lidar) or using a meteorological mast. A typical workflow of wind measurements often looks like the flowchart presented in Fig. 5.29.

The first step is to make a reliable, accurate and representative measurement campaign. A reliable (minimum one-year; multiple years is recommended) continuous measurement campaign must be carried out to avoid seasonal bias in the measurements. Since the mast height is very high, the measurement campaign needs to be carefully planned to include social acceptability.

After the measurement campaign has been completed, a long-term extrapolation of the measurements needs to be done, correcting for year-to-year variability in the

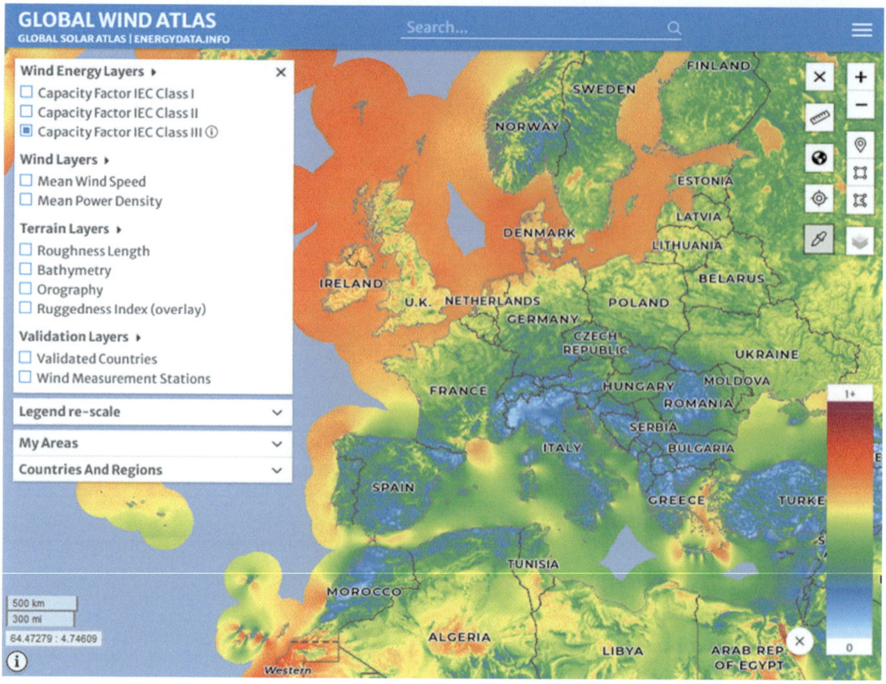

Fig. 5.28 Example of the global wind atlas. The current view presents the energy capacity factor using a turbine that meets the IEC Class III requirements [45], CC BY 4.0)

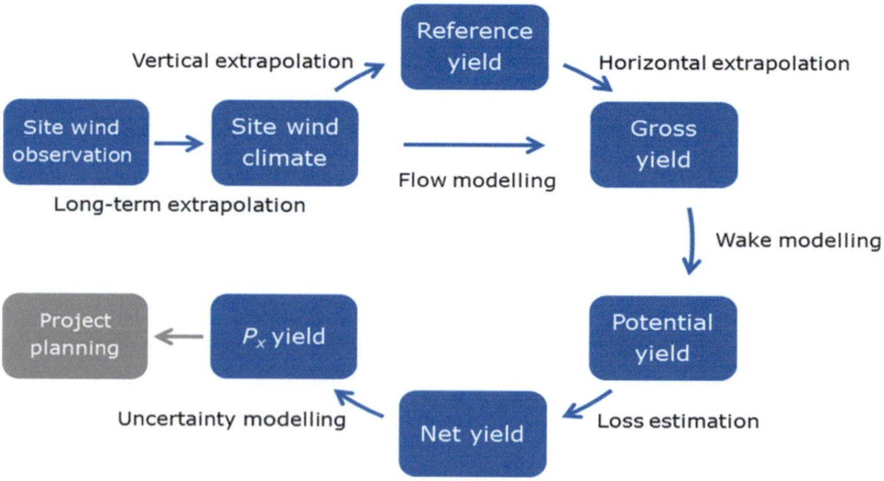

Fig. 5.29 Workflow to estimate the annual energy production of a wind power plant

wind resource and making the time series representative for the full lifetime of a turbine. Even after wind climate has been measured, flow modelling must still be done to obtain results for each turbine location. This is typically done using flow modelling and extrapolation. To minimize the vertical extrapolation model error, it is best practice to measure at heights close to the hub height.

To obtain the potential yield of the wind power plant we must still consider wake losses, for which linearized wake models (Sect. 5.7.2) are often used. Also, technical losses have to be estimated; for example, turbine unavailability due to maintenance, blade degradation, power grid curtailment or icing, after which the best estimate of the annual energy production can be obtained. Uncertainties must be included at all steps of the workflow. Investors often work with expected yields, P_{x}, which denote an estimate of the annual energy production using uncertainty.

Wind power plants are often referenced according to their rated power or capacity—here 500 MW. But if we want to design a system that would be able to produce 500 MW continuously for 8760 h annually, and to be able to compare it with "traditional" steam/gas turbine power plants running 95–98% of the year, the variability of wind needs to be considered. A good starting point is to use a wind atlas to get the local value of capacity factor. For simplicity, we assume that the power grid would be used as a 'virtual battery' and therefore there is no real energy storage required. In Chap. 7 the required energy storage for this system is discussed in more detail. A rough estimate of the actual capacity required for continuous production of 500 MW can be obtained by dividing the required power, 500 MW, by the capacity factor.

We have chosen two locations to discuss as examples: the Lestijärvi site in Finland and the Horns Rev 2 site in Denmark, situated in the North Sea (Fig. 5.30). Using the Global Wind Atlas, we can see that Lestijärvi and Horns Rev 2 have capacity factors of 0.5 and 0.66, respectively. This means that the required rated label power would be 500/0.5 = 1000 MW and 500/0.66 = 757 MW respectively, and the corresponding numbers of turbines would be 1000/6.6 = 152 for Lestijärvi and 757/15 = 51 for Horns Rev 2, assuming turbine sizes of 6.6 and 15 MW, respectively. However, if we also include the losses in the energy storage system a higher number of turbines is required to reach to 500 MW continuous production. Also, other sources of power may still be required when the wind speed is low.

The number and the size of the turbines are critical design parameters that effect the techno-economic performance of the project and also greatly affect the social acceptability, especially with regard to the impact on nature. As an example, (based on Wind Europe [48]) the average installed onshore turbine size was 4 MW in Europe in 2021, while the corresponding number for offshore installations was 8.5 MW. The turbine sizes ordered during 2021 for later installation for onshore and offshore were 4.9 MW and 11.2 MW, respectively.

To calculate the land and sea area needed, we use a grid arrangement and turbine grid spacing of $5D \times 7D$ (D = rotor diameter) in crosswind and downwind directions. Based on this information the total area in Lestijärvi would be $152 \times 5 \times 7 \times 0.170^2 = 157\ \mathrm{km}^2$, and in Horns Rev 2 $51 \times 5 \times 7 \times 0.236^2 = 99\ \mathrm{km}^2$. These would result in capacity densities of 500 MW/157 km^2 = 3.2 MW/km^2 for Lestijärvi and

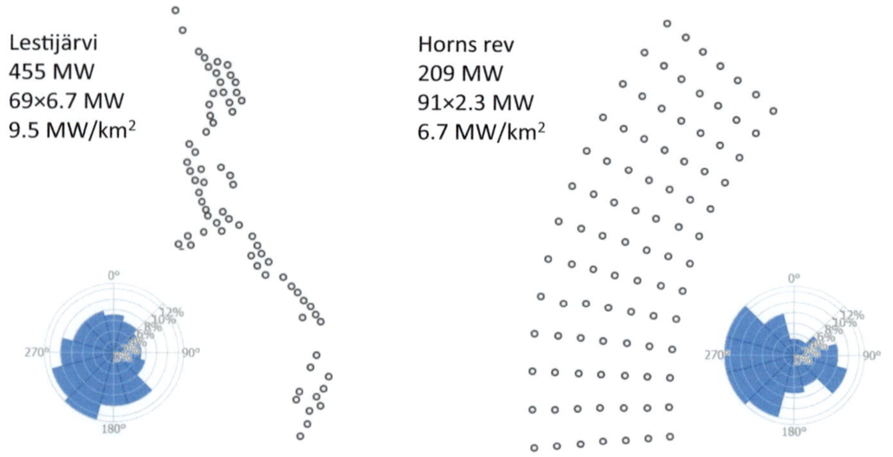

Fig. 5.30 Lestijärvi onshore and Horns Rev 2 offshore wind power plant' turbine arrangements. Arrangements are obtained from www.environment.fi for Lestijärvi, and from [47] for Horns Rev 2. The wind data has been obtained from the Global Wind Atlas [45]

5.1 MW/km^2 for Horns Rev 2. For comparison, the actual area of the Lestijärvi wind power plant is only 48 km^2, making the capacity density as high as 455/48 = 9.5 MW/km^2 meaning that the plant has a very compact construction. However, by checking the actual turbine arrangement (see Fig. 5.30) you can see that this 'dense' structure has been made possible by siting the turbines in narrow rows and aligning them perpendicular to the main wind direction, thereby avoiding array losses. The actual area on the Horns Rev 2 site is 31.4 km^2, with capacity 209.3 MW, meaning the capacity density of the Horns Rev 2 site is 6.7 MW/km^2 (Fig. 5.30).

Based on [49] average values in the United States were 3 MW/km^2 in 2009. [50] carried out a more recent analysis of wind power plants globally, and their average was 2.76 MW/km^2 for the United States. Meanwhile, [51] obtained value 3 MW/km^2 for the United States. Reference [52] found out that in Finland the capacity density is between 5.3 and 10.6 MW/km^2 (calculated using a grid of 5D × 7D).

5.7.2 *Interactions Between Turbines*

As a general rule the annual energy production of wind power plants containing N turbines can be calculated by summing up the production of all turbines and integrating this with respect to time, $\tau = 8760$ h.

$$E = \sum_{i=1}^{N} \int_{0}^{\tau} P_i(U_i(t))dt$$

The wind speed hitting a specific turbine, $U_i(t)$, is affected by the free wind site velocity before the first turbines (on the wind entering side), the wind direction and the interaction of the wakes with all the other turbines. These interactions are very complex. An isolated single turbine in an open area experiences the true free flow velocity. Regarding a Betz turbine, 59.3% of the kinetic energy of the wind passing the rotor disc is used and the velocity drops down to one third of the free wind value. In the wake the surrounding flow then starts to accelerate the flow exiting from the stream tube back up to the free wind value. This requires a long distance and the closer the next turbine is, the lower is the speed of air hitting its rotor. Therefore, sufficient spacing between turbines is essential. Typically, it is recommended to have 7–9 times the rotor diameter spacing if the turbines are assembled in rows.

Next the turbine wake and ways to estimate the wake impacts of several interacting turbines are discussed. First, we need to have a model for the expansion of the wake for a single turbine. It has been shown by Schlichting [32] that the wake diameter D is proportional to $x^{1/3}$ and the area is proportional to $x^{2/3}$, where x is the distance from the turbine. Instead of this, we use here the famous linear model developed by [53] that assumes linear dependency, $r = r_0 + kx$. The expansion coefficient k depends on the surface roughness, and is based upon [54].

$$k = 0.5/ln\left(\frac{z_h}{z_0}\right)$$

wherein z_h is the hub height and z_0 is the surface roughness length. Figure 5.31 shows the geometry of a single wake (a), but also the interaction with the following turbines and their wakes (b). In this chapter we focus only on an in-line configuration of turbines, or turbines fully inside each other's wakes. We neglect the cases where only part of the upstream wake interacts with the down-stream turbine(s). A more scattered orientation has been studied and an interested reader is recommended to study the works of [55, 56] and [57].

Based on Fig. 5.31 we can write the mass balance for the wake behind the first wind turbine, WT1 (a). Air enters with velocity U_∞ from the upper and lower wake borders, and the cross-sectional area of the expanding wake cone is $\pi\left(r_{w.2}^2 - r_0^2\right)$ where $r_{w.2}$ is the radius of the wake of turbine 1 at the location of turbine 2. For each turbine j, we assume that the pressure has already been equilibrated with the free flow value before the actual wake and then, we can write the ratio of outlet and inlet velocities as follows: (subscripts i and o refer to the inlet and outlet of the turbine control volumes, respectively).

$$\frac{U_{j.o}}{U_{j.i}} = 1 - 2a_j$$

The mass balance for the volume (a) is as follows.

$$U_\infty(1 - 2a_1)\pi r_1^2 + U_\infty\pi\left(r_{w.2}^2 - r_1^2\right) = U_{2.i}\pi r_{w.2}^2$$

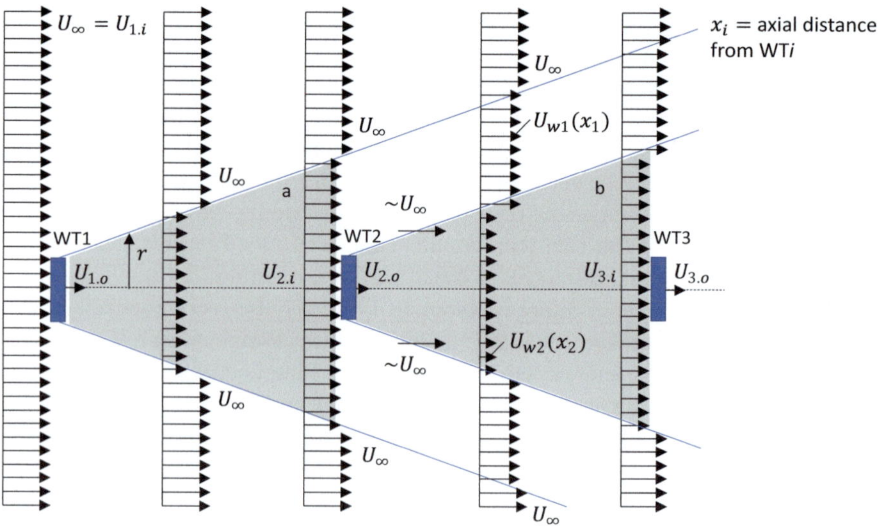

Fig. 5.31 Schematic of the wake model for a single (control volume a) and multiple wakes (control volume b)

From which we can solve the inlet velocity for the second wind turbine, WT2.

$$\frac{U_{2.i}}{U_\infty} = (1-2a_1)\frac{r_1^2}{r_{w.2}^2} + \left(1-\frac{r_1^2}{r_{w.2}^2}\right) = 1-2a_1\frac{r_1^2}{r_{w.2}^2} = 1-\frac{2a_1}{\left(1+\frac{kx_1}{r_1}\right)^2}$$

The axial induction factor can be solved from the overall thrust coefficient of the turbine. For example, when this wake hits the next turbine (as discussed for the optimal Betz turbine $a = 1/3$), the velocity again drops by a factor of $2a_2$. It was suggested by [53] that the velocities entering the control volume (b) from the upper and the lower boundaries approach U_∞, especially as the distance from the turbine increases. This means that the surrounding free flow has been able to accelerate the flow inside the previous inner wake already for the next turbine's wake. The mass balance for the control volume (b) is as follows:

$$U_{2.i}(1-2a_2)\pi r_2^2 + U_\infty \pi\left(r_{w.3}^2 - r_2^2\right) = U_{3.i}\pi r_{w.3}^2$$

From which we can solve the inlet velocity for the third wind turbine, WT3.

$$\frac{U_{3.i}}{U_\infty} = (1-2a_2)\frac{U_{2.i}}{U_\infty}\frac{r_2^2}{r_{w.3}^2} + \left(1-\frac{r_2^2}{r_{w.3}^2}\right) = 1-\frac{r_2^2}{r_{w.3}^2}\left(1-\frac{U_{2.i}}{U_\infty}(1-2a_2)\right)$$

This can also be written in general form for the inlet velocity for the turbine j as follows:

$$\frac{U_{j.i}}{U_\infty} = (1 - 2a_{j-1})\frac{U_{j-1.i}}{U_\infty}\frac{r_{j-1}^2}{r_{w.j}^2} + \left(1 - \frac{r_{j-1}^2}{r_{w.j}^2}\right)$$
$$= 1 - \frac{1}{\left(1 + \frac{kx_{j-1}}{r_{j-1}}\right)^2}\left(1 - \frac{U_{j-1.i}}{U_\infty}(1 - 2a_{j-1})\right)$$

The complete column of turbines can then be calculated by starting from the WT1 and continuing to the last turbine, considering each turbines' properties separately. To be able to demonstrate this in practice, we want to present a simplified case where all turbines are located at equal distances Δx and have equal rotors r_0 and all axial induction coefficients are equal to the optimal Betz value $a = 1/3$. In addition, we set $p = 1/\left(1 + \frac{k\Delta x}{r_0}\right)^2$ to simplify the calculation. Please note that $U_{1.i} = U_\infty$. This is then written for each turbine separately.

$$\frac{U_{j.i}}{U_\infty} = 1 - \frac{1}{\left(1 + \frac{kx}{r_0}\right)^2}\left(1 - \frac{U_{j-1.i}}{U_\infty}\frac{1}{3}\right) = 1 - p\left(1 - \frac{U_{j-1.i}}{U_\infty}\frac{1}{3}\right)$$

$$\frac{U_{2.i}}{U_\infty} = 1 - p\left(\frac{2}{3}\right)$$

$$\frac{U_{3.i}}{U_\infty} = 1 - p\left(\frac{2}{3}\right) - p^2\left(\frac{2}{3^2}\right)$$

$$\frac{U_{4.i}}{U_\infty} = 1 - p\left(\frac{2}{3}\right) - p^2\left(\frac{2}{3^2}\right) - p^3\left(\frac{2}{3^3}\right)$$

And in general, for equal turbines, this becomes for the jth turbine:

$$\psi_j = \frac{U_{j.i}}{U_\infty} = 1 - 2\sum_{i=2}^{j}\left(\frac{p}{3}\right)^{i-1} \rightarrow \frac{3(1-p)}{3-p}, \text{ when } j \rightarrow \infty$$

We are interested in predicting the efficiency of the whole plant. We could assume here that there are N rows directly in line with the flow and with each other, as defined above, and M columns sideways at a sufficient distance that they do not interact. We assume that the sideways distance is equal to the crosswind distance Δx for simplicity. We then need to consider only a single column of turbines.

The total efficiency of the plant is the actual power production of N turbines divided by the power from all N turbines that would not have any interactions. We also assume for now that the power coefficients are the same for each turbine.

$$\eta_{plant} = \frac{\frac{C_P}{2}\rho A_0 U_\infty^3}{N\frac{C_P}{2}\rho A_0 U_\infty^3}\sum_{i=j}^{N}\psi_j^3 = \frac{1}{N}\left(1+\sum_{j=2}^{N}\left(1-2\sum_{i=2}^{j}\left(\frac{p}{3}\right)^{i-1}\right)^3\right)$$

Here thew first value 1 in the outmost parenthesis refers to the first turbine that faces the free wind, and the summation is started from the second turbine onwards. We can also calculate the actual grid capacity density δ^* by dividing the total power produced by the area occupied, assuming a grid of $\Delta x \times \Delta x$. This has been written for the rated power and wind as follows:

$$\delta^* = \frac{\frac{C_P}{2}\rho A_0 U_\infty^3}{N\Delta x^2}\sum_{i=j}^{N}\psi_j^3 = \frac{P_r}{\Delta x^2}\frac{1}{N}\left(1+\sum_{j=2}^{N}\left(1-2\sum_{i=2}^{j}\left(\frac{p}{3}\right)^{i-1}\right)^3\right)$$

Example 5.6 Wind Power Plant Efficiency Analysis Analyze an offshore wind power plant of 10 equally spaced columns and 18 equally spaced rows of 100 m rotor diameter turbines in rated 12 m/s wind, corresponding to the rated power. Assume that the power coefficient is equal and constant with a value of 0.4, up to a rated power of 3.4 MW. The air density is 1.25 kg/m^3. Use a wake expansion coefficient of $k = 0.1$.

1. Calculate first how the dimensionless velocity ψ_j and power efficiency decreases from the first turbine to the three downwind turbines and discuss the findings, using spacing $\Delta x = 5.24D$.
2. Calculate the total plant efficiency as a function of the number of rows for the first 4 turbines.
3. Calculate the actual capacity density (with wake effects) and compare this with the label capacity efficiency. Discuss the results, again using a grid spacing of $\Delta x = 5.24D$.
4. Study the effect of the grid spacing on the power plant efficiency and capacity densities δ and δ^* by varying Δx from 4 to 8 and $16 \times D$.

1. The dimensionless velocity can be calculated easily using the model presented above, where the parameter p is as follows:

$$p = \frac{1}{\left(1+\frac{0.1\times 5.24\times D}{D/2}\right)^2} = \frac{1}{(1+2\times 0.1\times 5.24)^2} = 0.238$$

$$\psi_2 = \frac{U_{2.i}}{U_\infty} = 1 - 0.238\left(\frac{2}{3}\right) = 0.841$$

$$\psi_3 = 1 - 0.238\left(\frac{2}{3}\right) - 0.238^2\left(\frac{2}{3^2}\right) = 0.829$$

$$\psi_4 = 1 - 0.238\left(\frac{2}{3}\right) - 0.238^2\left(\frac{2}{3^2}\right) - 0.238^3\left(\frac{2}{3^3}\right) = 0.828$$

We can see that the value reaches equilibrium very quickly. The non-dimensional turbine specific powers can be calculated easily by raising these to the 3rd power. $\eta_1 = 1$, $\eta_2 = 0.841^3 = 0.596$, $\eta_3 = 0.829^3 = 0.569$, $\eta_2 = 0.828^3 = 0.567$. The turbine specific efficiencies were obtained from the equation below, where $\eta = \psi_j^3$

$$P_j = \frac{C_P}{2}\rho A_0\left(U_\infty \psi_j\right)^3 = \frac{C_P}{2}\rho A_0 U_\infty^3 \psi_j^3 = \frac{C_P}{2}\rho A_0 U_\infty^3 \eta_j$$

2. The total efficiency can be calculated by summing the specific efficiencies for all turbines and dividing this with the number of turbines with a turbine efficiency of 1 (which means that they always face the free wind conditions). For example, for the first 4 rows, we have:

$$\eta_{plant} = \frac{1 + 0.596 + 0.569 + 0.567}{4 \times 1} = 0.683$$

3. The label capacity density is equal to the rated power divided by the grid unit area.

$$\delta = \frac{P}{(5.24 \times D)^2} = \frac{\frac{C_P}{2}\rho A_0 U_\infty^3}{(5.24 \times D)^2} = \frac{0.4}{2} \times \frac{1.25 \times \pi \times 50^2 \times 12^3}{(5.24 \times 100)^2}$$
$$12.4\ \mathrm{MW/km^2}$$

This is purely a theoretical value. When we consider the actual capacity density reduction due to wake effects, we obtain the following (for 4 rows):

$$\delta^* = \frac{P^*}{(5.24 \times D)^2} = \frac{\frac{C_P}{2}\rho A_0 U_\infty^3 \eta_{park}}{(5.24 \times D)^2}$$
$$= 12.4\frac{\mathrm{MW}}{\mathrm{km}^2} \times 0.683 = 8.47\frac{\mathrm{MW}}{\mathrm{km}^2}$$

4. Finally, the power plant efficiencies are calculated for different turbine spacings of 4, 8 and 16*D*.

$$p = 1/(1 + 2 \times 0.1 \times 4)^2 = 0.309$$

$$p = 1/(1 + 2 \times 0.1 \times 8)^2 = 0.148$$

$$p = 1/(1 + 2 \times 0.1 \times 16)^2 = 0.057$$

$$\psi_2 = 1 - 0.309\left(\frac{2}{3}\right) = 0.794$$

$$\psi_3 = 1 - 0.309\left(\frac{2}{3}\right) - 0.309^2\left(\frac{2}{3^2}\right) = 0.773$$

$$\psi_4 = 1 - 0.309\left(\frac{2}{3}\right) - 0.309^2\left(\frac{2}{3^2}\right) - 0.309^3\left(\frac{2}{3^3}\right) = 0.771$$

And similarly for other spacings 8: $\psi_2 = 0.901$, and $\psi_4 = 0.896$ respectively and 16: $\psi_2 = 0.962$, $\psi_3 = 0.961$ and $\psi_4 = 0.961$ respectively. The plant efficiencies are then as follows:

$$\eta_{plant}(4) = \left(1 + 0.794^3 + 0.773^3 + 0.771^3\right)/4 = 0.61$$

$$\eta_{plant}(8) = \left(1 + 0.901^3 + 0.896^3 + 0.896^3\right)/4 = 0.79$$

$$\eta_{plant}(16) = \left(1 + 0.962^3 + 0.961^3 + 0.961^3\right)/4 = 0.92$$

By spreading the turbines out, the power plant efficiency significantly increases. Finally, the label and corrected capacity densities δ and δ are calculated. As above, the power in the 12 m/s free wind with power coefficient 0.4 is equal to 3.4 MW.

$$\delta(4) = \frac{3.4 \times 10^6}{(4 \times 100)^2} = 21.3\frac{\text{MW}}{\text{km}^2}$$

$$\delta(8) = \frac{3.4 \times 10^6}{(8 \times 100)^2} = 5.3\frac{\text{MW}}{\text{km}^2}$$

$$\delta(16) = \frac{3.4 \times 10^6}{(16 \times 100)^2} = 1.3\frac{\text{MW}}{\text{km}^2}$$

$$\delta^*(4) = 21.3 \times 0.61 = 13.0\frac{\text{MW}}{\text{km}^2}$$

$$\delta^*(8) = 5.3 \times 0.79 = 4.2\frac{\text{MW}}{\text{km}^2}$$

$$\delta^*(16) = 1.3 \times 0.92 = 1.2\frac{\text{MW}}{\text{km}^2}$$

These clearly show that a proper prediction of plant/array efficiency is essential in calculating the capacity density. Even though the plant efficiency increases significantly by increasing the turbine spacing, the impact of the larger area can be more significant on, for example environmental aspects and land use, and closer spacing enables more power per unit area to be squeezed out.

The simple model's results are compared next with computational fluid dynamics simulation-based results to investigate the differences. As an example of this kind of analysis, for an array of turbines arranged in squares or staggered triangles, [58] carried out detailed computational fluid dynamics simulations and demonstrated how wind speed and power decreases in these array structures. They simulated a wind power plant of $6 \times 18 = 108$ turbines; here only 6 rows are shown to save space. Figure 5.32 presents the squared and staggered arrangements.

Based on [58], the total plant efficiencies are re-calculated below and presented in Fig. 5.33. The row inlet velocities are also included. It is assumed that all power coefficients are equal in all rows. Results from the above simplified model are also presented for comparison.

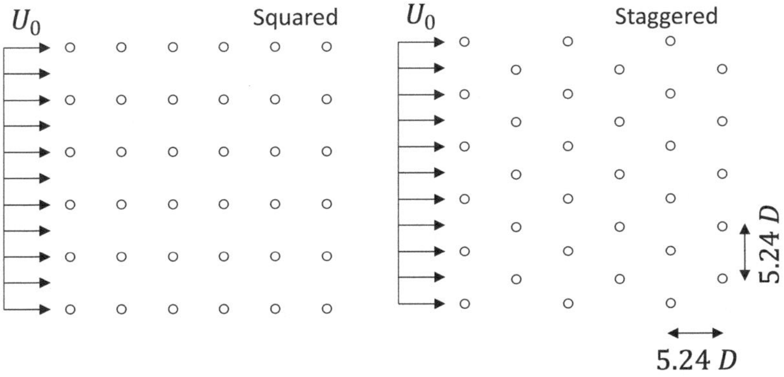

Fig. 5.32 Wind turbine arrangements where *D* is the rotor diameter. The hub height is 100 m and the rotor diameter is 100 m. Only the 6 first rows are presented with spacing $5.24 \times D$

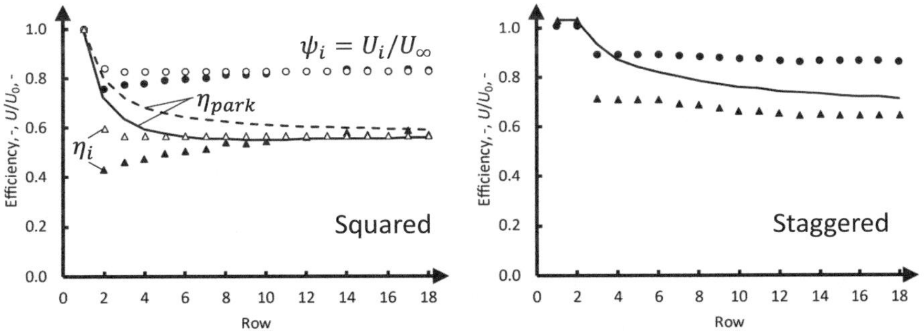

Fig. 5.33 Efficiency of each row turbines compared with the first row, plant efficiency (solid line) and row inlet velocities (filled dots and triangles), data from [58]. Open symbols and the dashed line present the simplified model

As can be seen, the differences between the models are surprisingly small. In the staggered arrangement, the efficiency is compared with the squared arrangement, making its efficiency ~ 3% better in the first row. In the squared arrangement, row-specific power dramatically drops by almost 60% from the first row to the second row and then the efficiency steadily increases to 60% of the first-row power towards the 18th row. For the staggered arrangement the row specific power does not drop for the second row, but then drops 30% for the third row and then gradually decreases to 40% of the first-row power. The total efficiency in the squared arrangement stabilizes at 55%, while in the staggered case it slowly decreases once efficiency rises above 70%. These results were obtained for a case where the wind blows perpendicular to the rows from the left. If the wind direction changes the results will immediately change, and the squared case approaches the staggered one and vice versa, since the distances between the turbines increase and decrease, respectively. This example is case specific, and these results cannot be directly applied to other spacings and turbine sizes. However, it clearly shows that in the case of wind turbine grids, the array efficiencies must be carefully studied.

Example 5.7 Four Turbines in Line with the Wind Analyze a system of four turbines placed in line with the wind, with three turbines behind the first one also in line. The rated power of each of them is 15 MW, the rotor diameter is 236 m, and the turbines are using a variable speed variable pitch control system. Assume that the rated wind speed is 10 m/s and the air density is 1.25 kg/m^3. Use a wake expansion coefficient $k = 0.1$ and spacing $\Delta x = 7D$.

1. Calculate the power coefficient at the rated power of the turbine at rated wind speed.
2. Analyze a case when the wind speed increases to 15 m/s. What is the wind power plant efficiency? Discuss the results.

1. The power coefficient at the rated power and wind speed is simply calculated from the power equation as follows:

$$C_P = \frac{P_r}{\frac{\pi}{8}\rho D^2 U_{\infty.r}^3} = \frac{15 \times 10^6}{\frac{\pi}{8} \times 1.25 \times 236^2 \times 10^3} = 0.549$$

2. When the wind speed is higher than the rated value, we need to limit the power by pitching the blades, to reduce the power coefficient. In the variable speed, variable pitch, VSVP system, C_P is maintained at the maximum constant value by allowing the rotor to increase the speed of rotation up to the rated power linearly with the wind speed, after which the blades are turned to limit the power production to the rated power when the wind speed increases further above this. Above the rated wind speed the power coefficient is simply calculated as follows (please note that the constant 548.7 is only for this case here and based on the above parameters):

$$C_P = \frac{P_r}{\frac{\pi}{8}\rho D^2 U_\infty^3} = \frac{15 \times 10^6}{\frac{\pi}{8} \times 1.25 \times 236^2} \frac{1}{U_\infty^3} = \frac{548.7}{U_\infty^3}$$

For example, in the case of a wind speed of 15 m/s, C_P is equal to 0.163. This means that the axial induction factor must be smaller and therefore, the velocity does not have to drop in the rotor as much as in the case of a turbine operating at the rated power. We can always write the power coefficient in terms of the axial induction factor as follows:

$$C_P = \frac{548.7}{U_\infty^3} = 4a(1-a)^2$$

This means that, depending on the power coefficient, the value of a varies. Due to the third order of the equation, we need to solve axial induction factor numerically. Please note that this might have three solutions; use the smallest positive value. For the first turbine, operating at a wind speed of 15 m/s, we obtain from the above equation a value of $a = 0.0445$. It means that the velocity decreases only by a fraction—$2 \times 0.0445 = 0.089$—before it enters the wake. Therefore, the velocity going into the first wake is equal to $U_{1.o} = (1 - 2 \times 0.0445) \times 15 = 13.7$ m/s. This is still above the rated wind, meaning that also the second turbine will produce full rated power. The general form of the velocity deficit was derived above.

$$\frac{U_{j.i}}{U_\infty} = 1 - \frac{1}{\left(1 + \frac{kx_{j-1}}{r_{j-1}}\right)^2}\left(1 - \frac{U_{j-1.i}}{U_\infty}(1 - 2a_{j-1})\right)$$

At the inlet of the second turbine, we have:

$$\frac{U_{2.i}}{U_\infty} = 1 - \frac{1}{(1 + 0.1 \times 2 \times 7)^2}\left(1 - \frac{15}{15}(1 - 2 \times 0.0445)\right) = 0.985$$

Note that $U_{1.i} = U_\infty$. The inlet velocity to the second turbine is equal to $0.985 \times 15 = 14.8$ m/s, and the corresponding axial induction factor is 0.0465. At the inlet of the third turbine, we have:

$$\frac{U_{3.i}}{U_\infty} = 1 - \frac{1}{(1 + 0.1 \times 2 \times 7)^2}(1 - 0.985(1 - 2 \times 0.0465)) = 0.981$$

The inlet velocity to the third turbine is equal to $0.981 \times 15 = 14.7$ m/s, and the corresponding axial induction factor is 0.0476. The velocity into the last turbine can be obtained as:

$$\frac{U_{4.i}}{U_\infty} = 1 - \frac{1}{(1 + 0.1 \times 2 \times 7)^2}(1 - 0.981(1 - 2 \times 0.0476)) = 0.980$$

The inlet velocity to the third turbine is equal to $0.98 \times 15 = 14.7$ m/s, and the corresponding axial induction factor is 0.0476. We can see that all the turbines therefore operate at the rated power, and the plant efficiency is equal to:

$$\eta_{park} = (1 + 1 + 1 + 1)/4 = 1$$

In conclusion, we can see that when the wind speed is higher than the rated wind speed, the array losses might not affect the plant operation. Of course, this requires that we use turbines that are dimensioned so that their rated wind speed is sufficiently low, and that there is reserve for producing constant power as the wind speed increases up to the cut-out value. This issue has been studied by others, for example recently by [59].

5.8 Impacts on Nature and Mitigation of Negative Effects

As with any energy production technology, wind energy has adverse environmental impacts, though the research to understand these effects has not developed as fast as the production of wind energy itself [7]. Planning and development of any large project, including wind energy, should address the potential impacts on wildlife and nature and include a plan to mitigate these. Reducing impacts on nature is important because while climate change accelerates nature loss, protecting and restoring nature is crucial for mitigating climate change itself, creating a vital feedback loop [60]. Wind power plants are often located away from human settlements due to noise and flicker effects. This may increase social acceptance but at the same time may increase the impact and disturbance on nature.

Wind power plants can affect nature and natural systems in unexpected and complex ways. The effects of wind energy projects on the environment occur throughout their lifecycle, starting from the mineral extraction and continues through the four main steps: characterization, construction, operation, and decommissioning. This section gives an overview of the environmental effects of onshore (land) and offshore (sea) wind power plants on the project site. It lists ways to mitigate these impacts and discusses the gaps in current knowledge. The main mechanisms of how on- and offshore wind power impact nature are the same, yet there are also some fundamental differences, and those are discussed separately below. The summary table at the end of the section outlines the main effects that wind power plants have on nature and wildlife, and the ways to mitigate those effects based on scientific literature. The focus of this section is on the planning/characterization, construction, and operation of wind power plants, excluding the impacts of mineral extraction.

The area impacted by a wind power project may be much larger than the area changed by the construction itself. The total area impacted by a wind power project consists of the direct impact area, the indirect impact area, and the cumulative impact area (Fig. 5.34). The direct impact area is the area where the existence of wind turbines directly harms nature. Changes in the environment due to clearance of forests and

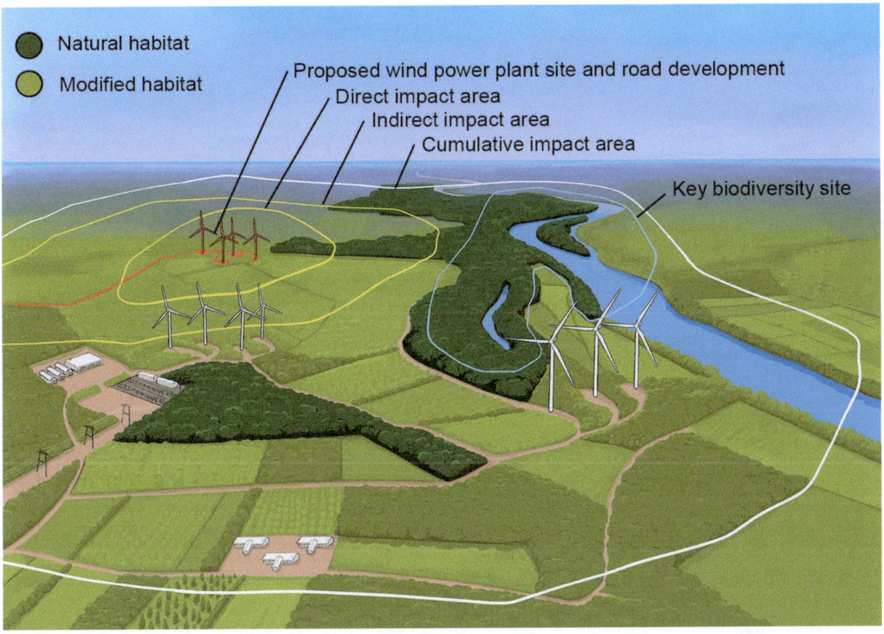

Fig. 5.34 The relationship between direct, indirect, and cumulative impacts on biodiversity—an illustrative example of an onshore wind development project © Maija Karala

other vegetation, for instance, has an indirect impact on nature outside of the direct impact area. The cumulative impact area consists of the combination of the direct and indirect impact areas, and also includes other human-caused disturbances in the area.

The construction of wind power plants in natural or semi-natural environments alters the landscape, reduces the quality of the natural environment, and often results in habitat loss, fragmentation, and degradation (Table 5.17). Species respond to these changes in numerous ways [61, 62], which can lead to wider population and ecosystem-level effects [63, 64] (Fig. 5.35). The impacts caused by wind power are, however, very site-dependent and may vary among species and between seasons. Good local knowledge and baseline data on nature and wildlife are needed to address potential impacts sufficiently during different phases of the project. Construction and operation of a wind power plant cause reduction or loss of available habitat (Fig. 5.35, Table 5.17). Turbines and the power plant may also form a movement barrier for some species (Table 5.17).

Displacement, reduction or disappearance of a certain species because of wind energy development can cause changes in the entire ecosystem; a process called trophic cascades (Fig. 5.36a, Table 5.17). A classic example of a top-down trophic cascade is removal/reduction of a top predator that results in changes in the prey population and, therefore, leading changes in the entire ecosystem structure. Several wind energy projects in the same area, or the development of wind power plants

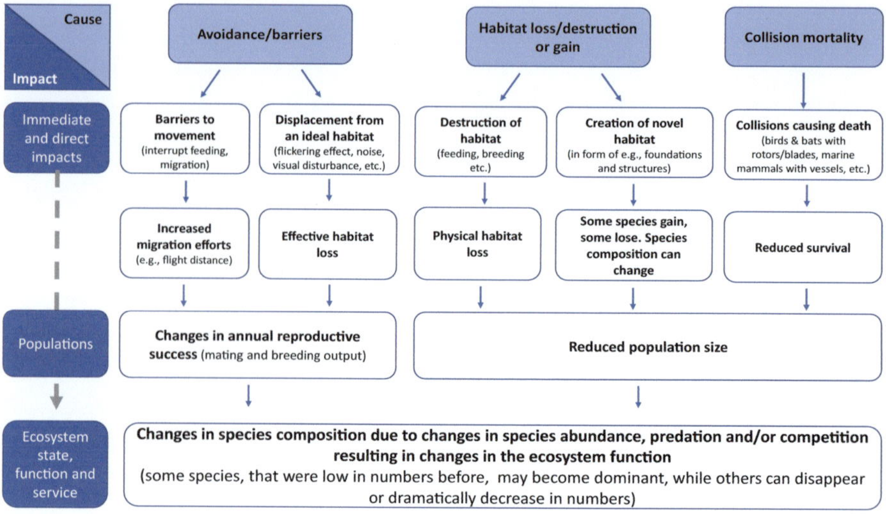

Fig. 5.35 Example of major risks caused by wind power plants (light blue boxes) and their potential effects on populations (dark blue boxes). The risks relate to behavioral changes caused by turbines, physical habitat changes (via loss or gain) and direct demographic changes resulting from mortality caused by collisions. These can result in changes in the overall population size and viability of the species inhabiting the wind power project area. Modified from [65]

in an already heavily occupied locality, increases the risk of cumulative impacts. Cumulative impacts are a result of many individual effects in the same area, none of which alone would result in serious effects on the species, ecosystems or their function (Fig. 5.36b). For species and ecosystems that are already threatened, or sensitive for natural reasons, these seemingly incremental impacts can together be significant.

The combined effects of wind power and other human activities, such as intensive fisheries, on the food web, are not so well understood [66]. Sometimes, effects can be seen only after a long period of time due to complex population-level interactions. For example, European golden plover (*Pluvialis apricaria*)—a species protected under European Union law—showed no significant effect during the construction of a wind power plant, but once the turbines were installed and in operation, the abundance of this bird species decreased by 79% in contrast to control areas [67]. The wind power plant was composed of thirty-five 2 MW turbines with a hub height of 69 m and a rotor diameter of 82 m [67]. This example illustrates how difficult it can be to predict impacts on nature even when the process is well-monitored throughout the planning and execution of the project.

The mortality of birds and bats from collisions with turbines or power transmission lines is the most studied direct impact [7, 68]. The construction and operation of wind power plants may cause more indirect disturbance to wildlife in the form of noise, artificial environmental light, vibrations, and increased human movement in the area. This may lead to avoidance of the area by some species, and the project can form

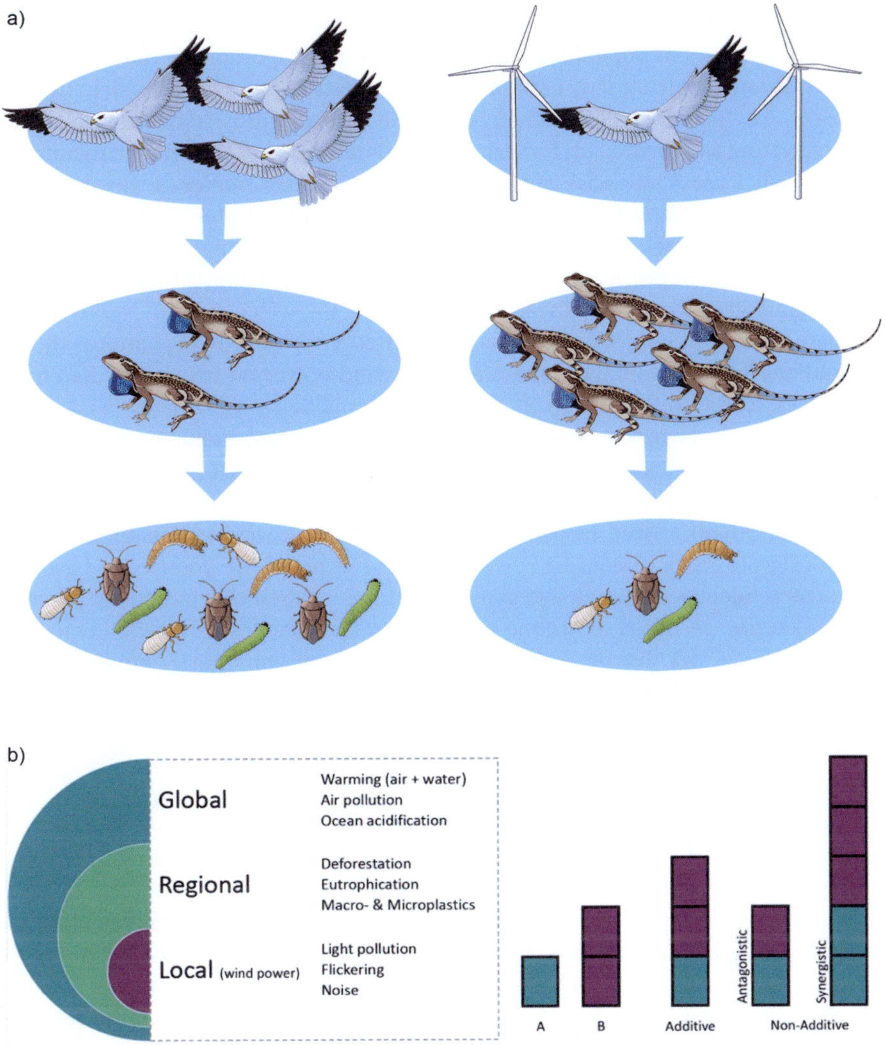

Fig. 5.36 **a**) A simplified example of a trophic cascade illustrating how a wind power development can change ecological communities in unexpected and complex ways via reduced predation and density-dependent competition. The wind power plants reduce the number of predatory birds (like buzzards, hawks, and kites) which increases the abundance of their prey (a lizard). Higher lizard densities change the behavior and physiology of the lizard, and this in turn impacts the abundance of the lower-trophic level prey (beetles and other arthropods), which may itself dramatically impact the vegetation of the ecosystem etc. [63]. This is an example of a top-down effect. An example of a bottom-up trophic cascade is given under the section on onshore wind power © Maija Karala. **b**) Multiple ecosystem effects that may be scaled from global (turquoise) to regional (green) and local (purple—the wind power plant scale). These stressors can act in an additive or non-additive (antagonistic, synergistic) way (exemplified as the interaction of one global environmental and one local wind power plant related stress effect)

a barrier to wildlife movement (Table 5.17), therefore reducing population connectivity. Connectivity of populations, in simple terms, is important for the dispersal of individuals and the genes they carry. Isolated populations have higher risk to decline and become extinct. Increased human activity on the site can also reduce the available area for nesting and feeding of, for example, birds and terrestrial or marine mammals, or larger fish. Case Studies 5.1 and 5.2 give two practical examples of ways to assess and minimize the impacts of wind energy projects on wildlife. The first example showcases an approach to minimize golden eagle collisions on wind turbine blades and the second example discusses the impacts of offshore energy projects on marine wildlife and ways to mitigate those impacts.

There is a lack of long-term datasets on the effects of wind power plants to wildlife [7], and any information about the effects of wind power on taxa other than birds and bats is scarce [69]. Though Strategic Environmental Assessments (SEAs) and Environmental Impact Assessments (EIAs) are used to manage and monitor the environmental impacts of wind power projects, and new guidelines to support sustainable wind power development are emerging in Europe and elsewhere (see for instance [70]), there is still a need to better understand multiple, and cumulative impacts. Monitoring programs should be actively used to improve the knowledge base and to overcome identified knowledge gaps [71]. If the location of wind power plants is mainly determined based on techno-economic rather than environmental considerations, there is a risk that the environmental impacts of wind power production are underestimated, with the potential for long-term consequences [63].

5.8.1 Avoiding and Mitigating Impacts on Nature in Wind Energy Projects

Most studies on mitigating the effects of wind power projects discuss reducing the impacts on wildlife during construction or operation; fewer studies list measures for avoiding the impacts [72]. Mitigation of potential adverse environmental impacts should be built into the project design early on. Good design and planning can help avoid costly surprises. For example, including realistic and comprehensive visualizations in permitting process documents facilitates dialogue between different stakeholders, officials, and the company working on the power plant. This process is iterative throughout the implementation of the project from planning to decommissioning. Measures should be adjusted as new information becomes available. Adaptive measures are important also, because sometimes identifying and proving cause/effect relationships can be difficult, or there might be a time-lag before the impacts are realized [73].

In accordance with the mitigation hierarchy, impacts on the environment should primarily be avoided, then mitigated, restored, and finally, impacts that cannot be avoided are offset, Fig. 5.37. Ideally, the target is that the project causes 'No Net

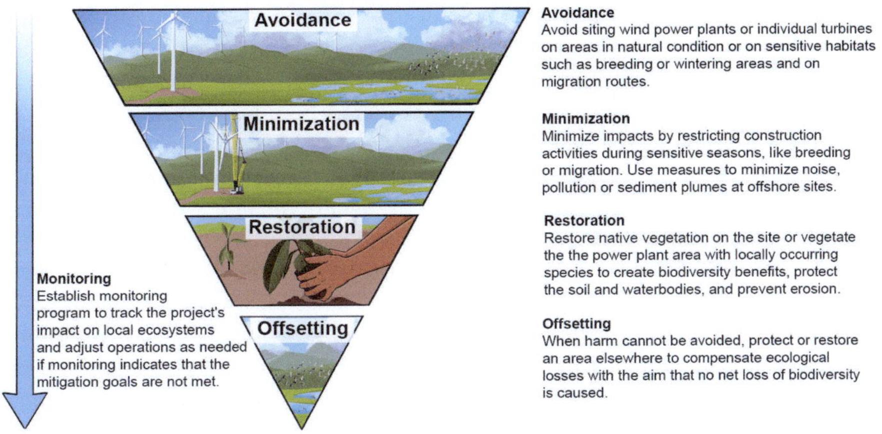

Fig. 5.37 Steps of mitigation hierarchy and example activities for mitigating potential adverse impacts of wind energy. The mitigation hierarchy is a tool to limit negative impacts of development projects on nature. The emphasis is on avoiding and minimizing negative impacts, and when those cannot be avoided, then compensating the residual impacts © Maija Karala

Loss' of biodiversity (NNL), or preferably, creates a 'Net Gain' (NG). (See more on mitigation hierarchy Sect. 2.3).

Managing the negative impacts of a wind energy site can cover the entire project area or individual turbines. These are called macro- and micro-siting, respectively. Usually, macro-siting options are discussed during the project planning phase and micro-siting is used when implementing the plan. Most of the mitigation actions have been applied to onshore wind power plants, yet there is evidence supporting the efficacy of managing impacts also at sea [7, 74].

Avoidance via careful placement of wind turbines, power lines and other related structures is the most important action. Areas always to be avoided include terrestrial and marine protected areas, sensitive areas with significant natural value such as wetlands, intact forests, reefs in marine environments, areas known to support threatened ecosystems or species, and migration corridors. Also, construction activities should be avoided during sensitive seasons such as during breeding and migration. Inside the wind power plant, micro-siting can include grouping individual turbines closer to each other to minimize the overall project footprint, or re-routing cables to mitigate (or reduce) their negative impacts. Bird collisions at onshore wind power plants are relatively well-studied (see also the golden eagle Case Study 5.1) and therefore there is a set of actions that can be taken to prevent collisions from happening. Turbines can, for instance, be aligned so that avian flight paths remain in-between turbines, thereby decreasing the overall risk of collision [61]. Noise and visual methods have also been proposed to mitigate collisions, though the efficacy of these is not always clear [12, 61, 75, 76]. Impacts of underwater noise on marine fauna (mammals, fish, sharks, turtles are also relatively well-understood and there are efficient ways to mitigate those impacts, for instance, via efficient monitoring and acoustic deterrent devices (ADD). Real-time and acoustic monitoring should be used prior the start of the construction activities to direct construction into areas and seasons when, for instance marine mammal activity is low [41]. Acoustic deterrents, also called pingers, are devices that emits pulses of high frequency sound to scare animals away from the construction site [77].

When adverse impacts cannot be avoided, they should be minimized. Again, there are several strategies proposed for mitigating avian collision with wind turbines. These include, among others, shutting down individual turbines, or the entire power plant during peak migration season, or feathering the turbine blades if the plant is not completely shut down, or make the blades more visible by painting those [7, 74, 75, 78]. Though shutting down turbines is an efficient way to prevent collisions from happening, it is not often used, probably because of reduced economic gain and lack of knowledge (Personal communication with Teemu Lehtiniemi, BirdLife Finland). Efficient application of this mitigation method could be done via a permit process. The operation of power plants could be prohibited during a specified migration season in the permit. At offshore construction sites, soft start piling, and bubble curtains can be used to minimize the impacts of noise. Soft start piling means that the construction power is elevated gradually leaving time for marine species to escape from the area and therefore minimizing the harmful exposure to noise [41]. A bubble curtain is a pneumatic barrier, where air bubbles are used to reduce the noise levels of construction [79]. Any unnecessary land work, soil sealing, or sediment disturbance should be avoided. Important areas for nature, e.g., nature conservation areas, areas in natural state or important areas for nature, like breeding, wintering, and migration routes should be protected with a buffer zone.

Restoration and re-vegetation of the site can also be used to minimize the overall project footprint. This can also provide great new business opportunities for improving the social acceptability of wind power. In general, minimizing and mitigating impacts of wind energy developments on- and offshore should be considered during the very early stages of project planning (See also Sect. 5.9 on social aspects on wind power). Examples of good on-site restoration (after [41, 73]) may include:

- Storing topsoil and sub-soil of the construction area and using later in restoration.
- Re-vegetating and restoring areas that are used only temporarily during construction. Soil, mulch, and vegetation debris (that contains natural seed stock) can be used to facilitate natural re-vegetation of disturbed areas.
- Use only locally occurring species in revegetation and monitor and manage potential spread of invasive alien species.
- At offshore sites, habitat restoration may include creation of seagrass meadows, or oyster reefs.
- Landscaping the project area by using site-specific, indigenous, and non-invasive species.
- Monitoring improves the knowledge base and helps to avoid and mitigate negative impacts.

There is always some level of residual damage left after large projects. Compensation is the last step in the mitigation hierarchy and can be used to compensate any remaining casualties. The compensation can be done through protecting or restoring areas outside the project area. More on compensation principles, see Sect. 2.3.

All human activities come with a cost of some sort to wildlife and there is no single solution that would enable to protect all species from all impacts stemming from renewable energy production, except the total avoidance. Mitigation measures are also typically species-specific and site-dependent, and it is not possible to address the entire affected wildlife community. Successful mitigation requires that enough time is scheduled in the project from the planning phase to the operation phase. Close cooperation between project planners, engineers, biologists, and contractors is needed to ensure that all possible mitigation steps are taken throughout the project (a successful team follows the T-profile introduced in Sect. 1.5).

5.8.2 Onshore Wind Power

The area impacted by an onshore wind project is relatively large, though 90% of the land is not directly used for energy production (See also Sect. 2.4 Renewable Energy and Land Use Change). Besides the turbines, the area consists of a collector sub-station, access roads, cabling between the substation and each turbine, and high-voltage power lines that connect the substation to the power grid. Intact and sensitive areas are more easily affected by wind power developments than already disturbed areas [80]. In forested areas, wind power projects lead to fragmentation [81], see also Table 5.17). Increased forest edge changes the micro-climate (temperature, humidity, light) of the area and consequently the species composition, with typical forest species displaced by species that can thrive under these new environmental conditions [82]. Negative effects can be minimized by directing constructions into already altered or marginalized landscapes [83]. To decrease the forest edge effect, and save as much natural habitat as possible, concentrated wind production areas are often better than multiple locations with a single or small number of turbines. As the Lestijärvi example, presented in Sect. 5.7.1, showed, the total area required by the wind power plant can differ considerably from the theoretical land use, depending on the local conditions. In the case of Lestijärvi, the area of the power plant area is bigger than the theoretical minimum area.

Most of the studies on the biodiversity impact of onshore wind power have focused on birds and bats, in particular their collisions with turbines [7, 68]. For example, prairie chickens [11, 84, 85], black kite [12], and migrating raptors [16, 86] have been shown to change their behavior and avoid locations near wind turbines, or are in danger of colliding with the turbine blades. Case Study 5.1 presents a study developing a method to site wind power plants in the vicinity of golden eagle (*Aquila chrysaetos*) territories without compromising their long-term survival.

Potential impacts on other taxa, or wider ecosystem functions are not so well-understood. However, there is evidence that the presence of wind turbines causes stress for at least some ungulates, like the roe deer (*Capreolus capreolus*) [13]. Also, European roe deer, hare [13] and wolf [14], have been shown to change their behavior and avoid locations near wind turbines. Findings from semi-domesticated reindeer (*Rangifer tarandus*) are case specific: in some areas, the reindeers have shown clear avoidance behavior [87], while in others, the impact has not been so clear [88]. Reference [89] found that studies that were limited in space and time often did not capture the disturbance of wind power on reindeers. For roe deer, the size of the wind power park should not exceed 18 turbines or 8.24 km^2 [13].

There is increasing evidence of the potential negative impacts of wind power on insects. For instance, species that congregate on hilltops, like some butterflies, could be attracted by the vertical structures of the turbines and be at risk of colliding with the spinning blades [90]. This risk is highest during summer and fall. Insects can also collide with blades in such large numbers, that the blades get covered by organic detritus. This layer disrupts the airflow and can even halve the power production [91–93] if the blades are not cleaned. Besides disrupting power production, poorly sited wind turbines can form a risk for some local, rare, or sensitive insect species via increased mortality, with potential bottom-up cascading effects like reduced food for higher trophic levels or other changes in the use of the environment (see for example Fig. 5.36a above, which presents an example of a top-down trophic cascade).

Some invasive insects also use wind turbine structures for overwintering, potentially helping opportunistic species to expand their range by facilitating winter survival [94]. See summary table (Table 5.17) for more information and examples of the impacts of wind power on nature.

Barotrauma—injury caused by sudden pressure change—has been hypothesized as a major source of bat mortality at wind power plants [95]. However, there is very little empirical evidence for this, and it may not represent a significant cause of bat mortality [95, 96].

Case Study 5.1: Golden Eagle (*Aquila chrysaetos*) and Onshore Wind Power

The golden eagle is globally the most widespread, large eagle (Fig. 5.38). Worldwide there are approximately 130,000 individuals distributed across the northern hemisphere [97]. Golden eagles are mostly local, but populations breeding at latitudes greater than 60° N are migratory. They inhabit open landscapes, plateaus and steppes; areas with little human activity [98, 99]. The extent of their predation area is large—approximately 300 km^2 [100]. Golden eagles are top predators and have no natural enemies. Therefore, almost all threats are caused by humans. Observations on ringed golden eagles in Finland showed that the most common causes of death include intentional killing (34%), disease (26%), traffic (11%) and electrocutions on power lines (20%) [101]. Golden eagles are known to collide with wind turbines as well [102–104]. They also show avoidance behavior around wind power production areas [105]. Their large predation area, low annual reproduction, and sensitivity to human activities make this species vulnerable to the effects of wind power production [106]. Even a slight increase in mortality rate can weaken the viability of the local population [107].

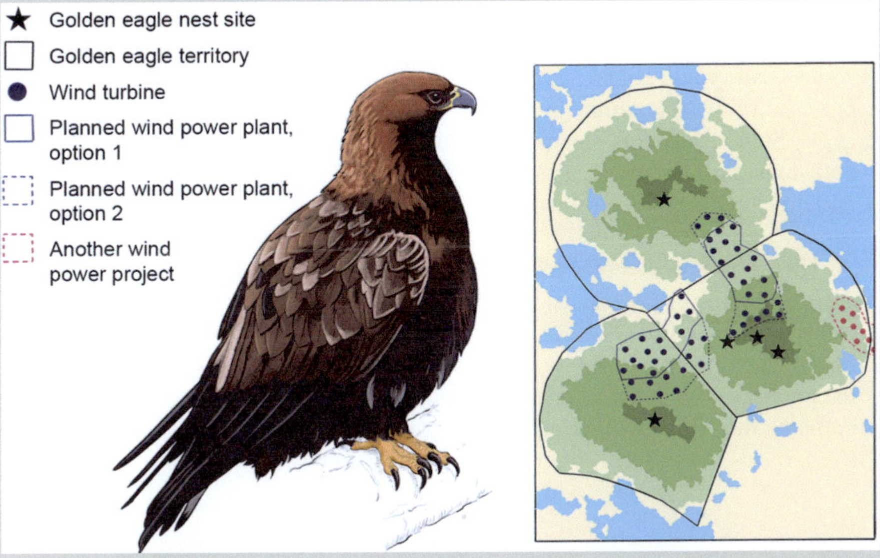

Fig. 5.38 Left: Adult male golden eagle. Right: wind power project options and golden eagle territories marked with lines (nests marked with a star). VE 1 (option 1) = 52 wind power turbines and VE 2 (option 2) = 28 wind power turbines. The green area is an existing nature conservation area (Natura 2000 site). The area marked with blue is an existing wind power production area. Right: the use of the three territories by golden eagles, based on results from the habitat model. The environments not used by the eagles are shown in blue, and the environments preferred by the eagles are in green, when compared to random points. Darker color signifies more intense use © Maija Karala

There are effective tools to analyze the risks of siting a wind power plant on golden eagle territories and breeding grounds. This case example introduces the main results of such an assessment [107] that used a habitat model which was created based on data from golden eagle populations in Finland [100, 108]. The model was used to calculate the movement and collision risk of golden eagles in three territories: Kokkoaapa, Selkäkangas, and Tulineva, under two siting alternatives (Fig. 5.38 right). Cumulative impacts (collisions with turbines and total stress level experienced by the eagles) with the nearby wind power production sites were evaluated as well, using the same habitat model that was used to assess the impact of the already existing wind power plant.

The assessment assumed that the unit power of the turbines is approximately 6–10 MW, the tip height up to 300 m, the hub height up to 200 m and the rotor circle diameter a maximum of 200 m. An increase in rotor size increases collision mortality. The collision probability was calculated using the "Band Collision Risk Model" [109].

The collision probability can be used to estimate the increased mortality risk for golden eagles, caused by collisions with the turbines. For long-lived species like the golden eagle, adult survival is important and even a small increase in mortality rate causes a drastic reduction in the population growth rate. The basis of the study was to demonstrate that the viability of the golden eagle territory will not be compromised, i.e., reproduction remains stable. This corresponds to a maximum increase in mortality risk of 3% per individual adult eagle. Statistically this means collision once every 16.6 years, i.e., about two collisions during the typical 30-year life cycle of a wind power project. If the additional mortality exceeds the 3%, there is a risk that the population will be lost on a long-term basis.

The study revealed that golden eagles mostly use the core area of the territory: the edges of swamps and sparsely wooded areas (Fig. 5.38, right). This information can further be used to compare the placement of individual turbines inside the wind power site, to profile potentially risky turbines and to compare the impact of removing or relocating individual turbines with the results of the collision modelling. The results of the study are summarized in Table 5.16.

Table 5.16 Summary of the assessment results (under alternative 1 and 2) across three golden eagle territories. See also Fig. 5.38

Territories	Alternative 1	Alternative 2	Conclusions and recommendations
Kokkoaapa	22 turbines	11 turbines	It is recommended that alternative 1 is not implemented. The combined impacts of alternative 1 and the current operating wind power plant are significant. By contrast the combined impacts of alternative 2 are assessed as moderate
Collision risk	Significant	Not significant	
Area disturbed (% of the territory)	13.5	6.7	
Natura 2000 area assessment—impacts on Golden eagle	Significant	Moderate, not significant	
Selkäkangas	17 turbines	10 turbines	The overall impact of alternative 1 are assessed to be moderate and the impact of alternative 2 minor. However, it is recommended that the two turbines closest to the nest are removed or relocated
Collision risk	Moderate but below threshold	Not significant	
Area disturbed	11.3	4.5	
Tulineva	10 turbines	7 turbines	The impact of alternative 1 are assessed as moderate and alternative 2 minor
Collision risk	Moderate but below threshold	Not significant	
Area disturbed	3.6	N/A	

The study used habitat, collision, and population models that are subject to some uncertainties. Though the models are based on best available data, there can be inaccuracies in the parameters used. For instance, it is difficult to predict how individual birds will respond to the turbines after two generations: will the avoidance behavior get stronger or weaker? How would this impact collision risk? Also, the flight observations can be prone to human error: the observability of the eagles varies, and there can be inaccuracies in flight heights and distances. Also, the impact of power transmission lines was not assessed. Therefore, the precautionary principle should always be applied when interpreting and making decisions based on modelling results.

5.8.3 Offshore Wind Power

Compared with the available knowledge of the impacts of onshore wind power plants, much less is known for offshore wind power plants. Whilst the impacts on migrating birds might be similar to those of onshore power, offshore wind energy construction can alter the pelagic (water column) and benthic (bottom) environments. The most appropriate areas for offshore wind energy construction from the techno-economical perspective are also often the most biodiverse and constitute important areas for natural marine environments, like offshore banks and reefs. These shallower grounds at open sea are vital breeding, feeding, wintering, and living spaces for many species. Near-shore areas also host diverse wildlife and are important fishing and recreational areas for coastal communities. In marine habitats the biggest change is the destruction of the seabed, whilst some of the other impacts may be more short-term. Construction of fixed foundations for turbines requires dredging bottom sediment and materials away so that the foundation and cables can be installed. This not only removes all species living in the patch, but also stirs-up material from the bottom, causing accumulation of material in the water column, with implications far beyond the direct impact area (see Fig. 5.34). Suspended particles increase turbidity and particles in the water column reduce light penetration, which in turn reduces photosynthesis of plants and algae and can disturb the whole trophic balance.

The construction itself creates an artificial reef, which can be a positive or negative development, via a so-called "reef effect". The structures create living spaces for hard-substrate (fouling) species, potentially increasing the overall diversity and density of species [110]. Offshore wind energy sites can also create a refuge for trawled fish species allowing fish to grow bigger and increase their numbers, resulting in better catches outside the area (spill-over effect, see for instance [111]. Thus, networks of offshore wind power plants can act like marine protected areas and therefore may be beneficial for the conservation of some threatened species [112, 113]. But created reefs can also create new living spaces for invasive alien species, or influence the native natural environment, if it replaces a former soft-bottom habitat with a hard-bottom community for instance, potentially resulting in replacing the original community with a totally new assembly [110].

Other marine traffic and activities such as fishing, recreational boats and cargo vessels are often displaced from the offshore wind energy area. While potentially benefitting some marine life via the reef-effect described above, displacement of activities might lead to increased pressure on biodiversity and nature elsewhere [114].

Underwater noise impacts on mammals are best studied in European waters. Studies show for instance that the harbor porpoise (*Phocoena phocoena*), harbor seal (*Phoca vitulina*), gray seal (*Halichoerus grypus*), and bottlenose dolphin (*Tursiops truncatus*) all avoid offshore wind project sites during construction because of the noise and vibration emitted [115–117], though seals seem to be less sensitive (or more adaptive) than cetaceans [116]. After construction has finished, some whales are attracted to the wind power plant area, most probably due to the reef effect and the increased availability of food in the area. The same is true for many fish and crustaceans (though effects vary from species to species): they tend to flee the sites during the construction because of turbidity, noise, and underwater vibration, and return once construction has ended [118, 119].

Many birds avoid offshore wind power plants during construction, and this can influence their migration, foraging, and resting behavior. During operation bird can collide with the turbine blades. Some bird species are also attracted by offshore wind power plants, potentially because of the availability of food (via the reef-effect), or long-distance migrators can use the pylons as resting sites.

Submarine power cables deliver the generated power from wind power plants to the onshore grid. Although underwater power cables have been in use since the mid nineteenth century, only recently have their environmental impacts been discussed. The temporary impacts on the underwater environment may include noise or chemical pollution, resuspension of sediment and pollutants, and damage or loss of underwater habitat during installation, maintenance, and decommissioning [120]. More permanent impacts include alteration of the habitat originating from the introduction of artificial substrates (which may allow new communities to form, replacing others) and risk of entanglement for larger fish and mammals. The long-term impacts from heat production, or through the emission of electromagnetic fields, leave the largest uncertainties regarding the potential impacts of submarine power cables on the underwater environment [120].

The onshore structures of the offshore wind energy production, like substations, grid connections, new road construction and expanding or building new harbors, can also alter formerly remote areas and make them more accessible for people. This might lead to new human settlements resulting in degradation of the natural environment (via changes in land use) and increased use of natural resources (via hunting, fishing, and other harvesting).

Impact of the wind wake of large offshore wind power plants on the hydrodynamic conditions in the ocean, and the resulting effects on marine ecosystems via primary production has received attention recently [20–22]. Primary production carried out mostly by plants and algae via photosynthesis form the basis of energy flow through any food chain in the ecosystem. It is kind of a chemical energy storage where living organisms form organic matter from inorganic carbon compounds. Increasing number of larger offshore wind power plants with larger turbines can change oceanographic and marine ecosystem conditions and change primary production not only just within the space required by the turbines, but the impacts can extend even hundreds of kilometers away [20, 22]. The detected consequences of wind wake include changes in stratification of the sea, reducing current velocities and circulation, decrease of bottom shear stress and potential to exacerbate oxygen deficiency in areas with low oxygen concentration [20]. Though still little is known about the exact ecosystem responses to wind wake this phenomenon has raised concerns because often the areas suitable for offshore wind power production overlap with highly productive marine ecosystems, hosting important feeding and nursery grounds for many marine species, such as at the coast of California [22], or the North Sea [20, 21]. Changes in the primary productivity in these kinds of locations can have severe, cascading ecosystem responses, including impacts on fish stocks. These research examples underline the importance of using precaution when planning large, offshore wind power plants. It is important to invest enough time on analyzing potential cumulative and cascading impacts and include monitoring programs into the plans to be able to avoid harmful impacts and build knowledge base and understanding.

Another little-known nature impact relates to chemical pollution stemming from the corrosion protection system of the turbine foundations. Corrosion protection is needed to protect the steel used in construction from the corroding effect of salt water so that the structures maintain their durability over the service life-time. Potential emissions from the corrosion protection include for example galvanic anodes emitting substantial amounts of metals, or organic compounds released from organic coatings (typically epoxy or polyurethan) [121]. These compounds can accumulate in marine sediments and filter feeding organisms like mussels. At current it is assumed that the environmental impacts of corrosion protection are low but there is not enough monitoring data to assess the impacts of pollution from this new source [121].

Case Study 5.2: Marine Mammals and Offshore Wind Power

Many species of marine mammal may interact with offshore wind power plants, although the nature and extent of these interactions will depend upon the location and characteristics of the development site. Harbour porpoises (*Phocoena phocoena*) are the most widespread and abundant species of marine mammal across European shelf seas [122]. Consequently, it is individuals of this species that have been most likely to be encountered within existing offshore wind power plants, requiring the development of measures to assess and minimize any risk of injury and disturbance to this European protected species. The harbour porpoise therefore provides a useful case study for understanding how we can assess the risk of wind power plant construction and operation to marine mammals and develop appropriate mitigation to reduce this risk under future development scenarios.

Harbour porpoises use high frequency sonar to navigate their environment and detect prey [123]. Although this species is considered particularly sensitive, potential injury or disturbance from the underwater construction noise has generally been the key concern for all marine mammals occurring in and around wind farm power plant [124].

In Europe, project level, assessments must demonstrate that the long-term viability of populations of these European Protected Species will not be impacted by the development. It's recognized that the key risks occur during construction, particularly during impulsive pile driving. In the near field, direct injuries may result from instantaneous pulses of extremely loud noise. Further afield, cumulative exposure to noise throughout piling cycles could also cause hearing damage. Finally, even in the far field, piling noise may disturb the animals' foraging patterns which, if this persists over long periods, could have longer term fitness effects [125, 126].

Typically, pre-consent assessments first require baseline surveys to estimate how many porpoises may be present in different parts of an offshore power plant site during construction [127]. Because piling noise may travel large distances, this also requires information on porpoise abundance and distribution across much larger areas surrounding power plant sites. Noise propagation modelling is then used to predict received levels of noise at different distances from the piling operation. Using available guidance on the noise thresholds that could lead to injury or disturbance, these two data sources can be integrated to estimate how many individuals may be affected. The consequences of these individual effects for protected populations can then be assessed using frameworks which model the population consequences of disturbance. These include frameworks such as DEPONS [126], developed specifically for porpoises, and others such as iPCOD [125], which can also be used for several other marine mammal species.

However, whilst straight forward in principle, there have been major uncertainties at every stage in this process. Data on variation in the porpoise population size and distribution is often based on very limited survey effort. Information on the source levels of piling noise and how this propagates through the water is also based on limited empirical data, and there is uncertainty over how existing measurements scale up when, for example, developers use increasingly large pile diameters and hammer energies. However, the largest data gaps, and most challenging to fill, exist in our understanding of how individual porpoises may respond to different noise levels, both in terms of their behavioural responses to noise, and in relation to potential cumulative damage to their hearing.

Following the precautionary principle, this uncertainty means that assessments of overall numbers of porpoises impacted by wind power plant development have been, intentionally, highly conservative. Despite this, when used within population modelling frameworks, short-term behavioural disturbance to relatively large numbers of individuals have not resulted in adverse impacts on the long-term conservation status of this species [125, 126]. However, given anticipated growth in offshore wind power plants, there is a need to reduce the level of conservatism in these assessments if we are to avoid unnecessary constraints on future developments. Research and monitoring carried out during the construction of recent North Sea windfarm developments has therefore aimed to reduce key uncertainties. For example, arrays of echolocation detectors and noise recorders were deployed on the seabed during the construction of the Beatrice and Moray East Offshore wind power plants in the Moray Firth, NE Scotland (Fig. 5.39). Occurrence of porpoises reduced by 8–17% during construction periods [128], but results indicated that fewer individuals were disturbed by individual piling events than had originally been predicted [129]. The behavioural dose–response curve developed from these data now provides more robust estimates of disturbance at different distances from piling vessels for environmental assessments required at future development sites.

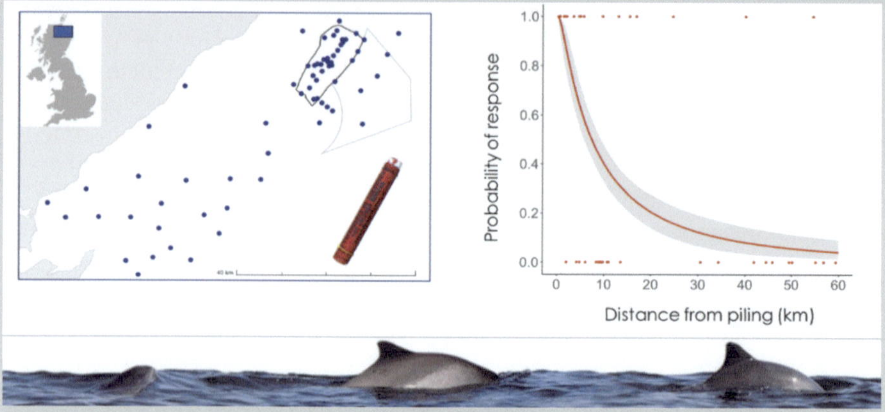

Fig. 5.39 Left: Locations of the CPOD echolocation detectors that were used to measure responses of harbour porpoises (Below) to pile-driving during the construction at offshore wind power plant sites in the Moray Firth NE Scotland. Right: Example of a dose–response curve base upon these data which can then be used in future assessments to predict how behavioural responses may vary at different distances from piling operations (from [129])

Whilst uncertainties over behavioural responses to piling noise can be reduced through new data collection, understanding of potential hearing damage remains dependent upon theory-based estimates [130]. Here, the focus has been on working with engineers to develop robust approaches to mitigating the risk of injury, particularly in the near field around piling vessels. In some countries such as Germany, legal limits for noise levels have encouraged the development of noise abatement systems, for example by using bubble curtains [131]. At the Moray Firth sites in Scottish waters, mitigation has involved a two-stage process [132] where a risk-based assessment was first made of the likelihood of porpoises being within injury zones at the start of piling. Once it had been demonstrated this this risk was extremely low, the residual risk could then be mitigated using acoustic deterrent devices prior to the start of piling. Although limits to piling noise levels may appear more precautionary given all these uncertainties, noise abatement methods may result in other environmental impacts, for example through additional vessels, higher resource use or extended construction timescales [133]. Indeed, monitoring of porpoises during construction of the Moray Firth wind power plants highlighted that porpoise occurrence and foraging was influenced by other vessel traffic as well as the piling noise that had been the focus of environmental assessments [128]. These findings highlight the need to explore these trade-offs when conducting environment assessment, and for engineers, regulators and environmental scientists to work closely together to optimize mitigation measures that can be applied practically and effectively in different contexts.

5.8.4 Summary Table

Table 5.17 summarizes the impacts on- and offshore wind power plants can have on nature and wildlife. Examples and the ways to mitigate negative impacts are drawn from scientific literature and discuss habitat loss (destruction, fragmentation, degradation), collision, electrocution, disturbance and pollution (noise, dust, light, vibration, solid and liquid waste), cumulative effects, trophic cascades and introduction of invasive alien species. The main purpose of the table is to summarize existing scientific literature of the impacts of wind power development on nature covering wide range of species, habitats and geographical locations. The list is not exhaustive so when starting to develop a wind power project, always visit the local information sources.

Table 5.17 Summary table of impacts, with examples and ways to mitigate for: habitat loss, collisions, avoidance and barrier effects, cumulative effects, trophic cascades and invasive alien species, respectively

Impact	Explanation	Onshore	Offshore	Ways to mitigate
Habitat loss: destruction, fragmentation. and degradation	Habitat is lost when it no longer can support its original species. Habitat can be lost via physical destruction, fragmentation, or degradation. The loss results in reduced biodiversity and changes in species abundance. The species can also be replaced by other, sometimes more generalist, species	Forest clearance can affect species through the loss of roosting and foraging habitats[1], or fragment a formerly continuous habitat into smaller patches with increasing edge[2]. The presence of vertical structures and movement of the blades can lower the quality of the habitat and cause avoidance or form a barrier. Clearance of vegetation increases run-off from land to nearby waterbodies and can change their status by making the water murky or euthrophied[3]	Bottom habitat is lost completely under a gravity base foundation. Former soft-bottom dominated habitats can be turned into a hard- bottom dominated habitat. Construction temporarily degrades the habitat via sediment plumes, smothering and noise[4]. During operation, changes in the water circulation can have an effect to the ecosystem, benthic communities and fish[5]	Avoid all constniction in sensitive areas and during sensitive seasons. Mitigate impacts during constniction by applying adaptive management and using micro-siting Limit clearance of the natural environment to a minimum, use restoration and offset residual damage. Implement measures to minimize soil erosion (drainage for access roads, avoid flooding land and/or damaging nearby water bodies)

(continued)

Table 5.17 (continued)

Impact	Explanation	Onshore	Offshore	Ways to mitigate
Collisions	Collisions with turbines (blades and towers) causing injury or death, or power distribution lines causing death via electrocution. At offshore wind power plants this can also include collisions with service vessels	Birds with low productivity and a late age of maturity (often large, carnivorous birds like eagles, hawks, hornbills, and herons) are more sensitive to the impact of increased mortality risk caused by collisions[6]. Conversely many wa- terbirds (e.g.. anseriformes) and songbirds do not collide with blades so often[7] Bats, especially long-distance disperses. have higher collision rates than birds[8]. Some bats are also attracted by the turbines	Migrating, foraging, and hunting birds and bats can be attracted by nacelle lights and are at risk of collision[9]. In Scotland gulls, white-tailed eagles, northern gan- nets and skuas were found to be most vulnerable to collision mortality[10] Marine mammals and sea turtles are at risk of collision with vessels when they surface to bask, breathe or eat[11]	Avoid siting turbines and transmission lines on migratory flyways, bat habitats and commuting routes. Fewer, large turbines can result in fewer collisions than many smaller ones[12]. Align turbines in parallel when there is a clear route for migration[13]. Shut-down, or slow-down ftirbines during specific seasons (like migration), or on demand. Paint blades and mark transmission lines to increase visibility[14] Collision with vessels: avoid sensitive seasons and reduce vessel speed[15]

(continued)

Table 5.17 (continued)

Impact	Explanation	Onshore	Offshore	Ways to mitigate
Avoidance and barrier effect	Species can be disturbed and avoid the wind power site during different phases of the project. Avoidance can be temporary (for instance due to noise, vibration, dust or sediment plumes during construction), or permanent, and occur of a radius of several kilometers[16]. The barrier effect occurs when the structures prevent an individual's movement from one area to another (feeding, wintering and breeding areas for example)	Species adapted to open landscapes can be disturbed by artificial, high structures such as wind turbines. Studies have found changes in habitat use by female greater sage- grouse and female greater prairie- chicken along with increased lek (= an area for mating display) abandonment among male greater prai- rie-chickens in locations near wind turbines[17] Black kites[18], European roe deer and hare[19] change their habitat use close to turbines, while wolves have been found to avoid breeding close to wind power plants[20]	At sea diver birds, common scoters and guillemots avoid offshore wind energy sites[21], and migrating raptors can adjust their flight trajectories to avoid wind power sites[22] Seismic surveys during the site characterization and construction can pose a risk to marine mammals, crustaceans, cephalopods. and fish[23]. During piling activities, harbor porpoises move as far as 20 km away[24]. This disturbance is temporary	Construction should be avoided during sensitive seasons. Mitigations should include understanding species-specific avoidance behavior and establishment of buffer zones around important areas, like breeding, resting, wintering, foraging and roosting areas, and along migration routes[25]. At sea important juvenile fish recruitment areas and seasons for marine mammals and fish should be avoided, and noise levels and sediment plumes reduced[26]

(continued)

Table 5.17 (continued)

Impact	Explanation	Onshore	Offshore	Ways to mitigate
Cumulative impacts	Cumulative impacts are changes in the environment that are caused by the combined effect of multiple human activities and natural processes. They are often difficult to predict and manage because of lack of sufficient environmental data, complexity of ecological processes, and the large scale of human activities	Long-distance migrants, like the Egyptian vulture, are exposed to human-induced mortality along their migration routes. Any additional mortality (like increased collisions with power transmission lines) can be significant[27]. Observations shows that hoofed mammals suffer reduced recruitment and change their habitat use in areas that have undergone progressive residential and energy development[28]	The cumulative impacts of offshore wind projects have not been studied extensively. Potential impacts include cumulative barrier impacts, habitat loss, underwater-noise impacts and collisions with vessels due to multiple wind power plant projects.[29]	Wind power production should be considered in relation to other human settlements in the area. It is important to have knowledge of the species in the area, particularly migrants and wide-ranging species[30]. Impacts can be mitigated by including corridors for species movement into the plans

(continued)

Table 5.17 (continued)

Impact	Explanation	Onshore	Offshore	Ways to mitigate
Trophic cascades	Indirect and strong interactions that can affect the entire ecosystem are called trophic cascades. Trophic cascades have an impact across at least tliree levels of the food web. via predator–prey interaction. In its simplest form predators limit the density of their prey and this in turn enhances the survival of the next lower trophic level. Human disturbance can convert these interactions resulting a change in the balance of the entire ecosystem	Reduced abundance of predatory birds led into increased density of lizards. The cascading effects included changes in the lizards' behavior, physiology and morphology that reflected a combination of a predation-free environment and density-dependent competition over food and other resources[31]	Changes in benthic habitat and hydrodynamic conditions can affect species composition and abundance and change predator–prey dynamics, altering the trophic interactions. Some fish can be attracted by the wind turbine structures, and this attracts foraging mammals and birds[32]. Structures also provide a substrate for mussels that play an important role in modifying the habitat and inducing ecological change through filtration[33]	After construction restore any areas no longer in use, to allow the original nature to return. Particularly important areas include grasslands, forests, and coastal ecosystems—such as dunes, mangroves, and oyster beds. Take preventive measures to protect waterbodies from erosion. Have enough knowledge of the local natural environment and biodiversity to be able to intervene appropriately to lessen or eliminate negative outcomes of the project

(continued)

Table 5.17 (continued)

Impact	Explanation	Onshore	Offshore	Ways to mitigate
Invasive alien species, AIS	Invasive alien species are plants, animals, pathogens or other organisms that are non-native to an ecosystem. They can have an adverse impact on biodiversity, ecosystem functioning, the economy or human health by causing a decline or loss of native species in local ecosystems	Creation of new open spaces, land disturbance and new access roads can facilitate the dispersal of IAS. Also, machinery, transferring soil and people may introduce IAS. Invasive alien species have been found to spread on wind power sites and maintenance roads enable the species to expand their range further along the road sides[34]	Invasive species can be introduced to a wind power site via vessel hulls and in ballast water, or with equipment[35]. Constructions can act as steppingstones for the dispersal of IAS[36]	Prevent introduction and movement of invasive species by washing vehicles before entering the area. Don't empty ballast waters in the offshore wind energy site. In landscaping, use soil from local or otherwise known sources to prevent spread of invasive species

1: [75, 134, 135], [136], [136]. 2: [17–19]. 3: [137]. 4: [41]. 5: [138, 139]. 6: [140–142]. 7: [142] 8: [143]. 9: [97, 144–147].10: [76].11: [148, 149]. 12: [142, 143, 150]. 13: [41, 151]. 14: [41, 78, 150]. 15: [152]. 16: [62, 80]. 17: [11, 84, 85]. 18: [12]. 19: [13]. 20: [14]. 21: [76, 153, 154]. 22: [16]. 23: [41, 155, 156]. 24: [116]. 25: [14, 85, 157]. 26: [41, 66]. 27: [158]. 28: [88]. 29: [41]. 30: [142]. 31: [63]. 32: [159]. 33: [113]. 34: [160]. 35: [161, 162]. 36: [113, 163]

5.9 Social Aspects of Wind Power

This section discusses the societal impacts and implications of wind energy projects and presents some tools which can be used to analyze societal impacts. The focus of this section is on how to create wind energy production that is regionally, economically, and socially equitable.

The main reason why we produce wind energy today is to provide societies with cleaner low-carbon energy; this is the starting point of the societal impacts of wind power. Wind power, as well as all other energy production concepts, have many different types of impact on surrounding communities. Most often, when discussing societal impacts of energy production, the focus is on how much gross domestic product (GDP)—that is how much added monetary value—the energy production generates and how much employment the energy project will bring. Together these measures impact human welfare at the national level. These are measurable and mostly positive societal impacts of wind energy production.

The International Renewable Energy Agency (IRENA) have measured different socio-economic impacts of renewable energy production. In 2021 wind energy production employed 1.3 million people worldwide [164]. The industry's employment effects have been rising since then, as wind power generation has increased

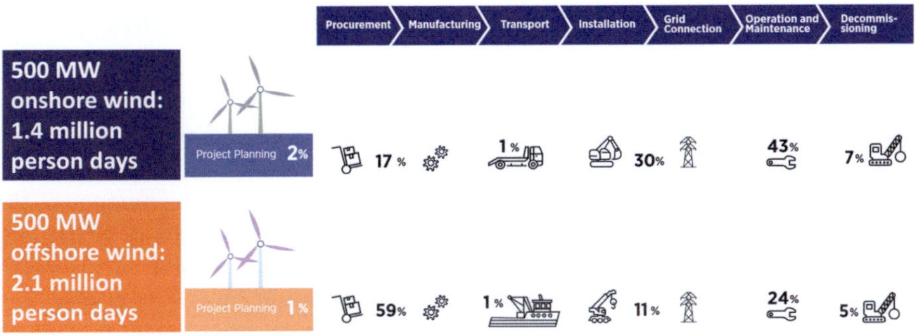

Fig. 5.40 Distribution of human resources and occupational requirement along the value chain for differ types of projects (After [166], based on data from IRENA)

globally from 1127 TWh (2017) to 1592 TWh (2020)—a 40% increase in three years [165]. Different phases of the value chain employ people differently. Based on IRENA's (2019) estimates, the most employee intensive stages are procurement along with operation and maintenance of facilities (see Fig. 5.40).

If we think about individual wind energy projects such as wind power plants, there are several local societal impacts that need to be considered. These include the benefit for local communities, welfare and health, the cultural importance of the region for local communities, livelihoods (both on land and at sea) and the social acceptability of the technology, which includes local communities' values and attitudes towards the project. Social acceptability has gained most attention although it is only one aspect of local impact. This is due to local resistance towards wind energy projects that regularly gain media and social media attention and the NIMBY (not-in-my-backyard) phenomenon, which means that although a person would accept the development of wind energy, it is not desired nearby one's home or region. The NIMBY argument is usually associated with unwanted land uses voiced by wealthy communities or other segments of society. The term is not new and has had scholarly attention for some decades [167–170]. The term NIMBY has been used widely in public discussions, especially in the early 2000s, but there has been less scientific support for the phenomena and in fact other issues such as policy and regulation has been found to slow down wind power projects more [171, 172]. Similarly, the acronym LULU has been used in the literature to denominate 'local unwanted land use'. The term LULU is usually used when homeowners oppose a project that would impact their neighborhood directly. They fear that such projects will have negative indirect consequences on their assets, such as devaluation of land or property value [173]. Both NIMBY and LULU are commonly framed as self-interest activism and driven by the motivation to protect their own turf [173].

5.9.1 Benefits and Health Impacts for Local Communities

One of the most important questions that needs to be answered while planning wind energy projects is how local communities will benefit the most. An example of local monetary impacts from Finland is where wind energy firms pay a rent to landowners for use of the land, and municipal land tax for using the site. The amount of the land tax is dependent on the land value and the size of the wind facility's buildings and power production. If the power plant capacity is less than 10 MW the tax varies between 0.93 to 1.80% and is the same as any building owner pays. If the power plant is larger than 10 MW then the power plant pays a special power plant tax, which can be as high as 3.1%. In Finland, local municipal authorities use the gathered tax to provide different public services, which benefits society. However, this is not always the case globally. Although, benefits and compensations are sometimes paid to people living near wind power plants, there are many other ways in which local communities can benefit from hosting wind energy facilities. For example, residents living near to a wind power plant could be offered discounted community electricity prices.

The noise and lighting associated with wind turbines not only disturbs wildlife and farm animals, but can also impact people living nearby, for example by causing sleep disturbance and noise annoyance [174, 175]. Several studies show that wind turbines located less than 1–2.5 km from the nearest human settlement increase sleeping disorders, chronic health conditions, impaired hearing, and noise sensitivity (see review on noise effects by [176]). In addition to distance from windmills, the level of the noise influence on living conditions. The World Health Organization (WHO) recommends an upper limit of 45 dB for wind turbine noise, but several countries have even lower-level recommendations [175]. At levels less than 40 dB the impacts are associated mostly with sound annoyance rather than health effects, such as stress or blood pressure levels [175].

Furthermore, if wind turbines are equipped with lighting to alleviate aviation collisions, the lights can impact on people's comfort if they are located too close to settlements, or on sites where visibility (for example over a lake) is very clear. Other health impacts can include the fear of blades coming lose, or even turbines collapsing, if the locations are not well chosen or the foundations poorly designed and constructed, as was the case in the Camocim municipality of Ceará state in Brazil in 2008 [177]. Wind power development projects such as Camocim in northern Brazil have gained the interest of scholars, particularly because of their harmful social impacts. A scoping review revealed that projects in that region were likely to cause socio-ecological problems, such as deforestation, expropriation of land, turbine noise, insignificant job generation, persistence of inequalities, migration to urban areas, extinction of cultural traditions, and food insecurity [178]. To overcome these issues, more emphasis should be given towards securing the land tenure for local and indigenous communities, providing royalty payments to these communities, establishing legal provisions to regulate the implementation of wind energy in the region, using good communication strategies to share information on the potential social benefits and possible harms associated with the project,and promoting public awareness and social participation [178].

5.9.2 Recognizing Local Livelihoods and Regional Justice

Different regions and areas have different meanings to local communities. When planning wind power plants, it is extremely important to understand the importance and the meaning of the land or waterbodies for local communities, because wind power plants do change the landscape and require a series of procedures and processes—for example cutting down forests, clearing the ground or sea bottom, installing alert lights on poles etc.

It is not uncommon to see evidence of local complaints about the way in which wind energy firms, in collaboration with government, have dealt with the regional impacts of wind power plants on the environment and people's livelihoods [177]. Most often the cases have led to local compensation or relocation of the local livelihoods and homes, especially in developing countries. Resettlements have a history in such areas, caused by other development projects such as dams and hydropower plants [179] leaving a legacy of conflict, fear, and suspicion in the impacted population.

The visual landscape effects of wind energy production should not be underestimated. Wind turbines are large facilities that can be detected from far away. Reference [180] compared different mast-like structures (wind turbines, chimneys, and transmission towers) and found that the visual image of wind power is largely tied to moral judgement of whether a respondent considers the production of wind energy a good thing. Paying careful attention to how to restore nearby landscapes after installing wind turbines, and how to reduce the visual landscape effects of wind turbines, may even foster positive attitudes towards regions with wind power areas and change the visual image of wind turbines within the landscape. But this aspect needs further research.

5.9.3 Social Acceptance and Social License to Operate

Renewable energy technologies such as solar, wind and green hydrogen are being deployed on a global scale. The integration of these technologies has significant impacts on different stakeholders and parts of society. Concepts related to low-carbon technology acceptance attempt to identify, map, predict, measure, and respond to the inherent risks involving stakeholders (such as local communities, consumers, non-governmental organizations (NGOs), and local authorities). Understanding the perception of these stakeholders is of paramount importance in the implementation of new energy solutions.

If the above-mentioned societal impacts are not considered when planning wind energy projects, the public acceptance of wind energy naturally reduces dramatically. This is especially the case if there are no legal requirements nor voluntary process to include members of the public and the local community and encourage them to participate in the planning phase. Different countries have different legal processes—such as the environmental impact assessment (EIA)—which include some societal impact assessment tools and the possibility for members of the public to give their opinion on the project.

The most prominent study on social acceptance of energy projects can be traced back to the framework proposed by [181], which introduced the 'triangle of social acceptance of renewable energy innovation' (Fig. 5.41), and accounted for socio-political, market and community acceptance. An empirical study carried out in Germany revealed that communities with strong identity, trust, social norms, and higher than average environmental concerns were positively associated with the willingness to participate in community energy projects [182]. Concepts such as energy justice are growing in relevance in the literature and are closely associated with social acceptance of projects. A previous study explored 71 renewable energy community cases in Europe and concluded that projects actively contributing to energy justice "by engaging with vulnerable and underrepresented groups and providing access to beneficial services to alleviate energy poverty remain the exception" [183]. Other more established industries, however, have been tackling risks and tensions involving societal actors with different tools and concepts. One such example is the social license to operate (SLO). With seminal literature coming from Canada [184], consolidated in Australia, and then reaching European and Latin American scholars, SLO has been used especially by the mining industry, with some relevance to the forestry industry and wind energy projects as well. The three core aspects integrating SLO are: trust, credibility, and legitimacy. Failing to comply with these principles can result in long-lasting and costly legal processes as occurred in Norway when the Norwegian Supreme Court ruled that an already constructed wind power park with 151 turbines violated the rights of the indigenous Sámi people, who have used the land for reindeer husbandry for centuries [185, 186].

In Finland, the most common impacts and perceptions associated with wind power and reported by local communities are (see, for example [106, 175, 187–190]):

- Impacts on health associated with wind turbines noise.
- Visual pollution (e.g., lights installed on the towers), landscape changes which can impact tourism.
- Wildlife and vegetation impacts (especially bats, birds and fish, when offshore).
- Impacts associated with decommissioning.
- Perception of wind power as an inefficient form of land use.
- Perception of wind power as a solution that is expensive to set up.
- Perception that there is an imbalance in the harm/benefit distribution in the rural–urban divides.
- Suspiciousness about the lack of local inclusion in wind power decisions.
- Suspiciousness about the fairness of local benefits.

Fig. 5.41 The 'triangle of social acceptance of renewable energy innovation' after [181]. The framework illustrates the social acceptance of energy projects which can be traced back to socio-political, market and community acceptance. © Maija Karala

- The need to educate local farmers with this alternative use of land.

In this section we have raised some critical societal concerns about wind power projects, but there are also good examples of how wind power has been negotiated, planned, and implemented, in a manner that considers local societal impacts. One such example is the establishment of a collaborative platform between reindeer farmers and wind power operators in Northern Finland (Case Study 5.3). Another good example is the Suolakangas wind power park, established in Finland in 2017. In this example, the conditions of the wind power project were agreed at the outset and included siting turbines at least 2 km from homes and limiting noise to 35 dB around holiday houses. There was a lot of interest in reconciling any disputes early in the project (including those associated with nature, holiday homes, nationally valuable landscape, recreation etc.). The key aspects that contributed to the success of this project were: (1) the trust that was built between the different parties in the process, (2) locals and other relevant actors got a feeling that their needs were considered, and changes were made to the process, and (3) the conditions the locals had set for the planning were obeyed. This shows that open negotiations, good planning and making compromises are very important success factors for establishing wind power plants.

Case Study 5.3: Wind Power Meets Reindeer Husbandry in Northern Finland

The northern regions across Fennoscandia are attractive for wind power development due to their sparse population and good wind conditions. In Finland, there is a clear strategic goal outlined by the government to increase the share of wind power considerably. In northern Finland wind power sites almost always overlap with reindeer husbandry areas. This co-occurrence generates encounters, or rather, clashes, between these two activities. Such encounters often result in bewilderment and frustration. The main reasons for this are the lack of timely communication and insufficient understanding of each other. This in turn results in mutual mistrust and an inability to negotiate and seek agreement. Here we present a case of building trust and mutual understanding between the two industries via assisted collaboration in Finland.

Studies have shown that the production of wind power can have a significant effect on reindeer herding [191] and that specific understanding and action regarding reindeer herder's rights is required within the project [192, 193]. Reindeer herding is based on pastures that provide areas for grazing at different phases of the herding cycle. In Finland the livelihood is protected by Reindeer Husbandry Act which protects the right to practice reindeer herding anywhere in the reindeer management area (RMA) regardless of land ownership (Fig. 5.42). Lands used as pastures are impacted by several forms of land use such as forestry, infrastructure construction, tourism, and mining. Increasing pressure from these various land uses shrink and weaken the quality of the area from the point of view of reindeer herding and reindeer husbandry [194, 195]. The effects of climate change are also most severe in the northern areas, which adds to the struggles of reindeer herding as a livelihood, and as a cultural heritage.

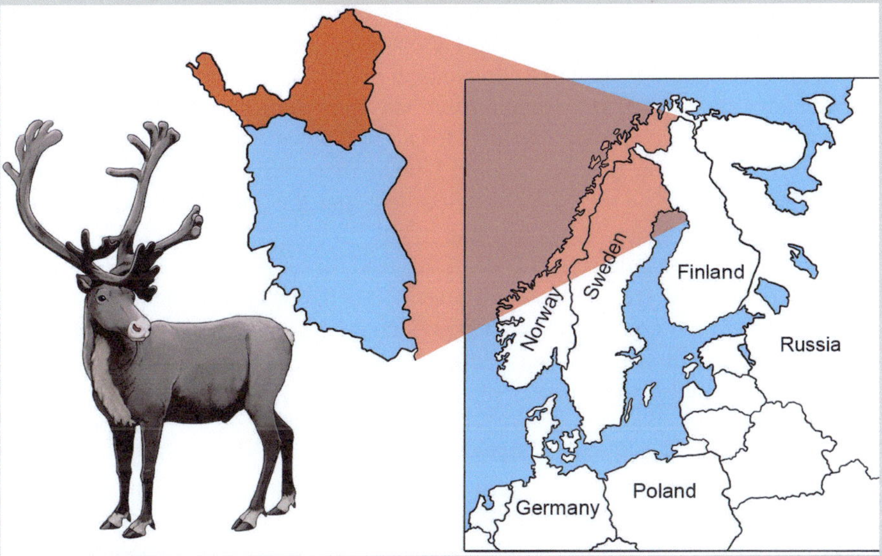

Fig. 5.42 The reindeer husbandry area in Finland (blue and red areas). A large portion of wind power development is currently located in the areas of RMA marked by blue. Area in red is the Sámi homeland where the Sámi people have constitutional self-government in the spheres of language and culture. Sámi people are the indigenous people of the Northern Europe, living in Sápmi, a cultural region stretching over four countries: Finland, Norway, Russia and Sweden © Maija Karala

The Step-by-Step Process Towards a Long-Standing Collaborative Platform

The Finnish Wind Power Association contacted Akordi—a neutral third-party actor and a not-for-profit company, specializing in environmental collaboration and conflict resolution—in the fall of 2019 to seek common ground with reindeer herders. Akordi initiated discussions with the Reindeer Herders' Association, who agreed to participate. The first step was to open a direct channel of communication, to approach the shared problem in a way that would be meaningful for both sides.

After several informal meetings mediated by Akordi, the two associations started to share an understanding of the current situation, and of each other's perspectives and needs. The idea of coming together at a joint summit gained support, and in September 2021 the members of the two associations, along with public officials and Akordi, convened to discuss their needs and deepen mutual understanding. Prior to the summit field visits to areas of reindeer herding cooperatives and wind power plants were organized, to promote dialogue and get to know each other.

During the summit participants engaged in small group dialogues where they shared their experiences and concerns about issues and voiced their needs. The wind power operators started to understand more about the nature of reindeer husbandry and the emerging boundary conditions, which need to be considered when looking for a suitable location for a wind power project. Similarly, the reindeer herders felt that they would benefit from being more aware of the different stages of wind power development and the steps that precede the actual decision making and construction.

A shared understanding emerged of the need for timely communication and information exchange, along with a sustained collaboration to identify where, and under which conditions wind power could be acceptable. The meeting resulted in a joint effort to draft recommendations for best practice for planning and operating wind power plants in the Finnish RMA (Fig. 5.42). For reindeer herders, learning about possible project plans as early as possible was considered important. Early engagement was seen as an indication of trust and respect. The commitment to minimize harm and compensation for residual damage were considered essential elements of an acceptable solution. The wind power operators on the other hand felt they would benefit from a better understanding of reindeer herding practices to be able to make informed decisions.

A working group was assembled to continue the dialogue, with the task of drafting guidelines for best practice for wind power projects in reindeer herding areas. Experts from the Reindeer Herders' Association, the Finnish Wind Power Association and Akordi worked together to finalize the draft and formulate: 'The guidelines for best practices for operating wind power plants in reindeer husbandry area'. The guidelines outline best practice at each stage of the wind power project, from the early stages, to planning, construction and operation.

The guidelines for best practice were presented and further discussed in the second joint summit between the reindeer herders, wind power developers and public officials, held in September 2022. The outcome of the summit was very action oriented, and furthermore, everyone was moving in the same direction. There was a clear understanding that what was needed next was to implement the new guidelines, and further strengthen the collaborative platform via regular meetings. The joint practical task was to seek resources to sustain the collaboration. The first steps towards building a longstanding collaborative platform are illustrated in Fig. 5.43.

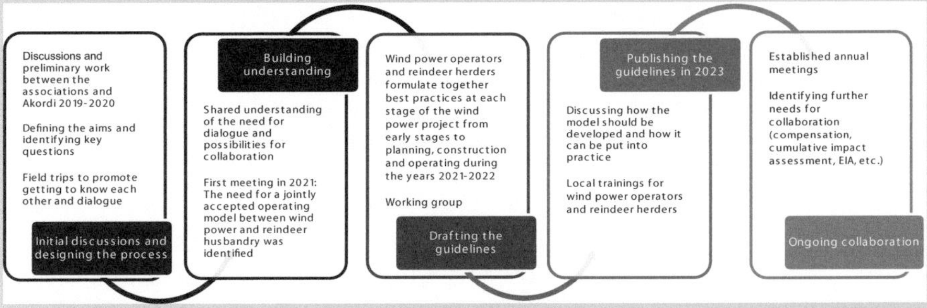

Fig. 5.43 The collaborative process of formulating the guidelines for good practice for operating wind power projects in a reindeer husbandry area

The best practice guidelines were published in January 2023 [196]. To turn the guidelines into reality a series training workshops was organized during spring 2023. The training workshops were designed to operationalize the new way of doing things, as outlined in the guidelines. This included the need to start to jointly understand and discuss complex and new issues, such as restoration or appropriate compensation for lost pasture areas (planting lichen for example). An important function of the training workshops was to bring together wind power operators, reindeer herders and the municipal land-use planners to learn skills and methods of negotiation with mutual gains. The training workshops had yet another important function, which was to help people get to know each other, and build social capital and trust which had previously been lacking.

The work described and the recommendations for best practice for wind power development apply to the reindeer husbandry area outside (south of) the Sámi homeland. The Sámi are the only indigenous people of the European Union inhabiting the northern parts of Norway, Sweden, Finland, and of the Murmansk Oblast, Russia (Fig. 5.42). Reindeer herding is one of the cornerstones of the Sámi culture. The Sámi community in Finland has clearly stated that wind power development is not acceptable in the Sámi homeland area due to the cumulative impacts of all land uses on the reindeer pastures; hence the homeland area was not part of the process described here.

Using a Neutral Third Party When Building Collaboration

What made it possible for the wind power operators and reindeer herders to come together and engage in a safe and trusting dialogue about the different needs and concerns regarding the two livelihoods? The foundations for the successful collaborative process described above were laid in previous encounters and experiences, resulting in trust and good working relationships between key personnel, enabling them to advance the collaborative endeavor together. This case demonstrates that even in situations where there is a desire to use land in a number of different ways, there are ways of seeking and finding mutually acceptable solutions. Involvement of a neutral third party and careful design of a step-by-step process enabled wind energy operators and reindeer herders to voice their concerns and build a missing platform for communication and problem solving, as well as form a basis for future collaboration (see for example [197–199]).

For Finnish wind power operators it is necessary to find ways of improving their social acceptance if they wish to operate in the RMA. In practice this means that wind power operators must be ready to collaborate with reindeer herders early on and study the local effects of the projects together with them while actively listening to the concerns relating to the construction of wind power plants. Engaging in collaboration beyond singular projects has enabled actors to learn how to anticipate possible contradictions and prevent conflicts.

The long history of disputes and lack of trust in the area emphasize the importance of having a neutral third party, whose role is to help design the process and mediate any potential conflicts [200]. In the process described here the neutral party helped to ensure the acceptability of the process by drafting goals for the collaboration in a way that would align with the interests of both groups.

This kind of joint problem solving has been found to result in sustainable outcomes and increased trust between the parties [201]. In the long run a well-organized and established collaboration between different parties (a collaborative platform) can help all parties anticipate and adapt to changes in the operating environment.

5.10 Summary

Traditionally wind power development has focused primarily on the technical and engineering sciences, such as wind conditions, atmospheric physics, fluid dynamics, composite structure design, the technical and material requirements of the turbines and electrical engineering [202]. However, as we have discussed in this book, overcoming the physical and technical challenges is just the first step. Expanding the use of wind power in accordance with the objectives of decarbonizing the energy sector and increasing the share of renewable energy sources requires better understanding, inclusion, and mitigation of ecological and social impacts. Wind energy production can impact terrestrial and marine life and ecosystems and local communities in many ways. Overlooking the impacts on nature or local groups' rights, can result in costly and long-lasting legal battles and prevent construction of energy projects [203], U.S. Department of Justice, Office of Public Affairs 2014, [204].

In this book we have decided to discuss the energy solutions in relation to three of the planetary boundaries: climate change, biosphere integrity and land system change. Regardless of the never-ending technical improvement, wind power is a mature and ready-to-use technology both on land and at sea. It is a pivotal source of energy on our journey towards carbon negative societies. And it is one of the solutions to bring the planetary boundaries associated with global heating back within their limits. Proper siting is the main question that determines if this technology works against or for the planetary boundaries that exist around biosphere and land use. Wind power can and should be harnessed to promote maintaining and restoring all planetary boundaries, preserving a safe operating space for humanity.

Wind power production relies on the wind conditions and therefore varies constantly. The capacity factor is used to describe the fraction of annual hours available for rated power production. The typical capacity factor is ~ 0.35 onshore, and ~ 0.45 offshore. These values mean that in order to produce the required energy amount of 4.38 TWh × (= 500 MW × 8760 h), we need to have a much larger fleet of installed turbines than 500 MW, and we need to utilize the flexibility of the electric power grid, use energy storage solutions and the demand side response to smooth the power output to 500 MW. Off course this constant 500 MW power requirement is artificial, but gives an idea of the challenge of connecting the large-scale variable production in the energy system. For onshore and offshore wind power, the required installed capacities are 1429 MW and 1111 MW, respectively. These numbers do not include the losses in the storage system; these are discussed in Chap. 7 Energy storage systems. To reduce the need for energy storage and stabilize the electrical grid, wind turbines can be designed to produce constant power over a larger range of wind speeds, which increases the efficiency of the interaction with the electrical network. 'Over-sized rotors', that are able to reach the rated constant power at low wind speeds, could be used.

Modern wind turbines use a variable speed system that maximizes power production by controlling the turbine rotational speed with the wind speed. This is made possible by the converter technology development that can produce output power with a constant grid frequency, even though the rotational speed of the rotor varies. To decrease the need for maintenance, state-of-the-art direct drive systems with no gearbox are used, especially in large offshore turbines sited far away from the shore. However, this creates a new challenge in terms of the demand for rare earth materials, such as Neodymium which is used in permanent magnets in direct drive generators. Anticipating the rapid expansion of wind power and the possibility of limited material resources, the optimum technical solution to maximize the sustainability of the increasing number of turbine fleets needs to be determined.

The above-mentioned themes form the technical framework for wind power plant development. But how do we overcome the associated ecological and social challenges? While the technical challenges are more universal in nature, most of the ecological and social challenges are local or regional, and case dependent. One limitation of the land use assessments typically used with wind power plants is that only directly impacted area is assessed meaning that the total impacts of the wind power plant are not necessarily properly captured. The total impacts, and the best siting options and shape of the wind power plant depends on the natural features of the site in question, along with distance from human settlements and local livelihoods. Typically, a distance of 5–7 rotor diameters is the required optimal spacing between the turbines to reduce wake losses. On the other hand, there is a need to minimize land use, to minimize the disturbance on nature. One suggestion is that instead of scattering turbines sparsely over large areas, it would be better to concentrate the turbines in larger gigawatt-sized power plants. This leaves a greater undisturbed area for nature and decreases the edge effect of the impact area (a larger continuous area has a smaller perimeter impacting the surrounding nature). Specific turbines can also be sited so that the length of new roads is minimized to minimize the loss and fragmentation of natural habitats. The ideal siting plan is a compromise of the area's natural features, topology, prevailing winds and the need to minimize the impact of wake on power generation. All in all, there is a growing need for robust tools to unify and assess the total impact area of wind power plants also giving a measure of the fragmentation area and decrease in habitat quality.

The direct impacts of a wind power plant come from the construction of the turbines, access roads and power lines and can include loss and fragmentation of natural habitats, including freshwater habitats, pollution and sedimentation of aquatic ecosystems, barriers to movement, dust, noise and vibrations. The indirect impacts or wind power development can be an end result of the changes in the environment, or resulting from new or increased human activity that is made possible by the improved road access. The cumulative impacts of several wind power projects, and combined effects with other human activities can create impacts that potentially threaten the long-term survival of some species in the project area, and if the same threat is repeated in many places, the combined impact can make the species even more vulnerable.

Avoiding harmful impacts (considering also cumulative effects), the first step of mitigation hierarchy, is crucial, and should be applied throughout the project, prior the construction (macro-siting) and in an adaptive manner during the construction of turbines (micro-siting). Protected areas, intact landscapes, and areas with important biological features should be left outside of wind power production areas. Where possible, wind power development on already disturbed lands should be favored, so as to reduce negative impacts on nature and potentially reduce the need for clear-up space for power lines and other support structures, like maintenance buildings. Adverse impacts can be minimized through applying good practices throughout the life cycle of the project. These practices include minimizing the clearance of natural vegetation during construction and locating any temporary activities on already disturbed land if possible, avoiding flooding and damaging natural streams, take measures to prevent soil erosion, handle any solid and liquid wastes carefully in particularly at offshore environments, prevent the transmission of invasive alien species via vehicles or soil, secure that locally sources materials (such as gravel) are legitimate and sustainably caught, restore temporary areas after construction, and set post-construction monitoring for the impacts on nature. Post-construction monitoring is important in improving the knowledge base of the impacts of wind power on nature. Understanding better the true extent of the impacts helps to improve the siting planning and develop practices for wind energy production.

Inclusion of local people and communities in the planning process is important. To find truly effective solutions, benefiting everyone—society as a whole, wind power companies and local residents—it is worth investing time in building processes that build and foster trust. One such example was presented and discussed in Case Study 5.3.

To summarize, continuous work, willpower and interdisciplinary cooperation is needed to address, understand and find solutions to execute wind power projects in a way that produces multiple benefits and does not seriously harm the nature or violate the rights of local communities. Mutual benefits should be sought, such as new business and job opportunities. Wind power also creates health benefits by emitting fewer pollutants and therefore improving air quality. On this path towards net zero and, eventually, carbon negative energy systems, we need innovation and the development of technologies that minimize material consumption and maximize materials recycling. We also need to understand how increasing the share of variable wind power production affects the stability of the electricity grid and how can we secure energy availability during times with no wind. Finally, we also need a better understanding of the long-term and cumulative impacts of wind power on natural and social systems, to identify gaps and develop a shared knowledge base with which to build best practices for a cleaner and just future.

Author contributions 5.1. Mika Järvinen and Hanna Paulomäki, 5.2 Mika Järvinen, 5.3. Mika Järvinen and Rogier Ralph Floors, 5.4, 5.5. and 5.6. Mika Järvinen, 5.7. Rogier Ralph Floors and Mika Järvinen, 5.8. Hanna Paulomäki and Christian Pansch, Case study 5.1. Hanna Paulomäki, Case study 5.2. Paul Thompson, 5.9. Eeva-Lotta Apajalahti and Mariana Galvão Lyra, Case study 5.3. Emma Luoma and Jonna Kangasoja, and 5.10. Mika Järvinen and Hanna Paulomäki. Subsections can be referenced accordingly.

References

1. Rockström J, Gaffney O (2021) Breaking boundaries—the science of our planet. Published by DK, 240 pp
2. Cleveland CJ (ed) (2009) Concise encyclopedia of the history of energy. Academic Press
3. Smil V (2017) Energy and civilization—a history. The MIT Press, Cambridge
4. Manwell JF, McGowan JG, Rogers AL (2002) Wind energy explained: theory, design and application. Wiley, Great Britain
5. Lundsager P, Frandser S, Christensen CJ (1980) Analysis of data from the Gedser wind turbine. Riso national laboratory, Danmark
6. United Nations (2021) Life cycle assessment of electricity generation options. United Nations economic commission for Europe, UNECE, GENEVA. https://unece.org/sites/default/files/2021-10/LCA-2.pdf
7. Dohm R, Drake D (2019) Wind energy effects on birds. In: Moorman CE, Grodsky SM, Rupp SP (eds) Renewable energy and wildlife conservation. John Hopkins University Press, pp 95–121
8. Fritsche U, Berndes G, Cowie A, Johnson F, Dale V, Langeveld H, Sharma N, Watson H, Woods J (2017) Energy and land use. Global land outlook working paper. https://doi.org/10.13140/RG.2.2.24905.44648
9. Ledec GC, Rapp KW, Aiello RG (2011) Greening the wind: environmental and social considerations for wind power development. World Bank Studies. http://hdl.handle.net/10986/2388

10. McDonald RI, Fargione J, Kiesecker J, Miller WM, Powell J (2009) Energy sprawl or energy efficiency: climate policy impacts on natural habitat for the United States of America. PLoS ONE 4(8):e6802. https://doi.org/10.1371/journal.pone.0006802
11. Lloyd JD, Cameron LA, Taber DA, LeBeau CW, McNew LB, Winder VL (2022) Prairie grouse and wind energy: the state of the science and implications for risk assessment. Wildl Soc Bull 46:e1305. https://doi.org/10.1002/wsb.1305
12. Marques AT, Batalha H, Rodriques S, Costa H, Pereire MJR, Fonseca C, Mascarenhas M, Bernardino J (2014) Understanding bird collisions at wind facilities: and updated review on the causes and possible mitigation strategies. Biol Conserv 179:40–52. https://doi.org/10.1016/j.biocon.2014.08.017
13. Łopucki R, Klich D, Gielarek S (2017) Do terrestrial animals avoid the areas close to turbines in the functioning wind farms in agricultural landscapes? Environ Monit Assess 189:343. https://doi.org/10.1007/s10661-017-6018-z
14. da Costa G, Petrucci-Fonseca F, Álvares F (2017) 15 years of wolf monitoring plans at wind farm areas in Portugal. What do we know? Where should we go? [Presentation]. Presented at 4th conference on wind energy and wildlife impacts (CWW 2017), Estoril, Portugal
15. Searle KR, O'Brien SH, Jones EL, Cook ASCP, Trinder MN, McGregor RM, Donovan C, McCluskie A, Daunt F, Butler A (2023) A framework for improving treatment of uncertainty in offshore wind assessments for protected marine birds. ICES J Mar Sci. https://doi.org/10.1093/icesjms/fsad025
16. Cabrera-Cruz SA, Villegas-Patraca R (2016) Response of migrating raptors to an increasing number of wind farms. J Appl Ecol 53(6):1667–1675. https://doi.org/10.1111/1365-2664.12673
17. Hanski I (2015) Habitat fragmentation and species richness. J Biogeogr 42:989–994. https://doi.org/10.1111/jbi.12478
18. Maxwell AE, Strager MP, Yuill CB, Petty JT (2012) Modeling critical forest habitat in the southern coal fields of west Virginia. Int J Ecol 2012:182683. https://doi.org/10.1155/2012/182683
19. Wood PB, Scott B, Bosworth SB, Dettmers R (2006) Cerulean warbler abundance and occurrence relative to large-scale edge and habitat characteristics. The Condor 108(1):154–165. https://doi.org/10.1093/condor/108.1.154
20. Daewel U, Akhtar N, Christiansen N, Schrum C (2022) Offshore wind farms are projected to impact primary production and bottom water deoxygenation in the North Sea. Commun Earth Environ 3:292. https://doi.org/10.1038/s43247-022-00625-0
21. Floeter J, Pohlmann T, Harmer A, Möllmann C (2022) Chasing the offshore wind farm wind-wake-induced upwelling/downwelling dipole. Front Marine Sci 9. https://doi.org/10.3389/fmars.2022.884943
22. Raghukumar K, Chartrand C, Chang G, Cheung L, Roberts J (2022) Effect of floating offshore wind turbines on atmospheric circulation in California. Front Energy Res 10:863995. https://doi.org/10.3389/fenrg.2022.863995
23. Susskind L, Chun J, Gant A, Hodgkins C, Cohen J, Lohmar S (2022) Sources of opposition to renewable energy projects in the United States. Energy Policy 165:112922. https://doi.org/10.1016/j.enpol.2022.112922
24. ENERCON (2015) Product overview. Enercon
25. Konstantinidis EI, Botsaris PN (2016) Wind turbines: current status, obstacles, trends and technologies. IOP Conf Ser Mater Sci Eng 161. 12079. https://doi.org/10.1088/1757-899x/161/1/012079
26. Horwath S, Hassrick J, Grismala R, Diller E (2020) Comparison of environmental effects from different offshore wind turbine foundations. Bureau of ocean energy management, department of the interior, office of renewable energy programs, USA
27. Vestas (2024) Material use in Vestas turbines. https://www.vestas.com/en/sustainability/environment/materials-and-rare-earths
28. Carrara S, Alves Dias P, Plazzotta B, Pavel C (2020) Raw materials demand for wind and solar PV technologies in the transition towards a decarbonised energy system. Publications Office of the European Union, Luxembourg

29. Gielen D (2021) Critical minerals for the energy transition. International Renewable Energy Agency, Abu Dhabi
30. Pfenninger S, Staffell I (2016) Long-term patterns of European PV output using 30 years of validated hourly reanalysis and satellite data. Energy 114:1251–1265. https://doi.org/10.1016/j.energy.2016.08.060
31. Staffell I, Pfenninger S (2016) Using bias-corrected reanalysis to simulate current and future wind power output. Energy 114:1224–1239. https://doi.org/10.1016/j.energy.2016.08.068
32. Schlichting H, Gersten K (2017) Boundary-layer theory, 9th edn. Springer Berlin, Heidelberg, 978-3-662-52917-1
33. Tennekes H (1973) The logarithmic wind profile. J Atmos Sci 30(2):234–238. https://doi.org/10.1175/1520-0469(1973)030%3c0234:TLWP%3e2.0.CO;2
34. Jenkins J, Ekanayake N (2017) Renewable energy engineering. Cambridge University Press
35. Tammelin B, Vihma T, Atlaskin E, Badger J, Fortelius C, Gregow H, Horttanainen M, Hyvönen R, Kilpinen J, Latikka J, Ljungberg K, Gylling Mortensen N, Niemelä S, Ruosteenoja K, Salonen K, Suomi I, Venäläinen A (2011) Production of the finnish wind atlas. Wind Energy. https://doi.org/10.1002/we.517
36. Betz A (1926) Wind-Energie und ihre ausnutzung durch Windmühlen. Vandenhoeck and Ruprecht, Göttingen, Germany (In German)
37. Glauert H (1935) Airplane propellers. In Durand WF (ed) Aerodynamic theory. Springer, Berlin, Heidelberg. https://doi.org/10.1007/978-3-642-91487-4_3
38. Eggleston DM, Stoddard F (1987) Wind turbine engineering design. Van Nostrand Reinhold, New York
39. Wilson RE, Lissaman P (1974) Applied aerodynamics of wind power machines. Oregon State University, Corvallis
40. De Vries O (1979) Fluid dynamics aspects of wind energy conversion. Advi-sory group for aerospace research and development neuilly sur seine, ADA076315
41. Bennun L, van Bochove J, Ng C, Fletcher C, Wilson D, Phair N, Carbone G (2021) Mitigating biodiversity impacts associated with solar and wind energy development. Guidelines for project developers. Gland, Switzerland, IUCN and Cambridge UK: The Biodiversity Consultancy, 260 pp. https://doi.org/10.2305/IUCN.CH.2021.04.en
42. Fu L, Wei Y, Fang S, Tian G, Zhou X (2018) A wind energy generation replication method with wind shear and tower shadow effects. Adv Mech Eng 10(3):1–11. https://doi.org/10.1177/1687814018759216
43. Hersbach H, Bell B, Berrisford P et al (2020) The ERA5 global reanalysis. Q J R Meteorol Soc 146:1999–2049. https://doi.org/10.1002/qj.3803
44. Saha S, Oorthi SM, Wu X et al (2014) The NCEP climate forecast system version 2. J Clim 27:2185–2208. https://doi.org/10.1175/JCLI-D-12-00823.1
45. Davis NN, Badger J, Hahmann AN, Hansen BO, Mortensen NG, Kelly M, Larsén XG, Olsen BT, Floors R, Lizcano G, Casso P, Lacave O, Bosch A, Bauwens I, Knight OJ, Potter van Loon A, Fox R, Parvanyan T, Krohn Hansen SB, Heathfield D, Onninen M, Drummond R (2023) The global wind atlas: a high-resolution dataset of climatologies and associated web-based application. B Am Meteorol Soc 104(8):e1507–e1525. https://doi.org/10.1175/BAMS-D-21-0075.1
46. IEC (2019) Wind energy generation systems—part 1: design requirements, international standard, 61400-1, edition 4.0 2019–02
47. Bormann R, Rehfeldt K, Wallasch A K and Lüers S (2018) Capacity densities of European offshore wind farms. Report SP18004A1, Deutsche WindGuard GmbH, Germany
48. Wind Europe (2022) Wind energy in Europe, Statistics and the outlook for 2022–2026. February 2022
49. Denholm P, Hand M, Jackson M, Ong S (2009) Land-use requirements of modern wind power plants in the United States. Technical Report NREL/TP-6A2–45834
50. Harrison-Atlas D, Maclaurin G, Lantz E (2021) Spatially explicit prediction of capacity density advances geographic characterization of wind power technical potential. Energies 14(12):3609. https://doi.org/10.3390/en14123609

51. Mai TT, Lantz EJ, Mowers M, Wiser R (2017) The value of wind technology innovation: implications for the U.S. power system, wind industry, electricity consumers, and environment. Office of scientific and technical information (OSTI), Golden, CO, USA
52. Rinne E, Holttinen H, Kiviluoma J, Rissanen S (2018) Effects of turbine technology and land use on wind power resource potential. Nat Energy 3:494–500. https://doi.org/10.1038/s41560-018-0137-9
53. Jensen NO (1983) A note on wind generator interaction. Risø national laboratory. Risø-M No. 2411. Denmark
54. Frandsen ST (1992) On the wind speed reduction in the center of large clusters of wind turbines. J Wind Eng Ind Aerod 39(1–3):251–265
55. Frandsen ST, Barthelmie R, Pryor S, Rathmann O, Larsen S, Højstrup J, Thøgersen M (2006) Analytical modelling of wind speed deficit in large offshore wind farms. Wind Energ 9:39–53. https://doi.org/10.1002/we.189
56. Hou P, Zhu J, Ma K (2019) A review of offshore wind farm layout optimization and electrical system design methods. J Mod Power Syst Clean Energy 7:975–986. https://doi.org/10.1007/s40565-019-0550-5
57. Kusiak A, Song Z (2010) Design of wind farm layout for maximum wind energy capture. Renew Energy 35:685–694. https://doi.org/10.1016/j.renene.2009.08.019
58. Dai X, Xu D, Zhang M, Stevens RJAM (2022) A three-dimensional dynamic mode decomposition analysis of wind farm flow aerodynamics. Renew Energy 191:608–624. https://doi.org/10.1016/j.renene.2022.03.160
59. Gupta T, Roy SB (2021) Recovery processes in a large offshore wind farm. Wind energy science discussions. Preprint. Discussion started 23 Feb. https://doi.org/10.5194/wes-2021-7
60. Pörtner HO, Scholes RJ, Agard J, Archer E, Arneth A, Bai X, Barnes D, Burrows M, Chan L, Cheung WL, Diamond S, Donatti C, Duarte C, Eisenhauer N, Foden W, Gasalla MA, Handa C, Hickler T, Hoegh-Guldberg O, Ichii K, Jacob U, Insarov G, Kiessling W, Leadley P, Leemans R, Levin L, Lim M, Maharaj S, Managi S, Marquet PA, McElwee P, Midgley G, Oberdorff T, Obura D, Osman E, Pandit R, Pascual U, Pires APF, Popp A, Reyes-García V, Sankaran M, Settele J, Shin YJ, Sintayehu DW, Smith P, Steiner N, Strassburg B, Sukumar R, Trisos C, Val AL, Wu J, Aldrian E, Parmesan C, Pichs-Madruga R, Roberts DC, Rogers AD, Díaz S, Fischer M, Hashimoto S, Lavorel S, Wu N, Ngo HT (2021) IPBES-IPCC co-sponsored workshop report on biodiversity and climate change. IPBES and IPCC. https://doi.org/10.5281/zenodo.47825382538
61. Drewitt AL, Langston RHW (2006) Assessing the impacts of wind farms on birds. Ibis 148:29–42. https://doi.org/10.1111/j.1474-919X.2006.00516.x
62. May RFA (2015) Unifying framework for the underlying mechanisms of avian avoidance of wind turbines. Biol Conserv 190:179–187. https://doi.org/10.1016/j.biocon.2015.06.004
63. Thaker M, Zambre A, Bhosale H (2018) Wind farms have cascading impacts on ecosystems across trophic levels. Nat Ecol Evol 2:1854–1858. https://doi.org/10.1038/s41559-018-0707-z
64. Wilson MW, Ridlon AD, Gaynor KM, Gaines SD, Stier AC, Halpern BS (2020) Ecological impacts of human-induced animal behaviour change. Ecol 23:1522–1536. https://doi.org/10.1111/ele.13571
65. Fox AD, Desholm M, Kahlert J, Christensen TK, Krag Petersen, I (2006) Information needs to support environmental impact assessment of the effects of European marine offshore wind farms on birds. Ibis 148(s1):129–144. https://doi.org/10.1111/j.1474-919X.2006.00510.x
66. Bergström L, Sundqvist F, Bergström U (2013) Effects of an offshore wind farm on temporal and spatial patterns in the demersal fish community. Mar Ecol Prog Ser 485:199–210. https://doi.org/10.3354/meps10344
67. Sansom A, Pearce-Higgins J, Douglas D (2016) Negative impact of wind energy development on a breeding shorebird assessed with a BACI study design. Ibis 158(3):541–555. https://doi.org/10.1111/ibi.12364
68. Hein CD, Hale AM (2019) Wind energy effects on bats. In: Moorman CE, Grodsky SM and Rupp SP (eds) Renewable energy and wildlife conservation. John Hopkins University Press, pp 122–145

69. Korfanta NM, Zero VH (2019) Effects of wind energy on wildlife. Emerging issues and underrepresented taxa. In: Moorman CE, Grodsky SM and Rupp SP (eds) Renewable energy and wildlife conservation. John Hopkins University Press, pp 146–173:267
70. European Commission, Directorate-General for Environment (2021) Guidance document on wind energy developments and EU nature legislation. Publications Office of the European Union. https://data.europa.eu/doi/10.2779/095188
71. Galparsoro I, Menchaca I, Garmendia JM, Borja Á, Maldonado AD, Iglesias G, Bald J (2022) Reviewing the ecological impacts of offshore wind farms. NPJ Sustain Ocean 1(1). https://doi.org/10.1038/s44183-022-00003-5
72. Vaissière A-C, Levrel H, Pioch S, Carlier A (2014) Biodiversity offsets for offshore wind farm projects: the current situation in Europe. Mar Policy 48:172–183. https://doi.org/10.1016/j.marpol.2014.03.023
73. Goodale MW, Milman A (2014) Cumulative adverse effects of offshore wind energy development on wildlife. J Environ Plan Manag 59(1):1–21. https://doi.org/10.1080/09640568.2014.973483
74. Langston RHW (2013) Birds and wind projects across the pond: a UK perspective. Wildl Soc Bull 37:5–18. https://doi.org/10.1002/wsb.262
75. Arnett EB, May RF (2016) Mitigating wind energy impacts on wildlife: approaches for multiple taxa. Hum-Wildl Interact 10:28–41. https://doi.org/10.26077/1jeg-7r13
76. Furness RW, Wade HM, Masden EA (2013) Assessing vulnerability of marine bird populations to offshore wind farms. J Environ Manage 15(119):56–66. https://doi.org/10.1016/j.jenvman.2013.01.025
77. McGarry T, Boisseau O, Stephenson S, Compton R (2017) Understanding the effectiveness of acoustic deterrent devices (ADDs) on minke whale (*Balaenoptera acutorostrata*), a low frequency cetacean. ORJIP Project 4, Phase 2. RPS Report EOR0692. Prepared on behalf of The Carbon Trust
78. May R, Nygård T, Falkdalen U, Åström J, Hamre Ø, Stokke BG (2020) Paint it black: efficacy of increased wind turbine rotor blade visibility to reduce avian fatalities. Ecol Evol 10(16):8927–8935. https://doi.org/10.1002/ece3.6592
79. Dähne M, Tougaard J, Carstensen J, Rose A, Nabe-Nielsen J (2017) Bubble curtains attenuate noise from offshore wind farm construction and reduce temporary habitat loss for harbour porpoises. Mar Ecol Prog Ser 580:221–237. https://doi.org/10.3354/meps12257
80. Gove B, Langston RHW, McCluskie A, Pullan JD and Scrase I (2013) Wind farms and birds: an updated analysis of the effects of wind farms on birds, and best practice guidance on integrated planning and impact assessment. Report T-PVS/Inf 15
81. Segers JL, Broders HG (2014) Interspecific effects of forest fragmentation on bats. Can J Zool 92:665–673. https://doi.org/10.1139/cjz-2014-0040
82. Aragón G, Abuja L, Belinchón R, Martínez I (2015) Edge type determines the intensity of forest edge effect on epiphytic communities. Eur J For Res 134:443–445. https://doi.org/10.1007/s10342-015-0863-5
83. Zimmerling JR, Pomeroy AC, d'Entremont MV, Francis CM (2013) Canadian estimate of bird mortality due to collisions and direct habitat loss associated with wind turbine developments. Avian Conserv Ecol 8:10. https://doi.org/10.5751/ACE-00609-080210
84. LeBeau CW, Johnson GD, Holloran MJ, Beck JL, Nielson RM, Kauffman ME, Rodemaker EJ, McDonald TL (2017) Greater sage-grouse habitat selection, survival, and wind energy infrastructure. J Wildl Manage 81:690–711. https://doi.org/10.1002/jwmg.21231
85. Winder VL, Carrlson KM, Gregory AJ, Hagen CA, Haukos DA, Kesler DC, Larsson LC, Matthews TW, McNew LB, Patten MA, Pitman JC, Powell LA, Smith JA, Thompson T, Wolfe DH, Sandercock BK (2015) Factors affecting female space use in ten populations of prairie chickens. Ecosphere 6(9):166. https://doi.org/10.1890/ES14-00536.1
86. Masden EA, Haydon DT, Fox AD, Furness RW (2010) Barriers to movement: modelling energetic costs of avoiding marine wind farms amongst breeding seabirds. Mar Pollut Bull 60(7):1085–1091. https://doi.org/10.1016/j.marpolbul.2010.01.016

87. Eftestøl S, Tsegaye D, Flydal K, Colman JE (2023) Effects of wind power development on reindeer: global positioning system monitoring and herders experience. Rangel Ecol Manag 87:55–68. https://doi.org/10.1016/j.rama.2022.11.011
88. Nellemann C, Vistnes I, Jordhøy P, Strand O, Newton A (2003) Progressive impact of piecemeal infrastructure development on wild reindeer. Biol Conserv 113:307–317. https://doi.org/10.1016/S0006-3207(03)00048-X
89. Vistnes I, Nellemann C (2008) The matter of spatial and temporal scales: a review of reindeer and caribou response to human activity. Polar Biol 31:399–407. https://doi.org/10.1007/s00300-007-0377-9
90. Voigt CC (2020) Insect fatalities at wind turbines as biodiversity sinks. Conserv Sci Pract 3:e366. https://doi.org/10.1111/csp2.366
91. Corten GP, Veldkamp HF (2001) Aerodynamics. Insects can halve wind-turbine power. Nature 412(6842):41–2. https://doi.org/10.1038/35083698. PMID: 11452297
92. Han W, Kim J, Kim B (2018) Effects of contamination and erosion at the leading edge of blade tip airfoils on the annual energy production of wind turbines. Renew Energy 115:817–823. https://doi.org/10.1016/j.renene.2017.09.002
93. Soltani M, Birjandi A, Seddighi Moorani M (2011) Effect of surface contamination on the performance of a section of a wind turbine blade. Scientia Iranica 18:349–357
94. Dudek K, Dudek M, Tryjanowski P (2015) Wind turbines as overwintering sites attractive to an invasive Lady Beetle, *Harmonia axyridis* Pallas (Coleoptera: Coccinellidae). Coleopt Bull 69(4):665–669. https://doi.org/10.1649/0010-065X-69.4.665
95. Baerwald EF, D'Amours GH, Klug BJ, Barclay RMR (2008) Barotrauma is a significant cause of bat fatalities at wind turbines. Curr Biol 18(16):R695–R696. https://doi.org/10.1016/j.cub.2008.06.029
96. AWWI (2019) Wind turbine Interactions with wildlife and their habitats—a summary of research results and priority questions. American Wind—Wildlife Institute
97. BirdLife International (2021) Aquila chrysaetos. The IUCN red list of threatened species. https://doi.org/10.2305/IUCN.UK.2021-3.RLTS.T22696060A202078899.en. Accessed 11 Dec 2022
98. Ponnikas S (2014) Establishing conservation management for avian threatened species. University of Oulu, doctoral thesis. ACTA Universitatis Oulensis, A Scientiae Rerum Naturalium 621. http://jultika.oulu.fi/files/isbn9789526203683.pdf
99. Watson J (1997) The golden eagle, 1st ed. Pub Poyser
100. Tikkanen H, Rytkönen S, Karlin O-P, Ollila T, Pakanen V-M, Tuohimaa H, Orell M (2018) Modelling golden eagle habitat selection and flight activity in their home ranges for safer wind farm planning. EIA Review 71:120–131. https://doi.org/10.1016/j.eiar.2018.04.006
101. Saurola P, Valkama J, Velmala W (2013) Suomen rengastusatlas I—the finnish bird ringing atlas. Vol. I. Luomus 552 pp
102. Katzner TE, Nelson DM, Brahan MA, Doyle JM, Fernandez NB, Duerr AE, Bloom PH, Fitzpatrick MC, Miller TA, Culver RCE, Baswell L, Dewoody JA (2017) Golden Eagle fatalities and the continental-scale consequences of local wind-energy generation. Conserv Biol 31:406–415. https://doi.org/10.1111/cobi.12836
103. Langgemach T, Dürr T (2020) Informationen über Einflüsse der Windenergie-nutzung auf Vögel.—Stand 07. Januar 2020, Aktualisierungen außer Fundzahlen hervorgehoben Verbraucherschutz. Staatliche Vogelschutzwarte. Brandenburg. https://lfu.brandenburg.de/lfu/de/aufgaben/natur/artenschutz/vogelschutzwarte/arbeitsschwerpunkt-entwicklung-und-umsetzung-von-schutzstrategien/auswirkungen-von-windenergieanlagen-auf-voegel-und-fledermaeuse/
104. Meller K (2017) Kirjallisuusselvitys tuulivoimaloiden vaikutuksista linnustoon ja lepakoihin. Työ- ja elinkeinoministeriön julkaisuja 27/2017 (in Finnish)
105. Fielding AH, Anderson D, Benn S, Dennis R, Geary M, Weston E, Whitfield P (2021) Non-territorial GPS-tagged golden eagles *Aquila chrysaetos* at two Scottish wind farms: avoidance influenced by preferred habitat distribution, wind speed and blade motion status. PLoS ONE 16(8):e0254159. https://doi.org/10.1371/journal.pone.0254159

106. Balotari-Chiebao F, Valkama J, Byholm P (2021) Assessing the vulnerability of breeding bird populations to onshore wind-energy developments in Finland. Ornis Fennica 98
107. Tikkanen H (ed) (2022) Hyvät käytännöt maakotkalle aiheutuvien vaikutusten arviointiin—esimerkkiraportti Nimettömänkankaan tuulivoimahankkeesta. Metsähallituksen luonnonsuojelujulkaisuja. Sarja A 241. (In Finnish)
108. Tikkanen H, Ekblad C, Karlin O-P, Leskinen J, Pakanen V-M (2022) Suomen maakotkien elinympäristöjen, lentokäyttäytymisen ja populaation mallintaminen tuulivoimasuunnittelun avuksi. Metsähallitus. (In Finnish)
109. Band W, Madders M, Whitfield DP (2007) Developing field and analytical methods to assess avian collision risk at wind farms. In: de Lucas M, Janss GFE, Ferrer M (eds) Birds and wind farms: risk assessment and mitigation. Quercus, Madrid, pp 259–275. https://www.natural-research.org/appliction/files/4114/9182/2839/Band_et_al_2007.pdf
110. Langhamer O (2012) Artificial reef effect in relation to offshore renewable energy conversion: state of the art. Sci World J 2012:1–8. https://doi.org/10.1100/2012/386713
111. Buxton CD, Hartmann K, Kearney R, Gardner C (2014) When is spillover from marine reserves likely to benefit fisheries? PLoS ONE 9(9):e107032. https://doi.org/10.1371/journal.pone.0107032
112. Hammar L, Perry D, Gullström M (2016) Offshore wind power for marine conservation. Open J Mar Sci 6:66–78. https://doi.org/10.4236/ojms.2016.61007
113. Slavik K, Lemmen C, Zhang W et al (2019) The large-scale impact of offshore wind farm structures on pelagic primary productivity in the southern North Sea. Hydrobiologia 845:35–53. https://doi.org/10.1007/s10750-018-3653-5
114. Zhang Y, Zhang C, Chang YC, Liu W-H, Zhang Y (2017) Offshore wind farm in marine spatial planning and the stakeholders engagement: opportunities and challenges for Taiwan. Ocean Coast Manag 149:69–80. https://doi.org/10.1016/j.ocecoaman.2017.09.014
115. Hastie GD, Russell DJF, McConnell B, Moss S, Thompson D, Janik VM (2015) Sound exposure in harbour seals during the installation of an offshore wind farm: predictions of auditory damage. J Appl Ecol 52(3):631–640. https://doi.org/10.1111/1365-2664.12403
116. Nehls G, Harwood AJP, Perrow MR, Pohlmann T (2019) Marine mammals. In: Perrow MR (ed) Wildlife and wind farms, conflicts and solutions. Vol 3 offshore: potential effects, chapter 6. Pelagic Publishing, Exeter, UK
117. Schaffeld T, Schnitzler J, Ruser A, Baltzer J, Schuster M, Siebert U (2022) A result of accidental noise pollution: acoustic flowmeters emit 28 kHz pulses that may affect harbor porpoise hearing. Front Mar Sci 9:892050. https://doi.org/10.3389/fmars.2022.892050
118. Harding H, Bruintjes R, Radford AN, Simpson SD (2016) Measurement of hearing in the Atlantic salmon (*Salmo salar*) using auditory evoked potentials, and effects of pile driving playback on salmon behaviour and physiology. Scott Mar Freshw Sci Report 7(11):51
119. Thomsen F, Luedemann K, Piper W, Judd A, Kafemann R (2008) Potential effects of offshore wind farm noise on fish. Bioacoustics 17:221–223. https://doi.org/10.1080/09524622.2008.9753825
120. Taormina B, Bald J, Want A, Thouzeau G, Lejart M, Desroy N, Carlier A (2018) A review of potential impacts of submarine power cables on the marine environment: knowledge gaps, recommendations and future directions. Renew Sust Energ Rev 96:380–391. https://doi.org/10.1016/j.rser.2018.07.026
121. Kirchgeorg T, Weinberg I, Hörnig M, Baier R, Schmid MJ, Brockmeyer B (2018) Emissions from corrosion protection systems of offshore wind farms: evaluation of the potential impact on the marine environment. Mar Pollut Bull 136:257–268. https://doi.org/10.1016/j.marpolbul.2018.08.058
122. Hammond P, Macleod K, Berggren P, Borchers D, Burtc L, Cañadas A, et al (2013) Cetacean abundance and distribution in European Atlantic shelf waters to inform conservation and management. Biol Conserv 164:107–122
123. Sørensen PM, Wisniewska DM, Jensen FH, Johnson M, Teilmann J, Madsen PT (2018) Click communication in wild harbour porpoises (Phocoena phocoena). Sci Rep 8(1):9702. https://doi.org/10.1038/s41598-018-28022-8

124. Bailey H, Brookes KL, Thompson PM (2014) Assessing environmental impacts of offshore wind farms: lessons learned and recommendations for the future. Aquat Biosyst 10(1):8. https://doi.org/10.1186/2046-9063-10-8
125. King SL, Schick RS, Donovan C, Booth CG, Burgman M, Thomas L, Harwood J (2015) An interim framework for assessing the population consequences of disturbance. Methods Ecol Evol. https://doi.org/10.1111/2041-210x.12411
126. Nabe-Nielsen J, van Beest FM, Grimm V, Sibly RM, Teilmann J, Thompson PM (2018) Predicting the impacts of anthropogenic disturbances on marine populations. Conserv Lett 11(5):e12563 https://doi.org/10.1111/conl.12563. UNSP
127. Faulkner RC, Farcas A, Merchant ND (2018) Guiding principles for assessing the impact of underwater noise. J Appl Ecol 55(6):2531–2536. https://doi.org/10.1111/1365-2664.13161
128. Benhemma-Le Gall A, Graham IM, Merchant ND, Thompson PM (2021) Broad-scale responses of harbor porpoises to pile-driving and vessel activities during offshore windfarm construction. Front Mar Sci 8. https://doi.org/10.3389/fmars.2021.664724
129. Graham IM, Merchant ND, Farcas A, Barton TR, Cheney B, Bono S, Thompson PM (2019) Harbour porpoise responses to pile-driving diminish over time. R Soc Open Sci 6(6):190335. https://doi.org/10.1098/rsos.190335
130. Southall B (2021) Evolutions in marine mammal noise exposure criteria. Acoust Today 17(2):52–60. https://doi.org/10.1121/AT.2021.17.2.52
131. Brandt MJ, Dragon AC, Diederichs A, Bellmann MA, Wahl V, Piper W, et al (2018) Disturbance of harbour porpoises during construction of the first seven offshore wind farms in Germany. Mar Ecol Prog Ser 596:213–232
132. Thompson PM, Graham IM, Cheney B, Barton TR, Farcas A, Merchant ND (2020) Balancing risks of injury and disturbance to marine mammals when pile driving at offshore windfarms. Ecol Solut Evid 1(2):e12034. https://doi.org/10.1002/2688-8319.12034
133. Merchant ND (2019) Underwater noise abatement: economic factors and policy options. Environ Sci Pol 92:116–123. https://doi.org/10.1016/j.envsci.2018.11.014
134. Cryan PM, Gorresen PM, Hein CD, Schirmacher MR, Diehl RH, Huso MM, Hayman DTS, Fricker PD, Bonaccorso FJ, Johnson DH, Heist K, Dalton DC (2014) Behavior of bats at wind turbines. PNAS 111(42):15126–15131. https://doi.org/10.1073/pnas.1406672111
135. Foo CF, Bennett VJ, Hale AM, Korstian JM, Schildt AJ, Williams DA (2017) Increasing evidence that bats actively forage at wind turbines. PeerJ 5:e3985. https://doi.org/10.7717/peerj.3985
136. Millon L, Colin C, Brescia F, Kerbiriou C (2018) Wind turbines impact bat activity, leading to high losses of habitat use in a biodiversity hotspot. Ecol Eng 112:51–54. https://doi.org/10.1016/j.ecoleng.2017.12.024
137. Ekholm P, Lehtoranta J (2012) Does control of soil erosion inhibit aquatic eutrophication? J Enviro Manage 93(1):140–146. https://doi.org/10.1016/j.jenvman.2011.09.010
138. ICES (2012) Report of the workshop on effects of offshore windfarms on marine benthos—facilitating a closer international collaboration throughout the North Atlantic Region (WKEOMB), 27–30 March 2012, Bremerhaven, Germany. ICES CM2012/SSGEF:13, p 57
139. Boström G, Ludwig E, Schneehorst A,. Pohlmann T (2019) Atmosphere and ocean dynamics. In: Perrow M (ed) Wildlife and wind farms—conflicts and solutions, vol 3. Offshore: Potential Effects. Exeter, UK
140. Angelov I, Hashim I, Oppel S (2013) Persistent electrocution mortality of Egyptian vultures *Neophron percnopterus* over 28 years in East Africa. Bird Conserv Int 23:1–6. https://doi.org/10.1017/S0959270912000123
141. Godino A, Garrido J, El Khamlichi R, Burón D, Machado C, Amezian M, Irizi A, Numa C, Barrios V (2016) Identificación de mortalidad por electrocución de aves rapaces en el sudoeste de Marruecos (Identification of mortality due to electrocution of birds of prey in southwestern Morocco). Malaga, Spain, IUCN. https://portals.iucn.org/library/node/46262
142. Thaxter CB, Buchanan GM, Carr J, Butchart SHM, Newbold T, Green RE, Tobias JA, Foden WB, O'Brien S, Pearce-Higgins JW (2017) Bird and bat species global vulnerability to collision mortality at wind farms revealed through a trait-based assessment. Proc Royal Soc B 284(1862):20170829. https://doi.org/10.1098/rspb.2017.0829

143. Roeleke M, Blohm T, Kramer-Schadt S, Yovel Y, Voigt CC (2016) Habitat use of bats in relation to wind turbines revealed by GPS tracking. Sci Rep 6:28961. https://doi.org/10.1038/srep28961
144. Hüppop O, Michalik B, Bach L, Hill R, Pelletier S (2019) Migratory birds and bats. In Perrow MR (ed) Wildlife and wind farms, conflicts and solutions, volume 3 offshore: potential effects, Chapter 7. Pelagic Publishing
145. Ahlén I, Baagøe HJ, Bach L (2009) Behavior of Scandinavian bats during migration and foraging at sea. J Mammal 90(6):1318–1323. https://doi.org/10.1644/09-MAMM-S-223R.1
146. Lagerveld S, Kooistra G, Otten G, Meesters L, Manshanden J, de Haan D, Gerla D, Verhoef H, Scholl M (2017) Bat flight analysis around wind turbines, a feasibility study. Wageningen Research Report C026/17. https://doi.org/10.18174/417091
147. Sjollema AL, Gates JE, Hilderbrand RH, Sherwell J (2014) Offshore activity of bats along the Mid-Atlantic coast. Northeast Nat 21(2):154–163. https://doi.org/10.1656/045.021.0201
148. Cates K, DeMaster DP, Brownell R, Silber G, Gende S, Leaper R, Ritter F, Panigada S (2017) Strategic plan to mitigate the impacts of ship strikes on cetacean populations 2017–2020. International Whaling commission (IWC)
149. NOAA Fisheries (2017) Understanding vessel strikes. NOAA Fisheries. https://www.fisheries.noaa.gov/insight/understanding-vessel-strikes
150. Bernardino J, Martins R, Bispo R, Moriera F (2019) Re-assessing the effectiveness of wire-marking to mitigate bird collisions with power lines: a meta-analysis and guidelines for field studies. J Environ Manage 252. https://doi.org/10.1016/j.jenvman.2019.109651
151. van der Winden J, van Vliet F, Patterson A, Lane B (2015) Renewable energy technologies and migratory species: guidelines for sustainable deployment. Bureau Waardenburg bv/CMS, AEWA, UNDP/GEF/Birdlife MSB Project. https://www.cms.int/sites/default/files/document/ETF1_Inf.1.pdf
152. Martin J, Sabatier Q, Gowan TA, Giraud C, Gurarie E, Calleson CS, Ortega-Ortiz JG, Deutsch CJ, Rycyk A, Koslovsky SM (2016) A quantitative framework for investigating risk of deadly collisions between marine wildlife and boats. Methods Ecol Evol 7(1):42–50. https://doi.org/10.1111/2041-210x.12447
153. Heinänen S, Žydelis R, Kleinschmidt B, Dorsch M, Burger C, Morkūnas J, Quillfeldt P, Nehls G (2020) Satellite telemetry and digital aerial surveys show strong displacement of red-throated divers (Gavia stellata) from offshore wind farms. Mar Environ Res 160. https://doi.org/10.1016/j.marenvres.2020.104989
154. Peschko V, Mendel B, Müller S, Markones N, Mercker M, Garthe S (2020) Effects of offshore windfarms on seabird abundance: strong effects in spring and in the breeding season, Mar Environ Res 162. https://doi.org/10.1016/j.marenvres.2020.105157
155. Kavanagh AS, Nykänen M, Hunt W, Richardson N, Jessopp MJ (2019) Seismic surveys reduce cetacean sightings across a large marine ecosystem. Sci Rep 9(1):19164. https://doi.org/10.1038/s41598-019-55500-4
156. Southall BL, Rowles T, Gulland F, Baird RW, Jepson PD (2013) Final report of the independent scientific review panel investigating potential contributing factors to a 2008 mass stranding of melon-headed whales (*Peponocephala electra*) in Antsohihy, Madagascar. Independent Scientific Review Panel. https://iwc.int/2008-mass-stranding-in-madagascar
157. Woodward I, Thaxter C, Owen E, Cook A (2019) Desk-based revision of seabird foraging ranges used for HRA screening. British Trust for Ornithology, Thetford, UK, p 139
158. Johnson HE, Sushinsky JR, Holland A, Bergman EJ, Balzer T, Garner J, Reed SJ (2016) Increases in residential and energy development are associated with reductions in recruitment for a large ungulate. Glob Change Biol 23:578–591. https://doi.org/10.1111/gcb.13385
159. Gill AB, Wilhelmsson D (2019) Fish. In: Perrow M (ed) Wildlife and wind farms—conflicts and solutions, vol 3. Offshore: Potential Effects. Exeter, UK: Pelagic Publishing
160. Silva M, Passos I (2017) Vegetation. In: Perrow MR (ed) Wildlife and wind farms—conflicts and solutions vol 1 Onshore: potential effects, Chapter 3. Pelagic Publishing, Exeter UK
161. Geburzi JC, McCarthy ML (2018) How do they do it?—Understanding the success of marine invasive species. In: Jungblut S, Liebich V, Bode M (eds) YOUMARES 8—oceans across

boundaries: learning from each other. Springer International Publishing, Cham, Switzerland, pp 109–124. http://link.springer.com/10.1007/978-3-319-93284-2_8
162. Iacarella JC, Davidson IC, Dunham A (2019) Biotic exchange from movement of static maritime structures. Biol Invasions 21(4):1131–1141. https://doi.org/10.1007/s10530-018-1888-8
163. De Mesel I, Kerckhof F, Norro A, Rumes B, Degraer S (2015) Succession and seasonal dynamics of the epifauna community on offshore wind farm foundations and their role as stepping stones for non-indigenous species. Hydrobiologia 756(1):37–50. https://doi.org/10.1007/s10750-014-2157-1
164. IRENA (2022) Renewable energy and jobs—annual review 2022. https://www.irena.org/publications/2022/Sep/Renewable-Energy-and-Jobs-Annual-Review-2022
165. IEA (2021) Wind power. IEA, Paris, https://www.iea.org/reports/wind-power
166. Renner M, Parajuli B, Ferroukhi R, Gehl Sampath P, Casals X, Khalid A, Hawila D, Feng J, Garcia-Banos C (2019) Measuring the socio-economic footprint of the energy transition: the role of supply chains. Analysis built on IRENA (2016) Renewable energy benefits: measuring the economics (IRENA 2016). https://doi.org/10.13140/RG.2.2.12085.70887
167. Brion DJ (1991) Essential industry and the NIMBY phenomenon. Quorum Books, New York
168. Dear M (1992) Understanding and overcoming the NIMBY syndrome. J Am plann assoc 58(3):288–300
169. Hunter S, Leyden KM (1995) Beyond NIMBY: explaining opposition to hazardous waste facilities. Policy Stud J 23(4):601–619
170. Siemens-Gamesa, environmental product declaration SG 8.0–167 DD
171. Rand J, Hoen B (2017) Thirty years of North American wind energy acceptance research: what have we learned? Energy Res Soc Sci 29:135–148
172. Wolsink M (2000) Wind power and the NIMBY-myth: institutional capacity and the limited significance of public support. Renew Energy 21:49–64
173. Schively C (2007) Understanding the NIMBY and LULU phenomena: reassessing our knowledge base and informing future research. J Plan Lit 21(3):255–266
174. Freiberg A, Schefter C, Hegewald J, Seidler A (2019) The influence of wind turbine visibility on the health of local residents: a systematic review. Int Arch Occup Environ Health 92:609–628. https://doi.org/10.1007/s00420-019-01403-w
175. Radun J, Maula H, Saarinen P, Keränen J, Alakoivu R, Hongisto V (2022) Health effects of wind turbine noise and road traffic noise on people living near wind turbines. Renew Sustain Energy Rev 157:112040
176. Karasmanaki E (2022) Is it safe to live near wind turbines? Reviewing the impacts of wind turbine noise. Energy Sustain Dev 69:87–102
177. Gorayed A, de Sousa MJ, de Andrade Meireles AJ, Brannstrom C, Vicente da Silva E, Ribeiro de Freitas AL (2016) Wind-energy development causes social impacts in coastal Ceará state, Brazil: the case of the Xavier community. J Coast Res 75:383–387
178. da Silva VP, Galvão MLDM (2022) Onshore wind power generation and sustainability challenges in Northeast Brazil: a quick scoping review. Wind 2(2):192–209
179. Urban F, Siciliano G, Wallbott L, Lederer M, Dang Nguyen A (2018) Green transformations in Vietnam's energy sector. Asia Pacific Policy Stud 5(3):558–582
180. Kirchhoff T, Ramisch K, Feucht T, Reif C, Suda M (2022) Visual evaluations of wind turbines: judgments of scenic beauty or of moral desirability? Landsc Urban Plan 226:104509
181. Wüstenhagen R, Wolsink M, Bürer MJ (2007) Social acceptance of renewable energy innovation: an introduction to the concept. Energy Policy 35(5):2683–2691
182. Kalkbrenner BJ, Roosen J (2016) Citizens willingness to participate in local renewable energy projects: the role of community and trust in Germany. Energy Res Soc Sci 13:60–70
183. Hanke F, Guyet R, Feenstra M (2021) Do renewable energy communities deliver energy justice? Exploring insights from 71 European cases. Energy Res Soc Sci 80:102244
184. Thomson I, Boutilier RG (2011) Social license to operate. SME Min Eng Handb 1:1779–1796
185. Engen S, Hausner VH, Fauchald P, Ruud A, Broderstad EG (2023) Small hydropower, large obstacle? Exploring land use conflict, indigenous opposition and acceptance in the Norwegian Arctic. Energy Res Soc Sci 95:102888

186. Norges Høyesterett (2021) HR-2021-1975-S, (sak nr. 20-143891SIV-HRET), (sak nr. 20-143892SIV-HRET) og (sak nr. 20-143893SIV-HRET), Anke over Frostating lagmannsretts overskjønn 8. juni 2020. https://www.domstol.no/globalassets/upload/hret/avgjorelser/2021/oktober-2021/hr-2021-1975-s.pdf (In Norwegian)
187. Balotari-Chiebao F, Brommer JE, Saurola P, Ijäs A, Laaksonen T (2018) Assessing space use by pre-breeding white-tailed eagles in the context of wind-energy development in Finland. Landsc Urban Plan 177:251–258
188. Koskinen A, Laitinen O (2010) Country report of Finland. In: Horbaty H, Huber S (2010) Task 28 social acceptance of wind energy projects, winning hearts and minds. In Conference Proceedings EWEC
189. Mäntymaa E, Pouta E, Hiedanpää J (2021) Forest owners interest in participation and their compensation claims in voluntary landscape value trading: the case of wind power parks in Finland. For Policy Econ 124:102382
190. Turunen AW, Tiittanen P, Yli-Tuomi T, Taimisto P, Lanki T (2021) Self-reported health in the vicinity of five wind power production areas in Finland. Environ Int 151:106419
191. Strand O, Colman JE, Eftestøl S, Sandström P, Skarin A, Thomassen J (2018) Vindkraft och renar: En kunskapssammanställning. Naturvårdsvärket Rapport 6799. (In Swedish)
192. Kløcker Larsen R, Staffansson J, Omma I-A, Lawrence R (2022) Paliskuntien ja hanketoimijoiden väliset sopimukset: vaikutukset porojen hyvinvointiin. UiT Norges arktiske universitet, Senter for Samiska Studier (In Finnish)
193. Nysten-Haarala S, Joona T, Hovila I (2021) Wind energy projects and reindeer herders rights in Finnish Lapland: a legal framework. Elementa 9(1):1–17
194. Forbes BC (2006) The challenges of modernity for reindeer management in northernmost Europe. Reindeer management in northernmost Europe. Springer, Berlin, Heidelberg, pp 11–25
195. Kumpula J, Jokinen M, Siitari J, Heikkinen J, Oinonen K, Shemeikka P, Kontio P, Niemelä A (2022) Poronhoidon muutokset ja sopeutumiskeinot eri maankäyttömuotojen ristipaineissa. Luonnonvara- ja biotalouden tutkimus 65/2022. (In Finnish)
196. Anttonen M, Kangasoja J, Luoma E, Paalatie H (2023) Tuulivoimahankkeiden suunnittelu ja operointi poronhoitoalueella—tuulivoima-alan ja poronhoidon näkemys hyvistä käytännöistä. Akordi Oy julkaisuja. (In Finnish)
197. Akordi (2018) Social acceptability of wind power—what was learned. https://akordi.fi/wp-content/uploads/SocialAcceptability_report2018.pdf
198. Fisher R, Ury W (1991) Getting to yes. Penguin Books, New York, NY
199. Foster-Fishman PG, Berkowitz SL, Lounsbury DW, Jacobson S, Allen NA (2001) Building collaborative capacity in community coalitions: a review and integrative framework. Am J Community Psychol 29(2):241–261
200. Elliot M (1999) The role of facilitators, mediators, and other consensus building practitioners. In: Susskind M, Thomas-Larmer (Eds.) The consensus building handbook. A comprehensive guide to reaching agreement. Thousand Oaks (Calif.), SAGE, pp 199–240
201. Walker GB, Daniels SE (2019) Collaboration in environmental conflict management and decision-making: comparing best practices with insights from collaborative learning work. Front Commun 4(2)
202. Veers P, Dykes K, Lantz E, Barth S, Bottasso CL, Carlson O, Clifton A, Green J, Green P, Holttinen H, Laird D, Lehtomäki V, Lundquist JK, Manwell J, Marquis M, Meneveau C, Moriarty P, Munduate X, Muskulus M, Naughton J, Pao L, Paquette J, Peinke J, Robertson A, Sanz Rodrigo J, Sempreviva AM, Smith JC, Tuohy A, Wiser R (2019) Grand challenges in the science of wind energy. Science 366:443. https://doi.org/10.1126/science.aau2027?src=getftr
203. Horsch R, Scavone A, Alexander T, Webber JK, McAliley N, Lumpkin CA (2009) Court halts construction and limits operation of wind project for failure to comply with Endangered Species Act. www.lexology.com/library/detail.aspx?g=8999cf47-9ffe-478a-837d-25c45280fd1e

204. U.S. Department of Justice, Office of Public Affairs (2014) Utility company sentenced in Wyoming for killing protected birds at wind projects. https://www.justice.gov/opa/pr/utility-company-sentenced-wyoming-killing-protected-birds-wind-projects-0

Chapter 6
Bioenergy and Waste

Mika Järvinen, Hanna Paulomäki, Han van Kasteren, Elina Oksanen, Iryna Herzon, Martin Forsius, Anu Akujärvi, Maria Holmberg, Virpi Junttila, Minna Pekkonen, Torsti Schulz, Heini Kujala, and Antti Arasto

What you make from a tree should be at least as miraculous as what you cut down.
—Richard Powers, The Overstory

Supplementary Information The online version contains supplementary material available at https://doi.org/10.1007/978-3-031-69856-9_6.

M. Järvinen (✉) · H. Paulomäki
Aalto University, Espoo, Finland
e-mail: mika.jarvinen@aalto.fi

H. van Kasteren
Eindhoven University of Technology, Eindhoven, Netherlands

E. Oksanen
University of Eastern Finland, Kuopio, Finland

I. Herzon
University of Helsinki, Helsinki, Finland

M. Forsius · A. Akujärvi · M. Holmberg · V. Junttila · M. Pekkonen · T. Schulz
Finnish Environment Institute, Helsinki, Finland

H. Kujala
University of Helsinki, Finnish Natural History Museum, Helsinki, Finland

A. Arasto
VTT, Espoo, Finland

M. Järvinen and H. Paulomäki (eds.), *Designing Renewable Energy Systems within Planetary Boundaries*, Green Energy and Technology,
https://doi.org/10.1007/978-3-031-69856-9_6

6.1 Introduction

Bioenergy is the oldest energy source employed by humans. It contains energy from the Sun, produced by plants and algae through photosynthesis and stored in chemical bonds. Biomass has many potential energy applications and is seen as one solution to climate change for its ability to substitute fossil fuels used for heating, electricity production and transport. Though renewable, its use releases greenhouse gases, and its sustainability depends on the type of products being substituted and the magnitude of use, since intense biomass production can induce negative land-use change, lead to land degradation, compete with food production and other land uses, and cause loss of biodiversity [35]. Biomass is also inherently limited in availability. Sustainable bioenergy production and use can happen only between ecological ceiling formed by the planetary boundaries and social foundation (see also Chap. 1). Climate change and other planetary boundaries are strongly interlinked, and this underlines the importance of, for example, solving climate change and biodiversity loss together, so that solutions in one area do not compromise those in another. While bioenergy can help societies to decarbonize, at the same time its use raises a number of risks around biodiversity, food supply, pollution, and wellbeing that must be addressed properly [65].

Since the domestication of fire, biomass has supplied almost all of our cumulative energy use. Fire has been used in a controlled manner since 800,000–1 600 000 years ago [86]. Cooking food with fire improved our ability to digest it and recover its energy, proteins and nutrients, and this was the major driver that enabled human survival in diverse environments and under challenging conditions [76]. Biomass also had a pivotal role at the dawn of industrialization, as it was extensively used in many applications, such as for heat and light, cooking food, and even metallurgy and shipbuilding. Due to growing populations and increased standard of living, biomass—especially forest biomass—was by the eighteenth and nineteenth centuries already overutilized in many regions [76]. Biomass remained the main energy source until the 1800s, when it was rapidly replaced by coal, oil and natural gas, hydropower, and eventually nuclear energy as the main energy sources (see also Chap. 1). Today, these fossil fuels are being replaced by solar, wind and other renewable energy sources to decarbonize our societies. Biomass retains its place on our energy palette, now and into the future. In this chapter we discuss the sustainable levels of biomass and its associated waste that support reaching climate and nature conservation goals.

Our planet Earth is a system based on carbon. When hydrogen, oxygen, and numerous other elements are combined with this carbon, a vast number of chemical species are created that form the raw materials and key building blocks of the Earth as we know it. In this chapter we focus on processes that strongly interact with the fast carbon cycle —that is, the carbon transfer between atmospheric CO_2 and the carbon-based species in biomass. The engine of the fast carbon cycle is the renewable energy from the Sun, which drives photosynthesis to convert CO_2 from the air and water into sugars, to be processed into more complex molecules that form the biomass. Along the way, it also produces oxygen. As discussed in Chap. 2, with time and under specific conditions, when biomass dies a very small fraction of it can be converted into fossil oil, gas, and coal. The Earth's current fossil resources, formed over millions of years, act now as a stable, long term chemical storage of CO_2. However, most of the carbon in living organisms, more than 90%, is released as CO_2 when biomass dies and decomposes— as nature's own version of "business and usual" [2].

Many types of biomass can be used for energy. Though wood is still the largest biomass energy resource, food crops, grasses, woody plants, residues from agriculture or forestry, and municipal and industrial wastes are also commonly used. Biomass can be converted into heat and power, or to other forms of energy carriers, in different ways, and there are several processes that can be used for this conversion. In *thermochemical processes*, heat transforms biomass (possibly with the aid of catalysts) into energy or energy carriers, by combustion, pyrolysis or gasification. Many of these processes are copied from the fossil fuel industry, and are mature and available at large scales. *Biochemical processes* use enzymes and micro-organisms to ferment biomass into alcohols, or anaerobic digestion to produce biogas. *Mechanical processes* use mechanical compression to press oil from seeds. In this book, we are interested in the stationary application of biomass for producing heat and power, so we have left out detailed discussion of transportation fuels and focus on the combustion of solid biomass and waste in the large-scale power plants. Sustainable transportation is, however, an important topic as there is also an urgent need to decarbonize the transportation sector. Biomass-based transportation fuels are adding to the pressure on biomass use, and raise several sustainability concerns, including loss of biodiversity, competition with food production, and water pollution, while failing to meet the greenhouse gas emission reduction targets. A good overview of the environmental impacts of the world's largest existing biofuel program, the US Renewable Fuel Standard, is given by Lark et al. [47].

When biomass is transported from across a local area and collected into a larger centralized reserve, we create a concentrated source of biogenic CO_2 that can be efficiently captured, utilized, or stored. This concept creates the basis for industrial bioenergy carbon capture and storage technology, BECCS—at the end of this chapter we discuss and analyze a range of BECCS technologies, focusing mainly on the Oxyfuel concept.

Bioenergy is one of the energy resources we require to replace fossil fuels for heating, electricity production, and transportation. As discussed in Chap. 2, approximately 10% of current total primary energy production is obtained from traditional biomass, mostly in developing countries for heating and cooking purposes. This significant share has negative impacts on the environment and societies, contributing to poor indoor air quality and causing significant health problems. The share of modern biomass use is far less significant.

While biomass is a renewable energy resource, there are limits to its sustainable use. The use of bioenergy always releases greenhouse gases, as carbon in biomass is oxidized. In addition, the impact of biomass use on land use and impacts on nature can be considerable, depending on how the biomass is grown and harvested. The sustainability of bioenergy depends to large extent on the magnitude and volume of the use; it is inherently limited in availability. Bioenergy will undoubtably form an important part of our future energy system, but sustainable bioenergy use can happen only within ecological and societal boundaries. The grand challenge is to learn to use biomass at sustainable levels, incorporating circular economy principles effectively. In this chapter we will critically discuss the benefits and limitations of biomass energy, from technical, ecological, and social viewpoints.

6.2 Net Primary Production and Biomass Availability

In a large-scale, bio-based economy, biomass would be used to replace fossil- carbon-based materials such as plastics, construction materials, chemicals, and fuels. This transition to the large-scale use of biomass resources raises the question of the amount of available biomass on Earth. In this section, we discuss the theoretical limitations and the potential for biomass production.

The total annual production of biomass by photosynthesis is known as the gross primary production, GPP. Typically, around 50% of GPP is consumed by the plants themselves to grow and to respire [2]. The remaining 50% forms the net primary production, NPP. Figure 6.1 shows the global annual net primary production of different biomass sources.

As Fig. 6.1 shows, tropical rainforest and open sea are by far the largest producers of global biomass, and satellite measurements by Field et al. [25] showed that the production at sea could be much higher, as much as 97 Gt per year. In addition to the rainforest ecosystem, other types of forest are also important for producing biomass, and therefore are important sinks for CO_2 capture from the atmosphere. As a whole, forests generate ~ 74 Gt of biomass annually, and (assuming 50% carbon content) capture $74 \times 0.5 \times 44/12 = 136$ Gt of CO_2. Regardless of the global biomass production value, all kinds of biodiversity, unique species, and species assemblages forming regional and local ecosystems, are important both for their own sake, and because of the multiple functions they produce and maintain.

With no anthropogenic disturbance, a significant share of the NPP, around 90%, will be consumed by heterotrophic respiration, mainly by decomposition by organisms in the soil, fungi and bacteria, with a significantly smaller fraction consumed by herbivores (plant eating animals, including humans). The remaining 10% is mostly consumed in natural fires, resulting in fluxes of volatile organic compounds such as methane. With these processes, the net carbon flow rate—also known as the net ecosystem exchange, NEE—can be zero, though more typically is slightly positive or negative. An example of a zero-balance or steady state ecosystem is an old forest,

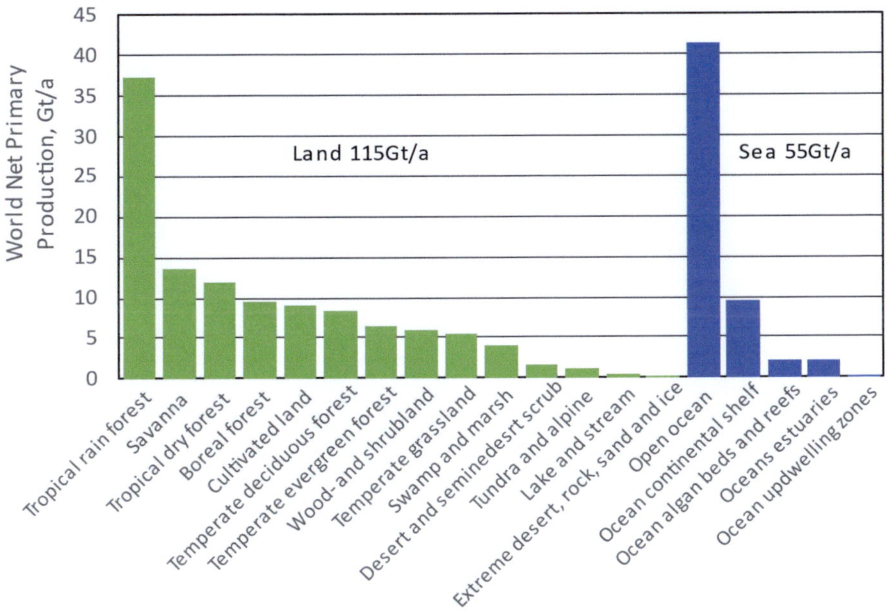

Fig. 6.1 Global net primary production (NPP). Amounts are dry mass (carbon content assumed 50%, [2]). Total combined production on land and sea is 170 Gt/a

where the death of old individuals can be compensated by the birth of new individuals. Old-growth forests also host high biodiversity [2].

Human activities and human-induced changes in the environment have strongly impacted the systems described above. We consume components of the net primary production for food, raw materials, and energy. The food and fuels are rapidly converted to CO_2, but if we build houses from wood materials, for example, their carbon is stored for longer periods, even hundreds of years. To better understand the magnitude of these different processes, Fig. 6.2 presents a simplified schematic of the global flows of bio-based materials used today [64].

As Fig. 6.2 shows, of the 115 Gt of dry land-based biomass produced annually, 12 Gt is used by humans. The fraction supplying all materials and chemicals is relatively low, ~ 1 Gt, compared to ~ 8.5 Gt used for food and energy. We might expect that all residual flows (currently losses) from food and feed could, in theory, provide raw inputs for bio-based material and chemical production. The current primary food supply uses almost twice as much carbon as needed for our current energy supply, as shown in Fig. 6.2. We expect, therefore, that increasing the volume of biomass for energy use to replace fossil energy sources will increase competition with biomass demand for food. For example, policies to replace gasoline and diesel with bioethanol and biodiesel have already led to higher food prices in countries such as Mexico [47].

Figure 6.3 presents a second scenario, in which we replace all fossil energy sources with bioenergy. In 2022 the primary energy consumption from fossil fuels

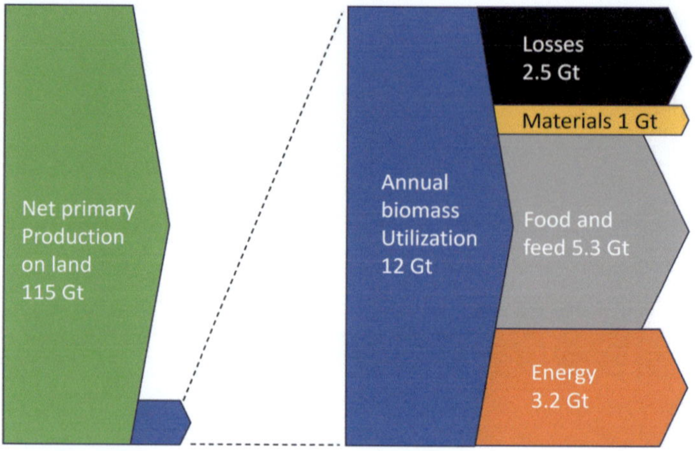

Fig. 6.2 Use of biomass for energy, food and feed, materials and losses [64]. The 12 Gt current annual biomass use (2015) is presented as a white-framed arrow for reference in Fig. 6.3

was 136,000 TWh. Using data from Table 2.1, if we assume the average heating value is 32.6 MJ/kg for fossil fuels and 13.8 MJ/kg for biomass, then the additional biomass required for energy use would be $32.6 \div 13.8 = 2.36$ times the amount of fossil fuels we use annually: $2.36 \times 15 = 35.4$ Gt. Of course, we assumed here that energy-efficient conversion technologies are available to convert biomass to green gas and liquid products to replace those already available from fossil sources, which is not the case. We also assumed in this scenario that all the other uses and losses (e.g., forest fires) would remain the same. As Fig. 6.3 shows, this means more than 40% of the current NPP would need to be used for energy—as a result, we would need to significantly increase the efficiency of biomass farming and fertilization. This scenario would present serious sustainability challenges around land use, biodiversity, and many social impacts, which are already significant at the current levels of bioenergy utilization shown in Fig. 6.2.

As discussed in Chap. 1, the global annual primary energy consumption is approximately 178,000 TWh, to which traditional biomass and modern biofuels contribute 11,100 TWh and 1140 TWh, respectively. Based on report by the Intergovernmental Panel on Biodiversity and Ecosystem Services [37], at this current utilization rate of bio- and other resources and land use, 75% of the land areas are severely altered, 85% of wetlands have been destroyed, and extinction threatens a greater number of species than ever before in human history.

A large number of studies have been carried out on the biomass resource requirements of a large-scale bioeconomy. For example, a literature review by Berndes et al. [5] covering 17 different studies found widely varied conclusions about the possible contribution of biomass to the future global energy supply to 2050. Estimates varied from 27,800 TWh (2 × the 2022 consumption levels) to 187,500 TWh/year (15 ×

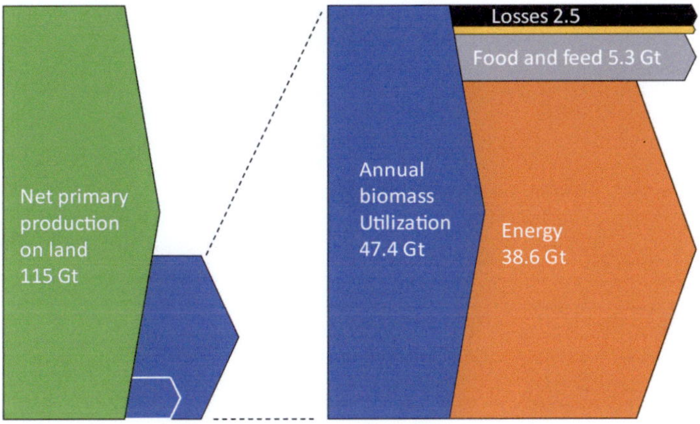

Fig. 6.3 For comparison with Fig. 6.2, a hypothetical scenario where all energy from fossil sources is replaced with bioenergy. This would require 35.4 Gt additional biomass, which would compromise other land-use needs, and ecological and social sustainability. The 12 Gt current annual biomass use (2015) from Fig. 6.2 is presented as a white-framed arrow for reference

), with the latter—more than the current global annual primary energy consumption—considered totally unrealistic, as discussed by Berndes et al. [5]. de Jong and van Ommen [13] presented a summary of resource estimation to 2030, where results varied from 23,300 TWh (IEA Current Policies) to 48,300 TWh (Shell's "Scramble" scenario). Compared with current use in 2022 (12,200 TWh), the above scenarios would already require significant increase in fertilization, harvesting, waste management, circular solutions, and demand for and changes to land use. The risk of pushing biodiversity loss and other planetary boundaries even further from their sustainable limits would necessarily increase.

As Fig. 6.2 showed, the required bio-based organic materials could be produced in theory from residues and agricultural waste streams. In addition, there may be great potential to replace some fossil-based materials using bio-based side and waste streams. For example, the annual total consumption of plastics is currently 0.4 Gt, from which 0.353 Gt forms plastic waste. Alongside increasing the recycled fraction of these plastics from 15%, we should aim to use as much as possible the side streams and waste fractions of bio-based materials as raw input materials for plastics [78].

However, several barriers must be solved to achieve this, both technically and economically. In particular, technologies need to be developed for extracting and converting fresh biomass and its residues into chemical building blocks for materials, and ultimately to products and fuels. The biorefinery concept aims to sustainable process biomass into food and feed, and a spectrum of other marketable products and energy. Figure 6.4 shows an example of the biorefinery concept for maximal utilization of lignocellulosic biomass.

Although not presented in Fig. 6.4 for lignocellulosic biomass, we also need to develop solutions which aim for making food production more sustainable and, even more importantly, to utilize all possible side streams from food and feed production

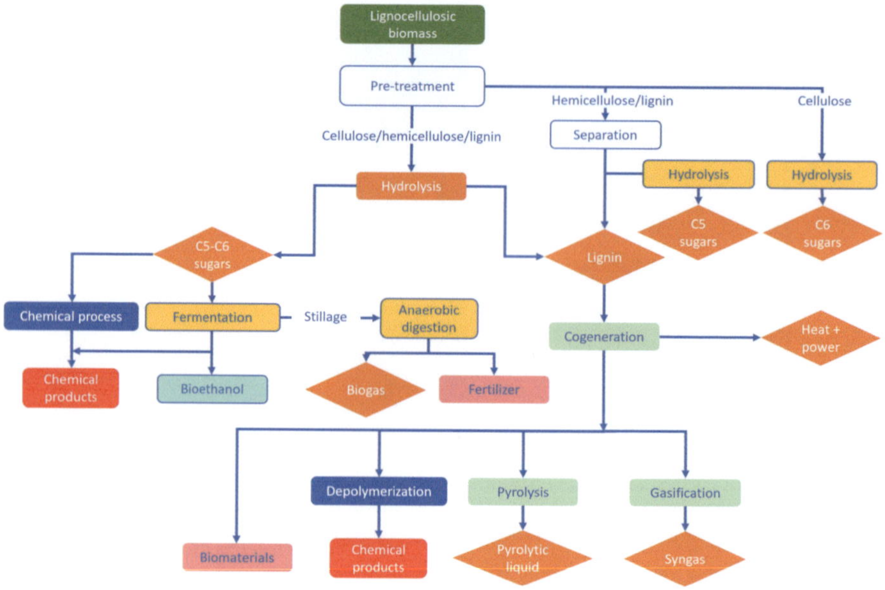

Fig. 6.4 Schematic representation of the concept of lignocellulosic biomass biorefining. Re-drawn from IEA [36]

as raw materials for chemicals, energy, and transportation fuels, as discussed above, see Fig. 6.2.

A biorefinery involves the disassembly, processing and refining of the compounds, chemicals, and minerals present in biomass and residues into valuable products, to achieve maximum value from the available biomass input. Biorefining processes the molecules in biomass through technologies such as separation via extraction, and chemical and thermal treatments. In this way, biorefining contributes to the production of chemicals and materials needed to build the bio-based material pyramid.

In short, aiming for a partially bio-based economy is attractive and feasible, to get the most out of valuable bio- and biowaste-based materials. In practice this is not easy to achieve. A great deal of effort must be focused on removing existing technical and economic barriers, and this will require practice-oriented research and innovations, including innovative business models based on fundamental research.

6.3 Principles of Photosynthesis

Plants, algae, and some photosynthetic bacteria (such as cyanobacteria and purple sulfur bacteria) are primary producers that fix atmospheric carbon (CO_2) during photosynthesis. The process of photosynthesis is initiated by chlorophyll and carotenoid pigments harvesting light in plant leaves, in special organelles called chloroplasts. During photosynthesis, plants containing chlorophyll use the energy of photons from the Sun, CO_2 from the atmosphere, and water and nutrients from the ground, to produce sugars (mainly glucose, but also sucrose, fructose, maltose, and galactose) and oxygen, O_2 (see Fig. 6.5). The reaction equation for photosynthesis can be written as

$$6\,CO_2 + 6\,H_2O = C_6H_{12}O_6 + 6\,O_2,\ \Delta H_p = +2808.1\,\text{kJ/mol}$$

Photosynthesis is an *endothermic* reaction (+ sign in the above reaction equation) and does not occur without energy from the Sun: the energy required to produce one mole of glucose is $\Delta H_p = +2808.1$ kJ/mol [54]. We can easily convert this to mass units, for example per kg of glucose (denoted with a subscript G), or per kg of carbon dioxide (subscript CO_2). The molecular masses are 0.044 kg/mol for CO_2, 0.018 kg/mol for H_2O, 0.18 kg/mol for glucose (written below as G), and 0.032 kg/mol for O_2. In the reaction equation below we multiply stoichiometric coefficients, ν_i, by the corresponding species' molar masses, $m_i = \nu_i M_i$, and then divide all terms by $\nu_i M_i$ for the species of interest (glucose or CO_2):

$$1.47\,\text{kg}\,CO_2 + 0.6\,\text{kg}\,H_2O = 1\,\text{kg}\,G + 1.07\,\text{kg}\,O_2,\ \Delta h_p = +15600\,\text{kJ/kg}_G$$

Fig. 6.5 Principle of photosynthesis. © Maija Karala

$$1\,\mathrm{kg\,CO_2} + 0.41\,\mathrm{kg\,H_2O} = 0.68\,\mathrm{kg}\,G + 0.73\,\mathrm{kg\,O_2},\ \Delta h_p = +10601\,\mathrm{kJ/kg_{CO_2}}$$

From this we see that the production of 1 kg of glucose requires 2808.1/0.180 = 15,601 kJ of energy and capturing 1 kg of CO_2 requires 2808.1/0.264 = 10,637 kJ. This reaction only utilizes Solar energy in the visible spectrum, with wavelengths 0.4–0.7 μm. At low levels of sunlight (irradiance less 100–200 W/m^2) the rate of photosynthesis is directly proportional to the irradiation, but levels off as irradiance increases. This means that plants can utilize only 10–20% of the maximum irradiance reaching the Earth's surface, which is in the order of 1000 W/m^2. The rate of photosynthesis can be increased by raising the concentration of CO_2 in the atmosphere, as is done in some greenhouses. In addition, sufficient water and nutrients are needed. In the end, the total reaction rate is limited by the mechanisms with the slowest rate or most limited availability [6]. A more detailed description of the reaction mechanisms and ways to increase the rate is presented in Info Box 6.1, and in the references provided.

Typical crops have an overall photosynthesis efficiency, η_{ps}, of just 0.5–1%. Higher 3% annual yields are reported for microalgae grown in bioreactors [6]. The efficiency is defined as the ratio of the chemical energy stored by the plant, E_c, defined on the lower heating value basis, to the solar insolation over a year, W (kWh/m^2),

$$\eta_{ps} = \frac{E_c}{WA} = \frac{E_c}{\int_0^{8760} G(t)A\,dt}$$

where the solar insolation is obtained by integrating the irradiance, G, over one year's time, and A is the horizontal area being considered, at ground level. The global horizontal irradiation values can be obtained from solar atlases—typical annual values for irradiation onto a horizontal plane are, for example: Southern Finland, 1000 kWh/m^2; Mogadishu, 2270 kWh/m^2; and Sydney, 1627 kWh/m^2 (https://globalsolaratlas.info).

Example 6.1 Approximate the land area needed to provide biomass-based energy for a country with annual electricity consumption ~ 90 TWh, and annual average insolation 1000 kWh/m^2. Assume that the produced biomass is burned in large-scale steam power plants with a total efficiency to electricity of $\eta_e = 0.3$. Assume that the efficiency of photosynthesis is $\eta_{ps} = 0.005$.

Solution: Total efficiency to electricity is the product of power plant and photosynthesis efficiencies, $\eta_{tot} = 0.3 \times 0.005 = 0.0015$. The required area can be solved from the above equation as,

$$A = \frac{E_c}{W\eta_{ps}} = \frac{E_e/\eta_e}{W\eta_{ps}} = \frac{E_e}{W\eta_{tot}} = \frac{90 \times 10^9\,\mathrm{kWh}}{1000\mathrm{kWh/m^2} \times 0.0015} = 60{,}000\,\mathrm{km}^2$$

Using Finland as an example with similar power consumption, and an arable area of 22,200 km^2, this would mean significant competition between food and energy demands, and result in new land-use changes with major biodiversity and social effects. In this calculation, losses in harvesting, logistics, and transportation were not included, which would all increase the required area even more.

Info Box 6.1

Improving photosynthesis and plant productivity.

During photosynthesis, energy from sunlight is converted into chemical energy in Adenosine triphosphate (ATP) and Nicotinamide adenine dinucleotide phosphate (NADP), which is used to fix and process atmospheric carbon into carbohydrates (sugars and starch). Photosynthesis has two main phases: (I) Light reactions, and (II) Carbon reactions. In the light reactions, photosynthetic organisms have a unique ability to split water (H_2O) into oxygen (O_2) and hydrogen (H^+), releasing electrons for the electron transport chain, producing energy from ATP and NADP for carbon processing. Carbon is fixed by the Rubisco (Ribulose-1,5-bisphosphate-carboxylase) enzyme—estimated to be the most abundant enzyme on Earth—through a process called the Calvin-Benson cycle (or C3 cycle because a 3-carbon compound, 3-phosphoglyserate, is produced). Carbon is taken up through special stomatal guard cells in the leaf surface, which form adjustable openings for parallel water transpiration and gas exchange.

In principle, light reactions, photosynthetic pigments, and Calvin-Benson cycle processes are very similar in all photosynthetic organisms, suggesting that photosynthesis has evolved only once in the very early phases of evolution. Plants, however, grow in almost every corner of the Earth, even under very harsh conditions. As a result, several modifications have evolved, in particular for the carbon fixation process, which is the 'weakest link' of photosynthesis. The Rubisco enzyme catalyzes not just carbon fixation, but also oxygen fixation in a competing process called photorespiration. Photorespiration limits photosynthetic production, but on the other hand plays an essential role in balancing energy flows and in nitrogen metabolism.

Plants have developed several carbon-enrichment mechanisms to increase the ratio of CO_2 to O_2 in the vicinity of the Rubisco enzyme, to increase carbon fixation efficiency. Plants growing in hot, dry areas need to restrict water loss (transpiration) by closing stomata during the daytime, which limits carbon uptake as well. Plants have therefore evolved to fix carbon from the atmosphere in two separate steps. C4 plants (such as maize) fix carbon first in the outer mesophyll cell layer via another carbon-fixing enzyme called PEP (Phosphoenolpyruvate) carboxylase, which does not react with oxygen. The first product is 4-carbon oxaloacetate (lending the 4 to the name for this photosynthesis type), which is then transported to inner plant tissues (i.e., bundle sheath cells) to be assimilated through the normal Calvin-Benson cycle into sugars. CAM (Crassulacean acid metabolism) plants (such as cacti) use the same two-step process and enzymes, but the steps are diurnally separated. The initial fixation of carbon occurs at night, with the PEP carboxylase enzyme, which has high affinity for CO_2, leading to accumulation of carbon in plant vacuoles (special organelles containing water and solutes) in malate form. The next day, carbon from the malates is transported to the Calvin-Benson cycle. Surprisingly, CAM metabolism can also be found in some aquatic plants. The impact of photorespiration is minimized in both C4 and CAM systems, but carbon fixation of C4 plants requires more ATP energy compared to C3 fixation. This is, however, not a limiting factor due to the high irradiation and sunlight energy in areas where these plants grow.

As described above, plants can be divided into three main groups according to their carbon fixation process: C3 plants, C4 plants, and CAM plants. This division also reflects different productivity properties, water-use efficiencies, and responses to changing climate. The highest productivity is usually seen in C4 crop plants, such as maize, sugarcane, and sorghum. There are about 8100 plant species in this group, mainly monocots (grasses). C4 plants mostly grow in southern and mid-latitudes, which are generally sunny and warm environments. They can retain high photosynthesis rates even in very high (> 40 °C) temperatures. C3 plants are the most abundant plant group and best adapted to northern latitudes, with less sunlight energy and cooler temperatures. Practically all boreal forest species are C3 plants. The productivity of C3 plants varies considerably among the species, taxa, and functional plant groups (trees, herbs, grasses, monocots, dicots, mosses, etc.), and depends also on the plants' environment. Several C3 plants, such as bamboos and rice, improve their photosynthetic efficiency by recapturing CO_2 from photorespiration. There are also attempts to breed beneficial C4 traits into rice and other important C3 crop plants. CAM plants, such as *Opuntia* and *Agave*, can have high production rates in optimal growing conditions, but the growth is usually restricted by limited water availability in their natural arid environment. CAM plants have the best water-use efficiency (WUE), the ratio of carbon assimilation or biomass production to the rate of water transpiration, while C3 plants are the least efficient by this measure.

Improving photosynthesis and plant productivity, particularly for crop plants, has been a key target in plant breeding for decades, with the aim of improving food security and increasing production of calories per hectare, known as land-use efficiency. The most promising targets in plant engineering have been improved kinetic properties of the Rubisco enzyme for carbon fixation, introduction of C4 processes into C3 plants, improved use of variable light intensities, a wider array of photosynthetic pigments (e.g., from cyanobacteria), more efficient carbohydrate and photosynthesis-related nitrogen metabolism to reduce the need for external fertilization, and improved structural properties (e.g., leaf size, position, canopy and leaf tissue density) allowing optimal exposure to light and carbon uptake. However, despite numerous breeding programs, improved molecular and genetic techniques, and ongoing research, real improvements in crop productivity have been very limited so far. Breeding for photosynthesis is challenging because carbon is an essential element and building material for many purposes and pathways, including plant growth, cell respiration (ATP formation), defense compounds, signaling, and reproduction. There are complex feedback loops and trade-off mechanisms between carbon fixation, growth and plant defense. In other words, high photosynthesis does not equate to high yield and productivity. Although increasing CO_2 concentrations in the atmosphere have principally increased photosynthetic rates of plants, the positive impact on growth and biomass accumulation is often transient and inherently down-regulated due to accumulation of CO_2 and sugars in leaf tissues, stomatal closure, and restrictions in water and nutrient supply. In many areas, climate change is increasing the presence of unfavorable growing conditions, including pressure from pests, pathogens, and extreme climatic conditions. Finally, the productivity and performance of plants is an outcome of carbon assimilation capacity, resistance and adaptation to changing conditions, and availability of resources necessary for the growth and reproduction, such as light energy, carbon, water, nutrients.

Ecologically, the productivity of ecosystems is measured and indicated by gross primary production (GPP), which is the amount of chemical energy, usually biomass, produced by the primary producers in a given time. Living organisms, however, always use a part of this energy for maintenance costs (i.e., cell respiration) leading to re-circulation of CO_2 to the atmosphere. This loss of energy is included (subtracted) in the other widely-used indicator, net primary production (NPP). At the ecosystem level, the most productive ecosystems are intensive agricultural and farming systems, fertile natural wetlands (such as marshes and estuaries), coral reefs, and moist terrestrial ecosystems such as tropical and temperate rainforests. Rainforests are located in constant supply of sunlight, warm temperatures and water, making them highly productive ecosystems both in GPP and NPP. However, it has been estimated that about 70% of fixed carbon is actually released back to the atmosphere due to long, warm nights maintaining high cell respiration levels. On the other hand, the least-productive ecosystems are deserts, arctic tundra, and deep-sea ecosystems, with harsh water, temperature and light conditions.

Despite the restrictions in engineering for improved photosynthesis at the plant level, there are several practical ways to increase photosynthesis at the ecosystem level [21]. For example, by substituting perennials for annual plants it is possible to utilize the length of growing seasons more efficiently in many areas, because annuals do not need time for nutrient retrieval (leaf yellowing and coloration) and winter-hardening process. On the other hand, this comes with trade-offs in other ecosystem properties, because annual cropping systems, due to the need for frequent mechanical and agrochemical operations, are prone to the degradation of soil structure and erosion risk, carbon and nutrient losses, and biodiversity declines. In the context of bioenergy production, perennial grasslands outperform annual crops on all ecosystem services (such as pest suppression, pollination, biodiversity conservation, carbon storage, water protection) but have lower productivity than crops with the highest yields, such as maize (e.g., Werling et al. [84]).

6.4 Biomass Types, Sustainable Availability Potential and Carbon Debt

This section discusses in more detail the sustainability aspects of different types of biomass feedstocks used in the energy sector, and potential conflicts and synergies with other types of land use. We examine the availability—and hence the potential and the limitations—of different biomass feedstocks, how the increased use of forests for bioenergy can compromise other values of forests, how growing biofuel feedstock crops can improve or degrade agricultural land, and how bioenergy production can lead to expansion of tree monoculture plantations, and to cultivated areas displacing forests, grasslands, peatlands, and wetlands.

The capacity of the Earth to produce biomass is limited by its biophysical boundaries [20, 76]. To be a viable alternative to fossil energy, bioenergy production should restrict direct and indirect competition with nature conservation or food security (see Chap. 2). Intensified harvesting of biomass should not lead to drastic declines of biodiversity. The many drivers to avoid biodiversity loss include economic factors, and the impacts of long-term productivity—that is, biomass regrowth—on ecosystems. For example, a global study with over 777,000 experimental plots across the globe containing 30 million trees revealed a positive biodiversity-production relationship, demonstrating that a continued loss of biodiversity will be followed by accelerating decline in forest productivity [48]. In other words, intensive management for maximal biomass output reduces the number of species, conversely, the more species a stand of forest contains, the higher its productivity is likely to be. A similar positive relationship between productivity and the number of *native* species is known from other ecosystems, including grasslands [14]. Liang et al. [48] also concluded that, while increases in native, dominant species raises productivity, increases in rare and non-native species reduces productivity. This suggests that the spread of non-native species should be restricted as much as possible.

The term bioenergy refers to organic materials derived from living or recently living organisms that can be used to produce energy. Bioenergy can be derived through various processes, and there are several biomass types used for bioenergy production, such as:

Wood and wood residues: This includes logs, sawdust, wood chips, and wood pellets. Branches, tree tops, and other residue from forestry activities are also utilized for bioenergy. Wood is one of the oldest—and still today most commonly used—biomass energy sources.

Energy crops: These are specifically grown for bioenergy purposes, and include grasses like switchgrass (*Panicum virgatum*) and silver grass (*Miscanthus* sp.), and fast-growing woody-crops like willow (*Salix* sp.), poplar (*Populus* sp.) and eucalyptus. Many edible crops are also used, such as sugar cane (*Saccharum officinarum*), sugar beet (*Beta vulgaris* subsp. *vulgaris*), corn (*Zea mays*), many grains, soybeans (*Glycine max*), and oil plants such as oil palm (*Elaeis guineensis*) and rapeseed (*Brassica napus*).

Aquatic plants: Plants like water hyacinth (*Eichhornia crassipes*) and duckweed (*Lemnoideae*) growing in water bodies can be harvested and used for bioenergy production.

Algae: Microalgae and macroalgae can be cultivated to produce biofuels such as biodiesel or biogas.

On top of wood residues there are also other waste streams, including from agriculture and the food industry, that form a key source of bioenergy supply in many scenarios. For example, in the net-zero emissions by 2050 scenarios (e.g., NZE Scenario by International Energy Agency, IEA), over 60% of the global bioenergy supply by 2050 should come from sustainable waste streams, alleviating the pressure on land use. These waste streams, discussed in detail under Sect. 6.4.2, include:

Agricultural residues: Byproducts from crop harvesting and processing, such as straw, corn stalks, rice husks, sugarcane bagasse, and other plant residues.

Animal manure: Organic waste from livestock and poultry can be used for biogas production through anaerobic digestion.

Food processing waste: Any byproducts and organic waste from food processing industries can be used to produce bioenergy.

Municipal solid waste (MSW): The specific type of biomass used for bioenergy depends on regional availability, technological capabilities, and environmental conditions. Various technologies exist to convert municipal solid waste biomass into useful forms of energy. Biomass can be combusted directly to produce heat and electricity, or it can be converted to gaseous or liquid biofuels.

Liquid fuels, or biofuels, are divided into three categories depending on their source and technology used. First-generation biofuels are processed directly from cultivated crops, such as oil seeds, sugar beets, or maize. Second-generation biofuels are produced from residues and waste. Third-generation biofuels are algae based and are still mostly under development. The use of first-generation biofuels has been questioned on multiple sustainability grounds, including the efficiency of net carbon gain and food security. Second-generation biofuels generally have a higher potential to reduce emissions, and minimal impacts on biodiversity because they mostly do not lead to land-use change. Third-generation algae-based biofuels have good prospects due to their high yields and versatility (algae can be converted to any kind of biofuels, such as biodiesel, gasoline, ethanol, propanol, and butanol) but their potential remains unrealized at a scale needed for environmental and economic gains. In Europe, there is a strong movement to limit the use of biomass as bioenergy to food waste and byproducts as localized solutions [19].

Biomass-based energy is often considered a sustainable alternative to reduce greenhouse gas emissions and mitigate climate change, as it can be used directly to substitute fossil fuels. However, while bioenergy replaces CO_2 emissions resulting from the use of fossil energy, the combustion and processes releases CO_2, other greenhouse gas emissions and other pollutants to the atmosphere and water from the soil and forest stands, as well as from use of inputs in cultivation, such as fertilizers. These emissions from the production and use of bioenergy can result in eutrophication of water bodies, acidification, and toxicity in the environment [32].

Habitat loss, fragmentation, and degradation are some of the major impacts caused by biomass energy production, especially when areas are converted to mono-cultural feedstock production [32]. In degraded or previously intensively cultivated lands, biomass energy production can also provide opportunities for the creation of biodiversity benefits and enhancement of ecosystem services [18, 30, 32]. In their review on the impacts of bioenergy crop production in temperate climates, Winberg et al. [85] confirmed that land-use changes due to bioenergy can have positive or negative effects on ecosystems, depending on the original land use, bioenergy crop type, and scale of bioenergy production. Perennial crop cultivation on arable land had the lowest negative impacts on biodiversity and ecosystem services. Using marginal or abandoned land for bioenergy crops cannot fully resolve these trade-offs. Some biomass crops, like eucalyptus or *Miscanthus,* can also be invasive [32].

When a large area of forest is harvested and converted to energy-crop plantation, a huge amount of biomass, carbon and nutrients are removed from the area. For example, tropical rainforest stores on average 45,000 ton/km^2 of dry biomass and 22,5000 ton/km^2 carbon, while tropical dry and temperate evergreen forests store 35,000 ton/km^2 dry biomass and 17,500 ton/km^2 carbon [2]. If the plant roots are removed, nutrients in the soil are easily drawn away by rain and erosion.

In addition to losing the existing carbon storage, the ongoing sink of carbon dioxide also drops for several years. The concept of carbon debt and its payback time has been developed to describe the impacts of removing natural vegetation from an area and replacing it with biomass production. The carbon debt concept was first introduced by Fargione et al. [24]: “We call the amount of CO_2 released during the first 50 years of this process the carbon debt of land conversion. Over time, biofuels from converted land can repay this carbon debt if their production and combustion have net GHG emissions that are less than the life-cycle emissions of the fossil fuels they displace. Until the carbon debt is repaid, biofuels from converted lands have greater GHG impacts than those of the fossil fuels they displace.”

Annual repayment rate means the rate of net carbon capture due to growing energy crops in plantations, and how much benefit the use of biofuel generates in comparison to using fossil fuels. The carbon debt concept is important when selecting energy crops to substitute fossil fuels, since this also determines the sustainability and climate-change mitigation potential of the crop. Table 6.1 presents some examples of carbon debt, repayment rate, and payback time.

Table 6.1 Carbon debt related to biomass use when the site area is cleared for energy crop farming, repayment rate, and carbon debt payback time [24, 32]

Biomass, fuel	Location	Carbon debt[a] (ton/km^{2})	Annual repayment rate[b] [ton/(km^2 y)]	Payback time[c] (y)
Palm biodiesel[b]	S-E Asia, peatland rainforest	N/A	N/A	629
Palm biodiesel	Indonesia, peatland rainforest	300,300	710	423
Soybean biodiesel	Brazil, tropical rainforest	28,700	90	319
Jatropha-based fuels[b]	Mature miombo woodland	N/A	N/A	187–966
Corn ethanol	USA, central grassland	11,100	120	93
Palm biodiesel	Indonesia, tropical rainforest	61,100	710	86
Corn ethanol	USA, abandoned cropland	5700	120	48
Soybean biodiesel	Brazil, Cerrado grassland	3300	90	37
Sugarcane ethanol	Brazil, Cerrado wooded	165	9.8	17
Prairie ethanol	USA, abandoned cropland	6	4.4	1
Prairie ethanol	USA, marginal cropland	0	7.8	No debt

[a]Includes CO_2 emission from the soils from above and below biomass resulting from land-use change
[b]Annual life cycle greenhouse gas reduction from biofuels, including the displacement of fossil fuels and soil carbon storage
[c]Number of years after conversion to biofuel production required for cumulative biofuel greenhouse gas reductions, relative to the fossil fuels they displace, to repay the biofuel's carbon debt
y refers to one year, N/A not reported

As Table 6.1 shows, the longest payback time is in the case of South Asian palm oil biodiesel in tropical peatland rainforest (629 years to break even), Indonesian peatland rainforest grown palm oil biodiesel (423 years), and Brazil tropical rainforest grown soybean biodiesel (319 years). In addition, the loss of biodiversity in these areas is enormous [75]. More precisely, oil palm use is associated with declines in biodiversity, while driving changes in ecosystem functioning and the provision of ecosystem services. Oil palm expansion has been estimated to affect 54% of all threatened mammals and 64% of all threatened birds globally [53, 75]. Based on Gasparatos et al. [32], the ethanol produced on already affected lands in the USA has a very minor impact on local carbon debt. In addition to this carbon debt and their effect on global climate, biofuels can also impact local micro-climates due to changes in surface albedo and evapotranspiration.

Bioenergy might wind up playing a much smaller role in climate change mitigation than suggested by many scenarios (such as the proposed ~ 28,000 TWh in IEA [36]), if all impacts—including those from the destruction of original ecosystems, loss of carbon stocks, and displacement of other land uses such as food production—are considered [55]. Based on the IPCC [40], large-scale land conversion for bioenergy, biochar (residue remaining after the pyrolysis of biomass), or afforestation (forestation of an area where there has not been recent tree cover) can increase risks to biodiversity, water and food security. They also suggested that, in contrast, restoring natural forests and drained peatlands, and improving sustainability of managed forests, enhance the resilience of carbon stocks and sinks and reduce ecosystem vulnerability to climate change. This means that bioenergy from biomass, though not negligible, can have only a relatively small contribution to sustainable climate change mitigation and energy transition. Though bioenergy can be obtained from many sources of biomass, its sustainably obtainable volumes are limited.

Example 6.2 Calculate how much CO_2 per kWh of heat is released from the combustion of high-quality coal, and from cold air-dried energy wood chips (alder, poplar, willow mixture). Lower heating value (water vapor released from the oxidation of hydrogen in the dry fuel is not condensed to liquid water to recover its heat) of the dry matter is 29.2 MJ/kg for coal and 17.5 MJ/kg for biomass, and moisture content is 0.04 and 0.3 on wet biomass basis, respectively. Carbon content is 0.73 for coal and 0.5 for biomass. Heat of water evaporation is 2.3 MJ/kg.

Solution: Let's first calculate the lower heating values as received (ar), which means that the water vapor released in drying escapes to the atmosphere and its latent heat is not recovered.

$$LHV_{ar} = (1 - 0.04) \times 29.2 - 0.04 \times 2.3 = 27.9\,\text{MJ/kg, coal}$$

$$LHV_{ar} = (1 - 0.30) \times 17.5 - 0.30 \times 2.3 = 11.6\,\text{MJ/kg, biomass}$$

The derivation of these will be discussed later. Only the combustion of dry matter produces heat, and the water component has to be evaporated. The energy released from 1 kg of burned biomass divided by that released from 1 kg of coal is 11.6/26.9 = 0.41, so to get the same amount of heat we must combust 1/0.41 = 2.42 times more wet biomass. 1 kg of coal contains 1 × 0.96 × 0.73 = 0.70 kg of carbon, while 1 kg biomass contains 1 × 0.7 × 0.5 = 0.35 kg of carbon. When these are oxidized to CO_2 via the reaction $C + O_2 = CO_2$, we get 0.7 × 44/12 = 2.57 kg of CO_2 from the coal, and 0.35 × 44/12 = 1.28 kg from the biomass. CO_2 emissions are then 2.57 kg from the coal, and 2.42 × 1.28 = 3.1 kg from the biomass, for same energy output. The emission coefficients (gCO_2/kWh) are then 2.57/27.9 kg/MJ = 331 g/kWh for coal, and 1.28/11.6 kg/MJ = 387 g/kWh for biomass.

This simple example shows that, in this case, the high-quality coal is actually the less CO_2-emitting fuel at the moment of combustion. However, if the wood chips are locally used, and sustainably produced fuels are used in the collecting and harvesting machinery, the fast-growing wood energy crops would be able to bind the released carbon dioxide and, over time, would be a much better feedstock option than the coal. As a local solution, this is seen as a sustainable option, but it might not be possible to scale-up this scenario to larger applications due to land-use and biodiversity-loss restrictions.

6.4.1 Forest-Based Biomass and Woody Crops

This section discusses the use and availability of forest-based biomass, deforestation, transformation of forests into plantations, and the impacts this energy source has on climate, nature, and land use. Globally, forests are the most important source for renewable bioenergy. Based on data from the World Bioenergy Association [87], solid biomass provided a supply of 13,700 TWh, accounting for ~ 86% of all biomasses used for energy production. From this wood fuel supplied 5820 TWh, black liquor in forest industry (the pulp and paper in making process fluid) supplied 560 TWh [23], char coal supplied 412 TWh, and wood pellets supplied 220 TWh. Traditional wood fuel was mostly used in Africa and Asia (both contributing by equal amounts, adding up to 73%) and char coal in mostly in Africa (70%) [87].

Forest ecosystems cover 40.2 million km^2 of the total global land area of 149 million km^2 [2, 69]. Figure 6.6 presents the historical development of land use for different purposes up to 2018.

As Fig. 6.6 shows, forest land area has globally decreased by 33% over the last 10,000 years, mostly due to grazing of domestic animals and growing of crops. Note also that the original 46 million km^2 of wild grasslands and shrubs has shrunk to just 17.4 million km^2, a decrease of 62%. Forest coverage in these land use statistics includes a considerable variety of ecosystem types.

Aside from the harvested trees, forest-based biomass includes wood residues, the byproducts and waste from forestry operations, with logging waste such as branches and tops of trees, damaged trees, and thinning materials used increasingly as energy sources. Stumps are also often harvested, against current recommendations—the removal of stumps decreases the growth of new tree generations as it disturbs the

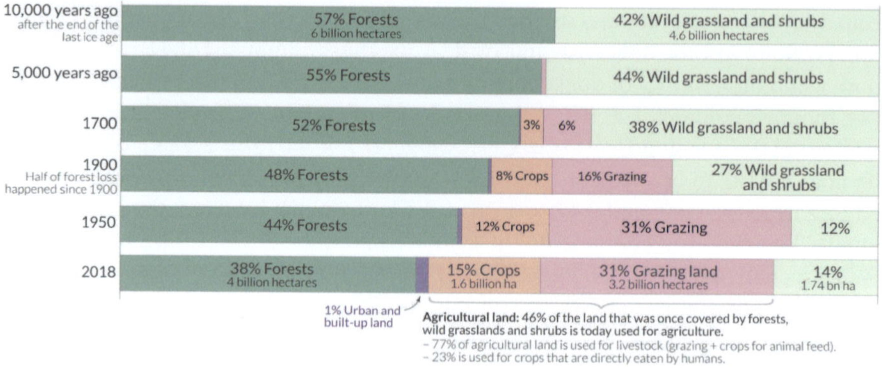

Fig. 6.6 Historical development of land area use. The total land area included is 105.8 million km^2; the remaining 43.2 million km^2 not included here is covered by deserts, glaciers, rocky terrain, and other barren land not suitable for growth ([69], CC BY 4.0). 100 ha = 1 km^2

soil, removes a significant amount of carbon and nutrients, and increases leaching [81]. Residues from the forest industry, such as sawdust, bark, and black liquor form another main source for energy use.

Forests can roughly be divided into primary forest (naturally occurring self-regenerating mixed stands), secondary forest (mixed stands that are replanted and managed, often in places of former forests), and plantations (monoculture stands of fast growing, often non-native, tree species, such as conifers at temperate latitudes and elevations, or eucalyptus or oil palms in sub-tropical and tropical regions). See Fig. 6.7 for examples of these forest types.

Fig. 6.7 Primary coniferous forest (left), secondary coniferous forest (middle), and plantation (right). © Maija Karala

Table 6.2 Forest area change from 1990–2020 in millions of km^2 [70]

	1990	2020	Change (%)
Europe	10.10	10.18	0.7
South America	9.74	8.44	−13.3
Africa	7.43	6.37	−14.4
North America	6.51	6.57	0.9
Asia	5.70	6.23	9.2
Rest of the world	2.90	2.81	−3.1
World total	42.40	40.60	−4.2

As shown in Fig. 6.7, global total forest area has been continuously decreasing: over the last 30 years, it has decreased from 42.4 to 40.6 million km^2. Table 6.2 breaks down this change across six main geographical areas.

Clearly, South America and Africa are the regions where the largest forest loss has occurred, by 13.3% and 14.4%, respectively. Forest resources have increased slightly in Europe between 1990 and 2020, with total forest area growing from 10.1 to 10.2 million km^2. Although there are major differences between the member states of Europe, the harvesting of roundwood has also been increasing [70]. In South America and Africa there is a gradual decline of forest resources, from 9.74 to 8.44 million km^2, and from 7.43 to 6.37 million km^2, respectively. With these changes in the forest area, the amount of stored carbon (carbon stock) increases or decreases accordingly. For example, in Europe carbon stock increased from 159 to 172 Gt, while in South America it declined from 162 to 145 Gt [22], Forest Europe [26].

The largest ongoing global trends are deforestation of naturally occurring mixed stands (primary forest), afforestation of former agricultural lands with monoculture plantations of conifers (at temperate latitudes and elevations) or eucalyptus and oil palms (in sub-tropical and tropical regions), and replacement of primary forests with plantations. The main reason for deforestation is the conversion of primary forests to agricultural land (pasture for beef production, fields for soya and palm oil), and to tree plantations for timber, pulp and bioenergy. The expansion of production of first-generation bioenergy crops (mainly oil palm and soya) drove 18% of deforestation between 2010–2014 [60].

Afforestation is driven mainly by increasing the land area for agriculture, and is mainly taking place in countries in tropical and sub-tropical areas [70]. Figure 6.8 presents drivers of afforestation.

As Fig. 6.8 shows, beef and oilseed production are the strongest drivers for loss of forest areas, with a combined share of 41%, and global deforestation rates of 21,000 km^2 and 9500 km^2 every year, respectively. As a very simple solution, changing to vegetarian diets and reducing meat consumption would have a significant positive impact on this problem. Large-scale production of biofuels from oilseed also presents a significant land-use change and threat to biodiversity. To expand plantations as sustainable sources of biomass, we should consider the impacts of land conversion

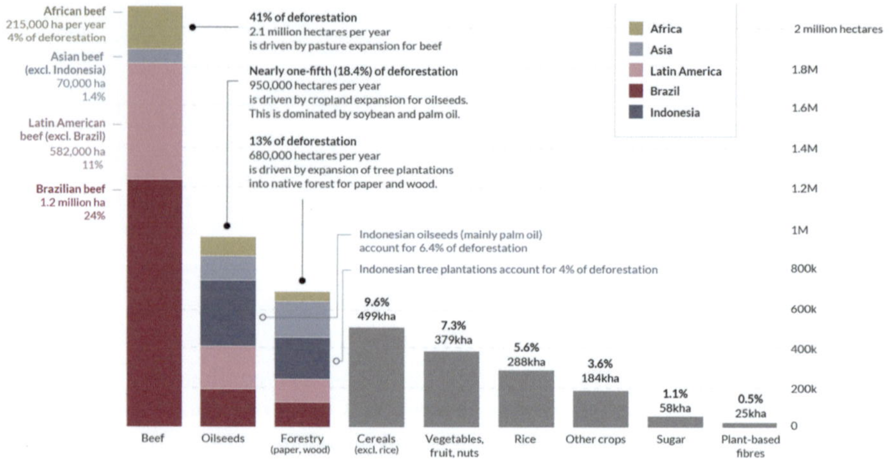

Fig. 6.8 Main contributors of deforestation ([60, 70], CC BY 4.0). 100 ha = km^2

and displacement of original ecosystems, with associated changes in carbon stocks and biodiversity.

Agricultural and forestry trade drives a large share of tropical deforestation emissions. The combustion of forest biomass has a considerable range of efficiencies, varying from very low efficiency for animal dung combusted in a simple open fireplace, to Oxyfuel combustion of dried wood chips in a state-of-the-art circulating fluidized bed boiler, producing a clean CO_2 stream that can be stored or used for chemical synthesis products. In some cases, using forest biomass for bioenergy can be sufficiently efficient and economic, but this is mostly due to subsidies [8]. It is argued that this form of bioenergy can buy time for developing more advanced forms of bioenergy. In some cases, however, the combustion of forest biomass can emit more CO_2 per unit of energy than burning fossil fuels [74].

The value of primary forests for biodiversity, climate regulation, water retention and cycling, and cultural and other livelihood services to humanity are irreplaceable [37]. As an example, tropical primary rainforests dominate as centers of species diversity: almost 30% of global vertebrate species are endemic to tropical forests and all in all tropical forests harbor over half of the world's vertebrates, many of which are experiencing high levels of extinction [61]. But all kinds of forests host locally and regionally important biodiversity and maintain important functions—for example, boreal forests store 30–40% of Earth's land-based carbon, making them critical to the fight against climate change [2].

The overall impacts of wide-scale removal of logging residues (branches, stumps, etc.) for energy are still poorly understood. The accumulated evidence points to considerable negative effects of this novel practice on biodiversity (and particularly species dependent on dead wood), water quality, reindeer herding, hunting and berry picking, and natural heritage values [67]. Positive benefits also exist, including fungal disease control and recreational values. This practice also poses risks for future biomass production, due to the loss of nutrients and impaired re-growth of trees (for example, nutrient losses have been reported up to 30% in Finland). Finally, the use of forestry residues at an economically attractive scale almost always implies clearcutting rather than non-clearcutting methods, because with the latter, wood residue is more scattered and difficult to collect [31].

There are unavoidable trade-offs in using land for competing purposes. Maximizing the harvest of wood biomass for bioenergy competes with maintaining or restoring forests to their natural state for biodiversity and long-term carbon storage. As an example, Finland provides a useful case study of multiple competing targets for their abundant, yet limited, forest resources. National chemical, forest, and energy sectors have each outlined targets for intensified forest biomass use, which combined (over 140 million m^3) are more than double that of the already-high 2019 logging rates from Finnish forests [50]. Excessive logging has already led to the decrease in the national carbon sink, jeopardizing the 2035 climate neutrality goal [49] and posing further risks to species already threatened by forestry.

Wood remains an affordable source of energy in the EU due to subsidies, and burning wood provided more than 60% of the EU's renewable energy in 2016 [8]. Only half of the wood pellets produced in the EU comes from byproducts and post-consumer wood, the rest being "primary woody biomass". The additional potential of bioenergy production from forests has been estimated to be small compared to current total energy consumption (for the Finland case see Antikainen et al. [3]).

In tropical and sub-tropical regions, the production of bioenergy plants is often linked to biodiversity loss of natural habitats. For example, palm oil production in Indonesia and Malaysia is the main driver for deforestation, causing loss of several endangered animal species, such as the Orangutan and the Sumatran Rhino [80]. In addition, the degradation of natural forest ecosystems causes these areas to shift from carbon sinks to carbon emitters, with negative consequences on climate. Another problem lies in increasing pollution of soil and water from exposure to fertilizers and pesticides, as well as residues due to field-burning practices. The protection of remaining tropical forests, especially peatlands, and the development and support of sustainable palm oil production and certified palm oil products are needed.

As a worrying example of what can go wrong, the utilization of the Amazon Rainforest has been far beyond the limits of sustainable development—see Fig. 6.9 as an example of increases in rainforest clearing in Mato Grosse, Brazil between 1992 and 2006 [58].

After clear-cutting the land is often turned over to monoculture, where ecological processes and species richness is poor when compared to the natural state. Areas are also converted into pastures for cattle farming, which can result in land degradation.

Plantations can be established in areas other than primary forests. In establishing fast-growing tree plantations on unforested land, it is important to start with a correct assessment of atmospheric carbon reduced, or net carbon gain. For example, if establishing the plantation requires draining of land (e.g., in bogs or wet areas), this leads to severe long-term release of soil carbon into the atmosphere, and these losses are not offset by the subsequent replacement of fossil fuels by bioenergy (see Chap. 2. 1). Examples include plantation expansion on present or former peatlands across

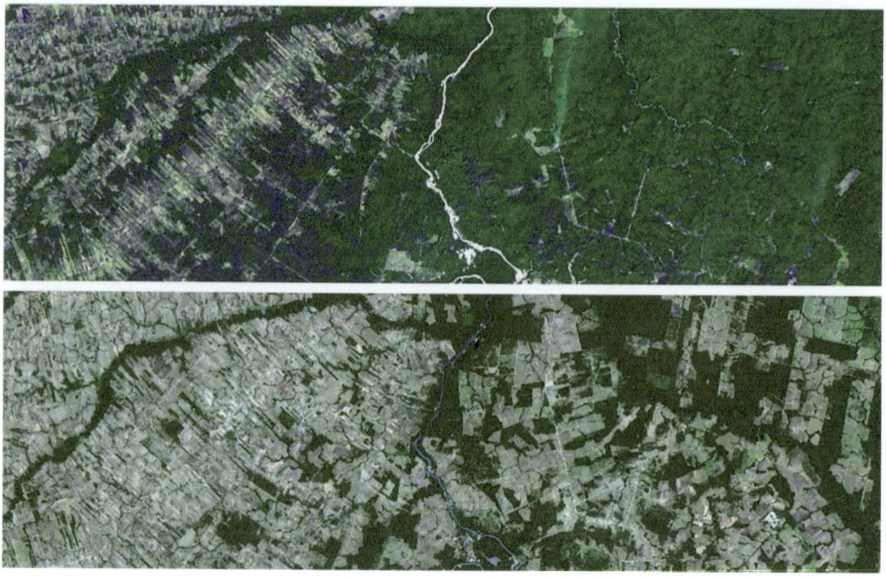

Fig. 6.9 Example of an extreme large-scale forest clearing in the Amazon: upper figure from 1992, 20% clearing; lower picture from 2006, same area, more than 80% clearing [58]. Actual land area size presented is 15 × 45 km. https://www.jpl.nasa.gov/images/pia11420-mato-grosso-brazil

Europe (e.g., in Ireland), and in tropical regions such as Indonesia [66]. Similarly, impacts on native biodiversity, water, soil, and livelihoods of rural communities should be assessed holistically. For example, rapid subsidized planting of monoculture of the radiata pine (*Pinus radiata D. Don*) on permanent grasslands in New Zealand (currently 1.6 million ha) has been shown to have detrimental impacts on biodiversity outside the plantations due to uncontrolled spread—radiata pine has become an invasive species [4].

Large differences exist between regions regarding technological readiness, management practices, harvesting potential, and utilization rate. For example, the EU Biodiversity Strategy proposes to increase the area of protected forest to cover at least 30% of the land in Europe. Intensive planting of trees in the EU area by 2030 will be carried out, along with stricter protections of European primary and old-growth forests. Climate change, however, is expected to increase the risk for forest disturbances—from storms, fires, pathogens and pests, for example—which can hamper the availability of biomass.

The challenge is to find ways of integrating multiple objectives in using forests. Transition from monocultures to mixed-species stands, and continuous forest growth-management practices are examples of measures that help to combine biomass harvesting with other commercial and non-commercial uses of forests, while also maintaining carbon stocks and biodiversity. Sustainable utilization of forest biomass also requires reassessment of available biomass volumes and multi-stakeholder discussion of the different values and uses of the forests. We need to increase the energy efficiency of all steps in the value chain from biomass to energy, to energy carriers. For example, if we could double or triple the primary energy utilization efficiency of traditional biomass in developing countries—responsible for 10% of global energy consumption—this would offer much more end-use energy; conversely, with higher efficiencies we could reduce the use of biomass, securing biodiversity increases. We could utilize the tools of technological innovations, increased resource efficiency (e.g., through digitalization), AI solutions, the circular economy for residuals and waste, as well as better management practices.

In Case Study 6.1 we present a spatial-modelling study of the current carbon balance, carbon conversion and storage processes, and biodiversity in boreal ecosystems in Finland. The tools and methods used here can be applied to other location as well.

Case study 6.1

Spatial modelling of carbon processes and biodiversity in boreal ecosystems: a Finnish case study

Finland, as part of the European union, is committed to the Paris Climate Change agreement, which aims to decrease net emissions of greenhouse gases to slow down climate change. To reach carbon neutrality, Finland should produce no more greenhouse gases (emissions: CH_4, CO_2, N_2O) than it can sequester (sinks). An area is considered carbon neutral when its emissions are at most as large as its sinks; if the area's emissions are lower than its sinks, the area is considered carbon negative. Actions towards carbon neutrality and carbon negativity aim to decrease climate warming. Estimates of national/regional-scale greenhouse gas (GHG) budgets for different land use classes and their future developments are therefore important for designing different management and mitigation measures, and for evaluating potential feedbacks to the climate system [27, 34]. Forests provide the largest boreal carbon sink, as well as key habitats for biodiversity protection, while forest harvesting causes emissions. Climate change has a positive impact on forest growth but also increases the uncertainty of the predictions [43, 46].

The basic equation for carbon neutrality is: Balance = emissions + sinks. The emissions are the fluxes of gases from land or water to the atmosphere, which are entered as positive numbers in the balance equation (Emissions > 0)—see Fig. 6.10. Sinks occur when carbon is bound into growing biomass in the process of photosynthesis, and are entered as negative numbers (sinks < 0). If the balance is positive, the area or land-use class acts as an emission source; if it is negative, the area or land-use class sequesters more carbon than it emits GHGs, and so acts as a sink.

Carbon dioxide equivalent (CO_2-eq) is a measure used to compare emissions of various GHGs based on their global warming potential (GWP). Compared with carbon dioxide (CO_2), with a GWP of 1, different values have been used over the years for the global warming potential of methane (CH_4) and nitrous oxide (N_2O). We used GWP values of 28 for CH_4 and 265 for N_2O for a 100-year time horizon in our example, taken from the IPCC AR5 [57].

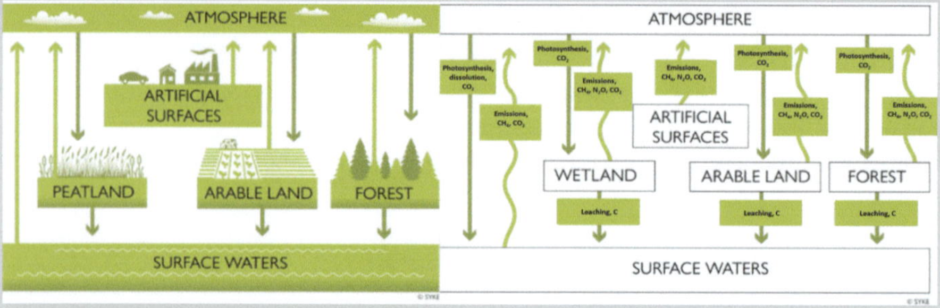

Fig. 6.10 Direction (left) and types (right) of fluxes of greenhouse gases between atmosphere and anthropogenic surfaces, peatland, arable land, forest and surface waters

Quantifying Sources and Sinks of Greenhouse Gases for Main Land-Use Categories

Following these principles, we calculated detailed, spatially-explicit estimates of current GHG emission sources and sinks for the primary land-use categories in mainland Finland [34]. We applied models for emissions from artificial surfaces (for CO_2, CH_4, N_2O), and for forest growth and carbon balance. Empirical emission coefficients were used to estimate emissions from drained forested peatland (CH_4, N_2O), cropland (CO_2), waterbodies (CH_4, CO_2), and undrained mires (CH_4, CO_2, N_2O).

We calculated gross emissions of 148.2 ± 7.6 $MtCO_{2.eq}$/a for 18 administrative units covering mainland Finland, using data representative of the period 2017–2025 (Fig. 6.11). Emissions from energy production, industrial processes, road traffic, and other sources in artificial surfaces amounted to 54.1 $MtCO_{2.eq}$/a. The loss of carbon in forest harvesting was the largest emission source in the land-use sector (59.8 $MtCO_{2.eq}$/a). Emissions from organic soils dominated cropland emissions (5.8 $MtCO_{2.eq}$/a). Rivers and lakes, and soils in undrained mires, added to the total GHG fluxes. The carbon sequestration from the atmosphere was 93.2 ± 13.7 $MtCO_{2.eq}$/a, with the main sink in forests on mineral soil (79.9 $MtCO_{2.eq}$/a). All sinks combined compensated for 63% of total emissions, and thus the net emissions were 55.0 ± 15.7 $MtCO_{2.eq}$/a. These results can be used for regional land use planning and resource management.

Dynamic Modelling of Carbon Processes in Forested Ecosystems

As Fig. 6.11 shows, forests totally dominate the carbon sink category in Finland, and forest harvesting is a major source of CO_2. However, the structural state of forests, their growth and economic use, as well as the type and condition of biodiversity, vary spatially and temporally. It is therefore important to also understand the dynamic aspects of these processes, for which process-based dynamic forest models are needed. Figure 6.12 shows the intensity of previous years' harvests by region as harvested biomass (measured as kgC) per hectare on the x-axis, and the area of each region on the y-axis—thus the area of the rectangles represents the total contribution of each region to the harvested biomass. The harvest intensity variability, combined with climatic conditions and soil fertility, make future scenario analysis and understanding the complex and correlated dynamics of forest carbon sequestration very challenging. Thus, careful modelling and analysis of the results are needed to draw a reliable picture of the regionally-varying potentials for carbon sinks in Finland.

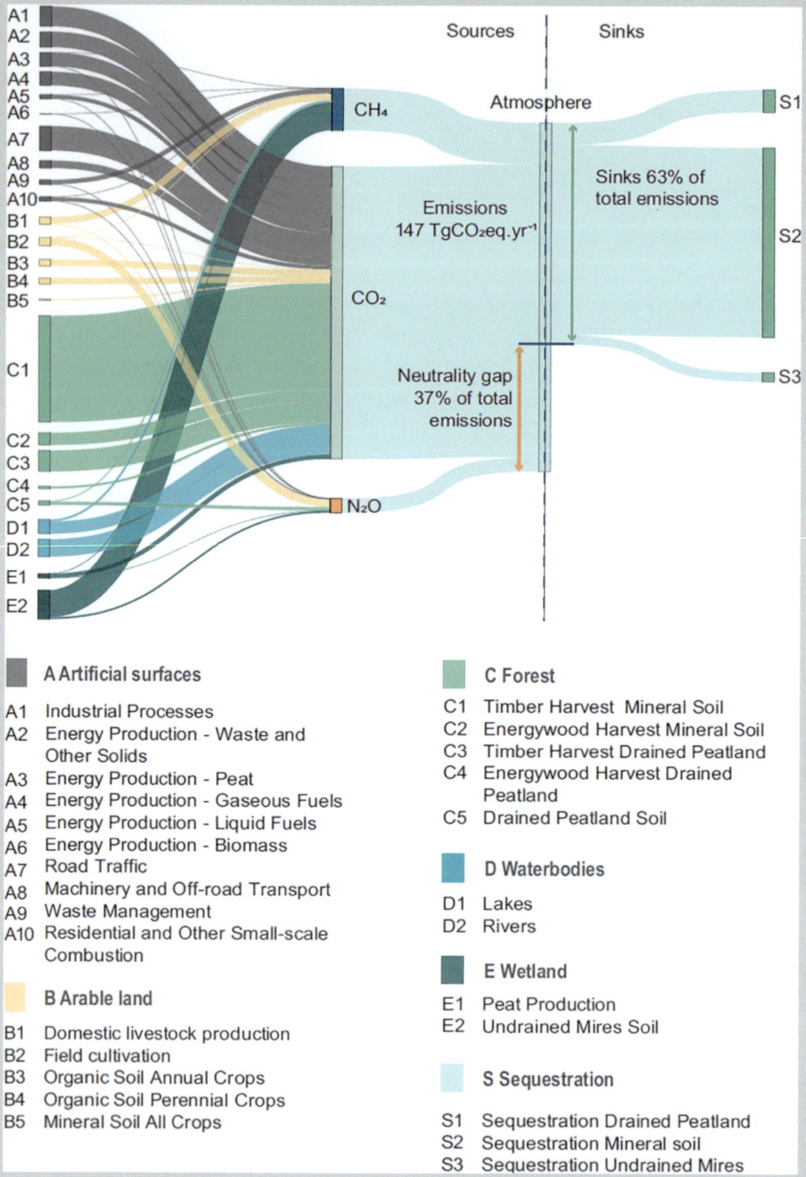

Fig. 6.11 Current GHG emission sources and sinks for main land-use categories in Finland. Current total emissions (left) are around 1.6 times larger than sinks (right). Total emissions of CH_4, CO_2, and N_2O from arable land, forest, undrained mires and water bodies are compensated by carbon sequestration in forests and in undrained mires (from Holmberg et al. [34]). Tg = Mt

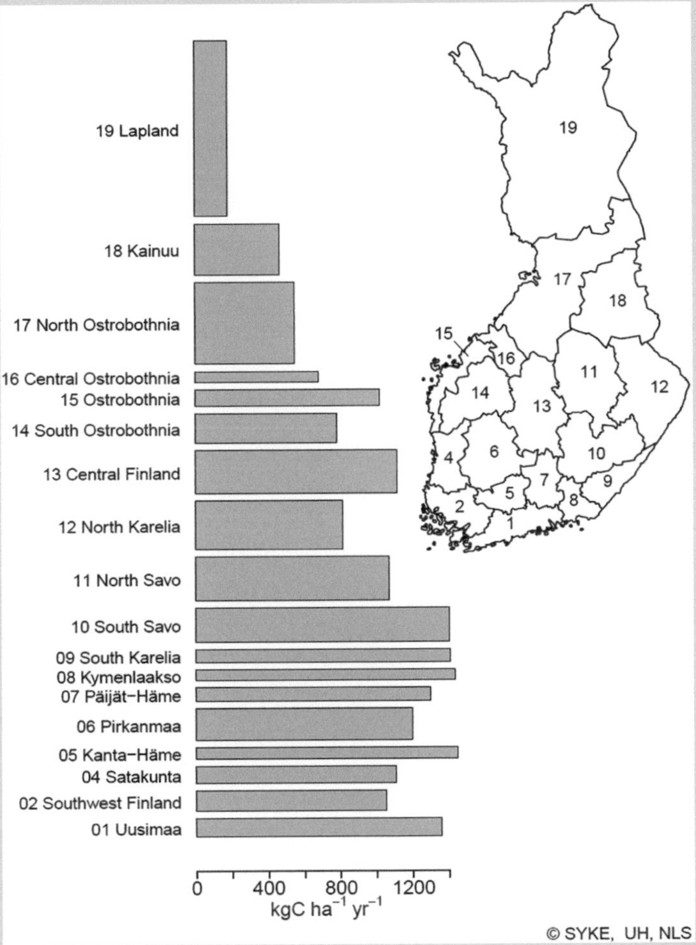

Fig. 6.12 Relative area of regions (y-axis) with respect to per area harvested biomass, for each region of Finland shown in the inserted map (redrawn from Junttila et al. [43])

The forest areas of Finland were modelled with a process-based model, PREBAS, which uses existing national forest inventory data from 2015 as the initial state of the forests, and various climate and harvest scenarios as drivers of the model [43]. There is a lot of uncertainty in models of natural processes, especially in carbon sequestration. Thus, the model's uncertainty was estimated with Monte-Carlo simulations using all existing information about the initial state and the model parameter uncertainties. The simulations were performed separately for combinations of harvest and climate scenarios, we used three climate scenarios, RCP2.6, RCP4.5 and RCP8.5, where RCP2.6 represents only small changes to current climate, and RCP8.5 represents the most severe climate change, with down-scaled weather data from five different climate models. Representative Concentration Pathways, RCPs, are often used to express the global heating impact of increased CO_2 concentration in terms of radiative forcing, in units of W/m^2. At the time of writing in 2024, the value is ~ 2 W/m^2, while the value reached by 2100 under the business-as-usual scenario is 8.5 W/m^2, corresponding to an increase in the average temperature of 3.5…4.5 °C. To return to a temperature increase of 1 °C, we should stop the increase of radiative heating at RCP2.5.

Four harvest scenarios (BaseHarv, MaxHarv, LowHarv and NoHarv) were used for an overview of situations where regional harvests: continue with the same intensity as before (BaseHarv); become even more intensive (MaxHarv); continue at a lower level (LowHarv); and cease after the year 2021 (NoHarv).

Forest carbon balance (Net Biome Exchange, NBE) is a sum of net ecosystem exchange (NEE), which represents carbon fixation in the ecosystem, and the yearly harvests, which remove carbon from the forests. Thus, the high level of carbon fixation does not automatically mean carbon sink—if the level of harvested carbon is higher, the region can even become a carbon source.

Figure 6.13 shows simulated cumulative distribution functions of average NBE values per region in Finland in the year 2035, under different climate and harvest scenarios. The NBE varies by region such that without harvests (NoHarv), the highest sinks (most negative values) are in the south, while under different intensities of harvests, the highest sinks are in the north, where the historical harvest intensity has been lower. For many regions in the south, the forests have become carbon sources with current and more intensive harvest levels by the year 2035. In Lapland, where the harvest intensity has been low, variations in harvest intensity levels do not have a big impact.

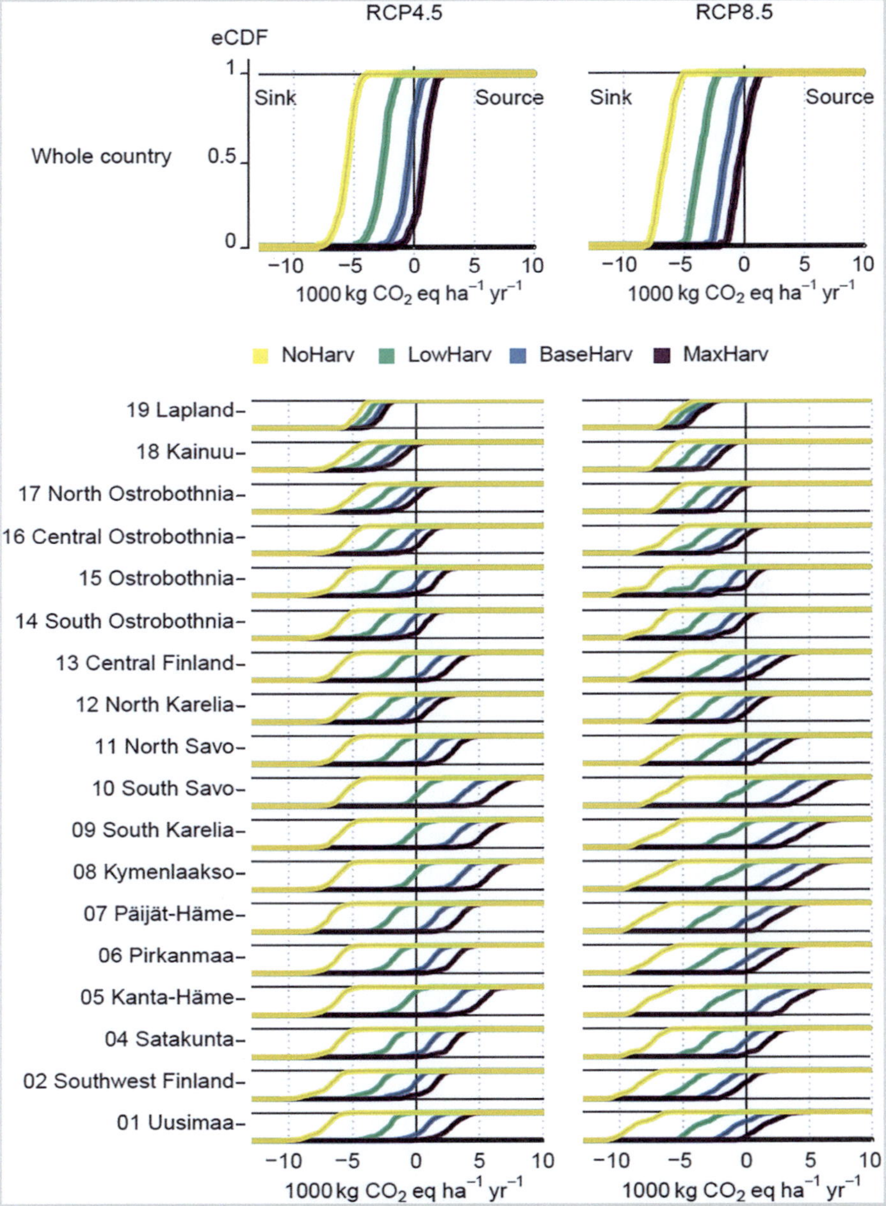

Fig. 6.13 Simulated empirical cumulative distribution functions (eCDF) of net biome exchange (kg $CO_{2.eq}$ ha^{-1} yr^{-1}) of forests per region in Finland in the year 2035, under different combinations of climate (RCP4.5, RCP8.5) and harvest scenarios. Negative value on the x-axis means a carbon sink, and positive value a carbon source (from Junttila et al. [43])

In this study, climate change has a positive impact on the growth of the forests, which can be seen in Fig. 6.13 as increasing sink from RCP2.6 to RCP8.5. However, the ecosystem model PREBAS does not include modules for biotic and abiotic disturbances, which are likely to increase significantly with warming climate. This introduces further uncertainty into these predictions.

Integrated Evaluation of Carbon Processes and Biodiversity in Forested Ecosystems

In addition to quantifying carbon and GHG fluxes and their changes in the landscape for climate change mitigation and use of natural resources, the protection of biodiversity (BD) is another important environmental issue. BD and nature's benefits to people (known as ecosystem services) underpin almost every aspect of human development: they help to produce food, clean water, regulate climate, and even control disease. Yet they are now being depleted and degraded faster than at any other point in human history [16]. The challenges posed by climate change, land use, and BD loss are also deeply interconnected problems [65]. Finland is committed to halting the loss of BD and increasing the protected areas and is implementing both EU-level and national strategies to reach this goal. For this reason, detailed, spatially distributed databases of BD values, which allow detection of valuable areas in the landscape and prioritization of potential new areas for BD protection, have been compiled [46]. By combining this with spatial information of carbon data as detailed above, it is possible to prioritize areas which are valuable for BD protection and carbon storage/sequestration both separately and jointly (Fig. 6.13). Such techniques thus allow for optimal land-use planning where both climate and BD related aspects are considered. It is also possible to give weights to the different values according to the priorities of decision makers. Moreover, the maps displayed in Fig. 6.14 give important information on the spatial distribution of the different values. Due to better environmental conditions, the carbon values are concentrated in more southern regions, whereas the BD values show a more equal spatial distribution. We have also shown by dynamic modelling that prioritized new forested areas for BD protection, according to an EU and national 10% protection target, would provide a significant carbon storage and sequestration potential by 2050, indicating complementarity of emission mitigation and conservation measures [27]. Our combined results can thus be utilized for integrating climate and BD policies, accounting of ecosystem services for climate regulation, and delimitation of areas for conservation. However, it is important to understand that there are large uncertainties in the spatial assessments of such complex topics, and therefore there is a continuous need to develop both the different databases and the evaluation methods.

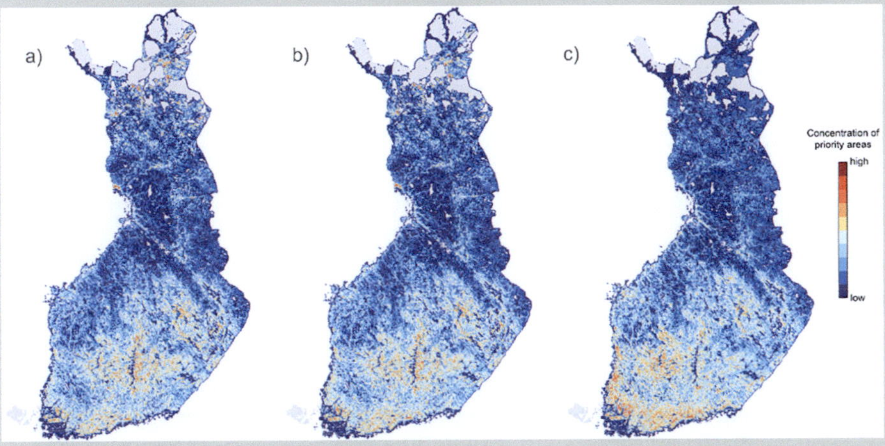

Fig. 6.14 Prioritization of forested areas important for biodiversity protection: **a** carbon stores and sinks, **c** biodiversity, and **b** both biodiversity and carbon sinks in Finland. The maps have been produced using the Zonation software (from Kujala et al. manuscript). The maps clearly indicate the different spatial distributions of the biodiversity and carbon values, indicating the need for careful landscape planning

6.4.2 Bioenergy from Agricultural Crops and Residues

Agricultural land covers approximately 46% of the global habitable land surface, corresponding to 48 million km^2 (Fig. 6.6). Approximately one-third of this, 16 million km^2, is used for crop production, while the rest, 32 million km^2, is mainly pastures and grasslands serving ruminant livestock, mostly for dairy and meat production. Looking at it another way, a staggering 77%, or 37 million km^2, of agricultural land is used for grazing and growing feed for of animals, while the remaining 23%, 11 million km^2, is used to grow food, materials and energy crops for humans. More specifically, 13% of croplands, 2.1 million km^2, is used for bioenergy and textile production, leaving 8.9 million km^2 for food production [62]. Around 30% of the cropland used to grow cereals, which humans could also consume, is actually for livestock feed, resulting in feed-food competition [56]. Figure 6.15 presents the summary of these figures.

It is very clear from these numbers that if we could move towards vegetarian diets, even cutting our meat consumption by 50%, the land use to graze domestic animal and grow food for them would decrease dramatically, and the area used for bioenergy could be significantly increased. This would also have positive impacts for direct and indirect emissions, and have positive impacts on human health.

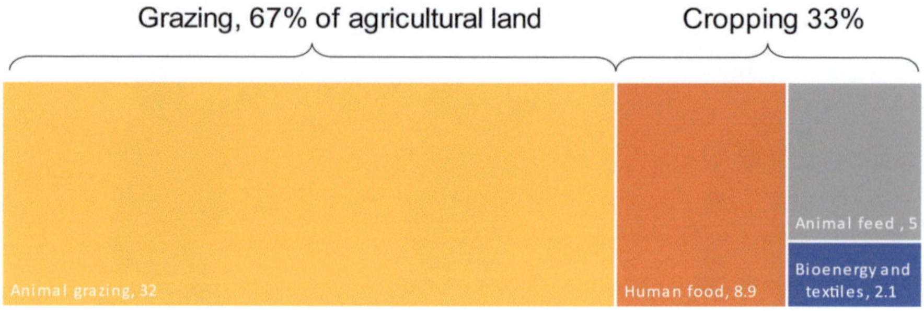

Fig. 6.15 The utilization of agricultural land area in total 48 million km^2

In terms of energy, agricultural crops account for 2.3 billion tons, or 6360 TWh, 46% of the solid biomass, offering a great opportunity for larger energy and material recovery, as discussed in Sect. 6.2. Liquid biofuels amount to 123 million tons, or 1130 TWh, ~ 7% of total bioenergy supply, mainly as biodiesel and bioethanol. Finally, biogas made by anaerobic digestion is responsible for 406 TWh, ~ 2% of total bioenergy supply [87].

The main food crops used for bioenergy production are starch-producing crops (such as maize), sugar-producing crops (such as sugar cane), and oil-producing crops (such as rapeseed/canola and oil palm). Other globally-important bioenergy plants are Jathropa, soybeans, cottonseed, sunflowers, switchgrass, and Miscanthus. Crop plants can be perennial, such as sugar cane and oil palms, or annual, such as maize and rapeseed. In general, starch- and sugar-containing crops are used for bioethanol production, while oil-containing crop species are used for biodiesel production; of these, only switchgrass and Miscanthus are classified as second-generation biofuel feedstocks because they are non-food crops.

Increasing the demand for agricultural crops—whether for food, feed, or biofuels—is connected to numerous environmental impacts regionally and worldwide. Among these are overuse of fresh water causing depletion of river flows and ground reservoirs, and pollution by fertilizers and pesticides. According to Jaureguiberry et al. [42], after land- and sea-use change, direct exploitation and pollution are the most dominant direct drivers of global biodiversity loss.

Production of bioenergy from agricultural land has been shown to have mixed effects on biodiversity, depending on the initial land use, production scale, and specific practices [41]. In temperate regions, negative impacts on biodiversity have been reported or forecasted from field-grown biomass: it drives further intensification of grassland use (for biogas production), resulting in, for example, loss of extensive grasslands, which are hotspots for biodiversity [41]. For Nordic regions, Framstad et al. [28] outlined only limited cases in which increased biomass harvesting for bioenergy has positive or neutral effects on biodiversity. One such example is harvesting of low-quality vegetation biomass from semi-natural grasslands in ways consistent with maintenance of their species-rich communities (ibid, [63])—harvesting is done only once or twice a season, and often late, after the flowering plants have had chance to seed and ground-nesting birds raised their young. However, the economic viability and relative net gain in energy from such extensive and low-scale harvesting remains unclear. Grasslands and other fields dedicated to bioenergy production tend to be harvested intensively to maximize the biomass output per hectare and harvest round. As a general principle, extracting "optimal rather than maximal quantities of biomass" may be necessary to maintain or enhance biodiversity on areas devoted to energy biomass production [29].

When the existing intensive agricultural practices are replaced with less intensive ones, and annual crops relaced with perennial plants, then bioenergy crops may have local benefits for biodiversity [41]. At the field scale, perennial bioenergy crops can create cropping mosaics, provide habitats and shelter for some species, and improve the connectivity of different sites. Using wider varieties of crops can also help to improve the status of degraded lands, thereby improving the quality of the soil [11, 52, 71].

Over 75% of global crops, including fruits and vegetables, coffee, cocoa, and many energy crops like Jathropa, are dependent on animal pollination mainly from pollinating insects such as bees, wasps, flies and butterflies. The decline in the number of wild pollinators is a global threat to food and energy crop production, and is linked to the increased use of chemicals in agriculture, and other anthropogenic factors like land-use change, climate change, intensive agricultural management practices, pollution, invasive alien species, and pathogens [77]. There are many sustainable practices in agricultural and forestry management, such as agroecological practices, agroforestry, and silvopastoral management (the practice of integrating trees, forage, and the grazing of domesticated animals in a mutually beneficial way), that prevent further negative impacts on biodiversity and reduce negative impacts of unsustainable land use, and that also benefits pollinator communities.

Agricultural residues have traditionally been used as energy sources across the world; their use for bioenergy production is still poorly developed, but has great potential as discussed previously. Because they require no additional land, they are regarded as the most promising win–win solution for producing bioenergy. The types and volumes of agricultural residues vary by region, and mostly include straw from cereals and non-cereal crops, corn stover, as well as wastes from processing of vegetables, fruit, nuts, cereal and oil crops. For example, in the European Union, the average amount of crop residues available for bioenergy was estimated at 425 TWh/year, which corresponds to about 3.2% of final energy consumption in the EU27 [72]. The limitations on their use for bioenergy stem from availability, logistics, and economics, as well as the associated removal of nutrients that would otherwise enter the soil, which are important for subsequent crops and soil carbon stock.

Finally, livestock manure and sludge can be a problematic waste product in regions with highly concentrated livestock production. If not properly treated, it causes GHG emissions, water pollution and eutrophication. On the other hand, this side stream provides an excellent opportunity for bioenergy production. Production of biogas from manure has been scaled up in many countries, producing methane for biofuel and fertilizer from the residue. Biogas can also be produced from rotational grasslands on farms without livestock, where such rotation is used to improve soil fertility, thus reducing the need for fossil fuels (e.g., [45]).

6.4.3 Bioenergy Crop Cultivation on Marginal Lands

Short-rotation plantations and short-rotation coppicing are developed on so-called marginal lands, typically defined as areas characterized by low productivity and reduced economic return [44]—although not all marginal lands are wasteland. On the contrary, many such areas are rich in biodiversity due to traditional multifunctional uses, these areas can host species specifically adapted to that environment. Marginal lands are especially targeted for growing second-generation bioenergy feedstocks, as an option for avoiding land competition between food and bioenergy (See Sect. 2.4 on land use). However, as with any cropping, multiple concerns exist over environmental impacts—such as erosion, land degradation, biodiversity, and climate change mitigation—and the overall sustainability of using marginal lands (see for example, [7]). In their review, Winberg et al. [85] suggested that in temperate climates, using marginal or abandoned land for bioenergy crops cannot fully resolve trade-offs between bioenergy crop production and biodiversity, conclusions that are complicated by a lack of empirical studies of the biodiversity value of marginal and abandoned land.

Some bioenergy crop production on marginal lands can have multiple benefits. For example, growing bioenergy crops on lands with low soil organic carbon content using appropriate practices can foster carbon sequestration, enhance soil and water quality, and support other ecosystem services [44]. In particular, the use of perennial grassland, switchgrass, and miscanthus could increase the soil carbon sink and reduce erosion.

Willows have been widely cultivated for bioenergy on marginal lands in northern latitudes (i.e., in Scandinavian countries) due to their good adaptation and resistance against cold and wet environment. Besides fast production of biomass, short-rotation willow plantations absorb excess nutrients, heavy metals, and toxic chemicals, which is useful in restoration of polluted or industrial areas, a process known as *phytoremediation* [51].

Development of short-rotation plantations and short-rotation coppicing is important for bioenergy production. Coppicing is an old method of farming trees, which involves cutting the branches of trees down to the stump—a tree that would grow normally grow 6–8 m high would, with regular coppicing, remain small and continually produce attractive new growth. The key tree species used in plantations across the world are willow (*Salix* spp), poplar (*Populus* spp), eucalyptus (*Eucalyptus* spp), Sitka spruce (*Picea sitchensis*), and other spruce species.

6.4.4 Peat as Energy

Peatlands are terrestrial wetland ecosystems with waterlogged conditions leading to peat accumulation. They form regionally important ecosystems with high biodiversity, and are important stores of carbon in the global carbon cycle [9, 88]. Peat is composed of dead materials, and *Sphagnum* mosses and other plant materials, decomposing slowly in low-oxygen and often cool conditions [73]. Natural peatlands are generally net sinks for carbon dioxide, and sources of methane emissions due to their anaerobic conditions [10, 68].

In addition to being carbon-dense landscapes, peatlands also play important roles in the provision of water resources and habitat, providing a range of rare, threatened, or declining habitats for plants and animals, and represent an important component of global biodiversity. Peatlands provide a valuable range of other nationally and internationally valuable ecosystem services, for example, natural flood regulation, and cultural services [88].

Peat has been used for energy mainly in Finland, Sweden, Ireland, Estonia, Latvia, Lithuania and Poland [73]. Due to environmental pressures, increasing carbon trade prices and climate targets the amount of peat for energy has declined dramatically in recent years. The IPCC [38] defined that, due to the very slow formation process, peat cannot be classified as a renewable energy source, as a biofuel or biomass fuel, despite its plant origins (remember that fossil fuels also have bio-based origin).

Peatlands cover 2.84% of the global land area, 4.23 million km^2, mainly in boreal and temperate zones across the Northern Hemisphere [88]. We should note that even though only slightly more than 10% of the world's peatlands are located in the tropics, their importance for the global carbon cycle is huge, as they are one of most carbon-dense ecosystems with a belowground carbon stock of around 105 Gt [59]. Tropical peatlands are not converted into peat production, instead they are threatened by drainage, deforestation, and conversion into agricultural lands either for food, feed, or energy crops like palm oil. Deforestation and land conversion leads to carbon loss and greenhouse gas emissions, land subsidence, loss of biodiversity and ecosystem services such as impaired water and nutrient recycling, and increased risk of fire as dry peat is highly flammable.

Peatlands that are actively accumulating peat are called "mires". Peatlands, especially mires, have been used for agriculture, forestry, raw material for energy and heat production, growth media for horticulture, agriculture and nurseries, and several other uses. For example, in Finland more than half of the natural 100,000 km^2 of mire area has been drained for forestry, agriculture, and peat production, which has resulted in biodiversity loss in mire habitats [73]. In addition, extensive drainage is leading to disturbed water fluxes, reduction of carbon stocks from mires, and impaired ecosystem services.

Peat extraction—the peeling of *Sphagnum* moss from the surface of the mire—causes serious damage to the mire ecosystem. Today, mires in some countries are being rewetted with the aim of restoring then to their natural state. One of the goals of the restoration is to promote carbon accumulation in the peat layer. Several studies on restoration indicate that, unfortunately, this is very challenging in some cases (e.g., [9, 68]). Instead of completely restoring industrial peat excavation sites, these areas could be used to implement large scale solar PV power plants. However, the hydrological conditions need to be adjusted to maximize intake of CO_2, and minimize release of CH_4.

6.5 Biomass and Waste to Energy

The majority of bioenergy—86%, 13,700 TWh—comes from the direct combustion of solid biomass: wood fuel, agricultural residues, wood char coal, and pellets. The word "direct" means here that they are burned in solid form. Liquid and gaseous biofuels comprise the remaining 14%, and biodiesel, bioethanol and biogas are all produced from solid biomass by thermochemical, mechanical or biochemical processes. The bio-oils inside seeds are naturally in liquid form within the solid plant matrix, but the seeds themselves are considered as solid biomass.

To properly burn, first the solid biomass must be sufficiently dried: exothermic oxidation reactions cannot sustain combustion if too large a fraction of the generated heat goes to endothermic evaporation of water. Evaporation consumes 2.442 MJ energy per 1 kg of evaporated water at 25 °C. The maximum allowed moisture content depends on the combustion technology used, and is typically between 50–60%.

By contrast, biochemical processes—anaerobic digestion to biogas and fermentation to alcohol—take place at low temperature, in an aqueous mixture with high water content, even above 95%. This saves energy, but the material and energy conversion efficiencies are not as high as those for high-temperature thermochemical processes. One potential benefit with fermentation and anaerobic digestion is that residue from the primary process can be directly utilized as a raw material for some other process. For example, the digestate from biogas production is an effective fertilizer, forming a very nice circular economy system: when all materials are used, and carbon and nutrients are ultimately returned to the soil. At the same time, the energy produced reduces the amount of energy required from other sources, and the bio-based fertilizer byproduct reduces the need for other fertilizers currently based on virgin fossil sources.

Before outlining the different processing pathways, the next section looks at the basic mass and energy balances of combustion applications.

6.5.1 Mass and Energy Balances in Combustion

Biomass sources are typically very heterogenous in their composition and properties. The water, oxygen and ash content are very important parameters as they affect the available energy yield in combustion. Water contained in the fuel requires heat to be evaporated, while the ash can be assumed mostly inert. Oxygen in the fuel does not generate any heat, but participates in the oxidation itself, degreasing the need for external O_2. Oxygen also "takes up space", which could instead be carbon, hydrogen, and sulfur, which all oxidize and generate heat.

The general (molar) oxidation reactions are:

$$C + O_2 = CO_2$$

$$H_2 + 0.5\ O_2 = H_2O$$

$$S + O_2 = SO_2$$

(Note that, though we discuss solid fuels here, these same equations can be used for liquids and gases by summing the quantities of C, H, S and O from all mixture components.) Converting from moles to masses we have:

$$12\,\text{kg}\,C + 32\,\text{kg}\,O_2 = 44\,\text{kg}\,CO_2$$

$$2\,\text{kg}\,H_2 + 16\,\text{kg}\,O_2 = 18\,\text{kg}\,H_2O$$

$$32\,\text{kg}\,S + 32\,\text{kg}\,O_2 = 64\,\text{kg}\,SO_2$$

or, presented as per 1 kg of combusted C, H_2 and S,

$$1\,\text{kg}\,C + 2.67\,\text{kg}\,O_2 = 3.67\,\text{kg}\,CO_2$$

$$1\,\text{kg}\,H_2 + 8\,\text{kg}\,O_2 = 9\,\text{kg}\,H_2O$$

$$1\,\text{kg}\,S + 1\,\text{kg}\,O_2 = 2\,\text{kg}\,SO_2$$

Based on the latter equations, we can calculate the required O_2 for complete oxidation and the subsequent amounts of product (flue) gases. Since air contains, by volume, 78% nitrogen, 21% oxygen and 1% other gases such as Argon and CO_2, this means 1 mol of oxygen is accompanied by 78/21 = 3.71 mol of nitrogen. This corresponds to 3.71 × 28/32 = 3.25 kg of N_2 per 1 kg of O_2. In the equations below: m_f is the mass of wet fuel; y_w is the mass fraction of water in the wet fuel; y_a is the mass fraction of ash in the dry fuel; y_C, y_H, y_S and y_O are the mass fractions of carbon, hydrogen, sulfur and oxygen in the dry, ash-free fuel; and F indicates combustible (dry, ash-free) fuel, respectively.

Input: From the above equations, we can work out the mass inputs to the combustion process:

$$m_{F.in} = m_f(1 - y_w)(1 - y_a) = a\,m_f$$

$$m_{W.in} = m_f\,y_w = b\,m_f$$

$$m_{A.in} = m_f(1 - y_w)y_a = c\,m_f$$

$$m_{O_2} = m_f\,a(2.67 \times y_C + 8 \times y_H + 1 \times y_S - y_O) = d\,m_f$$

$$m_{N_2} = m_{O_2} \times 3.25 = 3.25 \times d\,m_f = e\,m_f$$

The outputs are,

$$m_{CO_2,out} = m_f a \times 3.67 \times y_C = f\,m_f$$

$$m_{H_2O,out} = m_f(a \times 9 \times y_H + y_w) = g\,m_f$$

$$m_{SO_2,out} = m_f a \times 2 \times y_S = h\,m_f$$

$$m_{N_2} = m_f\,e$$

$$m_{F.out} = m_{W.out} = 0$$

$$m_{A.out} = m_f\,c$$

Since mixing with oxygen is often incomplete, reactions are not necessarily able to reach full equilibrium conversion. To burn the fuel completely, we need some excess air to make sure that there are no incombustibles left in the flue gases. The fuel-specific parameter λ, the air factor, is defined as the ratio of actual oxygen used to the theoretical amount needed. The actual required amount of oxygen, and the nitrogen coming in with that oxygen, are then.

$$m_{O_2,in} = m_{O_2}\,\lambda = d m_f\,\lambda$$

$$m_{N_2,in} = m_{N_2}\,\lambda = e\,m_f\,\lambda$$

The unused amounts of these in the flue gases (N_2 is not reacting) are simply.

$$m_{O_2,out} = m_{O_2}(\lambda - 1) = d\,m_f(\lambda - 1)$$

$$m_{N_2,out} = m_{N_2} = e\ m_f\ \lambda$$

In reality, the actual combustion reaction can be more precisely described with elementary reactions of radicals, since, for example, it is very unlikely that two hydrogen molecules would react with a single oxygen atom at the same time; nor there is such a thing as half a hydrogen molecule.

The energy balance for this combustion system can be simply presented as discussed in Sect. 3.1, with the essential difference that here we have N_S chemical species:

$$\sum_{i=1}^{N}\sum_{j=1}^{N_S}(\dot{m}_{i,j}h_j(T_i)) - \phi = \frac{dE}{dt}$$

Here, $\dot{m}_{i,j}$ refers to the mass flow rate at inlet or outlet connection i of species j, and $h_j(T_i)$ is the total enthalpy of species j at the temperature of location i, T_i. ϕ is the total heat loss flow rate (W). Total enthalpy is the sum of heat of formation at the reference state 0 and sensible heat needed to raise the temperature from the reference temperature, T_0, to observation temperature, T_i,

$$h_j(T_i) = h_j^0(T_0) + \int_{T_0}^{T_i} c_p(T)dT = h_j^0(T_0) + \overline{c}_p(T_i - T_0)$$

where $\overline{c}_p$ is the average specific heat capacity. In a steady state, adiabatic system, $\phi = 0\text{W}$, and the energy equation simplifies to

$$\sum_{i=1}^{N}\sum_{j=1}^{N_S}\left(\dot{m}_{i,j}h_j(T_i)\right) = 0$$

The heat of formation of two-atom molecules, such as oxygen and nitrogen, is equal to zero; ash is also assumed inert here, and so its heat of formation is also zero. For the combustion reaction discussed above, we can write (with W, F and A denoting liquid water, dry fuel, and ash-free fuel, respectively):

$$\begin{aligned}
&\dot{m}_{F.in}\left(h_F^0 + \overline{c}_{p.F}(T_{in} - T_0)\right) + \dot{m}_{W.in}\left(h_W^0 + \overline{c}_{p.W}(T_{in} - T_0)\right) \\
&+ \dot{m}_{A.in}\left(\overline{c}_{p.A}(T_{in} - T_0)\right) + \dot{m}_{O_2.in}\left(\overline{c}_{p.O_2}(T_{in} - T_0)\right) \\
&+ \dot{m}_{N_2.in}\left(\overline{c}_{p.N_2}(T_{in} - T_0)\right) - \dot{m}_{O_2.out}\left(\overline{c}_{p.O_2}(T_{out} - T_0)\right) \\
&- \dot{m}_{N_2.out}\left(\overline{c}_{p.N_2}(T_{out} - T_0)\right) \\
&- \dot{m}_{CO_2.out}\left(h_{CO_2}^0 + \overline{c}_{p.CO_2}(T_{out} - T_0)\right) \\
&- \dot{m}_{H_2O.out}\left(h_{H_2O}^0 + \overline{c}_{p.H_{2O}}(T_{out} - T_0)\right) \\
&- \dot{m}_{SO_2.out}\left(h_{SO_2}^0 + \overline{c}_{p.SO_2}(T_{out} - T_0)\right)
\end{aligned}$$

$$- \dot{m}_{A.in}\left(\overline{c}_{p.A}(T_{out} - T_0)\right) = 0$$

Note that the inlet temperature of the air is not necessarily the reference temperature—here, we assume that the fuel is preheated to T_{in}. This is justified since the combustion air is often used to transport fuel to the furnace.

Collecting all terms containing heat of formation on one side,

$$\begin{aligned}
&- \left(\dot{m}_{F.in}h_F^0 + \dot{m}_{W.in}h_W^0 - \dot{m}_{CO_2.out}h_{CO_2}^0 - \dot{m}_{H_2O.out}h_{H_2O}^0 - \dot{m}_{SO_2.out}h_{SO_2}^0\right) \\
&= \dot{m}_{F.in}\overline{c}_{p.F}(T_{in} - T_0) + \dot{m}_{W.in}\overline{c}_{p.W}(T_{in} - T_0) + \dot{m}_{A.in}\overline{c}_{p.A}(T_{in} - T_0) \\
&+ \dot{m}_{O_2.in}\overline{c}_{p.O_2}(T_{in} - T_0) + \dot{m}_{N_2.in}\overline{c}_{p.N_2}(T_{in} - T_0) - \dot{m}_{O_2.out}\overline{c}_{p.O_2}(T_{out} - T_0) \\
&- \dot{m}_{N_2.out}\overline{c}_{p.N_2}(T_{out} - T_0) - \dot{m}_{CO_2.out}\overline{c}_{p.CO_2}(T_{out} - T_0) \\
&- \dot{m}_{H_2O.out}\overline{c}_{p.H_2O}(T_{out} - T_0) - \dot{m}_{SO_2.out}\overline{c}_{p.SO_2}(T_{out} - T_0) \\
&- \dot{m}_{A.in}\overline{c}_{p.A}(T_{out} - T_0)
\end{aligned}$$

$$\begin{aligned}
&ah_F^0 + bh_W^0 - fh_{CO_2}^0 - gh_{H_2O}^0 - hh_{SO_2}^0 \\
&= \left(c\overline{c}_{p.A} + d(\lambda - 1)\overline{c}_{p.O_2} + e\lambda\overline{c}_{p.N_2} + f\overline{c}_{p.CO_2} + g\overline{c}_{p.H_2O} + h\overline{c}_{p.SO_2}\right)(T_{out} - T_0) \\
&- \left(a\overline{c}_{p.F} + b\overline{c}_{p.W} + c\overline{c}_{p.A} + d\lambda\overline{c}_{p.O_2} + e\lambda\overline{c}_{p.N_2}\right)(T_{in} - T_0) \\
&= \left(c\overline{c}_{p.A} - d\overline{c}_{p.O_2} + f\overline{c}_{p.CO_2} + g\overline{c}_{p.H_2O} + h\overline{c}_{p.SO_2} + \lambda\left(d\overline{c}_{p.O_2} + e\overline{c}_{p.N_2}\right)\right)(T_{out} - T_0) \\
&- \left(a\overline{c}_{p.F} + b\overline{c}_{p.W} + c\overline{c}_{p.A} + \lambda\left(d\overline{c}_{p.O_2} + e\overline{c}_{p.N_2}\right)\right)(T_{in} - T_0) \\
&= (T_{out} - T_0)(A + \lambda B) - (T_{in} - T_0)(C + \lambda D)
\end{aligned}$$

where

$A = c\overline{c}_{p.A} - d\overline{c}_{p.O_2} + f\overline{c}_{p.CO_2} + g\overline{c}_{p.H_2O} + h\overline{c}_{p.SO_2}$

$B = d\overline{c}_{p.O_2} + e\overline{c}_{p.N_2}$,

$C = a\overline{c}_{p.F} + b\overline{c}_{p.W} + c\overline{c}_{p.A}$

$D = d\overline{c}_{p.O_2} + e\overline{c}_{p.N_2}$

Note that A and B are calculated for the range $T_{out} - T_0$, and C and D for $T_{in} - T_0$. The left-hand side can be expanded to understand its meaning better:

$$\begin{aligned}
&a\,h_F^0 + b\,h_W^0 - f\,h_{CO_2}^0 - g\,h_{H_2O}^0 - h\,h_{SO_2}^0 \\
&= (1 - y_w)(1 - y_a)\left(h_F^0 + \frac{y_w}{(1 - y_w)(1 - y_a)}h_W^0 - 3.67 \times y_C\,h_{CO_2}^0\right. \\
&\left. - \left(9 \times y_H + \frac{y_w}{(1 - y_w)(1 - y_a)}\right)h_{H_2O}^0 - 2 \times y_S h_{SO_2}^0\right) \\
&= (1 - y_w)(1 - y_a)\left(h_F^0 - 3.67 \times y_C h_{CO_2}^0 - 9 \times y_H h_{H_2O}^0 - 2 \times y_S h_{SO_2}^0\right. \\
&\left. - \frac{y_w}{(1 - y_w)(1 - y_a)}\left(h_W^0 - h_{H_2O}^0\right)\right) \\
&= (1 - y_w)(1 - y_a) \times LHV_{daf} - y_w l_w^0 \\
&= (1 - y_w) \times LHV_{dry} - y_w l_w^0 \\
&= LHV_{ar}
\end{aligned}$$

This is simply the definition of the heating value of fuel as received (ar) and used. We then have an equation for the pre-heated general combustion system,

$$LHV_{ar} = (T_{out} - T_0)(A + \lambda B) - (T_{in} - T_0)(C + \lambda D)$$

This means that the net chemical energy released is used to heat up the combustion products plus the incoming inert nitrogen and the excess oxygen. The outlet or reaction temperature can be solved simply as

$$T_{out} = T_{ad} = T_0 + \frac{LHV_{ar}}{\lambda B + A} + \frac{\lambda D + C}{\lambda B + A}(T_{in} - T_0)$$

Since there were no heat losses considered, T_{ad} is the *adiabatic reaction temperature,* or *adiabatic combustion temperature.* We can see that: the higher the heating value, the higher the adiabatic temperature; the larger the air factor, the lower the temperature, since there is more gas (excess oxygen and nitrogen coming with the air) to be heated; and the higher the preheating temperature, the higher the adiabatic combustion temperature. We also note that to use this equation, unless the heat capacities are assumed as constant, iteration is required since the average heat capacities, and therefore A and B, vary with outlet temperature (C and D are cosntants since they are fixed to the range T_0 to T_{in}).

The average heat capacities are calculated from the definition of the sensible enthalpy,

$$\overline{c}_p = \frac{1}{(T - T_0)} \int_{T_0}^{T} c_p(T)dT$$

where T is T_{in} or T_{out} for inlet and outlet conditions, respectively. An Excel calculation sheet is available as an electronic supplementary material. This utilizes the polynomial correlations for heat capacities, see Chap. 3 for gas turbines. Table 6.3 presents compositions and lower heating values of dried biomasses as examples. More data can be found from the Phyllis2-database, https://phyllis.nl/.

Example 6.4 Calculate the composition of flue gases, their mass flow rates, lower heating value per wet mass of the fuel, and the adiabatic combustion temperature of energy willow as wet (50% moisture content). Its composition on dry ash free basis (daf) and lower heating value on dry basis are given in Table 6.4. Mass flow rate of the wet fuel is 1 kg/s, and the temperature of the air and fuel is 25 °C, the same as the reference temperature of the heating value. Heat capacities (assumed constant): $\overline{c}_{p.O_2} = 1.09$ kJ/kgK, $\overline{c}_{p.N_2} = 1.1672$ kJ/kgK, $\overline{c}_{p.CO_2} = 1.234$ kJ/kgK, $\overline{c}_{p.SO_2} = 0.888$ kJ/kgK, $\overline{c}_{p.H_2O} = 2.2877$ kJ/kgK, $\overline{c}_{p.A} = 1.0$ kJ/kgK.

Solution: Let's first calculate the constants a to h from the composition given in Table 6.4.

Table 6.3 Properties of example biomasses, composition and lower heating value on dry basis

	Proximate analysis, dry				Ultimate analysis, dry ash free					
Fuel	y_{vm}	y_{fc}	y_A	LHV (MJ/kg)	C,y_C	H,y_H	S,y_S	N,y_N	CI, y_{Cl}	O, y_O
Birch (#76)	83.8	14.8	1.4	18.45	50.90	7.60	0.50	0.50	–	40.5
Pine (#126)	85.7	14.1	0.2	19.30	51.70	5.79	0.01	0.06	0.02	42.42
Spruce (#2220)	89.6	10.2	0.2	17.93	47.40	6.30	–	0.07	–	46.2
Eucalyptus (#3120)	86.0	13.2	0.8	18.31	49.20	6.10	0.02	0.17	–	55.4
Forest residues (#846)	82.4	13.6	4.0	19.16	52.39	4.78	0.11	1.07	0.042	41.6
Anthracite coal (#1144)	7.7	89.4	2.9	33.86	91.60	3.50	0.83	1.60	0.085	2.39
Brown coal (#1519)	58.2	35.7	6.2	24.65	68.99	5.00	–	1.40	0.043	24.04
Horse manure (#3597)	67.0	23.2	9.8	18.00	54.33	6.81	0.16	1.54	–	37.15
Chicken manure (#1872)	61.4	11.2	27.4	12.68	49.46	6.86	1.06	4.90	–	37.71
Sewage sludge (#658)	53.5	11.5	35.0	14.03	52.31	7.54	2.00	7.23	0.123	30.78
Wheat straw (#459)	73.4	17.7	9.0	16.45	48.24	6.07	0.25	0.80	0.285	44.36
Rice straw (#702)	65.6	14.2	20.2	13.66	49.15	6.23	0.13	1.59	0.780	42.13
MSW (#2920)	72.6	10.6	16.8	18.70	59.19	9.80	0.29	2.19	–	28.53
Sugarcane bagasse (#894)	85.6	12.0	2.4	17.71	49.86	6.02	0.04	0.16	0.03	43.89
Miscanthus (#1743)	86.5	10.1	3.4	17.82	50.32	6.21	0.21	0.54	0.21	42.52
Willow (#345)	83.4	15.0	1.6	17.25	51.00	6.00	–	0.10		42.9
Energy willow (#3122)	–	–	1.8	18.42	50.71	6.21	0.05	0.72		42.31

O* by difference. Data obtained from Phyllis2-database for biomass, algae, feedstocks for biogas production, and biochar, https://phyllis.nl/, TNO Biobased and Circular Technologies, numbers refer to database ID. MSW refers to municipal solid waste

Table 6.4 Properties of energy willow chips (Phyllis database)

Moisture content, y_w	0.50
Ash content[a], y_a	0.018
Carbon[b], y_C	0.5071
Hydrogen[b], y_H	0.0621
Nitrogen[b]	0.0072
Sulphur[b], y_S	0.0005
Oxygen[b,*], y_O	0.4231
LHV[a], MJ/kg	18.42

[a]Dry basis, [b]dry ash free basis, [*]by difference

$$a = (1 - y_w)(1 - y_a) = (1 - 0.5)(1 - 0.018) = \frac{m_{F.in}}{m_f} = 0.491$$

$$b = y_w = \frac{m_{W.in}}{m_f} = 0.5$$

$$c = (1 - y_w)y_a = (1 - 0.5)0.018 = \frac{m_{A.in}}{m_f} = 0.009$$

$$\begin{aligned} d &= a(2.67 \times y_C + 8 \times y_H + 1 \times y_S - y_O) \\ &= 0.491 \times (2.67 \times 0.5071 + 8 \times 0.0621 + 1 \times 0.0005 - 0.4231) \\ &= \frac{m_{O_2.in}}{m_f} = 0.7015 \end{aligned}$$

$$e = d \times 3.25 = 0.7015 \times 3.25 = \frac{m_{N_2.in}}{m_f} = 2.280$$

$$f = a \times 3.67 \times y_C = 0.491 \times 3.67 \times 0.5071 = \frac{m_{CO_2.out}}{m_f} = 0.9138$$

$$g = a \times 9 \times y_H + y_w = 0.491 \times 9 \times 0.0621 + 0.5 = \frac{m_{H_2O.out}}{m_f} = 0.7744$$

$$h = a \times 2 \times y_S = 0.491 \times 2 \times 0.0005 = \frac{m_{SO_2.out}}{m_f} = 0.000491$$

With these and the average heat capacities defined above we get

$$B = d\bar{c}_{p.O_2} + e\bar{c}_{p.N_2} = 0.7015 \times 1.09 + 2.28 \times 1.1672 = 3.426\,\text{kJ/kg},$$

and

$$\begin{aligned} A &= -d\bar{c}_{p.O_2} + f\bar{c}_{p.CO_2} + g\bar{c}_{p.H_2O} + h\bar{c}_{p.SO_2} + c\bar{c}_{p.A} \\ &= -0.7015 \times 1.09 + 0.9138 \times 1.234 + 0.7744 \times 2.2877 \\ &\quad + 0.000491 \times 0.888 + 0.009 \times 1.0 \\ &= 2.135\,\text{kJ/kg} \end{aligned}$$

The lower heating value as received is $LHV_{ar} = (1 - 0.5) \times 18420 - 0.5 \times 2442 = 7989$ kJ/kg.

We can then write the reaction temperature equation,

$$T_{ad} = T_0 + \frac{LHV_{ar}}{\lambda B + A} = 25 + \frac{7989}{\lambda \times 3.426 + 2.144}$$

Note that the term containing C and D is not needed since inlet temperature is equal to the reference temperature.

With air factor values $\lambda = 1, 1.5$ and 2, we get adiabatic combustion temperatures of 1459, 1122 and 913 °C, respectively—the question then is, how accurate are these values, since the heat capacities were assumed as constants? See also the downloadable Excel sheet as electronic supplementary material that includes this and another more straightforward method of using enthalpies. The accurate values for the iterated solution are 1464, 1157 and 959 °C for $\lambda = 1, 1.5$ and 2, respectively.

Finally, Table 6.5 shows the mass flow rates and mass fractions of flue gas components.

As Table 6.5 shows, the main chemical species of the flue gases is nitrogen. If we want to capture CO_2 from the flue gases, this presents a problem, since N_2 dilutes CO_2 and decreases the rate of the capture processes. In Example 6.6 we discuss the Oxyfuel concept, which burns the fuel using only O_2, and as a result the CO_2 can be easily captured by condensing the water vapor.

Example 6.5 Calculate the adiabatic combustion temperature of raw biogas composed of 65% CH_4 and 35% CO_2 on volumetric (molar) basis. LHV of pure methane is 50 MJ/kg. Assume constant heat capacities: $\overline{c}_{p.O_2} = 1.09$ kJ/kgK, $\overline{c}_{p.N_2} = 1.1672$ kJ/kgK, $\overline{c}_{p.CO_2} = 1.234$ kJ/kgK, $\overline{c}_{p.H_2O} = 2.2877$ kJ/kgK.

To use the approach outlined above, we first need to convert the fuel composition into mass basis elemental composition of a single pseudo fuel. Assuming 1 kg of raw gas, we can easily calculate the amount of carbon, hydrogen and oxygen in the fuel. Mass and mole fraction are related by

Table 6.5 Mass flow rates of flue gas components per 1 kg/s of fuel

	Mass flow rates (kg/s)			Mass fractions (–)		
Species	$\lambda = 1$	$\lambda = 1.5$	$\lambda = 2$	$\lambda = 1$	$\lambda = 1.5$	$\lambda = 2$
CO_2	0.914	0.914	0.914	0.230	0.167	0.131
H_2O	0.774	0.774	0.774	0.195	0.142	0.111
Ash	0.009	0.009	0.009	0.002	0.002	0.001
SO_2	0.0005	0.0005	0.0005	0.00012	0.00012	0.00012
O_2	0.000	0.351	0.701	0.000	0.064	0.101
N_2	2.283	3.422	4.561	0.573	0.626	0.655
Total	3.980	5.470	6.960	1.000	1.000	1.000

$$y_i = \frac{m_i}{m} = \frac{N_i M_i}{NM} = \frac{x_i M_i}{M}$$

and the molar mass M of the gas mixture is then

$$M = \sum_{i=1}^{N} x_i M_i = 0.65 \times 0.016 + 0.35 \times 0.044 = 0.0258\,\text{kg/mol}$$

$$y_{\text{CH}_4} = 0.65 \times 0.016/0.0258 = 0.403$$

$$y_{\text{CO}_2} = 0.35 \times 0.044/0.0258 = 0.597$$

The lower heating value is therefore $LHV_{ar} = 0.403 \times 50 = 20.16\,\text{MJ/kg}$, since only methane will produce heat in combustion. There is no moisture or ash in the fuel, so the calculation is greatly simplified. We first calculate the mass fractions of carbon, hydrogen and oxygen in the raw gas:

$$y_C = \frac{y_{\text{CH}_4} M_C}{M_{\text{CH}_4}} + \frac{y_{\text{CO}_2} M_C}{M_{\text{CO}_2}} = \frac{0.403 \times 12}{16} + \frac{0.597 \times 12}{44} = 0.465$$

$$y_{H_2} = y_H = \frac{y_{\text{CH}_4} 4 M_H}{M_{\text{CH}_4}} = \frac{0.403 \times 4}{16} = 0.101$$

$$y_O = \frac{y_{\text{CO}_2} 2 M_O}{M_{\text{CO}_2}} = \frac{0.597 \times 32}{44} = 0.434$$

which (we check to make sure!) adds up to 100%. As we did for the willow calculation previously, we first calculate the constants a to h:

$$\begin{aligned}
a &= (1 - y_w)(1 - y_a) = 1 \\
b &= y_w = 0 \\
c &= (1 - y_w) y_a = 0 \\
d &= a(2.67 \times y_C + 8 \times y_H + 1 \times y_S - y_O) \\
&= (2.67 \times 0.465 + 8 \times 0.101 - 0.434) = 1.616 \\
e &= d \times 3.25 = 1.616 \times 3.25 = 5.251 \\
f &= a \times 3.67 \times y_C = 3.67 \times 0.465 = 1.707 \\
g &= a \times 9 \times y_H + y_w = 9 \times 0.101 = 0.909 \\
h &= a \times 2 \times y_S = 0
\end{aligned}$$

With these and the average heat capacities defined above we get

$$B = d\bar{c}_{p.\text{O}_2} + e\bar{c}_{p.\text{N}_2} = 1.616 \times 1.09 + 5.251 \times 1.1672 = 7.89\,\text{kJ/kg, and}$$

$$\begin{aligned} A &= -d\bar{c}_{p.O_2} + f\bar{c}_{p.CO_2} + g\bar{c}_{p.H_{2O}} + h\bar{c}_{p.SO_2} + c\bar{c}_{p.A} \\ &= -1.616 \times 1.09 + 1.707 \times 1.234 + 0.909 \times 2.2877 \\ &= 2.425\ \text{kJ/kg} \end{aligned}$$

We can then write the reaction temperature equation as

$$t_{ad} = t_0 + \frac{LHV_{ar}}{\lambda B + A} = 25 + \frac{20160}{\lambda \times 7.89 + 2.425}$$

With air factor values $\lambda = 1$, 1.5 and 2, we get adiabatic combustion temperatures of 1979 (1940), 1439 (1452) and 1132 °C (1169 °C), respectively—the values in parentheses correspond to accurate values using temperature-dependent heat capacities (see the same downloadable spreadsheet file used for Example 6.4).

6.5.2 *Combustion of Solid, Liquid and Gaseous Fuels*

In this section we will discuss the differences between, and dominating mechanisms of, combustion phenomena involving gaseous, liquid and solid fuels. Complete combustion requires fuel and oxidizer to meet each other to react. Temperature must be sufficient to have a high-enough chemical reaction rate, $r \propto \exp(-E_a/(R_u T))$, where E_a is the activation energy. Finally, there must be adequate time to burn all the fuel. These requirements can be reached through several mechanisms that we discussed below, illustrated with figures rather than mathematical models and equations.

Gaseous fuels can be combusted in a diffusion flame, or in a premixed flame. In addition, depending on the flow rate of the fuel and its Reynolds number, the flame can be laminar or turbulent. Here, Reynolds number is defined as $Re = uL_c/\nu$, where u is the average flow velocity in the nozzle opening, L_c is the characteristic dimension, (e.g., the nozzle diameter), and ν is the kinematic viscosity of the gas at the point where the velocity and diameter are defined. Different combustion regimes are nicely illustrated with a Bunsen burner (Fig. 6.16), which was developed in 1855 by Robert Bunsen, a German chemist (Britannica online 2024). The Bunsen burner was the forerunner of the gas-stove burner and the gas furnace.

The main parts of the Bunsen burner are the fuel flow control valve (a), adjustable air inlet holes (b), and the flame nozzle (c). When the air inlet holes are fully closed, the gaseous fuel and oxygen do not meet before the flame envelope—this results in the beautiful, candle-like diffusion flame. Fuel flows from inside towards the reaction front, which is the edge of the yellow visible flame (Fig. 6.17).

In the diffusion flame, there is no O_2 inside the flame envelope; instead, the oxygen diffuses inwards to the reaction front from the surrounding air. As the temperature of the gas increases towards the flame front inside the envelope, the hydrocarbon fuel rapidly decomposes into hydrogen and solid carbon (soot). The flame's yellow color

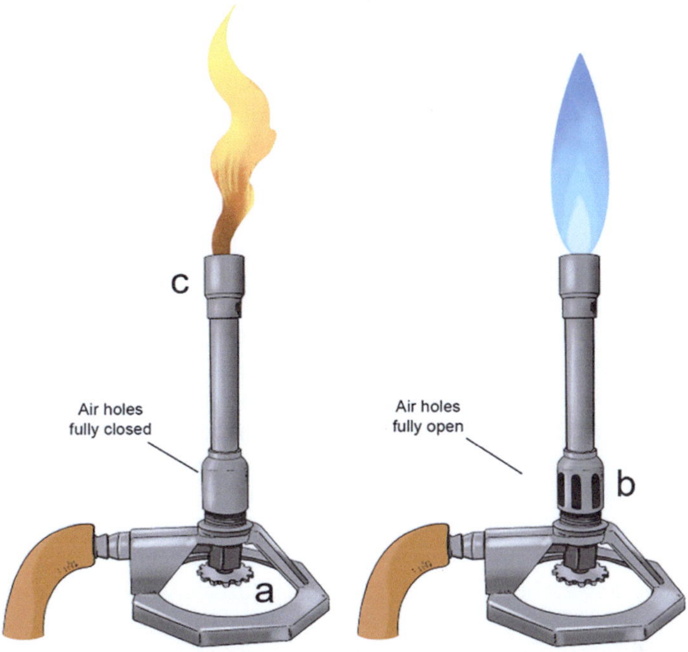

Fig. 6.16 Flame types in a Bunsen burner. Left figure is a diffusion flame and right figure presents a premixed flame. © Maija Karala

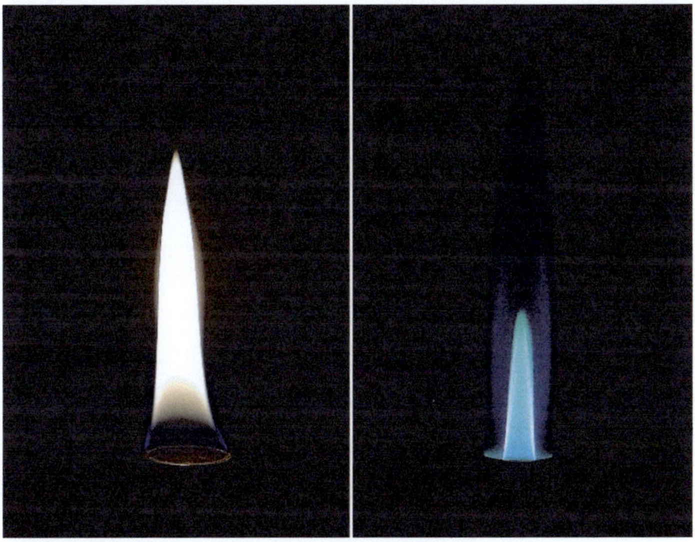

Fig. 6.17 Transition between a diffusion flame (left) to pre-mixed flame (right). Note how the sooty flame illuminates the brass nozzle, while the pre-mixed flame is dark at the nozzle

comes from the thermal radiation of the soot particles when they are oxidized in the flame front together with hydrogen.

When the air holes are fully opened, the fuel and air are fully pre-mixed before they enter the nozzle. This results in a blue, almost invisible flame since there is no soot formed. At the burner nozzle exit, the location and geometry of the flame is defined by the exit velocity of the mixture. If the velocity is too small, the flame will be drawn back into the nozzle tube and eventually quenched. If the velocity is too high, the flame will be blown off from the nozzle and quenched.

In contrast to gaseous fuels, liquid fuels first need to be atomized into small droplets, which then evaporate to gas, and this is then burned similarly to gaseous fuels. One application of spray combustion is the internal combustion diesel engine, where combustion takes place inside the cylinders—the fuel is directly sprayed into the cylinder and auto-ignited by compression. In the Otto engine, the fuel is atomized in the carburetor or by a spray nozzle, and then ignited by a spark plug. Another common example is gas turbines used in aviation, which use kerosine fuel—liquid fuels are more easily transported in planes, while gas fuels are more suited to stationary applications.

If we consider only a single fuel droplet, the combustion of fuel vapor inside the flame envelope delivers heat into the liquid droplet surface, causing more evaporation. The new vapor then flows outwards to the reaction front, where it is oxidized by the inwards diffusing oxygen. The combustion products diffuse outwards to the environment against the inwards diffusing oxygen. Since the incoming oxygen is preheated by the outgoing combustion products, the temperature of the flame can exceed the adiabatic combustion temperature without preheating. See Fig. 6.18 for an illustration of the droplet-flame principles.

When a cloud of droplets is burned, as in spray combustion applications, the oxygen concentration may be low inside the cloud of huge numbers of droplets. Since the evaporation keeps the temperature low, there is no reaction until the fuel is mostly evaporated, heats up, spreads, and meets with oxygen. This type of spray combustion is illustrated in Figs. 6.10a and 6.19b, which show the development of

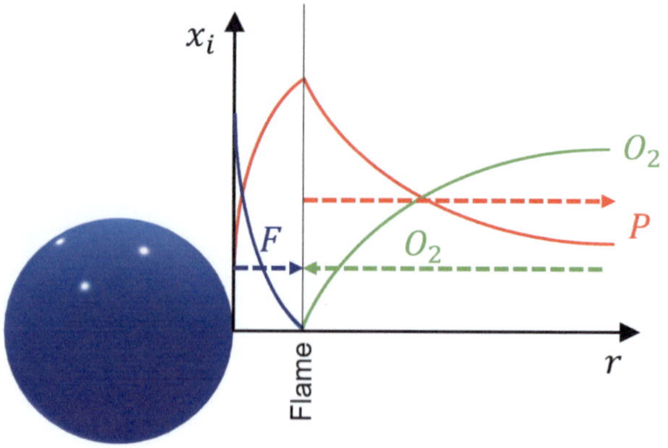

Fig. 6.18 Mechanism of liquid fuel combustion, F is the fuel and P refers to combustion products. The location of the flame front is presented as a thin vertical line, x_i is the mole fraction of gases

a n-dodecane ($C_{12}H_{26}$) liquid spray flame at an enclosure temperature of 627 °C, at pressure 60 bar and 15% O_2 concentration. These images used a computational fluid dynamics (CFD) simulation with the Large Eddy Simulation method (LES) [82]. This specific simulation took 3000 processor hours and run on 128 CPU units.

As the images show, the injected fuel spray must first heat up and evaporate before ignition can take place. The structure of the flame is very complex, and the simulation model is able to catch even the smallest details in the flame structure. CFD modeling with supercomputers can be used to analyze the processes of sprays, and is an especially useful tool when new types of sustainable fuels are developed for engine and gas turbine applications.

Finally, in contrast to gas and liquid droplet combustion, with solid fuel combustion there are several combustion phases, and different chemical and physical phenomena that control the rate of the combustion. Figure 6.20 shows a series of images taken from the combustion of a 20 × 20 × 20 mm cubic particle of spruce wood burned inside a wood stove on top of a glowing char layer. The particle was fixed on a 3 mm steel wire.

At the first stage of combustion of a solid fuel, the particle is heated, and the water it contains is evaporated (Fig. 6.20a). While there is free moisture in the particle, its surface temperature will stay at ~ 100 °C, the boiling temperature of water at atmospheric pressure. However, since part of the water is physically and chemically bonded, a slightly higher temperature is required to drive out the last remains of the water from the particle. The evaporation takes place at the particle's surface, as long as the water can flow freely—at the so-called critical water content, the liquid water is no longer mobile, and this results in the onset of the shrinking-core drying phase. The drying front starts to penetrate towards the center of the particle, and as the evaporation front moves inside the particle, the outer surface of the particle itself is able to heat up. When the surface temperature of the particle reaches the decomposition temperature of the organic material of the biomass, it starts to break down into gases, and begins to form char.

When the rate of this process is high enough, the gases will ignite outside of the particle and the devolatilization combustion stage begins (Fig. 6.20b). When the volatiles burn in the surrounding envelop flame, they radiate and conduct heat to the particle's surface, driving the particle processes to become autothermal and self-sustaining (Fig. 6.20c–d). Both the drying and devolatilization stages are heat-transfer-controlled processes.

As the temperature increases further, and after sufficient time has passed, all the volatiles are driven out from the particle. This leaves behind a porous char particle mostly formed from solid carbon and inorganic matter, such as metal oxides and other minerals in the biomass (Fig. 6.20e). The following stage, known as char combustion, is typically mass transfer limited by the flow of oxygen, as CO_2 or H_2O, from the environment to the particle surface (Fig. 6.20f). This is a heterogenous process taking place directly at the solid carbon surface. There are two parallel oxidation and two gasification reactions, and then both H_2 and CO can also be oxidized in the gas phase by reactions 6.5 and 6.6:

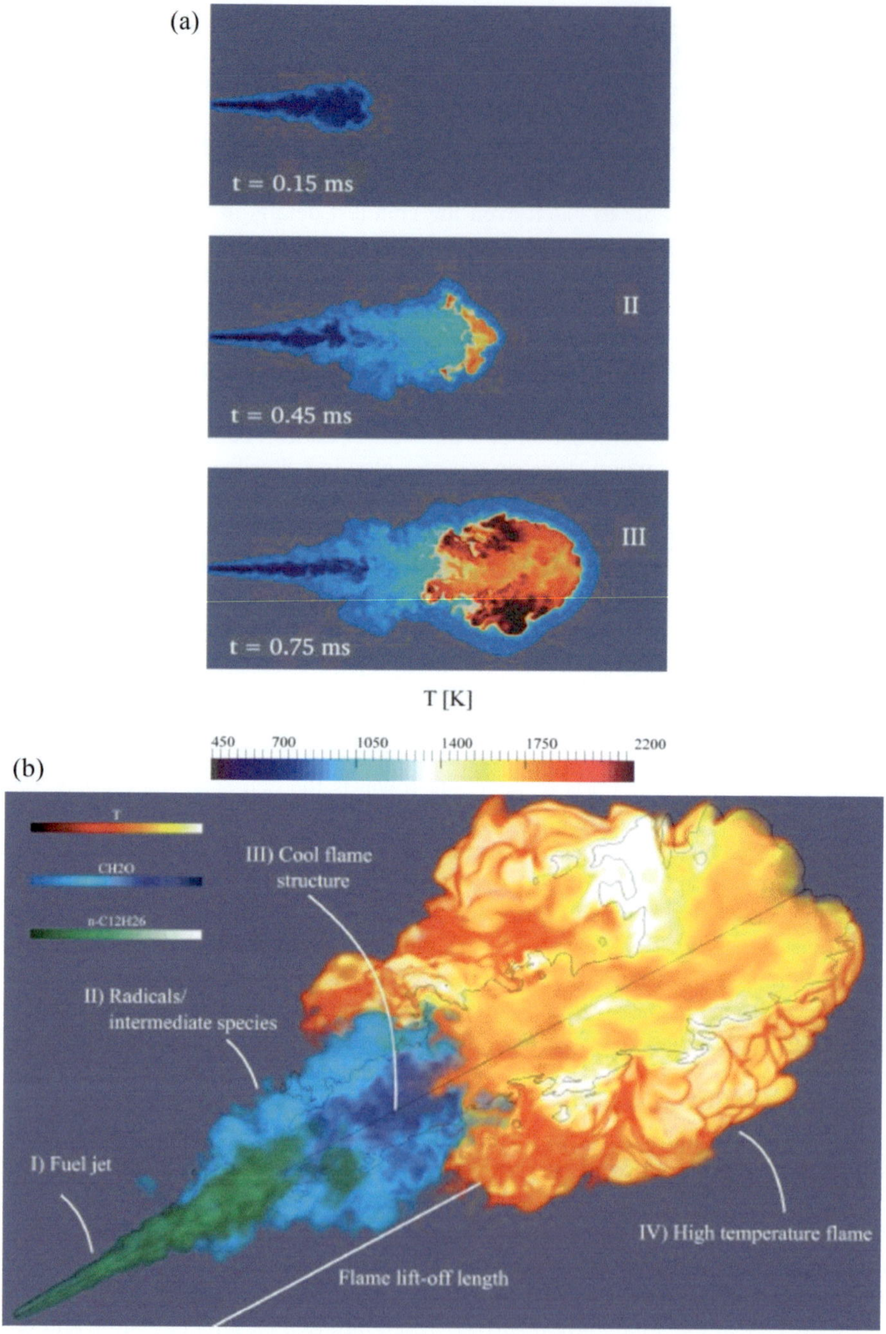

Fig. 6.19 **a** Development of a n-dodecane, $C_{12}H_{26}$, liquid spray flame at 627 °C, 60 bar and 15% of O_2 concentration [82]. The color scale represents temperature from 450 to 2200 K. **b** Development of a n-dodecane, $C_{12}H_{26}$, liquid spray flame [82], mechanisms explanations

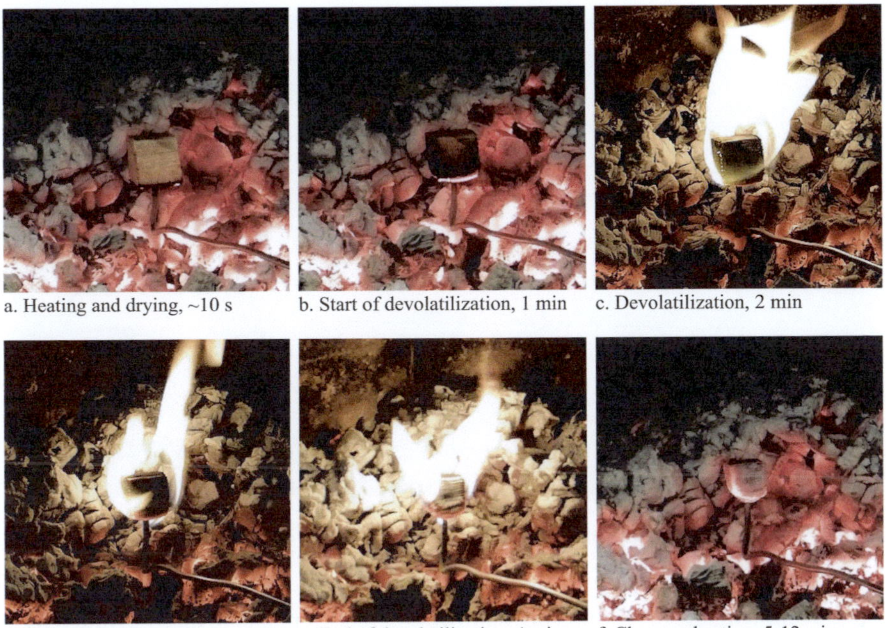

a. Heating and drying, ~10 s b. Start of devolatilization, 1 min c. Devolatilization, 2 min

d. Devolatilization, 3 min e. End of devolatilization, 4 min f. Char combustion, 5-12 min

Fig. 6.20 Combustion stages of solid fuel combustion. Times given are from the beginning of the experiment

$$C + O_2 \rightarrow CO_2 \tag{6.1}$$

$$C + 0.5\,O_2 \rightarrow CO \tag{6.2}$$

$$C + CO_2 \rightarrow 2CO \tag{6.3}$$

$$C + H_2O \rightarrow H_2 + CO \tag{6.4}$$

$$CO + 0.5\,O_2 \rightarrow CO_2 \tag{6.5}$$

$$H_2 + 0.5O_2 \rightarrow H_2O \tag{6.6}$$

If reactions 6.2–6.4 dominate, there can be again an envelope flame formed, pushing the flame front away from the surface, H_2 and CO formed oxidizing the surrounding gas phase. In this case the oxidized gas does not contain hydrocarbons, so there is no visible flame—the particle can actually be reacting via incomplete oxidation and gasification mechanisms. The char conversion step is typically much slower than the heat-transfer-limited stages, and can last tens of minutes, in

Fig. 6.21 Heterogenous combustion of wood logs

comparison with the drying and devolatilization phases that take place within tens of seconds to a few minutes. Of course, the duration greatly depends on the size of the particle—Fig. 2.21 shows wood logs burning, which typically takes tens of minutes (Fig. 6.21).

As Fig. 2.21 shows, these logs are already at the char combustion stage. The outer surface facing the surroundings is black and colder, and the downward side is glowing red hot. In this case it is important that the heat generation is sufficiently large in comparison with the heat losses, and that the system can maintain a high enough temperature to sustain the combustion reactions. The black charred wood block at the top acts as the shied for heat loss, keeping the temperature high below.

6.5.3 Large-Scale Stationary Combustion Systems and Processes

Technologies used to enable the green energy transition must have enough positive impact and up-scalability. For biomass-based energy sources this means, for example, that the biomass is grown, harvested and combusted in the most efficient and sustainable way. As we discussed earlier, around 10% of global primary energy is used in developing countries in simple three-stone-support campfires and low-quality stoves. Locally this increases deforestation and creates severe health problems when the particulate and gaseous emissions are inhaled by people cooking inside and heating their houses. In addition, since the energy efficiencies of these systems are very low, a lot of energy is wasted. As the combustion temperature remains low and mixing is very poor, only a small fraction of the heat of the flue gases can be utilized.

Simply by systematically improving these combustion setups, a huge improvement could be obtained.

Another option to maximize the energy in biomass is to burn it in larger, centralized power plants. Their size should be defined so that the fuel can be transported from an economically feasible and sustainable distance. Currently the largest units produce several hundred MW of electrical power—for example, the Alholmens power plant in Pietarsaari, Finland, which produces 240 MW electricity, 100 MW of process steam, and 60 MW of district heat, and the Fortum Värme power plant in Stockholm, Sweden, capable of producing 130 MW of electricity and 280 MW of district heat. These are both based on the circulating fluidized boiler technology. At the Alholmens power plant, the bottom area of the furnace is 8.5 × 24 m, and the boiler height is 40 m, live steam values are 165 bar and 545 °C, and the thermal input power of the boiler is 550 MW. Figure 6.22 shows a 3d model of the boiler.

In general, the objective of the boiler is to convert and capture the chemical energy of the fuel to thermal energy in water and steam, and minimize the emission of gases (CO_2, SO_2 and NO_x) and solids while doing so—see Fig. 6.23 for a schematic of the boiler components, and flows of water, fuel, ashes, flue gases and combustion air. The feedwater is first pre-heated to boiling temperature at the live steam pressure in the economizer (points 4–5 in Fig. 6.23). The evaporator converts the saturated liquid to saturated steam (5–6). The steam drum acts as the phase separator: saturated water flows down from the steam drum via the downcomer tubes to the evaporator, formed from tube-membrane walls. The walls receive efficient radiative heat flux

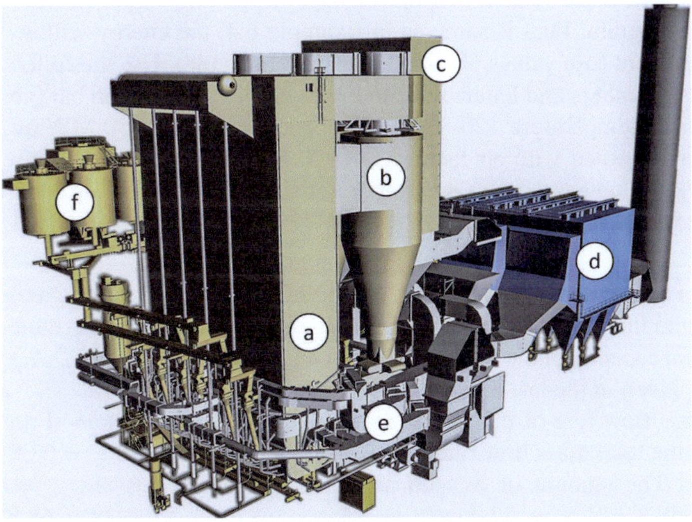

Fig. 6.22 The boiler and auxiliary components of the Alholmens kraft power plant. a = furnace, b = cyclone, c = heat transfer section, d = electrostatic precipitator, e = air feed system, and f = additives (e.g., CaO for SO_2 capture) feed system. © Ahlholmens Kraft, *Source* https://www.alholmenskraft.com/en/production/boiler

from the furnace, and the water is evaporated. Since the density of the two-phase flow containing steam bubbles and saturated water is smaller than the downcomer tubes' saturated water, this density difference creates a natural "lift" to circulate water in the evaporator surfaces. Saturated steam is then superheated to final live steam temperature, before it enters the high-pressure turbine (6–1). In some cases, reheating can be used, where the steam is taken back to the boiler after expansion in the high-pressure turbine—see Chap. 3 for more information on these processes.

As Fig. 6.23 shows, the fuel and the air enter the boiler from the furnace/evaporator section where combustion reactions take place, and which is the hottest part of the boiler. It is therefore important to have efficient phase change taking place at this location on the water side, to keep the metal surface temperatures low enough, in practice close to the boiling temperature, $t_{sat}(p_1)$.

There are several different types of boilers used in the combustion of solid fuels. Often the boiler type used for waste fuels is a grate furnace (GF), or a fluidized boiler—either a bubbling fluidized bed boiler (BFB), or a circulating fluidized bed boiler (CFB), as shown in Fig. 6.23. The key component of the CFB boiler is the *cyclone* that separates the solids from the flue gases. Solid particles are returned to the furnace, and gases are sent to the heat transfer section. In cases where the fuel is a very fine powder, a pulverized combustion (PC) is used. Although not shown in Fig. 6.23, these systems always include flue gas cleaning systems, for particulate emissions, SO_2, and often also for NO_x emissions.

Example 6.6 Calculate the heat loads in MW of all heat transfer surfaces of the CFB boiler—feedwater pre-heater, evaporator, superheater and air pre-heater—and draw the $\phi - T$ diagram. Fuel is same as in Example 6.4, the energy willow with 50% moisture content (use values obtained from that example). Use the following steam side values (numbers and letters refer to Fig. 6.23): live steam 166 bar (abs) (1), 545 °C; feedwater temperature 120 °C (4). Fuel energy input is 550 MW on *LHV* fuel basis. Fuel is burned with air factor $\lambda = 1.5$. *LHV* of the wet fuel, as received, is 7.989 MJ/kg. Flue gases leave the evaporator section at 1000 °C (b). Air enters the boiler at 25 °C (f).

Solution: From Example 6.3 we can get the mass flow rate of flue gases per 1 kg/s of fuel, and their mass fractions and average heat capacities—see Table 6.6. Total average heat capacity of the flue gases, $\overline{c}_{p.fg}$, can be calculated using $\overline{c}_p = \sum_{i=1}^{N} y_i \overline{c}_{p.i}$, and this is given in the last column of the table.

The mass flow rate of the fuel is $\dot{m}_f = 550/7.989 = 68.8\,\text{kg/s}$. From this and Table 6.6, the total mass flow rate of flue gas can be obtained, $\dot{m}_{fg} = 68.8 \times 5.47 = 376.6\,\text{kg/s}$. The amount of oxygen and nitrogen required is $\dot{m}_{O_2.in} = d\dot{m}_f\lambda = 0.7015 \times 68.8 \times 1.5 = 72.4\,\text{kg/s}$, and $\dot{m}_{N_2.in} = e\dot{m}_f\lambda = 2.28 \times 68.8 \times 1.5 = 235.4\,\text{kg/s}$, respectively, and these add up to $\dot{m}_{air.in} = 72.4 + 235.4 = 307.9\,\text{kg/s}$. Average heat capacity of the air is still needed. The mass fraction of oxygen in the air is $y_{O_2} = x_{O_2}M_{O_2}/M_{air} = 0.21 \times 32/28.964 = 0.232$, and the average of heat capacity of the air is then $\overline{c}_{p.air} = 0.232 \times 1.09 + (1 - 0.232) \times 1.167 = 1.149\text{kJ/kg}$.

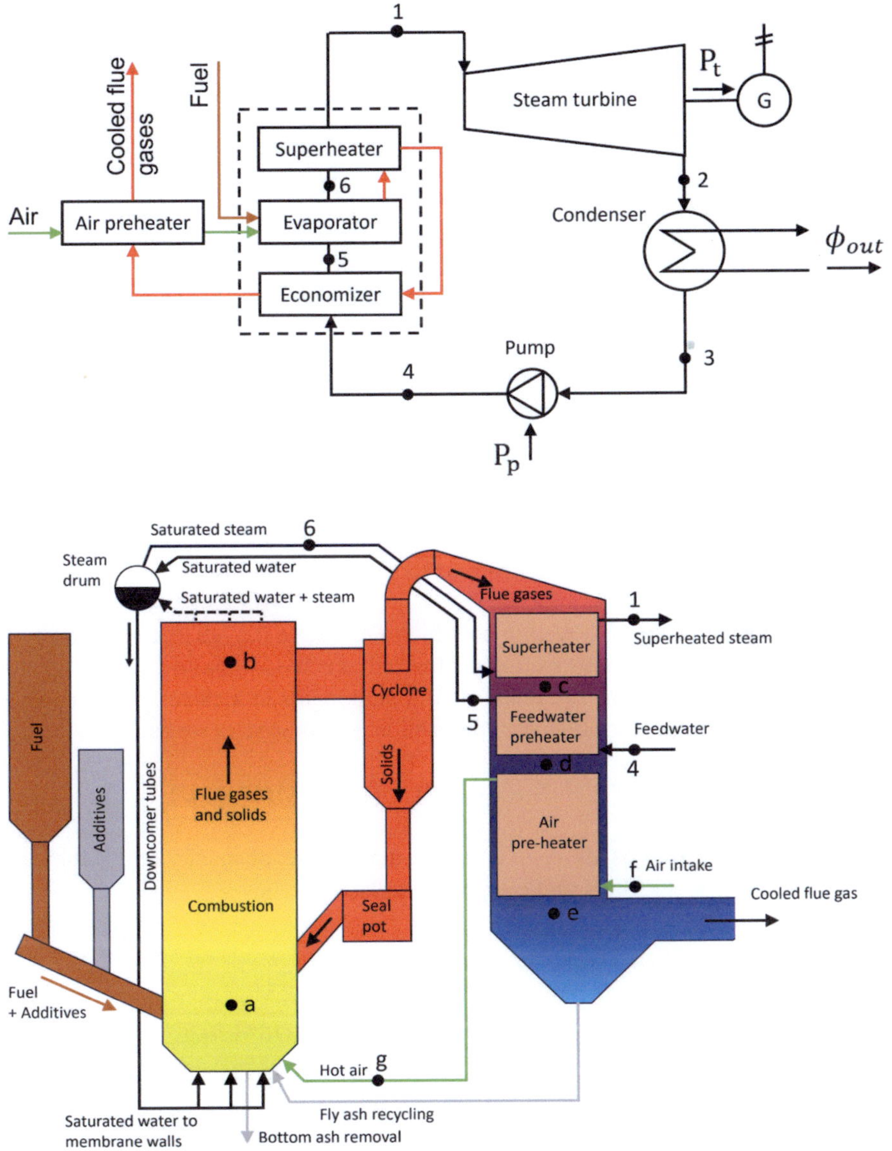

Fig. 6.23 Simplified flow schematic of the flows of air (green), fuel (purple), flue gases (red), water (black) and steam in the circulating fluidized boiler

Table 6.6 Mass of flue gas components per 1 kg/s of fuel, $\lambda = 1.5$

Species	m_i (kg/s)	y_i (–)	$\bar{c}_{p.i}$ (kJ/kgK)
CO_2	0.914	0.167	1.234
H_2O	0.774	0.142	2.288
Ash	0.009	0.002	1.000
SO_2	0.0005	0.00012	0.888
O_2	0.351	0.064	1.090
N_2	3.422	0.626	1.167
Total	5.470	1.000	1.333

On the steam side, we need to obtain the enthalpies, temperatures and the mass flow rate of the water to boiler. From the live steam pressure and temperature, and using the electronic supplementary materials Excel sheet, we can get the steam enthalpy as $h_1(545°C, 166\text{bar}) = 3419.5\,\text{kJ/kg}$. The evaporator temperature can be obtained from the saturation value, $t_5 = t_6 = t_{sat}(166\text{bar}) = 350.3°\text{C}$. The enthalpy of the saturated water and steam are $h_5 = h^L_{sat}(166\text{bar}) = 1673.8\,\text{kJ/kg}$ and $h_6 = h^G_{sat}(166\text{bar}) = 2561.2\,\text{kJ/kg}$, respectively. Enthalpy of the feedwater is $h_4(120°C, 166\text{bar}) = 515.4\,\text{kJ/kg}$.

Since the air is preheated, we are not able to calculate the adiabatic combustion temperature as in the previous examples. Instead, we need to build a model for the whole boiler and couple the operation of all components in a clever way. But first, let's study the model for the adiabatic combustion reactor system we built previously:

$$LHV_{ar} = (T_a - T_0)(A + \lambda B) - \left(T_g - T_0\right)(C + \lambda D)$$

With numeral values as above in Example 6.5, and heat capacity of dry fuel and liquid water, $c_{p.F} = 1\,\text{kJ/kgK}$ and $c_{p.W} = 4.2\,\text{kJ/kgK}$, we have

$$\begin{aligned}
A &= c\bar{c}_{p.A} - d\bar{c}_{p.O_2} + f\bar{c}_{p.CO_2} + g\bar{c}_{p.H_2O} + h\bar{c}_{p.SO_2} \\
&= 0.009 \times 1.0 - 0.7015 \times 1.09 + 0.9138 \times 1.234 \\
&\quad + 0.7744 \times 2.2877 + 0.000491 \times 0.888 = 2.135\,\text{kJ/kg} \\
B &= d\bar{c}_{p.O_2} + e\bar{c}_{p.N_2} = 0.7015 \times 1.09 + 2.28 \times 1.1672 = 3.426\,\text{kJ/kg} \\
C &= a\bar{c}_{p.F} + b\bar{c}_{p.W} + c\bar{c}_{p.A} \\
&= 0.491 \times 1.0 + 0.5 \times 4.2 + 0.009 \times 1.0 = 2.6\,\text{kJ/kg}
\end{aligned}$$

Putting these into the above equation we get $LHV_{ar} = (1 - 0.5) \times 18420 - 0.5 \times 2442 = 7989\,\text{kJ/kg}$. We can then write the equation for the reaction temperature as follows, with air factor $\lambda = 1.5$:

$$T_a = T_0 + \frac{LHV_{ar}}{\lambda B + A} + \frac{\lambda B + C}{\lambda B + A}\left(T_g - T_0\right)$$

$$= 25 + \frac{7989}{1.5 \times 3.426 + 2.135} + \frac{1.5 \times 3.426 + 2.6}{1.5 \times 3.426 + 2.135}\left(T_g - T_0\right)$$

$$T_a = 1123 + 1.064 \times \left(T_g - T_0\right)$$

This is a very interesting result—it means that adiabatic combustion temperature increases almost directly in proportion with pre-heating. To solve for the whole boiler, we need to link the water-steam side to flue gases and air preheating. The overall energy balance for the sections connected with water and steam is as follows: energy taken from the flue gases is equal to heat needed to preheat, evaporate and superheat the water (see Fig. 6.23 for process points),

$$\dot{m}_{fg} c_{p.fg} (t_a - t_d) = \dot{m}(h_1 - h_4)$$

For the remaining part, the air pre-heater, we can write

$$\dot{m}_{air} c_{p.air} \left(t_g - t_f\right) = \dot{m}_{fg} c_{p.fg} (t_d - t_e)$$

Note that t_f is the temperature of the surrounding environment, 25 °C. The air pre-heater is typically quite a large heat exchanger, and to keep its size reasonable, we accept that it is able to heat incoming air to 10 °C colder than the incoming flue gas. This temperature difference is called the pinch temperature:

$$t_d - t_g = \Delta t = 10^\circ C$$

We have now arrived at four equations, but they contain five unknowns, namely, t_a, t_d, t_e, t_g and $\dot{m}$,—to solve this, we need one more equation. We defined in the task description that the furnace exit temperature is $t_b = 1000\,^\circ\mathrm{C}$. The final equation should include this, so the furnace energy balance can be used:

$$\dot{m}_{fg} c_{p.fg} (t_a - t_b) = \dot{m}(h_6 - h_5)$$

which links the adiabatic combustion temperature t_a and mass flow rate of water $\dot{m}$. Now we have five equations and five unknowns, so this group of linear equations can be solved. Let's write this into linear form,

$$t_a - 1.064 \times t_g = 1123 - 1.064 \times t_0$$

$$t_d - t_g = \Delta t$$

$$\dot{m}_{fg} c_{p.fg} t_d - \dot{m}_{fg} c_{p.fg} t_e - \dot{m}_{air} c_{p.air} t_g = -\dot{m}_{air} c_{p.air} t_f$$

$$t_a - t_d - \dot{m}\frac{(h_1 - h_4)}{\dot{m}_{fg} c_{p.fg}} = 0$$

$$t_a - \dot{m}\frac{(h_6 - h_5)}{\dot{m}_{fg}c_{p.fg}} = t_b$$

Or in matrix form,

$$\boldsymbol{Q} \cdot \boldsymbol{x} = \boldsymbol{b}$$

$$\boldsymbol{Q} = \begin{pmatrix} 1 & 0 & 0 & -1.064 & 0 \\ 0 & 1 & 0 & -1 & 0 \\ 0 & \dot{m}_{fg}c_{p.fg} & -\dot{m}_{fg}c_{p.fg} & -\dot{m}_{air}c_{p.air} & 0 \\ 1 & -1 & 0 & 0 & \frac{(h_1-h_4)}{\dot{m}_{fg}c_{p.fg}} \\ 1 & 0 & 0 & 0 & \frac{(h_6-h_5)}{\dot{m}_{fg}c_{p.fg}} \end{pmatrix}, \boldsymbol{x} = \begin{pmatrix} t_a \\ t_d \\ t_e \\ t_g \\ \dot{m} \end{pmatrix}$$

$$\boldsymbol{b} = \begin{pmatrix} 1123 - 1.064 \times t_0 \\ \Delta t \\ -\dot{m}_{air}c_{p.air}t_f \\ 0 \\ t_b \end{pmatrix}$$

from which we find, with numeral values,

$$\boldsymbol{x} = \boldsymbol{Q}^{-1} \cdot \boldsymbol{b} = \begin{pmatrix} t_a \\ t_d \\ t_e \\ t_g \\ \dot{m} \end{pmatrix} = \begin{pmatrix} 1336.4\,°\text{C} \\ 235.5\,°\text{C} \\ 77.3\,°\text{C} \\ 225.5\,°\text{C} \\ 190.3\,\text{kg/s} \end{pmatrix}$$

When these are known, the rest of the process points and heat loads can be easily calculated. The heat load of the evaporator is $\phi_{EVA} = \dot{m}(h_6 - h_5) = 190.3 \times (2561.2 - 1673.8) = 168.9$ MW. The superheater is the next heat transfer surface after the evaporator. Since the mass flow rate on both flue gas and water sides is known, we can solve the flue gas temperature after the superheater from its energy balance,

$$\dot{m}(h_1 - h_6) = \dot{m}_{fg}c_{p.fg}(t_b - t_c)$$

$$t_c = t_b - \frac{\dot{m}(h_1 - h_6)}{\dot{m}_{fg}c_{p.fg}} = 1000 - \frac{190.3 \times (3419.5 - 2561.2)}{376.6 \times 1.333} = 674.6\,°\text{C}$$

The heat load of the superheater is $\phi_{SH} = \dot{m}(h_1 - h_6) = 190.3 \times (3419.5 - 2561.2) = 163.3$ MW. The economizer (i.e., the feedwater pre-heater) is the next heat transfer surface after the superheater in the flue gas channel. We solve the flue gas temperature after the economizer from its energy balance,

$$\dot{m}(h_5 - h_4) = \dot{m}_{fg} c_{p.fg} (t_c - t_d)$$

$$t_d = t_c - \frac{\dot{m}(h_5 - h_4)}{\dot{m}_{fg} c_{p.fg}} = 674.6 - \frac{190.3 \times (1673.8 - 515.4)}{376.6 \times 1.333} = 235.5\,^{\circ}\mathrm{C}$$

This is the same value already obtained above from the linear system of equations. The heat load of the economizer is $\phi_{ECO} = \dot{m}(h_5 - h_4) = 190.3 \times (1673.8 - 515.4) = 220.4\,\mathrm{MW}$.

The last heat transfer surface is the air pre-heater. Its heat load is readily available as $\phi_{APH} = \dot{m}_{fg} c_{p.fg} (t_g - t_f) = 376.6 \times 1.333 \times (235.5 - 77.3) = 79.4\,\mathrm{MW}$. However, the calculation of the air-preheater is not straightforward, since if the flue gas temperature decreases low enough, the water vapor of the flue gas will condense. For the sake of simplicity here, no condensation is allowed to take place. This means physically that the vapor pressure of the steam in the flue gases is not allowed to drop below the partial pressure (which is constant until condensation begins) of the water pressure in the flue gases. The molar mass of the wet flue gases is calculated from the data in Table 6.3, 0.0277 kg/mol.

$$p_{sat}(t_e) > p_0 x_{H_2O} = p_0 y_{H_2O} \frac{M_{H_2O}}{M_{fg}}$$
$$= 101325 \times 0.142 \times \frac{18}{27.7} = 9350\,\mathrm{Pa}$$

From this the temperature, t_e, can be iteratively solved to give $t_e > 44.2\,^{\circ}\mathrm{C}$. Since our temperature is 77.3 °C, there is no risk of condensation. Finally, we can draw the heat load—temperature diagram, as shown in Fig. 6.24.

6.6 Bioenergy Carbon Capture, Storage and Utilization

The objective of this section is first to discuss the principles and applications of carbon capture technologies based on biomass carbon. It is important to understand the big picture of how the natural CO_2 capture mechanism of photosynthesis acts as our carbon capture machine, and how the fast carbon cycle is being affected by these technologies. Then, we will discuss the magnitude of the technical potential of CO_2 capture technologies. After this, we review the products that can be produced from biomass, and how the biogenic carbon could be captured from these products. And finally, we take a closer look at one of the most promising carbon capture and storage (CCS) methods, Oxyfuel combustion, and carry out a simple analysis to demonstrate and better understand the potential and limitations of CCS. To be precise, we are focusing here only on the CC aspect of CCS—the carbon capture technologies—and not on the S, the storage part.

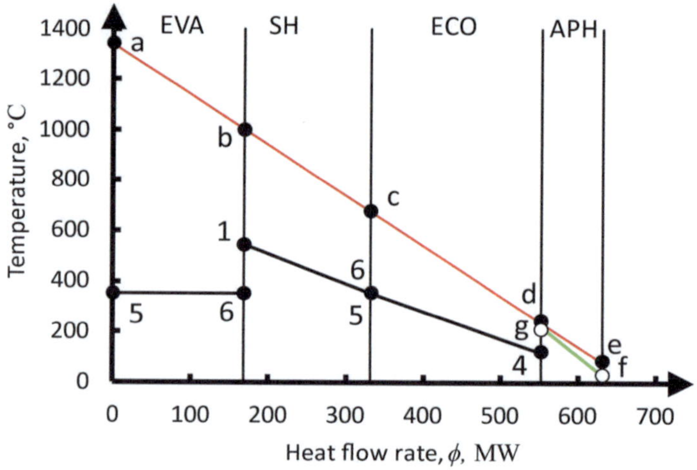

Fig. 6.24 Heat flow rate—temperature diagram of the boiler. SH = superheater, EVA = evaporator, ECO = feedwater preheater (economizer), and APH = air pre-heater. Red = flue gas, green = air, and black = water/steam. Inlet and outlet points of the air pre-heater are presented as open circles

Carbon capture methods can be divided into three kinds: pre-combustion methods, such as gasification combined with CO_2 separation before the gas combustion; post-combustion methods, such as the solvent-based methods (for example, the monoethanolamine (MEA) solvent process) or Calcium looping technology; and in-situ methods, such as the Oxyfuel combustion, in which the product gas contains only CO_2 and H_2O, which can be easily separated by condensation.

Well-functioning photosynthesis is the core of the bioenergy carbon capture pathway. This is maintained by various plants and algae, as part of healthy and functional ecosystems that support a resilient web of life. Natural ecosystems on the land and in the oceans are able to capture annually 58 and 49 Gt of carbon (corresponding to 213 Gt and 180 Gt of CO_2), respectively (see also Sect. 6.2, Fig. 6.1, masses are just multiplied here by 0.5 to get the carbon mass)—a total of 393 Gt of sequestrated CO_2. As we discussed earlier, if the Earth's fast carbon cycle system is not significantly disturbed by human uses, most of this will be released back to atmosphere, leaving a total $\pm$ 0 Gt CO_2 balance annually. In Sect. 2.2, we saw that the natural fast carbon cycle can bind ~ 5 Gt carbon from fossil emissions annually. For comparison, the current total anthropogenic CO_2 emissions and CCS capacity of human made systems is in the order of 40 Gt and 0.244 Gt of CO_2, respectively, indicating that nature is far more effective than our technical processes [79].

In their "Special Report: Global Warming of 1.5 °C", the IPCC presented four pathways, P1 to P4, to keep the global warming below 1.5 °C, as shown in Fig. 6.25 [39]. Different mitigation strategies can be used to achieve the net emissions reductions needed to limit global warming to 1.5 °C with no or limited overshoot.

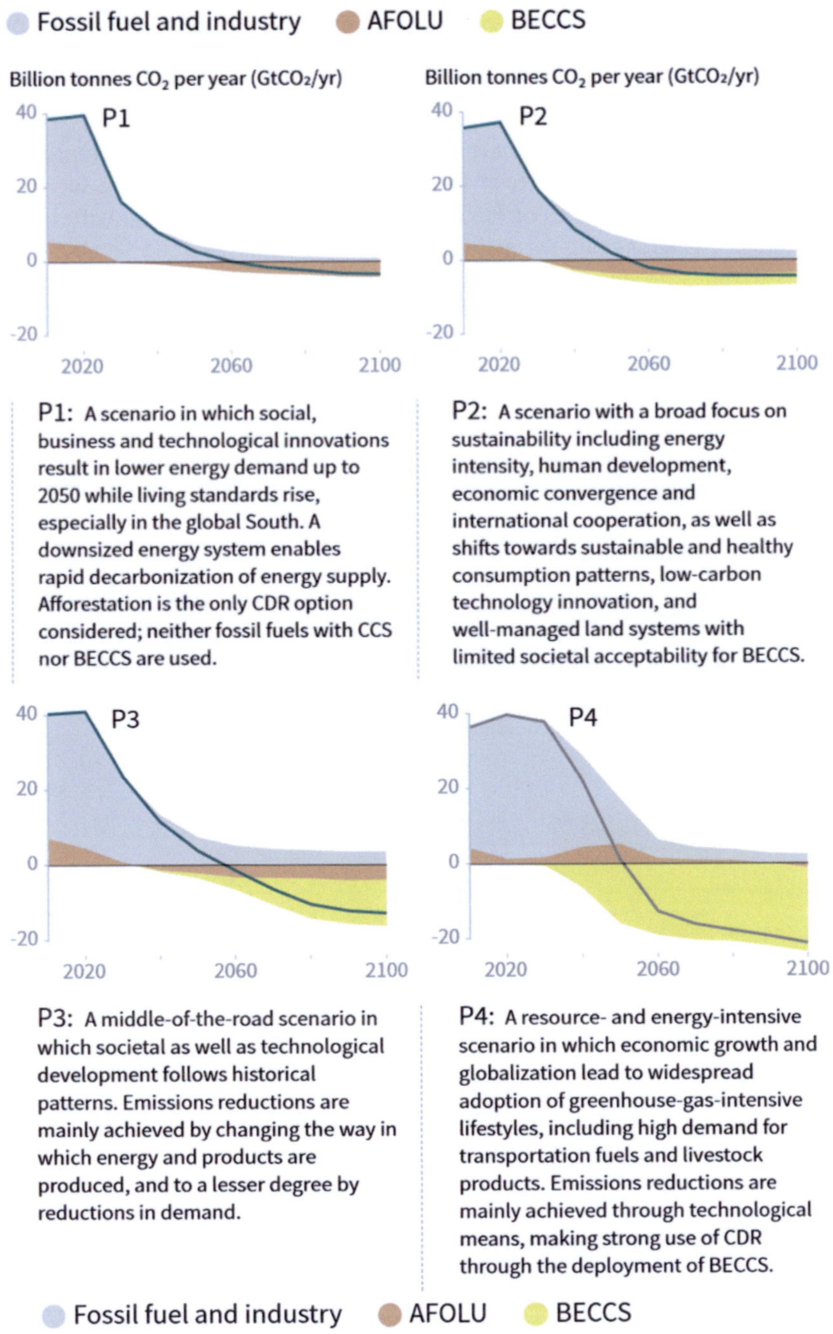

Fig. 6.25 Breakdown of contributions to global net CO_2 emissions in four illustrative model pathways of IPCC [39]. AFOLU = Agriculture, Forestry and Other Land Use, BECCS = Bioenergy Carbon Capture and Storage, CDR = Carbon Dioxide Removal [39]

In particular, scenario P4 is based on significant levels of bioenergy carbon capture and storage (BECCS) use—more than 20 Gt of annual CO_2 capture capacity—to stay below the 1.5 °C target. For example, the utilization of the annual growth of all boreal forests, which binds approximately 5 Gt of carbon (19 Gt of CO_2) annually (see Table 6.3), would never be sufficient without severely compromising biodiversity and the functionality of forests as a producer of ecosystem services. In addition, human-induced loss of biodiversity is already accelerating, with land use and land-use change among the major drivers of this loss ([37], see also Sect. 2.3).

Several routes are suitable for the conversion of biomass into final energy products or chemicals in combination with CCS. These can be divided into bio-chemical biofuels production, thermo-chemical production of biofuels and biochemicals, and biomass combustion for electricity and/or heat production (Fig. 6.26).

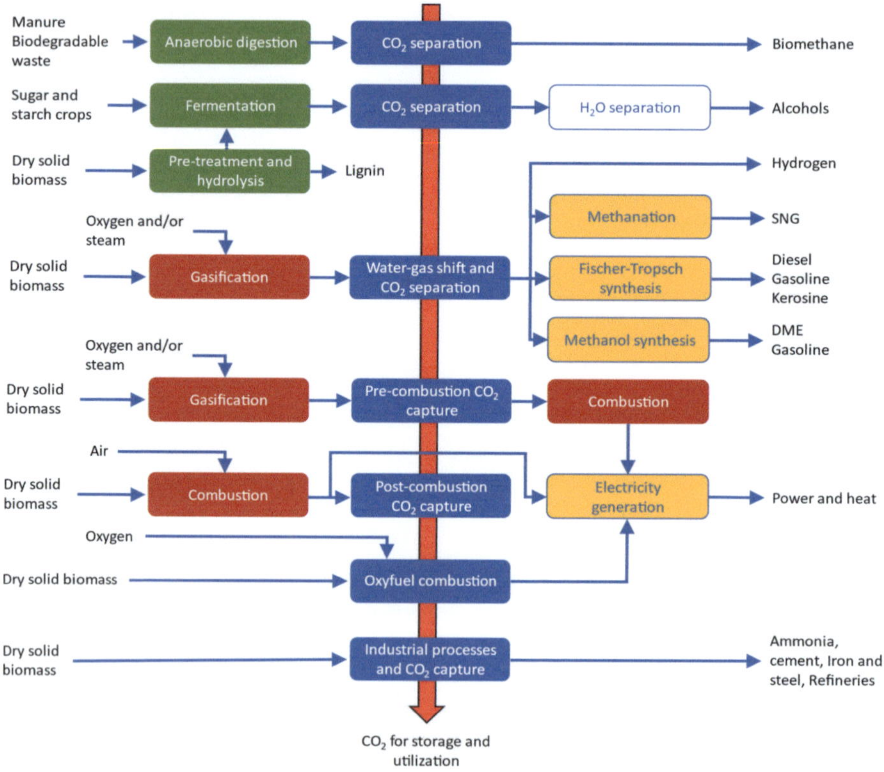

Fig. 6.26 Overview of biomass-based conversion routes with CCS (Redrawn based on ZEP [89])

6.6.1 Bio-Chemical Production of Biofuels and CO_2 Capture

Bio-chemical production of biofuels and CO_2 capture refers to processes where CO_2 is produced often in a highly concentrated point source for efficient capture. In the bio-chemical pathways, a significant amount of the carbon in the feedstock typically ends up in biofuels or biochemicals, resulting in smaller CO_2 emission side streams compared to heat and power generation. Note that, if liquid or gaseous biofuels are used for combustion, the total CO_2 emissions are larger than just burning the biomass directly, since the production of fuels also consumes energy, and all process steps contribute more losses. However, the impact of CO_2 capture on overall process yields is usually much smaller in the case of biofuels or biochemicals production—the CO_2 can either be separated, or in some cases, the fuel production process itself requires separation to ensure that downstream synthesis processes work properly.

Conventional biofuels from sugar or starch crops currently represent the largest capacity of all biofuel/biochemical production routes, while scale-up efforts are ongoing for thermo-chemical production routes. Below, we quickly review different production routes for biofuels.

Biomethane Production by Anaerobic Digestion

Anaerobic digestion of organic material to biogas is a process where biomass is broken down in several steps by different micro-organisms, in a reactor called a digester. Many organic raw materials can be used as feedstock, as long as they are biologically degradable. These include waste streams from animal, human, food, and organic sources, and green crops, excluding lignocellulosic woody biomasses. Process products include biogas containing 50–70% CH_4 and 30–50% CO_2, with trace amounts of sulfurous components, and a solid–liquid fraction called the digestate [13, 15].

Typical yields of biogas on dry biomass basis vary from 200 Nm^3/ton for cattle manure, to 500–600 Nm^3/ton for food remains, to as high as 800 Nm^3/ton for fish oil [83]. The water content of the digestate can be as high as 90–95%, and the liquid fraction of the digestate contains most of the nitrogen-based nutrients, such as ammonia. The solid fraction contains phosphorus, potassium and calcium nutrients—and, of course, the carbon. The digestate is typically treated in a centrifuge to separate the solid and liquid phases, allowing efficient utilization as fertilizers and ease of long-distance transportation [15].

Biogas can be upgraded into biomethane by separating the CO_2 and removing the sulfurous components. Since the properties of the product are similar to natural gas, it directly facilitates injection into the grid. CO_2 separation is a commercially-proven technology to produce biomethane, but faces certain challenges for the purposes of CCS, such as seasonal feedstock variability and relatively small available CO_2 streams.

Bioethanol Production by Fermentation

During ethanol fermentation, sugars from conventional biofuel feedstocks (for example, sugar cane, sugar beet, and the starch component of corn) are fermented into ethanol; the process also generates a CO_2 stream. Two-thirds of the carbon contained in the sugars ends up in the ethanol, and the remaining one-third forms nearly-pure CO_2. The CO_2 stream can be separated via gas liquid separation, while the ethanol–water mixture is typically separated via distillation. As an example, a typical US ethanol plant produces around 200 million liters per year (around 157,800 tons, using density of 0.789 ton/m^3), which corresponds to a pure CO_2 stream of 140,000 tons per year. The US is by far the largest producer of ethanol, with 59,121 Mm^3 of ethanol from 192 plants; Brazil is in second position with 31.26 Mm^3, and the EU is third with 5.45 Mm^3 (https://afdc.energy.gov/data).

Lignocellulosic feedstocks can also be used for ethanol production, although these require a pre-treatment step to isolate the cellulose from the lignin [13]. Subsequent chemical or enzymatic hydrolysis converts complex cellulose chains into simple sugars that can be fermented into ethanol. Around 62% of the carbon present in the feedstock ends up in the lignin stream, 25% in the ethanol product, and the rest in the pure CO_2 stream. Lignin can be used as a fuel to run the ethanol/water distillation, or as fuel for a combined heat and power plant, although post-combustion CO_2 capture should be implemented with these processes to significantly increase the CO_2 capture potential. However, all steps of the value chain in this process must remain on a sustainable level.

6.6.2 Thermo-Chemical Production of Biofuels and Biochemicals and BECCS

During thermo-chemical conversion, lignocellulosic and non-edible feedstocks are first dried and ground, and subsequently gasified using oxygen and/or steam as the reagent. The product from gasification is then cleaned and processed to form so-called synthesis gas (Syngas), which can be used in commercially available synthesis processes to form fuels and chemicals such as:

- Hydrogen, which can be further synthesized into ammonia, NH_3, and urea, CH_4N_2O.
- Synthetic natural gas (SNG) via methanation.
- Diesel, gasoline and kerosene (jet fuel) via fuel synthesis (e.g., Fischer–Tropsch) and via refining (often described as Biomass-to-Liquid, BtL, process).
- Methanol synthesis and upgrading to DME (dimethyl ether, a fuel additive) and gasoline, but also plastics, formaldehyde and acetic acid.

The required hydrogen-to-carbon monoxide ratio varies for different synthesis routes. Methanol or Fischer–Tropsch (FT) synthesis requires two hydrogen molecules per carbon monoxide molecule, while methane synthesis requires three

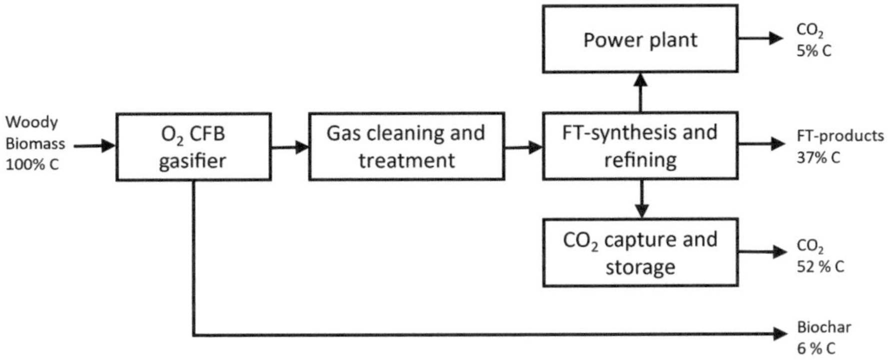

Fig. 6.27 Typical carbon distribution in the outlet streams of a Fischer–Tropsch Diesel plant [89]

hydrogen molecules per carbon monoxide molecule. This can be adjusted by the water–gas shift reaction and CO_2 separation, resulting in a relatively pure CO_2 stream. The capture technology is similar to pre-combustion CO_2 capture in Integrated Gasification Combined Cycle (IGCC) power plants, and is usually based on the use of physical absorption using solvents.

Figure 6.27 shows an example of the fate of carbon in a typical FT diesel plant, based on oxygen-fired Circulating Fluidized Bed (CFB) gasification. Slightly more than 50% of the carbon in the feedstock is released as relatively pure CO_2, while 37% ends up in the diesel stream. The CO_2 vented from the power island, and carbon in the char from the gasifier, makes up the balance.

Storage and Transport of CO_2

To reach net removal of CO2 from the atmosphere, captured CO_2 needs to be conditioned, transported to a storage site, and permanently stored in a way that is isolated from the atmosphere. Long-distance transport of carbon dioxide is needed unless a suitable geological formation for storage is located immediately below the recovery facility. CO_2 gas pipes are commercial technology, and the most common method in the transport of carbon dioxide. In the US, 40 Mt of CO_2 is transported annually, for the moment mainly for enhanced oil recovery (EOR). Carbon dioxide can also be transported as liquid on ships, trains or trucks, in thermally insulated tanks at temperatures below room temperature and pressures significantly lower than used in pipe transport. The most viable options for a full-scale CCS project are pipeline transport and/or ship transport.

After recovery, carbon dioxide must be stored in accordance with the European Union Directive on the Geological Storage of CO2 (known as the "CCS Directive", 2009/31/EC), permanently isolated from the atmosphere. For millions of tons of CO_2, there are only a few possible storage options that have been demonstrated at scale. These are underground storage locations, for example geological formations such as used oil and gas fields, saline aquifers, or sedimentary rock layers.

The geological storage methods of carbon dioxide have in common that the carbon dioxide is injected to supercritical pressures (> 74 bar). At these pressures, CO_2 behaves like steam, penetrating every crack in the rock, but its density is as high as a liquid. Carbon dioxide is injected into the pore spaces of the formation, typically at a depth of more than 800 m, so that the pressure created by overlying rock deposits matches the pressure of the injected carbon dioxide. The porosity and permeability of the formation determine the storage capacity in addition to the injection rate. The injection rate can be increased with additional injection wells. After carbon dioxide has been injected, several different physical and geochemical mechanisms prevent it from rising back to the surface—the most important of them is the presence of a cap rock or clay layer on top of the storage layer. The structure of the geological formation determines its suitability as a storage site: closed formations are safer reservoirs than open formations, but they are also significantly rarer.

6.6.3 BECCS in Biomass Combustion for Power and Heat Production

Biomass can be burned alone, or mixed with other feedstocks such as waste or industrial side streams. There are several technical routes for biomass co-firing which can be divided into indirect and direct processes. Indirect co-firing relies on the dedicated initial conversion of biomass to combustible gas in a fluidized bed gasifier. This produces a combustible gas that can be injected into an existing boiler. During direct co-firing, biomass is blended with other feedstocks such as waste or coal, milled and transported to the burners in the boiler. Biomass can also be ground in a dedicated biomass mill or modified coal mill—the ground biomass can be blended with pulverised coal and fed to the burners, fed via a dedicated biomass burner, or simply injected directly into the boiler.

Biomass co-firing ratios strongly depend on the characteristics of the biomass used and the layout of the power plant. Achieving elevated co-firing ratios has proved difficult as the untreated biomass is typically fibrous, fluffy, and inhomogeneous in nature. It has a lower volumetric energy density and different inorganic composition to coal; it is also hydrophilic and vulnerable to biodegradation. Significant plant modifications are therefore usually required, entailing additional, high investment costs.

An alternative is to use thermal pre-treatment technologies that increase the homogeneity, brittleness and/or the energy density of biomass. These take place at temperatures between 250–550 °C, with the torrefaction process occurring at the lower end of the range, and the pyrolysis process at the upper end [12]. The biomass co-firing share can be significantly increased with such pre-treatment technologies without major power plant modifications. As most methods do not affect the inorganic content of the biomass, boiler fouling, corrosion and emissions could still pose technical bottlenecks. However, high-temperature entrained flow gasification, as used in Integrated

Gasification Combined Cycle (IGCC) power plants, could reduce these limitations since part of the inorganic content of the pre-treated biomass is removed through the slag.

Biomass-only combustion is normally done in smaller, modified pulverized coal boilers, and can also be facilitated in existing, medium-sized circulating fluidized boiler (CFB) plants previously fired with coal or lignite. The latter are theoretically suited to fire up to 100% biomass and are generally based on CFB technology. CFB boilers are usually smaller than large utility boilers (ranging from 50 to 500 MW thermal power), more flexible around fuels, and typically located near urban areas or industrial facilities to supply heat.

Biomethane obtained from fermentation and upgraded by CO_2 separation or gasification-based bio-SNG could be used as fuel in natural-gas-fired combined gas turbine cycle power plants (NGCC), gas engines, or CHP plants. For gas-fired applications, there are in principle no co-firing ratio limitations, and biomethane and bio-SNG can be co-fired at any rate between 0 and 100%. There are also no fundamental restrictions to applying "conventional" post-combustion CO_2 capture technologies in these power plants (apart from economies of scale for CHP).

Gasification of biomass allows the use of a large variety of biomass feedstocks and, in theory, allows for pre-combustion CO_2 capture technologies that are also proposed for IGCC power plants. Technical improvements to future biomass-based IGCC (BIGCC) plants can therefore build on the experience and further development of gasification technology in the chemical sector, which produce base chemicals and transport fuels (e.g., FT-diesel or -kerosene).

Large industrial operations could present potential BECCS opportunities where a local heat or power requirement exists—particularly in the proximity of industrial clusters where CCS infrastructure can be shared, ensuring continuity and operational efficiency. Further synergies may be possible where systems are combined or shared, such as the application of low-grade heat to pre-drying the biomass, reducing the moisture content and improving the energy density of the feedstock.

The use of biomass in industry to replace fossil fuels includes a variety of potential applications: small- and medium-scale heat and power for industrial and domestic use (< 50 MW); as a fuel substitute in cement kilns; in the refining and chemicals industries as synthesis gas from gasification or pyrolysis oil; and via injection in blast furnace steel- and iron-making. However, as discussed above, the sustainability, land-use, and biodiversity impacts limit the up-scaling capacity of utilizing biomass significantly.

In the pulp and paper industry, most emissions originate from biogenic sources, since most of the on-site processes use biomass as a raw material. Although total site emissions are significant, they are scattered among different stacks, with the recovery boiler burning black liquor usually the largest source by far. Black liquor is a liquid process stream in chemical pulping processes that contains converted cooking chemicals, dissolved lignin, and other organics separated from the fibers used in pulp. In a recovery boiler, cooking chemicals are recovered for re-use, and lignin is burned to produce power and heat for the site [1]. Other sources in the pulp

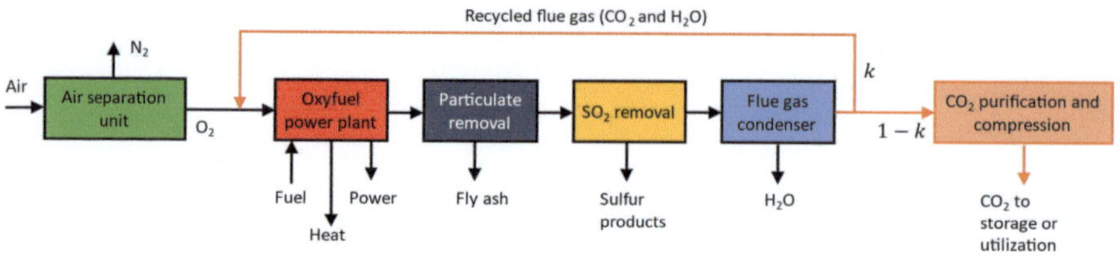

Fig. 6.28 Principle of the Oxyfuel power plant

and paper plant are the lime kiln, the bark boiler, and possibly other on-site heat and power production.

As for other industries, the potential for process integration could substantially reduce the energy penalty from CCS and associated capture costs. However, the size of single sources, as well as the heavily-integrated processes in modern pulp and paper mills, pose challenges in applying CCS to existing installations. Layout restrictions and impurities in flue gases also pose a challenge. Nevertheless, as aging recovery boilers are replaced in the future, a window of opportunity may open for the gasification of black liquor, or oxyfuel combustion, which present more feasible capture options.

Example 6.7 The boiler in Example 6.6 is converted into an oxygen-fired system as shown in Fig. 6.28. It is burning the same energy willow with 50% moisture content. The temperature after the flue gas condenser is set to 25 °C, the same as the inlet temperature of other streams. Air factor $\lambda = 1$, which means that there is no O_2 in the flue gases. Flue gas temperature after the evaporator is 1000 °C. What are the required mass flow rate of fuel and flue gas recirculation ratio, k, to generate the same amount of steam (168.9 MW evaporator surface heat load) as in Example 6.5 with the air fired boiler? For simplicity, exclude the particulate and sulfur removal blocks in your analysis by assuming that their efficiencies are 100%. (Note that in reality, Oxyfuel systems also require some of the electrical power generated to separate oxygen from the air, but this is neglected from this analysis.)

We first need to determine the dependency of the mass flow rates of species produced in combustion, and their recycled amounts. For example, for CO_2, the recycling ratio is defined as

$$\left(f\,\dot{m}_f + \dot{m}_{CO_2.r}\right)k = \dot{m}_{CO_2.r}$$

where f is the mass of CO_2 produced per 1 kg of wet fuel, and $\dot{m}_{CO_2.r}$ is the mass flow rate of recycled CO_2. From this we can find the recycled mass flow rate,

$$\dot{m}_{CO_2.r} = \frac{k}{1-k} f\,\dot{m}_f$$

For H_2O and SO_2 we can write

$$\dot{m}_{H_2O.r} = \frac{k}{1-k} g\dot{m}_f(1-\eta_W)$$

$$\dot{m}_{SO_2.r} = \frac{k}{1-k} h\dot{m}_f(1-\eta_S)$$

The total mass flow rates of these gases flowing in the furnace are therefore

$$\dot{m}_{CO_2,t} = \left(1 + \frac{k}{1-k}\right) f\dot{m}_f = \frac{1}{1-k} f\dot{m}_f$$

$$\dot{m}_{H_2O,t} = g\left(1 + \frac{k}{1-k}(1-\eta_W)\right)\dot{m}_f = g\left(\frac{1-k\eta_W}{1-k}\right)\dot{m}_f$$

$$\dot{m}_{SO_2,t} = h\left(1 + \frac{k}{1-k}(1-\eta_S)\right) = \dot{m}_f h\left(\frac{1-k\eta_S}{1-k}\right)\dot{m}_f$$

where η_S and η_W are the efficiencies of sulfur capture and water condensation, respectively, and g and h are the masses of H_2O and SO_2 produced per 1 kg of wet fuel. With these equations, using the average heat capacity approach, and recalling that heat of formation of two-atom molecules, such as oxygen and nitrogen, is equal to zero, the energy balance is

$$\begin{aligned}
&\dot{m}_{F.in}\left(h_F^0 + \overline{c}_{p.F}(T_{in} - T_0)\right) + \dot{m}_{W.in}\left(h_W^0 + \overline{c}_{p.W}(T_{in} - T_0)\right) \\
&+ \dot{m}_{A.in}\left(h_A^0 + \overline{c}_{p.A}(T_{in} - T_0)\right) + \dot{m}_{O_2.in}\left(\overline{c}_{p.O_2}(T_{in} - T_0)\right) \\
&- \dot{m}_{CO_2}\left(h_{CO_2}^0 + \overline{c}_{p.CO_2}(T_{out} - T_0)\right) \\
&- \dot{m}_{H_2O}\left(h_{H_2O}^0 + \overline{c}_{p.H_2O}(T_{out} - T_0)\right) \\
&- \dot{m}_{SO_2}\left(h_{SO_2}^0 + \overline{c}_{p.SO_2}(T_{out} - T_0)\right) \\
&- \dot{m}_{A.in}\left(h_A^0 + \overline{c}_{p.A}(T_{out} - T_0)\right) \\
&+ \dot{m}_{CO_2.r}\left(h_{CO_2}^0 + \overline{c}_{p.CO_2}(T_{in} - T_0)\right) \\
&+ \dot{m}_{H_2O.r}\left(h_{H_2O}^0 + \overline{c}_{p.H_2O}(T_{in} - T_0)\right) \\
&+ \dot{m}_{SO_2.r}\left(h_{SO_2}^0 + \overline{c}_{p.SO_2}(T_{in} - T_0)\right) \\
&- \dot{m}_{CO_2.r}\left(h_{CO_2}^0 + \overline{c}_{p.CO_2}(T_{out} - T_0)\right) \\
&- \dot{m}_{H_2O.r}\left(h_{H_2O}^0 + \overline{c}_{p.H_2O}(T_{out} - T_0)\right) \\
&- \dot{m}_{SO_2.r}\left(h_{SO_2}^0 + \overline{c}_{p.SO_2}(T_{out} - T_0)\right) = 0
\end{aligned}$$

We then collect like terms and present all mass flow rates in terms of the total fuel mass flow rate,

$$\begin{aligned}
&\dot{m}_{F.in}h_F^0 + \dot{m}_{W.in}h_W^0 - \dot{m}_{CO_2}h_{CO_2}^0 - h_{H_2O}^0\dot{m}_{H_2O} - \dot{m}_{SO_2}h_{SO_2}^0 + \dot{m}_{F.in}\overline{c}_{p.F}(T_{in} - T_0) \\
&+ \dot{m}_{W.in}\overline{c}_{p.W}(T_{in} - T_0) + \dot{m}_{A.in}\overline{c}_{p.A}(T_{in} - T_0)
\end{aligned}$$

$$
\begin{aligned}
&+\dot{m}_{O_2.in}\overline{c}_{p.O_2}(T_{in}-T_0)+\dot{m}_{CO_2.r}\overline{c}_{p.CO_2}(T_{in}-T_0)\\
&+\dot{m}_{H_2O.r}\overline{c}_{p.H_2O}(T_{in}-T_0)+\dot{m}_{SO_2.r}\overline{c}_{p.SO_2}(T_{in}-T_0)\\
&-\dot{m}_{CO_2}\overline{c}_{p.CO_2}(T_{out}-T_0)-\dot{m}_{H_2O}\overline{c}_{p.H_2O}(T_{out}-T_0)\\
&-\dot{m}_{SO_2}\overline{c}_{p.SO_2}(T_{out}-T_0)-\dot{m}_{A.in}\overline{c}_{p.A}(T_{out}-T_0)\\
&-\dot{m}_{CO_2.r}\overline{c}_{p.CO_2}(T_{out}-T_0)-\dot{m}_{H_2O.r}\overline{c}_{p.H_2O}(T_{out}-T_0)\\
&-\dot{m}_{SO_2.r}\overline{c}_{p.SO_2}(T_{out}-T_0)=0
\end{aligned}
$$

$$
\begin{aligned}
&a\,h_F^0+b\,h_W^0-f\,h_{CO_2}^0-g\dot{m}_{H_2O}-h\,h_{SO_2}^0+a\,\overline{c}_{p.F}(T_{in}-T_0)+b\,\overline{c}_{p.W}(T_{in}-T_0)\\
&\quad+c\,\overline{c}_{p.A}(T_{in}-T_0)+d\,\overline{c}_{p.O_2}(T_{in}-T_0)\\
&\quad+\frac{k}{1-k}f\,\overline{c}_{p.CO_2}(T_{in}-T_0)\\
&\quad+\frac{k}{1-k}g(1-\eta_W)\overline{c}_{p.H_2O}(T_{in}-T_0)\\
&\quad+\frac{k}{1-k}h(1-\eta_S)\overline{c}_{p.SO_2}(T_{in}-T_0)\\
&\quad-f\left(1+\frac{k}{1-k}\right)\overline{c}_{p.CO_2}(T_{out}-T_0)\\
&\quad-g\left(1+\frac{k}{1-k}(1-\eta_W)\right)\overline{c}_{p.H_2O}(T_{out}-T_0)\\
&\quad-h\left(1+\frac{k}{1-k}(1-\eta_S)\right)-c\overline{c}_{p.A}(T_{out}-T_0)=0
\end{aligned}
$$

$$
LHV_{ar}=A(T_{out}-T_0)-B(T_{in}-T_0)
$$

$$
T_{out}=T_0+\frac{LHV_{ar}}{A}+\frac{B(T_{in}-T_0)}{A}
$$

$$
\begin{aligned}
A=&\left(f\left(1+\frac{k}{1-k}\right)\overline{c}_{p.CO_2}+g\left(1+\frac{k}{1-k}(1-\eta_W)\right)\overline{c}_{p.H_2O}\right.\\
&\left.+h\left(1+\frac{k}{1-k}(1-\eta_S)\right)+c\,\overline{c}_{p.A}\right)
\end{aligned}
$$

$$
\begin{aligned}
B=&\left(a\overline{c}_{p.F}+b\overline{c}_{p.W}+c\overline{c}_{p.A}+d\overline{c}_{p.O_2}+\frac{k}{1-k}f\,\overline{c}_{p.CO_2}\right.\\
&\left.+\frac{k}{1-k}g(1-\eta_W)\overline{c}_{p.H_2O}+\frac{k}{1-k}h(1-\eta_S)\overline{c}_{p.SO_2}\right)
\end{aligned}
$$

$$
LHV_{ar}=a\,h_F^0+b\,h_W^0-f\,h_{CO_2}^0-g\,\dot{m}_{H_2O}-h\,h_{SO_2}^0
$$

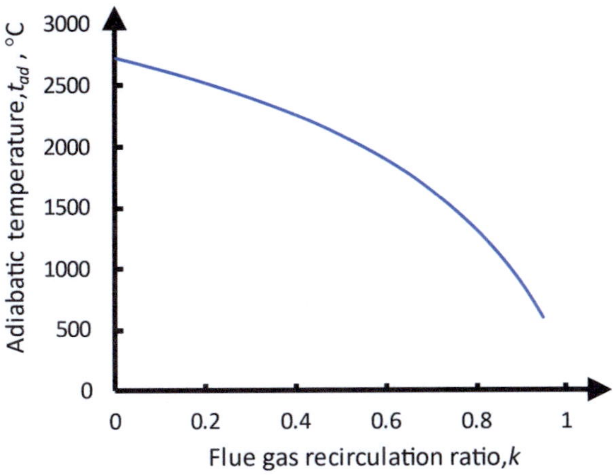

Fig. 6.29 Adiabatic combustion temperature in Oxyfuel combustion, $\eta_W = \eta_S = 1$

From these, we can solve the adiabatic combustion temperature of the Oxyfuel furnace as a function of the flue gas recirculation ratio, k (see Fig. 6.29). The values of all other parameters c, f, g and h, and the average heat capacities, are the same as in Example 6.5.

As Fig. 6.29 shows, the recirculation can be effectively used to control the adiabatic reaction temperature. To solve the mass flow rate of the fuel in Oxyfuel combustion that gives the same steam generation rate that was obtained in Example 6.5 with normal combustion systems, we need to use the energy balance of the evaporator that we already derived. The evaporator heat load needs to be set to the same value as the load for air-fired furnace, $\phi_{EVA} = 168900$ kW; we also require the gas temperature after the evaporator to be the same as with the air-fired furnace, $T_b = 1000$ °C, to avoid ash problems in the superheater section. We then get

$$\phi_{EVA} = 168900\,\text{kW} = \dot{m}_{fg} c_{p.fg}(T_a - T_b) = \dot{m}_{fg} c_{p.fg}(T_a - T_b)$$

where

$$\dot{m}_{fg} = \frac{\dot{m}_f}{1-k}(f + g(1 - k\eta_W) + h(1 - k\eta_S))$$

Since we have a single equation with two unknown variables, k and $\dot{m}_f$, we need an additional requirement to solve for a solution, similar to our calculations for the air-fired furnace. Figure 6.30 plots the fuel mass flow required as a function of the flue gas recycling ratio, for cases with full flue gas water condensation (solid lines) and with no water condensation (dashed lines). Black vertical lines show the point where the adiabatic combustion temperature drops to the furnace exit temperature, after which no steam generation is possible since $T_a - T_b = 0$.

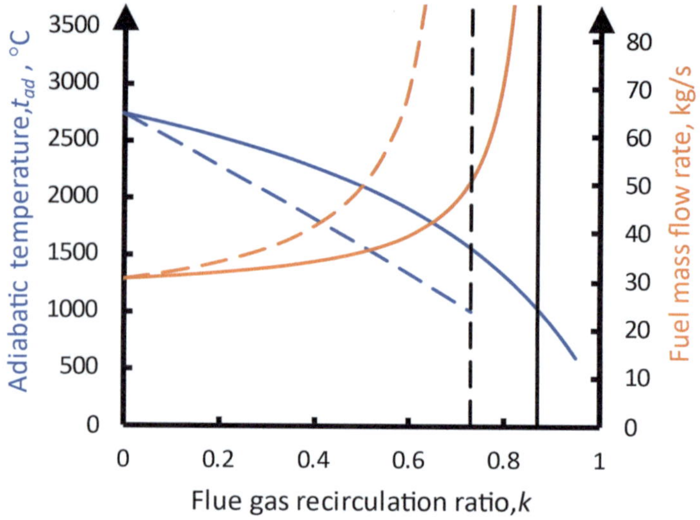

Fig. 6.30 Fuel mass flow rate and combustion temperature, as a function of flue gas recirculation ratio, to reach evaporator load of 168.9 MW. Oxygen and recycled flue gases are pre-heated to 225 °C, as in Example 6.5. Vertical black lines correspond the value of k when the adiabatic combustion temperature reaches the outlet temperature of the evaporator, 1000 °C. Dashed lines indicate no flue gas condensation, i.e., $\eta_W = 0$ and $\eta_S = 1$, and solid lines indicate full water vapor condensation, $\eta_W = \eta_S = 1$

As Fig. 6.30 shows, if the recycled flue gases contain water vapor, they cool the furnace more efficiently (dashed vs. solid blue lines) and the adiabatic flame temperature reaches the furnace exit temperature faster, at a lower flue gas recirculation ratio. Most importantly, the need for fuel is significantly smaller for Oxyfuel combustion (30–40 kg/s) than for air-fired combustion (~68.8 kg/s). At higher recirculation ratios, since the adiabatic temperature approaches the exit temperature, we need to increase the fuel input to compensate for the decrease in driving temperature difference in the evaporator.

The benefit of the Oxyfuel combustion is clear: the same amount of thermal energy can be produced by using (68.8–40)/68.8 = 42% less fuel. However, the production of pure oxygen consumes large amounts of electrical power. For example, a cryogenic air separation unit (ASU) consumes ~ 0.830 MJ/kg of oxygen [17]. If water alkaline electrolysis is used to generate the oxygen, it requires as much as 23.85 MJ/kg of oxygen. Clearly, oxygen production significantly reduces overall power plant efficiency. From these two technologies, only the ASU is able to operate with power generated solely from the combustion-driven steam power plant that it supplies with oxygen.

Oxyfuel combustion has several benefits over the other CCS methods, such as easier post-processing of flue gases due to their high CO_2 concentration. It does not require additional chemicals, such as the limestone required for Ca-looping CO_2 capture cycle. Since the mass flow rates are lower, much smaller systems are often sufficient. Finally, since significantly less nitrogen is used in the process, less thermal NOx is generated.

6.7 Towards Sustainable Use of Biomass: Mitigating Negative Impacts and Future Directions

In this chapter we have discussed the bioenergy production chain, from the principles of photosynthesis to the formation of different types of biomasses, their sustainable production limits, and different methods of biomass conversion to heat, power and energy carriers. We also briefly reviewed and analyzed different processes that could be used as bioenergy carbon capture and storage (BECCS) methods. The several calculation examples assist students to understand the application of the presented models.

Bioenergy will remain on the palette of energy resources needed to mitigate climate change in the future. As long as we keep eating, there will be bio-based side- and waste-streams generated, and these should be utilized as efficiently as possible. With biorefineries, we could significantly reduce the consumption of virgin fossil fuels, while also producing renewable raw materials for chemicals, energy and transportation fuels. Bioenergy will play roles in boosting local and national economies and job creation, and as a means to maintain energy independency. Bioenergy and biowaste also provide an opportunity to develop circular economy solutions, such as fertilizer production.

The reality is, however, that sustainable use of primary biomass sources is limited, and we will see more and more alternate, affordable renewable resources, such as solar and wind power, taking pressure away from the use of biomass; similarly, there will be true circular economy options available that decrease the combustion of waste. However, these other energy sources also come with trade-offs, such as variability; they also all have sustainability issues, such as the loss of biodiversity, land use and land-use change, and social acceptance concerns, which are discussed in Chaps. 4 and 5.

A sustainable bioeconomy is based on sustainability of the biomass, and several measures can be taken to mitigate the harmful impacts of the production and use of biomass. The following proposed solutions are based on Gasparatos et al. [32] and other references in this chapter:

- **Environmentally-friendly biomass production and utilization practices** which include avoiding monoculture plantations by promoting land-use practices that support understory vegetation, conserving riparian habitats and wetlands, and adopting wildlife-friendly management practices and pollution control mechanisms, including the use of crops that require less agrochemicals. Forestry should be further developed to improve solutions for balanced utilization of forest biomass, maintenance of biodiversity, carbon sequestration, carbon storage, and other ecosystem services. This includes, for example, promoting mixed-stand

forests, employing continuous growth and agro-forestry practices, increasing the amount of uncleared decayed and dead wood, and protection of old-growth and pristine forests.
- **Development of bioenergy production on already-degraded or marginal lands** helps to minimize negative impacts on nature, and can in some cases have ecological restoration effects. For example, some bioenergy crops, such as willow, can purify wastewater from sewage, and miscanthus and jatropha can be grown on contaminated land, and therefore have the potential for bioremediation and restoration of ecosystems. Note, though, that marginal lands with biodiversity values should be excluded from bioenergy production.
- **Designing of multi-functional bioenergy production areas** that combine several environmentally-friendly components supporting biodiversity better than single activities, including mixed-cropping (food, feed and bioenergy), crop rotation, wildlife corridors, and habitat patches left in their natural state in the energy production landscape. Such landscape approaches are beneficial not only for biodiversity; they can also result in higher biomass yields due to increased pollination and other ecosystem services, such as more stable microclimate and improved nutrient and water flow. However, such landscape approaches should incorporate a range of planning principles if they are to be successful.
- **Application of mitigation hierarchy steps** on single biomass production projects would help to minimize negative impacts on nature, and to move towards a mode of operation that is regenerative—the goal is not just to to achieve no overall negative impact on biodiversity (no net loss, NNL), but rather to achieve an overall positive outcome (net gain, NG). Net gain can be achieved by following the steps of the mitigation hierarchy, where negative impacts on nature are first avoided, then minimized and restored. Any residual damage that cannot be avoided is offset. See Sect. 2.3 on how to apply the mitigation hierarchy.

Though improved biomass cultivation practices require a transformation of the current business-as-usual approach, there are a number of effective practices that can be put in place immediately. Priority should be given to waste- and side-stream-based raw materials when developing sustainable bioenergy production. To facilitate the transformation, we should employ systematic land-use planning, and efficient development and implementation of policies and regulations, including improved monitoring and market-based mechanisms such as certification schemes.

There are also examples of technical solutions: innovative, smart, space-efficient farming and cropping systems could offer new opportunities to save valuable arable land, and can provide new pathways for decentralized production of bioenergy from waste- and side-streams. These innovative systems include: urban and peri-urban farming; vertical farming; soil-free farming techniques such as hydroponics, aquaponics and aeroponics; circular systems for water and nutrient management; and LED lightning technologies. While saving agricultural land, these systems also allow development and testing of new varieties and plant species, all-year cultivation, optimization of growing conditions, and lower costs for fertilizers, pest and disease management, and transportation. These indoor and potentially underground systems are relatively resistant to extreme weather conditions and changing climate. Development of intelligent sensors and controlling systems for environmental parameters, imaging systems (photonics) for plant performance, LED lightning technologies for selection of appropriate and optimum wavelengths for photosynthesis and productivity, and water purification technologies for the circular systems, all play a key role in this pathway. Naturally, these technological developments are also important and applicable in outdoor conditions, for the development of sustainable agricultural management and bioenergy production in conventional environments. Urban and peri-urban farming systems also include variable systems such as roof-top gardens, green walls, street landscaping, and many kinds of special gardens, which all have positive impacts on human health, wellbeing, aesthetic value, and local air temperatures.

Marine aquacultures offer new possibilities for the food industry, and alternatives to land-based agriculture, reducing the need for arable land. In particular, the cultivation of seaweeds (such as kelps and sea lettuce) is growing rapidly on a global scale. Seaweeds are rich in many valuable compounds, and can be used in food, feed, biomolecule industries, and energy production. They are also a potential source for still-unexplored biomolecules, which may be valuable for bio-based product development.

Seaweed cultivation also has great potential to promote the ongoing dietary protein shift away from red meat towards a more vegetarian diet. The environmental impacts of marine aquacultures can be smaller than conventional agriculture, due to their lower need for fertilizers, pesticides, and freshwater irrigation. Large seaweed beds are also very productive compared to terrestrial plants, forming effective carbon sinks. As a result, seaweed farming may help to mitigate ocean acidification and global warming. Further, seaweed farms may form attractive habitats for fish populations and other marine organisms, promoting biodiversity if harvested sustainably. The technical development of the necessary marine aquaculture infrastructure, and the creation of sustainability standards, are crucial for successful large-scale cultivation.

Development of innovative production systems based on microbial processes (yeast, cyanobacteria, microalgae) and genetic, molecular, and enzyme engineering also provide opportunities for a more sustainable bioeconomy. The most widely-used microbial food production systems are based on the function of yeast *(Saccharomyces cerevisiae*). Yeast has been successfully used for tailored biomolecule production (such as food oils and fatty acids), production of biofuels, and molecular engineering, even at industrial scales. Microbial production systems provide shorter, climate-independent operation cycles compared to plant-based systems. For example, local production of vegetable oils and proteins in European countries reduces the need for unsustainable cultivation of tropical plantations, associated with deforestation and biodiversity loss, as well as long-range transport. In particular, soyabean, palm and coconut oil production have major negative impacts on local and global environments.

Finally, molecular and genetic breeding techniques have opened new possibilities to use engineered short-rotation species for specific purposes, such as raw materials for growth media and bioreactors. For example, willow lines with altered wood composition (e.g., lignocellulose) have been generated. Modification of the wood's lignin content has been a key target in molecular breeding, resulting in engineered trees suited to food production chains—that is, fermentative conversion to food ingredients by microorganisms—or raw material for energy production.

These new and innovative solutions for producing biomass for bioenergy and other uses, including food and animal feed, are needed, but it must be noted that little is known of the efficiency of some of these technologies to reduce environmental pressures. Much depends on the scale and characteristics of the implementation. Development and use of new technologies should be coupled with careful monitoring of their impacts, and the implementation process adjusted if needed. Regardless, considering the scale of biomass and energy currently utilized by humans, the most urgent solution is to reduce the overall consumption of both biomass and energy in societies and social groups that have a disproportionate rate of overconsumption [33].

To summarize, we need bioenergy as an important part of the green transition, now and in the future, but the utilization of bioenergy has strict sustainability limitations that must be respected to bring human life back within planetary boundaries. As the efficiency of photosynthesis is very low, producing meaningful amounts of biomass requires large areas of land. We need to find enhanced circular economy solutions to reduce the use of virgin biomass, and to develop more area-efficient and sustainable ways to grow our food, for example by using vertical farming technologies. We need to secure biodiversity that is unique, and forms the foundation for peoples' livelihoods and sustainable development in all areas of activity, including energy, forestry or agriculture. We need to limit land use to secure sufficient food production for the continuously growing human population, and to protect freshwater bodies from eutrophication due to leakage of nutrients and soil carbon. We should heed the words of Carl Folke: *"Nature is not a sector in our society—it is a prerequisite for it."*

Author contributions 6.1 Mika Järvinen and Hanna Paulomäki, 6.2 Mika Järvinen and Han van Kasteren, 6.3 Elina Oksanen and Mika Järvinen, 6.4 Iryna Herzon, Mika Järvinen, Elina Oksanen and Hanna Paulomäki, 6.5 Mika Järvinen, 6.6 Mika Järvinen and Antti Arasto, 6.7 Hanna Paulomäki, Mika Järvinen, Antti Arasto, Iryna Herzon, Han van Kasteren and Elina Oksanen. Info box 6.1 Elina Oksanen, Case study 6.1 Martin Forsius, Anu Akujärvi, Maria Holmberg, Virpi Junttila, Minna Pekkonen, Torsti Schulz and Heini Kujala. Subsections can be referenced accordingly.

References

1. Adams TN, Frederick J, Grace, TM, Hupa M, Iisa K, Jones AK, Tran H (1997) Kraft recovery boilers. Tappi Press, ISBN 0-9625985-9-3, USA
2. Ågren GI, Andersson FO (2013) Terrestrial ecosystem ecology, principles and applications. Cambridge University Press. ISBN 978-1-107-64825-8
3. Antikainen R, Tenhunen J, Ilomäki M, Mickwitz P, Punttila P, Puustinen M, Seppälä J, Kauppi L (2007) Bioenergian uudet haasteet Suomessa ja niiden ympäristönäkökohdat. Nykytilakatsaus. In Finnish. 101 pp. Available at: https://helda.helsinki.fi/server/api/core/bitstreams/bc4d7435-9159-46aa-925f-dbca91b3e446/content

4. Bellingham PJ, Arnst EA, Clarkson BD, Forester LJ, Shaw WB, Sprague R, Wiser SK, Peltzer DA (2023) The right tree in the right place? A major economic tree species poses major ecological threats. Biol Invasions 25:39–60. https://doi.org/10.1007/s10530-022-02892-6
5. Berndes G, Hoogwijk M, Broek R (2003) The contribution of biomass in the future global energy supply: a review of 17 studies. Biomass Bioenerg 25:1–28. https://doi.org/10.1016/S0961-9534(02)00185-X
6. Blankenship et al (2011) Comparing photosynthetic and photovoltaic efficiencies and recognizing the potential for improvement. Science 332:805–809. https://doi.org/10.1126/science.1200165
7. Burland A, von Cossel M (2023) Towards managing biodiversity of European marginal agricultural land for biodiversity-friendly biomass production. Agronomy 13:1651. https://doi.org/10.3390/agronomy13061651
8. Camia A, Giuntoli J, Jonsson R. et al (2021) The use of woody biomass for energy production in the EU. European Commission, Joint Research Centre, Publications Office. https://data.europa.eu/doi/10.2760/831621
9. Chapman S et al (2003) Exploitation of northern peatlands and biodiversity maintenance: a conflict between economy and ecology. Front Ecol Environ 1(10):525–532
10. Christensen TR, Jackowicz-Korczyński M, Aurela M, Crill P, Heliasz M, Mastepanov M, Friborg T (2012) Monitoring the multi-year carbon balance of a subarctic palsa mire with micrometeorological techniques. Ambio 41(Suppl 3):207–217. https://doi.org/10.1007/s13280-012-0302-5
11. Dauber J, Jones MB, Stout JC (2010) The impact of biomass crop cultivation on temperate biodiversity. GCB Bioenergy 2(6):289–309. https://doi.org/10.1111/j.1757-1707.2010.01058.x
12. de Jong W, van Ommen JR (2014) Biomass as a sustainable energy source for the future. John Wiley Sons Inc. ISBN: 9781118304914
13. de Jong W and van Ommen JR (2015) Biomass as a sustainable energy source for the future: fundamentals of conversion processes. John Wiley Sons Inc. ISBN 978-1-118-30491-4
14. Dee LE, Ferraro PJ, Severen CN et al (2023) Clarifying the effect of biodiversity on productivity in natural ecosystems with longitudinal data and methods for causal inference. Nat Commun 14:2607. https://doi.org/10.1038/s41467-023-37194-5
15. Deublein D, Steinhauser A (eds) (2008) Biogas from waste and renewable resources: An Introduction. Wiley-VCH Verlag GmbH & Co. KGaA. https://doi.org/10.1002/9783527621705
16. Diaz S et al (2019) Pervasive human-driven decline of life on Earth points to the need for transformative change. Science 366:eaax3100. https://doi.org/10.1126/science.aax3100
17. Ding G, He B, Cao Y, Wang C, Su L, Duan Z, Song J, Tong W, Li X (2018) Process simulation and optimization of municipal solid waste fired power plant with oxygen/carbon dioxide combustion for near zero carbon dioxide emission. Energy Convers Manag 157:157–168. https://doi.org/10.1016/j.enconman.2017.11.087
18. Dotta G, Verdade LM (2011) Medium to large-sized mammals in agricultural landscapes of south-eastern Brazil. Mammalia 75:345–352. https://doi.org/10.1515/MAMM.2011.049
19. EEAC Network (2022) Towards a sustainable food system—a position paper on the framework law. https://eeac.eu/wp-content/uploads/2022/10/Towards-a-sustainable-food-system-_-An-EEAC-Network-Position-Paper-PV.pdf
20. Erb KH, Lauk C, Kastner T, Mayer A, Their MC, Haber H (2016) Exploring the biophysical option space for feeding the world without deforestation. Nat Commun 7:11382. https://doi.org/10.1038/ncomms11382
21. Evans JR (2013) Improving photosynthesis. Plant Physiol 162:1780–1793. https://doi.org/10.1104/pp.113.219006
22. FAO (2020) Global forest resource assessment. Food and Agriculture Organization (FAO), Rome. https://fra-data.fao.org/
23. FAO (2023) FAOSTAT, Forestry production and trade. https://www.fao.org
24. Fargione J, Hill J, Tilman D, Polasky S, Hawthorne P (2008) Land clearing and the biofuel carbon debt. Science 319:1235–1238. https://doi.org/10.1126/science.1152747

25. Field CB, Behrenfeld MJ, Randerson JT, Falkowski P (1998) Primary production of the biosphere: integrating terrestrial and oceanic components. Science 281:237–240
26. Forest Europe (2020) State of Europe's Forests 2020. https://foresteurope.org/wp-content/uploads/2016/08/SoEF_2020.pdf
27. Forsius M, Holmberg M, Junttila V, Kujala H, Schulz T, Paunu VV, Savolahti M, Minunno F, Akujärvi A, Bäck J, Grönroos J, Heikkinen RK, Karvosenoja N, Mäkelä A, Mikkonen N, Pekkonen M, Rankinen K, Virkkala R (2023) Modelling the regional potential for reaching carbon neutrality in Finland: sustainable forestry, energy use and biodiversity protection. Ambio 52:1757–1776. https://doi.org/10.1007/s13280-023-01860-1
28. Framstad E, Berglund H, Gundersen V, Heikkilä R, Lankinen N, Peltola T, Risbøl O, Weih M (2009) Increased biomass harvesting for bioenergy: effects on biodiversity, landscape amenities and cultural heritage values. Nordic Council of Ministers p 161
29. French KE (2018) Assessing the bioenergy potential of grassland biomass from conservation areas in England. Land Use Policy. https://doi.org/10.1016/j.landusepol.2018.12.001
30. Fry DA, Slater F (2008) The effect on plant communities and associated taxa of planting short rotation willow coppice in Wales. Asp Appl Biol 90:287–293
31. Gałęzia T (2016) Economic methods for the utilisation of logging residues. For Res Pap 77(1):50–55. https://doi.org/10.1515/frp-2016-0006
32. Gasparatos A, Doll CNH, Esteban M, Ahmed A, Olang TA (2017) Renewable energy and biodiversity: implications for transitioning to a green economy. Renew Sustain Energy Rev 70:161–184. https://doi.org/10.1016/j.rser.2016.08.030
33. Hickel J, O'Neill DW, Fanning AL, Zoomkawala H (2022) National responsibility for ecological breakdown: a fair-shares assessment of resource use, 1970–2017. Lancet Planet Health 6:e342–49
34. Holmberg M, Junttila ST, Grönroos J, Paunu VV, Savolahti M, Minunno F, Ojanen P, Akujärvi A, Karvosenoja N, Kortelainen P, Mäkelä A, Peltoniemi M, Petäjä J, Vanhala P, Forsius M (2023) Role of land use in Finland's greenhouse gas emissions. Ambio 52:1697–1715. https://doi.org/10.1007/s13280-023-01910-8
35. Hurmekoski E, Smyth CE, Stern T, Verkerk PJ, Asada R (2021) Substitution impacts of wood use at the market level: a systematic review. Environ Res Lett 16:123004. https://doi.org/10.1088/1748-9326/ac386f
36. IEA (2023) World energy outlook. International Energy Association
37. IPBES (2019) Global assessment report on biodiversity and ecosystem services of the intergovernmental science-policy platform on biodiversity and ecosystem services. In: Brondizio ES, Settele J, Díaz S, Ngo HT (eds) IPBES secretariat, Bonn, Germany, 1148 pp. https://doi.org/10.5281/zenodo.3831673
38. IPCC (2006) 2006 IPCC guidelines for national greenhouse gas inventories: glossary. www.ipcc-nggip.iges.or.jp/public/2006gl/pdf/0_Overview/V0_2_Glossary.pdf
39. IPCC (2018) Summary for Policymakers. In: Masson-Delmotte V, Zhai P, Pörtner H-O, Roberts D, Skea J, Shukla PR, Pirani A, Moufouma-Okia W, Péan C, Pidcock R, Connors S, Matthews JBR, Chen Y, Zhou X, Gomis MI, Lonnoy E, Maycock T, Tignor M, Waterfield T (eds) Global warming of 1.5°C. An IPCC special report on the impacts of global warming of 1.5°C above pre-industrial levels and related global greenhouse gas emission pathways, in the context of strengthening the global response to the threat of climate change, sustainable development, and efforts to eradicate poverty. Cambridge University Press, Cambridge, UK and New York, NY, USA, pp 3–24. https://doi.org/10.1017/9781009157940.001
40. IPCC (2023) Climate change 2023: synthesis report. In: Lee H, Romero J (eds) Contribution of working groups I, II and III to the sixth assessment report of the intergovernmental panel on climate change, editors, IPCC, Geneva, Switzerland, 184 pp. https://doi.org/10.59327/IPCC/AR6-9789291691647
41. Immerzeel DJ, Verweij RA, Van Der Hilst F, Faaij APC (2014) Biodiversity impacts of bioenergy crop production: a state-of-the-art review. GCB Bioenergy 6:183–209. https://doi.org/10.1111/gcbb.12067

42. Jaureguiberry P, Titeux N, Wiemers M, Bowler DE, Coscieme L, Golden AS, Guerra CA, Jacob U, Takahashi Y, Settele J, Díaz S, Molnár Z, Purvis A (2022) The direct drivers of recent global anthropogenic biodiversity loss. Sci Adv 8. https://doi.org/10.1126/sciadv.abm9982
43. Junttila V, Minunno F, Peltoniemi M, Forsius M, Akujärvi A, Ojanen P, Mäkelä A (2023) Quantification of forest carbon flux and stock uncertainties under climate change and their use in regionally explicit decision making: case study in Finland. Ambio 52:1716–1733. https://doi.org/10.1007/s13280-023-01906-4
44. Kang S, Post WM, Nichols JA, Wang D, West TO, Bandaru V, Izaurralde RC (2013) Marginal lands: concept, assessment and management. J Agric Sci 5(5):129–139. https://doi.org/10.5539/jas.v5n5p129
45. Koppelmäki K, Parviainen T, Virkkunen E, Winquist E, Schulte RPO, Helenius J (2019) Ecological intensification by integrating biogas production into nutrient cycling: modeling the case of agroecological symbiosis. Agric Syst 170:39–48. https://doi.org/10.1016/j.agsy.2018.12.007
46. Kujala H, Minunno F, Junttila V, Mikkonen N, Mäkelä A, Virkkala R, Akujärvi A, Leikola, Heikkinen RK (2023) Role of data uncertainty when identifying important areas for biodiversity and carbon in boreal forests. Ambio 52:1804–1818. https://doi.org/10.1007/s13280-023-01908-2
47. Lark TJ, Hendricks NP, Smith A, Pates N, Spawn-Lee SA, Bougie M, Booth EG, Kucharik CJ, Gibbs HK (2022) Environmental outcomes of the US renewable fuel standard. PNAS 119(9):e2101084119. https://doi.org/10.1073/pnas.2101084119
48. Liang J et al (2016) Positive biodiversity-productivity relationship predominant in global forests. Science 354:aaf8957. https://doi.org/10.1126/science.aaf8957
49. Luke (2022) Sustainability report 2022, case study 13.2 tackling climate change: agriculture and forestry support schemes must be developed boldly to meet climate targets. https://www.luke.fi/en/about-luke/responsibility/sustainability-report-2022/sdg-13-take-urgent-action-to-combat-climate-change-and-its-impacts/132-tackling-climate-change-agriculture-and-forestry-support-schemes-must-be-developed-boldly-to-meet-climate-targets
50. Majava A, Vadén T, Toivanen T, Järvensivu P, Lähde V, Eronen JT (2022) Sectoral low-carbon roadmaps and the role of forest biomass in Finland's carbon neutrality 2035 target. Energy Strategy Rev 41:100836. https://doi.org/10.1016/j.esr.2022.100836
51. Marron N (2015) Agronomic and environmental effects of land application of residues in short-rotation tree plantations: a literature review. Biomass Bioenergy 81:78–400. https://doi.org/10.1016/j.biombioe.2015.07.025
52. Meehan TD, Gratton T, Diehl E, Hunt ND, Mooney DF, Ventura SJ, Barham BL, Jackson RD (2013) Ecosystem-service tradeoffs associated with switching from annual to perennial energy crops in riparian zones of the US midwest. PLoS ONE 8(11):e80093. https://doi.org/10.1371/journal.pone.0080093
53. Meijaard E, Garcia-Ulloa J, Sheil D, Wich SA, Carlson KM, Juffe-Bignoli D (2012) Mire with micrometeorological techniques. Ambio 41(Suppl 3):207–217. https://doi.org/10.1007/s13280-012-0302-5
54. Mekonen EA, Mekonnen YT, Fatoba SO (2023) Thermodynamic prediction of biogas production and combustion: the spontaneity and energy conversion efficiency from photosynthesis to combustion. Scientific African 21:e01776. https://doi.org/10.1016/j.sciaf.2023.e01776
55. Merfort L, Bauer N, Humpenöder F, Klein D, Strefler J, Popp A, Luderer G, Kriegler E (2023) Bioenergy-induced land-use-change emissions with sectorally fragmented policies. Nat Clim Change 13:685–692. https://doi.org/10.1038/s41558-023-01697-2
56. Mottet A, Haan C, Falcucci A, Tempio G, Opio C, Gerber P (2017) Livestock: on our plates or eating at our table? A new analysis of the feed/food debate. Glob Food Secur 14:1–8. https://doi.org/10.1016/j.gfs.2017.01.001
57. Myhre GD et al. (2013) Anthropogenic and natural radiative forcing. In: Climate change 2013: the physical science basis. contribution of working group I to the Fifth assessment report of the intergovernmental panel on climate change, Cambridge University Press, Cambridge, United Kingdom and New York, NY, USA.
58. NASA (2008) https://www.jpl.nasa.gov/images/pia11420-mato-grosso-brazil

59. Page S, Mishra S, Agus F, Anshari G, Dargie G, Evers S, Jauhiainen J, Jaya A, Jovani-Sancho AJ, Laurén A, Sjögersten S, Suspense IA, Wijedasa LS, Evans CD (2022) Anthropogenic impacts on lowland tropical peatland biogeochemistry. Nat Rev Earth Environ 3:426–443. https://doi.org/10.1038/s43017-022-00289-6
60. Pendrill F, Persson UM, Godar J, Kastner T, Moran D, Schmidt S, Wood R (2019) Agricultural and forestry trade drives large share of tropical deforestation emissions. Glob Environ Change 56:1–10. https://doi.org/10.1016/j.gloenvcha.2019.03.002
61. Pillay R, Venter M, Aragon-Osejo J, González-del-Pliego P, Hansen A, Watson E, Venter O (2021) Tropical forests are home to over half of the world's vertebrate species. Front Ecol Environ 20(1):10–15
62. Poore J, Nemecek T (2018) Reducing food's environmental impacts through producers and consumers. Science 360:987–992. https://doi.org/10.1126/science.aaq0216
63. Popp D, Von Gillhaussen P, Weidlich EWA, Sträuber H, Harms H, Temperton VM (2017) Methane yield of biomass from extensive grassland is affected by compositional changes induced by order of arrival. GCB Bioenergy 9:1555–1562. https://doi.org/10.1111/gcbb.12441
64. Popp J, Kovács S, Oláh J, Divéki Z, Balázs E (2021) Bioeconomy: biomass and biomass-based energy supply and demand. New Biotechnol 60:76–84. https://doi.org/10.1016/j.nbt.2020.10.004
65. Pörtner HO, Scholes RJ, Agard J, Archer E, Arneth A, Bai X, Barnes D, Burrows M, Chan L, Cheung WL, Diamond S, Donatti C, Duarte C, Eisenhauer N, Foden W, Gasalla MA, Handa C, Hickler T, Hoegh-Guldberg O, Ichii K, Jacob U, Insarov G, Kiessling W, Leadley P, Leemans R, Levin L, Lim M, Maharaj S, Managi S, Marquet PA, McElwee P, Midgley G, Oberdorff T, Obura D, Osman E, Pandit R, Pascual U, Pires APF, Popp A, Reyes-García V, Sankaran M, Settele J, Shin YJ, Sintayehu DW, Smith P, Steiner N, Strassburg B, Sukumar R, Trisos C, Val AL, Wu J, Aldrian E, Parmesan C, Pichs-Madruga R, Roberts DC, Rogers AD, Díaz S, Fischer M, Hashimoto S, Lavorel S, Wu N, Ngo HT (2021) IPBES-IPCC co-sponsored workshop report on biodiversity and climate change. IPBES and IPCC. https://doi.org/10.5281/zenodo.47825382538
66. Ramdani R, Lounela AK (2020) Palm oil expansion in tropical peatland: distrust between advocacy and service environmental NGOs. For Policy Econ 118:102242. https://doi.org/10.1016/j.forpol.2020.102242
67. Ranius T, Hämäläinen A, Egnell G, Olsson B, Eklöf K, Stendahl J, Rudolphi J, Sténs A, Felton A (2018) The effects of logging residue extraction for energy on ecosystem services and biodiversity: a synthesis. J Environ Manage 209:409–425. https://doi.org/10.1016/j.jenvman.2017.12.048
68. Renou-Wilson F, Moser G, Fallon D, Farrell CA, Müller C, Wilson D (2019) Rewetting degraded peatlands for climate and biodiversity benefits: results from two raised bogs. Ecol Eng 127:547–560. https://doi.org/10.1016/j.ecoleng.2018.02.014
69. Ritchie H (2021) The world has lost one-third of its forest, but an end of deforestation is possible. Published online at OurWorldInData.org. Retrieved from: 'https://ourworldindata.org/world-lost-one-third-forests'
70. Ritchie H (2021) Cutting down forests: what are the drivers of deforestation? Published online at OurWorldInData.org. https://ourworldindata.org/what-are-drivers-deforestation
71. Robertson B, Doran P, Loomis L, Robertson J, Schemske D (2011) Perennial feedstocks enhance avian diversity. GCB Bioenergy 3:235–246. https://doi.org/10.1111/j.1757-1707.2010.01080.x
72. Scarlat N, Martinov M, Dallemand J (2010) Assessment of the availability of agricultural crop residues in the European Union: potential and limitations for bioenergy use. Waste Manage 30(10):1889–1897. https://doi.org/10.1016/j.wasman.2010.04.016
73. Schilstra AJ (2001) How sustainable is the use of peat for commercial energy production? Ecol Econ 39:285–293
74. Schlesinger WH (2018) Are wood pellets a green fuel? Science 359:1328–1329. https://doi.org/10.1126/science.aat2305

75. Silva KCL, Tabarelli M, Vieira ICG (2023) Oil palm plantations in an aging agricultural landscape in the eastern Amazon: pushing Amazon forests farther from biodiversity-friendly landscapes. Biol Conserv 283. https://doi.org/10.1016/j.biocon.2023.110095
76. Smil V (2017) Energy and civilization—a history. The MIT Press, Cambridge Massachusetts, London, England. ISBN 9780262035774
77. Smith MR, Mueller ND, Springmann M, Sulser TB, Garibaldi LA, Gerber J, Wiebe K, Myers SS (2022) Pollinator deficits, food consumption, and consequences for human health: a modeling study. Environ Health Perspect 130(12) 127003:1–12. https://doi.org/10.1289/EHP10947
78. Statista (2024) Plastic waste worldwide, statistics & facts. https://www.statista.com/topics/5401/global-plastic-waste/#topicOverview
79. Steyn M, Oglesby J, Turan G, Zapantis A, Gebremedhin R, Zapantis A, Al Amer N, Havercroft I, Ivory-Moore R, Yang X, Abu Zahra M, Pinto E, Rassool D, Williams E, Consoli C, Minervini J (2023) Global status of CCS 2022. Global CCS Institute
80. Tan KT, Lee KT, Mohamed AR, Bhatia S (2009) Palm oil: addressing issues and towards sustainable development. Renew Sustain Energy Rev 13(2009):420–427
81. Walmsley JD, Godbold DL (2010) Stump harvesting for bioenergy—a review of the environmental impacts. Forest Int J Forest Res 83(1):17–38. https://doi.org/10.1093/forestry/cpp028
82. Wehrfritz A, Kaario O, Vuorinen V, Somers B (2016) Large eddy simulation of n-dodecane spray flames using flamelet generated manifolds. Combust Flame 167:113–131. https://doi.org/10.1016/j.combustflame.2016.02.019
83. Wellinger A, Murphy J and Baxter D (2013) The biogas handbook—science, production and applications. Woodhead Publishing Limited. ISBN 978-0-85709-741-5
84. Werling BP, Dickson TL, Isaacs R, Gaines H, Gratton C, Gross KL, Liere H, Malmstrom CM, Meehan TD, Ruan L, Robertson BA, Robertson GP, Schmidt TM, Schrotenboer AC, Teal TK, Wilson JK, Landis DA (2014) Perennial grasslands enhance biodiversity and multiple ecosystem services in bioenergy landscapes. PNAS 111(4):1652–1657. https://doi.org/10.1073/pnas.1309492111
85. Winberg J, Smith HG, Ekroos J (2023) Bioenergy crops, biodiversity and ecosystem services in temperate agricultural landscapes—a review of synergies and trade-offs. GCB Bioenergy 15:1204–1220. https://doi.org/10.1111/gcbb.13092
86. Wolfson R (2018) Energy, environment and climate. W.W. Norton & Company, New York. ISBN 9780393622911
87. World Bioenergy Association (2023) Global bioenergy statistics report
88. Xu J, Morris PJ, Liu J, Holden J (2018) PEATMAP: refining estimates of global peatland distribution based on a meta-analysis. CATENA 160:134–140. https://doi.org/10.1016/j.catena.2017.09.010
89. ZEP (2012) Biomass with CO_2 Capture and Storage (Bio-CCS)—the way forward for Europe. European technology platform for zero emission fossil fuel power plants. http://www.biofuelstp.eu/downloads/bioccsjtf/EBTP-ZEP-Report-Bio-CCS-The-Way-Forward.pdf

Chapter 7
Energy Storage Systems

Mika Järvinen, Konsta Turunen, Ari Seppälä, Janne Hirvonen, Neha Garg, and Annukka Santasalo-Aarnio

No free energy device will ever be allowed to reach the market.
—Nikola Tesla

7.1 Introduction

Energy storage is everywhere. Mobile electronics like phones and laptops use Li-ion batteries, which can be charged and discharged even up to several times a day. All electronic devices contain numerous capacitors that take and release energy continuously. In mechanical clocks and watches, the spring needs to be wound, charging it with potential energy which is then released to rotate the clock components. Cars have an endless number of different types of energy storage—the need for storage is clear for electrical and hybrid vehicles, but combustion engine cars also have several

M. Järvinen (✉) · K. Turunen · A. Seppälä · N. Garg · A. Santasalo-Aarnio
Aalto University, Espoo, Finland
e-mail: mika.jarvinen@aalto.fi

J. Hirvonen
University of Tampere, Tampere, Finland

M. Järvinen and H. Paulomäki (eds.), *Designing Renewable Energy Systems within Planetary Boundaries*, Green Energy and Technology,
https://doi.org/10.1007/978-3-031-69856-9_7

kinds of energy storage. For example, the fuel tank is a very simple kind of chemical energy storage filled with liquid or gaseous fuel, and the engine uses acid-lead batteries and capacitors that store electrical energy. In addition, all rotating masses in the car store mechanical energy, especially the flywheel that keeps the engine running smoothly. Driving uphill charges the car with gravitational potential energy that is released when coming back down. Even people need to charge their energy storage by eating food and converting its chemical energy to mechanical work and heat, or storing it for later use. In our bodies, excess energy is stored as fats that can be consumed later as an emergency, long-term energy storage.

The operation of our societies is also based on large energy systems supported by large energy storage. For example, hydropower uses stored potential energy by accumulating water in lakes at high altitudes, from rain or by damming rivers. Combustion-based power plants use solar energy stored in chemical form in biofuels and fossil fuels, which is released in combustion as heat, and then converted to mechanical and finally electrical energy. The hot water storage tank used in many houses is an example of a thermal storage system. These are charged by burning gas or oil, or using electrical resistive heating, and discharged when the water is consumed. Power plants that burn fossil fuels, biomass or waste can run continuously since their energy stores are already charged with fuels (e.g., huge piles of coal or wood chips), or because fuel is continuously transported to the power plant. This logistical transport step requires significant amounts of energy, and there are further losses in the delivery chain. The benefit of these systems is that we have precise control of the primary energy delivery, and weather conditions do not affect the production.

For renewable energy systems such as wind and solar power, on the other hand, the production does depend on weather conditions. To meet demand and secure stable energy flow, we need energy storage systems that are charged when excess renewable energy is available, and discharged during periods of no wind or sunshine. It is also possible to use demand-side response/management and economic control, which means that consumption is shifted to times of energy surplus, and energy is saved when production is low. In addition, industrial systems typically require a constant feed of raw materials, such as hydrogen or its derivatives, and large-scale energy storage can be used to connect variable renewables with different plants and industries.

Energy cannot be created, but it can be converted from one form to another. Conversion processes almost always have losses in the form of heat—to increase the efficiency of energy systems, heat losses should be utilized in other processes. For example, we could use low temperature heat for district heating, by upgrading it with heat pumps; or we could use it to make power with an organic Rankine cycle. Different energy forms and conversion pathways are illustrated in Fig. 7.1. Bioenergy for power and solar energy stored in chemical form are presented as green and purple arrows, respectively.

Chemical energy can be easily converted to heat by combustion. However, due to thermodynamic and kinetic limitations, these reactions are often incomplete and generally irreversible. Electrical energy can be completely converted to heat by resistive heating. The bioenergy route to electricity begins with solar irradiance (Fig. 7.1,

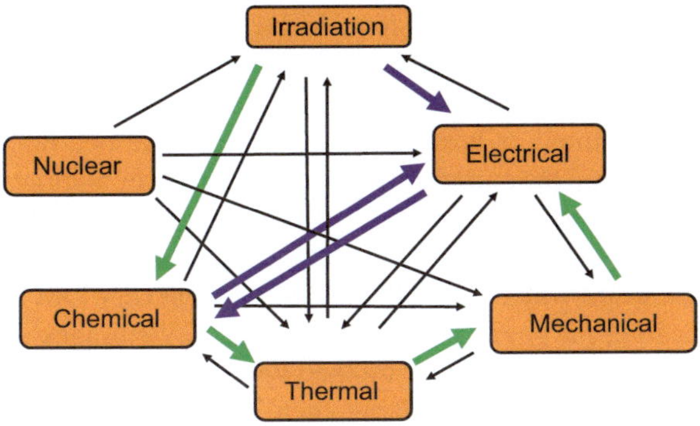

Fig. 7.1 Examples of energy forms and some conversion pathways. As examples, green arrows indicate bioenergy routes and purple arrows represent solar PV cell battery conversions

green arrows). At first, solar irradiance is converted and stored by photosynthesis as chemical energy in sugars and more complex products (efficiency of converting solar irradiance to stored chemical energy, $\eta \sim 0.001-0.02$). Chemical energy stored by plants and algae can be converted to thermal energy through combustion ($\eta \sim 0.9$). This can be converted in a boiler to internal energy of the steam ($\eta \sim 1$) that is then converted to mechanical energy in a steam turbine ($\eta \sim 0.4$), and finally, electrical energy in a generator ($\eta \sim 0.98$). Each step has losses, and typically, the greater the number of conversion steps, the lower the total efficiency becomes. The PV-panel-to-battery route to electricity also begins with irradiance (Fig. 7.1, purple arrows). Irradiance is first converted to electrical energy in the PV panel ($\eta \sim 0.20$), which is then stored in a battery ($\eta_{charge} \sim 0.95$) to be used when needed ($\eta_{discharge} \sim 0.95$).

If we compare the PV with bioenergy, bioenergy is itself an energy storage, while with PV, we need a separate storage—for example, a battery. Combustion of biomass creates CO_2 emissions, while the operation of the of the PV panel emits no CO_2.—however, their operation can result in other kinds of environmental impacts (see Chap. 4 on solar power and Chap. 6 on sustainability of biomass), and manufacturing of solar panels can be CO_2 intensive if fossil energy is used in production and mining of the materials. As the total electrical energy produced from solar irradiance depends on the product of all conversion step efficiencies, the total efficiency of bioenergy is far smaller than the PV panel route.

The form of energy required at the final stage determines the kind of energy storage and conversion path needed, to minimize losses and environmental impacts, for the lowest combined cost. Balancing the cost is difficult at times, as there are no generally-agreed principles, for example, for monetizing the value of nature, which in turn directs economic screening towards techno-economic optimum and externalization of the impacts on nature [20]. Another important factor is the availability of storage capacity. For example, in 2022 the largest single battery energy storage, in Moss Landing, California, had a storage capacity of 1600 MWh and power rating of 400 MW, and used lithium-ion battery technology. In our 500 MW load system used throughout this book, this battery system could be effectively used to balance hourly variation for 3–4 h, but it would be far too modest to handle monthly or annual variation. To cover annual storage requirements, we would need hundreds of these battery-based systems in parallel, which with current technologies would not be economically feasible or sustainable given their enormous material requirements.

When you compare 1600 MWh stored as electrical energy in a battery, and 1600 MWh stored as heat in 100 ºC water, there is a huge difference in their ability to do work—the superiority of one system over another depends ultimately on the use application. Heating can be provided using the energy stored in either system, but doing other kinds of work (such as powering machines) is just not possible by using the warm water—the efficiencies become very low. If we want to use the energy for heating purposes at 100 °C, both energy storages are equally good— however, if the battery system is combined with a heat pump, and there is a suitable large-scale waste heat source available, then 3–5 times more heat can be generated.

Moreover, storage time affects the suitability of each storage system, and the cost of the energy storage is also an important factor. To date, large scale batteries are still very expensive, while storing hot water needs only a large, insulated tank or pool. Different energy storage systems and their selection principles are discussed in more detail below.

Different energy storage technologies are shown in Fig. 7.2, where they are grouped according to the form of energy stored. The figure does not consider the form of energy prior the storage—for example, electrical energy can be stored as thermal, mechanical, or chemical energy. The schematic in Fig. 7.1 is useful in considering the couplings and conversion steps, if we also add conversion efficiencies between the different forms of energy.

This chapter discusses and analyzes several sensible energy storage technologies, including water pool and bore hole storage related to the heating of buildings, and molten salt storages in concentrated solar power plants (see also Chap. 3). For latent heat storage, we only focus on phase-change materials and steam accumulators. Hydrogen energy storage and various Power to X (P2X, where X stands for heat

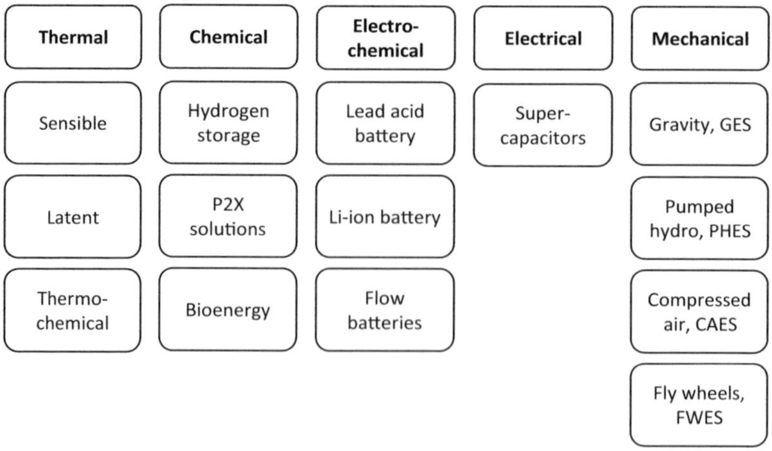

Fig. 7.2 Energy storage technologies

or chemicals and fuels) products such as e-methane, e-methanol and e-ammonia, will be discussed. Biomass, as nature's own chemical energy storage option, is more broadly discussed in Chap. 6 on Bioenergy and Waste. We will examine electrochemical energy storage in Lithium-ion batteries, and pure electrical energy storage in supercapacitors. For mechanical energy storage, we focus on gravity, pumped hydro, compressed air, and flywheel energy storage systems.

In addition to the increasing electrical energy needs of decarbonized societies, there is also a growing need for heat in buildings (heating water temperature has to be 70–90 °C, to avoid growth of algae). Heat is also required by many industrial processes, usually in the range of 150–500 °C, but some industrial processes, for example in the metallurgical industry, require very high temperatures from 700 to 2000 °C [4].

In an energy system based on a large capacity of variable renewable energy sources, the main function of energy storage is to stabilize the system by filling energy gaps when there is no production available. All storage technologies have their optimal application ranges and key performance indicators.

Before delving further into the details, let's look at the status of the technologies discussed above to gain a general understanding of their maturity and available potential. Table 7.1 presents storage capacities, power rating and case examples of the existing systems. Note that for several cases, such as the CSP power plants, only the hours that the plant can run without sunlight are given, and actual storage capacity is not indicated. The actual power output with the storage can often be much smaller than given for the design case production. From the real industrial examples given in the Table 7.1 we see that the largest storage capacities in 2023 are 90 GWh and 40 GWh for heat and electricity storage, respectively.

The challenge is to design a storage system that can stabilize and optimize the variable-energy-based systems with the existing electrical grid and load infrastructures. However, we must not forget that the most important priorities are to reduce the consumption of energy overall, and to shift energy use to times of excess of production and reduce consumption during times of lower production.

7.2 Energy Storage as a Key Component in Large-Scale Energy Systems

To understand the role of energy storage, let us first consider a practical example to explain the rationale for and operation of a large-scale energy storage system. Figure 7.3 shows the main components of an example system: wind power plant, energy storage, load, and power grid. Note that both the energy storage and the grid can act as a sink and a source of energy; in the figure, arrows indicate positive direction. We can write the energy balance of the energy transported between the system components simply as

Table 7.1 Storage technologies in renewable energy systems, with available maximum capacities, time scales, and suitable applications

Technology and stored energy form	Case example	Application
Thermal energy storage, TES		
Sensible heat pool storage	Vantaa Energy, Finland, 2023[a]. 90 GWh (H), 140 °C water, rock cavern of 10^6 m^3	District heating, in construction phase
Sensible heat bore-hole storage	Tianjin Cultural Centre, China, 2018[b]. 3789 × 200 m deep bore holes	Ground-source heat pumps sink and source
Sensible heat thermal oil storage	NOOR I, Morocco, 2015[c]. 160 MW_e, 3 h storage, 293–393 °C , (H → E)	CPS, parabolic trough collector
Sensible heat molten salt storage	Crescent Dunes, US, 2015[c]. 110 MW_e, 10 h storage, 289–565 °C , (H → E)	CPS, solar tower collector
Phase change steam accumulator	Khi Solar One, S. Africa, 2016[c]. 50 MW_e, 2 h storage, 270–530 °C , (H → E)	CPS, solar tower collector
Chemical energy storage		
Bioenergy	Alholmens Kraft 2, Finland, 2001[d]. 240 MW_e, 160 MW. (C → H + E)	Combined heat and power production
Hydrogen energy storage	LKAB Hybrit, Sweden, 2023[e]. Pilot cavern storage is 100 m^3 ~ 100 MWh, at 250 bar. (C)	Steel industry
Mechanical energy storage		
Gravity storage	EnergyVault, Switzerland, 2020[f]. Lifted weight, 32 t concrete blocks, 120 m height, 20 MWh_e (M → E)	RES balancing
Flywheel energy storage	Beacon Power—Hazle Spindle, US, 2021[g]. 200 × 0.1 MW = 20 MW_e, 15 min, 5 MWh_e. (M → E)	RES balancing
Pumped hydro storage	Fengning, China, 2021[h]. 12 × 300 MW = 3.6 GW_e, 40 GWh_e, 16 h. (M → E)	RES balancing
A-CAES, isothermal advanced energy storage, liquid piston	Willow Rock, Hydrostor, Canada, ongoing 2023[i]. 500 MW, 4 GWh_e, 8 h. (M → E)	RES balancing
Electrical and electrochemical energy storages		

(continued)

Table 7.1 (continued)

Technology and stored energy form	Case example	Application
Supercapacitor energy storage	Several applications and manufacturers, such as AVX, Cornell Dubilier, Eaton, Maxwell, Panasonic	RES, wind turbine converters
Battery energy storage	Moss landing, US, 2021[j]. 400 MW_e, 1.6 GWh, 4 h. Li-ion. (E)	RES balancing

H heat, *E* electricity, C = chemical energy, M = mechanical energy
[a] Vantaan [105] under construction phase, [b] Yin and Wu [112], [c] Khan et al. [58], [d] Pohjolan [81], [e] LKAB [68], [f] EnergyVault [30], [g] Beacon [8], [h] Andritz [2], [i] Hydro [51], [j] Vistra [106]. *CSP* concentrated solar power plants, *RES* renewable energy systems

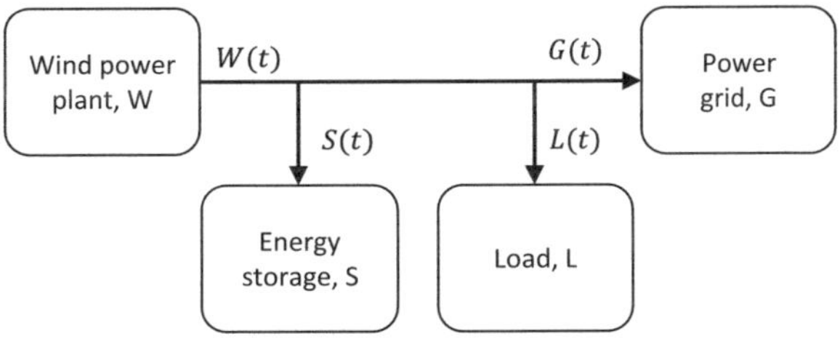

Fig. 7.3 A system based on renewable wind power production, energy storage, load, and grid connection

$$W(t) - L(t) - G(t) - S(t) = 0$$

If there is excess production of wind power, it can be stored, and when the storage is full, the power can be fed to the grid. If there is a deficit of production, energy can be taken from the storage, and if the voltage gets too low, the required energy can be taken from the grid.

We can calculate the charging rate $S_{in}(t)$, discharging rate $S_{out}(t)$, total energy flow rate with the storage (from energy system side, not in the storage itself) $S(t)$ and rate of energy accumulation in the storage $\frac{dE}{dt}$, as

$$S_{in}(t) = \min\left(\max(W(t) - L_w(t), 0), \frac{S_{\max}}{\eta_{in}}, \frac{E_{\max} - E(t)}{\Delta t \times \eta_{in}}\right)$$

$$S_{out}(t) = \min\left(\max(L_w(t) - W(t), 0), S_{\max} \times \eta_{out}, \frac{E(t) - E_{\min}}{\Delta t} \times \eta_{out}\right)$$

$$S(t) = S_{in}(t) - S_{out}(t)$$

$$\frac{dE}{dt} = S_{in}(t) \times \eta_{in} - \frac{S_{out}(t)}{\eta_{out}}$$

Here $S_{\max}$ is the maximum rate of charging and discharging; η_{in} and η_{out} are efficiencies of charging and discharging, respectively; $E_{\max}$ and $E_{\min}$ are the maximum and minimum available storage capacities; L_w is the load target set to be filled by wind power—say, 30% on annual consumption—with the remainder taken from the grid. For a system rated to meet 100% energy production using variable renewables, $L_w(t) = L(t)$

The first term in the $S_{in}(t)$ equation represents the energy available from excess production to be stored, the second term ensures the actual charging rate does not exceed the maximum rate, and the last term ensures the storage is not filled above its maximum capacity. By using the min function, we select the limiting case from these three options. Similarly, the first term in the equation for $S_{out}(t)$ gives the energy removed from the storage to load, the second term makes sure that the de-charge rate does not exceed the maximum rate, and the last term ensures the storage is not emptied below the allowed minimum charge. $S(t)$ is the total energy transfer between the overall energy system, and the final equation represents the rate of accumulation of stored energy after all losses. Note that when charging, less energy ends up to being stored than is taken from the energy system, and when draining the storage, less energy is obtained from the storage than the reduction in stored capacity. Finally, the energy transfer required between the system and the grid can be simply solved from the overall system energy equation,

$$G(t) = W(t) - L(t) - S(t)$$

As an example, we consider Helsinki, the capital city of Finland. The hourly load curve is available from Helen [41]. Power output was obtained from www.renewables.ninja [95] to calculate wind power production of an imaginary offshore wind power plant in the Baltic Sea off-shore from Helsinki. Figure 7.4. shows the hourly production of wind power and consumption for the city of Helsinki and 1020 MW (68 × 15 MW) of turbines in the plant. The lower figure is a magnification of the first 1000 h.

By integrating over hours in a year, the annual production, $P = 4{,}375{,}000$ MWh, and consumption, $C = 4{,}269{,}000$ MWh, can be obtained. This means that we are theoretically able to fill the annual energy consumption, but it is produced at wrong times to meet the demand—therefore, storage is needed. The capacity factor is $P/(1020\text{MW} \times 8760\text{h}) = 0.49$, so the average production rate is ~ 500 MW. The amount of stored energy must remain between the size of the storage and minimum charge, $E_{\max}$. Here we use 0 MWh for the minimum and then vary the maximum size as a free parameter, defined from the fraction of annual consumption, $s = E_{\max}/C$. We set the maximum charge and de-charge rates both at $S_{\max} = 700$ MW, and assume the charging and discharging efficiencies are $\eta_{in} = \eta_{out} = 0.9$. Initial storage is set to 0 MWh.

Let's first examine how the storage behaves when its size is 3% of annual consumption, $s = 0.03$—see Fig. 7.5.

The results show that the energy storage is filled up very fast, and then excess is fed to the power grid; then, when the storage is fully emptied at 2700 h, power

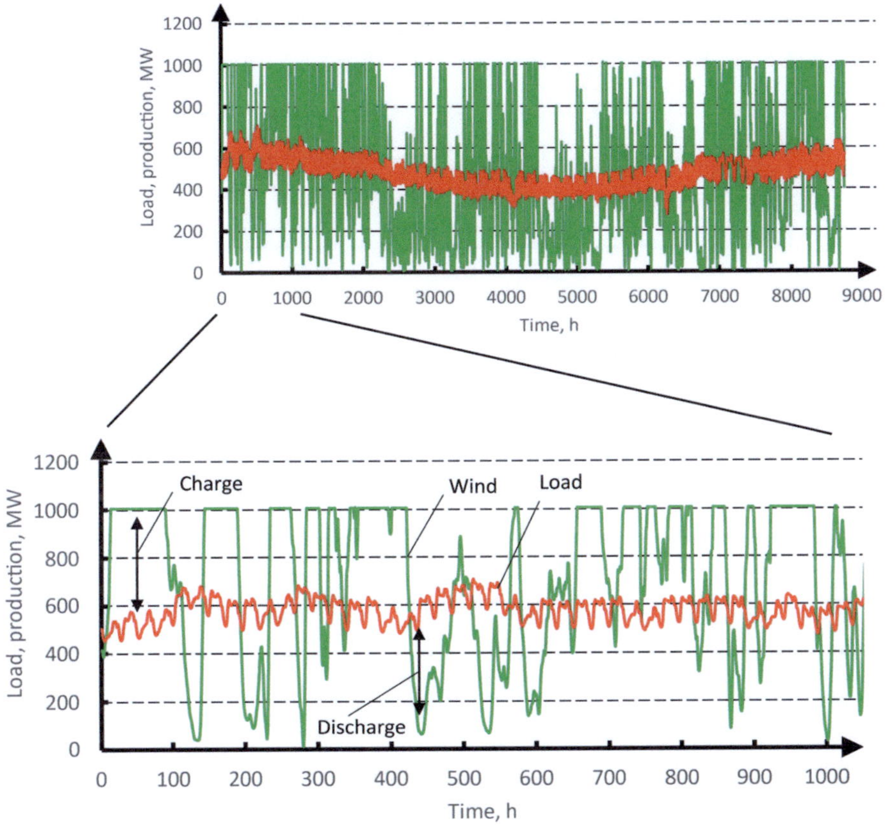

Fig. 7.4 Power demand curve and production from the wind power plant (rated 1020 MW) for the city of Helsinki, Finland, in 2015. Red line represents consumption, green line shows wind power generation

must be taken from the grid. Now let's look at the case where the storage is very large—20% of annual consumption ($s = 0.2$)—see Fig. 7.6.

There are still some periods where grid power is needed, at the end of summer and during fall, but in comparison with the smaller storage in Fig. 7.5, the system is now more stable. The storage is clearly too large, since the maximum state of charge, SOC, is only ~ 25%, but it is still not sufficient to make the system grid-independent. There is no need to feed power to the grid in this scenario, so it is not shown.

The above simple analysis shows that it is very challenging to create a storage system that makes the energy system grid-independent across a year with entirely variable renewable energy sources. The storage sizes needed for these scenarios—131 and 876 GWh—are not even currently available (see Table 7.1).

Next, let's consider system operation during a one-month period, set the renewables' share and target value to just 30% of annual consumption, and simulate three cases: first with no storage, and then storage of 0.1% and 1% of annual energy

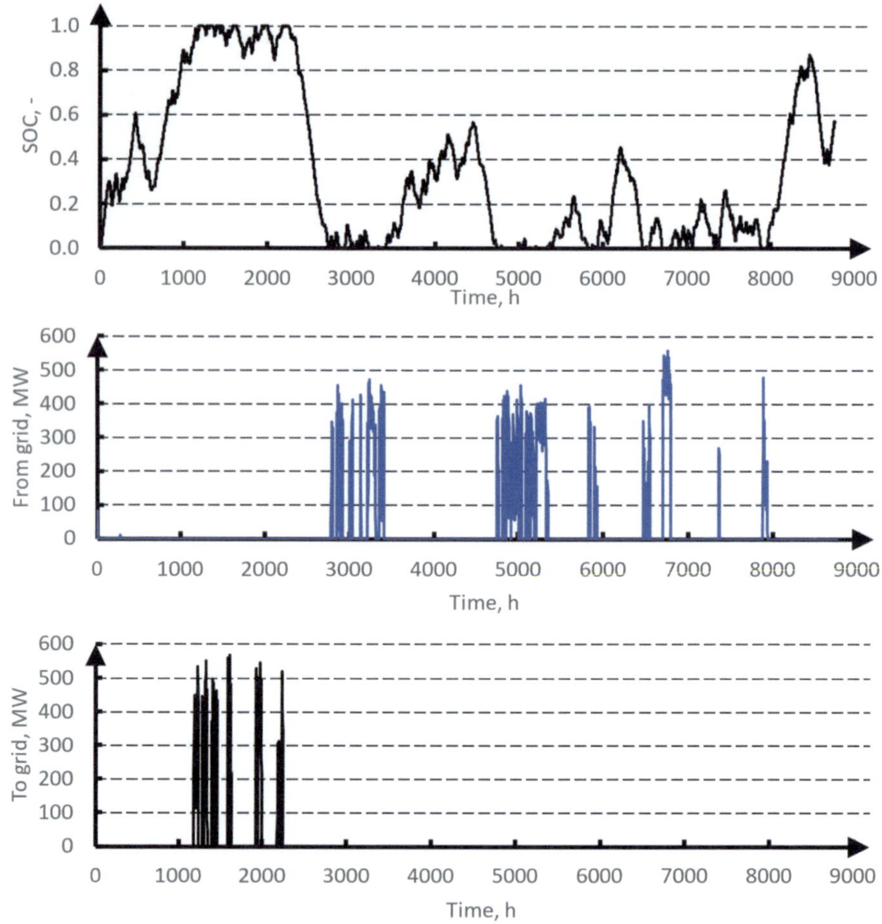

Fig. 7.5 Stored energy (state of charge, SOC, top), and power taken from (middle) and fed to (bottom) the grid. Wind power plant (1020 MW) for the city of Helsinki, Finland, in 2015. Storage size is 131,400 MWh, 3% of annual consumption

consumption. We set the rated capacity of the wind power plant at 306 MW, and we also assume that the load is constant at 500 MW. The results are presented in Figs. 7.7 and 7.8.

As can be seen from Fig. 7.7, there are large variations in grid use as wind power varies. The minimum grid power of 194 MW corresponds to the times, when the full rated power of wind is generated. Let us then study how we can smooth the grid use by implementing storage of 0.1 and 1% of annual consumption, corresponding to 4.38 GWh and 43.8 GWh, respectively. These are presented in Fig. 7.8.

As shown in Fig. 7.8, the grid use profile with 0.1% storage capacity (4.38 GWh) is already much more stable, but varies still between two levels. The lower level of 194 MW corresponds to the situation where the storage is fully charged (wind

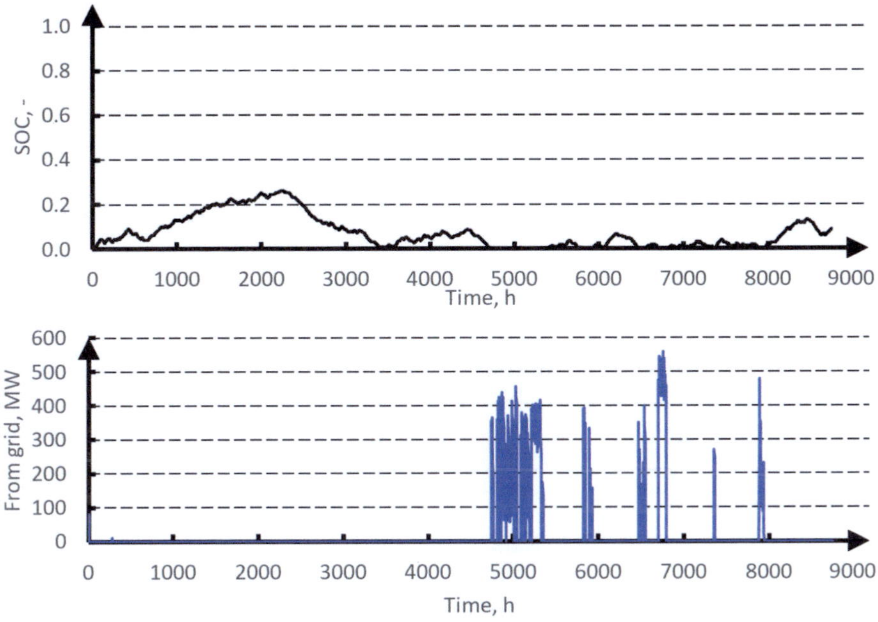

Fig. 7.6 State of charge (top) and power taken from the grid (bottom). Wind power plant (1005 MW) for the city of Helsinki, Finland, in 2015. Storage size 876,000 MWh, 20% of annual consumption

production at rated power), and the upper level of 334 MW corresponds to the rated wind power to reach 30% annual production. With ten times the storage capacity, 43.8 GWh, as Fig. 7.8 shows, the monthly grid-use profile is stabilized— however, towards the end of the January the storage approaches capacity (SOC > 0.8), and if wind power production continues into February, the storage would soon be full.

Nonetheless, clearly on monthly level 1% storage capacity is sufficient in January. There are several energy storage options that could be used; these are discussed in more detail in the next sections. We will first focus on different energy forms and conversion efficiencies, and then discuss in more detail different energy storage technologies and their limitations.

7.3 Thermal Energy Storage Systems

One major issue in the utilization of renewable energy is the seasonal mismatch of energy and demand. For example, in the most Northern and Southern areas, there is plenty of solar energy available in summer, but the greatest energy consumption is in the cold and dark of winter, when solar energy output falls dramatically. Similarly, waste heat from industrial processes, or from the water and ventilation systems of buildings, cannot be fully utilized during the summer period. Therefore, we need

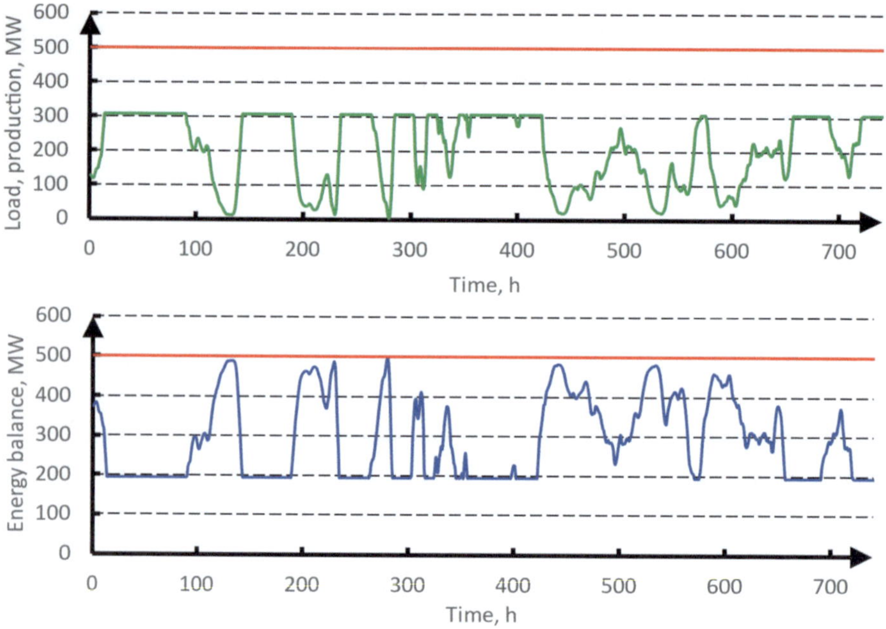

Fig. 7.7 Wind power (306 MW, 30% of annual consumption) with no storage. Constant monthly load 500 MW (red line). Wind power production (green line, top), and grid electricity consumption (blue line, bottom)—the difference between the red and blue line is the share of wind, same as in the top figure

seasonal energy storage systems that allow storage not just for days and weeks, but over many months. Figure 7.9 presents a range of different thermal energy storage technologies.

Sensible thermal energy storage is the most familiar type of thermal energy storage, as it functions through temperature changes of materials: feeding energy into the storage increases the temperature of the system. A common example is a residential hot water storage tank that is kept at around 60 °C to meet the hot water demand of residents. Due to its high heat capacity, low cost and good availability, water is the most-used sensible heat storage material—other materials include air, sand, and rock. Sensible heat storage could also be utilized by storing heat in the walls and other structures of a building, or into rocks underground. The most common seasonal energy storage systems utilize sensible heat.

Latent thermal energy storage exploits the energy associated with phase changes of materials, for example from solid to liquid, or liquid to gas, and vice versa. A common example of a phase-change energy storage is frozen water, which is often used to cool drinks as ice cubes. A lot of energy is released or contained in the phase-change process, which makes latent heat a great way to store energy in large quantities. The problem is that the phase change process is always tied to some specific temperature, such as water freezing and melting at 0 °C. To use latent heat

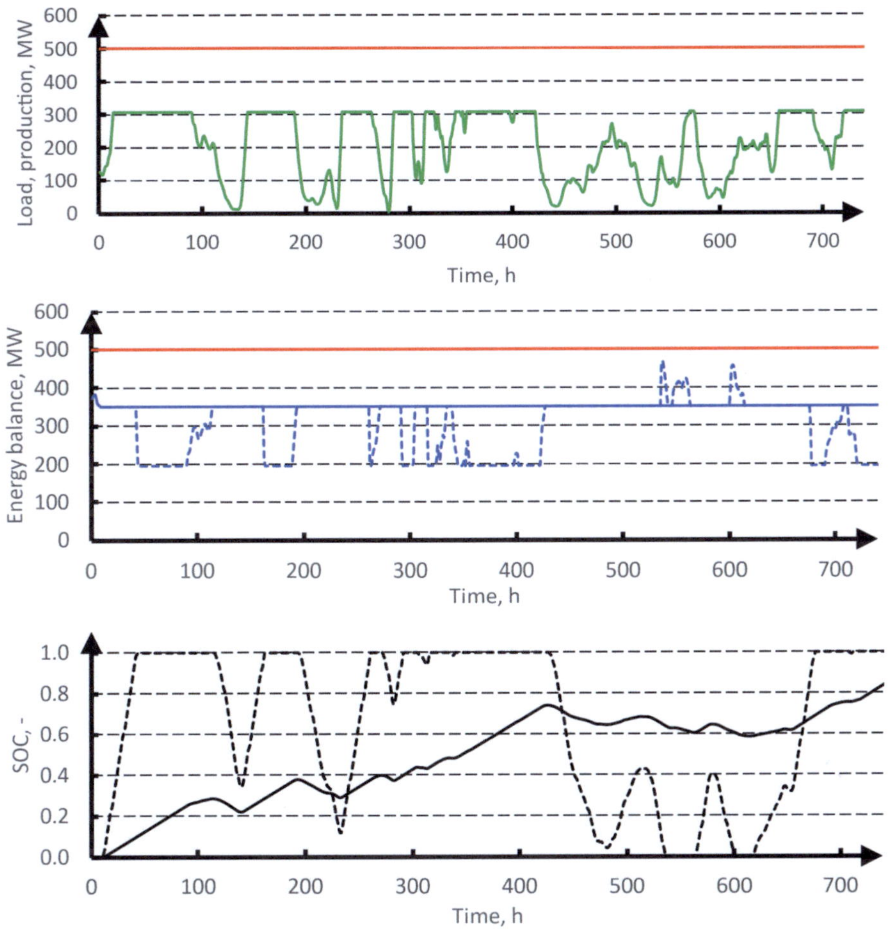

Fig. 7.8 Monthly (January 2015) energy balance for constant 500 MW consumption (red line), wind power production (green line, top), grid use (blue, middle), and state of charge (bottom), with 0.1% storage (dashed lines) and 1% storage (solid lines). Wind power covers 30% of annual consumption

for specific purposes, such as the heating of buildings, requires very specific phase-change materials (PCM) with phase-change temperatures close to those of the use case. For example, PCM could be used for passive cooling of houses: a layer of PCM on the ceiling could absorb excess heat from the building during a hot day, which would reduce the temperature in the room. At night-time, cold air could be fed into the room through the windows to cool down the PCM and discharge the stored heat, allowing for more heat absorption the next day.

Finally, thermochemical thermal energy storage uses reversible chemical reactions to store and release heat. This could mean using thermal energy to separate two chemicals from each other to store energy, and releasing that energy by combining

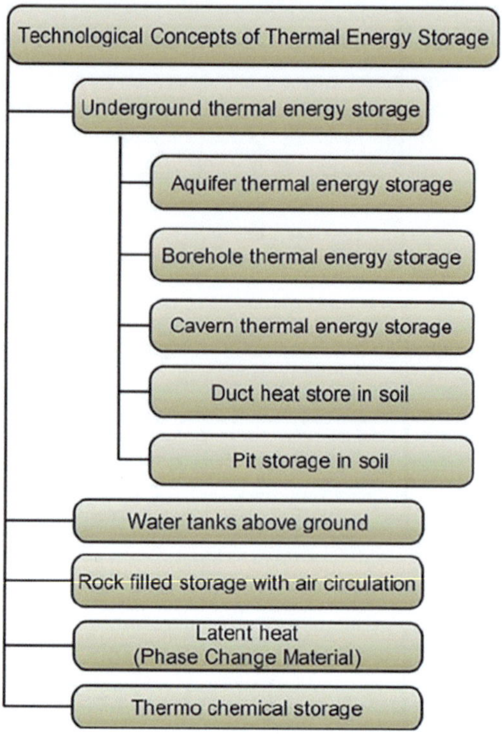

Fig. 7.9 Thermal energy storage technologies

the materials again. Since the energy is not released unless the chemicals are allowed to react, it is possible to store energy with practically no losses, making it ideal for long-term, seasonal thermal energy storage. Chemical energy storage also provides a very high energy density, which minimizes space requirements. The challenge with thermochemical energy storage is in finding durable low-cost materials that can be utilized at available temperatures.

7.3.1 Seasonal Pool Storages

Hot water is the simplest way to store thermal energy, and is commonly used in hot water storage tanks. There are several reviews on this technology available, for example [110]. This principle is also used in the pit/pool thermal energy storage (PTES). In PTES, a large pit is dug into the ground, covered with a watertight insulation lining to prevent water from leaking into the environment, and covered with a floating, insulating lid. Water in the storage pit becomes stratified in layers of different temperatures based on density, with high temperatures on top and low temperatures at the bottom. Energy is stored by feeding high temperature water into the top layers through a diffuser that minimizes mixing of the temperature layers.

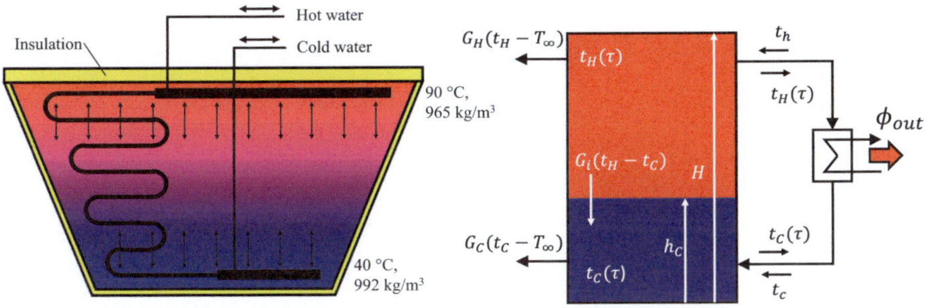

Fig. 7.10 Cross section of the pit thermal energy storage PTES (left), and schematic of a simplified model (right). Thin vertical arrows represent the movement of different thermoclines. Note the difference in density with temperature as shown. Positive direction is hot water inwards

Energy is then extracted by feeding low temperature water through another diffuser at the bottom, and extracting through the diffuser at the top—see Fig. 7.10.

PTES may be used for both seasonal energy storage and short-term storage, depending on the storage capacity compared to energy generation and consumption levels and patterns. The temperature of an unpressurized system is limited by the boiling point of water and the durability of the lining materials. Typically, the high temperature limit is 90 °C.

A variant of PTES is cavern thermal energy storage, where the water is stored in underground caverns. This has been employed by district heating companies that have repurposed old oil storage into large-scale thermal energy storage. A cavern lined with solid rock or metal can also be pressurized, allowing for storage temperature exceeding 100 °C.

The energy storage capacity of a PTES is determined by the water volume and temperature of the system. The higher the temperature, the more energy is contained in the water. The temperature of the stored water can be compared to the temperature needed for the connected heating system, and the temperature of the cold inlet flow that is being heated up by the storage—in a real system, there is a broad distribution of water at different temperatures and volumes, both vertically and horizontally.

To simplify the analysis, we build here a simplified model for pool thermal energy storage. We assume that there are only two temperatures: the cold temperature at the bottom $t_C(\tau)$, and the temperature of the hot water $t_H(\tau)$, that may both vary with time τ. The simplified schematic of the system is presented in Fig. 7.10 (right). We assume that the incoming temperatures of the hot and cold water are constant with time, t_c and t_h. If we assume that the density and heat capacity are constants, and define the positive direction of $\dot{C}$ as inwards for the hot water (corresponding to negative sign of the cold water out from the bottom at the same time), the energy conservation equation for the hot and cold water can be written as follows:

$$\max(\dot{C},0)t_h - \max(-\dot{C},0)t_H - G_H(t_H - t_\infty) - G_i(t_H - t_C) = \dot{C}t_H + C\frac{dt_H}{d\tau}$$

$$-\max(\dot{C},0)t_C+\max(-\dot{C},0)t_c-G_C(t_C-t_\infty)+G_i(t_H-t_C)=-\dot{C}t_C+C\frac{dt_C}{d\tau}$$

$$\frac{dm_H}{d\tau}=\dot{m}$$

$$\frac{dm_C}{d\tau}=-\dot{m}$$

where $\dot{C}=\dot{m}c_p$ and $C=mc_p$ for simplicity, the reference temperature, t_0, is set equal to 0 °C, and $G=hA$ are the conductances: G_o and G_i are heat-transfer conductances between hot water and environment, and hot water and cold water, respectively. $\dot{m}$ is the mass flow rate of water in and out from the pool. The surface areas and total heat transfer coefficients need to be calculated individually for every case.

In the first equation, the first and second term on the left correspond to energy of water in and out from the pool's hot water side—in the second equation, these have opposite signs since the incoming mass flow rate at the top is outgoing from the bottom to keep the pool total mass constant, and vice versa. The third terms correspond to the water losing heat to the environment, and fourth terms present the heat transfer between the hot and cold sections. On the right of the first two equations, the first terms are the accumulation of energy by mass change, and second terms are the accumulation of energy by changing temperature.

Since the above four equations form a non-linear system of ordinary differential equations, numerical solution methods must be used:

$$C_H\frac{dt_H}{d\tau}=a_Ht_H+b_Ht_C+c_H$$

$$C_C\frac{dt_C}{d\tau}=a_Ct_H+b_Ct_C+c_C$$

$$\frac{dm_H}{d\tau}=\dot{m}$$

$$\frac{dm_C}{d\tau}=-\dot{m}$$

where $a_H=-\left(\max\left(-\dot{C},0\right)+G_H+G_i+\dot{C}\right)$
$b_H=G_i$, $c_H=\max\left(\dot{C},0\right)t_h+G_Ht_\infty$,

$$a_C=G_i,$$

$b_C=-\left(\max\left(\dot{C},0\right)+G_C+G_i-\dot{C}\right)$, and

$$c_C=\max\left(-\dot{C},0\right)t_c+G_Ct_\infty.$$

Note that constants a and b must be updated at every time step, since geometry and temperatures are changing. Geometrical parameters and conductances can be calculated from the masses of cold and hot water—see Example 7.1 below.

The efficiency of a thermal energy storage system is defined as the annual ratio of energy discharged and energy charged (both in kWh), or the fraction of stored energy that can be utilized after storage. We consider the heat lost from the hot water as losses. The heat in the cold water cannot be utilized. We get,

$$\eta = \frac{E_{out}}{E_{in}} = \frac{E_{in} - E_{losses}}{E_{in}} = 1 - \frac{E_{losses}}{E_{in}}$$

$$E_{in} = E_0 + \int_0^{8760\,\mathrm{h}} \max\big(\dot{C}(\tau), 0\big) t_h d\tau$$

$$E_{losses} = \int_0^{8760\,\mathrm{h}} (G_H(t_H - t_\infty) + G_i(t_H - t_C))d\tau$$

Generally, the larger the thermal storage system, the better the efficiency. A sensible thermal energy storage is constantly losing energy to its surrounding environment due to the temperature difference between the storage and the environment. These losses happen through the exterior surface of the storage system, partly in the ground and partly through the air. When the volume of a regular-shaped object increases, the external surface also increases, but not quite as much as the volume—this means that the surface area to volume ratio goes down with increasing volume. Thus, relative to the stored energy, the heat losses are reduced. Figure 7.11 presents the surface area to volume ratios and fraction of annually lost heat E_{losses}/E_{in} as a function of the storage volume and insulation thickness.

As Fig. 7.11 shows, large storage size is extremely important to minimize losses. For example, if the insulation thickness is 1 m and storage volume is 1000 m^3, annual losses are ~ 12.5% (0.125), resulting in efficiency of 87.5%.

Example 7.1 Conversion of an Old Oil Tank to Hot Water Pool Storage
Analyze the operation of a 6280 m^3 old oil tank converted into a hot water pool storage—see Fig. 7.10. The tank has the shape of a cylinder with a flat top and bottom, with diameter and height both 20 m. The tank is insulated with mineral wool, either $s = 0.05$ or 0.2 m thickness, with thermal conductivity $\lambda = 0.06$ W/mK. Internal and external heat transfer coefficients between water and air are 300 and 26 W/m^2K, respectively, and overall wall conductance between water and ground is 0.16 W/m^2K.

Solution: The tank is initially completely filled with 90 °C hot water before a cold period of – 25 °C outside temperature, lasting seven days. During discharge, cold water of 45 °C is fed from the bottom—the same water that came from the upper part to an external heat exchanger, wherein the temperature drops 45 °C. Storage can

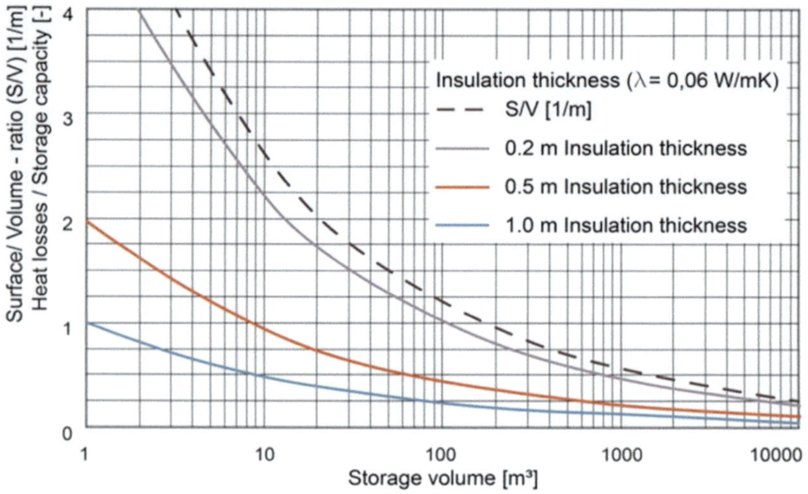

Fig. 7.11 Surface area to volume ratio and annual heat loss fraction of pool thermal energy storage, PTES [69]

be fully discharged by the end. Using the simple model developed above, simulate the temperature of the hot and cold water for seven days and plot these with time, together with the heat flow rate utilized. How much energy could be stored in total? Water density, thermal conductivity, and heat capacity can be assumed to be constant, as 970 kg/m^3, 0.024 W/mK, and 4200 J/kgK, respectively.

Since the stored water must be sufficient for seven days, we can first calculate the mass flow rate, 6280 m^3/(7 × 24 × 3600 s) = 0.0104 m^3/s, corresponding to 0.0104 m^3/s × 970 kg/m^3 = 10.1 kg/s. The heat capacity flow rate is then 10.1 × 4200 J/kg K = 42,324 W/K. The initial mass of hot water is $m_{H.0} = 6{,}091{,}600$ kg, and for numerical stability reasons we also need to assume a small amount of cold water, $m_{C.0} = 1000$ kg. Masses are calculated as $m_H = m_{H.0} - \dot{m}t$, $m_C = m_{C.0} + \dot{m}t$, and the areas between the hot water and the air, and between the cold water and the air, are

$$A_H = \frac{\pi}{4}D^2 + \pi DH \times \left(\frac{m_{H.0} - \dot{m}t}{m_{H.0}}\right) = \frac{\pi}{4}D^2 + \pi DH \times \left(1 - \frac{\dot{m}t}{m_{H.0}}\right)$$

$$A_C = \pi DH \times \frac{m_{C.0} + \dot{m}t}{m_{H.0}}$$

We assume that the ground under the tank remains well insulated and heat loss from the bottom can be neglected. Overall heat transfer coefficients (W/m^2 K) from water through the wall (w) to outside and from hot water to cold water vertically (i), are

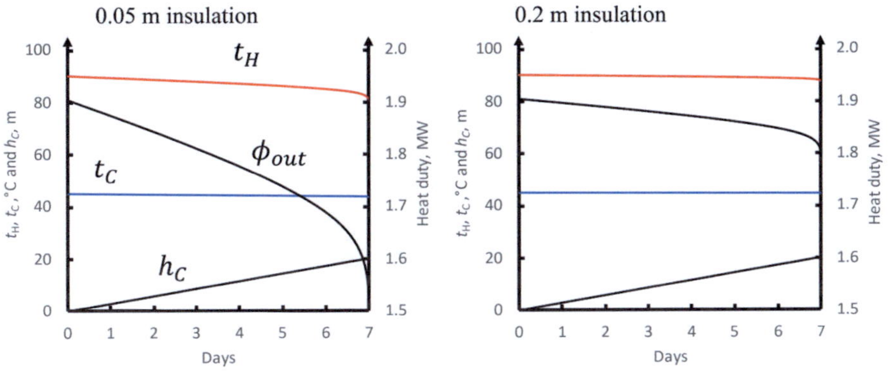

Fig. 7.12 Discharging a hot water pool thermal storage. h_C is the vertical location of the cold-water surface

$$h_w = \frac{1}{\frac{1}{\alpha_w} + \frac{s}{\lambda_i} + \frac{1}{\alpha_a}} = \frac{1}{\frac{1}{300} + \frac{0.4}{0.04} + \frac{1}{26}} = 0.1$$

$$h_i = \frac{2\lambda_w}{H} = \frac{2 \times 0.024}{20} = 0.0024$$

respectively (we assume a linear vertical temperature profile). The area between hot and cold regions is equal to $A_i = \pi D^2/4$. With these all defined, we can implement the system of equations into numerical solution software to find the results—here, we used a simple Excel spreadsheet calculation using Euler's 1st order method. Figure 7.12 presents the results for 0.05 and 0.2 m insulation thicknesses.

Total energy output is 301 and 316 MWh for 0.05 and 0.2 m insulation thicknesses, respectively, for 45 °C cold inlet temperature. These result in storage efficiencies of 94% and 99% for a period of one week.

In Finland, large thermal energy storages have been used as buffer storages for large district heating grids. For example, energy company Helen, the municipal energy utility in Helsinki, operates a 320,000 m^3 cavern thermal energy storage system, built in an old oil storage facility. It can supply heating power at 120 MW, which allows for four days of full-power operation. A similar repurposing project has been implemented in Vaasa, Finland, where a 210,000 m^3 hot water cavern has about 8 GWh of energy capacity and can supply 100 MW of heat flow rate.

7.3.2 Seasonal Bore Hole Thermal Storages

Borehole thermal energy storage (BTES) is an energy storage system based on vertical heat exchanger tubes drilled into the ground. While borehole fields are already commonly used by ground-source heat pump systems, in a BTES the heat transfer is bidirectional. In BTES, the ground itself serves as the medium of energy storage.

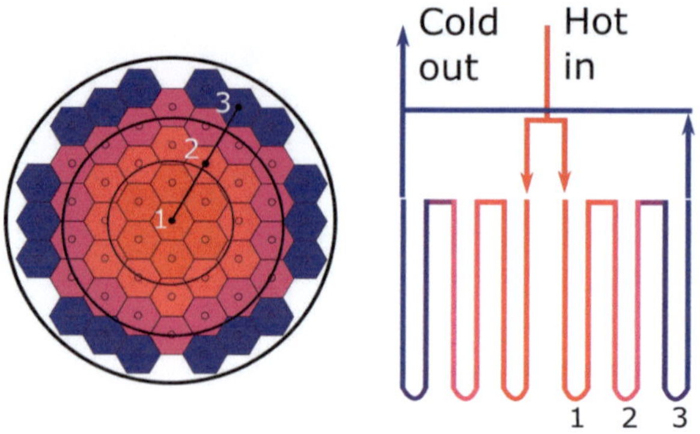

Fig. 7.13 A borehole field with a series connection between boreholes [45]

Thus, the properties of the ground itself affects how much energy can be stored and how fast the heat can be injected to or extracted from the storage. Figure 7.13 represents the operational principle of the system.

Heat flow rate of a BTES is determined by the number of boreholes, the flowrate of the heat transfer fluid, the temperature difference between the heat transfer fluid and the storage, as well as the heat transfer coefficient between the pipes and ground. To increase the temperature change of inlet and outlet streams, several boreholes can be connected in series to allow for a longer period of heat transfer—as a trade-off, this will reduce the total flow rate of the system, as all the series-connected pipes effectively become one single pipe.

Modelling of a bore hole thermal energy storage system requires the solution of the multidimensional heat conduction equation and the energy equation for the pipe flow, with proper boundary conditions (see Hirvijoki and Hirvonen [44] as one example of this):

$$\rho_R c_R \frac{\partial T_R}{\partial t} = k_R \nabla^2 T_R$$

$$\frac{c_w \dot{m}}{2\pi r_{hole}} \frac{dT_R}{dz} = h_w (T_R - T_w), \quad z \in [0, L]$$

$-k_R \frac{\partial T_R}{\partial r} = h_w (T_R(z) - T_w(z))$, at the outer pipe wall, $r = r_p$,

$-k_R \frac{\partial T_R}{\partial r}(z) = 0$, $z \in [0, 2L]$ zero heat flux in radial direction at the radial distance of $r = r_B$. This describes an infinite periodic lattice of installed boreholes with small distances, or one borehole with a sufficiently large domain reserved.

$-k_R \frac{\partial T_R}{\partial z} = -q_{Earth}$, vertical direction at the depth of $z = 2L$, heat flow from the center of the Earth.

$-k_R \frac{\partial T_R}{\partial z} = h_A (T_A(r) - T_R(r))$, $z = 0$, $r \in [0, r_B]$, vertical direction at the Earth's surface.

where T_R, ρ_R, c_R, k_R are the temperature (K), density (kg/m^3), heat capacity (J/kgK) and thermal conductivity of the soil (W/mK), respectively; $c_w \dot{m} dT$ is the change rate of energy content in the water in the vertical direction in the outer pipe; and $h_w(T_R - T_w)2\pi r_p dz$ is the corresponding convective rate of heat transfer between the surface and water. Figure 7.14 shows the principle of the coaxial bore hole heat exchanger operation, where the pipe between the inner and outer sections is assumed to be well insulated.

The above model requires an effective numerical solution method, for example the finite element method used by Hirvijoki and Hirvonen [44]. The energy storage capacity of a BTES is determined by the volume of soil and rock surrounding the boreholes, and the density and heat capacity of the ground material. As a result, the distance between the boreholes, or the borehole density, influences the energy storage capacity. A dense borehole field has can provide more power, but has correspondingly reduced energy storage capacity.

The energy storage efficiency of a BTES depends on the ground material and the temperature level of the storage. High temperature storage in a very conductive environment can incur high losses; on the other hand, heat losses will also gradually heat the environment, which tends to reduce losses. Thus it can take several years until a BTES system reaches its final operational efficiency. As an example, Fig. 7.15 shows the input, output, and average temperature of a BTES system across five years of operation.

As the figure demonstrates, the system stabilizes after 2 years of operation. Figure 7.16 shows annual energy balance of three BTES systems (cases 1, 4 and 22 in a series of case studies in [45]), including the amount of energy fed from the collectors during the summertime, energy discharged from the storage, and the required district heating.

As Fig. 7.16 shows, case 1 required a significant share of district heating. In cases 4 and 22 the size of the heat collectors and size of the storage are larger, and much less district heating is then required.

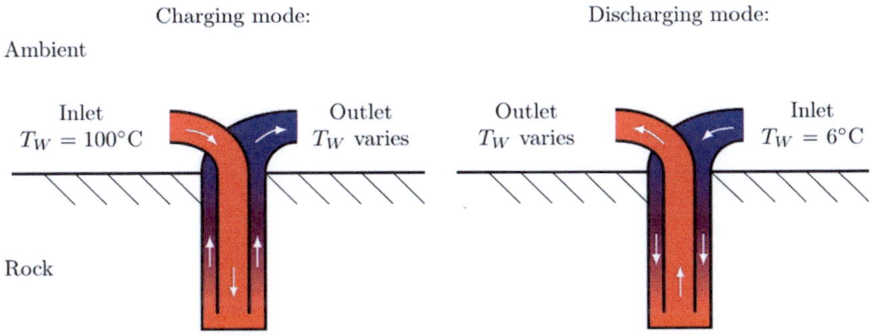

Fig. 7.14 An illustration of the borehole operation in charging and discharging modes. The inner pipe is assumed insulated (Hirvijoki and Hirvonen [44], CC. 4.0)

Fig. 7.15 Temperature development in a BTES system over five years, in case 4 in the Fig. 7.16 [45]

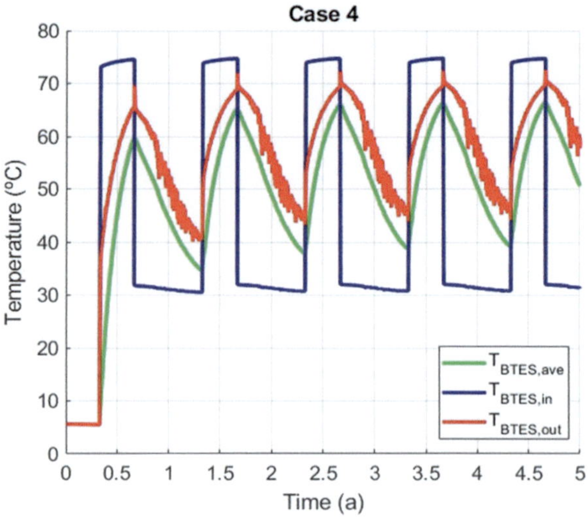

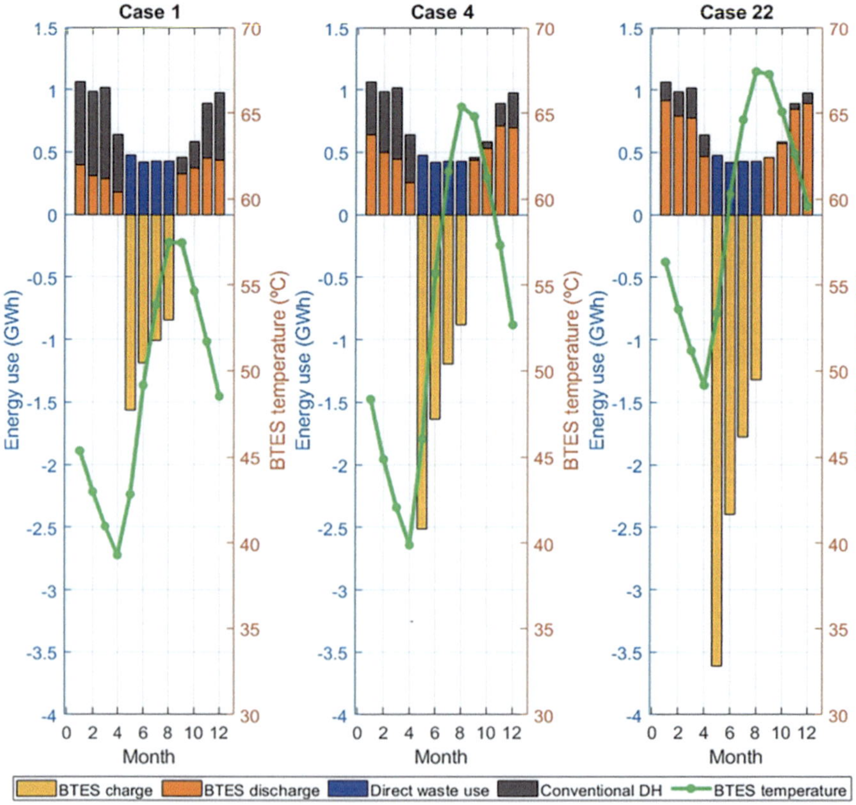

Fig. 7.16 Seasonal thermal storage cycle of three different district heating and BTES systems [45]

The Drake Landing Solar Community (DLSC) in Canada is perhaps the most famous example of seasonal solar energy storage [57]. The community has 52 energy-efficient houses, with solar thermal collectors installed on garage rooftops. The community contains a 30,000 m^3 BTES system, which is charged by the solar collectors and can supply 98% of annual space heating demand.

7.3.3 Steam Accumulator

The steam accumulator is a well-insulated pressure vessel that is charged by feeding in steam at times when there is excess production in the steam generator of a power plant, and discharged when additional steam is required in the turbine or other processes. It therefore balances generation and/or consumption, and the system is able to operate with a smaller boiler as generation peaks are avoided. Alternatively, the steam accumulator could be used to secure overnight operation, as will be shown with a simple example below.

The incoming steam could be superheated or saturated, at the pressure p_{in}, while outgoing steam from the steam battery is always saturated at pressure p_2, which rules out the application at higher temperatures above the critical point. To be able to control the outflow, the accumulator pressure $p(t)$ must be higher than the utilization pressure. The steam is used at the pressure p_2, and normally the pressure difference between the accumulator and use-point has to be kept above $p(t) - p_2 > 1$ bar to secure proper control valve operation [94]. When the steam accumulator pressure is much higher than the actual pressure needed, a significant share of the energy is wasted in the throttling valve, instead of using it to do work in the steam turbine.

Steam accumulators are used in the process industry to balance the need for variable consumption. In connection to renewable energy, as an example, steam accumulators are used with smaller concentrated solar power plants to smooth the power generation in the turbine when the solar conditions vary. There are numerous ways to implement a steam accumulator into a power plant—we study here only a simple case, with steam inlet and outlet joints, and an additional water connection to replace water lost through evaporation. We also assume that the steam accumulator is adiabatic and there are no heat losses, and the mixing is assumed perfect inside the vapor and the liquid phases. In addition, we assume that when charging, there is no flow out from the tank, and when discharging, there is no inflow. Water is added between charge–discharge cycles to replace any loss by evaporation. See Fig. 7.17 for a description of the system operation.

The cycle starts with a discharged tank in a saturated condition (a). We then add an incremental volume element of vapor into the system (b). We assume that all the vapor remains first below the water surface, expanding the total combined liquid–gas volume. Then, we let the pressure do work, forcing the vapor into the liquid while condensing it, and after this we enter the stage (c). These incremental steps are then repeated a sufficient number of times (only two are shown in the figure).

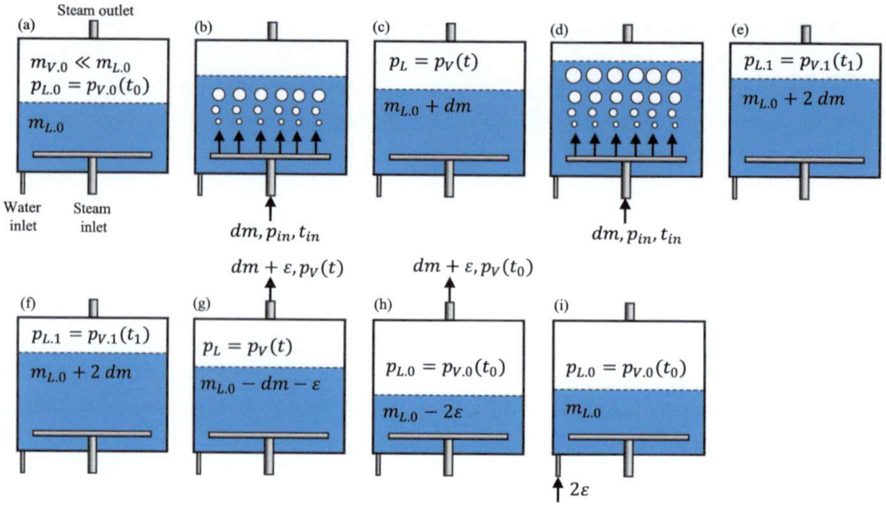

Fig. 7.17 Charging (**a–e**) and discharging (**f–h**) and adding make-up water (**i**) to the steam accumulator

During the compression and condensation of the vapor, the change of the total volume below the liquid surface is calculated as follows [55, 74]:

$$dV = (m + dm)(v_L + dv_L) - mv_L - dmv_{in} = mv_L + dm(v_L - v_{in})$$

Note that the last term in the middle of the equality is the volume occupied by the bubble of vapor fed into the liquid. The vapor pressure does work on the system and presses the vapor into the liquid. Based on the first law of thermodynamics, in a closed and adiabatic system, for the liquid in this work phase (the importance of the vapor is negligible) we can write

$$\begin{aligned} dU &= (m + dm)(u_L + du_L) - mu - dmu_{in} = mu + dm(u_L - u_{in}) \\ &= -pdV = -pmv_L - pdm(v_L - v_{in}) \end{aligned}$$

From this we can solve by integration to get an equation for the mass added due to pressure increase,

$$\frac{dm}{m} = \frac{du_L + pdv_L}{h_{in} - h_L} = \frac{Tds_L}{h_{in} - h_L}$$

$$\frac{\Delta m_{in}}{m_0} = exp\left(\frac{\Delta u_L + \overline{p}\Delta v_L}{h_{in} - \overline{h_L}}\right) - 1$$

where $\Delta u_L = u_L(p_1) - u_L(p_0)$, $\Delta v_L = v_L(p_1) - v_L(p_0)$, $\overline{p} = (p_0 + p_1)/2$ and $\overline{h_L} = (h_L(p_0) + h_L(p_1))/2$. Δu_L and Δv_L are simply calculated as a difference between start and end states, 0 and 1, respectively. During the discharging step, noting that Δu_L and Δv_L are now calculated from higher to lower pressure, the equation becomes,

$$\frac{\Delta m_{out}}{m_0} = exp\left(\frac{\Delta u_L + \overline{p}\Delta v_L}{\overline{h_V} - \overline{h_L}}\right) - 1$$

where $\Delta u_L = u_L(p_0) - u_L(p_1)$, $\Delta v_L = v_L(p_0) - v_L(p_1)$, $\overline{p} = (p_0 + p_1)/2$ and $\overline{h_L} = (h_L(p_0) + h_L(p_1))/2$ and $\overline{h_V} = (h_V(p_0) + h_V(p_1))/2$. Remember that in reality, p_0 must remain typically ~ 1 bar above the utilization pressure for the throttling valve to operate properly.

Tables 7.2, 7.3 and 7.4 present calculated mass changes for 0, 50 and 100 °C superheats (temperature different to boiling point) between different pressure levels. Δm is positive for charging and negative for discharging phases.

How do we read the results in Tables 7.2, 7.3, and 7.4? As an example, let's look at the case of $\Delta t = 50°C$, Table 7.2, and consider a steam battery that operates between 120 and 80 bars. During the charging step, we use superheated steam at 120 bar and 375 °C. This steam condenses and heats the material until the pressure in the storage reaches in input pressure of 120 bar—then the flow stops, and the temperature corresponds to saturated value. We can read from the Table 7.3 the value of the corresponding mass increase in the storage, $\Delta m/m_0 = 0.115$ (the intersection of the 80 bar row and 120 bar column)—that is, the new mass is equal to $m = 1.115 \times m_0$.

Table 7.2 Charging (+) and discharging (−) mass changes, $\Delta m/m$, between pressure levels p_0 and p_1. $\Delta t = 0$ °C

$\Delta t = 0°C$		t_{in}	180	212	250	276	295	311	325
		$t(p_{in})$	180	212	250	276	295	311	325
		p_{in}	10	20	40	60	80	100	120
$t(p_0)$ °C	p_0 bar		Input and full storage conditions, for output only p_{in} is relevant						
180	10	Initial accumulator state	0.000	0.077	0.191	0.292	0.396	0.509	0.639
212	20		−0.072	0.000	0.103	0.193	0.284	0.382	0.494
250	40		−0.160	−0.094	0.000	0.079	0.156	0.239	0.332
276	60		−0.224	−0.161	−0.073	0.000	0.070	0.143	0.226
295	80		−0.278	−0.217	−0.133	−0.065	0.000	0.067	0.141
311	100		−0.326	−0.267	−0.187	−0.123	−0.062	0.000	0.067
325	120		−0.371	−0.315	−0.239	−0.177	−0.120	−0.062	0.000

Charging steam and discharging always saturated

Table 7.3 Mass changes, $\Delta m/m$, charging steam is super-heated, $\Delta t = 50$ °C, discharging always saturated

$\Delta t = 50°C$		t_{in}	230	262	300	326	345	361	375
		$t(p_{in})$	180	212	250	276	295	311	325
		p_{in}	10	20	40	60	80	100	120
$t(p_0)$ °C	p_0 bar		Input and full storage conditions, for output only p_{in} is relevant						
180	10	Initial accumulator state	0.000	0.072	0.174	0.261	0.343	0.427	0.516
212	20		–0.072	0.000	0.094	0.173	0.247	0.322	0.400
250	40		–0.160	–0.094	0.000	0.070	0.136	0.202	0.270
276	60		–0.224	–0.161	–0.073	0.000	0.061	0.121	0.184
295	80		–0.278	–0.217	–0.133	–0.065	0.000	0.056	0.115
311	100		–0.326	–0.267	–0.187	–0.123	–0.062	0.000	0.055
325	120		–0.371	–0.315	–0.239	–0.177	–0.120	–0.062	0.000

Table 7.4 Mass changes, $\Delta m/m$, charging steam is super-heated, $\Delta t = 100$ °C, discharging always saturated

$\Delta t = 100°C$		t_{in}	280	312	350	376	395	411	425
		$t(p_{in})$	180	212	250	276	295	311	325
		p_{in}	10	20	40	60	80	100	120
$t(p_0)$ °C	p_0 bar		Input and full storage conditions, for output only p_{in} is relevant						
180	10	Initial accumulator state	0.000	0.068	0.163	0.240	0.313	0.385	0.459
212	20		–0.072	0.000	0.088	0.160	0.226	0.290	0.357
250	40		–0.160	–0.094	0.000	0.065	0.125	0.182	0.241
276	60		–0.224	–0.161	–0.073	0.000	0.056	0.110	0.164
295	80		–0.278	–0.217	–0.133	–0.065	0.000	0.051	0.103
311	100		–0.326	–0.267	–0.187	–0.123	–0.062	0.000	0.049
325	120		–0.371	–0.315	–0.239	–0.177	–0.120	–0.062	0.000

We then discharge this storage and release the pressure down to 80 bar. This is equal to $\Delta m/m_0 = -0.120$ (the intersection of the 120 bar row and 80 bar column). The water mass in the storage after discharge is simply $m = 1.115 \times (1 - 0.120) \times m_0 = 0.981 \times m_0$. This means that due to the super-heated steam, we are losing a small fraction of the steam by evaporation, which needs to be compensated by adding more water.

Without going into the details, the above model only gives the theoretical maximum amount of steam that can be stored and released; it does not consider the actual mass flow rates of input and output steam. These can be estimated, for example, by the simple pressure loss model of a valve [94]:

$$\dot{m}_{in} = \frac{K_v p_{in}}{300}\sqrt{1 - 5.67 \times \left(0.41996 - \frac{\max(p_{in} - p(t), 0)}{p_{in}}\right)^2}$$

$$\dot{m}_{out} = \frac{K_v p_{in}}{300}\sqrt{1 - 5.67 \times \left(0.41996 - \frac{\max(p(t) - p_2 - \Delta p_V, 0)}{p(t)}\right)^2}$$

The valve coefficient K_v can be read from the chart provided in Appendix 7.1. During discharging, when the pressure in the storage, $p(t)$, decreases from p_1 to p_0, we need to adjust the valve opening β to maintain constant flow rate and power production in the downflow Rankine cycle. As already mentioned above, this also means significant energy losses at higher pressure differences.

The energy storage capacity of the steam battery varies significantly based on what the energy is stored for. If we are storing heat to be used in a steam turbine, there is much less energy available compared with storing steam for heating purposes only, since the heat-to-power efficiency of the Rankine cycle is typically less than 45%. Energy fed in (change from $p_0 = p_2 + \Delta p_v \rightarrow p_1\ p$) and taken out (at the utilization pressure, change from $p_1 \rightarrow p_2 + \Delta p_v$) from the adiabatic (perfectly insulated) steam battery is equal to

$$E_{in} = \Delta m_{in} h_{in} = m_0\left(\exp\left(\frac{\Delta u_L + \overline{p}\Delta v_L}{h_{in} - \overline{h_L}}\right) - 1\right) \times h_{in}$$

and

$$E_{out} = \Delta m_{out} h_V(p_2)$$

$$= m_0 \exp\left(\frac{\Delta u_L + \overline{p}\Delta v_L}{h_{in} - \overline{h_L}}\right)\left(1 - \exp\left(-\frac{\Delta u_L + \overline{p}\Delta v_L}{\overline{h_V} - \overline{h_L}}\right)\right) \times h_V(p_2)$$

After each cycle's discharge step, the liquid water mass has to returned to the nominal level. We can calculate the required mass of refill/reject water, $\varepsilon = \Delta m_{out} - \Delta m_{in}$, from the equations derived above.

The volume of the steam accumulator, V_{SA}, can be approximated based on the required energy storage capacity, E_{SA}, the maximum fill fraction at the beginning of the discharge step, φ, and the equation derived above for mass release during the discharging step. In the beginning of the discharge step, the contents are at the maximum level, and during discharge the water level drops as water is flashed due to pressure reduction. Defining Δu_L and Δv_L for the discharging process direction, from p_{in} to $p_0 = p_2 + \Delta p_v$, including the losses in the control valve, and only considering the energy that enters the use system at pressure p_2, we have

$$E_{SA} = \frac{V_L h_V(p_2)}{v_L(p_1)}\left(1 - \exp\left(\frac{\Delta u_L + \bar{p}\Delta v_L}{\overline{h_V} - \overline{h_L}}\right)\right)$$
$$= \frac{V_{SA}\varphi h_V(p_2)}{v_L(p_1)}\left(1 - \exp\left(\frac{\Delta u_L + \bar{p}\Delta v_L}{\overline{h_V} - \overline{h_L}}\right)\right)$$

$$V_{SA} = \frac{E_{SA} v_L(p_1)}{\varphi h_V(p_2)\left(1 - \exp\left(\frac{\Delta u_L + \bar{p}\Delta v_L}{\overline{h_V} - \overline{h_L}}\right)\right)}$$

Energy density can be also calculated as

$$\frac{E_{SA}}{V_{SA}} = \frac{\varphi h_V(p_2)}{v_L(p_1)}\left(1 - \exp\left(\frac{\Delta u_L + \bar{p}\Delta v_L}{\overline{h_V} - \overline{h_L}}\right)\right)$$

The volume of the storage pressure vessel and the volumetric storage density are calculated for different pressure ratios in Tables 7.5 and 7.6, respectively.

Table 7.5 Volume of the storage tank in m^3, storage capacity of 1000 kWh of heat at the average pressure between p_{in} and $p_0 = p_2 + \Delta p_v$. $\Delta p_v = 1$bar is the minimum pressure difference in the control valve

		$t(p_{in})$	180	212	250	276	295	311	325
		p_{in}	10	20	40	60	80	100	120
$t(p_0)$ °C	p_2 bar		Upper pressure level						
180	10	Steam state at exit	–	26.9	11.9	8.8	7.4	6.6	6.0
212	20		–	–	20.3	12.1	9.3	7.9	7.0
250	40		–	–	–	27.4	15.3	11.3	9.3
276	60		–	–	–	–	32.4	17.5	12.6
295	80		–	–	–	–	–	35.9	18.9
311	100		–	–	–	–	–	–	37.9
325	120		–	–	–	–	–	–	–

Table 7.6 Storage density (MJ/m^3) for different pressure ratios

		$t(p_{in})$	180	212	250	276	295	311	325
		p_{in}	10	20	40	60	80	100	120
$t(p_2)$ °C	p_2 bar		Upper pressure level						
180	10	Steam state at exit	–	134	303	410	488	549	597
212	20		–	–	177	297	385	455	512
250	40		–	–	–	132	236	319	388
276	60		–	–	–	-	111	206	286
295	80		–	–	–	–	–	100	191
311	100		–	–	–	–	–	–	95

Example 7.2 Dimensions of a steam accumulator for the overnight operation of a concentrated solar power plant

A small 50 kW$_e$ concentrated solar power plant needs a steam battery to cover overnight operation for 11 h (see [26]). The basic saturated Rankine cycle operates between 9 bar and 0.2 bar, and has turbine isentropic efficiency of 0.85 and generator efficiency of 0.95. Steam accumulator maximum operation pressure is 40 bar. Assume throttling valve pressure loss of $\Delta p_v = 1$bar. Calculate the size of the steam accumulator.

Solution: To estimate the size of the storage, we first need to determine the amount of steam required. From the electrical power production of 50 kW, we can simply calculate the mass flow rate of steam required. Enthalpies can be obtained from generally available steam tables, $h_1 = 2773$ and $h_{2s} = 2180$ kJ/kg, see Chap. 3 for physical properties in electronic supplement, and we find

$$\dot{m} = \frac{P_e}{\eta_g \eta_s (h_1 - h_{2s})} = \frac{50}{0.95 \times 0.85 \times (2773 - 2180)} = 0.093 \text{ kg/s}$$

The overnight steam needed for 11 h is therefore $m_s = 3683$ kg. We need a steam accumulator able to deliver this while pressure decreases from 40 to 10 bars (to secure proper valve operation). This can be solved simply from the discharge equation,

$$V_{SA} = \frac{m_s v_L(p_1)}{\varphi\left(1 - \exp\left(\frac{\Delta u_L + \overline{p}\Delta v_L}{\overline{h}_V - \overline{h}_L}\right)\right)}$$

where $v_{L.1}(40 \text{ bar}) = 0.0013 m^3/kg$, $v_{L.2}(10 \text{ bar}) = 0.0011 m^3/kg$, $\varphi = 0.9$, $u_{L.1}(40 \text{ bar}) = 1082$ kJ/kg, $u_{L.2}(10 \text{ bar}) = 762$ kJ/kg, $\overline{h}_V(25 \text{ bar}) = 2802$ kJ/kg, $\overline{h}_L(25 \text{ bar}) = 962$ kJ/kg. From this we obtain the minimum required tank size of $V_{SA} = 30.7$ m^3. For comparison, Ehtiwesh et al. [26] used an accumulator tank of 160 m^3 in their system, securing a broader range of operation parameters.

7.3.4 Latent Phase Change Energy Storage

Latent heat storage exploits the thermal energy involved in a phase transition, such as melting and solidification. This method enables high energy storage densities at a narrow temperature range, because the phase transition typically absorbs and releases thermal energy at a constant temperature. The storage materials, knows as phase change materials (PCMs), often utilize solid–liquid transition, as it offers relatively large latent heat of melting and a small change in density between the solid and liquid state. Nevertheless, other transitions can be used as well: the liquid–gas phase change provides large latent heat, but the gaseous state requires large storage volumes. On the other hand, solid–solid transitions typically have small changes of volume, but they possess relatively small latent heat.

The operational principle of latent heat storage can be demonstrated by employing a solid–liquid PCM for heat storage or thermal regulation. In this case, the storage is charged by increasing the temperature of the PCM in a solid state, which eventually melts the material. The charging process is illustrated by the red line in Fig. 7.18, which plots the stored thermal energy in palmitic acid as a function of the material temperature.

Initially, stored energy increases steadily as the solid material is heated. When the melting temperature, T_m, is reached, melting begins, and heat is absorbed in the phase transition at a constant temperature. After melting is complete the temperature of the liquid PCM increases once more. Ideally, the PCM remains in the liquid state for the time the thermal energy is stored. However, heat loss to ambient gradually

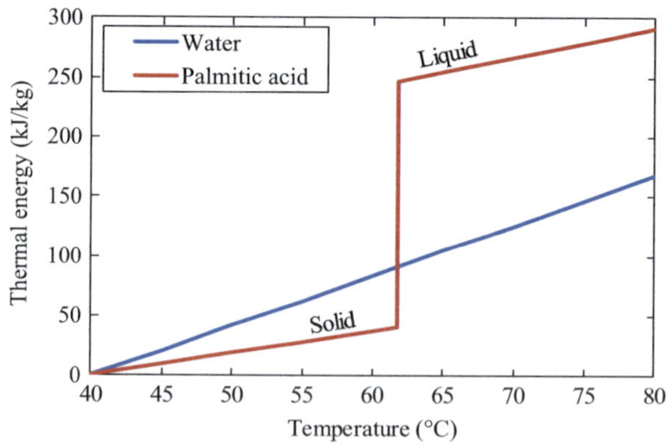

Fig. 7.18 A graph showing thermal energy versus temperature for water and palmitic acid. A blue line for water shows a steady increase in thermal energy as temperature rises. A red line for palmitic acid shows a sharp increase in thermal energy at approximately 62 °C, indicating a phase change from solid to liquid. Reference point is set at 40 °C, 1 bar. The graph includes labels "Solid" and "Liquid" for palmitic acid

decreases the temperature of the liquid, and if the PCM is not re-heated, the temperature eventually reduces below T_m, initiating solidification. This decreases the amount of dischargeable thermal energy and the efficiency of the storage.

During discharging, thermal energy is extracted from the storage material and the PCM solidifies, releasing the latent heat. The discharging process follows the same red line shown in Fig. 7.18, in ideal cases at least. The same principle of thermal energy storage can be used to provide cooling—for example, ice is used in cooling packs.

Figure 7.18 also illustrates the advantage of the latent heat storage. The slopes of the lines are determined by the specific heat capacity (kJ/kgK) of the material, which is the basis for the sensible heat storages (shown for water as the blue line). In comparison, the latent heat of the phase transition stores thermal energy at a narrow temperature range, indicated by the vertical jump in the red line in Fig. 7.18. The stored thermal energy, q_{sto} (kJ/kg), can be determined as the sum of heat transfer during the temperature changes and phase transitions of the storage material,

$$q_{sto} = \int_{T_1}^{T_m} c_{p,s} dT + \Delta H_m(T_m) + \int_{T_m}^{T_b} c_{p,l} dT + \Delta H_b(T_b) + \int_{T_b}^{T_2} c_{p,g} dT$$

where c_p is the specific heat capacity (kJ/kgK) at a constant pressure, $\Delta H_m(T_m)$ is the latent heat of melting (kJ/kg) at T_m, and $\Delta H_b(T_b)$ is the latent heat of boiling (kJ/kg) at the equilibrium boiling temperature T_b and constant pressure. The subscripts s, l, g, 1 and 2 refer to the solid, liquid, gas, start and end points, respectively. If the solid–liquid phase change is used to store thermal energy, the above equation would comprise only the sensible heat of heating solid material from the start temperature to the melting temperature (1 $\longrightarrow$ m), the latent heat of melting at T_m, and then the sensible heat of heating the liquid material from melting temperature to the end temperature T_2. Often the sensible heat parts are omitted because their share is insignificant over a small temperature range—however, the magnitude of the sensible heat depends on the applied temperature range and should always be assessed.

PCMs are typically categorized by their composition into organic and inorganic materials, and further into other classes, such as paraffins, alcohols and carboxylic acids (organic materials). Another convenient method of categorization is by phase-transition temperature, into high-temperature (> 300 °C), medium temperature (100–300 °C), and low-temperature (< 100 °C) materials. This categorization eases the selection of the PCM, as the phase-transition temperature should match the temperature levels of the application. (The above temperature ranges are approximate, and variance can be found in the literature.)

When selecting solid–liquid PCMs, their suitability is mainly determined by the latent heat of melting (J/g) and melting temperature (°C). Table 7.7 lists a few typical PCMs that have been suggested for latent heat storages. A potential PCM should have a suitable T_m for the application, as the majority of the thermal energy is absorbed and released at T_m. Even though the melting temperature limits the selection of PCMs for different applications, a large array of PCMs have been studied over a wide temperature range. A PCM should also have a large latent heat of melting because it determines the amount of thermal energy that can be stored. The latent heat of melting typically ranges from 100 J/g to 400 J/g, but high-temperature PCMs can achieve even higher values [78, 80, 109].

PCMs have different properties and characteristics, which can manifest as physical, chemical, technical, and economic limitations. To illustrate the variety of challenges related to PCMs, we highlight a few examples. Supercooling is a phenomenon where a PCM remains in the liquid state below the melting temperature—this is undesirable in conventional short-term thermal energy storage (TES) because it delays the release of heat energy and reduces the amount of dischargeable heat at T_m. A PCM may also crystallize to a different crystal structure than the thermodynamically-stable solid structure, a phenomenon known as polymorphism. This is also undesirable, as the properties of the material, such as melting temperature, latent heat of melting, and crystal growth rate, change, as shown in Table 7.7 by the properties of the metastable polymorph of erythritol. PCMs should also be chemically stable, non-toxic, and non-corrosive to prevent undesired reactions. Furthermore, because PCMs typically have low thermal conductivity, heat exchanger system should be carefully designed for adequate heat transfer rates. In technical terms, the existence of the liquid phase often requires precautions against leakage, such as envelopment or encapsulation of the PCM.

The above-mentioned examples highlight the importance of characterizing the PCM to understand and predict its behavior in applications—otherwise, the storage may perform worse than expected. There are several methods for characterizing the different properties, and obtaining reliable results can be time-consuming—see the available databases [37, 39]. Similarly, different characteristics of PCMs require individual assessments of the advantages and shortcomings for each case. A more thorough list and recommended screening practices can be found, for example, in Ding [22] and Cabeza [14].

Latent heat storage can be applied as passive systems, where the ambient temperature fluctuates (passively) above and below the melting temperature. These systems operate as thermal buffers that reduce temperature fluctuations, with applications in electronics, clothing, or building materials, for example. Active systems are charged and discharged by actively transferring heat to and from the storage. Typically, latent heat storage is recommended for short-term storage applications, ranging from hours to days. Small-scale commercial examples include building materials, heating and cooling pads, and small-scale active storage systems in buildings, similar to hot-water tanks. For example, paraffin based PCM panels were installed to the walls of test cubicles in Spain for passive cooling on a hot day [15]. Addition of the panels reduced the internal temperature of the cubicles on average by 1 °C, as shown in Fig. 7.19. In addition, the PCM also decreased fluctuations in temperature.

In long-term storage over months, heat loss gradually reduces the dischargeable energy content during the storage period. Nevertheless, if the storage is large, the efficient storage time can be prolonged, since more energy is stored, and the ratio of storage volume to surface area increases, which reduces heat loss to the environment.

Table 7.7 Thermophysical properties of selected phase-change materials from different classes

Material	T_m (°C)	ΔH_m (J/g)	$C_{p,s}$ (J/g K)	$C_{p,l}$ (J/g K)	ρ_s (kg/dm^3)	ρ_l (kg/dm^3)	λ_s (W/m K)	λ_l (W/m K)	Ref.
Organics									
Water	0	333	2.02 (– 10 °C)	4.18 (20 °C)	0.92 (– 10 °C)	0.998 (20 °C)	2.26 (– 10 °C)	0.60 (20 °C)	b
Capric acid	32	145	2.0 (15 °C)	2.1 (50 °C)	0.85 (23 °C)	0.89 (40 “C >	0.21 (9 °C)	0.17 (60 °C)	c
Palmitic acid	62	206	1.9 (25 °C)	2.4 (80 °C)	0.90 (23 °C)	0.85 (70 °C)	0.30 (19 °C)	0.22 (75 °C)	c
Paraffin vax 46–50[a]	41–43	228	2.48 (8 °C)	2.76 (80 °C)	0.83 (8 °C)	0.77 (80 °C)	0.21 (8 °C)		d
Paraffin vax 52–54[a]	49–53	220	1.97 (8 °C)	2.10 (80 °C)	0.87 (8 °C)	0.77 (80 °C)	0.19(8 °C)		d
Erythritol	118 105^{ms}	330–353 316^{ms}	1.4 (20 °C)	2.8 (120 °C)	1.44 (20 °C)	1.29 (120 °C)	0.7…0.9 (RT)	0.34 (120 °C)	e, f, g
Mannitol	166–168 157^{ms}	230–340	1.3(25 °C)	2.8 (187 °C)	1.49		0.6–1.3 (RT)	0.37(178 °C)	e, h, i
Salt hydrates									
Calcium chloride hexahvdrate	29.5	198	1.66 (25 °C)	2.3 (30 °C)			0.5–1.1 (30 °C)	0.52 (40 °C)	j
Sodium acetate trihydrate (SAT)	58	278	1.69 (25 °C)	2.98 (90 °C)	1.45 (20 °C)	1.28 (90 °C)	0.5–0.7 (30 °C)	0.40 (70 °C)	k
Salts									
$LiNQ_3$	254	360–390	1.67 (230 °C)	2.04 (270 °C)					b,l
NaCl	802	480	0.95 (330 °C)			1.56 (330 °C)			b

(continued)

Table 7.7 (continued)

Material	T_m (°C)	ΔH_m (J/g)	$C_{p,s}$ (J/g K)	$C_{p,l}$ (J/g K)	ρ_s (kg/dm^3)	ρ_l (kg/dm^3)	λ_s (W/m K)	λ_l (W/m K)	Ref.
Metal alloys									
Aluminum	660	397	1.03 (330 °C)		2.7 (25 °C)	2.38 (660 °C)	218 (530 °C)		b

ms metastable polymorph. *RT* Room temperature
[a] Commercial product, [b] Haynes et al. [40], [c] Kahwaji et al. [56], [d] Ukrainczyk [104], [e] Seppälä et al. [90], [f] Höhlein et al. [47], [g] Puupponen et al. [82], [h] Lian et al. [64], [i] Jian et al. [54], [j] Meisingset and Grønvold [71], [k] Meisingset and Gronvold [70]. and [l] Takahashi et al. [98]

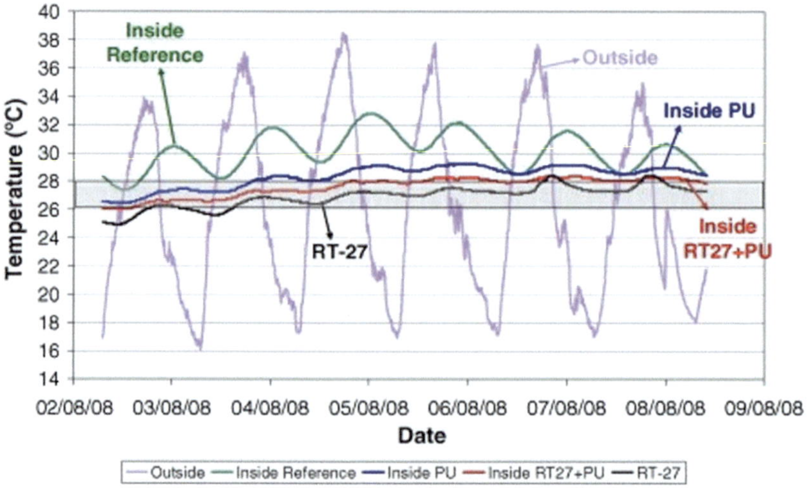

Fig. 7.19 A line graph showing temperature variations during one week from 02/08/08 to 08/08/08. Green, blue and red lines represent temperatures inside differently insulated cubicles: without insulation (Inside Reference, green), polyurethane insulation (Inside PU, blue) and polyurethane & RT-27 PCM insulation (Inside RT27+PU, red). Outside and RT-27 PCM temperatures are shown in purple and black, respectively. As shown by the red line, addition of RT-27 can reduce the inside temperature and its fluctuations

For instance, a snow storage system has supplied 90% of the hospital's cooling demand in Sundsvall, Sweden, by storing 30,000 m^3 of snow (capacity of 60,000 m^3) [92, 97]. The snow is gathered from the area into a pit and stored until summer, under a 0.2 m layer of wood chips.

While there are commercial PCM suppliers, latent heat storage has not yet been widely applied on an industrial scale. One of the main challenges is to reduce the cost of PCMs and their associated latent heat storage systems, to make them competitive against well-established water- and ground-based sensible heat storage systems. Furthermore, latent heat systems cannot offer a one-size-fits-all solution, because the properties of PCMs vary and must be assessed for different applications case by case. Therefore, more knowledge is needed on material behavior and long-term operation of large-scale pilot projects for the reliable use of PCMs and latent heat storage systems.

Indeed, current work on latent heat storage have been attempting to overcome the challenges outlined above, as the following examples show. Improvement in the thermal conductivity of PCMs has been attempted by inserting highly conductive materials, such as metals and carbon, into to the PCM. For passive storage, phase-stabilized PCMs have been synthesized by adding polymers—these help the material to maintain its structure like a solid material, even in a liquid state, preventing leakage. Supercooling has been suggested as a mechanism for long-term storage of the latent heat of melting. In this method, the melted PCM is purposely supercooled in a metastable liquid state to the storage temperature, which is below PCM's melting temperature. At this state, the latent heat of melting can be stored indefinitely, as long as solidification does not occur—and when heat is needed, the material is solidified by, for example, inserting a seed crystal, which releases the latent heat. One of the largest challenges of this method is to prevent premature solidification, while maintaining an adequate solidification rate during heat release [48, 102].

There are also several applications emerging for latent heat TES. High-temperature PCMs may be useful in Carnot batteries, where electricity is transformed to thermal energy, and then, after a storage period, back to electricity. Electrical appliances could benefit from a thermally regulating PCM, which would prevent overheating while keeping the source temperature constant. In addition, textiles, food preservation, and the life sciences could take advantage of PCMs' characteristic property of maintaining a constant temperature.

Example 7.3 Household Latent Heat Thermal Storage

Consider a latent heat storage system for heating warm water for a household. We assume that the household consumes 400 L of 40 °C water in a day, while the incoming temperature of the water is 10 °C. What size of latent heat storage could supply the warm water consumption for the household for one day?

Solution: First, the PCM should have a melting temperature above 40 °C and below the maximum operating temperature to ensure feasible operation—for this example, let's assume a maximum operating temperature of 70 °C for the storage. There are several options for PCM, but here we use sodium acetate trihydrate (SAT), because it has high latent heat of melting, and a melting point between 40 and 70 °C (58 °C, see Table 7.7).

In one day, the household uses heat for warm water as follows:

$$Q_w = V_w \rho_w \int_{T_{in}}^{T_{out}} c_{p,w} dT = 400\text{L} \times 0.998\,\frac{\text{kg}}{\text{L}} \times 4.18\,\frac{\text{kJ}}{\text{kg}^\circ\text{C}} \times (40\,^\circ\text{C} - 10\,^\circ\text{C}) = 50.06\ \text{MJ}$$

The SAT storage should supply at least the same amount heat ($Q_{sto} = Q_w$) when it is cooled from the maximum operating temperature. If we assume that, at the end of the discharge, the storage is at 40 °C, the specific heat capacities are approximately constant across the operating temperature range, and using the properties of SAT listed in Table 7.7, the mass of SAT can be calculated:

$$Q_{sto} = m_{sat} q_{sto}$$

$$q_{sto} = \int_{T_{\min}}^{T_m} c_{p,s} dT + \Delta H_m(T_m) + \int_{T_m}^{T_{\max}} c_{p,l} dT$$

$$\begin{aligned} m_{sat} &= \frac{Q_{sto}}{\left(\int_{T_{min}}^{T_m} c_{p,s} dT + \Delta H_m(T_m) + \int_{T_m}^{T_{max}} c_{p,l} dT\right)} \\ &= 50.06 \ \text{MJ} / \left(1.69 \frac{\text{J}}{\text{g °C}} \times (58 \ \text{°C} - 40 \ \text{°C}) + 278 \frac{\text{J}}{\text{g}} \right. \\ &\quad \left. + 2.98 \frac{\text{J}}{\text{g °C}} \times (70 \ \text{°C} - 58 \ \text{°C})\right) \\ &= 145.4 \ \text{kg} \end{aligned}$$

$$V_{sat} = m_{sat} / \rho_{sat,l} = 145.7 \, \text{kg} / 1.28 \, \frac{\text{kg}}{\text{L}} = 113.6 \, \text{L}$$

At least 114 L of SAT is required for the storage; in reality, the heat exchanger will increase the overall size of the storage, and heat loss during storage reduces the amount of dischargeable heat. SAT also suffers from extensive supercooling, which can inhibit the discharge process. For comparison, the size of a hot water tank with similar capacity typically ranges from 250 to 300 L.

Additionally, we can determine the share of the overall heat transferred to and from the material from sensible heat,

$$\frac{\left(\int_{T_{\min}}^{T_m} c_{p,s} dT + \int_{T_m}^{T_{\max}} c_{p,l} dT\right)}{q_{sto}}$$

In this example case, sensible heat accounts for 19% of the heat transferred to and from the material at the selected temperature range. Therefore, it should be accounted in dimensioning such a storage system.

7.4 Mechanical Energy Storage Technologies

Mechanical energy storage can be divided into: gravitational potential energy storage technologies, such as solid gravity energy storage (SGES) and pumped hydro energy storage (PHES); kinetic energy storage technologies, such as flywheel energy storage (FWES); and elastic potential energy storage technologies, such as compressed air energy storage (CAES). Below, we briefly discuss the operational principles and theoretical limitations of these four categories.

7.4.1 Solid Gravity Energy Storage Systems, SGES

Solid gravity energy storage systems are in principle formed from large weights that are lifted with electric-motor-driven pulleys when excess energy is available, and dropped when energy is needed, turning generators with pulleys. There are numerous technologies available, which are broadly discussed by Tong et al. [99] in their review, for example. As an example, Fig. 7.20 shows the largest tower-based system in the world in 2023, in Rudong, China. The system is around 122 m tall, and uses concrete blocks each weighing 32 tons—their density, which should be maximized, is more than 2000 kg/m^3.

Weights can be made from waste materials, such as mine tailings or waste concrete. Round-trip efficiencies for these systems can be between 80 and 90%. As a simple calculation, each 32-ton block can store 10.1 kWh when lifted to 122 m ($\eta_g = 0.95$)—therefore to meet the storage capacity of 100 MWh, more than 9900 blocks are needed. In Fig. 7.20 we can see there are approximately $25 \times 25 = 625$ vertical lifting channels, which means that the system requires intelligent weight logistics and storage systems that can pack the weights efficiently at the upper level; however, the details of the system are not available.

Figure 7.21 shows the dynamics of the block-pulley system for different conditions.

Note that the upwards-accelerating and -decelerating cases can be similarly solved from Newton's laws ("+" direction downwards):

$$\sum F = ma$$

To keep the forces on the cable finite, we need to allow a finite acceleration period, assumed to be a constant linear acceleration, a. In case (a), with fixed position or constant velocity upwards or downwards, the force in the cable is equal to the gravitational force. When accelerating downwards, the force on the cable, F_w, is smaller than gravity, and when decelerating at the bottom, the force is larger than gravity. In addition to mechanical and electrical losses, both of these transition periods affect the system's efficiency.

Fig. 7.20 Gravity energy storage system under construction in Rudong, China, 2023. Contractor Energy Vault, Customer China Tianying Group. 25 MW power, 100 MWh storage capacity. *Source* https://www.energyvault.com © EnergyVault

The amount of potential energy charged in the system is equal to $E = mgz$. For the constant-velocity period of motion, the electrical power and energy needed in the charging phase, P_m and E_{in}, and the generator power and energy, P_g and E_{out} in the discharging phase, are given by

$$E_{in} = P_m \Delta t = \frac{mgu\Delta t}{\eta_m \eta_{mec}} = \frac{mg\Delta z}{\eta_m \eta_{mec}}$$

$$E_{out} = P_g \Delta t = mgu\Delta t \eta_g \eta_{mec} = mg\Delta z \eta_g \eta_{mec}$$

The maximum power that can be produced from a lowering mass m at constant speed u can be calculated from

$$P_g = mgu\eta_g \eta_{mec}$$

For example, in the case of Rudong, to generate 25 MW with a dropping speed of 1 m/s, we would need a mass of 2800 tons; if the speed is 5 m/s, 655 tons is sufficient ($\eta_g \eta_{mec} = 0.9$). These correspond to a total of 2800/32 = 88 and 2800/655 = 20 of the 32 ton blocks, all moving simultaneously. If the vertical height is 120 m, power would be available for 120 s for the 1 m/s dropping speed, and 24 s for the 5 m/s dropping speeds. If the entire stored mass of ~ 320,000 tons is dropped together, for

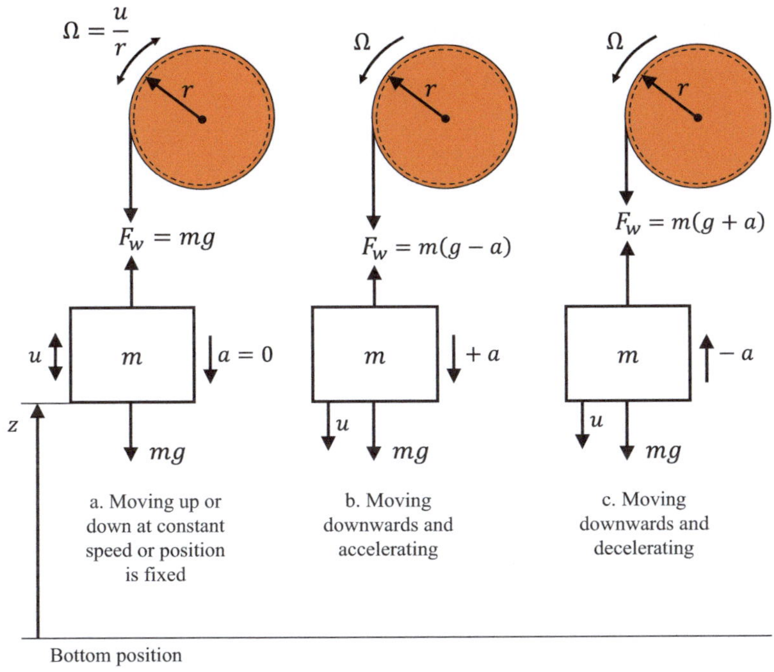

Fig. 7.21 Dynamics of moving weight and force on the wire cable: **a** fixed position or constant velocity, **b** accelerating downwards, and **c** decelerating downwards

the 100 MWh capacity, the velocity needed is only 0.0085 m/s to provide 25 MW of power, for a duration of four hours—of course, using the weights all at once is not practical, and the system has to be divided into smaller units.

If we also consider the acceleration and deceleration phases and their decreased efficiencies, the power fed in and obtained are as follows:

$$E_{in} = \frac{m(g+a)\Delta z_a}{\eta_{m.a}\eta_{mec}} + \frac{mg(\Delta z - \Delta z_a - \Delta z_d)}{\eta_m \eta_{mec}} + \frac{m(g-a)\Delta z_d}{\eta_{m.d}\eta_{mec}}$$

$$E_{out} = m(g-a)\Delta z_a \eta_{g.a}\eta_{mec} + mg(\Delta z - \Delta z_a - \Delta z_d)\eta_g \eta_{mec} + m(g+a)\Delta z_d \eta_{g.d}\eta_{mec}$$

If a and Δz are the same for the acceleration and deceleration phases, and if the efficiency is also the same at smaller velocities, the equations reduce back to their simplified versions shown above. The energy loss in acceleration is recovered in the breaking phase, a phenomenon also employed in hybrid vehicles, for example. In reality, the efficiency will decrease, as the efficiency is typically an increasing function of the generator rpm and reaches a maximum value at the design point.

Finally, the round-trip efficiency of an ideal system can be calculated as follows.

Table 7.8 Solid gravity energy storage materials' properties [99]

Material	Density, t/m^3	Price, \$/t (2022)
Lead	11.3	251.1
Brass	8.8	1177.2
Iron	7.9	15.7
Tin	7.3	5399.5
Zinc	7.1	439.5
Aluminum	2.7	337.5
Concrete	2.5	7.4
Sand	1.5	1.6

$$\eta = \frac{E_{out}}{E_{in}} = \frac{mg\Delta z \eta_g \eta_{mec}}{\frac{mg\Delta z}{\eta_m \eta_{mec}}} = \eta_g \eta_m \eta_{mec}^2$$

Example 7.4 Design of solid gravity storage, system dimensions, and materials selection

We wish to design a solid gravity-based energy storage system able to generate 500 MW using the Energy Vault technology presented above. As discussed, the system is based on 32 t solid blocks, and the height of the system is 122 m. Based on https://www.energyvault.com, we could assume that the horizontal area needed for a 100 MWh/25 MW system is approximately 130 m × 130 m, that a horizontal distance between the blocks of 0.5 m is needed for supporting structures, and that the round-trip efficiency is equal to 0.9 as above.

Calculate: 1. The horizontal area needed for a 500 MW system and its storage capacity; 2. The size of the blocks, assuming they are cubes of concrete; and 3. Estimate the size of the storage if lead blocks are used instead, and discuss the technical, economic, and sustainability benefits of using concrete or waste materials as storage material—see Table 7.8 for material properties.

Solution

1. To produce 500 MW, we need to have 500/25 = 20 of the 25 MW units. One 25 MW unit requires a land area of approximately 130 × 130 = 16,900 m^2, so 20 units would require an area of 16,900 × 20 = 340,000 m^2, corresponding to a square of around 580 m × 580 m. Storage capacity is also 20 times larger, 0.1 GWh × 20 = 2 GWh, and this would be sufficient for 4 h at full power as above.
2. The size of the blocks is simply calculated from their weight, 32,000 kg, and density, 2500 kg/m^3, as $L = \sqrt[3]{32{,}000/2500} = 2.34$ m. Having 25 of these side by side, plus the required 0.5 m spacing, gives 26 × 0.5 m + 25 × 2.34 m = 71.5 m, fitting into the estimated width.
3. Using lead as the block material would decrease the size significantly as the density is much larger, 11,300 kg/m^3. The size would be only $L = \sqrt[3]{32{,}000/11{,}300} = 1.41$ m. Having 25 of these side by side, plus the 0.5

m spacing, gives 26 × 0.5 m + 25 × 1.41 m = 48.3 m, decreasing the required land area by a fraction of $1 - 48.3^2/71.5^2 = 0.54$. The total mass of the blocks needed in the 500 MW/2 GWh system would be $m = E_{out}/(gz\eta_g) = 6.3 \times 10^6$ t. The price of the lead blocks is therefore 6.3×10^6 t × 251.1 \$/t = 1510 M\$, and of the concrete blocks is 6.3×10^6 t × 7.4 \$/t = 47 M\$—just on a cost-basis of the block materials, concrete is much better. However, the other building and frame materials required for the smaller lead-based system should be considered in detail also. From a sustainability point of view, using concrete or even waste-concrete materials would be much better than using virgin lead, steel, or other valuable metals for this purpose.

Up-scaling the solid gravity energy storage technology requires huge amounts of materials for the weights, and for a sustainable solution we need to find a maximally-circular-economy-based solution—for example, using mine tailings from the increasing excavation of rare earth elements.

7.4.2 Pumped Hydro Energy Storage, PHES

Globally, pumped hydro is the most-used energy storage technology, with more than 160 GW of installed capacity [84]. This mature technology represented more than 90% of global stationary storage capacity, including 93% of the storage capacity in the United States at the end of 2020, and an estimated 97% of the total storage capacity installed in the European Union (EU). Pumped storage capacity is heavily concentrated, with more than 86% of the capacity in 2021 installed in just four regions: China (36 GW), Japan (27.6 GW), the United States (21.9 GW), and the European Union (52.2 GW) [84]. Figure 7.22 shows an example of a pumped hydro energy storage system in Venda Nova II, Portugal.

In conventional hydropower and pumped hydro power plants built on flatter landscapes, the dam can create a large negative environmental impact by flooding and degrading the land area within the reservoir—see for example Fig. 7.23 from the Sobradinho reservoir in Brazil [49].

As Fig. 7.23 shows, the distance that the "shore" location moves, is 15–20 km between fully charged and discharged states. This type of reservoir has a huge impact on the local ecosystems because, during the dry season, flora and fauna adapted to life close to the water find themselves several kilometers from the river, with wasteland in between. For this reason, droughts can be particularly devastating. The land-use impacts of pumped hydro storage are discussed in Sect. 2.4.4.

In contrast to these, steep valley locations are preferable since they allow large vertical height variation, they occupy less land, evaporation losses are smaller due to the smaller water surface area, and impacts on the area's nature, and methane and CO_2 emission from the soil, are far less dramatic. In these steep valley reservoirs, the water surface can move vertically as much as 60–150 m; the highest variation, 164 m, is reported in the Limberg II site in Austria [49]. As the power of the pump/

Fig. 7.22 Pumped hydro energy storage in Venda Nova II, Portugal. The system uses two parallel reversible Francis pump turbines, with total average power 192 MW. The synchronous speed of the generators is 600 rpm, average net head is 410 m, and discharge rate is 50 m^3/s ([108], © Voith)

Fig. 7.23 Conventional seasonal pumped hydro energy storage in Sobradinho, Brazil [49]. Gray and black areas refer to full and empty storages, respectively (vertical height difference of 12 m). Flooded grey area is 4214 km^2 and useful water volume is 28.7 km^3

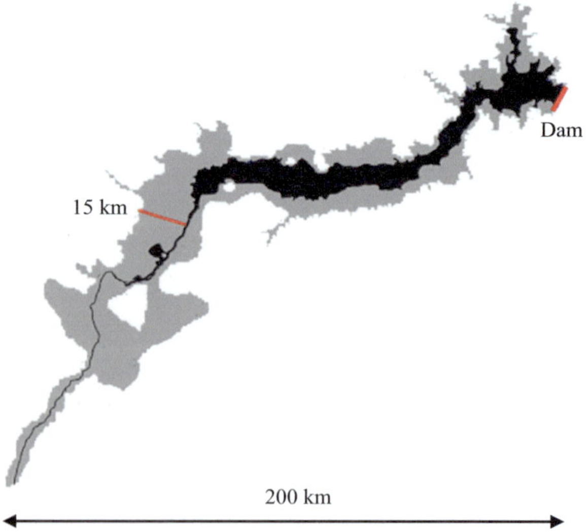

turbine is directly proportional to head and volumetric flow rate, $P \propto \dot{V}\Delta z$, and if the head changes significantly, the flow rate must change inversely to maintain constant production. Changing the flow rate means changing the flow velocity in the pump turbine—and to operate the turbine at optimal condition, its rpm must be changed, in analogy to what we learned in Chap. 5 for wind turbines (optimal tip-speed ratio). This variable speed design was made possible by the introduction of variable speed generators and converter technologies, as with wind power plants.

Traditional pumped hydro energy storage relies on high heads—that is, large vertical distance, Δz, between the surfaces of the upper and lower reservoirs—to reach the required power and storage capacity. Round-trip efficiencies, from pumping to production, are commonly around 80% [83]. The largest units are already in the range of 3600 MW [2]. However, suitable topography is not always available, and low-head technologies are also being developed for flatter landscapes [46, 86]. These low-head systems have vertical heights of 2–30 m, meaning that the area of land impacted is large, and therefore the potential negative impacts on nature caused by the hydro energy storage will be higher as well.

Pumped hydro energy storage systems use two reservoirs at different vertical heights. During excess power production from other renewable sources such as wind and solar, water is pumped to the upper reservoir, and during peak consumption hours in the grid, water can be released back to the lower reservoir through the turbines to produce power. Since the 1950s, most systems rely on combined pump-turbine technologies that greatly simplify the system configuration [83].

Energy balances for pumped hydro energy storage systems can be derived from the open flow system energy equation derived in Chap. 3. Assuming that the system is momentarily at steady state, for generation and pumping directions, with subscripts u and l refering to upper and lower reservoir surfaces respectively (see also Fig. 7.24 below), we get

$$p_u + \rho g z_u + \frac{\rho u_u^2}{2} - \Delta p_t - \sum_{i=1}^{N} \left(\xi_i \frac{L_i}{d_i} + C_i \right) \frac{\rho u_i^2}{2} = p_l + \rho g z_l + \frac{\rho u_l^2}{2}$$

$$p_l + \rho g z_l + \frac{\rho u_l^2}{2} + \Delta p_p - \sum_{i=1}^{N} \left(\xi_i \frac{L_i}{d_i} + C_i \right) \frac{\rho u_i^2}{2} = p_u + \rho g z_u + \frac{\rho u_u^2}{2}$$

In the upper equation the terms on the left are static pressure, potential energy and kinetic energy at the upper reservoir surface, turbine pressure drop, friction and minor flow losses in the piping system, and on the right they are static pressure, potential energy and kinetic energy at the lower reservoir surface. The terms in the lower equation on the left are static pressure, potential energy and kinetic energy at lower reservoir surface, pump pressure increase, friction and minor flow losses in the piping system, and on the right they are static pressure, potential energy and kinetic energy at upper reservoir surface. L_i and d_i are pipe section lengths and diameters, ξ_i is the friction loss coefficient obtained from the Moody diagram, and C_i is the minor

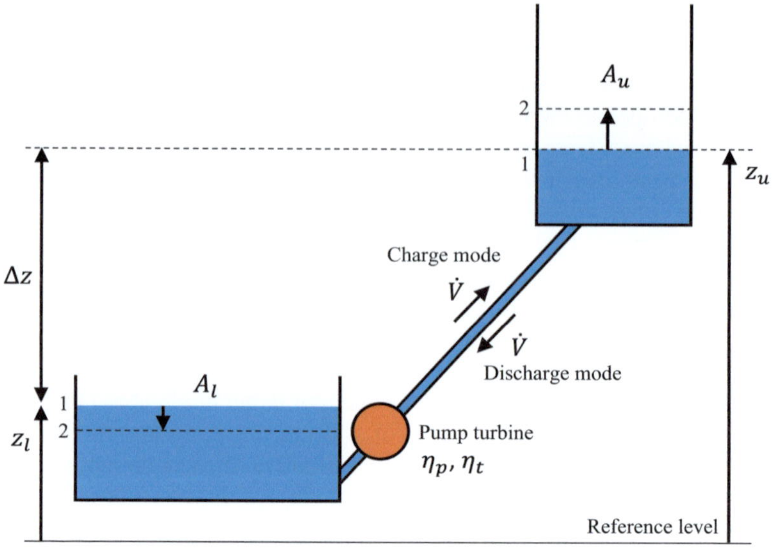

Fig. 7.24 Pumped hydro energy storage principle

loss coefficient for valves, contractions, expansions, bends, etc. (see for example, Streeter and Whiley [96]).

Since $p_u = p_l$ and $u_u = u_l$ and we can write the local velocity in terms of volumetric flow rate, these equations simplify to

$$\Delta p_t = \rho g \Delta z - \sum_{i=1}^{N} \left(\xi_i \frac{L_i}{d_i} + C_i \right) \rho \frac{8\dot{V}^2}{\pi^2 d_i^2}$$

$$\Delta p_p = \rho g \Delta z + \sum_{i=1}^{N} \left(\xi_i \frac{L_i}{d_i} + C_i \right) \rho \frac{8\dot{V}^2}{\pi^2 d_i^2}$$

For accurate calculations the loss terms must be considered, but for now we assume that they are negligible for simplicity. From the energy balances for the power of the pump and turbine, we can then write

$$P_p = \frac{\dot{m}\Delta h}{\eta_p} = \frac{\dot{m} v \Delta p}{\eta_p} = \frac{\dot{V} \rho g \Delta z}{\eta_p}$$

$$P_t = \dot{m}\Delta h \eta_t = \dot{V} \rho g \Delta z \eta_t$$

Enthalpy change is defined as

$$\Delta h = \Delta u + p_2 v_2 - p_1 v_1 + \frac{u_2^2}{2} - \frac{u_1^2}{2} + g\Delta z \sim v\Delta p$$

where Δp is across the pump or turbine. Since liquid is assumed incompressible, $v_2 = v_1$, and flow area is assumed to stay constant within the pump and turbine, so $u_1 = u_2$. We see that the same power can be obtained by a high head and low volume flow rate, or vice versa. The head, $z_u - z_l = \Delta z$, is a function of time—it decreases when discharging, and increases when charging. Figure 7.24 shows a schematic of this system.

We can write a combined mass-conservation equation for both reservoirs—assuming the total mass of water remains constant, neglecting precipitation and evaporation terms, we get.

$$\frac{dV}{dt} = \frac{dV_u}{dt} + \frac{dV_l}{dt} = \frac{A_u dz_u}{dt} + \frac{A_l dz_l}{dt} = 0$$

$$dz_u - dz_l = d\Delta z$$

where u and l refer to upper and lower reservoirs. We can then solve for the derivative of head with time, taking the positive direction of the volumetric flow rate outwards from the upper reservoir:

$$\frac{d\Delta z}{dt} = \frac{dz_u}{dt}\left(1 - \frac{A_u}{A_l}\right) = \frac{\dot{V}}{A_u}\left(1 + \frac{A_u}{A_l}\right)$$

From this we see that if the lower reservoir is a large body of mass of water, such as a sea or a river, only the upper reservoir level will change; or, if the reservoirs have equal surface areas, for example, the rate of change of the head is twice the rate of upper-level change. We can calculate the energy stored in charging, and the energy released in discharging, as

$$E_{in} = \int_0^{\tau} P_p dt = \int_0^{\tau} \frac{\rho g \Delta z}{\eta_p} \frac{A_u}{\left(1 + \frac{A_u}{A_l}\right)} \frac{d\Delta z}{dt} dt$$

$$= \int_{\Delta z_1}^{\Delta z_2} \frac{\rho g \Delta z}{\eta_p} \frac{A_u}{\left(1 + \frac{A_u}{A_l}\right)} d\Delta z = \frac{\rho g}{\eta_p} \frac{A_u}{\left(1 + \frac{A_u}{A_l}\right)} \frac{1}{2}\left(\Delta z_2^2 - \Delta z_1^2\right)$$

$$E_{out} = \int_0^{\tau} P_t dt = \int_{\Delta z_1}^{\Delta z_2} \rho g \Delta z \eta_p \frac{A_u}{\left(1 + \frac{A_u}{A_l}\right)} d\Delta z = \rho g \eta_p \frac{A_u}{\left(1 + \frac{A_u}{A_l}\right)} \frac{1}{2}\left(\Delta z_2^2 - \Delta z_1^2\right)$$

For a system where the lower reservoir is significantly larger in area, $A_l \gg A_u$, these reduce to

$$E_{in} = \frac{\rho g A_u}{\eta_p} \frac{1}{2}\left(z_{u.2}^2 - z_{u.1}^2\right) = \frac{\rho g A_u (z_{u.2} - z_{u.1})}{\eta_p} \frac{(z_{u.2} + z_{u.1})}{2} = \frac{mg\bar{z}}{\eta_p}$$

$$E_{out} = \rho g A_u (z_{u.2} - z_{u.1}) \eta_t \frac{(z_{u.2} + z_{u.1})}{2} = mg\bar{z}\eta_t$$

where $\bar{z}$ corresponds to average height difference between start and end water surface levels. In the case of solid storage, the stored mass was constant but its vertical location changes, while with pumped hydro, the vertical reservoir bottom level is fixed but the storage water mass changes above this fixed bottom level.

Example 7.5 Comparison of 500 MW/2 GWh Pumped Hydro Energy and Solid Gravity Storage Systems

Compare two pumped hydro energy storage and the solid gravity energy systems each providing 500 MW and 2 GWh. They both have an area of 340,000 m^2, and in each case the base level of the upper-most storage location is 122 m from the lower position (that is, all water and weight blocks are lifted above this limit). We assume that the lower reservoir for the pumped hydro is a large river, and its surface level remains constant.

1. Calculate the required active vertical height difference, $(z_{u.2} - z_{u.1})$, in the upper reservoir for the PHES. Assume that pump and turbine efficiencies are $\eta_t = \eta_p = 0.9$.
2. Estimate the required upper-level storage height for the solid gravity storage, assuming a 0.5 m spacing in all directions between the weights.

Solution:

1. The upper reservoir height can be calculated from the required storage capacity,

$$E_{out} = \rho g \eta_t A_u \frac{\left(z_{u.2}^2 - z_{u.1}^2\right)}{2}$$

$$H = z_{u.2} - z_{u.1} = \sqrt{z_{u.2}^2 + \frac{2E_{out}}{\rho g \eta_t A_u}} - z_{u.1}$$

$$= \sqrt{122^2 + \frac{2 \times 2 \times 10^9 \times 3600}{1000 \times 9.81 \times 0.9 \times 340000}} - 122 = 18.3 \text{ m}$$

   So the water level of fully charged reservoir is at 140.3 m from the river's surface.

2. In the case of solid gravity storage with concrete blocks, the upper-level storage height requirement can be estimated as follows. Since we assume that the system is horizontally scalable, it is sufficient to analyze a single 25 MW unit. We can pack (130 m – 0.5 m)/(2.34 m + 0.5 m) = 46 blocks along a side, and so we can fit 46 × 46 = 2116 blocks in each storage layer. Since 9900 blocks were needed for a 25 MW unit, we require 9900/2116 ~ 5 storage layers. And since we assume a 0.5 m vertical spacing between layers, the height of the upper level is equal to 5 × (2.34 + 0.5) m = 14.2 m. This would make the solid storage slightly

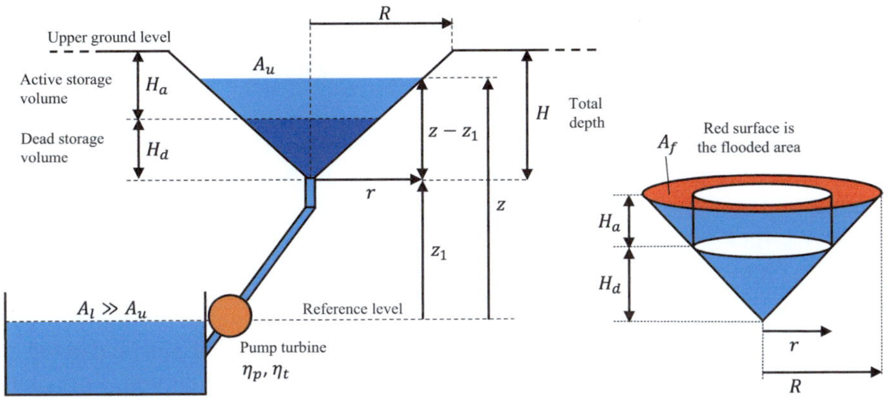

Fig. 7.25 Pumped hydro energy storage system with a conical upper reservoir, simplified model of the Muquém site. The Red surface is the flooded area. Note that dimensions are not to scale

smaller than the pumped hydro technology, however without the detailed system configuration this is only a rough estimate.

Example 7.6
Analyze the planned Muquém seasonal pumped hydro energy storage in Brazil. Higher and lower levels of the upper reservoir from the reference level of the river surface are 289 and 139 m, respectively. The flooded area is 52 km^2 and active storage volume is 7.8 km^3. Average turbine flow rate is 958 m^3/s, and generator efficiency is assumed to be 0.9. Calculate: 1. Land use for the upper reservoir in km^2; 2. The storage capacity in TWh; 3. The turbine power in MW; and 4. The storage density in km^2/TWh. Figure 7.25 shows the system dimensions.

Solution:

1. Land requirement for the upper reservoir can be estimated from the cut-cone geometry. We assume that the upper reservoir has circular cross section, and the side wall of the cone is straight. The acceptable active volume height variation given is $H_a = 150$ m. The cross section's outer ring is equal to a flooded area of $A_f = 52$ km^2, we can write this as follows and solve total radius of the reservoir:

$$A_f = \pi R^2\left(1 - \left(\frac{r}{R}\right)^2\right) = \pi R^2\left(1 - \left(1 - \frac{H_a}{H}\right)^2\right)$$

$$R^2 = \frac{A_f}{\pi\left(1 - \left(1 - \frac{H_a}{H}\right)^2\right)}$$

We also know the active volume, 7.8 km^3, for which we can write

$$V_a = \frac{\pi}{3}\left(R^2H - r^2(H - H_a)\right) = \frac{HA_f}{3}\frac{\left(1 - \left(1 - \frac{H_a}{H}\right)^3\right)}{\left(1 - \left(1 - \frac{H_a}{H}\right)^2\right)}$$

We can iteratively solve this equation for the total reservoir height, and with the given values we get $H = 368.6$ m and $R = 5053$ m. Note that H is only a shape parameter describing the active volume, and the dead volume might be a different shape. We then get $z_1 = 289 - 368.6 = -79.6$ m (also a shape parameter). The total area requirement, including the area occupied by the dead volume, is therefore $A = \pi \times 5.053^2 = 80.2\ \text{km}^2$.

2. The equation for the stored energy in a cone-shaped reservoir can be derived as in the discussion above, integrating the turbine power equation with time (note the negative sign):

$$E_{out} = -\int_0^t P_t dt = -\int_0^t \dot{V}(t)\rho g z(t)\eta_t dt$$

$$= -\int_0^t \pi r^2 \frac{dz}{dt}\rho g z(t)\eta_t dt$$

$$= -\int_{z_1+H}^{z_1+H_d} \pi r^2 \rho g z \eta_t dz$$

Note that the local radius of the cone is a function of the height from the cone bottom, $(z - z_1)$. From Fig. 7.25 we see that $r/(z - z_1) = R/H$, where R and H are the total height and radius of the reservoir cone, respectively. Inserting these into the equation for E_{out} and integrating gives

$$E_{out} = -\int_{z_1+H}^{z_1+H_d} \pi\left(\frac{R}{H}\right)^2 (z - z_1)^2 \rho g z \eta_t dz$$

$$= -\pi\left(\frac{R}{H}\right)^2 \rho\, g\, \eta_t \int_{z_1+H}^{z_1+H_d} (z - z_1)^2 z\, dz$$

$$= -\pi\left(\frac{R}{H}\right)^2 \rho\, g\, \eta_t$$

$$\left(\frac{(z_1 + H_d)^4 - (z_1 + H)^4}{4} - \frac{2\left((z_1 + H_d)^3 - (z_1 + H)^3\right)z_1}{3} + \frac{\left((z_1 + H_d)^2 - (z_1 + H)^2\right)z_1^2}{2}\right)$$

With the values $\eta_t = 0.9$, $\rho = 1000$ kg/m^3, $z_1 + H_d = 139$ m and $z_1 + H = 289$ m, the storage capacity is equal to 4.63 TWh. A more detailed analysis by Hunt et al. [49], including a tributary river for the upper reservoir, estimates the storage capacity to be 10 TWh.

3. Turbine power at average head of 214 m is $P_t = 958 \times 214 \times 1000 \times 9.81 \times 0.9 =$ 1.8GW.
4. Land use intensity is the ratio of land requirement to storage capacity: 80.2 km^2/ 4.63 TWh = 17.3 km^2/TWh.

If the pumped storage system uses a river as the water resource, instead of damming the river, as in the case in Muquém site, it would be better to have the pumped storage system installed in parallel so that the ecosystem of the river is minimally disturbed. This concept, illustrated in Fig. 7.26, is referred as Run-of-the-River Seasonal Pumped-Storage (RRS-PS), and is also planned for Muquém site [49]. The characteristics of the buffer reservoir are not presented for simplicity.

This system is used to extract continuous amounts of water from the river during periods of high river flow, and then flexible amounts of water back to the river during low-flow periods. The buffer storage is used as a standard lower reservoir, enabling daily to weekly operation—its main purpose is to avoid harmful variation in the main river ecosystems. However, even this system is not free from problems, since pumping large amounts of water to the upper reservoir can harm fish, and variable flow velocities and heights in the main river affects habitats on the shoreline. In

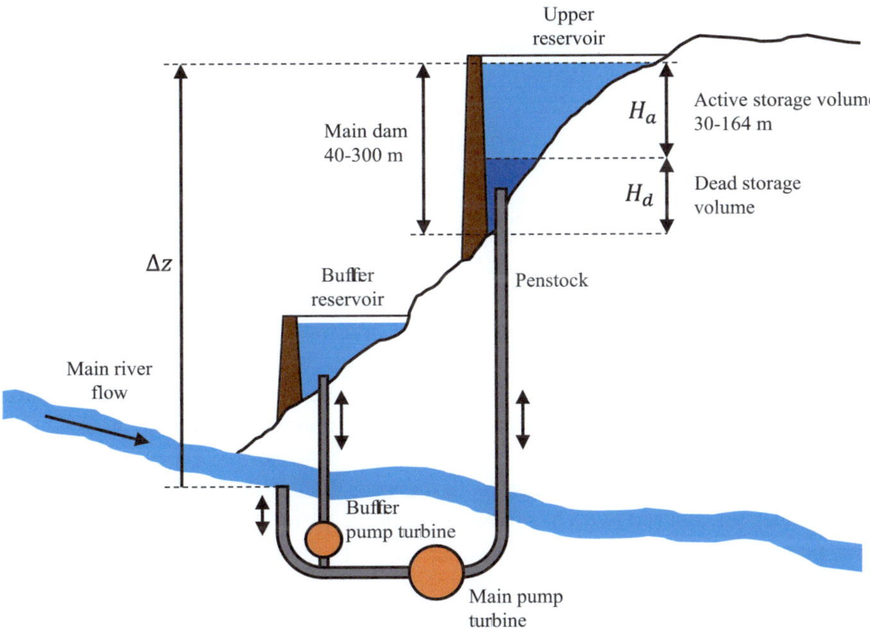

Fig. 7.26 Principle of the Run-of-the-River Seasonal Pumped-Storage system [49]

addition, as the Muquém site would utilize a tributary river for the upper reservoir, its nature impacts should also be considered.

7.4.3 *Flywheel Energy Storage, FESS*

In principle, the core of the flywheel energy storage system, FESS, is simply a rotating mass containing kinetic energy. Flywheel energy storage can be charged by using a motor to increase the rotational speed, and discharged by connecting the rotating mass to a generator that decreases the rotational speed through breaking torque and power, while converting the kinetic energy into electrical energy.

There are several reviews written on flywheels—for example, Arani et al. [3], Mousavi et al. [75], Choudhury [17], and and Li and Palazzolo [63]—and these are excellent sources for scientific papers on the topic. In this section we will review the basic operating principles of FESSs, discuss some physical limitations, and try to understand the applicable scale.

Flywheels are everywhere. For example, in a car engine the purpose of the flywheel is to smooth the oscillating torque from the crankshaft, driven by the forces from linearly moving pistons. FESS can be used in power systems to store energy, for example to create a real inertia for a system that is otherwise based on variable speed, such as in converter-coupled wind turbines. In contrast to traditional power production that relies on constant-speed grid-forming steam- and gas-turbine power generators, variable-speed grid-following systems require a stabilizing component that maintains real momentum in the system.

Figure 7.27 (left) shows the main components of a basic flywheel energy storage system (modified from Choudhury [17] and Li and Palazzolo [63]), including the flywheel rotor and the generator/motor on the axel, where most of the energy is stored. There are several ways to arrange this rotating unit—for example, the flywheel rotor could serve as the generator-motor, placed outside the fixed stator in the center [17]. Radial and thrust bearings at both ends of the axel are often magnetic, which reduces losses by avoiding all physical contact between the fixed components and the rotating mass. In the system presented here, the generator is located outside the sealed vacuum housing, to make cooling more effective. The pressure in the sealed casing is kept as low as possible with a vacuum pump, to avoid viscous friction between the rotor and gas inside the casing. A 3-phase generator is connected to a bi-directional motor/generator to allow connection of the storage with a constant-frequency grid.

In a system used to stabilize the grid the flywheel is constantly rotating synchronously with the grid frequency—if there are fluctuations in the grid frequency, stability can be assured with the storage system. Combining this storage with a wind-power-driven system can increase the capacity factor significantly (e.g., [50]).

One of the world's largest flywheel energy storage systems, constructed by Siemens Energy in Ireland in 2023, uses a single 120 tonne flywheel rotating at 3000 rpm. The system also has a grid stabilizer (synchronous condenser) and provides an inertia of 4 GWs (www.siemens-energy.com). As a smaller-scale example, Beacon

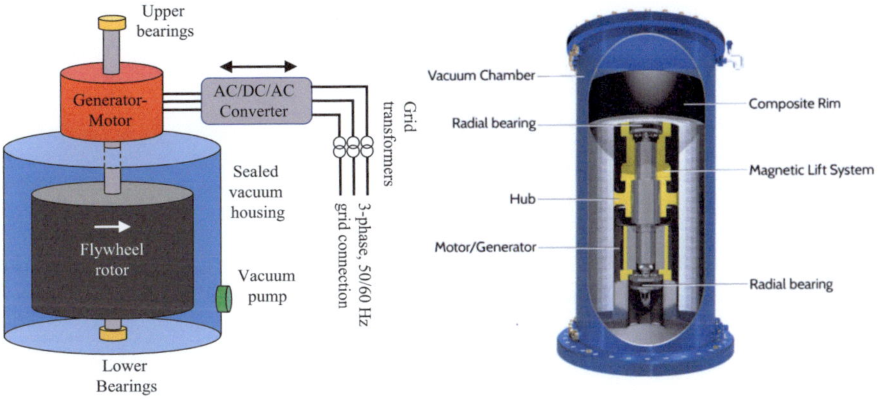

Fig. 7.27 Components of the flywheel energy storage (left, modified from Choudhury [17]) and example of a commercial flywheel storage system (Beacon Power)

Power offers several different sizes of modular systems up to 25 kWh/300 kW or 36 kWh/145 kW, with maximum rotational speed of 16,000 rpm. These units can be combined to reach the desired power and capacity of the overall energy storage system.

To examine the operation of these systems, let us first analyze the amount of energy stored in a rotating mass, m, with angular velocity of ω around a pivot point p—see Fig. 7.28.

We consider the system to be composed of infinitesimal masses, m_i, rotating at speed u_i at a fixed distance r_i from the pivot point. Kinetic energy of a single mass is equal to $1/2 m_i u_i^2$. Summing over the solid object,

$$E_{kin} = \sum_{i=1}^{N} \frac{1}{2} m_i u_i^2 = \sum_{i=1}^{N} \frac{1}{2} m_i \omega^2 r_i^2$$

This can be presented in integral form as

Fig. 7.28 Schematic of the flywheel parameters

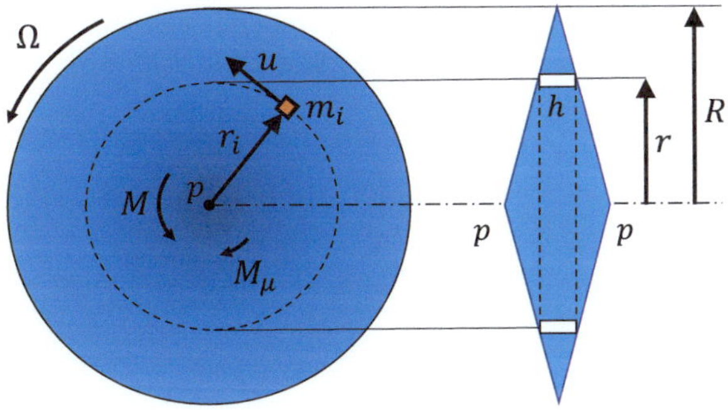

$$E_{kin} = \int_{R_0}^{R} \frac{1}{2} 2\pi r^2 h(r)\rho(r)\omega^2 r^2 dr$$

$$= \frac{\omega^2}{2} \int_{R_0}^{R} 2\pi r^3 h(r)\rho(r)dr = \frac{1}{2} I\omega^2$$

where ω is the angular velocity of the system, $h(r)$ is the radial thickness profile of the rotationally symmetric volume, $\rho(r)$ is the density at distance r from the rotational axis, and the I is called the inertial moment, given by

$$I = \int_{R_0}^{R} 2\pi r^3 h(r)\rho(r)dr$$

In reality, not all of the kinetic energy is available for use. Typically, the rotational speed is allowed to drop to a certain fraction of the maximum speed, to avoid dynamic s. For example, Bernard et al. [10] suggested a ratio of maximum to minimum rotational speeds of 3—that is, $\omega_{\min} = \omega_{\max}/3$. Then, the energy storage capacity equation becomes

$$E_{kin} = \frac{1}{2} I\left(\omega_{\max}^2 - \omega_{\min}{}^2\right) = \frac{1}{2} I\omega_{\max}^2 \left(1 - \left(\frac{\omega_{\min}}{\omega_{\max}}\right)^2\right) = \frac{1}{2} I\omega_{\max}^2 f_\omega$$

Based on a set ratio of rotational speeds of 3, we can utilize ~ 89% of the stored maximum energy ($f_\omega = 0.89$).

Example 7.7
Calculate the inertial moment of a homogenous constant-thickness steel disk, where radius $R = 1$ m, $R_0 = 0.05$ m, axial disk thickness $h = 1$ m, density ρ =7800 kg/m^3. Find the stored energy when the disk rotates at 10,000 rpm. Assume that $\omega_{\min} = \omega_{\max}/3$.

Solution: The inertial moment can be calculated from the integral equation derived above. For a constant-thickness and -density disk, we can write

$$I = 2\pi h\rho \int_{R_0}^{R} r^3 dr$$

$$= \frac{2\pi h\rho(R^4 - R_0^4)}{4} = \frac{m\left(R^2 + R_0^2\right)(R^2 - R_0^2)}{2\left(R^2 - R_0^2\right)} = \frac{m(R^2 + R_0^2)}{2}$$

where m is the total mass of the flywheel, and we use $a = R^2$ and $b = R_0^2$, and note that $\left(a^2 - b^2\right) = (a + b)(a - b)$ to get the final form.

With the numeral values provided we get $I = 12{,}252$ kg m^2, and the mass of the disk is 24.4 tons. Angular velocity is equal to 10,000/60 × 2π = 1047 rad/s. Energy storage capacity is equal to

$$E_{kin} = \frac{1}{2}12,252 \times 1047^2 \times 0.89 = 5979\ \text{MJ} = 1.66\ \text{MWh}$$

However, the mechanical strength of the material and flywheel structure sets the maximum rotational speed. Based on Sorensen [93], we can calculate the maximum stress $\sigma_{\max}$ in a cylindrical disk-shaped flywheel, made from isentropic homogenous material. For a disk with a hole, we get

$$\sigma_{\max} = \frac{\rho\omega^2 R^2}{4}\left((3 + \nu) + (1 - \nu)\left(\frac{R_0}{R}\right)^2\right)$$

and for a disk without a hole,

$$\sigma_{\max} = \frac{\rho\omega^2 R^2}{8}(3 + \nu)$$

Here ν refers to Poisson's ratio (for steel $\nu = 0.3$, aluminum 0.33) and R_0 and R are the hole and outer disk diameters, respectively. This maximum stress occurs in tangential direction, and is located at the inner surface of the hole at R_0. For example, the maximum stress in the steel flywheel analyzed previously would be 7061 MPa, which significantly exceeds the tensile strength of the material, 250 MPa for normal structural steel, and even exceeds the strength of 2617 MPa for 2800 Maraging steel. The above construction would therefore not be possible.

Turning the question the other way around, what is the maximum rotational speed that above system can take? From the equations above we can solve for rotational speeds of 1882 rpm and 6088 rpm for structural and Maraging steels, respectively. The corresponding energy storage potentials are 0.059 and 0.62 MWh, and energy storage densities are 19 and 196 kWh/m^3.

On the other hand, we can also combine the equations for kinetic energy and maximum stress to obtain

$$\omega^2 = \frac{4\sigma_{\max}}{\rho R^2\left((3 + \nu) + (1 - \nu)\left(\frac{R_0}{R}\right)^2\right)} = \frac{u^2}{R^2}$$

$$E_{kin} = \frac{1}{2}I\omega^2 f_\omega = \frac{2I\sigma_{\max} f_\omega}{\rho R^2\left((3 + \nu) + (1 - \nu)\left(\frac{R_0}{R}\right)^2\right)}$$

$$= \frac{m(R^2 + R_0^2)\sigma_{\max} f_\omega}{\rho R^2 \left((3+\nu) + (1-\nu)\left(\frac{R_0}{R}\right)^2 \right)}$$
$$= \frac{V(R^2 + R_0^2)\sigma_{\max} f_\omega}{R^2 \left((3+\nu) + (1-\nu)\left(\frac{R_0}{R}\right)^2 \right)}$$

where V refers to the volume of the solid material. We can solve for the energy density, J/m^3, for the disk with and without a hole, using the relevant equation from above:

$$\frac{E_{kin}}{V} = \sigma_{\max} f_\omega \frac{\left(1 + \left(\frac{R_0}{R}\right)^2\right)}{\left((3+\nu) + (1-\nu)\left(\frac{R_0}{R}\right)^2\right)} = \sigma_{\max} f_\omega K_m$$
$$\frac{E_{kin}}{V} = \sigma_{\max} f_\omega \frac{2}{(3+\nu)} = \sigma_{\max} f_\omega K_m$$

(When using these equations, please observe the correct unit conversions: Pa = J/m^3.) This shows that the maximum energy storage density is not actually a function of the rotational speed. For a solid cylinder and a cylinder with hole ($R_0 = 0.05$m)we get $K_m = 0.606$ and $K_m = 0.303$, respectively.

The energy storage capacity is

$$\begin{aligned} E_{kin} &= \sigma_{\max} f_\omega K_m = 2617\text{MPa} \times 0.89 \times 0.303 \\ &= 2617\text{MNm/m}^3 \times 0.89 \times 0.303 = 706\text{MJ/m}^3 \\ &= 706\text{MWh}/3600/\text{m}^3 = 196\text{kWh/m}^3. \end{aligned}$$

We can also solve for the maximum circumferential speed from the tension equation (see Larsonneur [61]),

$$u_{\max} = \sqrt{\frac{4\sigma_{\max}}{\rho\left((3+\nu) + (1-\nu)\left(\frac{R_0}{R}\right)^2\right)}}$$

For K_m values for other shapes, please see e.g., Sorensen [93] or Choudhury [17]. Table 7.9 presents some properties of flywheel materials, based on tensile strength. These values should only be used as very rough estimates.

The models discussed above describe the storage potential of the flywheel in the steady state; to understand the charging and discharging phases, we need to analyze the time-dependent conservation of the angular momentum of the system,

$$\frac{d(I\omega)}{dt} = M - M_\mu = \left(\frac{P_m \eta_m}{\omega}, -\frac{P_g}{\omega \eta_g}\right) - f\omega$$

Table 7.9 Flywheel material properties. For maximum circumferential speed, a hole size of $R_0/R = 0.05$ was used

Material	Density, kg/m^3	Tensile strength, MPa	Design stress, MPa	Poisson's ratio ν, –	u_{max}, m/s	References
Al alloy	2700	500		0.33	~ 500	[93]
Steel 4340	7700	1520	–	0.3	~ 490	Li and Palazzolo [63
Maraging steel	8000	2700	900	0.3	~ 640	[93]
Carbon fiber (40% epoxy)	1550	1500	750	–	~ 1100	[93]
Kevlar Fiber (40% epoxy)	1400	1800	1000	–	~ 1250	[93]
Carbon fiber T1000	1520	1950	–	–	~ 1250	Li and Palazzolo [63
Carbon fiber AS4C	1510	1650	–	–	~ 1150	Li and Palazzolo [63

The left side is the change rate of angular momentum, while on the right, the first term is the torque applied (positive for motor and negative for generator use, P_m and P_g are electrical powers of motor and the generator) and the second term is the friction moment (f is the friction factor). As discussed previously, the storage enclosure is typically a vacuum, and active magnetic bearings are used to minimize friction losses. If we assume there are no friction losses and that the inertial moment is constant, the above equation can be integrated with time to give

$$\frac{1}{2}I\omega^2 - \frac{1}{2}I\omega_{t=0}^2 = \left(P_m\eta_m, -\frac{P_g}{\eta_g}\right)t$$

Comparing this with the discussion above, the left side of the equation represents the change in stored kinetic energy, and the right side is the work done to/by the flywheel during time t. The efficiencies are not constant with power and ω, and power will also vary, and the friction moment should be included—in general, numerical integration is required.

Example 7.8

A flywheel energy storage system has a rotating mass of 200,000 kg, which initially rotates at 3000 rpm (50 Hz). Estimate the available energy until the minimum allowed rotation speed (47.5 Hz, 2850 rpm) is reached. The radius and length of the flywheel are 1 m and 4 m, respectively. The flywheel is made from ultra-heavy tungsten steel (18,000 kg/m^3) and has a center hole of $R_0 = a \times R$, and the ratio b of flywheel length to radius is 4. Generator efficiency (including mechanical efficiency) is 95%.

Solution: To calculate the inertial moment, we need to know the flywheel dimensions. From $b = h/R = 4$ and the mass provided, we can calculate the center hole radius. The mass is equal to

$$m = \pi\left(R^2 - R_0^2\right)h\rho = \pi\left(R^2 - R_0^2\right)bR\rho = \pi R^3\left(1 - a^2\right)b\rho$$

from which we can find the inner flywheel radius,

$$a = \frac{R_0}{R} = \sqrt{1 - \frac{m}{4\pi R^3 \rho}} = \sqrt{1 - \frac{200,000}{4 \times \pi \times 1^3 \times 18,000}} = 0.34$$

This then gives $R_0 = 0.671$m. We can then calculate the inertial moment,

$$I = \frac{mR^2\left(1 + a^2\right)}{2} = 200,000 \times \frac{1^2\left(1 + 0.34^2\right)}{2} = 111,581 \text{ kg m}^2$$

With the initial angular velocity $\omega = \frac{3000}{60} \times 2\pi = 314.2 \text{ s}^{-1}$, final angular velocity $\omega = 2850/60 \times 2\pi = 298.5\text{s}^{-1}$, and generator efficiency 95%, the available electrical energy can be found:

$$E = \frac{1}{2}I\left(\omega^2 - \omega_0^2\right)\eta_g = \frac{1}{2} \times 111,581 \times \left(314.2^2 - 298.5^2\right) \times 0.95 = 537 \text{ MWs}$$

To achieve a storage capacity of several GWs, the system either needs more complex mechanisms to allow much lower rpm, or it must be able to change the rotating inertial moment to keep the rpm above the allowed minimum value. Off course, with the AC-DC-AC converter frequency fixed to the grid the rpm can drop much more, but then the benefit of the true inertia of the rotating mass is lost.

7.4.4 *Compressed Air Energy Storage Systems, CAES*

Compressed air/gas energy storage systems are one option for storing electrical energy as mechanical energy at medium- and large-scales. Several technology reviews are available, including Budt et al. [11], Gouda et al. [36], Bazdar et al. [7], and Adip et al. [1].

The operational principle of CAES is simple: excess power supply available in the grid is used to compress air into a storage tank; then, when demand is larger than supply, pressurized gas from the storage tank is used to generate the required extra power. As with other storage technologies, numerous technical solutions are already available for CAES, and it is not possible to study them all within this book—see the above-mentioned reviews for a broader view. Here, we focus on two interesting technological solutions: advanced adiabatic compressed air energy storage, AA-CAES, and liquid piston compressed air energy storage, LP-CAES.

The operating principle of the advanced adiabatic CAES is: a compressor pressurizes the gas to the storage pressure p_c (100–200 bars); the heat generated by the compression is transferred to the thermal energy storage; and finally, the gas is fed into an adiabatic storage tank or cavern. When discharging the storage, the gas is heated using the thermal storage, and fed into the gas turbine to generate power. The gas can simply be air, but this technology can also be combined with the storage of hydrogen to have extra benefits.

When the pressure is released in the turbine, the temperature will significantly decrease, and if the initial temperature is too low this will cause freezing and icing of the turbine and pipelines. With the connection to thermal energy storage, this problem can be avoided. Figure 7.29 presents the a schematic of an AA-CAES system—note that the compressor and turbine are not mechanically connected, as is the case with normal gas turbines.

One of the first models for AA-CAES was presented by Glendenning [32], and we use this idea to illustrate the principles of this type of storage first. We also draw on the theory for gas turbines presented in Chap. 3. During steady state operation, the power needed in compression $P_{e.c}$ and the power obtained in the turbine $P_{e.t}$ can be calculated as follows:

$$P_{e.c} = \frac{\dot{m} c_p T_1}{\eta_m \eta_{mec} \eta_{s.c}} \left(\left(\frac{p_c}{p_{atm}} \right)^{\frac{\gamma-1}{\gamma}} - 1 \right)$$

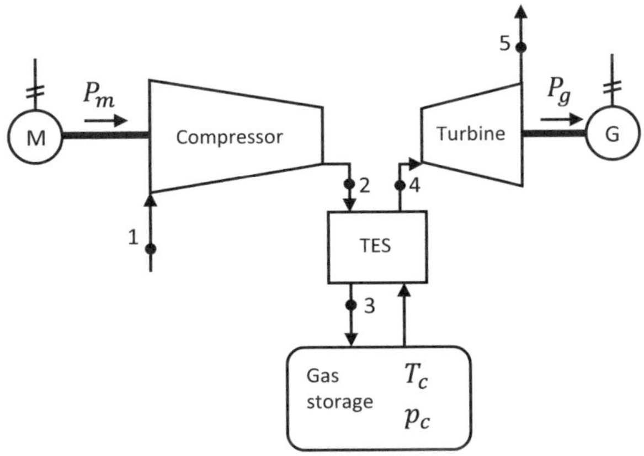

Fig. 7.29 Schematic of the advanced adiabatic compressed air (gas) storage

$$P_{e.t} = \eta_g \eta_{mec} \eta_{s.t} \dot{m} c_p T_4 \left(1 - \frac{1}{\left(\frac{p_c}{p_{atm}}\right)^{\frac{\gamma-1}{\gamma}}} \right) = \eta_m \eta_{mec} \eta_{s.t} \dot{m} c_p T_4 \left(\frac{\left(\frac{p_c}{p_{atm}}\right)^{\frac{\gamma-1}{\gamma}} - 1}{\left(\frac{p_c}{p_{atm}}\right)^{\frac{\gamma-1}{\gamma}}} \right)$$

The heat generated in the compression is transferred into the thermal energy storage at temperature T_2. If we assume that the storage tank or cavern is adiabatic (well-insulated) and the thermal storage is ideal, the temperature fed into the turbine is same as the temperature after the compression : $T_4 = T_2$.: Also, the pressures are $p_2 = p_3 = p_c = p_4$ and $p_1 = p_5$. As a result, we can solve for T_4:

$$T_4 = T_1 \left(1 + \left(\left(\frac{p_c}{p_{atm}} \right)^{\frac{\gamma-1}{\gamma}} - 1 \right) \frac{1}{\eta_{s.c}} \right)$$

We assume that charging and discharging times are equal, $\Delta t_c = \Delta t_d = \Delta t$, compressor and turbine power are constant with time, and the system operates at the average cavern pressure p_c. With this, we can calculate the storage efficiency,

$$\eta = \frac{E_{e.t}}{E_{e.c}} = \frac{\int_0^{\Delta t_d} P_t(t)dt}{\int_0^{\Delta t_c} P_c(t)dt} = \frac{P_t \Delta t}{P_c \Delta t}$$

$$= \frac{\dot{m} \Delta t c_p T_1 \eta_g \eta_{mec} \eta_{s.t} \left(1 + \left(\left(\frac{p_c}{p_{atm}} \right)^{\frac{\gamma-1}{\gamma}} - 1 \right) \frac{1}{\eta_{s.c}} \right) \left(\frac{\left(\frac{p_c}{p_{atm}}\right)^{\frac{\gamma-1}{\gamma}} - 1}{\left(\frac{p_c}{p_{atm}}\right)^{\frac{\gamma-1}{\gamma}}} \right)}{\frac{\dot{m} \Delta t c_p T_1}{\eta_m \eta_{mec} \eta_{s.c}} \left(\left(\frac{p_c}{p_{atm}} \right)^{\frac{\gamma-1}{\gamma}} - 1 \right)}$$

$$= \eta_g \eta_m \eta_{mec}^2 \eta_{s.t} \left(1 + (1 - \eta_{s.c}) \left(\frac{p_c}{p_{atm}} \right)^{-\frac{\gamma-1}{\gamma}} \right)$$

We have assumed that same mass that was compressed in the storage comes out again—that is, $\dot{m}\Delta t$ is constant.

The drawback of this simplified model is that it does not give a way to estimate the storage capacity. As discussed by Roos and Haselbacher [85], even though the compression has losses as heat, this heat can be recovered by the TES, and then it can be partly utilized when energy is fed to the turbine.

An example of a more detailed model that considers the actual change in the cavern pressure, and incorporates filling and emptying the cavern, was presented by Roos and Haselbacher [85]. To calculate the efficiency more accurately, and to estimate the storage capacity, the power, cavern mass, and energy balance equations were integrated with time, between minimum and maximum operation pressures, $p_{c.in}$ and $p_{c,\max}$. The fundamental assumption of the model is that the *time average* temperature during charging and discharging of the gas flowing in and out from the TES is the same, and assumed constant when integrating the power equations:

$T_4(t) \approx \overline{T}_2 = \overline{T}_4$. In addition, temperature out from the TES during charging, T_3, is assumed constant and is treated as a model parameter. During discharging, the exit cavern temperature $T_3(t)$ is equal to the cavern temperature, $T_c(t)$.

Total roundtrip efficiency of the storage can be calculated as follows.

$$\eta = \eta_m \eta_g \eta_{mec}^2 f_\eta$$

where η_m and η_g are electrical efficiencies of motor and generator, η_{mec} is the mechanical efficiency of both the compressor and turbine (not to be confused with isentropic efficiency), f_η is a complex function describing the plant efficiency, neglecting electrical and mechanical losses but including the compression, operation of the thermal energy storage, and expansion in the turbine. These can be derived from the thermodynamic principles already discussed in the Chapter 3 for gas turbines (see details in [85]). Without discussing the detailed derivation of the model, Fig. 7.30 presents a calculated example of the f_η values for different ratios of maximum storage pressure and atmospheric pressure, $p_{c.\max}/p_{atm}$, along the x-axis. The storage operates between maximum (charged) and minimum (discharged) pressures, $p_{c.max}$ and $p_{c.min}$, and the ratio $p_{c.\max}/p_{c.\min}$ forms the x-axis. Other values used in the calculations were $\eta_m = \eta_g = \eta_{mec} = 0.98$, the heat capacity ratio is equal to $\gamma = 1.4$, and the isentropic efficiency of the compressor and turbine are $\eta_{s.c} = \eta_{s.t} = 0.85$.

Other cases need to be calculated following the model presented by Roos and Haselbacher [85]. We see from the graph that the higher the storage pressure and the lower the difference between min and max operation pressures, the higher the efficiency becomes. To analyze the storage capacity, the following non-dimensional storage capacity function can be derived:

$$\frac{E_{el}}{p_{c.\max} V_c} = \frac{\gamma}{\gamma - 1} \eta_m \eta_g \eta_{s.t} \frac{T_{atm}}{T_3} \frac{T_3}{T_{c.\max}} f_t$$

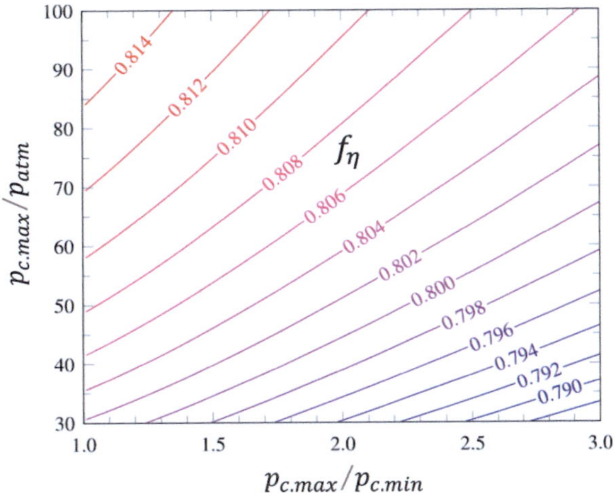

Fig. 7.30 Plant efficiency, f_η, without electrical and mechanical losses, $\eta_m = \eta_g = \eta_{mec} = 0.98$, $\gamma = 1.4$, $\eta_{s.c} = \eta_{s.t} = 0.85$ [85]

where we have

$$\frac{T_3}{T_{c.\max}} = \frac{1 - \frac{p_{c.\min}}{p_{c.\max}}}{\gamma\left(1 - \left(\frac{p_{c.\min}}{p_{c.\max}}\right)^{1/\gamma}\right)}$$

$$f_t = \frac{\overline{T}_2}{T_{atm}}\left(1 - \left(\frac{p_{c.\min}}{p_{c.\max}}\right)^{\frac{1}{\gamma}} - \frac{1}{2-\gamma}\left(\frac{p_{atm}}{p_{c.\max}}\right)^{\frac{\gamma-1}{\gamma}}\left(1 - \left(\frac{p_{c.\min}}{p_{c.\max}}\right)^{\frac{2-\gamma}{\gamma}}\right)\right)$$

$$\frac{\overline{T}_2}{T_{atm}} = 1 + \frac{1}{\eta_{s.t}}\left(\frac{\gamma}{2\gamma-1}\left(\frac{p_{c.\min}}{p_{atm}}\right)^{\frac{\gamma-1}{\gamma}}\left(\frac{p_{c.\max}}{p_{c.\min}} - 1\right)^{-1}\left(\left(\frac{p_{c.\max}}{p_{c.\min}}\right)^{\frac{2\gamma-1}{\gamma}} - 1\right) - 1\right)$$

Figure 7.31 shows $E_{el}/p_{c.\max}V_c$ for different ratios of maximum storage pressure to atmospheric pressure, $p_{c.max}/p_{atm}$, along the y-axis, and the ratio $p_{c.\max}/p_{c.\min}$ along the x-axis.

Example 7.9
Compare the roundtrip efficiency predictions from the models of Glendenning [32] and Roos and Haselbacher [85] for a 500 MWh AA-CAES with electricity storage capacity, with maximum and minimum pressures of 100 and 50 bars, (so $p_{c.\max}/p_{atm} = 100$ and $p_{c.\max}/p_{c.\min} = 2$). Assume that $\eta_m = \eta_g = \eta_{mec} = 0.98$, $\gamma = 1.4$, $\eta_{s.c} = \eta_{s.t} = 0.85$, and $T_3/T_{atm} = 1$. Also estimate the volume of the storage cavern.

Solution: The average storage pressure needed for Glendinning's model can be simply calculated as $p_c = 0.5 \times (100 + 50) = 75$bar. The round-trip efficiency is then

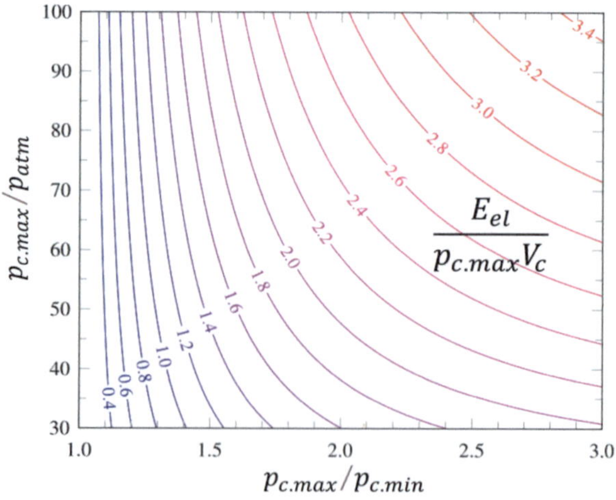

Fig. 7.31 Non-dimensional storage capacity. $\eta_m = \eta_g = \eta_{mec} = 0.98$, $\gamma = 1.4$, $\eta_{s.c} = \eta_{s.t} = 0.85$, and $T_3/T_{atm} = 1$ [85]

$$\eta = 0.98 \times 0.98 \times 0.98^2 \times 0.85 \times \left(1 + (1 - 0.85)\left(\frac{75}{1}\right)^{-\frac{0.4}{1.4}}\right) = 0.81$$

To use the model from Roos and Haselbacher [85], f_η must be first read from Fig. 7.30, from the intersection of $p_{c.\max}/p_{atm} = 100$ and $p_{c.\max}/p_{c.\min} = 2$, giving $f_\eta = 0.8105$. The round-trip efficiency is then

$$\eta = \eta_m \eta_g \eta_{mec}^2 f_\eta = 0.98 \times 0.98 \times 0.0.98^2 \times 0.8105 = 0.748$$

Energy storage capacity for a 500 MWh system can be obtained for Roos's model by first reading the non-dimensional storage capacity from Fig. 7.31, $E_{el}/(p_{c.\max} V_c) = 2.675$, from which we can solve the storage volume, $V_c = E_{el}/(p_{c.\max} \times 2.675)$. With $p_{c.\max} = 100 \times 10^5 Pa$ and $E_{el} = 500 \times 10^6 \times 3600$ Ws $= 1.8 \times 10^{12}$ Ws, we find the cavern volume

$$V_c = 1.8 \times 10^{12}/\left(100 \times 10^5 \times 2.675\right) = 67290 \text{ m}^3$$

This is equivalent to a cube 40.7 m along each side. Note that for other values of η_m, η_g, η_m, γ, $\eta_{s.c}$, $\eta_{s.t}$ and T_3/T_{atm}, Figs. 7.30 and 7.31 are no longer applicable and the actual model equations must be used.

Next, we discuss the liquid piston compressed air energy storage system, LP-CAES—for technology reviews, see for example Gouda et al. [36], Patil and Ro [77], and Bazdar et al. [7]. In a LP-CAES system, during the charging stage a liquid piston—a vertical column of liquid, normally water—is pumped from the bottom into a "cylinder" to compress a fixed amount of gas or air. When discharging, the compressed air is allowed to push the water out, to then do work in a pump-turbine.

The main advantages in comparison with traditional piston systems is that sealing is automatically secured by the liquid surface and there are no leaks, as is the case when using piston rings. In addition, friction losses between the piston and cylinder wall are mostly avoided, and wear is avoided completely. Another benefit is that, since the work is done by a pump-turbine, with incompressible liquid, the efficiency is higher in comparison with gas systems, as these features all reduce losses. The highest efficiency can be obtained from an isothermal process, but this is very challenging to achieve in reality. One option used in the LP-CAES systems is to spray water as small droplets from the top of the gas cylinder to cool the gas efficiently. LP-CAES is also suitable for installation underwater or on the seabed, which offer the benefit of large amounts of available natural cooling, as well as the hydrostatic pressure allowing higher density of the gas in the discharged state.

Here we discuss the operational principle of this type of sub-sea system as an example—Fig. 7.32 shows a schematic of the charging phase of an LP-CAES system installed on the seabed, and Fig. 7.33 presents the compression process as (p, V) diagrams for isothermal, polytropic and adiabatic processes.

The first step is the compression from initial pressure and volume (p_0, T_0, V_0) to pressure p_c by pumping liquid into the cylinder. If this is a real polytropic process,

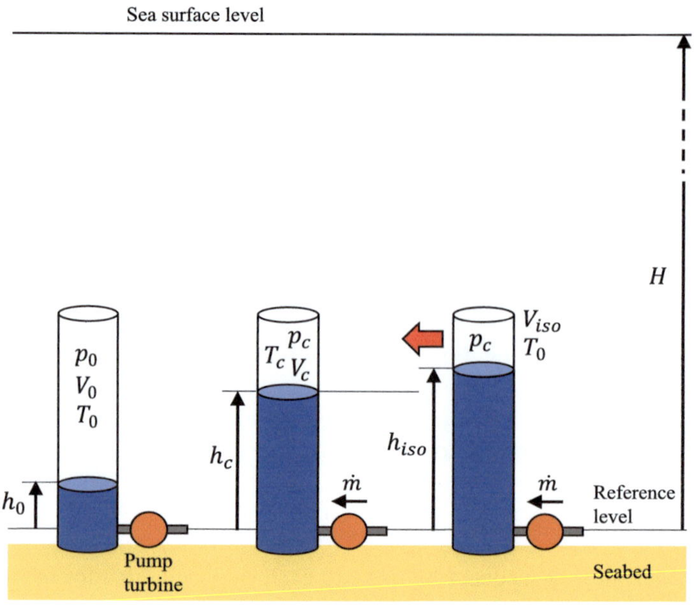

Fig. 7.32 Principle of the sub-sea LP-CAES system charging phase

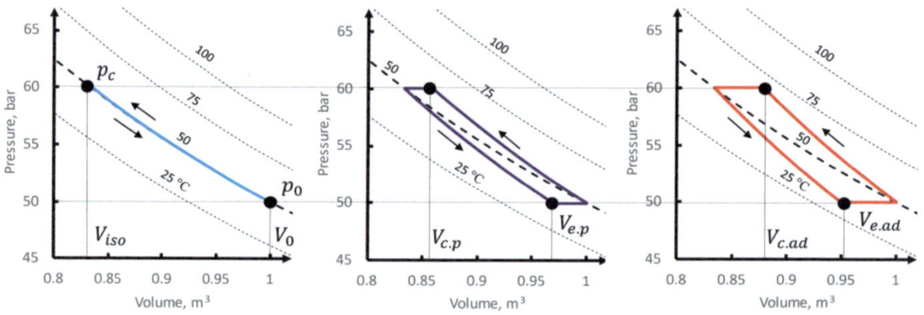

Fig. 7.33 Storage cycle of the LP-CAES process as $(p - V)$ diagrams, for iso-thermal (left), poly-tropic (middle, $pV^{1.2}$) and adiabatic (right, $pV^{1.4}$) processes. Depth 500 m (50 bar), maximum pressure 60 bar. Black dots indicate the end points of compression and expansion

the end state is $(p_c, T_{c.p}, V_{c.p})$, see Fig. 7.33. After this, the gas is cooled back to the original temperature at constant pressure—which means that the volume of the gas decreases and more water needs to be pumped in, and the end state is then (p_c, T_0, V_{iso}). V_{iso} corresponds to the volume equal to the one reached in isothermal compression, see the light blue line on the Fig. 7.33. It is clear that unless the process can be kept isothermal, we would need an efficient thermal storage system for heating and cooling. As shown in the graphs, the closer we are to an adiabatic process, the larger the area inside the cycle borders becomes, and the lower the process efficiency becomes.

The compression as a whole can be presented mathematically as follows—note that $p_c V_{iso} = p_0 V_0$ by definition of an isothermal process.

$$\begin{aligned} W_c &= \int_{V_0}^{V_{c.p}} (p(t) - p_0)dV + \int_{V_{c.p}}^{V_{iso}} (p_c - p_0)dV \\ &= \int_{V_0}^{V_{c.p}} (p(t) - p_0)dV + (p_c - p_0)\big(V_{iso} - V_{c.p}\big) \\ &= \int_{V_0}^{V_{c.p}} (p(t) - p_0)dV - \frac{(p_c - p_0)}{p_c}(p_c V_{c.p} - p_0 V_0) \\ &= \int_{V_0}^{V_{c.p}} (p(t) - p_0)dV - \left(1 - \frac{1}{\pi}\right)(p_c V_{c.p} - p_0 V_0) \end{aligned}$$

Note that the sign of the first term is negative and that the work is a function of time through the time dependence of the gas pressure. After the compression step is complete, the theoretical amount of energy stored, assuming isothermal ideal compression, can be calculated. Note again that $p_c V_{iso} = p_0 V_0$ by definition of an isothermal process, which would be the optimal path to discharge the storage. The stored energy is

$$\begin{aligned} E &= \int_{V_0}^{V_{iso}} (p - p_0)dV = \int_{V_0}^{V_{iso}} \left(\frac{p_0 V_0}{V} - p_0\right)dV \\ &= p_0 V_0 \ln\left(\frac{V_{iso}}{V_0}\right) - p_0(V_{iso} - V_0) = -p_0 V_0\left(\ln(\pi) + \frac{1}{\pi} - 1\right) \end{aligned}$$

The efficiency of compression is then

$$\eta_c = \frac{p_0 V_0\big(\ln(\pi) + \frac{1}{\pi} - 1\big)}{-\int_{V_0}^{V_{c.p}} (p(t) - p_0)dV + \big(1 - \frac{1}{\pi}\big)(p_c V_{c.p} - p_0 V_0)}$$

During expansion in the polytropic process, as the gas expands, its temperature decreases, and the volume ends up as $V_{e.p}$. To reach the start temperature, the gas must be heated at constant pressure to T_0. Similarly, for the expansion state, keeping the isothermal expansion as the reference available energy, we find

$$\eta_e = \frac{\int_{V_{iso}}^{V_{e.p}} (p(t) - p_0) dV}{p_c V_{iso} \ln(\pi) - p_0 (V_0 - V_{iso})}$$

Example 7.10
Integrate the compression and expansion equations for the polytropic process, using $pV^n = p_0 V_0^n$ during compression and $pV^n = p_c V_{iso}^n$ during expansion, and find the equation for the round-trip efficiency. Compare the round-trip efficiency of isothermal and polytropic process ($n = 1.2$) operating between 50 and 60 bars, with $p_0 = 50$ bar, $p_c = 60$ bar, and $\eta_m = \eta_g = \eta_{mec} = 0.98$. Estimate the required volume for a 500 MWh electricity storage capacity.

Solution: Note that thermal efficiency of the isothermal process is 1, the compression and expansion efficiencies for the polytropic process are

$$\eta_c = \frac{p_0 V_0 \left(\ln(\pi) + \frac{1}{\pi} - 1\right)}{-\int_{V_0}^{V_{c.p}} \left(p_0 \left(\frac{V_0}{V}\right)^n - p_0\right) dV + p_0 V_0 \left(1 - \frac{1}{\pi}\right)\left(\frac{p_c V_{c.p}}{p_0 V_0} - 1\right)}$$
$$= \frac{\ln(\pi) + \frac{1}{\pi} - 1}{\frac{\pi^{\frac{n-1}{n}} - 1}{n-1} + \frac{1}{\pi^{\frac{1}{n}}} - 1 + (\pi - 1)\left(\frac{1}{\pi^{\frac{1}{n}}} - \frac{1}{\pi}\right)}$$

$$\eta_e = \frac{\int_{V_{iso}}^{V_{e.p}} \left(p_c \left(\frac{V_{iso}}{V}\right)^n - p_0\right) dV}{p_c V_{iso} \ln(\pi) - p_c V_{iso} \left(1 - \frac{V_{iso}}{V_0}\right)} = \frac{\frac{p_c V_{iso}}{n-1} \left(\frac{\pi^{\frac{n-1}{n}} - 1}{\pi^{\frac{n-1}{n}}}\right) - p_c V_{iso} \left(\frac{1}{\pi^{\frac{n-1}{n}}} - \frac{1}{\pi}\right)}{p_c V_{iso} \left(\ln(\pi) + \frac{1}{\pi} - 1\right)}$$
$$= \frac{\frac{1}{n-1}\left(1 - \frac{1}{\pi^{\frac{n-1}{n}}}\right) - \frac{1}{\pi^{\frac{n-1}{n}}} + \frac{1}{\pi}}{\ln(\pi) - \left(1 - \frac{1}{\pi}\right)}$$

Round-trip efficiency of the isothermal process is simply $\eta = \eta_m \eta_g \eta_{mec}^2 = 0.98^4 = 0.92$, and for the real process the efficiency is

$$\eta = \eta_m \eta_g \eta_{mec}^2 \eta_c \eta_e = \eta_m \eta_g \eta_{mec}^2 \frac{\frac{1}{n-1}\left(1 - \frac{1}{\pi^{\frac{n-1}{n}}}\right) - \frac{1}{\pi^{\frac{n-1}{n}}} + \frac{1}{\pi}}{\frac{\pi^{\frac{n-1}{n}} - 1}{n-1} + \frac{1}{\pi^{\frac{1}{n}}} - 1 + (\pi - 1)\left(\frac{1}{\pi^{\frac{1}{n}}} - \frac{1}{\pi}\right)}$$
$$= 0.98^4 \frac{\frac{1}{1.2-1}\left(1 - \frac{1}{1.2^{\frac{1.2-1}{1.2}}}\right) - \frac{1}{1.2^{\frac{1.2-1}{1.2}}} + \frac{1}{1.2}}{\frac{1.2^{\frac{1.2-1}{1.2}} - 1}{1.2-1} + \frac{1}{1.2^{\frac{1}{1.2}}} - 1 + (1.2 - 1)\left(\frac{1}{1.2^{\frac{1}{1.2}}} - \frac{1}{1.2}\right)}$$
$$= 0.922 \times 0.70 = 0.65$$

The theoretical volume of the required air tank can be calculated from the stored energy, $500\text{MWh} = 1.8 \times 10^{12}$ Ws:

$$E = p_0 V_0 \left(\ln(\pi) + \frac{1}{\pi} - 1 \right) = p_c V_{iso} \left(\ln(\pi) + \frac{1}{\pi} - 1 \right)$$

$$V_{iso} = \frac{1.8 \times 10^{12}\ \text{Ws}}{60 \times 10^6 \text{Pa}\left(\ln(1.2) + \frac{1}{1.2} - 1\right)} = 1,920,000\text{m}^2$$

This is equivalent to a cube 124 m along each side—of course, this would be obtained by combining several smaller tanks together. In comparison with Example 7.7, the LP-CAES system is far larger than the AA-CAES operating between 50 and 100 bars.

7.5 Electrical and Electrochemical Energy Storages

This section gives a short development history of capacitors, super-capacitors and batteries, followed by the current status and existing best available examples of these technologies, along with a basic understanding of the operational and design principles. We also give few examples of the largest existing units in 2024.

As discussed previously, over the last two decades renewable energy has gained increasing prominence in response to the exponential growth in global energy demand. For example, in the US a staggering 834 TWh of electric power—approximately 21% of the total electricity generated—was derived from renewable energy sources [103]. For comparison, according to the European Environment Agency [25], ~ 23% of the energy consumed in the EU in 2022 was generated from renewable sources. This is a slight increase from 2021 figures, largely driven by strong growth in solar power. The share is also amplified by a 2022 reduction in non-renewable energy consumption linked to high energy prices.

The share of renewables in Europe is expected to keep growing. However, meeting the new target of 42.5% for 2030 will require more than doubling the rates of renewables deployment seen over the past decade, and requires a deep transformation of the European energy system [25]. The main challenge associated with renewable energy lies in the lack of energy storage capacity, which requires new systems to be sustainable, cost-effective, high efficiency, scalable, and safe. Energy storage devices are key components in the field of energy preservation and its utilization.

Electrochemical energy storage systems convert electrical energy into chemical potential energy, and vice versa [66]. Among the different energy storage devices, electrochemical energy storage, with the help of capacitors and batteries, plays a significant role, especially in electric vehicles and portable electronics, but also increasingly in grid ancillary services on a large scale [23].

7.5.1 Supercapacitors

A capacitor is device used to store an electric charge Q (its unit is 1 C, 1 Coulomb). Capacitors are made, for example, from two conducting plates separated by a dielectric material. Figure 7.34 shows the operating principle of a basic capacitor when charged and discharged.

Initially, we have a neutral capacitor that has both plates at neutral charge, and so their potential difference is equal to 0 V, as shown in Fig. 7.34a. For charging, we connect the battery terminals to the capacitor (b). The current will flow, and transport charge carriers into the capacitor until its voltage is same as the battery voltage. Note that there is no flow of current through the capacitor at any moment (unless the voltage is too high and the capacitor breaks). The battery is then disconnected, and the charge remains in the capacitor, keeping its voltage constant (c). The capacitor is discharged by connecting it in parallel with a load, or by short-circuiting the terminals as shown in (d). Current will flow in the reverse direction to that in step (b) until the potential is again 0 V.

During the charging and discharging stages, the flow of current can be calculated as

$$I(t) = \frac{dQ}{dt} = \frac{d(CV)}{dt} = C\frac{dV}{dt}$$

Assuming the initial voltage is 0 V, the charge in the capacitor is simply

$$Q = CV = \int_0^{\tau} I(t)dt$$

If the voltage is increased from V_1 to V_2 at a constant rate of $R_V = dV/dt$ (V/s), the charge accumulated in the capacitor is then

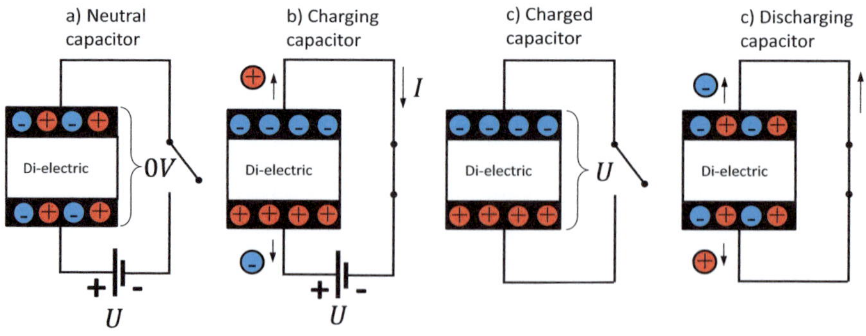

Fig. 7.34 A capacitor, and its charging and discharging steps

$$Q = C(V_2 - V_1) = \frac{1}{R_V} \int_{V_1}^{V_2} I(V)dV$$

The energy stored in the capacitor, E_c (J), can be calculated from the power fed into the capacitor within time τ:

$$E = \Delta W = \int_0^{\tau} P(t)dt = \int_0^{\tau} V(t)I(t)dt = \int_0^{\tau} V(t)C\frac{dV}{dt}dt$$
$$= \int_{V_1}^{V_2} VCdV = \frac{C}{2}\left(V_2^2 - V_1^2\right)$$

It can also be found from the work required to push the charge Q into the capacitor against the potential difference,

$$E = \Delta W = \int_{V_1}^{V_2} Q(V)dV = \int_{V_1}^{V_2} CVdV = \frac{C}{2}\left(V_2^2 - V_1^2\right)$$

As an example, the capacitance of a system composed of two parallel plates is

$$C = \varepsilon\varepsilon_0\frac{A}{d}$$

where A is the area of the plates, d is the distance between the plates, ε is the relative permittivity (dielectric constant), and $\varepsilon_0 = 8.854 \times 10^{-12}$F/m is the permittivity of the vacuum. To give a few examples, the relative permittivity of air is $\varepsilon \sim 1$, paper $\varepsilon \sim 5$, Mica $\varepsilon \sim 3$–6, and glass $\varepsilon \sim 7$. The unit of the capacitance is the Farad (F).

To increase the capacitance of a system, capacitors can be connected in parallel or in series. Parallel connection (p) increases the current and series connection (s) increases the voltage, and the total capacitance for these are

$$\frac{1}{C_p} = \sum_{i=1}^{N} \frac{1}{C_i}$$

$$C_s = \sum_{i=1}^{N} C_i$$

The first capacitor was invented by E. G. von Kleist in 1745, and at the same time independently by P. van Musschenbroek of the University of Leiden [29]. Electrochemical capacitors (supercapacitors) have garnered more attention due to their higher power density and ultra-cyclic stability [113]. In the middle of the eighteenth century, the electrical double layer (EDL) structure and the formation of electrical potential at the solid–electrolyte interface were put forward by H. von Helmholtz, which seeded the fundamentals for a charge storage mechanism and electrical double layer capacitance (EDLC) [42]. A century later, the EDLC and the charge–discharge processes were demonstrated in an experimental electrochemical cell constructed by H. I. Becker for the General Electrical Company [9]. This led to the first commercial supercapacitor based on EDLC, constructed by Ohio State's Standard Oil Company and subsequently licensed to the Nippon Electrical Corporation (NEC) in 1971 for computer technologies [100].

The term 'supercapacitor' was first coined by NEC, who also marketed the device by that name. Afterwards, the term 'supercapacitor' became familiar in Asia. Soon after, in 1982, the Pinnacle Research Institute (PRI) developed a high-power electrochemical cell based on ruthenium oxide, and commercialized it under the name 'PRI ultracapacitor, which popularized the alternate term 'ultracapacitor' in North America [12].

In the past several decades, electrochemical energy storage systems have seen enormous growth, introducing the new concepts of pseudo capacitance [16], battery-type behavior [114], and asymmetric and hybrid device [107] architectures towards high-performance and next-generation energy storage devices (Fig. 7.35).

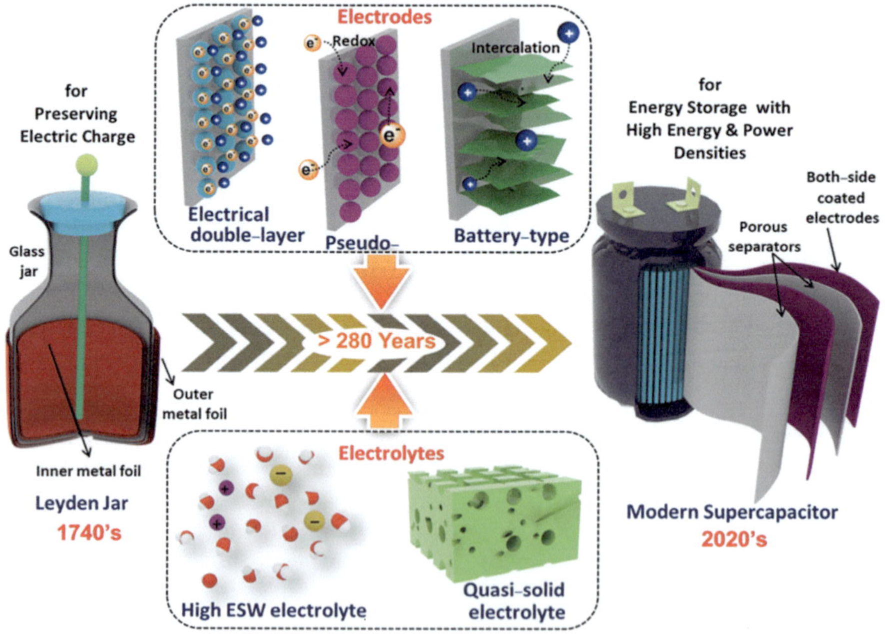

Fig. 7.35 Evolution of modern supercapacitor devices through various developments in their potential components [60]

The development of high-performing electrode and electrolyte materials is crucial to achieve improved electrochemical energy storage. The introduction of multi-role nanostructured materials has enormously boosted the electrochemical performances of supercapacitors [38]. Furthermore, the specific energy—energy stored per unit mass of electrode material—for supercapacitors is typically less than 30 Wh/L, which is usually inferior to that of batteries (~ 500 Wh/L), but is often increased by utilizing electrolytes that can operate in wider potential windows [66]. Ionic liquids, unconventional electrolytes (water in salt, WIS), and deep-eutectic solvents (DES) are examples of high-performing electrolyte systems for supercapacitors [65].

The basic components of a typical supercapacitor device, such as electrode materials, binder and conductive additives, current collectors, electrolytes, and separators, will be discussed shortly. The most common electrode materials are generally categorized into five sub-sections: carbon-based, transition-metal compounds, two-dimensional (2D) layered materials, metal–organic materials, and covalent-organic frameworks. Further, the electrolyte systems are grouped into three sub-sections: conventional aqueous, conventional nonaqueous, and unconventional electrolytes.

To evaluate most of the electrochemical characteristics of electrode and electrolyte materials, we can use various electrochemical techniques such as cyclic voltammetry (CV), galvanostatic charge–discharge (GCD), electrochemical impedance (EIS), and linear-sweep voltametric (LSV) measurements [5]. The main purpose of this electrochemical characterization in the three-electrode configuration is to identify the reversibility of the individual electrode processes, the stable operational potential window, and the charge-transfer kinetics at electrode–electrolyte interfaces. After the successful electrochemical characterization, the actual supercapacitor device is fabricated in a two-electrode configuration for laboratory testing. In this configuration, the device has positive and negative electrodes, and a separator soaked in the electrolyte solution is sandwiched between the electrodes to complete the cell. A given electrode material can act as either a positive or negative electrode depending upon the material property, which is identified with the help of the CV experiment. The charge–discharge behavior, cyclic stability, and device resistance can be determined using the above-mentioned techniques for the two-electrode cell.

Without going into details, in a typical cyclic voltammetry technique the current response of electrode material is measured when a potential is applied, with the help of a reference electrode [28]. The CV measured by monitoring the current response during two linear potential sweeps in a cyclic manner is called a cyclic voltammogram. At first, the potential is swept at a rate, R_V (V/s), from an initial voltage V_1 (0 V) to a final voltage V_2 (~1 V) in the first linear potential sweep, and subsequently reversed back to V_2 from V_1 in the second potential sweep. In general, the CV is rectangular or quasi-rectangular in shape—see Fig. 7.36. Mathematically, the current response $I(t)$ is related to the capacitance $C = Q/V$ and the potential sweep at a particular rate $R_V = dV(t)/dt$,

$$I(t) = C\frac{dV(t)}{dt} = CR_V$$

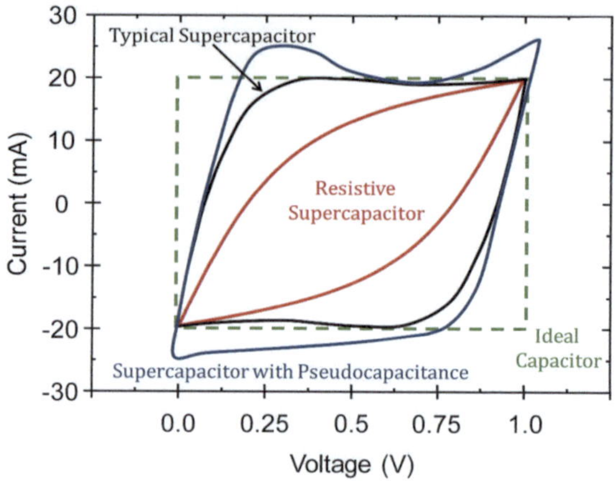

Fig. 7.36 Comparison of CV responses of ideal capacitor, typical EDLC capacitor, resistive supercapacitor, and supercapacitor with pseudo-capacitance. Giannakou [31], CC BY 4.0

From the cyclic voltammogram, the specific capacitance C_{sp} (F/g) of the electrode material can be calculated by modifying the above equation. When a cyclic voltammogram is measured for an electrode material with a mass m (g), at a scan rate of R_V (V/s) over a voltage range of $\Delta V = V_2 - V_1$, the C_{sp} is

$$C_{sp} = \frac{C}{m} = \frac{1}{mR_V(V_2 - V_1)} \int_{V_1}^{V_2} I(V)dV$$

where $\int_{V_1}^{V_2} I(V)dV$ is the area between $I(V)$ and the horizontal V-axis in Fig. 7.36. We should note that the above equation is valid only for the capacitance calculation if the CV is rectangular, or very close to rectangular.

The specific energy storage capacity, E_{sp} (Wh/kg), can also be calculated,

$$E_{sp} = \frac{E_c}{3600 \times m} = \frac{1}{3600 \times m} \int_{V_1}^{V_2} Q(V)dV$$

$$= \frac{C}{3600 \times m} \int_{V_1}^{V_2} VdV = \frac{C_{sp}}{3600 \times 2}\left(V_2^2 - V_1^2\right)$$

Another typical method is the charge–discharge process. Supercapacitors have such large capacitance values and currents that standard measuring equipment cannot be used to measure their capacity—instead, capacitance is measured as follows [52]: (1) Charge capacitor for 30 min at the rated voltage, (2) Discharge capacitor through a *constant current load system*. Constant discharge rate (A) is typically defined as $I_d = aC(\text{F})U(\text{V})$, where $a = 0.0004\text{A/F}$ and $a = 0.004\text{A/F}$ for energy storage and power applications, respectively; (3) Measure the time, Δt, taken for voltage to drop

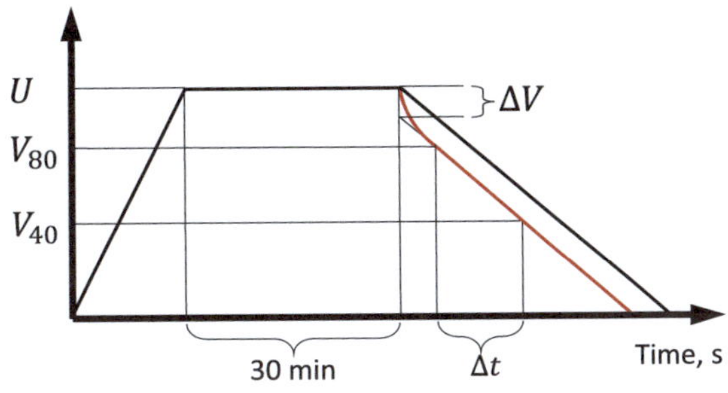

Fig. 7.37 Capacitance measurement using the charge–discharge method [52]

from $V_{80} = 0.8U$ to $V_{40} = 0.4U$; and finally, (4) Calculate the capacitance using:

$$C = \frac{I_d \Delta t}{(V_{80} - V_{40})}$$

The internal resistance can also be determined from the same measurement, using the initial voltage drop ΔV (see Fig. 7.37):

$$R_i = \frac{\Delta V}{I_d}$$

In practice, the maximum power is limited by the internal resistance of the capacitor [72]. The current flow through the resistance creates losses that dissipate as heat—too-high peak temperature results in thermal stresses and shorten the capacitor lifetime, and should be therefore avoided. Based on Liu et al. [67] and Burke [13], the power density (W/kg) is inversely proportional to internal resistance,

$$P_{sp} = \frac{U^2}{4mR_i}$$

where U and m are the nominal voltage and mass, respectively. Typical values for $P_{sp}$are 10–20 kW/kg [13].

Example 7.11
Table 7.10 presents some properties of a sample of Eaton XL60 electric double layer capacitors (EDLC). Calculate the energy and charge storage capacities, specific energy and charge storage capacities, maximum power, and 1 s pulse current for the XL60-2R7308T-R model.

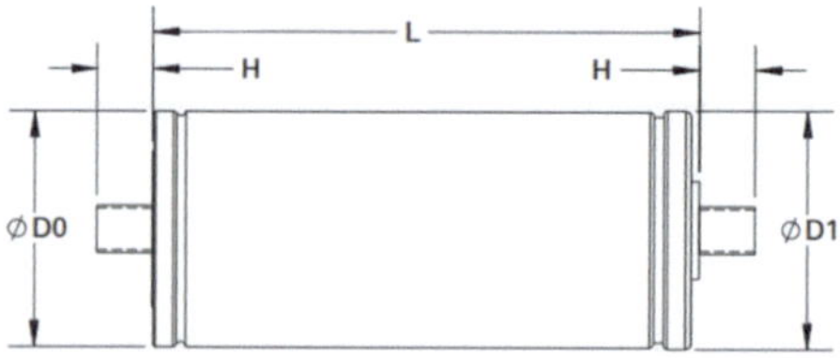

Solution: Energy and charge storage capacities and specific energy and charge storage capacities are calculated from the capacitance, voltage, and mass provided:

$$E = \frac{C}{2}U^2 = \frac{3000}{2}2.7^2 = 10,935\text{J} = 3.04\ \text{Wh}$$

$$E_{sp} = \frac{E}{m} = \frac{3.04}{0.515} = 5.90\ \text{Wh/kg}$$

$$Q = CU = 3000 \times 2.7 = 8100\text{C} = 8100\text{As} = 2.25\text{Ah}$$

$$Q_{sp} = \frac{Q}{m} = \frac{2.25}{0.515} = 4.37\text{Ah/kg}$$

Maximum power and specific power are

$$P_{\max} = \frac{U^2}{4R_i} = \frac{2.7^2}{4 \times 0.00023} = 7924\text{W} = 7.9\ \text{kW}$$

$$P_{sp} = \frac{P_{\max}}{m} = \frac{7.8}{0.515} = 15.4\text{kW/kg}$$

Note that average voltage during discharge is $U/2$, which is where the constant 4 comes from here.

Pulse current is defined as the current for 1 s duration from full rate voltage to half voltage. Time constant of the capacitor is $R_iC = 0.00023 \times 3000 = 0.69\text{s}$. Based on Eaton [24], the pulse current is

$$I_p = \frac{0.5 \times VC}{1 + R_iC} = \frac{0.5 \times 2.7\text{V} \times 3000\text{F}}{1\text{s} + 0.00023 \times 3000\text{s}} = 2396\text{A}$$

Table 7.10 Specifications of Eaton © supercapacitors [24]

Capacitance[1] (F)	Part number	Maximum working voltage (V)	Maximum initial ESR[1] (mΩ)	Nominal leakage current[2] (mA)	Stored energy[3] (Wh)	Peak power[4] (W)	Pulse current
3000	XL60-2R7308W-R	2.70	0.23	5.0	3.0	7.900	2400
3000	XL60-2R7308T-R	2.70	0.23	5.0	3.0	7.900	2400

L = 138 mm, D0 = 60.3 mm and mass 515 g. [24] © Eaton)

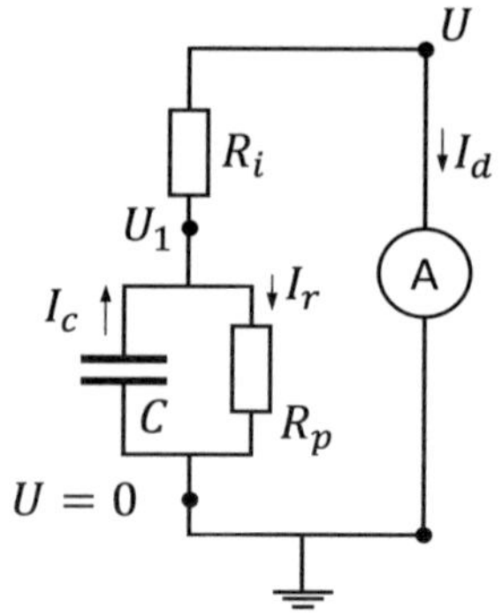

Fig. 7.38 Supercapacitor equivalent circuit, A is the constant current load system

Example 7.12

Figure 7.38 shows the equivalent circuit of a supercapacitor. Build a model to calculate capacitor voltage as a function of time, and draw the time voltage profile for a test procedure as outlined above. Use the Eaton XL60-2R7308T-R model from the previous example, $C = 3000\text{F}$, $R_i = 0.00023\Omega$, $U_0 = 2.7\text{V}$ and use $R_p = 1000\Omega$.

Solution: We first need to write the current balance at the point U_1, and then substitute the definitions of currents coming from the capacitor and resistor current. We have

$$I_C - I_R - I_d = 0$$

$$-C\frac{dU_1}{dt} - \frac{U_1}{R_p} - I_d = 0$$

From these we get the ordinary differential equation,

$$\frac{dU_1}{dt} = -\frac{U_1}{R_pC} - \frac{I_d}{C}$$

which can be integrated to obtain the following model for the voltage U_1:

$$U_1(t) = (U_0 + I_dR_p)\exp\left(-\frac{t}{R_pC}\right) - I_dR_p$$

Finally, writing the overall voltage of the capacitor block we get,

$$\begin{aligned} U(t) = U_1(t) - I_dR_i &= (U_0 + I_dR_p)\exp\left(-\frac{t}{R_pC}\right) - I_d(R_p + R_i) \\ &= U_0\exp\left(-\frac{t}{R_pC}\right) + I_dR_p\left(\exp\left(-\frac{t}{R_pC}\right) - 1\right) - I_dR_i \end{aligned}$$

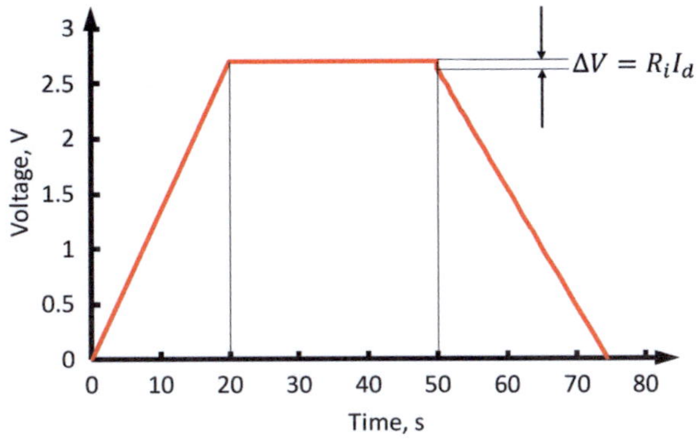

Fig. 7.39 Charge–discharge curve of a supercapacitor

From this we can see that just before connecting the current load the voltage is U_0, but it immediately drops to $U = U_0 - I_d R_i$ when load is connected. Since the current load system also has an internal time constant, we can see the non-linear delay in the measurement curve, as seen in Fig. 7.37. The calculated charge–discharge curve is presented in Fig. 7.39, comprising a linear voltage increase from 0 to 20 s, then constant voltage charging for 30 s, and then discharge at constant current with 40 mA/F—that is, 324 A current for 3000 F capacitance.

If this was a measured curve, internal resistance could be determined from the initial voltage drop $I_d R_i$ shown in the figure—but since this value was provided in the example, it was simply included in the calculation.

7.5.2 *Battery Energy Storage Systems, BESS*

A battery is a device that stores electrical energy in chemical potential energy. In contrast to capacitors, batteries keep the voltage level almost constant until the stored energy is completely used; as noted above, a capacitor's potential varies with the level of charge. As a result, their applications are clearly different.

A battery contains one or several identical cells that each store electrical power as chemical potential energy in two electrodes separated by an electrolyte [34]. Battery energy storage offers some benefits over, for example, pumped hydro and compressed gas energy storage systems, such as higher round-trip efficiency, and flexibility in installation, location, and applications across several sectors. Several excellent reviews are available for battery energy storage systems, see for example Deguenon et al. [21] and Zhao et al. [116].

Batteries are everywhere, and it is easy to forget how dependent our societies are on these devices. Laptops, mobile phones and a vast number of other mobile systems, as well as both traditional combustion engine and electric vehicles all rely on batteries. Batteries are also a key part of our energy systems already, especially off-grid systems running solely on renewable energy. Several types of batteries are

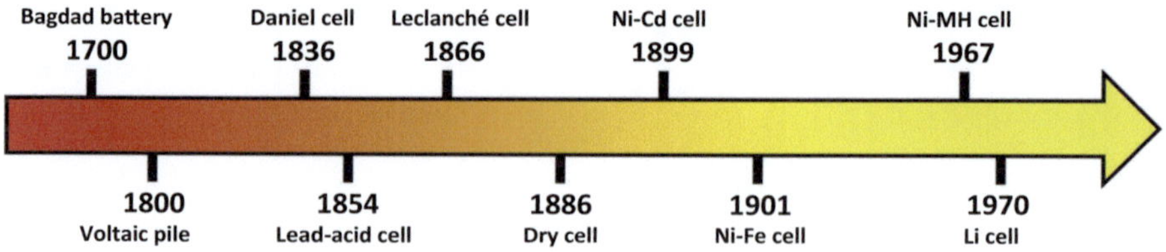

Fig. 7.40 The historical timeline of the development of battery technologies

currently used, for example, the traditional lead-acid battery is still widely used in cars and off-grid PV systems due its low cost. Lead-acid batteries used in cars, the so-called SLI (Starting, Lightning and Ignition) batteries, are designed to provide the large currents needed for short durations in starting the combustion engine, and only allow discharging by 20%—they have very short lifetimes if they are heavily discharged, and only ten 80–90% discharges will reduce the lifetime significantly [53]. In contrast, the batteries used with off-grid PV systems, for example, need to sustain continuous charge–discharge cycles without affecting their lifetime.

Along the timeline of technological development, several notable advancements have played a significant role in shaping and evolving modern batteries. Figure 7.40 illustrates the timeline of introduction of the common types of batteries.

Perhaps the earliest example of the technology was between 250 and 224 A.D. when the inhabitants of the Parthian civilization electroplated gold onto silver using jars comprised of an iron rod within a copper cylinder, an assemblage known as the “Bagdad Battery” [29]. However, Alexander Volta is considered the real discoverer of the battery [29]. He made the first successful demonstration of a modern battery in 1800, commonly referred to as the Voltaic pile. Other developments include the Daniel cell in 1836, and the first rechargeable battery, the lead-acid battery, in 1854. Lithium-based batteries, introduced in the 1970s, were the last to emerge in the progression of battery technologies.

There has been a lot of research on electrochemical batteries. A recent study identified lithium-ion, sodium sulfur, lead acid, and vanadium redox flow batteries as the most suited for grid support [88], see also Hesse et al. [43]. Lead-acid batteries have been on the market for a long time, however the most popular technology available today is lithium-ion batteries. Vanadium redox batteries have been recently proposed as a more sustainable alternative for grid support, but the technology has not yet established as strong a market as lithium-ion batteries. A lot of research has focused on comparing battery technologies, however there is no perfect battery for all applications.

Batteries can be basically divided into two categories: primary batteries, and secondary batteries. Primary batteries are single-use batteries, typically used and then discarded—or, better yet, recycled. We will not discuss these further in this book, but will focus instead on the large-scale secondary battery storage systems.

Secondary batteries contain electrochemical cells whose chemical reactions can be reversed by applying a certain voltage to the battery, to make the current flow in

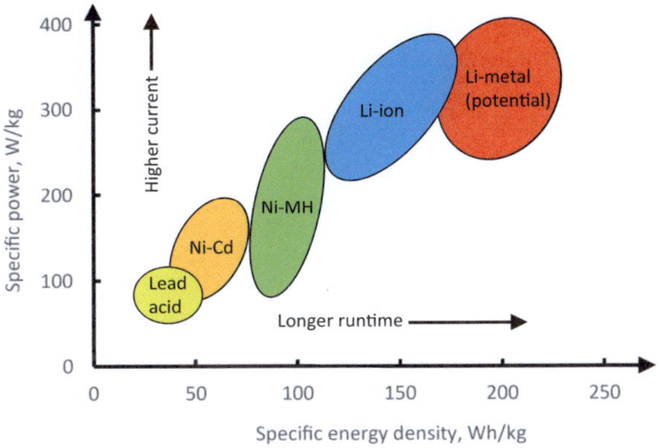

Fig. 7.41 Comparison of secondary battery types (Redrawn from Miao et al. [73])

the reverse direction with respect to discharging. Small-capacity secondary batteries are used to power portable electronic devices, like mobile phones and other gadgets and appliances, while heavy-duty secondary batteries are used in powering various electric vehicles and other high-drain applications, such as load levelling in electricity generation. They are also used as stand-alone power sources, combined with DC-AC inverters to supply AC electricity. Although the initial cost of acquiring rechargeable batteries is always higher than that of primary batteries, they are more cost-effective over the long-term.

Secondary batteries can be further classified based on their chemistry—this is very important because the chemistry determines some of the attributes of the battery, including its specific energy, cycle-life, shelf-life, and price. Rechargeable batteries in common use include Nickel–Cadmium (Ni–Cd), Nickel-Metal Hydride (Ni-MH), Lead-Acid, and Lithium-ion (LIBs) batteries. The specific power and energy of the most prevalent secondary battery types are compared in Fig. 7.41.

LIBs offer high voltage, high energy density, and good efficiency, making them attractive for various storage applications. In 2023, LIB technology was the clear global market leader in grid-connected systems, with a greater than 90% share [21]. A LIB consists of one or more individual electrochemical cells, usually, an ion-porous separator is placed in the electrolyte between the two electrodes to prevent short circuiting [35]. The cells can be connected in series or in parallel depending on the desired characteristics of the system, often called a battery pack [117]. While fully charged, the Li-ions in the cell are located on the anode (negative electrode), which is made of the graphite. During discharge, the Li-ions travel through the electrolyte via a separator, to the positive cathode. The negative electrode reaction at the graphite anode is [79]

$$\mathrm{Li}_x\mathrm{C}_6(s) \leftrightarrow \mathrm{C}_6(s) + x\mathrm{Li}^+ + xe^-$$

We have here assumed that graphite forms a hexagonal grid in which Li ions are able to fit between the layers. The negative electron continues to the external circuit, and the positive lithium ion diffuses via the separator to positive electrode, where it combines with cobalt oxide and takes the electrode from the external circuit.

$$x\,\mathrm{Li}^{+} + xe^{-} + \mathrm{Li}_{(1-x)}\mathrm{CoO}_2(s) \leftrightarrow \mathrm{LiCoO}_2(s)$$

These two equations combine to give the overall reaction from charged to discharged cell:

$$\mathrm{Li}_x C_6(s) + \mathrm{Li}_{(1-x)}\mathrm{CoO}_2(s) \leftrightarrow \mathrm{C}_6(s) + \mathrm{LiCoO}_2(s)$$

The parameter x can take values between 0 and 1. For $x = 1$ we get

$$\mathrm{LiC}_6(s) + \mathrm{CoO}_2 \leftrightarrow \mathrm{C}_6(s) + \mathrm{LiCoO}_2(s)$$

The work required to make this reaction take place, and to exchange x electrons, is equal to Gibbs free energy,

$$\Delta G^0 = xFE^0$$

which is the work required to move the charge xF over the potential difference E^0.

Figure 7.42 shows the charge–discharge cycle of a Li-ion battery. The separator is used to prevent the electrodes short circuiting. The electrodes are connected externally with a conductor, allowing electrons to flow to produce an electric current. When all the ions have moved to the cathode, a current no longer flows through the conductor, and the cell is fully discharged.

LIBs are secondary batteries, meaning that the reactions described above can be reversed by applying an electric current to the electrodes, causing the ions to return to the other electrode and recharge the cell—and so the reactions taking place on the electrodes are reversed. Different types of LIBs are typically characterized by their cathode material, which also affects the operational parameters of the battery.

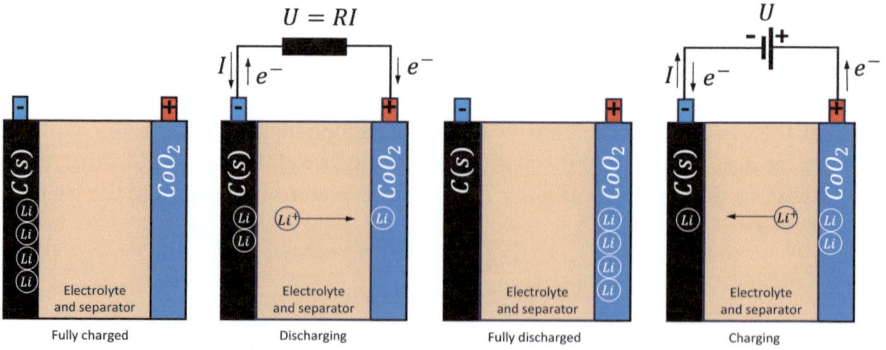

Fig. 7.42 Charge–discharge cycle of a Li-ion battery, $x = 1$

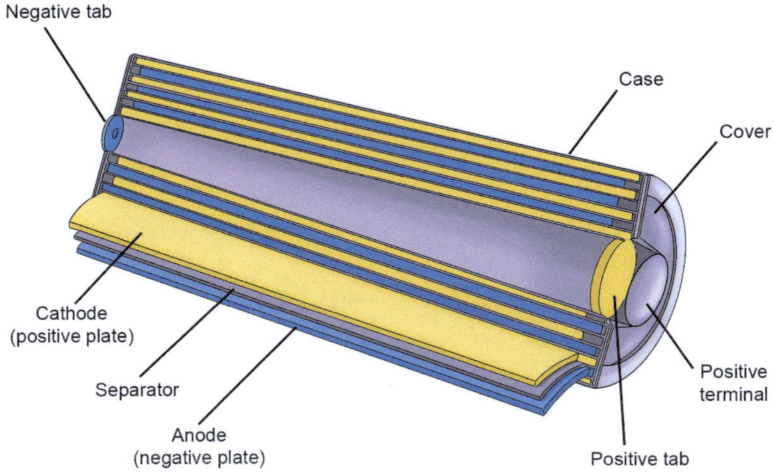

Fig. 7.43 Structure of a cylindrical Li-ion cell. © Maija Karala

The most common LIB types by cathode material [117] are Lithium iron phosphate ($LiFePO_4$), Lithium Nickel Manganese Cobalt Oxide (NMC), Lithium Nickel Cobalt Aluminum Oxide (NCA), Lithium Cobalt Oxide (LCO, also Fig. 7.48), Lithium Manganese Oxide (LMO), and Lithium Titanate (LTO).

Lithium-ion cells are produced in different shapes and styles, including coin cells and button cells, cylindrical or pouch forms, and prismatic cells, depending on the desired end use. The structure of a cylindrical cell is presented in Fig. 7.43.

Other than the differing cathode materials, the composition of the other components are quite standardized. The anode is usually made from graphite-coated copper foil, sometimes with added silicon [59], although other materials such as lithium titanate can also be used. The electrolyte is a liquid mixture of an organic solvent, usually multiple carbonate compounds such as ethylene carbonate, and a lithium salt, often lithium hexafluorophosphate. The separator is a porous polymer membrane that allows lithium ions to pass through [117], sometimes also including a thin layer of ceramic material to improve mechanical integrity [59]. The cathode is made from a lithium-based compound layered onto aluminum foil. As noted above, the cathode material greatly affects the performance characteristics of the battery, such as the specific energy and specific power of the battery. The copper and aluminum bases for the electrodes also act as current collectors for the cell, and are connected to the poles of the battery [59].

Example 7.13

Rechargeable lithium-ion batteries are among today's most successful battery systems. Consider a typical lithium-ion battery found in phones and laptops, with $LiCoO_2$ as the positive electrode material and graphite C_6 as the negative electrode material. Use $\Delta G^0 = -405.2$ kJ/mol for the full cell, and Faraday's number F =

96,485 C/mol. Calculate (1) the reversible cell voltage, and (2) the specific charge Q_{sp} (Ah/kg) and specific energy E_{sp} (Wh/kg). Assume that $x = z = 1$.

Solution: (1) As shown above, $z = 1$ if one electron is transferred per one Li ion. To calculate the reversible cell potential, we can use $\Delta G^0 = -zF\Delta E^0$, from which we can solve the potential, $\Delta E^0 = -\Delta G^0/zF$. The overall reaction is

$$LiC_6(s) + CoO_2(s) \leftrightarrow C_6(s) + LiCoO_2(s)$$

$$\Delta E^0 = -\frac{\Delta G^0}{zF} = -\frac{-405,200}{1 \times 96,485} = 4.2\text{V}$$

(2) To calculate the specific charge Q_{sp} (C/g or Ah/kg) for a fully charged battery, a system composed of 1 mol of LiC(s) and 1 mol of CoO_2 (the reaction products are fully on the left side when fully charged), we use Faraday's law (1 C = 1 C = 1 As) to get

$$\begin{aligned} Q_{sp} &= \frac{zF}{1 \times M_{LiC_6}(s) + 1 \times M_{CoO_2}(s)} \\ &= \frac{1\text{mol} \times 96485\text{C/mol}}{1\text{mol} \times 79.001\text{g/mol} + 1\text{mol} \times 90.933\text{g/mol}} \\ &= 567.8\text{C/g} = 157.7\text{Ah/kg} \end{aligned}$$

The specific energy is therefore

$$E_{sp} = Q_{sp}E_0 = 157.7 \times 4.2 = 662.4\text{Wh/kg}$$

We should note that this includes only the active electrode materials; the actual mass of the battery would be much higher.

Requirements and monitoring of large-scale battery energy storage systems

For a large-scale battery energy storage system, both the power in MW and energy in MWh must be specified. The power-to-energy ratio is normally higher in situations where a large discharge of energy is required within a short time, such as for frequency regulation applications. For pricing purposes, however, the energy rating is usually the quoted measure.

Lithium-ion battery life expectancy has been studied widely, but there is no fixed agreement on real-world lifetime as this depends on various factors, including how well the battery has been maintained. With industrial data we can assume that a well-kept battery can perform for up to 15 years. The stress factors that most impact battery performance and life expectancy, as suggested by Xu [111], are:

1. Number of Cycles. Every battery is rated for a number of full charge–discharge cycles across its lifetime. How frequently the charge/discharge state changes dictate how long we can expect the battery to serve.

2. State of Charge (SOC). The available active charge in the battery can be expressed as SOC: if the SOC is 0% the battery is completely empty, and if SOC is 100% the battery is completely full. For each application and battery system there is an optimum SOC which should be maintained during operation.
3. Depth of Discharge (DOD). Depth of discharge means the amount that is charged and discharged over a period of time. It is complementary to SOC, which means to maintain an optimum SOC requires an optimum DOD as well.
4. Current Rate (C-rate). Current rate expresses how fast a battery can be charged and discharged to maximum capacity. C/1rate means that a battery will be discharged to its full capacity within one hour. The C-rate can be determined based on the application and business case.
5. Battery Cell Temperature. Battery performance is highly dependent on temperature; degradation increases with higher temperatures. Accordingly, an optimum temperature should be be maintained around the battery cell.
6. Battery Life. Besides the use-case and all other factors described above, a battery's performance also depends on aging and the passage of time. As the battery ages a natural degradation occurs which reduces the state of health (SOH).

Some of these stress factors do not have as high an impact on redox flow batteries as on lithium-ion batteries. For example, lithium-ion batteries have a maximum life of around 3500 cycles, while flow batteries can last around 13,000 cycles [115]. The main reason for this is that the design of the flow battery allows the electrolyte, the main source of energy storage, to be easily refilled. However, a life cycle performance analysis comparing lithium-ion and flow batteries for ancillary services has not yet been carried out.

Battery Management Systems (BMS) are commonly used to monitor and protect lithium-ion batteries. The purpose of a BMS is to provide battery safety and longevity, monitor state-of-function in the form of state-of-charge, state-of-energy and state-of-health, prompt caution and indicate need for service including warnings of high temperature, cell imbalance or calibration, and indicate end-of-life when the capacity falls below the user-set target threshold. They are an integral part of lithium-ion battery storage systems, and play a crucial role in ensuring their safe and efficient operation [6]. High-quality lithium-ion batteries, when operated within their specified voltage and temperature ranges, do not typically pose a fire risk—however, deviations from safe operating parameters can lead to damage and, in extreme cases, fires. It is also important to note that a faulty BMS can be a major cause of battery faults.

One of the most common and fundamental uses of a BMS is to monitor cell voltage. By monitoring the voltages of individual cells, the BMS can prevent overcharging or over-discharging situations, which can be detrimental to the batteries. Charging and discharging processes generate heat, and different battery types have varying sensitivities to high and low temperatures. Therefore, the BMS also monitors the temperature of the cells to prevent potential damage.

In addition to monitoring functions, BMSs can have computational capabilities. These can include calculating internal impedance, tracking charge–discharge cycles, determining operational time, and utilizing coulomb counting (a method to estimate the battery's State of Charge (SOC), but which may become less accurate over time with constant use). Advanced BMS systems may incorporate active cell voltage balancing, ensuring that the voltages of individual battery cells remain within safe limits and preventing localized over-voltages or under-voltages. Below, we discuss these most important key-performance indicators.

The rate of charge and discharge, the so-called C-rate, is a very important parameter of battery energy storage. Maximum C-rate describes the system's capability of charging and discharging, which is used for hardware specifications broadly—for example, a C-rate of C/20 means a current that discharges the battery in 20 h; for a maximum *charge* storage capacity of $Q_s = 240$ Ah battery, the a C/20 rate is 12 A.

Battery capacities are not constant, they depend on the rate of charge and discharge. For example, with a lead-acid battery and very high discharge rate, due to limited mass transfer rate, only the surface of the plates is active—this significantly reduces the capacity since the inner parts of the porous plates are not utilized.

The units of C-rate are A/Ah (or W/Wh), and it is mathematically defined as.

$$C_{rate} = \frac{I}{Q_s} \approx E_{rate} = \frac{P}{E_s} = \frac{UI}{\int_0^{\tau} I(t)U(t)dt}$$

where the "$\approx$" means we assume that the voltage of the battery remains constant [116], and P and E_s are the power transfer rate and the *energy* storage capacity of a battery.

To maximize the battery lifetime, its condition must be continuously monitored and controlled: state of charge (SOC), state of energy (SOE), and state of health (SOH) are the key performance indicators to be monitored, as defined below. In the equations below, Q_C is the energy charge and E_C is the stored energy at the present state, and $Q_{\max}$ is the maximum charge and $E_{\max}$ is the maximum energy storage capacity at the present state of the battery. Q_0 and E_0 are the values of these for a new system [116]. The state of charge and state of energy are related by

$$SOC = \frac{Q_C}{Q_{\max}} = \frac{\int_0^{Q_C} dQ}{\int_0^{Q_{\max}} dQ} \approx SOE = \frac{E_C}{E_{\max}} = \frac{\int_0^{Q_C} U(Q)dQ}{\int_0^{Q_{\max}} U(Q)dQ}$$

The state of health is defined as performance compared with the original stage of a new system, both for charge and energy stored:

$$SOH = \frac{Q_{\max}}{Q_0} = \frac{\int_0^{Q_{\max}} dQ}{\int_0^{Q_0} dQ} \approx \frac{E_{\max}}{E_0} = \frac{\int_0^{Q_{\max}} U(Q)dQ}{\int_0^{Q_0} U(Q)dQ}$$

Monitoring these indicators and the thermal load of the battery is essential for maximizing the life of the battery energy storage system. Too much charging and discharging contributes to degradation, as high current rates can induce side reaction and degradation of the electrodes. Degradation depends strongly on the temperature,

Fig. 7.44 Part of the battery energy storage at Moss Landing [19, 62, 106]. This horizontal row contains two vertical levels of 18 racks. © LG

and managing and maintaining the temperature below the recommend value is crucial to prevent excessive thermal stress and damage.

In 2024, the largest battery energy storage systems are located at Vistra Energy's Moss Landing energy storage facility in California, USA, with a capacity of 400 MW/1600 MWh—see Fig. 7.44.

The system is based on LG Energy's TR1300 racks that have 327.7 kWh nominal power, maximum charge and discharge powers and voltages of 92.4 kW and 87 A, respectively, and DC voltage range of 924–1294 V (nominal voltage of 1131 V). Each of the 22 battery packs in each rack has an energy charge capacity of 290 Ah [62]. Other large installations include the Manatee Energy Storage Center project in Florida, US, with 409 MW/900 MWh, the Victorian Big Battery in Geelong, Australia, with 300 MW/450 MWh, the Mccoy Solar energy project BESS in California, US, with 230 MW/920 MWh, and the Elkhorn battery in California, US, with 183 MW/730 MWh [91].

7.6 Chemical Energy Storage

When energy is stored, efficiency is of the essence: we want to obtain as much of the stored energy back as possible, in a useful form. We have seen previously that electricity storage, mechanical and electrochemical storage, and thermal storage systems all have quite high storage efficiencies—but when electrical energy is converted into a chemical compound, the losses are higher.

Regardless of these higher losses, chemical storage has a role to play in the energy storage strategies, as it offers entirely different branches of chemical products that often are not converted back into energy. Additionally, these chemical components offer significant energy density and stability, and therefore they fit well with long-term storage, and provide additional logistics pathways for transferring energy from one place to another. We should also note that chemical products are high value commodities with numerous uses serving a multitude of needs, and therefore their primary use should *not* be combustion back to energy. On the contrary, chemical storage provides a decarbonization pathway for those sectors that are difficult to electrify directly.

Remember that fossil fuels—hydrocarbons—are an excellent form of chemical storage, and are the foundation of our current energy, transportation, and chemical industries. Fossil fuels are readily transported—we have very efficient logistic chains for them—and they are easy to store, providing flexibility for the energy system. When we phase out fossil energy sources, we will require new kinds of flexibility for the energy system of the future. One evident need is seasonal electricity storage, where current electricity storage systems cannot offer cost efficient solutions. These are also mainly fixed in storage capacity, whereas chemical storage is independent of the production, and can be used as commodities instead of purely as a temporary store of energy.

Chemical compounds are also the means for decarbonization for the chemical and metallurgical industries, and for heavy transport applications, where the technologies for direct electrification do not enable high-enough energy density. If we want to fully decarbonize these sectors, we will need help from chemical storage—specifically, hydrogen and its derivatives. With synthetic compounds, clearly the carbon and hydrogen sources must be sustainable, otherwise synthetic fuel production does not make any sense. The advantage is that, as the chemical itself does not know how it was produced, we can utilize synthetic chemicals from renewable sources with the same infrastructure as their fossil-based counterparts.

7.6.1 Production of Green Hydrogen: Water Electrolysis

We have chemically produced Hydrogen for centuries. The main production method has been from natural gas via steam reforming reactions (SMR), with ~ 98% of H_2 made by this pathway [33]:

$$CH_4 + H_2O \rightarrow 3H_2 + CO, \Delta H = +206\text{kJ/mol}$$

$$CH_4 + 2H_2O \rightarrow 4H_2 + CO_2, \Delta H = +165\text{kJ/mol}$$

If we write this SMR reaction in mass units, we see that per one kg of hydrogen produced, we produce 44/8 = 5.5 kg of CO_2, which is therefore the main product of the reaction. Unless this CO_2 is captured (as is the current mainstream processing path), this process forms a significant emissions source.

One alternative is to utilize methane pyrolysis, where methane (fossil or synthetic) is converted into hydrogen and solid carbon:

$$CH_4 \rightarrow 2\ H_2 + C(s),\ \Delta H = +74.85\text{kJ/mol}$$

This process produces very small CO_2 emissions and could therefore be an alternative to electrolysis processes where electricity is used (and this electricity is not always, or in every country, fully CO_2-free). This solid carbon can be then collected and used in carbon-demanding industries—for example, in carbon filtering, or as a raw material for lithium-ion battery negative electrodes (which presently uses mined mineral-based graphite). Surely, this methane pyrolysis process still leans on the fossil products and delay the phase out of fossil sources.

Water electrolysis has been in commercial use for several decades—in fact, the basic idea of the alkaline electrolysis cell technology goes back to 1789 [101]. Further developments throughout the nineteenth century culminated in Schmidt's 1900 patent [89] for an industrial system. This was commercialized by Maschinenfabrik Oerlikon in Switzerland in 1902, and the technology has been used commercially since then. For many years its main product was high-quality oxygen for medical and other applications—but of course, splitting water molecules creates both oxygen and hydrogen molecules:

$$H_2O(l) \rightarrow H_2(g) + 0.5\ O_2(g)$$

This reaction requires a total energy input of $\Delta H_l^0 = +285,840$ J/mol (subscript l refers to liquid water), and this energy can be provided by heat or electrical energy. The change in entropy in the reaction is $\Delta S_l^0 = +163.15\text{J/molK}$, and the free Gibbs energy can be calculated from these at the reference state,

$$\Delta G_l^0 = \Delta H_l^0 - T_0 \Delta S_l^0 = 285{,}840 - 298.15 \times 163.15 = +237{,}200\ \text{J/mol}$$

The positive sign indicates that the reaction is not spontaneous and needs additional energy. Note that, in standard thermodynamic reference conditions ($T_0 = 298.15\text{K}$ and $p_0 = 1\text{bar}$), water is in liquid form and hydrogen and oxygen are gases. Electrolysis can also be carried out with water vapor, and the reaction then becomes

$$H_2O(g) \rightarrow H_2(g) + 0.5O_2(g)$$

requiring a total energy input of $\Delta H_l^0 = +241,800\text{J/mol}$. The entropy change is $\Delta S_l^0 = +44.10\text{J/molK}$, and the Gibbs free energy is then

$$\Delta G_g^0 = \Delta H_g^0 - T_0 \Delta S_g^0 = 241,800 - 298.15 \times 44.10 = +228{,}651\ \text{J/mol}$$

But what do these terms in the equations mean? By definition, the Gibbs free energy, ΔG, is the minimum work required to run the reaction. In electrolysis, this work is done electrically by moving charge carriers across a voltage potential E (unit

V), corresponding to

$$\Delta G = nFE$$

where n is the number of electrons exchanged per water molecule converted to hydrogen, and $F = 96,485\text{C/mol}$ is Faraday's constant, the charge of one mole of electrons. We will examine this equation more below and discuss how it relates to the efficiency and power requirements of hydrogen production.

The term $T\Delta S$ represents the heat, ΔQ, in the reaction per mole of water, in addition to the electrical energy. This comes directly from the definition of entropy, $\Delta S = \Delta Q/T$. Finally, ΔH is the total energy needed to make the reaction happen, and is equal to the reaction enthalpy. Let's put these equations to use, by calculating the energies needed at different temperatures—in the following, we keep pressure constant at 1 bar.

To calculate enthalpy, entropy and Gibb's free energy at different temperatures, we need to find expressions for them as functions of temperature. We have heat capacity available as a polynomial function of temperature T,

$$C_{p.i}(T) = a_i + b_i T + c_i T^2 + d_i T^3 + e_i T^4$$

and from this and basic thermodynamic relations, we can calculate all other required values. Enthalpy H and entropy S at constant pressure and below the boiling point of water, 100 °C, are defined as

$$H_i(T) = H_i^0 + \int_{T_0}^{T} C_{p,i}\, dT = H_i^0 + \left(a_i(T - T_0) + \frac{b_i}{2}(T^2 - T_0^2) + \frac{c_i}{3}(T^3 - T_0^3)\right.$$
$$\left. + \frac{d_i}{4}(T^4 - T_0^4) + \frac{e_i}{5}(T^5 - T_0^5)\right)$$

$$S(T) = S^0 + \int_{0}^{\mathrm{T}} \frac{C_p}{T} dT = S_i^0 + \left(a_i ln\left(\frac{T}{T_0}\right) + b_i(T - T_0) + \frac{c_i}{2}(T^2 - T_0^2)\right.$$
$$\left. + \frac{d_i}{3}(T^3 - T_0^3) + \frac{e_i}{4}(T^4 - T_0^4)\right)$$

This is simple for H_2 and O_2, but for water we also need to consider evaporation. Below 100 °C, the above equations can be used with liquid water data. When we move above the boiling point, there is a step in enthalpy and entropy equal to the heat of evaporation at 100 °C, 40,615 J/mol. Above 100 °C we need to use data for water vapor (note we must also use H_i^0 and S_i^0 values for vapor). The above equations are then

$$H_i(T) = H_i^0 + \int_{T_0}^{T_b} C_{p,i}^l \, dT + \Delta H(T_b)_{l-g} + \int_{T_b}^{T} C_{p,i}^g \, dT = H_i^0 + \int_{T_0}^{T} C_{p,i}^g \, dT$$

$$= H_i^0 + \left(a_i^g (T - T_0) + \frac{b_i^g}{2}\left(T^2 - T_0^2\right) + \frac{c_i^g}{3}\left(T^3 - T_0^3\right) + \frac{d_i^g}{4}\left(T^4 - T_0^4\right) + \frac{e_i^g}{5}\left(T^5 - T_0^5\right) \right)$$

$$S(T) = S^0 + \int_{0}^{T_b} \frac{C_p}{T} dT + \Delta(T_b)S_{l-g} + \int_{T_b}^{T} \frac{C_{p \cdot g}}{T} \, dT = S^0 + \int_{T_0}^{T} \frac{C_{p \cdot g}}{T} \, dT$$

$$= S_i^0 + \left(a_i^g \ln\left(\frac{T}{T_0}\right) + b_i^g (T - T_0) + \frac{c_i^g}{2}\left(T^2 - T_0^2\right) + \frac{d_i^g}{3}\left(T^3 - T_0^3\right) + \frac{e_i^g}{4}\left(T^4 - T_0^4\right) \right)$$

with data to be used as shown in Table 7.11.

In both equations, integration to vapor can be carried out via two paths, either: (1) heating from 25 °C liquid to 100 °C liquid, evaporation at boiling temperature t_b, and finally heating from boiling temperature to observed temperature; or (2) starting directly from vapor at 25 °C and heating directly to the observed temperature (which is above the boiling temperature).

Figure 7.45 presents values for $\Delta G(T)$, $\Delta H(T)$ and $T\Delta S(T)$ as functions of temperature:

$$\Delta H(T) = H_{H_2} + 0.5H_{O_2} - H_{H_2O},$$

$$\Delta S(T) = S_{H_2} + 0.5S_{O_2} - S_{H_2O}$$

Table 7.11 Thermodynamic properties of water liquid, water vapor, hydrogen and oxygen and, $T_0 = 298.15$K (data from Petrucci et al. [79]), J/mol, kg/mol

	H_2O(l)	H_2O(g)	H_2(g)	O_2(g)
a_i	7.3458E+02	3.2566E+01	2.7324E+01	2.6538E+01
b_i	– 7.8147E+00	3.2297E–04	4.6799E–03	1.0671E–02
c_i	3.4746E–02	1.2262E–05	– 3.0046E–06	– 2.1470E–06
d_i	– 6.8743E–05	– 4.0074E–09	1.3345E–09	– 2.9078E–10
e_i	5.1120E–08	0.0000E+00	0.0000E+00	0.0000E+00
H_i^0	– 2.8580E+05	– 2.4180E+05	0.0000E+00	0.0000E+00
S_i^0	6.9910E+01	1.8880E+02	1.3070E+02	2.0510E+02
G_i^0	– 2.3710E+05	– 2.2860E+05	0.0000E+00	0.0000E+00

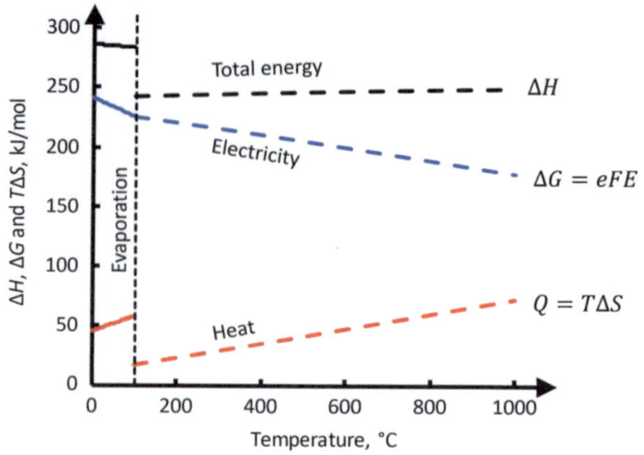

Fig. 7.45 $\Delta G(T)$, $\Delta H(T)$ and $T\Delta S(T)$ as functions of temperature for atmospheric water electrolysis, per mole of water

and

$$\Delta G(T) = \Delta H(T) - T\Delta S(T)$$

We see from Fig. 7.45 that the total energy required for the electrolysis of water vapor is smaller than for liquid water—of course, to get this "benefit" we need to have the steam already available. In general, the total energy needed is fairly constant as temperature increases, but the contribution from electrical power significantly decreases, while more heat is required correspondingly to fill the total energy need.

Can we identify an optimum temperature? This depends on the available technology—and since high temperature electrolysis (such as SOEC, Solid Oxide Electrolysis Cell) is not yet available on large industrial scales, we still need to rely on lower temperature processes. If high temperature processes were available, then one criteria for technology solutions would the price of energy: which is cheaper, electricity or heat? In combination with a system that produces high-temperature waste heat, this would offer significantly lower power consumption than low-temperature processes. With AEC (Alkaline Electrolysis Cell) and PEMEC (Proton Exchange Membrane Electrolysis Cell) temperature is often limited by the durability of the membrane and electrolyte materials to 80–90 °C [33].

Electrolysers are electrochemical devices comprising two electrodes and a semipermeable separator. Oxidation and reduction reactions take place at the electrodes—the reactions depend on the conditions of the electrolyzer, and are therefore discussed separately below with each technology. In general, all electrolyzers have a positive anode, where the O_2 evolution reaction occurs, and a negative cathode where the H_2 evolution reaction takes place. In the case of an alkaline electrolyzer cell, this takes two electrons from the external circuit and breaks the water into one hydrogen molecule and two hydroxide ions:

$$2H_2O + 2e^- \rightarrow H_2 + 2OH^-$$

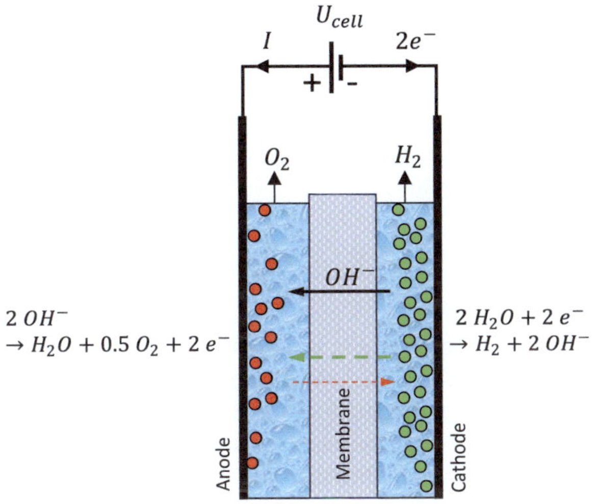

Fig. 7.46 Schematic of the alkaline electrolysis cell. Dashed lines represent the diffusion of gases—more H_2 diffuses to the oxygen side due to its smaller molecular size

Note that the electrical charge is equal on both sides. The hydroxide ions then diffuse through the alkaline solution and the semi-permeable membrane (which separates the H_2 and O_2 on their own sides) to the positive anode, where the O_2 evolution reaction releases a half oxygen molecule (atom) and two electrons back to the external circuit per H_2 molecule released at the cathode. The charge from the hydroxide ions continues through the external electrical circuit as two electrons,

$$2OH^- \rightarrow H_2O + 0.5O_2 + 2e^-$$

These reactions add to the overall process discussed above, giving

$$H_2O \rightarrow H_2 + 0.5O_2$$

See Fig. 7.46 for a schematic of the alkaline electrolysis cell.

The driving force of the reaction is the electrical potential difference across the cell, U_{cell}, while the minimum work required is defined by the Gibb's free energy ΔG,

$$\Delta G = nFE$$

From this we can solve the free-energy voltage E, and also calculate its value at the reference temperature,

$$E^0 = \frac{\Delta G^0}{nF} = \frac{+237100\ \text{J/mol}}{2 \times 96485 \frac{\text{C}}{\text{mol}}} = 1.229\text{V}$$

However, we need to have an amount of energy at least equal to the reaction enthalpy to continuously run the electrolysis reaction—free-energy voltage is just able to start the reaction, but then extra heat would be required to maintain a continuous reaction without external heating. We can also define another limit, the thermoneutral voltage, based on the overall reaction enthalpy. At the reference temperature of 25 °C its value is

$$V^0 = \frac{\Delta H^0}{nF} = \frac{+285800\ \mathrm{J/mol}}{2 \times 96485\frac{\mathrm{C}}{\mathrm{mol}}} = 1.481\mathrm{V}$$

As a summary, there is no reaction taking place below or at 25 °C, at the voltage of $U_{cell} \leq E^0 = 1.229\mathrm{V}$. If the cell voltage is $1.229\mathrm{V} \leq U_{cell} \leq V^0 = 1.481\mathrm{V}$ the reaction will start but it not able to sustain its operation, cooling the system. Therefore, to run the electrolysis continuously without external heating, the cell voltage must be higher than the thermoneutral voltage, $U_{cell} > V^0 = 1.481\mathrm{V}$. The value of voltages V and E at other temperatures can be solved directly from data in Fig. 7.45.

When no additional heat is provided, the heat needed to meet the required reaction enthalpy is taken from the dissipation of the electrical energy losses from the excess voltage used. In this case, we define the overvoltage as $\Delta U = U_{cell} - E$. In addition, the system has electrical resistive losses, lumped into a single resistance, IR_e, that creates an additional voltage drop. If $nF(\Delta U + IR_e) = T\Delta S$, the system does not need any additional heat, and this condition is known as the thermo-neutral electrolysis voltage of the cell, $\Delta U + IR_e = V$.

If the electrolyte is externally heated by the amount $Q > T\Delta S$ before it enters the electrolyzer, the free-energy voltage is sufficient to run the process. In real situations, to achieve a maximum rate of H_2 generation, the flow of current through the electrolyzer cells is increased. This in turn increases the ohmic and mass-transfer losses in the system, which means that the cell voltage must be increased to maintain the reaction. All excess voltage will dissipate to heat.

As discussed, the theoretical amount of energy required can be estimated from thermodynamics. By using the minimum theoretical power requirement, just above the ΔH, the rate of the process is very low, and a large system size would be needed to produce the gases at a meaningful rate. To reach higher generation rates, higher voltage and current are needed—and since losses are proportional to current, they increase as current increases. We define the cell efficiency as the ratio of theoretical work required to the actual work done,

$$\eta_{cell}(T, p, i) = \frac{E(T, p)It}{U_{cell}(T, p, i)It} = \frac{E(T, p)}{U_{cell}(T, p, i)}$$

noting that the transferred charge (C, coulomb) is the electrical current I (A = C/s) multiplied by time, the cell voltage is a function of the current density of the cell, $i = I/A$, pressure and temperature, and $\eta_{cell}(T, p, i)$ is a function of the temperature, pressure and current density. It is decreased by resistance along the path of the current, first in the electrodes, then in ohmic and mass transfer resistances in the electrolyte and membrane. For example, gas bubbles in the electrolyte create a diffusion resistance, since ions are not able to travel in gases. Figure 7.47 shows an illustration of an electrolysis cell's $U_{cell}(i)$ curve.

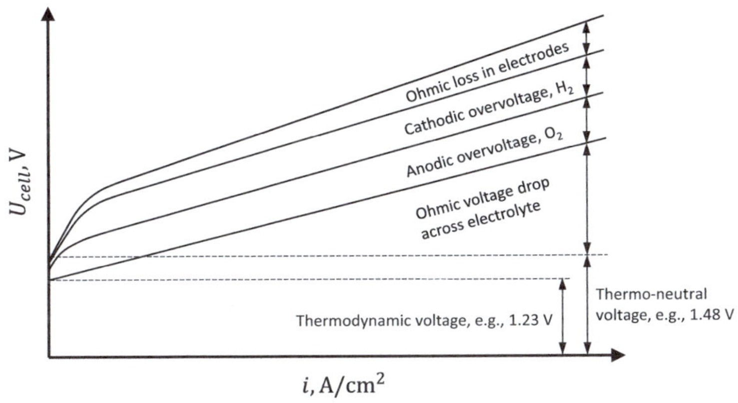

Fig. 7.47 Electrochemical $U_{cell}(i)$ characteristic of electrolyzer cell and contributions of different mechanisms (not to scale)

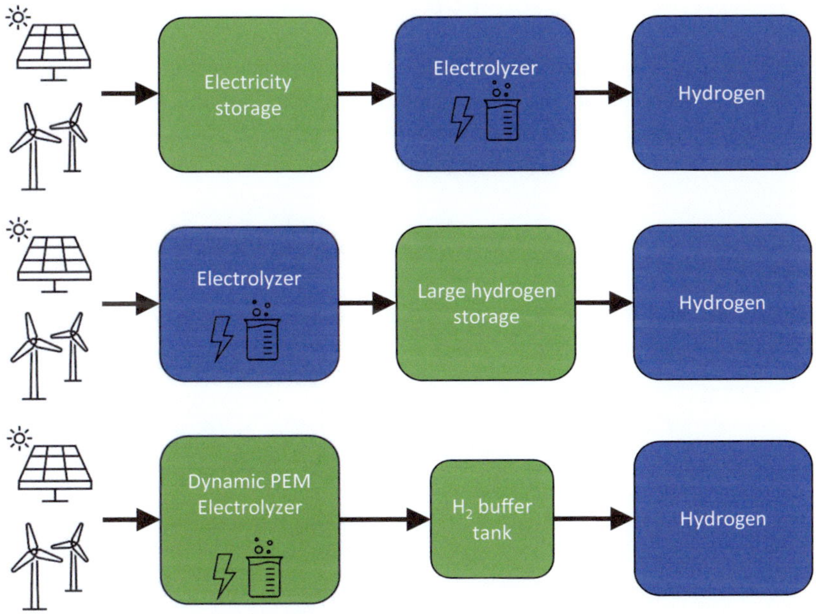

Fig. 7.48 Renewable energy storage strategies for continuous H_2 delivery, including hydrogen production for preparing the energy carrier. Flexible components is indicated by green color

Another useful characteristic is the Faradaic efficiency, defined separately for the anode and the cathode. This measures the efficiency of electrical current's conversion to gases:

$$\eta_{f.anode} = \frac{\dot{N}_{O_2}}{I/4F}$$

$$\eta_{f.cathode} = \frac{\dot{N}_{H_2}}{I/2F}$$

In a perfectly operating electrolyzer, both gases would be kept to their own sides, separated by the membrane. However, there are some internal leakages, especially due to small size of the hydrogen molecule, and these can be monitored by the Faradaic efficiency. If the gases are able to mix they react back to water, which decreases the process efficiency, since part of the work done is lost by reverse reaction. In addition, hydrogen is a safe gas as a pure gas, but if it forms mixture with oxygen/air that is between 3.9 to 98.5 mol%, it is very explosive [33]. This must be taken into account when planning the infrastructure.

In practice, the electrolyzer efficiency is calculated as the ratio of energy into the hydrogen produced to the electrical energy used. With the number of cells in series in the electrolyzer as n_{cell}, the energy supplied to the electrolyzer by a DC source is

$$E_{in} = UIt = n_{cell}U_{cell}It$$

where $U = n_{cell}U_{cell}$ is the total voltage at the terminals of the electrolyzer stack, assuming no connection losses. The current flowing through the electrolyzer is I (the same for all cells in series), and if all the cells all produce equal amounts of hydrogen, the total amount of hydrogen produced during time t is

$$N_{H_2} = n_{cell}\frac{\dot{N}_e}{2}\eta_F t = n_{cell}\frac{I}{2F}\eta_F t$$

where $F = 96,485\ \mathrm{A/mol}$ is Faraday's constant. For an electrolyzer operating with liquid water (l) and steam (g), the energy converted into hydrogen is as follows.

$$E_{\mathrm{H_2}.l} = n_{cell}\frac{I}{2F}t\eta_F\Delta H_l = n_{cell}\frac{I}{2F}t\eta_F \times 285{,}800$$

$$E_{\mathrm{H_2}.g} = n_{cell}\frac{I}{2F}t\eta_F\Delta H_g = n_{cell}\frac{I}{2F}t\eta_F \times 241{,}800$$

The efficiencies are then

$$\eta_l = \frac{E_{\mathrm{H_2}.l}}{E_{in}} = \frac{n_{cell}\frac{I}{2F}t\eta_F \times 285,800}{n_{cell}U_{cell}It} = \frac{\eta_F}{2F} \times \frac{285{,}800}{U_{cell}(i,T,p)}$$

$$\eta_g = \frac{E_{H_2 \cdot g}}{E_{in}} = \frac{n_{cell}\frac{I}{2F}t\eta_F \times 241,800}{n_{cell}U_{cell}It} = \frac{\eta_F}{2F} \times \frac{241{,}800}{U_{cell}(i, T, p)}$$

Based on these, the efficiency is independent of the generation rate of the hydrogen, but depends on the cell voltage, U_{cell}, which is a strong function of the cell current density.

More often than efficiencies, specific energy consumption is used to compare different electrolyzers. It is defined as the energy required to produce one normal cubic meter, m^3, or 1 kg of hydrogen. For electrolyzer operation with liquid water, reaction enthalpy is $\Delta H_l^0 = 285,800\text{J/mol} = 285,800/0.002\text{J/kg} = 39.694\text{kWh/kg} = 3.245\text{kWh/Nm}^3$, and for steam systems the reaction enthalpy is $\Delta H_g^0 = 241,800\text{J/mol} = 241,800/0.002 = 33.583 \text{ kWh/kg} = 2.746 \text{ kWh/Nm}^3$. Using ΔH^0 based efficiency (power + heat), the energy required can be calculated as follows, for liquid and vapor systems, respectively:

$$e_l = \frac{\Delta H^0}{\eta_{\Delta H}} = \frac{39.694}{\eta_{\Delta H}}\frac{\text{kWh}}{\text{kg}} = \frac{3.245}{\eta_{\Delta H}}\frac{\text{kWh}}{\text{Nm}^3}$$

$$e_g = \frac{\Delta H^0}{\eta_{\Delta H}} = \frac{33.583}{\eta_{\Delta H}}\frac{\text{kWh}}{\text{kg}} = \frac{2.746}{\eta_{\Delta H}}\frac{\text{kWh}}{\text{Nm}^3}$$

For typical AEC ($\eta_{\Delta H} = 0.7$) and PEM ($\eta_{\Delta H} = 0.8$) with liquid, and SOEC ($\eta_{\Delta H} = 1.0$) for steam, specific energy consumptions are

$$e_{AEC} = 56.7\frac{\text{kWh}}{\text{kg}} = 4.6\frac{\text{kWh}}{\text{Nm}^3}$$

$$e_{PEMEC} = 49.618\frac{\text{kWh}}{\text{kg}} = 4.1\frac{\text{kWh}}{\text{Nm}^3}$$

$$e_{SOEC} = 33.6\frac{\text{kWh}}{\text{kg}} = 2.7\frac{\text{kWh}}{\text{Nm}^3}$$

At the time of writing this book, there are two main industrial-scale electrolyzer systems to perform the water splitting reaction with electricity: alkaline electrolyzers, and proton exchange electrolyzers. Alkaline electrolysis cells (AEC) are a mature technology and can be purchased and deployed at scale; however, they are not very efficient and have low current densities. They operate at a constant temperature on liquid electrolyte and benefit from continuous operation. Since they are a mature technology, they have also been the first to be implemented at large scale for green hydrogen production. As they operate at alkaline conditions, they can use Nickel as a catalyst material, which is not critical and fairly low cost. For this reason, AEC will play an important role in the future, helping to widen the use of different materials used in electrolyzers.

Polymer electrolyte membrane electrolysis cells (PEMEC) have recently come to market, and are now at the large demonstration stage. Their history goes back to the fuel cell production and space missions, where PEM fuel cells are still currently being used. In PEMEC, the membrane material is negatively charged and therefore repels electrons while attracting protons (H^+). After water molecules are split at the anode to form oxygen molecules, protons and electrons, the protons cross though the membrane and the electrons go to outer circuit—they meet at the cathode and reduce to become hydrogen molecules. The advantage of the membrane is also that there is a clear separator between the hydrogen and oxygen gases, and therefore both gases are very pure. This electrolyzer can also be operated with more flexibility at different loads in dynamic operation.

PEMEC are one prominent technology for green hydrogen production. With the high voltage requirements for water splitting in acidic conditions, there are very limited options for materials suited to these systems. Currently, the most common materials are Pt for the cathode and InO_2 for the anode. Future development is directed towards minimizing these components by adding other metals to bi- and tri-catalytic alloys. As the alkaline conditions are very appealing and provide more options for electrode materials, alkaline-ion exchange membranes have also been developed, however they still suffer from slower ion conductivity and require more research [33].

The solid oxide electrolysis cell (SOEC) is an electrolyzer that operates at higher temperatures (600–800 °C) and converts steam into oxygen and hydrogen. This technology is built on solid oxide fuel cell research, and at high temperature the same device can be used as a reverse solid oxide system, and operate as a fuel cell and electrolyzer, since the electrode materials are the same. SOEC technology uses ceramic cell materials that have metallic oxide components such as Nickel, Yttrium, and Lanthanum [33]. In these systems, the electrolyte is a solid, oxygen ion (O^{2-}) conductive ceramic, and this only operates at higher temperatures. A special feature of this technology is that it can be also operated in co-electrolysis mode, where a simultaneous feed of steam and CO_2 produces synthetic gas ($CO + H_2$), and possibly other hydrocarbons as well, that can be directly utilized as fuels. Table 7.12 gives a comparison of the different characteristics of the main electrolyzer types.

7.6.2 *Hydrogen Logistics*

With the variability of renewable energy generation, flexibility must be integrated somewhere along the value chain. Hydrogen is often discussed as a flexibility factor, but there are still several alternatives on where the flexibility can exist in the system—some scenarios are presented in Fig. 7.48. The best use-case would be to pressurize the hydrogen to a required pressure and transport it directly to a large, industrial-scale customer. This is the main way that hydrogen systems have been implemented in the past, but the hydrogen source has been natural gas, which itself is an energy storage—therefore, the flexibility has actually been provided by the fossil fuel.

Table 7.12 Comparison of electrolyzer technologies [33]

	AEC	PEMEC	SOEC
Maturity, 1–10	10	9	7–8
Price, €/kW (2050)	800–1300 (150)	1200–2000 (150)	1000–2000 (250)
Charge carrier	OH^-	H^+	O^{2-}
Current density, A/cm^2	0.2–0.5	0.8–2.5	1–2
Operation temperature, °C	25–120	50–80	700–850
Operation pressure, bar	1–200	1–350	1–5
Ramping rates (cold/warm), min	50/6	10–20/2	600/10
Specific energy consumption, kWh/kg H_2	46	55	40
Efficiency, % (LHV)	70	60	81
Lifetime, h	100,000	10,000–50,000	500…2000
Capacity range, kW	– 5000	0–200	–
Positive features	Low cost, technically mature, high reliability, elevated pressure (10 MPa has been tested)	No corrosive electrolytes, high pressures, high current densities, suitable for transients	If high temperature waste heat is available, very high efficiency can be obtained Co-electrolysis option of H_2O and CO_2 to syngas
Negative features	Large size, moderate current densities, not suitable for transient operation	Catalyst materials expensive and limited, high price	Still under development, material durability issues. Not suitable for transient operation

As presented in Fig. 7.48, one option is to combine an electrolyzer plant with an electricity storage system, for instance a hydro power plant, battery storage, or a flywheel, that provides the flexibility for the renewable energy plant (upper path). In this way, the electrolyzer plant could operate in optimal conditions and provide a constant hydrogen stream for the end user—but this would cause additional costs for the energy system. A second option is to directly connect the electrolyzer to the renewable energy system, and it then operates in optimal conditions when there is renewable energy available (mid path); when renewable energy production is low, the electrolyzer is entirely shut down. To provide a constant hydrogen stream, the system has a hydrogen storage system that will balance the production need. In this scenario the electrolyzer needs to be oversized so that part of the hydrogen produced is not provided immediately to the customer, but stored for periods of lower renewable energy input.

The third option is to operate the electrolyzer in dynamic mode, which means that the electrolyzer operation will be modified according to electricity input (lower path). A PEMEC system is the most tolerant for dynamic operation, and the operation can vary between 5 and 100% capacity [87]. In this case, only a very small buffer tank of hydrogen would be required. There is currently insufficient knowledge about this mode of dynamic operation and its impact on materials and electrolyzer lifetime; however, this method has attracted a large interest in the recent years.

Renewable energy production is not always located where the energy will be needed. We know that, in the future, we will need to build a large quantity of electricity transmission lines, and this may become technically unfeasible or very costly. An alternative way to transport the renewable energy is to first convert electricity into hydrogen and then transfer the hydrogen through a pipeline. This would offer an alternative transmission network, which would also provide additional flexibility for the total energy system. The larger the hydrogen production capacity, the lower the transfer cost per kg of hydrogen.

A hydrogen pipeline can also operate as its own hydrogen storage, and as a buffer, as the pressure of the hydrogen can be varied (with certain limits), and therefore can change the mass of the gas within the pipeline. Many places around the globe have an existing natural gas pipeline available. Hydrogen can also be blended and transported with methanol (fossil, or possibly synthetic in the future), since these pipelines already exist, and so the related capex investments are lower.

Safety is a very important topic when we consider hydrogen production, storage, and logistics—and so it is useful to remember that hydrogen has been the most-used commodity in the chemical industry for over a century, and there are very detailed standards for hydrogen logistics. Hydrogen by itself does not ignite or explode, but with oxygen (volume fraction from 3.9 to 98.5%) it creates an explosive gas mixture. Additionally, hydrogen is the smallest molecule on earth, and therefore its storage and logistics need special attention. For instance, old natural gas pipes can also transport some fraction of hydrogen [33], but 100% hydrogen—therefore the direct use of old infrastructure needs to be carefully analyzed when implementing more hydrogen streams for industrial processes.

7.6.3 Hydrogen Storage

When we discuss industrially-relevant hydrogen production, especially if connected with the renewable energy sources, large hydrogen storage systems must be applied. As discussed above, fossil fuels provided us with large storage capacities, and this flexibility must be now located in other parts of the energy system. At small scales a variety of different hydrogen storage systems are available—for instance, hydrogen can be connected to various chemical components, and then later released. But for grid-stabilizing levels, only large-scale options are presented here.

Hydrogen can be stored underground in geological reserves, especially in salt caverns, or any other existing built or natural caverns [27]. However, this depends on the geological properties of the area—some countries have these available, making them particularly appealing locations for hydrogen storage. For regions where suitable salt caverns do not exist, it is possible to prepare lined rock caverns (LRC), where human-made or natural rock caverns are strengthened with steel and concreate liners. In 2024, the first commercial LRC demonstration is being built in Sweden as a part of the HYBRIT-project [68], and this will bring additional expertise on LRC construction and highlight the issues and challenges that need to be explored when storing hydrogen in large quantities. As just one example, hydrogen has different physical behavior to natural gas—for instance, it releases heat when depressurized—and this needs to be considered in the storage design [33, 74].

Steel tanks have also been proposed for hydrogen storage, as they are a mature, commercially-available technology; they are also very costly if the quantity of hydrogen is large. These will most likely be used for buffer storage, especially for industrial purposes. As noted above, pipelines have also been proposed for storing hydrogen, and they will be part of the total hydrogen logistics chain—however, since the pipeline pressure is not very high, they cannot entirely solve the hydrogen storage problem. Larger systems, most likely underground, will offer the most cost-efficient solutions for industrial-scale hydrogen storage.

7.6.4 Energy Carriers

As discussed in the beginning of this section, there is a large variety of applications where either the energy density of direct electrification or the use of hydrogen is not enough. As our transportation sector has been optimized for fossil-based fuels, high energy–density fuels have dictated the development of aviation, shipping and heavy-duty vehicles for the past century. As a result, we do have the potential to convert green hydrogen into higher energy–density molecules that we can call energy carries. In this process, hydrogen reacts with N_2 or CO_2, which can both be captured from the atmosphere (or CO_2 can also be captured from industrial point sources). By this route, it is possible to produce any hydrocarbons that can be used in industrial processes or for powering machinery with existing technologies. This route is often referred as Power to X (P_2X), or power to fuels, as this is a possibility for the electrification of sectors that are difficult to directly electrify in other ways. The fuels produced with this method are often referred as synthetic fuels or e-fuels. The different routes are presented in Fig. 7.49.

Technically, with this method it is possible to produce hydrocarbons with any molecular length, but the longer the carbon chain, the more we need to add hydrogen in the process, which in turn means higher consumption of electrical energy. Therefore, it is preferable to produce molecules that are as simple as possible—methane, methanol and ammonia are the most preferred energy carriers in the current scenarios. These fuels are also easier and cheaper to store and transfer than pure hydrogen. Methanol is a liquid at room temperature, and the current logistics have a lot of experience of handling liquid fuels. The marine sector has existing infrastructure and

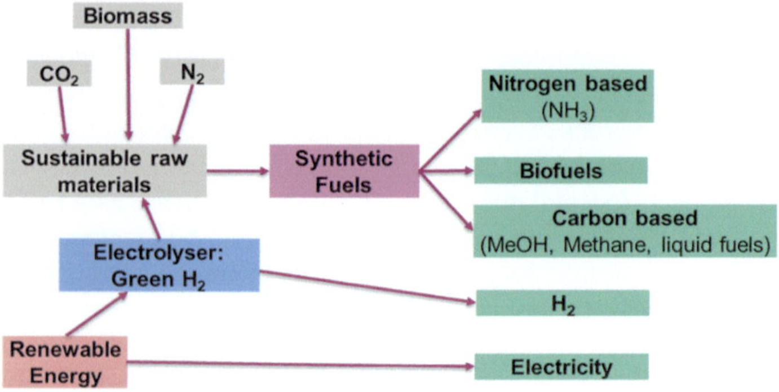

Fig. 7.49 Renewable energy storage strategies including hydrogen production for preparing different energy carriers

knowledge of handling natural gas (fossil methane), and since its synthetic counterpart has the same chemical and physical properties, the marine industry can easily be transformed to work with synthetic methane. Ammonia is a gas at room temperature, and has not yet been used in the transport sector; however, the changes required for ammonia are smaller than would be needed for pure hydrogen.

Longer hydrocarbons are even closer to the fossil fuels used in the transportation sector, and it is possible to produce synthetic diesel and naphtha (suitable for the highly-regulated aviation market). At the moment, bio-based diesel and naphtha are available, and they will also have an important role in the transportation sector in the future. However, if we want to entirely get rid of fossil fuels in transportation, we do not have enough bio-based raw materials to cover the need. For this reason, we will need synthetic fuels to support the phase-out of the fossil components.

This all sounds like a dream come true—why is it, then, that we have not been able to get rid of fossil fuels over a short time-frame, if we can produce all the energy carriers though these biofuel or synthetic fuel routes? The first challenge is that CO_2 is a very different starting molecule than the crude oil that we are used to. Crude oil is ancient biomass, where CO_2 was captured from the atmosphere by plants and then converted to hydrocarbons through photosynthesis within the plant. After the plant died, the biomass was pressurized underground, and the carbon chain further increased. As a result, crude oil is composed of very long hydrocarbons with high energy density. In a current fossil oil refinery, this long carbon chain is cracked into smaller pieces to produce different hydrocarbon products.

By contrast, along the synthetic route, we use the CO_2 molecule as the first step. As presented in the Fig. 7.50, CO_2 itself has zero energy content, which does not make it an appealing energy carrier. We need to boost the energy content with hydrogen, which in turn requires a lot of electrical energy. The molecule produced, in this case methanol, has a mid-energy content.

Sometimes synthetic fuels have been criticized for the inefficiency of their process routes. Figure 7.51 visualizes some of the challenge connected with synthetic fuel production [18]. In the synthetic routes, we will need to first produce both CO_2 and

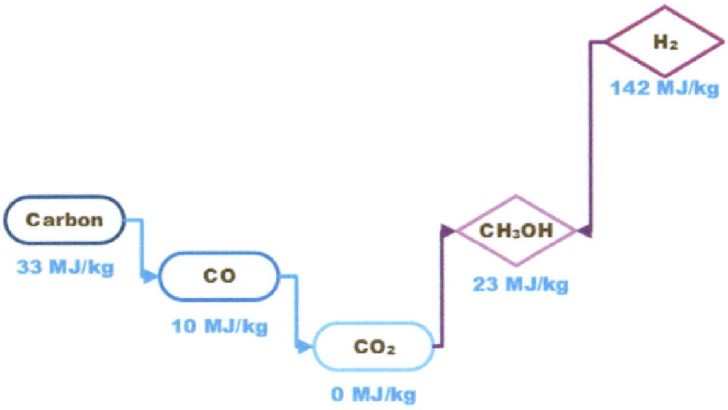

Fig. 7.50 Energy content for different carbon compounds and hydrogen. CO_2 has no energy content, but when it reacts with hydrogen, the derivative energy carriers have higher energy content

H_2. CO_2 can be obtained directly from the atmosphere (blue line), but as the concentration of CO_2 in the air is very low, there is larger energy input for this process. We can also utilize point sources of CO_2, for instance from a power-plant emissions, and preferably from biogenic CO_2 combustion from a bio-process plant (green line)—in which case the energy efficiency of the process is higher. Additionally, we will need to produce the hydrogen, and depending on the electrolyzer technology, the efficiency can vary. The work referenced compared the three most-used electrolyzer technologies—alkaline electrolysis cell (AEC), Polymer electrolyte membrane electrolysis cell (PEMEC), Solid oxide electrolysis cell (SOEC)—and their reported efficiencies. For simplicity, an average value of the electrolyzer efficiency is used in Fig. 7.51.

When comparing to the fossil route (red) it is good to keep in mind that the CO_2 capture and the carbon-chain build up was completed by natural photosynthesis processes, millions of years ago. The fossil route has higher efficiency because part of the work has been completed by nature, and so a direct comparison in not possible. We have been utilizing this gift from nature for a century, and now we must be able

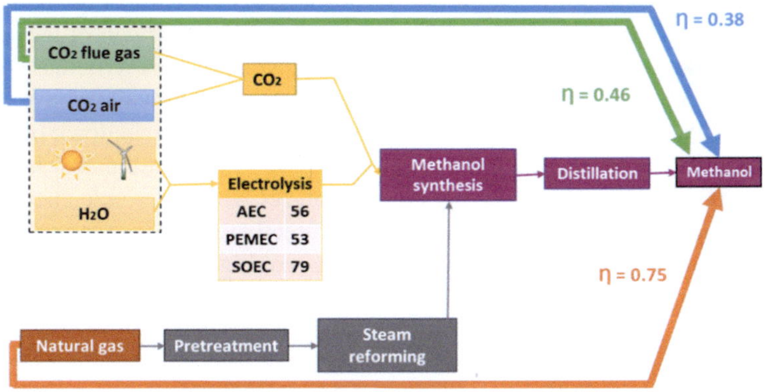

Fig. 7.51 Different process routes for methanol production from natural gas and from renewable energy sources. Electrolyzer types are: AEC, alkaline electrolysis cell; PEMEC, polymer membrane electrolysis cell; and SOEC, solid oxide electrolysis cell. Number values are obtained from Cifre and Badr [18]

to produce carbon-based products by utilizing the current biomass (that has taken the CO_2 from the atmosphere), or we need to capture the CO_2 ourselves.

The second problem for synthetic fuels are the current production costs and increase in volumes. There are not yet large-enough, concentrated CO_2 streams available for processing at industrial scales. Processing CO_2 into long hydrocarbons is a chemical process for which we do not possess strong expertise, and this transition will require new, more-efficient catalyst materials that are not sensitive to the water product created in CO_2 hydrogenation process:

$$\mathrm{CO_2 + 3H_2 \rightarrow CH_3OH + H_2O},\ \Delta H = -49.16\mathrm{kJ/mol}$$

Compare this to methanol production from syngas used in the traditional methanol processing:

$$CO + 2H_2 \rightarrow CH_3OH,\ \Delta H = -90.77\mathrm{kJ/mol}$$

Additionally, notice that in the latter reaction, only two hydrogen molecules are required, in comparison to the three molecules required for the CO_2 hydrogenation. This third molecule of hydrogen is then “lost” in the process, in the production of the water molecule (that was initially broken in the water-splitting step with renewable energy), and this decreases the overall conversion from renewable energy to methanol. One interesting future solution to overcome this challenge in chemical processing is the co-electrolysis of water (steam) and CO_2 in a solid oxide electrolysis cell (SOEC), to produce syngas without producing a significant quantity of water [33]. As water is also a reactant in this process, it can be redirected for that purpose.

The third issue for synthetic fuels is associated with their cost. Since they are chemically entirely similar, if they were financially competitive with their fossil counterparts, synthetic fuels would have already been implemented widely across the world, and would easily phase out fossil fuels. As shown in Fig. 7.51, one of the issues is that fossil raw materials are more energy dense, as part of the work has been provided by nature in creating both CO_2 and hydrocarbons. Whereas, when we produce the same components synthetically, we need to perform all the process steps ourselves. This contributes to the price of synthetic fuels—a breakdown of the costs is presented in Fig. 7.52. The most important thing to note in the cost breakdown is that the H_2 (hydrogen produced in an electrolyzer with renewable energy) is the largest contribution to the costs overall. H_2 cost directly reflects the renewable energy prices; therefore, synthetic energy fuel products are highly dependent on low-cost renewable energy availability.

Fig. 7.52 The cost breakdown of synthetic methanol production with addition of the Methanex price (cost for the current fossil-based methanol) [76]

7.7 Summary

As the world transitions to an energy system significantly based on variable wind and solar power, the demand for energy storage systems on all scales increases. Today, in 2025, the greatest challenges are in the lack of large-scale, long-term energy storage systems. As discussed in this chapter, many of these systems are technically achievable, but they would need very large commitments in terms of physical size, land-use impact, and materials requirements to make a meaningful impact at the system level.

The first major solution to this challenge is to cut our consumption, in every aspect. As discussed in Chap. 1, today we are over-using the biocapacity of the planet—actually, a single planet is not enough, we need 1.75 planets to fulfill our current consumption demands. Of course, it will be very difficult to make this change, but just think about it: cutting our consumption by the required amount of more than 50% would make finding the overall energy production and storage solutions much, much easier.

Another solution is to rethink the way we are using energy. Could we synchronize our energy-intensive activities with the variability in our energy generation? With smart demand-side response systems and automation, we could significantly decrease the need for peak power and, at the same time, cut down the need for energy storage and avoid the need for peak power generation systems that result in capital expenditures, material requirements, land-use impacts, and many other challenges—all to make our living more sustainable.

Finally, we need to explore all existing energy storage technologies, and optimize the energy system so that there is a perfect alignment with storage technologies and applications. In designing and developing energy storage solutions, it is important to consider the whole energy value chain and its overall efficiency. For example, although the individual efficiencies of the components in an electrolysis-hydrogen storage fuel cell system are quite high, the overall efficiency drops after multiplying the component efficiencies with each other—it the electrolyzer efficiency is 65%, storage 80% and fuel cell 65%, the overall efficiency is only $0.65 \times 0.8 \times 0.65 = 0.338$, and we loose more than 60% of our precious energy. We must always try to select the system with the highest efficiency; of course, we also need to make the solution economically profitable, and we must also ensure that sustainability requirements are satisfied. Even though the efficiency of the hydrogen storage cycle, i.e., water electrolysis-storage-fuel cell, is low, we must remember that the overall efficiency of the internal combustion engine pathway, from the production of fossil gasoline from crude oil, or from production of e-fuels to move our cars is much lower as a whole.

We have discussed in this chapter many existing, mature energy storage solutions, as well as many that are still emerging. Because this is only a single book, it has not been possible to cover everything. We hope that the discussions here give readers and students inspiration to study, learn and discover more. We also hope that building a solid theoretical basis in these technologies and sciences will provide a solid foundation for new discoveries.

Author contributions Mika Järvinen wrote the sections 7.1 and 7.2. The section 7.3.1 was written by Janne Hirvonen and Mika Järvinen, 7.3.2 by Janne Hirvonen, 7.3.3 by Mika Järvinen, and the section 7.3.4 by Konsta Turunen and Ari Seppälä. Mika Järvinen wrote the section 7.4. Neha Garg, Annukka Santasalo-Aarnio and Mika Järvinen wrote the section 7.5. Section 7.6 was written by Annukka Santasalo-Aarnio and Mika Järvinen. Mika Järvinen and Annukka Santasalo-Aarnio wrote the Section 7.7. When citing, refer to section number, include all author names, and use Chapter DOI number.

Appendix 7.1: Valve sizing chart for the saturated steam © Spirax-Sarco

Based on the maximum inlet pressure, move horizontally to the critical pressure drop line. Move then vertically downwards to the required maximum flow rate and read the value of valve coefficient from the K_{VS} lines. Critical pressure drop line corresponds to choking flow condition, giving the maximum flow through the nozzle.

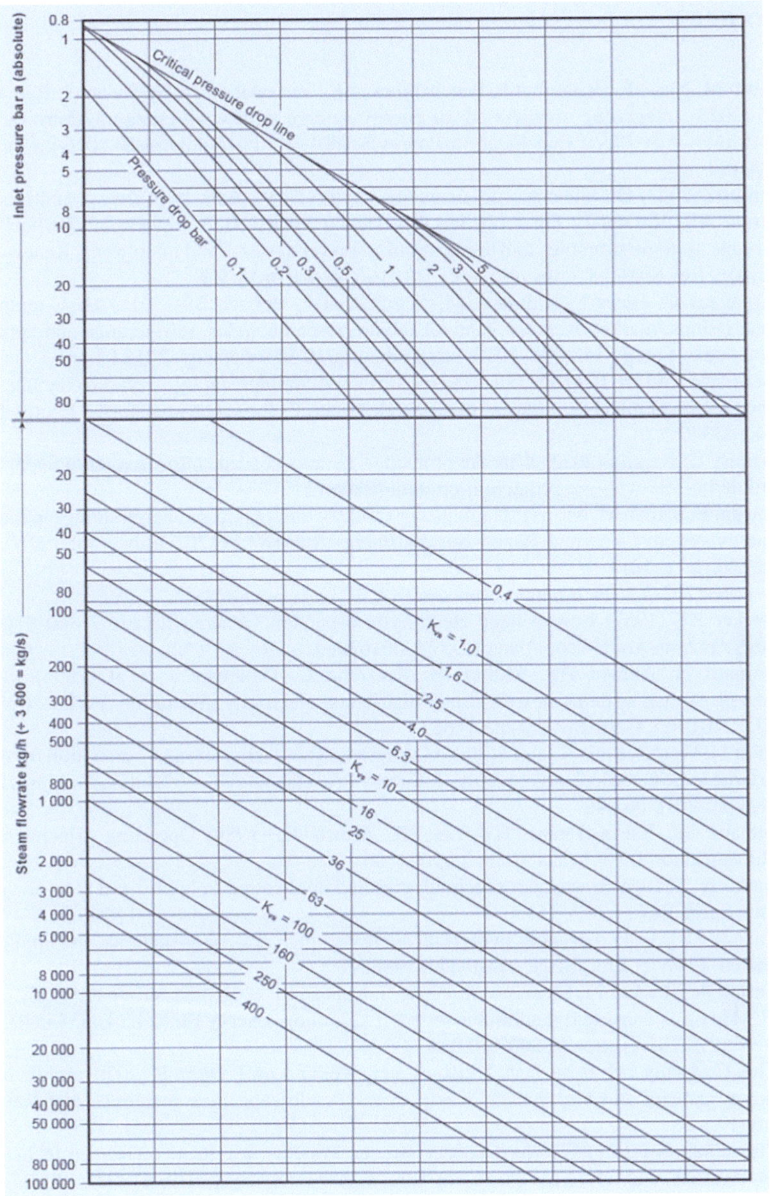
Inlet pressure bar a (absolute)
Critical pressure drop line
Pressure drop bar
0.1
0.2
0.3
0.5
1
2
3
5
0.8
1
2
3
4
5
8
10
20
30
40
50
80
Steam flowrate kg/h (÷ 3 600 = kg/s)
20
30
40
50
80
100
200
300
400
500
800
1 000
2 000
3 000
4 000
5 000
8 000
10 000
20 000
30 000
40 000
50 000
80 000
100 000
0.4
K_{vs} = 1.0
1.6
2.5
4.0
6.3
K_{vs} = 10
16
25
36
63
K_{vs} = 100
160
250
400

References

1. Adib M, Nasir F, Haghighat F, Panchabikesan K, Venkataramani G, Tiwari S, Ramalingam V (2023) Integrating compressed air energy storage with wind energy system—a review. e-Prime—Adv Electr Eng Electron Energy 5:100194. https://doi.org/10.1016/j.prime.2023.100194
2. Andritz (2023) On-line customer magazine, Hydro News, Asia. https://www.andritz.com
3. Arani AK, Karami H, Gharehpetian GB, Hejazi MSA (2017) Review of flywheel energy storage systems structures and applications in power systems and microgrids. Renew Sustain Energy Rev 69:9–18. https://doi.org/10.1016/j.rser.2016.11.166
4. Arpagaus C, Bless F, Uhlmann M, Schiffmann J, Bertsch SS (2018) High temperature heat pumps: market overview, state of the art, research status, refrigerants, and application potentials. Energy 152:985–1010. https://doi.org/10.1016/j.energy.2018.03.166
5. Baig MM, Gul IH, Baig SM, Shahzad F (2021) The complementary advanced characterization and electrochemical techniques for electrode materials for supercapacitors. J Energy Storage 44:103370
6. Battery University (2023) Why are protection circuits needed? https://batteryuniversity.com/article/bu-304-why-are-protection-circuits-needed
7. Bazdar E, Sameti M, Nasiri F, Haghighat F (2022) Compressed air energy storage in integrated energy systems: a review. Renew Sustain Energy Rev 167:112701. https://doi.org/10.1016/j.rser.2022.112701
8. Beacon (2024) www.beaconpower.com
9. Becker HE (1957) Low voltage electrolytic capacitor. General-Electric 2 800 616. 1957. https://patents.google.com/patent/US2800616A/en
10. Bernard N, Ahmed HB, Multon B, Kerzreho C, Delamare J et al (2003) Flywheel energy storage systems in hybrid and distributed electricity generation. PCIM 2003, May, NURNBERG, Germany, hal-00674687
11. Budt M, Wolf D, Span R, Yan J (2016) Compressed air energy storage—an option for medium to large scale electrical-energy storage. Energy Procedia 88:698–702. https://doi.org/10.1016/j.egypro.2016.06.046
12. Bullard GL, Sierra-Alcazar HB, Lee HL, Morris JL (1989) Operating principles of the ultracapacitor. IEEE Trans Magn 25:102–106
13. Burke A (2010) Ultracapacitor technologies and application in hybrid and electric vehicles. Int J Energ Res 34:133–151
14. Cabeza LF (2021) Advances in thermal energy storage systems—methods and applications, 2nd ed. Elsevier Ltd. ISBN: 978-0-12-819888-9
15. Castell A, Martorell I, Medrano M, Perez G, Cabeza LF (2010) Experimental study of using PCM in brick constructive solutions for passive cooling. Energy Build 42(4):534–540. https://doi.org/10.1016/j.enbuild.2009.10.022
16. Choi C, Ashby DS, Butts DM, DeBlock RH, Wei Q, Lau J, Dunn B (2020) Achieving high energy density and high-power density with pseudo-capacitive materials. Nat Rev Mater 5:5–19
17. Choudhury S (2021) Flywheel energy storage systems: a critical review on technologies, applications, and future prospects. Int Trans Electr Energ Syst 31(9):e13024. https://doi.org/10.1002/2050-7038.13024
18. Cifre PG, Badr O (2007) Renewable hydrogen utilization for the production of methanol. Energy Convers Manag 48:519–527
19. Colthorpe A (2023) Moss landing: world's biggest battery storage project is now 3 GWh capacity. https://www.energy-storage.news/moss-landing-worlds-biggest-battery-storage-project-is-now-3gwh-capacity/
20. Dasgupta P (2021) The economics of biodiversity: the Dasgupta review. HM Treasury, London
21. Deguenon L, Yamegueu D, Kadri SM, Gomna A (2023) Overcoming the challenges of integrating variable renewable energy to the grid: a comprehensive review of electrochemical

battery storage systems. J Power Sources 580:233343. https://doi.org/10.1016/j.jpowsour.2023.233343
22. Ding Y (2021) Thermal energy storage materials, devices, systems and applications. United Kingdom, Croydon: Royal Society of Chemistry. ISBN: 978-1-78801-717-6
23. Dutta A, Mitra S, Basak M, Banerjee T (2023) A comprehensive review on batteries and supercapacitors: development and challenges since their inception. Energy Storage 5(1). https://doi.org/10.1002/est2.339
24. Eaton (2020) Technical Data 10339
25. EEA (2023) Share of energy consumption from renewable sources in Europe. https://www.eea.europa.eu/en/analysis/indicators/share-of-energy-consumption-from?activeAccordion=ecdb3bcf-bbe9-4978-b5cf-0b136399d9f8
26. Ehtiwesh A, Kutlu C, Su Y, Riffat S (2023) Modelling and performance evaluation of a direct steam generation solar power system coupled with steam accumulator to meet electricity demands for a hospital under typical climate conditions in Libya. Renew Energy 206:795–807
27. Elberry AM, Thakur J, Santasalo-Aarnio A, Larmi M (2021) Large-scale compressed hydrogen storage as part of renewable electricity storage systems. Int J Hydrogen Energy 46(29):15671–15690. https://doi.org/10.1016/j.ijhydene.2021.02.080
28. Elgrishi N, Rountree KJ, McCarthy BD, Rountree ES, Eisenhart TT, Dempsey JL (2018) A practical beginner's guide to cyclic voltammetry. J Chem Educ 95:197–206
29. Encyclopedia Britannica (1966) Universities of Chicago, Oxford, Cambridge and London
30. EnergyVault (2024) www.energyvault.com
31. Giannakou P (2022). Comparison of CV responses of ideal capacitor, typical EDLC capacitor, resistive super-capacitor and supercapacitor with pseudo-capacitance. https://doi.org/10.6084/m9.figshare.20186636.v1
32. Glendenning I (1976) Long-term prospects for compressed air storage. Appl Energy 2:39–56
33. Godula-Jopek (2015) Hydrogen production by electrolysis. Wiley-VCH. ISBN 978-3-527-33342-4
34. Goodenough JB (2018) How we made the Li-ion rechargeable battery. Nat Electron 1(3):204. https://doi.org/10.1038/s41928-018-0048-6
35. Goodenough JB, Park KS (2013) The li-ion rechargeable battery: a perspective. J Am Chem Soc 135(4):1167–1176. https://doi.org/10.1021/ja3091438
36. Gouda EM, Fan Y, Benaouicha M, Neu T, Luo L (2021) Review on liquid piston technology for compressed air energy storage. J Energy Storage 43:103111. https://doi.org/10.1016/j.est.2021.103111
37. Gunasekara SN, Barreneche C, Fernández AI, Calderón A, Ravotti R, Ristić A, Weinberger P, Paksoy HÖ, Koçak B, Rathgeber C, Chiu JN, Stamatiou A (2021) Thermal energy storage materials (TESMs)—what does it take to make them fly? Crystals 11(11):1276. https://doi.org/10.3390/cryst11111276
38. Guo X, Zheng S, Zhang G, Xiao X, Li X, Xu Y, Xue H, Pang H (2017) Nanostructured graphene-based materials for flexible energy storage. Energy Storage Mater 9:150–169
39. Hauer A, Fumey B, Gschwander S, Lager D, Lázaro A, Rathgeber C, Ristić A, van Helden W (2020) Material and component development for thermal energy storage executive summary of the final report. Task 58 Annex 33. https://iea-es.org/wp-content/uploads/public/IEA-ES-TCP-Annex-33-Executive-Summary_revised.pdf
40. Haynes WM, Lide DR, Bruno TJ (2017) CRC Handbook of chemistry and physics, 97th ed. eBook, CRC Press, Boca Raton. ISBN: 78-1-4987-5429-3
41. Helen (2015) www.helen.fi
42. Helmholtz H (1853) (1853) On some laws of the distribution of electric currents in physical conductors with application to animal electrical experiments. Ann Phys 165:211–233
43. Hesse HC, Schimpe M, Kucevic D, Jossen A (2017) Lithium-ion battery storage for the grid—a review of stationary battery storage system design tailored for applications in modern power grids. Energies 10(12):2107. https://doi.org/10.3390/en10122107
44. Hirvijoki E, Hirvonen J (2022) The potential of intermediate-to-deep geothermal boreholes for seasonal storage of district heat. Renewable Energy 198:825–832. https://doi.org/10.1016/j.renene.2022.08.070

45. Hirvonen J, Kosonen R (2020) Waste incineration heat and seasonal thermal energy storage for promoting economically optimal net-zero energy districts in Finland. Buildings 10(11):205. https://doi.org/10.3390/buildings10110205
46. Hoffstaedt JP, Truijen DPK, Fahlbeck J, Gans LHA, Qudaih M, Laguna AJ, De Kooning JDM, Stockman K, Nilsson H, Storli P-T, Engel B, Marence M, Bricker JD (2022) Low-head pumped hydro storage: a review of applicable technologies for design, grid integration, control and modelling. Renew Sustain Energy Rev 158:112119
47. Höhlein S, König-Haagen A, Brüggemann D (2017) Thermophysical characterization of $MgCl_2 \cdot 6H_2O$, xylitol and erythritol as phase change materials (PCM) for latent heat thermal energy storage (LHTES). Materials 10(4):444. https://doi.org/10.3390/ma10040444
48. Hua W, Lv X, Zhang X, Ji Z, Zhu J (2023) Research progress of seasonal thermal energy storage technology based on supercooled phase change materials. J Energy storage 67:107378. https://doi.org/10.1016/j.est.2023.107378
49. Hunt JD, Byers E, Riahi K, Langan S (2018) Comparison between seasonal pumped-storage and conventional reservoir dams from the water, energy and land nexus perspective. Energy Convers Manag 166:385–401. https://doi.org/10.1016/j.enconman.2018.04.044
50. Hutchinson AJ, Gladwin DT (2023) Capacity factor enhancement for an export limited wind generation site utilizing a novel flywheel energy storage strategy. J Energy Storage. 68:107832. https://doi.org/10.1016/j.est.2023.107832
51. Hydro (2024) www.hydrostor.ca
52. IEC (2022) Fixed electric double-layer capacitors for use in electric and electronic equipment. IEC 62391-1 and IEC 62391-2. International Electrochemical Commission
53. Jenkins J, Ekanayake N (2017) Renewable energy engineering. Cambridge University Press, Cambridge. ISBN: 139781107028487
54. Jian R, Sun K, Li R, Zhang Y, Wang W, Yin H, Fang D, Shi Q, Tan Z (2017) Heat capacities of some sugar alcohols as phase change materials for thermal energy storage applications. J Chem Thermodyn 115:233–248
55. Jotuni, P, Ryti H, Pöyhönen O (1966) Handbook of technology 2, E. Thermodynamics (In Finnish: Tekniikan käsikirja 2, E. Termodynamiikka), Gummerus, Jyväskylä, Finland
56. Kahwaji S et al (2017) Fatty acids and related phase change materials for re-liable thermal energy storage at moderate temperatures. Sol Energy Solar Cells 167:109–120. https://doi.org/10.1016/j.solmat.2017.03.038
57. Karasu H, Dincer I (2020) Life cycle assessment of integrated thermal energy storage systems in buildings: a case study in Canada. Energy Build 217:109940. https://doi.org/10.1016/j.enbuild.2020.109940
58. Khan KI, Asfand F, Al-Ghamdi SG (2022) Progress in research and technological advancements of thermal energy storage systems for concentrated solar power. J Energy Storage 55:105860. https://doi.org/10.1016/j.est.2022.105860
59. Kwade A, Haselrieder W, Leithoff R, Modlinger A, Dietrich F, Droeder K (2018) Current status and challenges for automotive battery production technologies. Nat Energy 3(4):290–300. https://doi.org/10.1038/s41560-018-0130-3
60. Lakshmi KCS, Vedhanarayanan B (2023) High-Performance supercapacitors: a comprehensive review on paradigm shift of conventional energy storage devices. Batteries 9(4):202. https://doi.org/10.3390/batteries9040202
61. Larsonneur (1990) Design and control of active magnetic bearing systems for high-speed rotation. Dissertation, ETH 9140 Zurich
62. LG (2024) www.lgessbattery.com
63. Li X, Palazzolo A (2022) A review of flywheel energy storage systems: state of the art and opportunities. J Energy Storage 46:103576. https://doi.org/10.1016/j.est.2021.103576
64. Lian YN, Chen AT, Suurkuusk J, Wadsoe I (1982) Polyol-water interactions as reflected by aqueous heat capacity values. Acta Chem Scand A36(9):735–739
65. Liang T, Hou R, Dou Q, Zhang H, Yan X (2021) The applications of-water-in-salt electrolytes in electrochemical energy storage devices. Adv Funct Mater 31:2006749

66. Liu J, Zhang JG, Yang Z, Lemmon JP, Imhoff C, Graff GL, Li L, Hu J, Wang C, Xiao J et al (2013) Materials science and materials chemistry for large scale electrochemical energy storage: from transportation to electrical grid. Adv Funct Mater 23:929–946
67. Liu X, Dai X, Wei G et al (2017) Experimental and theoretical studies of nonlinear dependence of the internal resistance and electrode thickness for high performance supercapacitor. Sci Rep 7:45934. https://doi.org/10.1038/srep45934
68. LKAB (2024) www.lkab.com
69. Mangold D, Deschaintre L (2012) Seasonal thermal energy storage. Solar heating and cooling program IEA. https://www.iea-shc.org/Data/Sites/1/publications/IEA_SHC_Task45_B_Report.pdf
70. Meisingset KK, Grønvold F (1984) Thermodynamic properties and phase transitions of salt hydrates between 270 and 400 K III. $CH_3CO_2Na{\cdot}3H_2O$, $CH_3CO_2Li{\cdot}2H_2O$, and $(CH_3CO_2)2Mg{\cdot}4H_2O$. J Chem Thermodyn. https://doi.org/10.1016/0021-9614(84)90003-X
71. Meisingset KK, Grønvold F (1986) Thermodynamic properties and phase transitions of salt hydrates between 270 and 400 K IV. $CaCl_2{\cdot}6H_2O$, $CaCl_2{\cdot}4H_2O$, $CaCl_2{\cdot}2H_2O$, and $FeCl_3{\cdot}6H_2O$. J Chem Thermodyn. https://doi.org/10.1016/0021-9614(86)90130-8
72. Meng T, Ramasubramanian B, Sundarrajan S, Xuan Y, Ramakrishna S (2024) Unleashing capabilities of supercapacitors: Strategies to reduce internal resistances. J Power Sources 596:234068. https://doi.org/10.1016/j.jpowsour.2024.234068
73. Miao Yu, Hynan P, Jouanne AV, Yokochi A (2019) Current li-ion battery technologies in electric vehicles and opportunities for advancements. Energies 12:1074
74. Moran, MJ, Shapiro HN (1993) Fundamental of engineering thermodynamics. , Wiley, New York. ISBN: 0-471–59275-7
75. Mousavi SMG, Faraji F, Majazi A, Al-Haddad K (2017) A comprehensive review of flywheel energy storage system technology. Renew Sustain Energy Rev 67:477–490
76. Nyári J, Magdeldin M, Larmi M, Järvinen M, Santasalo-Aarnio A (2020) Techno-economic barriers of an industrial-scale methanol CCU-plant. J CO_2 Util 39:101166. https://doi.org/10.1016/j.jcou.2020.101166
77. Patil VC, Ro PI (2020) Modeling of liquid-piston based design for isothermal ocean compressed air storage system. J Energy Storage 31:101449. https://doi.org/10.1016/j.est.2020.101449
78. Pereira J, Eames P (2016) Thermal energy storage for low and medium temperature applications using phase change materials—a review. Appl Energy 177:227–238. https://doi.org/10.1016/j.apenergy.2016.05.097
79. Petrucci RH, Herring FG, Madura JD, Bissonnette C (2010) General chemistry—principles and modern applications. Pearson ISBN: 978–0–13–206452–1
80. Pielichowska K, Pielichowski K (2014) Phase change materials for thermal energy storage. Prog Mater Sci 65:67–123. https://doi.org/10.1016/j.pmatsci.2014.03.005
81. Pohjolan Voima (2024) www.pohjolanvoima.fi
82. Puupponen S, Mikkola V, Ala-Nissilä T, Seppälä A (2016) Novel micro structured polyol–polystyrene composites for seasonal heat storage. Appl Energy 172:96–106. https://doi.org/10.1016/J.APENERGY.2016.03.023
83. Rehman S, Al-Hadhrami LM, Alam MM (2015) Pumped hydro energy storage system: a technological review. Renew Sustain Energy Rev 44:586–598
84. REN21 (2022) Renewables 2022. Global Status Report REN21 Secretariat, Paris
85. Roos P, Haselbacher A (2022) Analytical modeling of advanced adiabatic compressed air energy storage: literature review and new models. Renew Sustain Energy Rev 163:112464. https://doi.org/10.1016/j.rser.2022.112464
86. Ruiz RA, de Vilder LH, Prasasti EB, Aouad M, De Luca A, Geisseler B, Terheiden K, Scanu S, Miccoli A, Roeber V, Marence M, Moll R, Bricker JD, Goseberg N (2022) Low-head pumped hydro storage: a review on civil structure designs, legal and environmental aspects to make its realization feasible in seawater. Renew Sustain Energy Rev 160:112281
87. Sayed-Ahmed H, Toldy AI, Santasalo-Aarnio A (2024) Dynamic operation of proton exchange membrane electrolyzers—critical review. Renew Sustain Energy Rev 189:113883. https://doi.org/10.1016/j.rser.2023.113883

88. Schmidt O, Melchior S, Hawkes A, Staffell I (2019) Projecting the future levelized cost of electricity storage technologies. Joule 16. www.cell.com/joule/fulltext/S2542-4351(18)30583-X
89. Schmidt O (1899) Apparat zur Elektrolyse von Wasser. Patent DE 111 131 Jun. 13.
90. Seppälä A, Turunen K, Yazdani M (2022) Thermal conductivity of sugar alcohols. Sol Energy Mater Sol Cells 243:111796. https://doi.org/10.1016/j.solmat.2022.111796
91. Shah J (2022) Sauer energy international. https://www.saurenergy.com/solar-energy-news/the-top-5-largest-battery-energy-storage-systems-worldwide
92. Skogsberg K, Nordell B (2001) The Sundsvall hospital snow storage. Cold Reg Sci Technol 32(1):63–70. https://doi.org/10.1016/S0165-232X(00)00021-5
93. Sorensen B (2017) Renewable energy. Academic, New York. ISBN: 978-0-12-804567-1
94. Spirax-Sarco (2014) The steam and condensate loop. Gloucestershire, UK, ISBN 978–0–9550691–5–4
95. Staffell I, Pfenninger S (2016) Using bias-corrected reanalysis to simulate current and future wind power output. Energy 114:1224–1239
96. Streeter VL, Wylie EB (1983) Fluid mechanics. Singapore: McGraw-Hill ISBN 0-07-548015-8
97. Swedish Environmental Protection Agency (2011) Snötipp blir kylanläggning i Sundsvall. Information Fakta, Goda Exempel Energieffektivisering Klimp – Klimatinvesteringsprogram, Naturvårdsverket, Stockholm, Sweden. http://naturvardsverket.divaportal.org/smash/get/diva2:1773260/FULLTEXT01.pdf
98. Takahashi Y, Sakamoto R, Kamimoto M (1988) Heat capacities and latent heats of $LiNO_3$, $NaNO_3$, and KNO_3. Int J Thermophys 9(6):1081–1090. https://doi.org/10.1007/bf01133275
99. Tong W, Lu Z, Chen W, Han M, Zhao G, Wang X, Deng X (2022) Solid gravity energy storage: a review. J Energy Storage 53:105226
100. Treacy EA (2020) NEC Corporation. Encyclopedia Britannica. https://www.britannica.com/topic/NECCorporation
101. Troostwijk A, Deiman J (1789) Sur une manière de decomposer l'Eau en air inflammable et en air vital. Obs sur la Phys 35:369–378
102. Turunen K, Mikkola V, Laukkanen T, Seppälä A (2023) Long-term thermal energy storage prototype of cold-crystallizing erythritol-polyelectrolyte. Appl Energy 332:120530. https://doi.org/10.1016/j.apenergy.2022.120530
103. UEIA (2021) Independent statistics and analysis. Monthly Energy review, US Energy Information Administration-EIA
104. Ukrainczyk N (2010) Thermophysical comparison of five commercial paraffin waxes as latent heat storage materials. Chem Biochem Eng Q 24(2):129–137. https://doi.org/10.15255/CABEQ.2014.240
105. Vantaan Energia (2024) www.vantaanenergia.fi
106. Vistra (2024) https://investor.vistracorp.com
107. Vlad A, Singh N, Rolland J, Melinte S, Ajayan PM, Gohy JF (2014) Hybrid supercapacitor-battery materials for fast electrochemical charge storage. Sci Rep 4:4315
108. Voith (2024) https://voith.com
109. Wei G, Wang G, Xu C, Ju X, Xing L, Du X, Yang Y (2017) Selection principles and thermophysical properties of high temperature phase change materials for thermal energy storage: a review. Renew Sustain Energy Rev 81:1771–1786. https://doi.org/10.1016/j.rser.2017.05.271
110. Xiang Y, Xie Z, Furbo S, Wang D, Gao M, Fan M (2022) A comprehensive review on pit thermal energy storage: technical elements, numerical approaches and recent applications. J Energy Storage 55C:105716. https://doi.org/10.1016/j.est.2022.105716
111. Xu B (2013) Degradation-limiting optimization of battery energy storage systems operation. Academia Edu. http://www.aca-emia.edu/6390087/http_citeseerx_ist_psu_edu_viewdoc_down-load_doi_10_1_1_401_8095_and_rep_rep1_and_type_pdf
112. Yin B, Wu X (2018) Performance analysis on a large-scale borehole ground source heat pump in Tianjin cultural centre. IOP Conf Ser Earth Environ Sci 121(5):052071. https://doi.org/10.1088/1755-1315/121/5/052071

113. Yu Z, Tetard L, Zhai L, Thomas J (2015) Supercapacitor electrode materials: nanostructures from 0 to 3 dimensions. Energy Environ Sci 8:702–730
114. Yu F, Huang T, Zhang P, Tao Y, Cui FZ, Xie Q, Yao S, Wang F (2019) Design and synthesis of electrode materials with both battery-type and capacitive charge storage. Energy Storage Mater 22:235–255
115. Zakeri B, Syri S (2015) Electrical energy storage systems: a comparative life cycle cost analysis. Renew Sustain Energy Rev 42:569–596. https://www.sciencedirect.com/science/article/abs/pii/S1364032114008284
116. Zhao C, Andersen BP, Træholt C, Hashemi S (2023) Grid-connected battery energy storage system: a review on application and integration. Renew Sustain Energy Rev 182:113400. https://doi.org/10.1016/j.rser.2023.113400
117. Zubi R, Dufo-López R, Carvalho M, Pasaoglu G (2018) The lithium-ion battery: state of the art and future perspectives. Renew Sustain Energy Rev 89:292–308. https://doi.org/10.1016/j.rser.2018.03.002

Chapter 8
Electrical Power Generation, Conversion and Transmission Systems

Matti Lehtonen, Edris Pouresmaeil, Mika Järvinen, and Hanna Paulomäki

Electric power is everywhere present in unlimited quantities and can drive the world's machinery without the need of coal, oil, gas, or any other of the common fuels.
—Nikola Tesla

The objective of this chapter to give students a basic understanding of the electrical power systems required in renewable energy systems. This includes generators, motors, frequency inverters, AC/DC and DC/AC converters and control systems, as well as the interconnected networks needed to integrate all these components and resources as a working entity. The impacts of power transmission lines on nature are also briefly discussed in Sect. 8.1.

8.1 Introduction to AC and DC Systems

An electric power system consists of electrical networks, power generation units, and electrical loads connected into the system. In large generating units, such as thermal power plants and wind power plants, the power is usually generated at medium

M. Lehtonen (✉) · E. Pouresmaeil · M. Järvinen · H. Paulomäki
Aalto University, Espoo, Finland
e-mail: matti.lehtonen@aalto.fi

M. Järvinen and H. Paulomäki (eds.), *Designing Renewable Energy Systems within Planetary Boundaries*, Green Energy and Technology,
https://doi.org/10.1007/978-3-031-69856-9_8

voltage (6–20 kV) and the voltage is then elevated using transformers to the transmission voltage level (typically 220 or 400 kV) before feeding it into the transmission grid system. High voltage is required to achieve high transmission power and long transmission distance. The transmitted power depends on the product of voltage and current; on the other hand, the current defines the required cross section of the power line conductor. As a result, the most feasible way to increase the transmission capacity is to increase the voltage. Another factor favoring high voltages is the voltage drop caused by the load current in the phase conductors: the higher the operation voltage, the smaller the relative impact of this voltage drop.

In a transmission network, the power can be transmitted for hundreds, even thousands of kilometers before it enters the load centers, such as major industrial consumers or large cities. Here the voltage is stepped down to a lower level for local transmission and distribution. First the voltage is reduced to 110 kV for the subtransmission network; this network level is also often called a high voltage distribution network. The role of the 110 kV system is as an intermediate network between transmission and distribution levels. Some larger loads, like industrial plants, are directly connected to the 110 kV network, as are many medium-size power-generation units. The transmission distances for a 110 kV network are typically at most a few hundred kilometers. The local distribution networks are connected in subtransmission systems using high- or medium-voltage transformers, typically 110/20 or 110/10 kV. Just a few hundred meters before the customers, the voltage is then stepped down to 0.4 kV, which is the standard low voltage delivered to small loads. See Fig. 8.1 for a description of this power transmission system.

The power distribution system consists of medium- and low-voltage networks. In major cities, the reach of a medium-voltage network is only a few kilometers, but in lightly-loaded rural networks, it can be up to 40–60 km. In cities, the length of low-voltage lines is typically 100–200 m, and in rural networks around 500 m, but not much more. The line voltage—the voltage between the two phases of a three-phase system—is 0.4 kV, and the voltage delivered to the single-phase loads is the voltage between phase and neutral, which is 230 V. In communities, larger loads, such as small and medium size industries, or commercial and office buildings, often connect to the network directly at the 20 kV level using their own 20/0.4 kV transformers.

The benefit of a large integrated power system is the possibility of locating the generating units in places of favorable conditions. These locations are often distant from the load centers. For example, large nuclear plants are located in places where cooling water is available, and where it is possible to arrange large vacant land areas around the plants. Large hydro-power plants naturally require significant water flow over a large vertical distance, and preferably a reservoir for better power control. Wind-power plants are located in places where vacant land areas are combined with favorable wind conditions. A large, interconnected power system also makes it possible to utilize large generation units, which in turn bring economies of scale. And finally, one of the most important benefits of large, interconnected power systems is the security of supply: a diverse mix of generation units will make sure that there is power available should any single unit or resource fail. This makes it easier and more cost-effective to arrange back-up power.

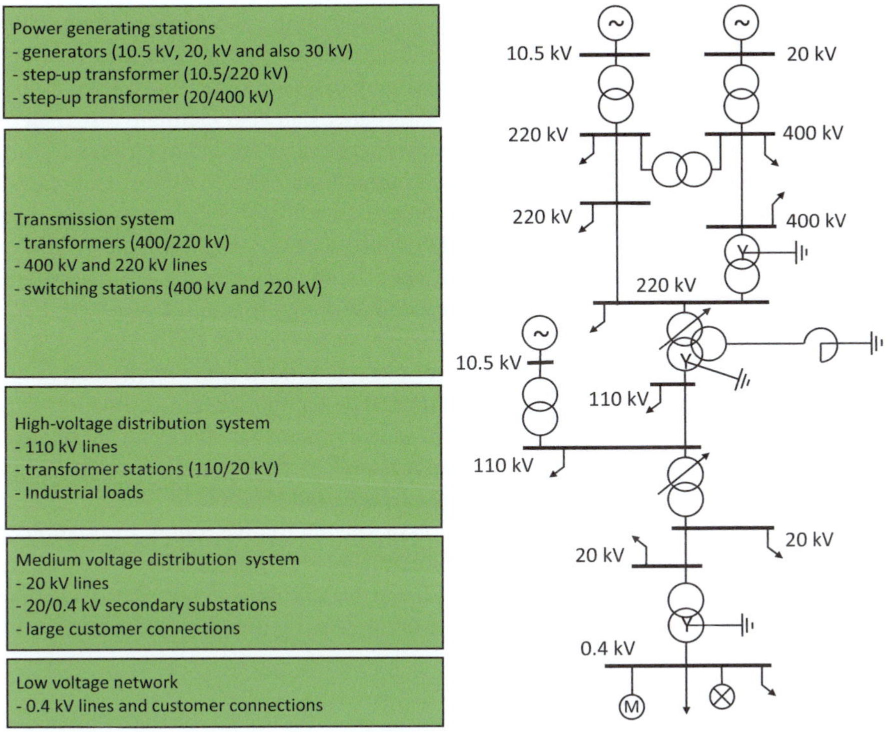

Fig. 8.1 An integrated electric power system (Y represents a star connection with a neutral point)

Transmission networks are typically connected to the networks of neighboring countries by strong connections, which enable power exchange and larger power markets. For instance, Finland is connected to Sweden using two 400 kV AC lines in the north, and two submarine DC cables between Olkiluoto and Forsmark nuclear power plant sites in the south. In addition, there are two submarine DC cables between Finland and Estonia. The common Nordic power system enables shared power markets, where the electricity is produced where it is most cost efficient at a given time. Also, a large power system enables larger generating unit size, since the reserve capacity to account for any single failure can be shared between neighboring countries.

Electrification of our societies means that, in addition to renewable energy production sites, transmission infrastructure also needs to be expanded. Until now this land-use impact has been negligible compared to other human land uses. But as decarbonization efforts expand further, larger areas of land are needed to produce and transport renewable electricity, increasing the potential negative impacts on nature and wildlife [4]. The main impacts caused by the construction of power transmission line networks include habitat loss, fragmentation, and degradation via vegetation removal [3]. The impacts are site specific and depend on the status of the land prior to construction, land-use history, topography, productivity, and vegetation type that is removed. For example, in forested areas fragmentation created by power transmission corridors increases edge effects, and may displace some forest species and disrupt animal movement (see Fig. 8.2). Disturbance can also facilitate dispersal of invasive alien species. Concentration of many energy projects and other human uses in one area can also result in significant and cumulating negative impacts on nature [2]. Moreover, power transmission lines can create collision or electrocution risks for birds and bats. Electrocution is associated with low- or medium-voltage power lines for birds that use pylons as perches when hunting or nesting. Collisions and electrocutions can create a significant increase in mortality for some large, long-lived species, including bustards, cranes, storks, geese, swans, eagles and vultures ([1, 2]; see also case study 5.1 on siting wind power close to golden eagle territories).

Negative impacts can be effectively minimized by applying a mitigation hierarchy throughout the project planning phases. The starting point for successful mitigation is a robust knowledge of the species and ecosystems in the planning area. In a mitigation hierarchy, negative impacts are first avoided and minimized through project planning and design, including identification of site alternative routes, evaluation of measures and improvement of practices. When negative impacts cannot be avoided or minimized, restoration can be used to bring back nature values that were lost during construction. Finally, any residual damage can be offset in accordance with

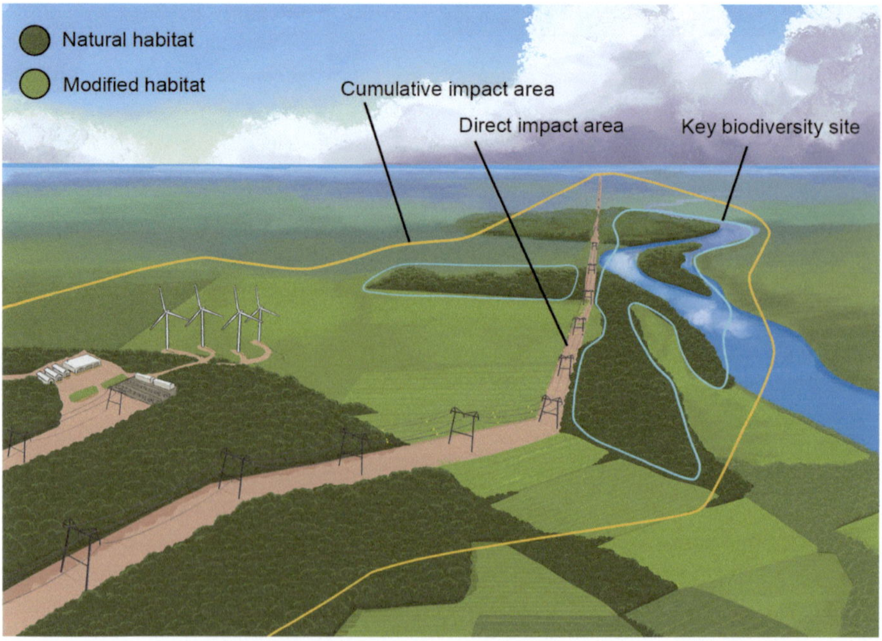

Fig. 8.2 Land use impact and habitat fragmentation by power lines. © Maija Karala

offset criteria (see more on mitigation hierarchies in Sect. 2.2, and Chaps. 4 and 5 on solar and wind power).

Though nature impacts—and therefore mitigation measures—are always site specific, some generalizations can be made to identify areas that should always be avoided. These include ecologically important areas, such sensitive breeding sites, important species migration routes, feeding and wintering areas, wetlands, and protected areas [2]. Routing power transmission lines through these areas should be avoided, or the effectiveness of mitigation measures, such as burying the power lines, should be carefully evaluated.

The most important measure to avoid bird and bat collisions is routing. In the early planning phase, important areas for birds and bats, like migration routes or wetlands, should be identified and avoided as far as possible. Marking transmission lines with bird diverters helps to increase the visibility of the power lines, and have become a standard practice which has proven to significantly reduce the numbers of collisions; on average, marking the lines has been shown to halve the numbers of bird collisions [2]. Risk of bird electrocution can be almost eliminated through construction of safe distribution lines that include insulation and spacing of conductors [2]. Such measures have proven to be efficient and cost-effective. Further work is needed, along with the development of ecological tools to aid the evaluation of the additional land-use impacts resulting from renewable energy production, including the additional transmission power line networks, to avoid and mitigate all unnecessary impacts.

Let us next consider basics of an alternating current (AC) power system. When a loop of wire is rotated in a constant magnetic field, the voltage generated in the loop, $u(t)$, can be determined according to the Faraday's induction law (Fig. 8.3).

$$\begin{aligned} u(t) &= -\frac{d\phi(\mathrm{t})}{dt} = -\frac{d(BA(t))}{dt} \\ &= -\frac{d(BA\cos(\omega t))}{dt} = \omega BA\sin(\omega t) \end{aligned}$$

where B is the magnetic flux density (unit is tesla, $\mathrm{T} = \mathrm{kg/(s^2A)}$), $A(t)$ is the area of the loop, which surrounds the magnetic flux, $\phi = BA(t)$, and $\theta = \omega t$ is the angle between the wire loop and the plane perpendicular to the magnetic field. When rotating a full cycle, the loop produces one cycle of sinusoidal voltage, as shown in Fig. 8.4.

The amplitude of the generated voltage is $\widehat{U} = \omega BA$, the same as the peak voltage. In power systems the convention is to use RMS (root mean square) voltages. This gives the equivalent energy, E, that the sinusoidal curve produces over a half cycle (180° or π radians), corresponding to time $t = \pi/(2\pi f)$, as

$$\begin{aligned} E &= \overline{P}t = \overline{P}\frac{\pi}{2\pi f} = \overline{UI}\frac{\pi}{2\pi f} \\ &= \frac{\overline{U}^2}{R}\frac{\pi}{2\pi f} = \frac{\widehat{U}^2}{R}\frac{1}{2\pi f}\int_0^{\pi}\sin^2(\alpha)d\alpha \end{aligned}$$

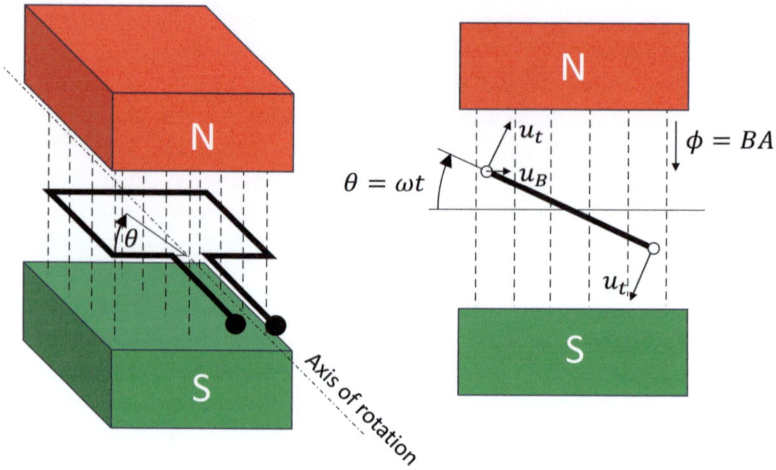

Fig. 8.3 Basics of an AC generator—a rotating winding loop in a magnetic field

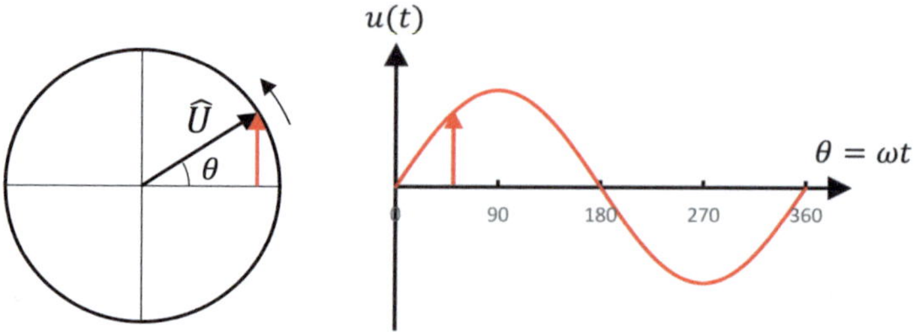

Fig. 8.4 Sinusoidal voltage produced by the loop of Fig. 8.2 and a rotating phasor

For $\overline{U}^2$ we then get

$$\overline{U}^2 = \widehat{U}^2 \frac{1}{\pi} \int_0^{\pi} \sin^2(\alpha) d\alpha$$
$$= \widehat{U}^2 \frac{1}{\pi} \left(\left(\frac{\pi}{2} - \frac{1}{4} \sin(2\pi) \right) - \left(\frac{0}{2} - \frac{1}{4} \sin(0) \right) \right) = 0.5 \times \widehat{U}^2$$

from which we can solve the root mean square voltage, $\overline{U} = U_{RMS} = U = \sqrt{0.5} \times \widehat{U} = 0.7071 \times \widehat{U}$. From here onwards we will write U_{RMS} as U for simplicity; similarly for the current we will write I_{RMS} as I.

Another important parameter, the average voltage, is defined as

$$U_{ave}\pi = \widehat{U}\int_{0}^{\pi}\sin(\alpha)d\alpha$$

$$U_{ave} = \frac{\widehat{U}}{\pi}(\cos(0) - \cos(\pi)) = \frac{2}{\pi} \times \widehat{U} = 0.637 \times \widehat{U}$$

In a practical generator, the loop in Fig. 8.3. is constructed from tens or hundreds of winding turns, which is then connected to one phase; real AC power systems then adopts a three-phase system, arranged so that the three windings of different phases have a 120-degree phase angle between them. In AC power systems, the sinusoidal voltages and currents are usually modelled using phasors. These are rotating vectors, the projection of which along the real axis gives the instantaneous value of the quantity concerned. We represent vectors (e.g., $\boldsymbol{U}$) with bold symbols, in contrast to scalars (e.g., U) that have no direction. The voltages and currents in a three-phase system can thus be modelled using three phasors, with phase angles of 120° between them—see Fig. 8.5. In this figure, the blue phasors are phase voltages, which represent the voltages between phase and neutral (which in normal operating situations are at zero potential); the red lines represent line voltages, that is, phase-to-phase voltages. Usually, when speaking about power system voltages, we mean these line voltages. Using a three-phase system has two benefits. The first is that a return wire is not needed, since when arriving at the neutral side of the load, the three phase currents sum up to zero, cancelling each other. The second benefit is that the three phases can produce a rotating magnetic field. This field can be used directly in producing rotation and mechanical torque in electrical motors.

Let us next consider the power transmitted in a three phase AC system. This power can be modelled as apparent power $\boldsymbol{S}$, real power $\boldsymbol{P}$, and reactive power $\boldsymbol{Q}$. Their magnitudes are as follows (please note that they also have directions):

$$S = 3U_P I = \sqrt{3}U_L I$$

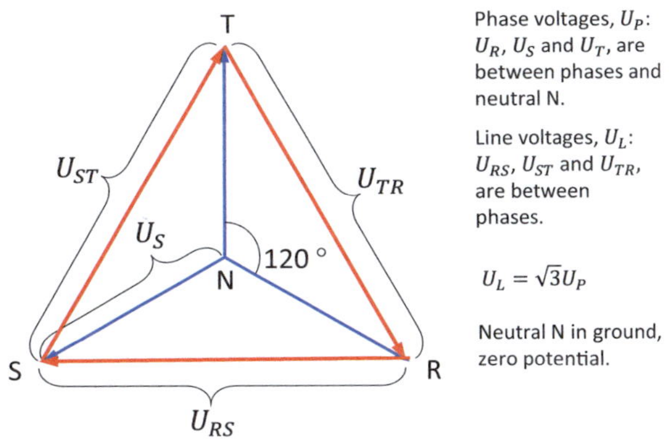

Fig. 8.5 The basic concepts of a three-phase system

$$P = S\cos(\varphi)$$

$$Q = S\sin(\varphi)$$

$$S = \sqrt{P^2 + Q^2}$$

The units of real power, reactive power and apparent power are W, VAr, and VA, respectively. The amount of real power transmitted depends on the phase voltage U_P, phase current I, and the phase angle difference φ between these two quantities. When this phase angle is zero, such as in the case of a purely resistive load, $\cos(\varphi) = 1$, and all the power transmitted is real power—that is, it is producing useful forms of energy such as heat or mechanical work. The term $\cos(\varphi)$ is also known as the power factor. When the phase angle φ is increased, the power factor is reduced, real power becomes smaller, and there will be more reactive power. Reactive power oscillates between the source and load, and does not provide useful work, since when integrated over one full power system cycle its energy will be zero. However, reactive power is useful for the load, since it is needed to create the magnetic field, which enables induction and hence the operation of electrical machines. Apparent power is the square sum of real and reactive power; it represents the overall power flow in the power system, and defines the capacity requirements for power transmission and power system components.

Resistances, inductances, and capacitances are basic building components of power systems. However, to model a rectifier, for example, we would also need to add models for all used semiconductors. When a voltage is applied to a resistance, it produces a current in phase with the voltage; only the real power is consumed in the resistive component. When voltage is applied to an inductive component, the component resists the rise of current until the magnetic field of the inductance has been built up—in a physical sense, the inductive component stores electrical energy in the magnetic field. During this process the current lags the voltage, and in a fully inductive component it lags by 90°. Hence, the inductive component consumes reactive power. The behavior of a capacitance is the opposite: when a capacitor is connected to a power supply, the voltage across the component builds up with a delay which depends on how quickly the current is able charge the electric field of the capacitor. In this case, the electrical energy is stored in the electric field of the capacitor, and since this process takes time, the voltage lags the current—in a fully capacitive component, the current leads the voltage by 90°, and the component produces reactive power. In practice, we model these components as AC impedances, which we call the reactance, X—see Fig. 8.6, where j is an imaginary unit corresponding to a 90-degree phase shift, and angular velocity $\omega = 2\pi f$, where f is the power system frequency (50 Hz in European countries). The use of these is discussed in the following examples for a single-phase system.

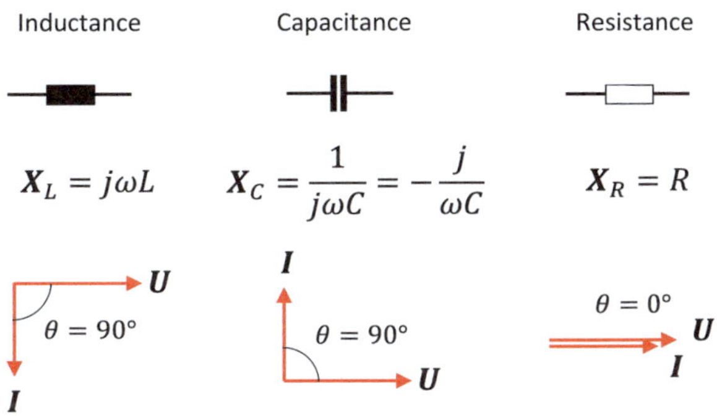

Fig. 8.6 Impact of inductance, capacitance and resistance on reactive power. Positive rotation is anti-clockwise

Example 8.1 An AC voltage of $U = 230$ V (rms) is applied across a 100 Ω resistive load (Fig. 8.7). Calculate: peak values $\hat{I}$ and $\widehat{U}$; rms value of the current, I; peak-to-peak values I_{p-p} and U_{p-p}; and real power P, reactive power Q and apparent power S.

Solution: The peak, peak-to-peak, and average (over the positive half cycle) voltages are

$$\widehat{U} = U/\sqrt{2} = 1.414 \times 230V = 325.2 \text{ V}$$

$$U_{p-p} = 2 \times \widehat{U} = 650.4 \text{ V}$$

$$U_{ave} = 0.637 \times \widehat{U} = 0.637 \times 325.2 = 207.2 \text{ V}$$

The rms current can be calculated from the Ohm's law since we have only resistive load—the peak, peak-to-peak and average currents are

$$I = U/R = 230 \text{ V}/100 \ \Omega = 2.3 \text{ A}$$

$$\hat{I} = 1.414 \times 2.3 \text{ A} = 3.25 \text{ A}$$

$$I_{p-p} = 2 \times \hat{I} = 4.6 \text{ A}$$

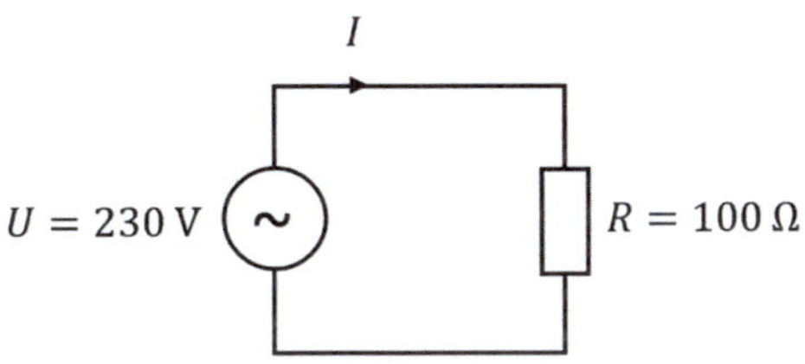

Fig. 8.7 AC source with a resistive load

$$I_{ave} = 0.637 \times \hat{I} = 0.637 \times 3.25\ \text{A} = 2.07\ \text{A}$$

For a resistive load, the real power is equal to apparent power since the voltage and the current are in the same phase—simply the product of rms voltage and currents, $P = S = UI = 230 \times 2.3 = 529$ W. For a resistive load, the reactive power is equal to 0 W.

Example 8.2 An AC voltage is applied over a 100 Ω resistive load and an inductive reactance of 100 Ω (Fig. 8.7). The rms current is 2 A. Calculate the power factor, applied voltage, and the real, reactive and apparent powers, P, Q and S, respectively. Draw the voltage-current phasor, impedance, and power diagrams.

Solution: The phase difference is

$$\theta = \text{atan}(X_L/R) = \text{atan}(100/100) = 45°$$

and the power factor is then $\cos(45°) = 0.7071$. The impedance—the total opposition to alternating current—is

$$Z = \sqrt{100^2 + 100^2} = 141.4\ \Omega$$

as shown in Fig. 8.7. The applied voltage can be calculated from the rms current (the current is set at a phase angle of 0°) and impedance (Fig. 8.8):

$$U = ZI = 141.4 \times 2 = 282.8\ \text{V}$$

Alternatively, this problem can also be formulated using vector forms,

$$\boldsymbol{Z} = 141.4\ \Omega\angle 45° = 100\ \Omega + j100\ \Omega$$

$$\boldsymbol{I} = 2\ \text{A}\ \angle 0° = 2\ \text{A} + j0$$

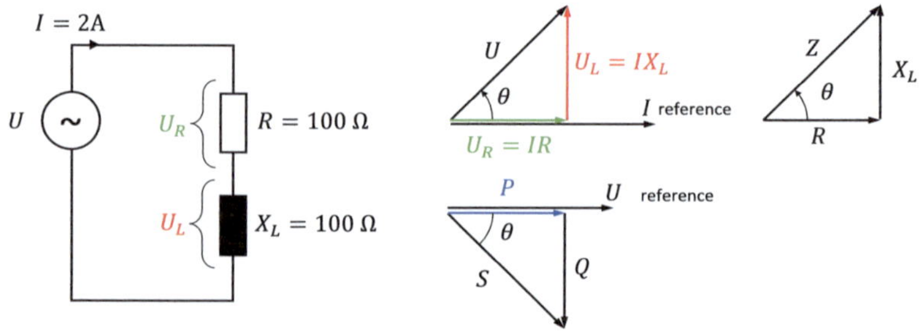

Fig. 8.8 AC source with resistive and inductive loads (left). Voltage, impedance, and power diagrams (right)

and so we get

$$\boldsymbol{U} = \boldsymbol{ZI} = (100 + j100)(2 + j0) = 200 + j200 = 282.8 \text{ V } \angle 45°$$

The real power can be calculated in two ways:

$$P = I^2 R = 2^2 \times 100 = 400 \text{ W}$$

or

$$P = UI \cos(\theta) = 282.8 \times 2 \times 0.7071 = 400 \text{ W}$$

Reactive power is the component perpendicular to real power,

$$Q = UI \sin(\theta) = 282.8 \times 2 \times 0.7071 = 400\text{VAR}$$

This is lagging by 45° since, in an RC circuit, current lags the voltage. Finally, apparent power is the hypotenuse of the power triangle,

$$S = \sqrt{P^2 + Q^2} = \sqrt{400^2 + 400^2} = 565.7 \text{ VA}$$

Example 8.3 An AC voltage is applied over a 100 Ω resistive load, an inductive reactance of 100 Ω, and a capacitive load of a) 50 Ω, and b) 100 Ω (Fig. 8.9). The rms current is 2 A. Calculate the power factor, applied voltage, and real, reactive and apparent powers, P, Q and S. Draw the voltage-current phasor, impedance and power diagrams.

Solution: The phase differences are

$$\theta_a = \text{atan}((X_L - X_C)/R) = \text{atan}((100 - 50)/100) = 26.6°$$

$$\theta_b = \text{atan}((X_L - X_C)/R) = \text{atan}((100 - 150)/100) = -26.6°$$

and power factor is then $\cos(26.57°) = \cos(-26.57°) = 0.894$. The impedance is

$$Z_a = \sqrt{100^2 + (100 - 50)^2} = 111.8 \ \Omega$$

$$Z_b = \sqrt{100^2 + (100 - 150)^2} = 111.8 \ \Omega$$

as shown in Fig. 8.9. The applied voltage can be calculated from the rms current and impedance (current at phase angle 0°),

$$U_T = ZI = 111.8 \times 2 = 223.6 \text{ V}$$

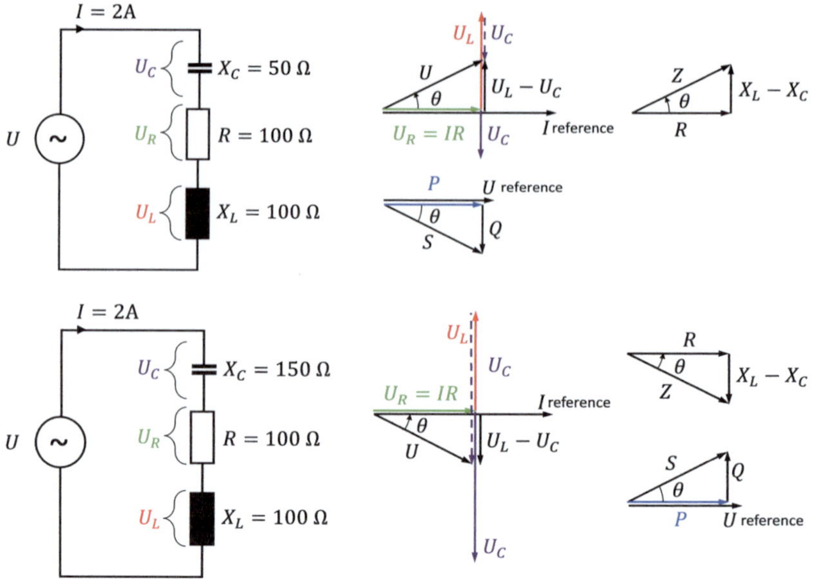

Fig. 8.9 AC source with resistive, capacitive, and inductive loads, and the corresponding voltage, impedance, and power diagrams. $X_C = 50\ \Omega$ (top) and $X_C = 150\ \Omega$ (bottom)

Alternatively, using vectors,

$$\boldsymbol{Z_a} = 111.8\ \Omega\angle 26.6^\circ = 100\ \Omega + j50\ \Omega$$

$$\boldsymbol{Z_b} = 111.8\ \Omega\angle - 26.6^\circ = 100\ \Omega - j50\ \Omega$$

$$\boldsymbol{I} = 2\ \text{A}\ \angle 0^\circ = 2\ \text{A} + j0$$

and so we get

$$\boldsymbol{U_a} = \boldsymbol{Z_a I} = (100 + j50)(2 + j0) = 200 + j100 = 223.6\ \text{V}\ \angle 26.6^\circ$$

$$\boldsymbol{U_b} = \boldsymbol{Z_b I} = (100 - j50)(2 + j0) = 200 - j100 = 223.6\ \text{V}\ \angle - 26.6^\circ$$

The real power is equal to the voltage multiplied only by the component of the current that is in phase with the voltage, and can be calculated in two ways:

$$P = I^2 R = 2^2 \times 100 = 400\ \text{W}$$

or

$$P = UI\cos(\theta) = 223.6 \times 2 \times 0.894 = 400 \text{ W}$$

Reactive power is the component perpendicular to real power,

$$Q_a = Q_b = UI\sin(26.6) = 223.6 \times 2 \times 0.894 = 200 \text{ VAR}$$

In case (a) the reactive power lags the real power by 26.6° since the current is lagging the voltage. In case (b) the reactive power leads the real power by 26.6° since, as the system is capacitive, voltage lags current. Apparent power the hypotenuse of the power triangle,

$$S_a = S_b = \sqrt{P^2 + Q^2} = \sqrt{200^2 + 400^2} = 447.2 \text{ VA}$$

The real power can be considered dissipated heat, which only occurs in the resistive load.

Example 8.4 Compare the transmission efficiencies of a one-phase and a three-phase system. Both systems need to deliver a total power of 3 kW to a load using a 230 V source. Resistance and reactance in both systems in a single wire is $R_w = 1\ \Omega$ and $X_w = 1\ \Omega$, respectively. We assume that the three-phase system has a fully symmetrical load, and that it can be described by a single-phase equivalent circuit. Figure 8.10 illustrates both systems.

Solution: The overall impedance is defined as $\boldsymbol{Z} = \boldsymbol{U}/\boldsymbol{I}$, from which we can solve for the current,

$$\boldsymbol{I} = \frac{\boldsymbol{U}}{\boldsymbol{Z}} = \frac{230 \text{ V}\angle 0°}{R_w + R_L + jX_w} = \frac{230 \text{ V}\angle 0°}{\sqrt{(R_w + R_L)^2 + X_w^2}\angle \text{atan}\left(\frac{X_w}{R_w + R_L}\right)}$$

The absolute value of the current is then

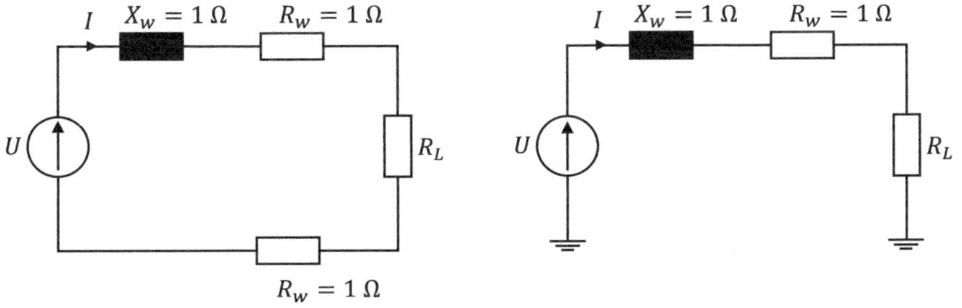

Fig. 8.10 Single-phase system (left), and equivalent circuit of a symmetric three-phase system (right)

$$I = \frac{230}{\sqrt{(R_w + R_L)^2 + X_w^2}}$$

The power used in the loads is $P_L = R_L I^2$, from which we can find the required load resistance,

$$R_L = \frac{P_L}{I^2} = \frac{P_L\big((R_w + R_L)^2 + X_w^2\big)}{230^2}$$

After some cleaning up we have

$$R_L \frac{230^2}{P_L} = (R_w + R_L)^2 + X_w^2 = R_w^2 + 2R_w R_L + R_L^2 + X_w^2$$

$$R_L^2 + R_L\left(2R_w - \frac{230^2}{P_L}\right) + R_w^2 + X_w^2 = 0$$

We then solve for the load resistance,

$$R_L = \frac{-\left(2R_w - \frac{230^2}{P_L}\right) \pm \sqrt{\left(2R_w - \frac{230^2}{P_L}\right)^2 - 4\left(R_w^2 + X_w^2\right)}}{2}$$

For the single-phase system, $R_w = 2\ \Omega$ and $P_L = 3000$ W, and for the three-phase system, $R_w = 1\ \Omega$, and $P_L = 1000$ W. In the three-phase system, only one-third of the power is transported per a phase. The corresponding single- and three-phase load resistances are 13.3 Ω and 50.9 Ω, respectively. The currents are then

$$\boldsymbol{I}_{1p} = \frac{230\ \text{V}\angle 0°}{\sqrt{(2 + 13.3)^2 + 1^2}\text{atan}\left(\frac{1}{2+13.3}\right)} = \frac{230}{15.33}\frac{\angle 0°}{\angle 3.74°} = 15.0\ \text{A} \angle - 3.74°$$

$$\boldsymbol{I}_{3p} = \frac{230\ \text{V}\angle 0°}{\sqrt{(1 + 50.9)^2 + 1^2}\text{atan}\left(\frac{1}{1+50.9}\right)} = \frac{230}{51.9}\frac{\angle 0°}{\angle 1.10°} = 4.4\ \text{A} \angle - 1.10°$$

Transfer losses, P_w, can then be calculated from these,

$$P_{w.1p} = R_w I_{1p}^2 = 2 \times 15^2 = 450\ \text{W}$$

$$P_{w.3p} = R_w I_{3p}^2 = 1 \times 4.4^2 = 19.4\ \text{W}$$

Finally, we get the transmission efficiencies,

$$\eta_{1p} = \frac{3000}{3000 + 450} = 87\%$$

$$\eta_{3p} = \frac{1000}{1000 + 19.4} = 98\%$$

As we can see from these calculations, the losses in three-phase systems are significantly smaller—but of course, the three-phase system uses 50% more cable (or 100% more if it also includes a neutral wire of same thickness). This means that if we accept the same losses for the three-phase system as for the single-phase system, we could save materials by using much smaller cable cross-sectional areas. In the single-phase system, the return cable has the same thickness as the feed-in cable. Also, as the three-phase loads are often never fully symmetrical, the fourth neutral return cable is typically the same thickness as the line cables. However, many three-phase loads do not have the neutral wire—for example, three-phase motors.

8.2 Components of the Electrical Power System

8.2.1 Powerlines

In this section, we present electrical models of various power system components, starting with power transmission lines. The structure of a typical overhead line used in transmission networks is illustrated in Fig. 8.11. The towers are typically made of steel, and the phase conductors are bare metal wires attached to the tower cross arms using suspension-type insulations. In addition, two additional wires typically run along the top of the transmission line—these wires are connected to ground, and their purpose is to protect the lines from direct lightning strikes by leading the lightning surge current to the ground, keeping impulse voltages within acceptable limits, and protecting the network.

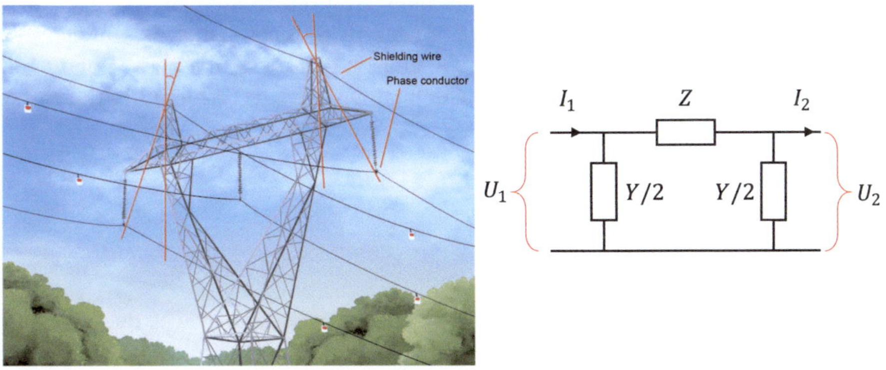

Fig. 8.11 Power transmission line and it's electrical "π-circuit"

Power transmission lines are usually represented by a π-circuit model, as shown in Fig. 8.11, consisting of a series impedance Z connected between the line ends, 1 and 2, and shunt admittance Y, divided in two with each half connected between the line ends and neutral (ground). These two elements Z and Y are

$$Z = (r + j\omega l)s$$

$$Y = (g + j\omega c)s$$

where r and l are the resistance and inductance per line kilometer, respectively, s is the line length in kilometers, and $\omega = 2\pi f$ is the angular frequency. The resistance is due to the internal resistance of the conductor wire and its magnitude is usually very low. The inductance is caused by the magnetic field in, around, and between the phase conductors, and its value becomes larger as the spacing of the conductors increases—in practical power transmission lines, the inductive impedance is the dominant part of the series impedance.

The shunt admittance is comprised of leakage conductance g, and phase-to-neutral capacitance c. The conductance is very small and can in most cases be omitted from the line model. The capacitance is caused by the electric field stored between the phases and neutral—this is important to the reactive power balance and thus to the voltage control of the line. Capacitances produce reactive power, whereas load current flowing in the series reactance consumes reactive power. In a lightly-loaded line, the reactive power consumption is smaller than the reactive power generated by the line capacitances, and the voltage of the line tends to rise; conversely, in a heavily-loaded line, the reactive power consumption is larger than the reactive power generated by the line capacitances, and the voltage tends to decrease. In these situations, the reactive power balance of the line must be controlled by feeding either inductive or capacitive reactive power into the network, to maintain the voltages at an acceptable level.

This kind of bare conductor overhead lines are used in high AC voltage transmission grids when transmitting large power over long distances. In densely-populated areas, underground cables are also used—in these, the phase conductors are surrounded by an insulation layer, which allows lines to be buried in the soil or in underground ducts or tunnels.

8.2.2 Transformers

Inductive power transformers are used to connect parts of a power network with different voltage levels. The transformer's main components are an iron core, and primary and secondary windings. When the transformer is energized, a magnetic flux is generated in the iron core, as shown in Fig. 8.12.

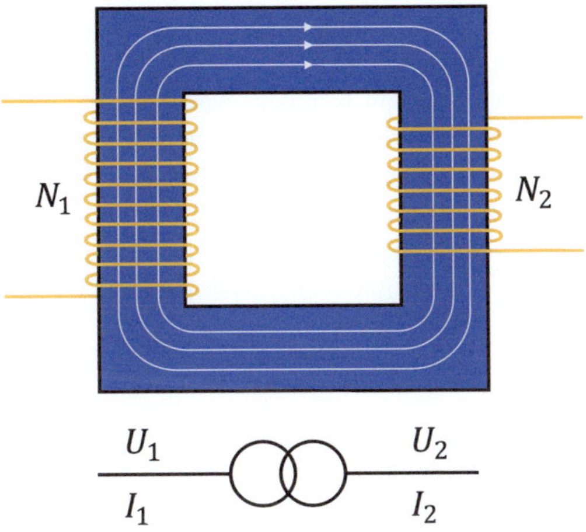

Fig. 8.12 Inductive power transformer (top) and its circuit symbol (bottom). N_1 and N_2 are the number of turns in the primary and secondary windings, and U and I are the corresponding voltages and currents

This flux creates, due to the induction law, a similar voltage for every winding turn. The transforming ratio is therefore defined by the ratio of winding turns in primary and secondary coils,

$$\frac{U_1}{U_2} = \frac{N_1}{N_2}$$

Since, in an ideal transformer, the power into the primary should equal to the power drawn from the secondary, the changes in voltage and current must be inversely related:

$$S = \sqrt{3}U_2 I_2 = \sqrt{3}U_1 I_1$$

$$\frac{I_1}{I_2} = \frac{U_2}{U_1} = \frac{N_2}{N_1}$$

In practical large power transformers, the windings are covered by paper tape insulation between adjacent winding turns. The main insulation between the windings and the grounded parts of the transformer is based on mineral oil. Oil also acts as a coolant, moving heat from the windings to a set of radiators.

8.2.3 Generators

Generators are fundamental components of electrical power systems, responsible for converting mechanical energy into electrical energy for use in external circuits such as electrical loads and devices. They play a key role in various power generation methods, from conventional fossil-fuel-based systems to renewable energy sources (RESs).

Generators are typically based on the principle of electromagnetic induction, as described by Faraday's law: when a conductor, normally in the form of copper coil, is exposed to a changing magnetic flux, a voltage (known as the electromotive force, *EMF*) is generated in the conductor. The amount of *EMF* (in volts) is proportional to the rate of change of the magnetic flux with respect to time, $d\phi/dt$, (in webers per second or volts), and so Faraday's law can be simply stated as

$$EMF = -\frac{d\phi}{dt}$$

When connected with a load, this *EMF* causes a flow of electric current, resulting in the generation of electrical power. The power generated through this process is transmitted by transmission lines to supply the electricity needed for electrical appliances.

There are different types of power generators working in electrical power systems, each with different characteristics and applications. The three main types of power generators are synchronous generators, asynchronous generators (induction generators), and permanent magnet generators. The general principle of these generators is discussed briefly below.

Synchronous Generators

Synchronous generators, also known as alternators, are a type of electrical machine used widely in traditional power generation applications, particularly in conventional power plants such as nuclear, hydroelectric, and thermal power plants, to generate a stable and synchronized power output. They play an important role in converting mechanical energy into electrical energy by using the principle of electromagnetic induction. The operation of synchronous generators is closely linked to the synchronization of their rotational speed with the frequency of the electrical grid to which they are connected. The synchronous speed, N_s, for a synchronous generator can be calculated from the number of poles, P, and the frequency of the grid, f, as

$$N_s = \frac{120 \times f}{P}$$

For example, the N_s of a four-pole synchronous generator in an electrical grid operating at a frequency of 50 Hz is 1500 revolutions per minute (rpm), while the N_s of a two-pole generator for the same grid would be 3000 rpm.

Figure 8.13. shows the main components of a simplified synchronous generator.

As shown in Fig. 8.13, the stator and rotor are the two main components of a synchronous generator. The stator consists of a stationary outer frame in which multiple sets of three-phase windings are arranged. These windings create a rotating magnetic field when supplied with alternating current (AC) from the power grid. The rotor is the rotating part of the synchronous generator, which is fixed onto a

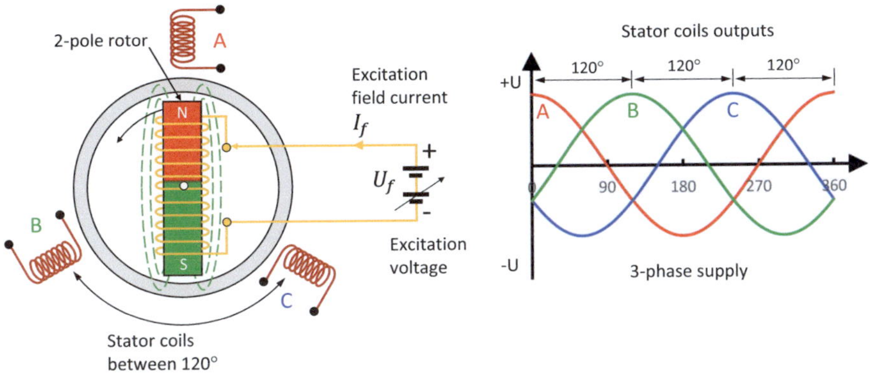

Fig. 8.13 Structure and operation of a synchronous generator

shaft. The rotor's magnetic field is generated by a direct current (DC) applied to the rotor windings, creating a set of poles with a magnetic field that interacts with the stator's magnetic field. The interaction between the rotating magnetic field of the rotor and the stationary stator windings creates an electromotive force (EMF) in the stator windings, which leads to the flow of electric current in the stator windings. This current flow produces the electrical output of the synchronous generator.

Synchronous generators have several important advantages, including their ability to provide stable and synchronized electrical power to different levels of power grids. This synchronization is fundamental to maintaining the stability of the grid frequency and ensuring the reliability of the overall power system, including power generation, transmission lines, and distribution networks. In addition, synchronous generators offer high efficiency in converting mechanical energy into electrical energy, which makes them suitable for large-scale power generation in various conventional power plants.

Inverter-based power generators are a type of synchronous generator equipped with power electronics, such as inverters, to facilitate grid integration of variable renewable energy sources or energy storage systems. By using power inverters, the output frequency of these generators can be adjusted to match the grid frequency, facilitating seamless power transfer from sources to power grids. Figure 8.14 shows a wind-power system (based on a permanent magnet synchronous generator, PMSG, or an asynchronous squirrel cage induction generator, SCIG) using a full power converter.

Permanent-Magnet Generators (PMGs)

Permanent magnet synchronous generators (PMSGs) are a type of electrical synchronous generator that use permanent magnets to directly induce electrical currents in the generator's windings. They can be used in combination with gear boxes, or as direct-drive systems with no gearbox. With a direct-drive system, since the rotor axis rotates very slowly, the number of poles must be increased to increase

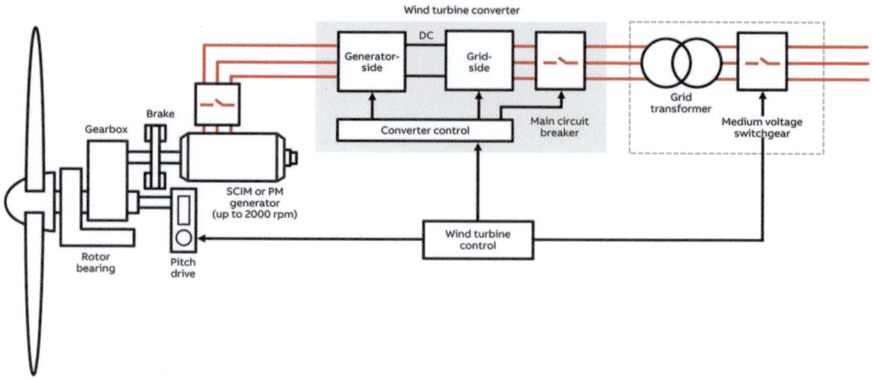

Fig. 8.14 Wind power train based on generator combined with full power converter (Courtesy of ABB)

the electrical output frequency. This is only possible by increasing the diameter, and hence the perimeter on which the permanent magnets are installed.

PMGs in general consist of stator and rotor components. The permanent magnets in the rotor create a constant magnetic field. As the rotor turns relative to the stator windings, the changing magnetic field induces an EMF in the copper coils of the stator. This induced EMF produces an alternating current in the stator windings. One of the key advantages of PMGs is their high energy efficiency relative to other generator types, due to the lack of an excitation system, which reduces energy losses.

The use of permanent magnets also enhances reliability and reduces maintenance requirements, as there are no brushes or slip rings to wear over time. If used as a direct-drive system, there is also no gearbox needed, which significantly simplifies the system configuration and reduces maintenance. As a result, PMGs can operate smoothly for longer periods without needing frequent repairs or replacements. Figure 8.15 shows a schematic of a permanent magnet induction generator.

PMGs are widely used in renewable energy systems, especially in wind turbines. Their compact design and high efficiency make them ideal for small-scale wind turbines used in residential sectors and off-grid settings. In addition, PMGs are extensively used in hydroelectric power plants, which provide a reliable and efficient means of converting mechanical energy from flowing water into electrical power.

However, PMGs also have limitations. The magnetic strength of permanent magnets can decrease over time due to temperature variations and other factors, which may affect the performance of the generator. PMGs may carry higher initial costs due to the price of high-quality permanent magnets. In addition, as discussed in Chap. 5 on wind power, the materials needed for the production of permanent magnets are limited.

Despite these challenges, advances in material science and engineering techniques have significantly improved the performance and cost-effectiveness of PMGs. As the demand for renewable energy sources continues to grow, PMGs play an ever more

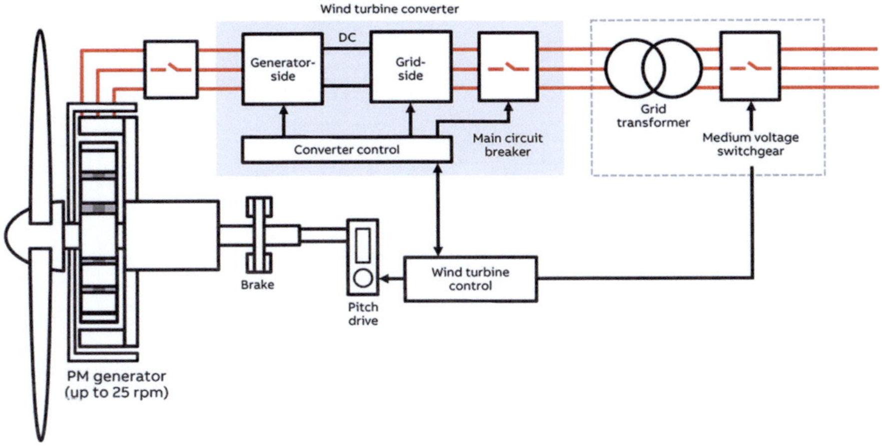

Fig. 8.15 PMG in wind power application with a full power converter– note the relatively large diameter of the rotor and the absence of a gearbox (Courtesy of ABB)

important role in sustainable power generation, and will contribute to a greener and more sustainable energy future.

Asynchronous Generators

Asynchronous generators, also known as induction generators, were in the past a fundamental component in electricity production from renewable energy systems, such as wind turbines and hydroelectric power plants. Figure 8.14 presented one example of a system containing a squirrel cage induction generator with a full power converter. The working principle of asynchronous generators is electromagnetic induction, where a rotating magnetic field is created in the stator by applying a three-phase voltage to the stator windings. If used as a motor, this rotating magnetic field induces voltage and current in the rotor windings—this causes the rotor to rotate at a lower speed than the synchronous speed of the rotating magnetic field, resulting in what is known as slip frequency.

This slip frequency is a key characteristic of asynchronous generators, as it allows them to operate at variable speeds. If the rotor speed is faster than the synchronous magnetic field, they generate power; if the speed is slower, they act as electrical motors. The input power and the frequency vary with the wind conditions, leading to fluctuations in the speed of the rotating elements. Asynchronous generators can adapt to these changes by adjusting their rotational speed to match the varying input sources, thus efficiently converting mechanical energy into electrical energy to meet the demands. They therefore are also able to tolerate wind load fluctuations better, and decrease the mechanical loads created on the turbine components.

Asynchronous generators offer unique advantages, such as automatic start-up capability and the flexibility to operate in either grid-connected or stand-alone

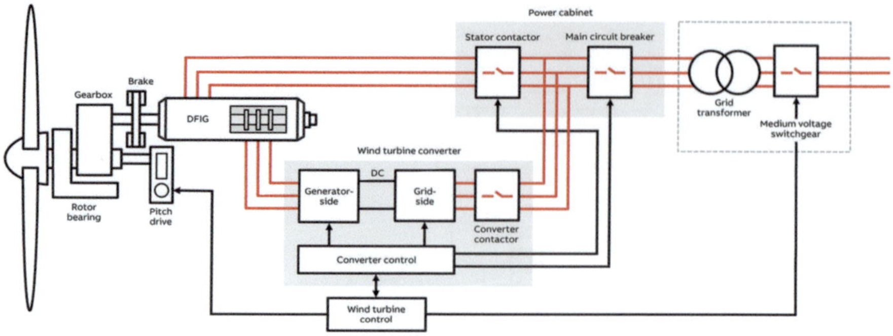

Fig. 8.16 Doubly-fed induction generator (DFIG) in wind power application (courtesy of ABB)

modes—in grid-connected operation, they deliver power to the grid, while in stand-alone mode, they can independently supply electricity to local or remote loads.

Design of the control systems of wind turbines, using two fully rated power converters (AC-DC on the turbine side and DC-AC on the grid side), offers flexibility, since both synchronous or asynchronous generators, and gear-box-based or direct-drive systems, can be used. The drawback of fully-rated converters is that all produced power passes through both converters, resulting in losses and thermal loads in the components. One solution to this is the doubly-fed induction generator, DFIG (Fig. 8.16).

As its name suggests, only part of the produced power will go through the converter, reducing losses as well as size and cost—typically, 70% of the power produced from the stator does not pass through the converters. In the DFIG, a variable-frequency voltage is fed into the rotor circuit of a wound rotor induction generator. Turbines with a DFIG have a limited speed range, typically $\pm$ 30% of the rated speed. This restricted speed range captures much of the energy obtained by the variable-speed operation of fully-rated power converters, but more importantly, it allows the rotor to change speed in response to wind-speed fluctuations, reducing mechanical load peaks. A special feature on the DFIG is that, at full speeds, both stator and rotor are feeding power to the grid, but at lower speeds, a fraction of the power is fed back to the rotor.

The link between asynchronous generators and inverter-based generators lies in their role within renewable energy systems. While asynchronous generators are well suited for stand-alone applications due to their self-starting capability, inverter-based generators excel in grid-connected scenarios, where they can actively control their output frequency and synchronize with the power grid. This capability allows inverter-based generators to enhance grid stability, contribute to the provision of ancillary services, and increase the integration of renewable energy sources and energy storage systems into the existing power grid. Together, asynchronous generators and inverter-based generators form a comprehensive solution for harnessing renewable energy sources and ensuring a more sustainable and resilient energy landscape.

8.2.4 Frequency Converters and Storage Applications

Frequency converters play an important role in modern power systems, enabling the efficient and seamless integration of renewable energy sources such as wind and solar, as well as facilitating energy storage through high-power and high-energy storage systems. This section explores the basic concepts, working principles, and various applications of frequency converters in the context of wind, solar, and battery storage technologies.

Frequency converters play a crucial role in modern power systems by enabling the efficient conversion of electrical power between different frequencies and voltage levels. These power electronic devices are essential components for the integration of renewable energy sources, such as wind and solar, into power grids, as well as for enabling energy storage through battery systems. The basic principle of frequency converters lies in their ability to convert direct current (DC) into alternating current (AC) or vice versa, depending on the application's requirements. This conversion is achieved through the use of power semiconductor devices, such as insulated gate bipolar transistors (IGBTs), which facilitate high frequency switching operations. As a result, frequency converters can mimic the desired frequency and voltage characteristics of the output power. Figure 8.17. shows the basic model of an AC-DC-AC frequency converter.

The generated variable three-phase power is first rectified in the generator side converter; this is stabilized with a capacitor prior to the grid side converter. In the grid side converter, insulated gate bipolar transistors (IGBTs) are used to construct a sinusoidal AC output from the constant rectangular DC-voltage pulses, by turning them

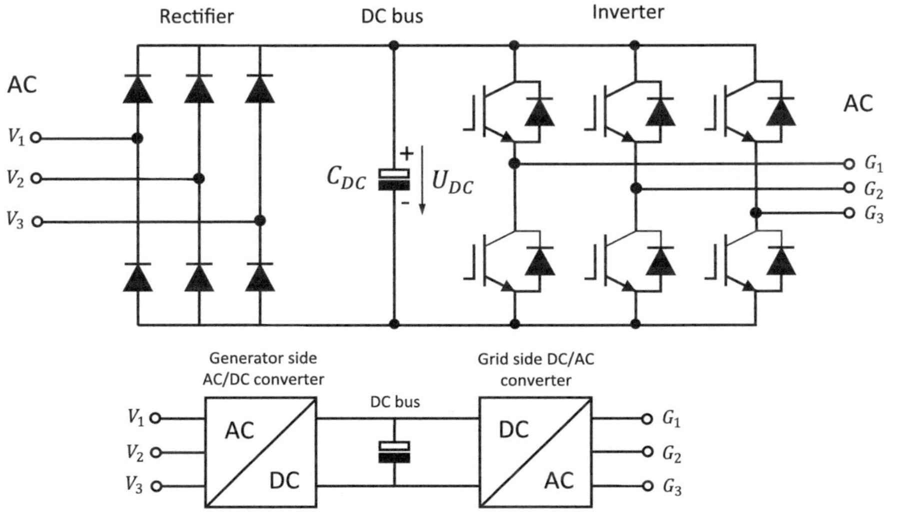

Fig. 8.17 Schematic diagram of a three-phase AC-DC-AC converter. V and G refer to variable generator production and fixed frequency grid feeding, respectively

on and off at precisely-set intervals—see Fig. 8.18 for a simplified representation of a single-phase system.

In Fig. 8.18, when switch S_1 is on and switch S_2 is off, output voltage is $+U_{DC}$, and when S_2 is on and S_1 is off, output voltage is $-U_{DC}$. The area under the modulation curve is set equal to the target sine wave area, for numeral interval widths of $\pi/12$.

In the context of renewable energy applications, frequency converters are widely used in wind-energy systems. They enable the implementation of variable speed wind turbines, allowing efficient power capture over a wide range of wind speeds. By adjusting the turbine's rotational speed through frequency conversion, wind turbines can maximize power output while minimizing mechanical stress on their components. In addition, frequency converters are essential for the grid integration of wind-power plants, as they convert the turbine's variable AC output power and frequency into stable, grid-compatible AC power. Several examples for wind-power systems were presented in Figs. 8.14, 8.15 and 8.16.

As shown in Fig. 8.19, frequency converters, specifically solar inverters, are fundamental components in solar energy systems. They convert the direct current generated by solar panels into AC power, making it suitable for grid connection or local consumption. Solar inverters also incorporate maximum power point tracking (MPPT) algorithms, ensuring that solar panels operate at their maximum power output by continuously adapting to changing solar irradiance and temperature conditions.

Frequency converters also play a crucial role in battery energy storage systems (BESS). In this context, they are essential for grid stabilization, as they control the flow of power between the battery and the grid, injecting or absorbing power as needed to maintain grid frequency within acceptable limits. Additionally, frequency converters enable energy arbitrage in BESS, allowing the storage of excess energy during periods of low demand, and discharging it during peak demand, optimizing energy costs and grid operation.

Bidirectional Converters

AC/DC-to-DC/AC converters, also known as back-to-back converters, play a key role in modern power systems, enabling an efficient, bidirectional power flow between

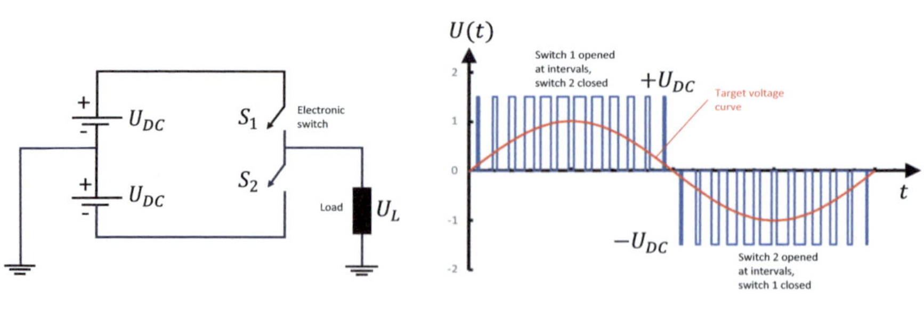

Fig. 8.18 Operation of a single-phase inverter system using pulse-width modulation to create a sine wave voltage, $U_{DC} = \pm 1.5$ V and $\widehat{U} = 1$ V

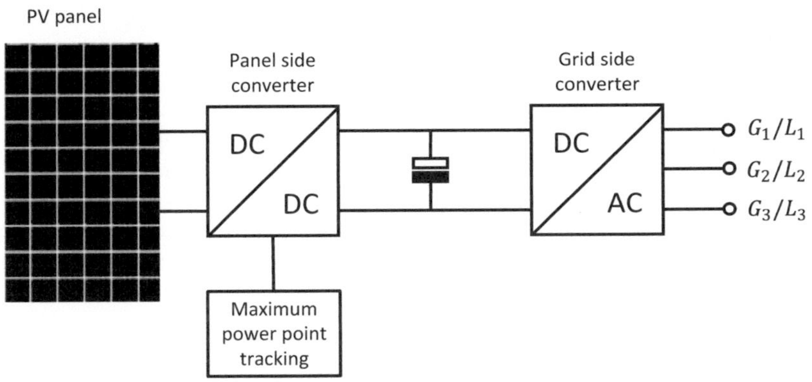

Fig. 8.19 Application of frequency converters for the integration of solar power into power grids (*G*) or load centers (*L*)

two AC grids with different voltage and frequency levels. These converters act as intermediaries, bridging the gap between two distinct power systems, and facilitating seamless bidirectional energy exchange.

Back-to-back converters are a specialized configuration of two connected power-electronic converters. The first, referred to as the AC/DC converter, converts alternating current from one AC grid or power source, such as a wind power plant with a constant or variable frequency, into direct current. The second, known as the DC/AC converter or inverter, then converts the DC power back to AC and feeds it into the second grid or load with a specific voltage and frequency. This arrangement enables a bidirectional power flow between two sectors and enables various power system functionalities.

The back-to-back topology provides the unique advantage of decoupling the voltage and frequency of two interconnected AC grids, making it highly suitable for interconnecting grids with different characteristics. This decoupling capability allows for flexible power transfer between grids with varying voltage levels and facilitates the integration of renewable energy sources, which often exhibit variable output characteristics. Moreover, the power exchange through back-to-back converters can be precisely controlled and regulated, making them indispensable components in a wide range of power applications.

Notably, back-to-back converters find essential applications in renewable energy integration, electrical vehicle charging, and high-voltage direct current (HVDC) systems. For instance, back-to-back converters are employed in wind farms to convert the variable-frequency output from wind turbines into stable and synchronized AC power for seamless integration into the main grid. Similarly, in electrical vehicle charging networks, these converters enable bidirectional charging systems, allowing electric vehicles to supply power back to the grid during peak demand or emergencies, thus promoting grid flexibility and energy storage capabilities.

8.3 Connecting Variable AC and DC Power Sources to the Grid

Variable renewable energy resources (VRES) such as wind and solar power are dependent on the weather, and are hence volatile and variable. This feature challenges the power system balance, since in the electrical network the generation and demand must always be equal. In this section we first present the conditions for power transmission in electric power systems, and then discuss the theoretical principles of power balance.

8.3.1 Limits of Power Transmission

In high voltage transmission lines the resistance in typically negligible in comparison with the inductive reactance, and the static power transmission maximum value P is limited by the power angle equation,

$$P = \frac{U_1 U_2}{X} \sin(\delta)$$

where U_1 is the sending-end line voltage, U_2 is the receiving-end line voltage, X is the reactance of the power line, and δ is the power angle, i.e., the phase angle between the voltages in line ends— Fig. 8.20 shows the simplified circuit for this system.

When the transmission power is increased, the angle δ also increases, until at the angle 90° the maximum transmission power is reached. If we try to increase power beyond this point, the power transmitted will decrease and the system will collapse—Fig. 8.21 illustrates the safe operation range.

The static limit is theoretical, since in practical power systems we must leave some margin for disturbance and power oscillations. Hence the practical maximum power transmission corresponds to some lower power angle value, for instance 45°. The power angle usually defines the maximum transmission capacity at higher voltage levels, such as 400 kV; when entering the subtransmission (110 kV) and distribution network (20, 0.4 kV) levels, the current and voltage limits of the lines and transformers are more restricting.

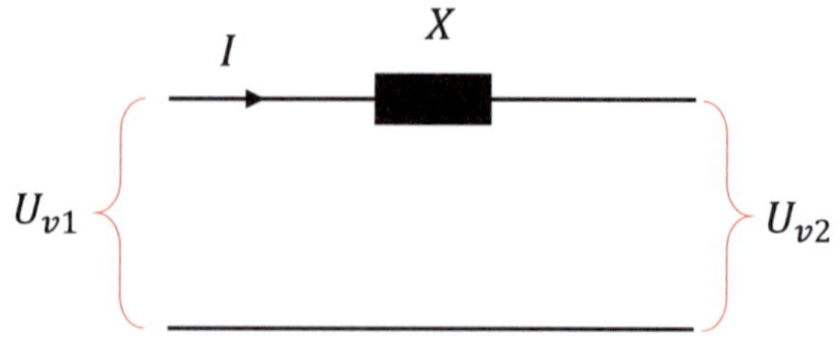

Fig. 8.20 Circuit of a high-voltage power line with reactance only

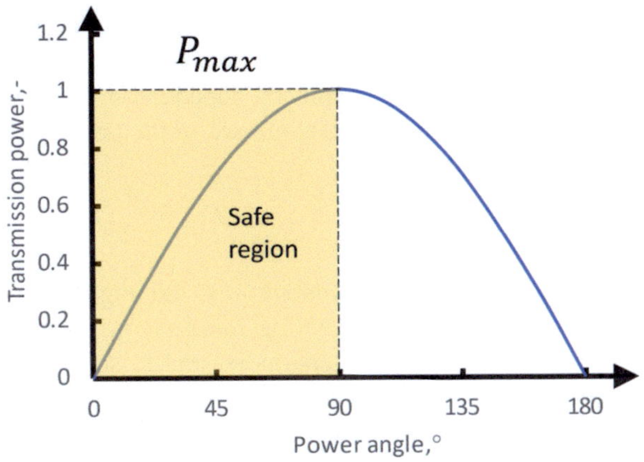

Fig. 8.21 The power angle and the stable area of static power transmission

Due to the load current in lines and transformers, power losses appear in component resistances. These losses appear as heat in the conductive and insulation materials, and this excess heat must be removed to prevent accelerated aging or even immediate damage to materials. Thermal limits usually set the loading limits of transformers and short lines, especially power cables, in terms of maximum allowed load currents.

In the case of overhead lines, voltage drop can also be a limiting factor for power transmission. Here the decisive parameter is the voltage received by the electricity customers, and power quality standards set limits to the acceptable voltage levels. Eventually these limits dictate the variation band of voltages in the distribution networks.

Voltage and current are also limiting factors when defining the local renewable energy hosting capacity (HC) of distribution systems. HC is defined as the maximum infeed of local (renewable) generation that can be transmitted upstream in the power distribution network without violating the technical limits of its component tolerances. HC is usually important when considering local photovoltaic generation—at the time of writing this book, 2024, approximately 70–80% of PV generation is installed at the customers' premises at the lower power system levels. However, the share of larger, utility-scale PV installations seems to be increasing. The most common limiting factors of HC in distribution networks are load current capacity in densely populated areas, and voltage rise due to the excessive PV infeed in suburbs and rural areas. Other possible limiting factors are imbalance and other power quality parameters, such as flicker and harmonics.

8.3.2 Power Balance

Sudden losses of generating units, and steep generation ramps of wind or solar, challenge the power balance in a power system. The imbalance of generation and demand is seen in frequency variations. In the case of excessive generation, the frequency tends to rise, while generation deficiency leads to a fall in frequency. The rate of change of the frequency depends on the inertia of the power system. The kinetic energy in the rotating masses of the power system is

$$E_{kin} = \frac{1}{2} J \omega^2$$

where J is the inertial moment, ω is the angular velocity, $\omega = 2\pi n$, and n is the rotational speed of the rotating mass (s^{-1}). Inertial energy is mostly tied to the rotating masses of turbogenerators, and to some extent in the motor drives of the loads. The inertia of a generator is often measured as the inertia constant H, which describes for how many seconds the kinetic energy can supply the generator's rated power.

The inertia of power systems has become an issue, as converter-connected wind and solar power push conventional synchronous generators out of the power market, thus reducing the system inertia overall. Converter-connected variable renewable energy systems (VRES) in their present form do not provide inertia into the power system; at the same time, the generation ramps of VRES increase power balance deviations. This combination makes the frequency deviations faster, larger, and more challenging from the power system security point of view.

The behavior of the power system frequency in the case of a sudden loss of a power generation unit is depicted in Fig. 8.22. Note that the time scale is in the order of seconds or tens of seconds. Initially, the rate of change of frequency depends on the size of the power deficiency compared to the total system inertia. The missing power is first taken from the system's rotating masses: when the frequency falls, the primary reserves are activated, and they follow the power control characteristics, also called droop. These primary reserves are also known as spinning reserves, since they are obtained from already running power units.

The role of primary reserves is to stabilize the frequency. After the primary reserves have stabilized the situation, the frequency still stays under the normal value (Fig. 8.22, left). This is because the new operation point corresponds to a higher power setting, which is found at a lower frequency (Fig. 8.22, right).

After the deployment of primary reserves, the secondary reserves are activated. Their role is to remove the frequency deviation, and at the same time to release the primary reserves so that they can cope with any new, similar power loss. A common practice has been to activate secondary reserves within in 15 min. In case of lower inertia in the system, faster responding primary reserves are needed. The primary reserves are dimensioned such that they can handle the sudden loss of the largest generation unit in the power system in question.

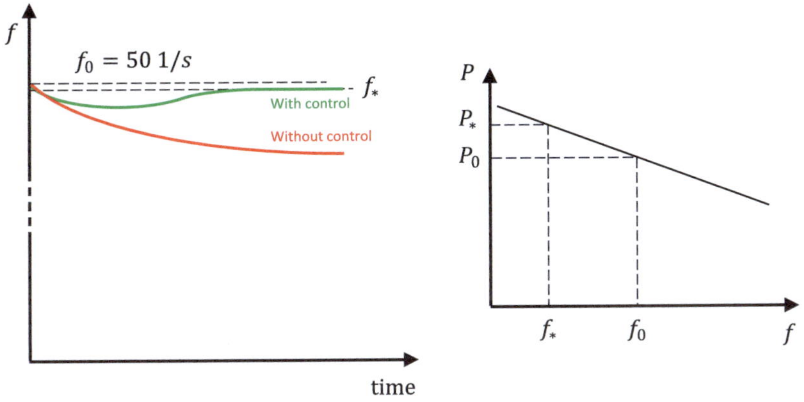

Fig. 8.22 The behavior of power system frequency in the case of a sudden power deficiency, and the control characteristics (droop) of the primary reserves

8.4 Operation of the Power System as a Whole

The operation of power and energy systems is divided between the reserve markets, operated by the local transmission system operators (TSO), and the commercial energy markets, operating typically at one-hour or 15-min resolution. In general, we talk about ancillary services, essential reliability services, or grid support services—these services are procured in European power systems by reserve markets described below. Let us first consider the reserve markets operated by the TSOs. Starting from the lowest time resolution (fastest response time), in the European context the reserve market is typically divided as follows.

FFR, Fast Frequency Reserves. These are activated in approximately one second. Their purpose is to slow down the rate of change of the frequency (RoCoF) and thus compensate the lack of inertia in the power system.

FCR, Frequency Containment Reserves. These primary reserves stabilize the power system frequency during disturbances and large generation ramps. They are divided into normal state reserves, FCR-N, which cope with the small frequency variations and try to keep the power system frequency close to the rated value (50 Hz), and FCR-D for disturbances, which is the type of primary reserves described in the previous section.

FRR, Frequency Restoration Reserves. These are secondary reserves, the purpose of which is to return the frequency back to nominal after larger disturbances. They are further divided into two types: automatic and manual reserves.

The energy markets are typically operated in the framework of electricity exchange. For instance, in the Nordic area, electricity is traded in the Nord Pool, where the main market is divided into the day-ahead (DA) market called Elspot, and the intra-day market (ID) called Elbas. The trading period has previously been one hour, but will be deceased to 15 min in spring 2024.

In the Elspot market, the DA selling and buying offers are submitted by noon the previous day, Scandinavian time. The generation offers are summed in the order of rising price, and the buying offers in descending price order. The point where these two summations intersect defines the price of the electricity for all energy at that time. In the ID market, the power balances of the trading units are then fine-tuned by the Elbas trade. Here the price of trades varies since the prices are fixed whenever sell and buy offers match.

In the power market a very important actor is the balance responsible party (BRP). This is a stakeholder who is responsible for the commercial power balance being as agreed—that is, the sold and purchased energy being equal. BRP is typically some large power company who can master the commercial power balance through its resources. In the event that this power balance does not match, the BRP must buy up- or down-regulation from the TSO. In practice, the manually-operated FRR forms the up-regulation market, and hence the price of negative balance deviation to the BRP and for down-regulation the price for positive balance, respectively. With the increase of VRES in power systems, the volume of balancing power—that is, the up and down regulation—is expected to increase.

8.5 Conclusions and Future Directions

For the time being, the volumes of variable renewable energy resources like wind and solar power are rapidly growing within power and energy systems. The electricity networks tend to form bottlenecks for the increased utilization of these energy sources: in high voltage transmission grids, due to the problems with power balance, and in distribution networks, because of the limited solar PV hosting capacity.

For transmission system operators, the lack of inertia can be compensated by procuring more fast reserves, and by installing synchronous compensators. These are electrical machines, running on idle, with the sole purpose of providing inertia in case of rapid balance deviations. In the longer term, inertia can probably be found from different types of energy storage solutions, and some of the inverter-based resources (wind, solar, batteries) can be operated with grid-forming controls to support the grid when there are no, or few, synchronous machines online. Other stability issues beyond frequency stability (lack of inertia) will also emerge in wind- and solar-dominated power systems, where grid-forming inverters will play a crucial role.

For distribution systems, the lack of hosting capacity will lead to increased need for network investments. The load flows of distribution networks will become larger, and much more volatile, for several reasons:

- The increase in local solar power generation will occasionally turn consumers into producers, when local generation exceeds their own demand. This situation has given rise to a modern practice of calling this kind of electrical customers "prosumers".

- Electric vehicles will increase the load on distribution networks, especially at the lower system levels. On the other hand, EV charging offers some flexibility that can be utilized for optimal power system operation.
- Volatile and variable power prices motivate customers to develop demand-side flexibility. Demand response (DR) solutions enable, for example, timely changing of power demand to benefit from cheap market hours.

At the higher-power and -energy system level, the electrification of district heating sectors will increase power demand, but also offer some centralized sources of flexibility, which can serve balance management in the power system. Sector coupling between power and heat makes these large heat storage systems available for balancing purposes.

Finally, the decarbonization of industries and societies requires extensive electrification and transition to a hydrogen economy: sectors and industries that are not feasibly decarbonized by electrification will be made sustainable through hydrogen. Examples of this development include fertilizers (for example, NH_3), derivatives of hydrogen as carbon neutral fuels in aviation and maritime traffic, and replacing oil by hydrogen derivatives as raw materials for plastics. Hydrogen derivatives can also provide storage buffers for power systems, as well as the flexible operation of electrolyzers—not forgetting, of course, the direct use of hydrogen for industrial processes and reserve power generation.

Author contributions: Lead author: Matti Lehtonen in collaboration with Edris Pouresmaeil and Mika Järvinen. Hanna Paulomäki contributed for Sect. 8.1.

References

1. Angelov I, Hashim I, Oppel S (2013) Persistent electrocution mortality of Egyptian vultures *Neophron percnopterus* over 28 years in East Africa. Bird Conserv Int 23:1–6. https://doi.org/10.1017/S0959270912000123
2. Bennun L, van Bochove J, Ng C, Fletcher C, Wilson D, Phair N, Carbone G (2021) Mitigating biodiversity impacts associated with solar and wind energy development. Guidelines for project developers. Gland, Switzerland, IUCN and Cambridge UK: The Biodiversity Consultancy 260 pp. https://doi.org/10.2305/IUCN.CH.2021.04.en
3. Hernandez RR, Easter SB, Murphy-Mariscal ML, Maestre FT, Tavassoli M, Allen EB, Barrows CW, Belnap J, Ochoa-Hueso R, Ravi R, Allen MF (2014) Environmental impacts of utility-scale solar energy. Renew Sustain Energy Rev 29:766–779. https://doi.org/10.1016/j.rser.2013.08.041
4. van de Ven DJ, Capellan-Peréz I, Arto I, Cazcarro I, de Castro C, Patel P, Gonzalez-Eguino M (2021) The potential land requirements and related land use change emissions of solar energy. Sci Rep 11:2907. https://doi.org/10.1038/s41598-021-82042-5

Further Reading

5. Gussow M (2012) Basic electricity, Schaums's easy outlines. McGrawHill. ISBN 978-0-07-178068-1
6. Kundur PS, Malik OP (2022) Power system stability and control, 2nd Edn. McGraw Hill, ISBN-13: 978-1260473544
7. Mohan N, Undeland TM, Robbins WP (2002) Power electronics: converters, applications, and design, 3rd edn, Wiley. ISBN-13:978-0471226932

Chapter 9
Securing a Sustainable Energy Transition: Summary and Conclusions

Hanna Paulomäki and Mika Järvinen

Although time is not on our side, it is not too late to make the conscious decision to change paths. Our descendants deserve nothing less.
—Partha Dasgupta

In this book, we have covered the technical and sustainability aspects of some selected renewable energy technologies and storage solutions, introduced the concept of planetary boundaries and doughnut economies, and discussed the importance of biodiversity for our existence on Earth. We've stressed the importance of addressing both ecological and social sustainability when developing solutions to climate change mitigation. Additionally, we have examined the economic principles, bottlenecks and solutions for a transition towards renewable energy technologies. As the planetary boundaries and key aspects defining the social foundation are strongly decoupled and interlinked, these should be solved simultaneously. Solutions to climate change should not have unacceptable, negative impacts on reaching, for example, other UN

H. Paulomäki (✉) · M. Järvinen
Aalto University, Espoo, Finland
e-mail: hanna.paulomaki@lut.fi

M. Järvinen
e-mail: mika.jarvinen@aalto.fi

M. Järvinen and H. Paulomäki (eds.), *Designing Renewable Energy Systems within Planetary Boundaries*, Green Energy and Technology,
https://doi.org/10.1007/978-3-031-69856-9_9

Sustainable Development Goals aimed at building and securing an ecologically and socially sustainable and resilient future [21].

The transition away from fossil-fuel-based systems is necessary to halt climate change. Our focus on the sustainability challenges associated with developing and constructing renewable energy installations is not intended to suggest that renewable energy systems are somehow worse than those based on fossil fuels; at the same time, we must not overlook the potential challenges inherent in the transition. For example, some of the solutions to climate change can contradict nature conservation targets, yet many nature restoration efforts help to curb climate change by enhancing and maintaining carbon capture and storage from the atmosphere. Also, failure to include local communities in planning processes can delay or halt renewable-energy projects, negating the benefits they may otherwise bring. The ongoing transition towards cleaner energy sources is necessary, but this is also an opportunity to focus on building a truly sustainable system. In this sense, we are at an extremely important moment in history, one that we should not pass by without a detailed analysis of all the available energy generation, storage, transmission, and end-use options. We should design and build an energy system that leads to decarbonization, minimizes land-use and biodiversity losses. Also, the transition brings about opportunities to create more just energy systems, enabling more equal distribution of benefits and costs across different groups in our societies.

Addressing climate change by implementing the energy transition—for example through the construction of wind and solar power, as well as use of Bioenergy Carbon Capture and Storage (BECCS)—will induce land-use changes that impact the environment and local communities. These impacts can be neutral, negative or positive—or even all these at the same time—depending on perspective. For example, if a natural or semi-natural area is cleared for renewable-energy development, species that require untouched environments can be lost, while species that favor or tolerate human-changed landscapes can benefit from the changes in the environment. Human activities have already impacted over 75% of terrestrial landscapes and 65% of marine areas and this threatens a growing number of species with extinction—which is then also a threat to economies, food production and security. In affected communities, some groups of people may benefit from a new development, while others may be impacted negatively. It is therefore important to identify any unacceptable impacts on nature and on local communities, and to set clear and measurable goals, indicators and monitoring programs for the mitigation of negative ecological and social outcomes.

The definition of a sustainable economy can also be problematic. Measured in purely economic terms, economic sustainability can undermine the other pillars of sustainability, and lead to the externalization of costs imposed on the environment, and consequently on societies. Partha Dasgupta has emphasized in his report [4] that economic benefits resulting from a continued depletion of natural capital are not just unsustainable, they are short-sighted since they jeopardize the opportunities of future generations.

The objective of this final chapter is to first review and highlight the key issues from each of the prior chapters. We also raise some additional issues for further consideration, such as the sufficiency of critical materials, that can pose threats to successful large-scale implementation of renewables—by acknowledging these early in the energy transition, hopefully we can avoid surprises later. Finally, we present some action points and ways forward.

9.1 Key Messages of the Chapters

In the first and second introductory chapters we formulated the objectives of the book, and discussed briefly three relevant planetary boundaries [16, 17] in the ongoing green-energy transition: climate change, biodiversity loss, and land use. The nine planetary boundaries (climate change, biosphere integrity, land-system change, freshwater use, biochemical flows, ocean acidification, atmospheric aerosol loading, stratosphere ozone depletion, and novel entities) are all tightly interconnected and affected by various human activities, including the expansion of large-scale renewable-energy installations. However, in this book we limited our focus on the three mentioned boundaries. The main message of the first two chapters was that there is a need to broaden the scope of traditional techno-economical optimization of renewable-energy projects. Large-scale renewable-energy projects, needed to tackle climate change, can impact nature and societies in complex ways. Their implementation requires several transformations, for example, in businesses and earning models (including accounting for the value of nature in economic assessment), subsidy systems, energy and material value chains, land-use, or in developing circular economy solutions.

Though replacing fossil-energy systems with renewable energy reduces CO_2 emissions, climate change cannot be solved in isolation from the surrounding natural environment and societies—which we illustrated as "carbon tunnel vision" (see Fig. 1.12), where our narrow focus on solving climate change turns our attention away from other, equally-grave challenges. For this reason, we discussed broadly the importance of biodiversity to humanity, and gave an example of how to measure the biodiversity footprint of a large-scale energy project. We also emphasized the importance of considering the social sustainability of large-scale energy projects, and discussed how social acceptability is formed, and how power relations influence energy policies and systems. Finally, we explained the basics of economic evaluation of renewable energy projects.

Another aspect from the introductory chapters we want to highlight is the level of energy and material consumption. On average, at the moment we are using the biocapacity of almost two planets—and if we zoom in to examine consumption in just the most developed countries, the Earth's annual biocapacity is consumed in only a few months. To squeeze our way of life back to the limits of one planet requires transformations and new innovations, not only technical innovations in our energy systems, but—perhaps even more importantly—social innovations fostering much needed changes in our attitudes, habits, and levels of consumption. This relates to the strong sustainability framework that we illustrated with the Stockholm Resilience

Centre's wedding cake model on UN Sustainable Development goals (see Fig. 1.5), which emphasizes that societies and economies should be seen as an embedded parts of the biosphere, so that the spheres of sustainability are not sectoral but built on well-functioning, diverse and healthy ecosystems. A strong sustainable solution to climate change needs to fit within all planetary boundaries and support social well-being, as explained by the doughnut economy model [13, 14].

Chapter 3 discussed thermal power plant technologies and ways to maximize their efficiency, including the Rankine cycle (steam power plant), the gas turbine cycle, the combined gas turbine cycle, combined heat and power generation, and the Sterling Engine. These robust systems can be used efficiently with renewable energy sources, such as solar and geothermal energy, and with sustainably-produced e-Fuels made from green hydrogen and captured CO_2. Steam and gas turbines and internal combustion engines are made from basic metals, cast iron, different steel grades, aluminum and copper, which are already available at large scales and can be fully recycled, making their use in a circular economy possible.

The final part of Chap. 3 discussed heat pumps, which are the most efficient way of generating heat, even up to 180 °C, within an energy system based on electricity instead of combustion processes. Heat pumps can utilize the low-value waste heat from several industrial processes and, for example, from wastewater treatment plant effluents, maximizing these processes' energy efficiency and minimizing heat pollution to water bodies. As heat pumps are in principle inverse heat engines, they bring the same benefits regarding the technical maturity and materials.

Chapter 4 on solar power covered an analysis of solar resources and availability, the operational principles and application of solar PV panels, currently-available and emerging technologies, and their impacts on nature and communities. Solar energy input to the Earth is the largest available renewable energy source, and the power produced by solar PV panels is currently the cheapest form of renewable energy. Due to significant developments in solar PV technology, efficiencies are continually increasing and—most importantly—the production cost of silicon PV panels has come down significantly. In addition to being the cheapest energy production option, solar PV is also often socially more acceptable than wind power.

PV panels can be installed on rooftops and walls of buildings or other man-made structures, which reduces their direct land-use impact close to zero. However, in this book our focus is on large-scale ground-mounted solar PV power plants, which are being constructed at an accelerating rate. The technological requirements are the same regardless of the type of installation, but the impacts of ground-mounted solar power plants on land use, on the natural environment, and on local communities become a greater concern than for smaller-scale PV installations in urban areas.

Siting is the single most important decision when developing industrial-scale, ground-mounted solar power. Areas in a natural state, areas of ecological importance (such as migration routes, feeding, breeding, and wintering sites), and areas next to protected locations, should always be avoided. Instead, construction should be directed to locations already changed by human activities, and to disused industrial sites such as old peat production areas, mines, and other marginal or less valuable lands for biodiversity and local societies. In fact, solar power development on degraded lands has been shown to bring biodiversity benefits, and combining land rehabilitation with solar energy production looks promising—similarly, there is a growth in projects combining solar power with other land uses, like agriculture. Agrivoltaics have been shown to increase economic stability for farmers, as income is generated from both food and energy production. Shading from solar arrays can also help crops to survive better during heatwaves. Usually, co-location options require a larger spacing of the solar panels than what would be required for energy production alone, but overall the land use can become more efficient.

Chapter 4 also discussed important social aspects tied to solar energy projects. On the one hand, they can bring positive outcomes, such as the creation of diverse jobs, from design and research to installation and maintenance, enhancing livelihoods and social welfare. Particularly when such projects are community-driven, they can empower local communities, allowing citizens to become investors, innovators, or energy prosumers. On the other hand, however, large-scale solar projects can disrupt local communities' traditional ties to land, impacting livelihoods, or causing forced relocations. Often, powerful actors control decision-making processes regarding the distribution of costs and benefits, marginalizing vulnerable groups and excluding diverse perspectives, which can lead to increased social inequalities and opposition to solar energy projects. Moreover, global solar supply chains can involve human rights violations, conflict financing, unmet safety standards, or even child labor, raising significant ethical concerns. All these social aspects should be carefully analyzed and addressed when planning and developing renewable energy projects.

Chapter 5 on wind power started with the technical operating principles, basic design of turbine blades, an analysis of wind as an energy resource, and some aspects of turbine siting and wake effects in the design of wind-power plants. Since the amount of installed wind power is rapidly increasing, the land-use impact resulting from this expansion is growing as well—we need to understand and address the subsequent negative impacts on nature and wildlife, as well as on local communities. Ways to avoid and mitigate these negative impacts exist, and the sustainable expansion of renewable energy production are built on the foundations of understanding and addressing these potentially harmful impacts.

Similar to solar energy, wind power is relatively dilute; harnessing it in meaningful quantities requires large rotors and considerable areas of land. Therefore, site selection is crucial for sustainable large-scale implementation of wind power. Wind turbines require considerable spacing to reduce the wake effect, in contrast to solar panels which can be installed much closer together. The space between the turbines—often forests, grasslands, or fields—remains largely untouched, and this allows the land to be used for other purposes. However, the visual impacts and associated noise can lower the quality of the area for many species as well as local residents. The area required for installation and supporting infrastructure, such as maintenance roads, transmission lines, and grid connections, creates environmental changes that result in habitat loss, degradation and fragmentation. Many species are known to avoid wind power production sites, and some bird and bat species collide easily with wind turbines blades (which can have tip speeds around 290 km/h) and powerlines. The subsequent increases in mortality can be significant, for example on large, slowly reproducing bird species such as the white-tailed eagle. Similarly, offshore wind-power development creates changes in aquatic ecosystems, and some of these are poorly understood—such as the impact of wind-wake on the hydrodynamic conditions in the ocean, and the resulting effects on marine ecosystems via primary production.

Like solar power development, protected areas and locations with high ecological importance should be excluded from consideration for wind power development, and during the planning process important environmental features, such as migration routes, breeding, feeding and wintering grounds, should be recognized and protected, either by timing construction activities outside sensitive seasons or limiting activities altogether in the most vulnerable sites and apply buffer zones that protect these sites from the impacts of the construction and operation of the power plant. Environmental concerns, including impacts on wildlife, are also a matter of social acceptability; concerns over nature have been identified as a growing source of opposition to and barrier for renewable energy development.

In Chap. 6 we focused on the use of biomass and waste for energy, and material and chemical production. We concluded that the sustainable use of biomass is limited by its large land-area requirement, and subsequent negative impacts on nature. Meeting the needs of the growing population (the current prediction is that our population will stabilize for around 10–11 billion, from 8 billion in 2024) challenges current systems of production for food, energy, and materials. As most of the terrestrial world is already impacted by human uses, there is not much space left for further expansion. We need to find more cost-efficient land uses to meet our needs, and improve circular economy solutions to reduce the demand for virgin materials. The single, most efficient way to alleviate land use pressure is to favor vegetarian diets and minimize meat consumption. As discussed in Sect. 6.4.2, around 77% of agricultural land, or 37 million km^2, is used for grazing and growing feed for domestic animals. With a smaller consumption of meat, much more land could be used for sustainable energy generation. Moreover, meat production requires significant amounts of water, which is yet another depleted resource globally. As an example, the production of 1 kg of beef requires 8 kg of feed and 14 500 l of water, while 1 kg of pork takes 3 kg of feed and 6000 l of water.

Even with such a change in our diets, with the great improvement in climate change mitigation it would bring, the production of our food and the feed required for the remaining domestic animals would still generate large amounts of side- and waste-streams. These include woody parts, branches and leaves of plants, seeds and peelings. Beyond the consumption of food and feed by animals and humans, we have urine, manure and human excrement—all rich in valuable nutrients. In fact, as discussed, these side-streams could act as the raw materials for our chemicals, fertilizers, and even sustainable fuels. We need to develop and up-scale waste-based biorefineries to utilize these resources as efficiently as possible.

Since the efficiency of the photosynthesis is very low, typically less than 1%, much less energy can be obtained per land area through the growth of biomass as compared with PV panels. However, biomass is a natural chemical storage of solar energy, and biomass and derived biofuels can be used, for example, to balance variability in wind-

and solar-power generation. Even more important than the storage of solar energy, growing and existing biomass is nature's own carbon capture and storage machine, with enormous capacity. Sustainably produced and harvested biomass can be very efficiently used as a fuel, exploiting the same combustion and conversion systems designed for fossil fuels. We must ensure that our use of biofuels does not exceed the rate that plants are able to bind the CO_2 released in their harvesting, production and combustion.

In fact, to decrease the atmospheric CO_2 content, the release of carbon to the atmosphere should be less than the amount removed by plants. This feature of biomass can be utilized directly as a carbon capture method, referred as the BioEnergy Carbon Capture and Storage, or simply the BECCS. The principle of BECCS is that we let nature to do the capture part of CO_2 from the atmosphere, then efficiently burn the biomass, and finally capture the concentrated point-stream of CO_2 from the power plant and store it permanently, for example in geological underground storage. The capacity of this process is again limited by the amount of sustainably-available biomass, as well as the availability of suitable safe and large-scale CO_2 storage locations. As an essential component of our future energy systems, all the sustainable potential for BECCS should utilized.

Chapter 7 looked at the various energy storage solutions needed to stabilize the variability of wind- and solar-power production. The storage of energy in large quantities over several months is extremely challenging—in particular, the size of seasonal energy-storage systems becomes very large, requiring considerable areas of land and a lot of construction materials. To limit the required capacity of the largest storage systems, we need to re-think the energy system as a whole—for example, do we need to store energy as electricity, or is it better to store it as low-temperature heat? We need to consider which components of energy generation and load are already, or can be made, flexible. Technically both wind and solar energy can be curtailed rapidly if needed. Unfortunately, this has negative impacts on economic performance; if generation stops, no income is produced. Hydropower can be adjusted up or down rapidly within design limits. The concept of demand-side response is also a very important tool, making consumption economically more attractive at times of over-production, and less attractive in periods of low production. Through peak-shaving and valley-filling we can manage with a smaller installed capacity, less storage, and lower capital investments. We should determine how to make power loads, including electrolyzers, more tolerant to variability in power generation.

In addition to storing energy for stationary heat and power applications, there are so-called hard-to-abate sectors that are difficult to electrify, including heavy road transportation, marine sector, aviation and chemical, cement and metallurgical industries. These have been built entirely on fossil-based raw materials, and so as we move away from fossil resources, we must find alternative solutions for these sectors. For transportation, e-fuels made from hydrogen and captured CO_2 is a promising solution—these would allow the continued use of internal combustion engine and gas turbine systems, which is a great benefit in terms of up-scaling. For chemical and metallurgical industries based on coal, oil and natural gas, we are also able to replace these with green hydrogen and CO_2 as raw materials.

Finally, in Chap. 8 we discussed electrical power systems, which are playing an increasingly important role with the decarbonization and electrification of power systems. As we install variable energy production systems in greater numbers, the need for both energy storage and large-scale power transmission systems increases. Often the best generation sites are far from where consumers are located, requiring efficient transmission systems—off-shore wind power is a good example. To reduce land-use impacts resulting from power transmission lines, cooperation between various project developers and grid operators is required to construct an optimal power transmission system.

9.2 Challenges

There are several hurdles along the way to success in the energy transition. We next take a critical look at the availability of materials needed for the planned expansion of renewable energy systems. After this, we discuss the synergies and trade-offs between renewable energy systems and sustainable development goals.

9.2.1 Materials Availability for the Green Energy Transition

In comparison with the mature energy and material value chains in the fossil economy (including both traditional metals and fuels), a new decarbonized energy system will require materials that are not yet mined at large scales. We might even ask if we are converting the fossil economy into a minerals' economy—are critical materials becoming the "new oil"? We understand well the negative impacts of coal mining, oil drilling and natural-gas use, but what are the impacts of large-scale mining of rare and critical materials for building components for wind turbines, electric motors, solar panels, electrolyzers, fuels cells, and batteries?

As an example, the manufacture of silicon solar cells, the most common PV technology, requires a lot of materials and energy, to melt sand and several other energy-intensive steps. With several emerging new technologies, such as the perovskite solar cells, the amounts of energy and materials can be significantly decreased; unfortunately, harmful chemicals are still used in their production, which must be safely managed. Wind turbines, on the other hand, require metals such as neodymium for the permanent magnets, and submarine cables add a significant material requirement to offshore power plants. As an illustration, Fig. 9.1. Summarizes the materials required for 500 MW solar PV, onshore wind, and offshore wind power plants.

To better understand the status of the available materials, [5] carried out a broad survey of the most important materials needed for wind turbines, solar panels, and batteries. Figures 9.2, 9.3 and 9.4 present a summary of this study.

It often takes 20–30 years to start a new mine, sometimes even longer to minimize the negative impacts on nature and societies. It is evident from Fig. 9.2, 9.3

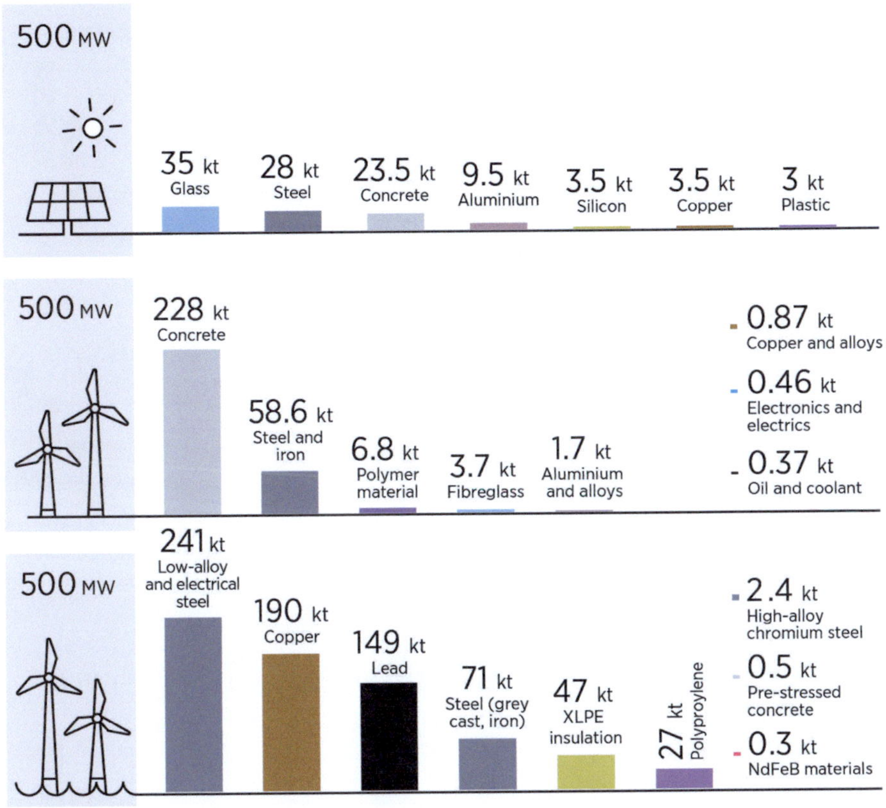

Fig. 9.1 Materials requirement for a 500 MW PV solar, onshore wind and offshore wind power plants [8]. Note that for offshore wind power, the distance from land affects the required submarine cable materials (Copper, Lead, XLPE, polypropylene)

and 9.4 that the rate of energy transition will already be limited by the availability of some materials from 2025 onwards with the current mining and production activity, shown as the red dashed line. As an example, how to interpret the figures, the need for Dysprosium (Dy), needed in permanent magnets of the electric motors and generators, is already 50% larger in the high demand scenario in 2025 in comparison with its current production, see Fig. 9.3. To stay on track to reach our climate goals, we will need to multiply our mining activity well above the existing capacity in 2024 and invest heavily in circular economies. How this can be done sustainably, minimizing negative impacts on nature, water resources and surrounding societies, is the key question.

Figure 9.5 illustrates the way circular economy helps the energy transition. The green curve presents the increasing amount of materials in use. For example, when a wind power plant reaches its lifetime, it is decommissioned. To be able to minimize the need for new materials to manufacture new replacing turbines, the materials from

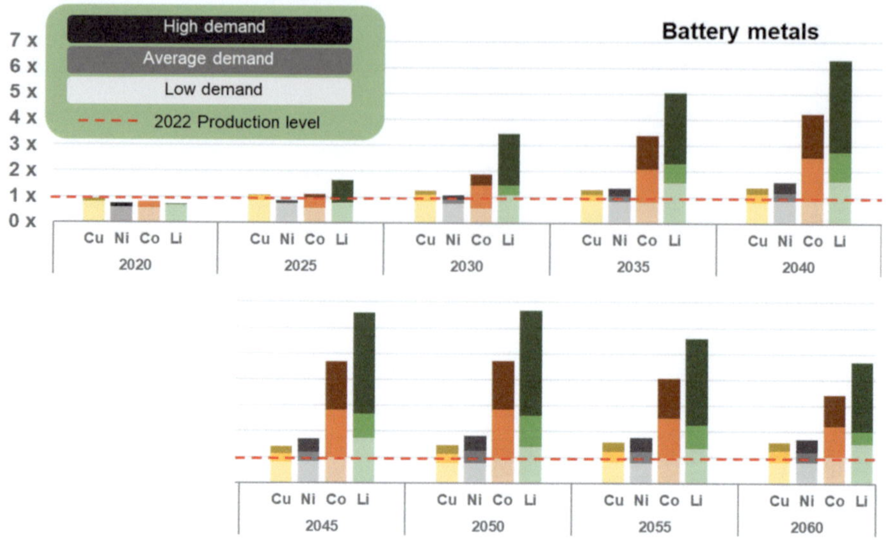

Fig. 9.2 Required mining volumes to reach climate targets: battery metals. The y-axis is multiples of 2022 production levels [5]

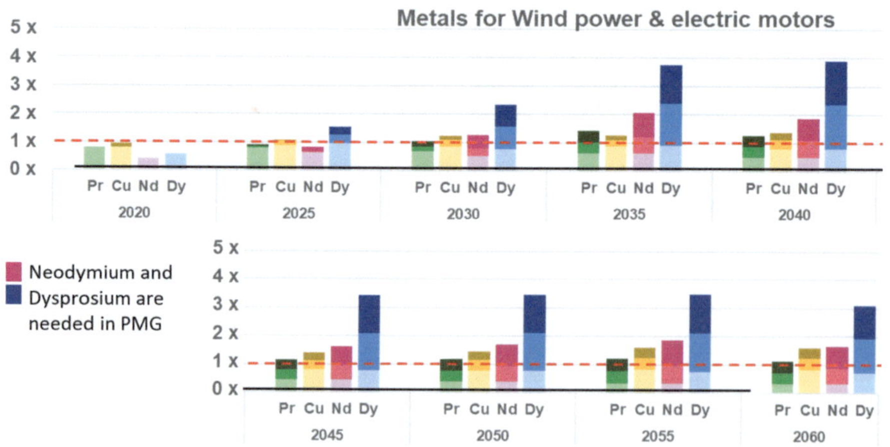

Fig. 9.3 Required mining volumes to reach climate targets: metals for wind turbines, motors and generators (PGM = permanent magnet generators). The y-axis is multiples of 2022 production levels [5]

the old turbines should be recycled and utilized as much as possible. The losses in the recycling are presented as short arrows upwards from the recycling loops. Materials to replace these losses, plus the materials needed to increase the installed capacity is presented as the arrows entering the loop from below. To minimize the time to reach the required level of materials in use, our material recovery processes should be as

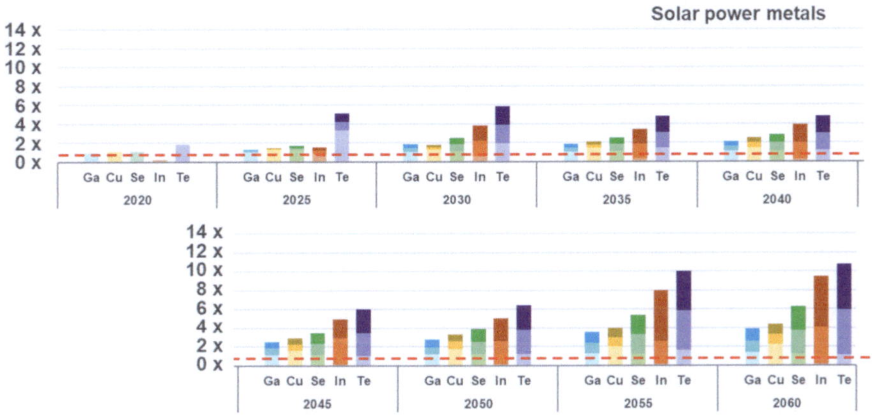

Fig. 9.4 Required mining volumes to reach climate targets: metals for solar panels. The y-axis is multiples of 2022 production levels [5]

efficient as possible, to avoid losses of the valuable elements in the waste fraction, the mine tailings. In addition, when carrying out mining activities, different actors should collaborate to utilize all metals in the mined rock, not just a single element by one actor.

To reach renewable energy implementation targets, the mined metals and mine tailings should be maximally utilized and recycled, to avoid and mitigate material losses. The available amount of metallic materials is finite, but in contrast to fossil fuels, they are in theory indestructible and indefinitely reusable, as they are retained mostly in solid or liquid form.

Fig. 9.5 Circular economy in upscaling the amount of materials needed for the new energy system. Not to scale; the fraction of losses should minimal

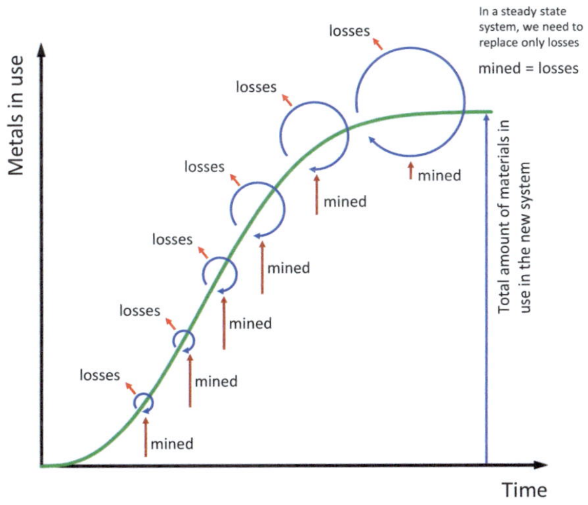

9.2.2 *Balancing Different Sustainability Goals*

Every energy technology we use has an impact on nature and society; conversely, energy technologies are impacted by societies, by existing policy frameworks, by availability of resources and finance, and by prevailing power dynamics. Limiting global warming and bringing solutions to sustainability challenges requires a transformative change in energy production technologies, industries and related infrastructure, and will result in significant changes in society and impacts on land-use.

As renewable energy sources are more dilute, and are therefore collected from a broader spatial distribution than some other forms of energy production (see Fig. 9.6), it is important to manage land-use impacts via thorough planning processes to identify areas with the least conflict with both nature and communities. Fortunately, there is flexibility in siting opportunities for solar and wind power, and it is possible to direct construction to areas already altered by human uses. [1], for example, estimated that globally there is 19 times more such altered lands that could be used for renewable energy production to meet the Paris agreement goals.

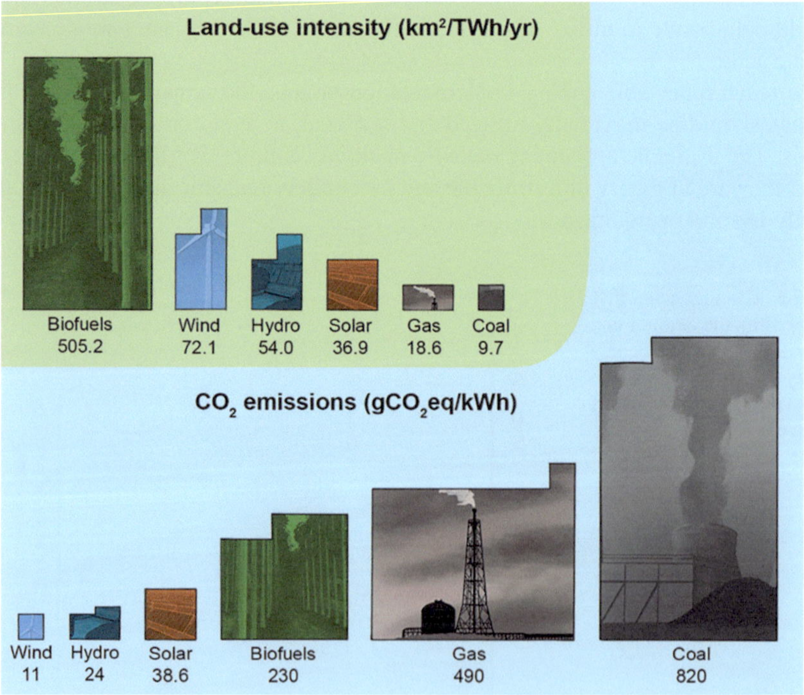

Fig. 9.6 Land-use intensity (km^2/TWh/yr) for fossil fuels and renewable energy sources, adapted from [9]. © Maija Karala

On the other hand, poorly-sited renewable power plants can lead to land-use conflicts, or to the release of carbon stocks through loss of biodiversity. Kiesecker et al. [10] estimated that unsuccessful siting of renewable energy could release around 4 billion tonnes of CO_2 from land and soils, which would equal around 8% of the overall emission reduction needed to meet the Paris Agreement goals. Today, over 17% of wind, solar and hydropower globally operate within important nature conservation areas [15]. In addition, solar energy and—in particular—bioenergy production can compete with land needed for food crops, potentially impacting food security. Biofuel production, as well as keeping vast arrays of solar panels clean, can use significant amounts of water, which impacts local water availability and can pollute waterways with fertilizers and chemicals.

Therefore, during planning it is important to invest time and effort in recognizing trade-offs, understanding local conditions and state of nature, and identifying areas with least conflict. This is a prerequisite to avoid building infrastructure on ecologically vulnerable or important areas, and during sensitive seasons, such breeding or migration. The use of mitigation hierarchies to avoid, minimize, restore, and offset negative ecological or social impacts provides a pathway for action that contributes not just to the goal of achieving net-zero, but also enables positive ecological and societal outcomes.

One way to make sustainability synergies and tradeoffs more understandable is to compare the impacts of different climate mitigation measures on other sustainable development goals (Fig. 9.7) [21]. These interconnected goals provide a widely-accepted framework for assessing different sides of sustainability measures and can be used to illustrate synergies and potential trade-offs of actions aimed at reaching the goals. Climate-change mitigation and measures to adapt to climate change are crucial for sustainable development in many ways. Climate change poses a threat to ecosystem health and biodiversity, which affect human health, livelihoods, food production and security, and other factors. However, despite many clear synergies, there can also be significant trade-offs that need to be addressed, especially at a local level, to secure a just and sustainable energy transition.

A successful energy transition should be planned and implemented so that, alongside halting climate change (Goal 13), actions contribute to achieving other SDGs that secure sustainable economic growth, social well-being, and ecological sustainability. Figure 9.7 illustrates the connections between different renewable energy solutions that can be used to reach Goal 13 on Climate Action, and other sustainable development goals. There are several synergies, especially for economies and societies, but also potential trade-offs, and the way these are addressed in the energy transition will eventually determine how sustainably we manage the transition towards clean energy sources.

Trade-offs on nature can include, for example, destruction or degradation of protected natural areas, or areas with ecological importance, and local loss of biodiversity as we discussed earlier. Trade-offs for local communities can relate to value of land, access to important provisioning services such as agriculture or fisheries, degradation of important cultural or recreational areas, job insecurity, and resource depletion, with significant gendered impacts. Renewable energy projects can negatively

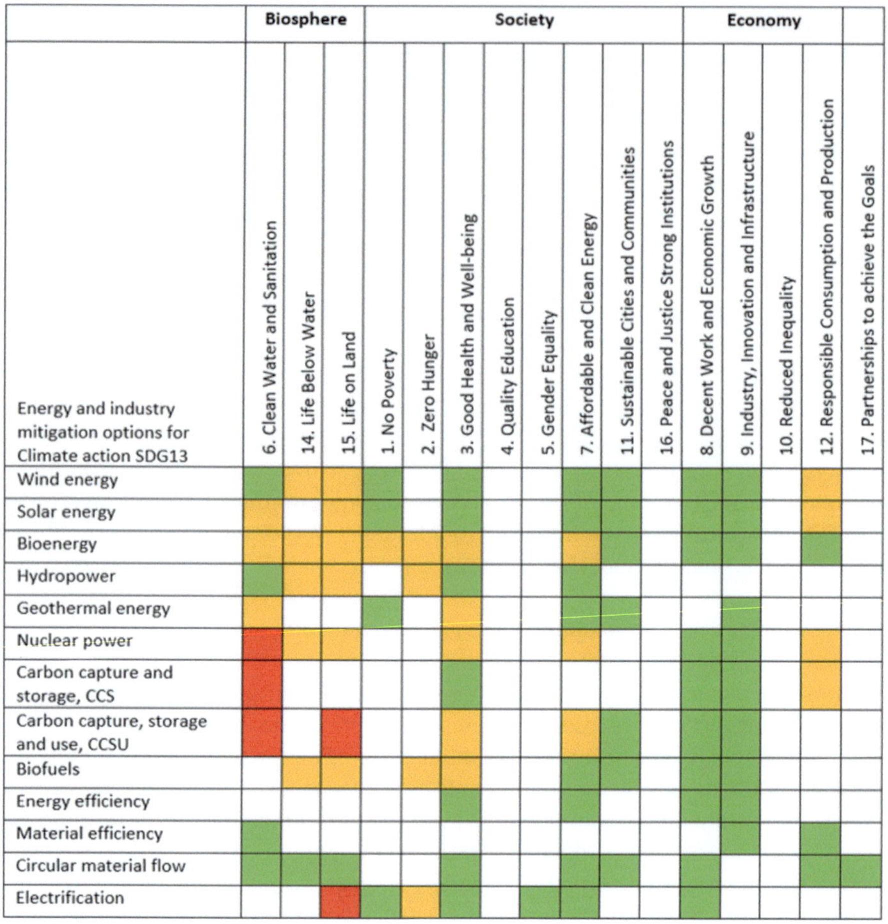

Fig. 9.7 Connections between climate mitigation options (Sustainable Development Goal, SDG 13) on other SDGs. The data is derived from the [6] report on climate change mitigation, and the different SDGs are grouped together in accordance with the SDG wedding cake model (see also Chap. 1, Fig. 1.5) to highlight how social and economic sustainability are built on a healthy and well-functioning biosphere. Green = synergy, red = trade-offs, yellow = both synergies and trade-offs, blank = not assessed

impact social justice or gender dynamics and exacerbate existing inequalities by disproportionately affecting the rights and livelihoods of vulnerable groups, such as rural communities, women, indigenous people, and minorities. Achieving synergies between renewable energy and local communities' well-being depends on addressing power imbalances where more resourceful actors, such as large energy corporations, use their power to fulfill their interests, while local communities, in more vulnerable positions, might lack the opportunity to participate in decision-making processes. Failure to address ecological impacts and impacts on local communities can result in opposition and delays to renewable energy development, and therefore undermine reaching climate targets.

Balancing these trade-offs requires a comprehensive understanding of the local conditions, and an inclusive approach built on technological innovation, policy support, community engagement, and environmental stewardship.

9.3 Concluding Remarks, a Way Forward

Transition to renewable energy technologies, driven mostly by solar and wind energy, provide many benefits over fossil-fuel-based energy production. However, it is equally important to acknowledge that this transition, while necessary, can induce impacts at a local level that compromise sustainable development. Addressing, avoiding and mitigating these negative ecological and social impacts locally form the cornerstones of sustainable energy transition. The objective of this last section is to present ideas to analyze the magnitude of change required, and different scenarios for reducing carbon dioxide emissions.

We'll start this discussion by estimating the needed capacity of renewable energy for decarbonization and how this would impact land use. We will then expand the discussion to current carbon footprints of select countries, and show how these should be changed to reach decarbonization. We hope that this perspective makes it easier to understand the scale of the challenge. Finally, we provide some action points for making the energy transition.

9.3.1 Emission Reduction Wedges

Our first tool for this discussion is the Stabilization Wedges approach (Pacala and Socolow 2004), which aims to stabilize total annual emissions at a constant level of ~ 7 Gt carbon (26 Gt of CO_2) between 2003 and 2053. This level was estimated to keep atmospheric CO_2 concentration below 500 ppm until 2054. During this period, we would have time to develop and scale up new technologies, such as larger wind turbines and more efficient solar panels, implement CCS, build more nuclear power, and maximize afforestation and reforestation to support natural carbon sinks and storage, and secure biodiversity. This wedge tool was presented in 2004, and as shown in Fig. 9.8, almost twenty years later emissions have increased from 26 to 37.4 Gt (using 2023 data) while energy consumption has increased from 130 000 TWh to 176 000 TWh.

Figure 9.8 shows the emissions from fossil energy sources until 2019, and then several emission reduction options, including the implementation of onshore and offshore wind, electric vehicles, and implementation of land-use-intensive options. In this model we use simple linear wedges, so that the analysis is simplified, and the solutions and their challenges can be more easily understood. We defined our target

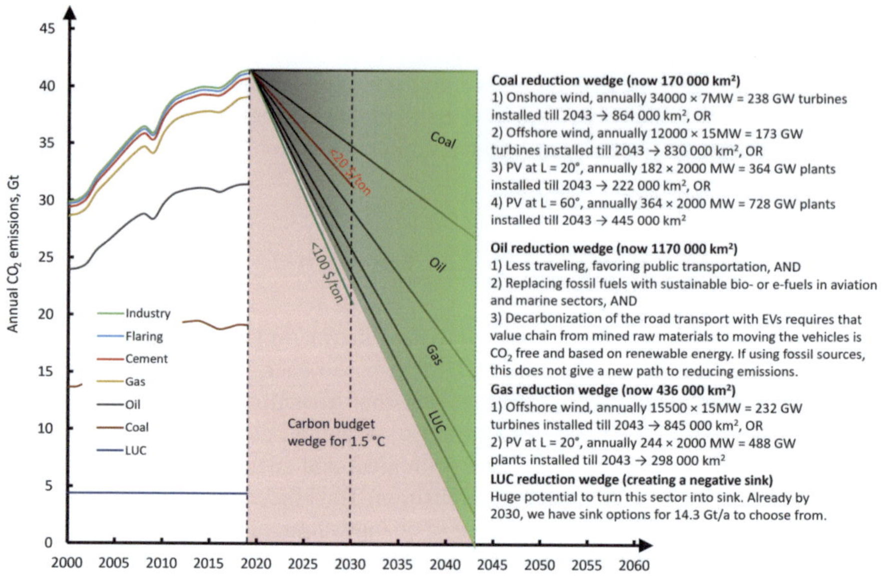

Fig. 9.8 Approximations of CO_2 emission reduction wedges, annual renewable energy capacity additions, and total required land areas. Red and green decreasing lines starting from 2019 present the estimates for obtainable CO_2 emission reduction with technology prices below 20 and 100 UDS/t of CO_2 emissions reduced, respectively [7]. The area under both pink and green wedges is 500 Gt, corresponding to the assumed CO_2 budget to stay below 1.5 °C [7]. As a reference, the surface area of the USA is 9 834 000 km^2. The main assumption is that our power consumption would not increase—if it does increase, the installed capacities need to be increased accordingly. Estimations of current area needed for coal, oil and gas are based on [12]. See Appendix 9.1 for calculation details

to limit temperature increase to 1.5 °C, giving us a carbon budget of approximately 500 Gt CO_2 from 2019. With the annual emission in 2019 of 41.7 Gt, this gives us around 24 years (2019 to 2043) to make the transition to a net-zero energy system ($500/41.7 \times 2 = 24$, see Chap. 2.1 for details of this approach). Our carbon budget corresponds to the area of the pink wedge in Fig. 9.8; the equal-sized triangular green area is the reduction wedge. The principal idea of the wedge approach is that we do not have make the complete system transformation all at once; instead, the additions to capacity can be split over the reduction period evenly. The starting point is the primary energy consumption and CO_2 emissions in 2019 of 43 700 TWh/14.7 Gt (coal), 53 400 TWh/12.3 Gt (oil), and 39 100 TWh/7.6 Gt (natural gas). CO_2 emissions from land-use change we assume as 4.4 Gt.

Figure 9.8 presents several options to decarbonize our coal and natural gas use. In most cases the land use increases, but, for example, installing solar PV at southern latitudes can decrease the required land area in comparison with natural gas use. In particular, the decarbonization of road transport significantly depends on the electricity generation technologies used, and it is difficult to estimate area changes and CO_2 reduction potential. If we just compare the energy needed to move cars with renewable electricity sources, this significantly reduces the required land area. Unfortunately, the main CO_2 emissions of EVs typically come from the mining and manufacturing steps; in addition, the amount of materials required might be not available from currently-operating mines. The land use change wedge (to reduce the current land use emissions to zero) is shown but is not analysed in more detail; there are several tools to create a negative sink from this sector. Based on the synthesis report of IPCC [7], there are potential emission reduction options of 14.3 Gt per year from land-use related sectors—for example: reducing conversion of natural ecosystems, 4 Gt/a, carbon sequestration in agriculture, 3.4 Gt/a; ecosystem restoration, afforestation, reforestation, 2.8 Gt/a; shifting to sustainable diets, 1.8 Gt/a; and sustainable forest management, 1.4 Gt/a.

9.3.2 National Emission Reduction Targets for a just Energy Transition

We discussed above the global emission reduction targets required to remain below 1.5 °C—but how do we make this happen in reality? What does this requirement mean for different countries? We next discuss the concept of defining a target value for per capita emissions (see, for example, [19]). As a starting point, Fig. 9.9 shows the current per capita emissions for several countries. The red curve indicates the global average per capita emissions needed to stay within our carbon budget. Predicted population increases [22] are included in the calculations.

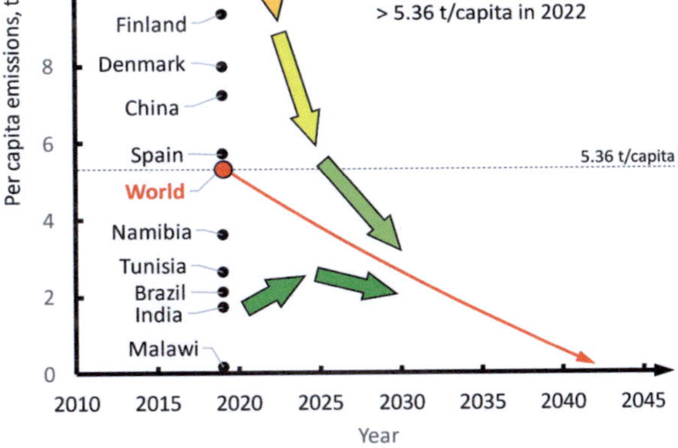

Fig. 9.9 Per capita consumption-based emission of CO_2 in 2019, and the required decrease (red curve) to remain within the carbon budget of 500 Gt until 2043. Note that 41% of the global population has a carbon footprint larger than 5.36 t, while the other 59% are already well below the target

As Fig. 9.9 shows, more than 40% of the Earth's population has carbon footprint larger than the average 5.36 tons of CO_2—therefore, most of the population lives in countries with lower carbon footprint than the average. The large red dot indicates the starting point from which we should reduce the global average footprint to zero by 2043 to keep the average temperature increase below 1.5 °C, as presented in Fig. 9.8. To have a just transition, countries below the current average value should be allowed to develop their economies while aiming to keep their per capita emissions below the global average. This would be achievable by utilizing renewable energy technologies with very low emissions (already cheaper than fossil options), preferably without passing through the fossil economy. It can be argued that developing countries should have a right to increase their CO_2 emissions to bridge the economic gap with developed countries who have benefitted from the fossil fuel era. If this happens, and India and countries in Africa increase their standard of living with fossil fuels, we are not able to reach climate targets. Therefore, enabling a just and sustainable energy transition, for example via financial and other incentives, should be one of the key priorities of developed countries. On the other hand, countries already with carbon footprint higher than the global average should reduce their emissions rapidly and increase their natural sinks as much as possible. As discussed with Fig. 9.8, there is a large selection of low-price technologies we could choose from to achieve this outcome.

In overall, weather we succeed in this huge challenge depends not only on available technologies, but rather it's a complex political process, demanding profound changes in how we think, what we do, and how we cooperate as societies in a global scale.

9.3.3 *Action Points for Energy Engineers*

In this book we have discussed the need for a 'transformative change' to bring human society back within planetary boundaries. The question of how to bring greenhouse gas emissions close to zero is pivotal, and energy transition has a central role in the ongoing sustainability transformation. In its simplest form, transformative change means doing things differently: getting rid of practices that maintain and enforce negative trajectories, and inventing and strengthening practices that secure and sustain sustainable solutions. Transformative change includes several parallel, long-term, and often complex processes that can also be characterized by risks and uncertainty. In this energy transition, risks can arise, for example, through changes in geopolitical situations that result in reduced availability and increased prices of critical materials. On the other hand, a transition towards renewable energy sources will also help to improve energy independency and security of supply for different nations and regions, by reducing dependency on fossil-fuel producers.

Enforcing a sustainable energy transition can be visualized with an X-curve, shown in Fig. 9.10 [18], that starts with converting old practices, such as harmful policies and subsidies, and building new practices, including both innovating new solutions as well as enforcing existing, good practices such as circular economy solutions and sustainable land-use practices. The conversion of dominant practices will culminate in breaking down and phasing out harmful practices, such as over-consumption, unsustainable land use, and eventually use of fossil fuels. Building new practices will lead to scaling up and institutionalizing successful practices such as sustainable finance mechanisms, sustainable lifestyles, and clean and sustainable energy systems.

Note that Fig. 9.10 is just an example; also, as we learned earlier, different countries are in different stages of energy transition, and so the challenges that need to be addressed vary as well. Regardless of the stage, some action points along the way towards a sustainable energy system include:

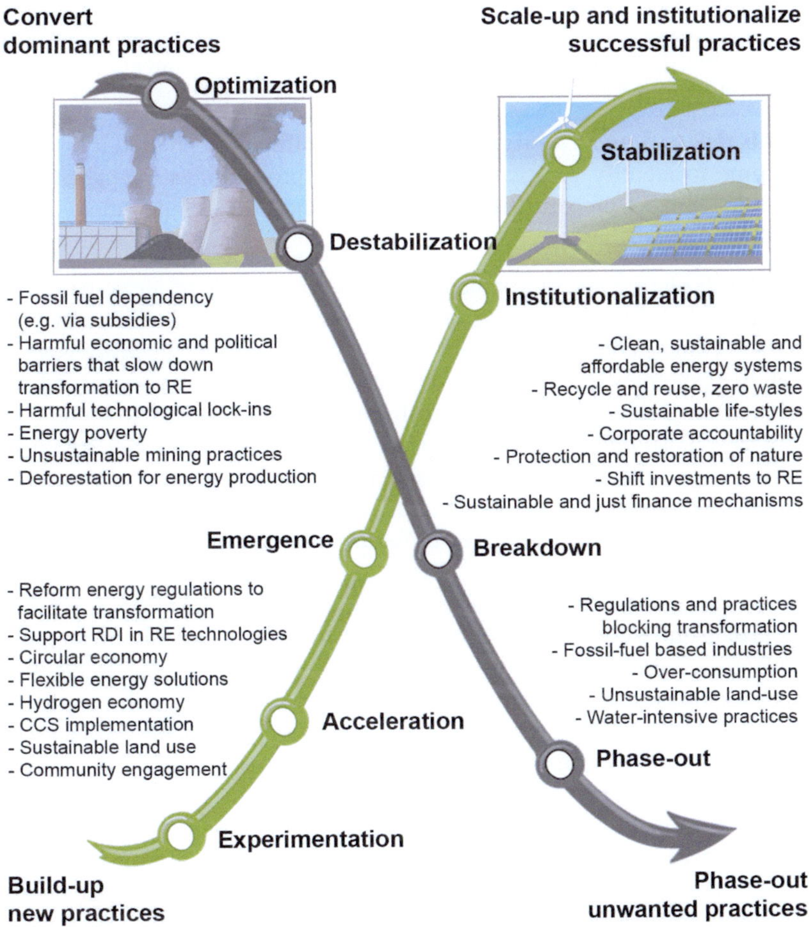

Fig. 9.10 X-curve representing the steps needed for the transition to a clean and sustainable energy system. RE = renewable energy, RDI = research, development and innovation, CCS = carbon capture and storage (inc. direct air capture, DAC). Modified from [18] © Maija Karala

Setting a goal that guides further action. What needs to happen in your country to keep up with the Paris agreement goal? How can you build new social norms or legal changes that enable enforcing the transformation?

Being systemic and future oriented. It might be difficult to predict the future, but with thorough preparation and planning it is possible to direct the range of possible futures. This involves broadening the scope of planning to incorporate long-term thinking, including also addressing the interconnectedness of techno-economic-socio-ecological systems. What are the necessary steps and key intervention points having an impact across systems towards the goal? How can you help in designing a more sustainable future?

Address path dependencies. Past decisions, investments and credits tie us to current energy systems and direct what is possible during the coming decades. How can we ensure that the decisions taken today will not lock us into unexpected, harmful systems in the future?

Deal with risk and uncertainty. As the future is difficult to predict, it is important to take time to identify potential risks and their probabilities (so-called "known unknowns"), and to prepare for things that cannot yet be identified (so-called "unknown unknowns").

Innovation. Fostering an environment that encourages flow of ideas, new ways of thinking and operating, and innovation. There is an urgent need to identify scalable solutions to climate change that do not conflict with other sustainability goals.

From the examples in the X-curve in Fig. 9.10 we see that a successful energy transition to sustainability requires a transformative change across all sectors. In this book we have mostly focused on the impacts of renewable-energy systems on nature and societies, and the interdependencies of this techno-economic-socio-ecological system. Energy transition therefore requires coordination across different scales and sectors of society. Coordination across different value chains is important as impacts may vary from one place to another—for example, activities in one place can affect biodiversity or human livelihoods in another far away, via global trade, markets, or land displacement. Researchers have also raised concerns that institutional inertia—the tendency to resist adoption of new practices and maintain the status quo—is holding back the progressive adoption of 100% renewable-energy solutions [3]. They believe the energy transition could be faster, and result in lower costs and higher sustainability, if international institutions adopt the latest insights from renewable energy systems research.

High-income countries bear larger responsibility for global ecological crises, and their ecological debt to the rest of the world is larger than that of lower-income countries, as illustrated for example with the Earth overshoot data. This places a greater responsibility on high-income countries to radically reduce their consumption of resources, to take all available measures to avoid further ecological degradation, and to invest in and implement measures that make maximum use of circular economies and low- to zero-emission solutions.

Halting the loss of biodiversity is crucial alongside actions to curb climate change, and jointly solving these two crises is about respecting the planetary boundaries that maintain all life on this spaceship called Earth. These boundaries are non-negotiable, and for that reason, apart from finding solutions to the climate change, we need to learn a new narrative for sustainability. We can use the energy transition as one tool in transformative change towards a transgenerational just and sustainable future. We can ask ourselves: what prevents us from living in ways that serve not just ourselves, but also future generations, other species, and the entire, diverse, interconnected world?

Appendix 9.1 Land use and emission calculations

For the coal and gas reduction wedges, we have assumed that the electricity consumption remains constant, which is not the case in reality, since it would increase with a system where no waste heat is available from combustion-based processes. The rough approximation principle for replacing coal and gas as power producing technologies is quite simple. By multiplying the primary energy consumption of coal and natural gas by total electrical efficiency of the fossil energy systems, we can estimate the produced electrical energy (TWh). From this, the corresponding power of the renewable energy system can be obtained by dividing this with the term $C_f \times 8760$, the effective rated power hours. Capacity factors C_f used were 0.35, 0.48, 0.23, 0.11 for onshore, offshore wind, and PV at 20° and 60° latitudes, respectively. Efficiency of coal and gas turbine combined cycle power plants are assumed 0.4 and 0.6, respectively. Primary energy-based emission factors are 0.337 t/MWh and 0.196 t/MWh per heat and 0.843 t/MWh and 0.326 t/MWh per electricity for coal and natural gas, respectively. Emissions coefficients for onshore wind, offshore wind and solar PV is 0.012, 0.013 and 0.037 t/MWh, respectively (see Table 2.3). Capacity densities were assumed 6.6 and 39.3 MW/km^2, for wind and solar PV.

As an example, the annual capacity additions in the case of converting coal with solar PV at latitude 20°, we did the calculations as follows. Electrical energy generated from the primary energy in coal is, $E_e = 43700\text{TWh} \times 0.4 = 17480\,\text{TWh}$. Since even solar PV creates CO_2 emissions, its emission factor being 0.037 t/MWh, we are able to reduce emissions by $17480 \times 10^6\text{MWh} \times (0.843 - 0.037)\,\text{t/MWh} = 14.1$ Gt from the current emission from coal combustion of 14.7 Gt, still leaving 0.6 Gt unavoided emissions coming from the manufacturing of solar panels. Total required capacity of solar PV is $P = 17480\text{TWh}/(0.228 \times 8760\text{h}) = 8740\,\text{GW}$, and required surface area is $8740000\text{MW}/\left(39.3\text{MW/km}^2\right) = 222391\,\text{km}^2$. Annually, 2019–2043, we need to add $8740/24 = 364\,\text{GW}$ of panels, corresponding to 182 solar PV power plants of 2 GW rated power (the scale of world's largest plant size in 2024). As for comparison, the added PV capacity 2023 was 444 GW. The annual installation rate is predicted to increase even as high as 1 TW per year by 2030. If we could globally decrease our energy consumption, reaching our targets would be much easier.

The calculation of the emissions reduction from the oil reduction wedge is very complicated. Based on [20] a fraction of 0.49 of oil is used in road transport. We assume that this is also the fraction from the primary energy consumption. Therefore, energy consumed by road traffic is $54300\text{TWh} \times 0.49 = 26151$ TWh. As an example, based on McDonald (2009) land use intensity of petroleum products production onshore (~70% of oil is produced onshore) is 44.7 km^2/TWh (per primary energy), so the current land requirement is $26151\text{TWh} \times 44.7\ \text{km}^2/\text{TWh} = 1170000\ \text{km}^2$. In average, the yield of gasoline and diesel from the crude oil is 0.71, and based on Ritchie (2023), only a fraction of 0.2 of the fuel's energy is used to move the car, so the annual energy used to move cars is $26151\ \text{TWh} \times 0.71 \times 0.2 = 3713\ \text{TWh}$. Based on [11] total efficiency on electric cars from electricity to movement is 0.89, and therefore, the amount annual renewable electricity required is $3713\ \text{TWh}/0.89 =$ 4172 TWh. The amount of needed new wind or solar power can be estimated from this as above. However, the main source of CO_2 emissions in the value chain of electrical cars is in the mining of raw materials, for example for batteries and cars manufacturing [2]. Emissions related to build the renewable energy systems to generate the power needed to move the cars can be approximated by multiplying the needed energy by the emission factor, for example for solar PV, $4172\text{TWh} \times 0.037\text{t}/\text{MWh} = 154\text{Mt}$.

References

1. Baruch-Mordo S, Kiesecker JM, Kennedy CM, Oakleaf JR, Opperman JJ (2019) From Paris to practice: sustainable implementation of renewable energy goals. Environ Res Lett 14:024013
2. Bickert S, Kampker A, Greger D (2015) Developments of CO_2-emissions and costs for small electric and combustion engine vehicles in Germany. Transp Res Part D: Transp Environ 36:138–151. https://doi.org/10.1016/j.trd.2015.02.004
3. Breyer C, Khalili S, Bogdanov D et al (2022) On the history and future of 100% renewable energy systems research. IEEE Access 10:78176–78218. https://doi.org/10.1109/ACCESS.2022.3193402
4. Dasgupta P (2021) The economics of biodiversity: the Dasgupta review. London: HM Treasury pp. 610. https://www.gov.uk/government/publications/final-report-the-economics-of-biodiversity-the-dasgupta-review
5. Granvik P, Jokilaakso A (2021) Metals and critical factors for the green transition. Aalto University, International Process Metallurgy Symposium
6. IPCC (2022) Climate change 2022: Mitigation of Climate change. contribution of working group III to the Sixth assessment report of the intergovernmental panel on climate change. Shukla PR, Skea J, Slade R, Khourdajie AA, van Diemen R, McCollum D, Pathak M, Some S, Vyas P, Fradera R, Belkacemi M, Hasija A, Lisboa G, Luz S, Malley J (eds.). Cambridge University Press, Cambridge, UK and New York, NY, USA. https://doi.org/10.1017/9781009157926
7. IPCC (2023) Summary for policymakers. In: Climate change 2023: Synthesis report. contribution of working groups I, II and III to the sixth assessment report of the intergovernmental panel on climate change, IPCC, Geneva, Switzerland, pp. 1–34, https://doi.org/10.59327/IPCC/AR6-9789291691647.001
8. IRENA (2022) Green hydrogen for industry: a guide to policy making, Int Renew Energy Agency, Abu Dhabi. https://www.irena.org/-/media/Files/IRENA/Agency/Publication/2022/Mar/IRENA_Green_Hydrogen_Industry_2022_.pdf?rev=720f138dbfc44e30a2224b476b6dfb77

9. Kiesecker J & Naugle D (2017) Energy Sprawl Solutions: Balancing Global Development and Conservation. Island Press, Washington. https://doi.org/10.5822/978-1-61091-723-0
10. Kiesecker JM, Baruch-Mordo S, Kennedy CM, Oakleaf JR, Baccini A, Griscom BW (2019) Hitting the target but missing the mark: unintended environmental consequences of the paris climate agreement. Front Environ Sci 7. https://doi.org/10.3389/fenvs.2019.00151
11. Kirk K (2024) Electric vehicles use half the energy of gas-powered vehicles. Yale Clim Connect. https://yaleclimateconnections.org/2024/01/electric-vehicles-use-half-the-energy-of-gas-powered-vehicles/
12. McDonald RI, Fargione J, Kiesecker J, Miller WM, Powell J (2009) Energy sprawl or energy efficiency: climate policy impacts on natural habitat for the United States of America. PLoS ONE 4(8):e6802. https://doi.org/10.1371/journal.pone.0006802
13. Raworth K (2012) A safe and just space for humanity. Oxfam Discussion Paper 26
14. Raworth K (2017) Doughnut economics: seven ways to think like a 21st-century economist. Random House UK, pp 384. ISBN 9781847941398
15. Rehbein J, Watson J, Lane J, Sonter L, Venter O, Atkinson S, Allan J (2020) Renewable energy development threatens many globally important biodiversity areas. Glob Change Biol 26:3040–3051
16. Rockström J, Steffen W, Noone K, Persson Å, Chapin FS, Lambin E, Lenton TM, Scheffer M, Folke C, Schellnhuber H, Nykvist B, De Wit CA, Hughes T, van der Leeuw S, Rodhe H, Sörlin S, Snyder PK, Costanza R, Svedin U, Falkenmark M, Karlberg L, Corell RW, Fabry VJ, Hansen J, Walker B, Liverman D, Richardson K, Crutzen P, Foley J (2009) Planetary boundaries: exploring the safe operating space for humanity. Ecol Soc 14(2):32. http://www.ecologyandsociety.org/vol14/iss2/art32/
17. Rockström J, Gupta J, Qin D et al (2023) Safe and just Earth system boundaries. Nature 619:102–111. https://doi.org/10.1038/s41586-023-06083-8
18. Silvestri G, Diercks G, Matti C (2022) X-curve—a sensemaking tool to foster collective narratives on system change, EIT Climate-KIC, https://drift.eur.nl/app/uploads/2023/08/X-Curve-booklet-DRIFT-EIT-Climate-KIC-2022-1.pdf
19. Sorensen (2008) Pathways to climate stabilization. Energy policy 36:3505–3509. https://doi.org/10.1016/j.enpol.2008.05.028
20. Statista (2024) Statista research department. https://www.statista.com/statistics/307194/top-oil-consuming-sectors-worldwide/
21. United Nations (2015) Transforming our world: the 2030 agenda for sustainable development. Resolution adopted by the General Assembly on 25 September 2015, A/RES/70/1
22. United Nations (2022) World population prospects. DESA, Population Division

Correction to: A Sustainable Path for Renewable Energy Technologies

Mika Järvinen, Hanna Paulomäki, Elina Oksanen, Dilip Khatiwada, Sami El Geneidy, Maiju Peura, Janne S. Kotiaho, Eeva-Lotta Apajalahti, Alicja Dańkowska, Mariana Galvão Lyra, Ayesha Sadiqa, Roosa Talala, Samuel Cross, and Antti Arasto

Correction to:
Chapter 2 in: M. Järvinen and H. Paulomäki (eds.), *Designing Renewable Energy Systems within Planetary Boundaries*, Green Energy and Technology, https://doi.org/10.1007/978-3-031-69856-9_2

In the original version of the book, the following belated corrections have been incorporated: The author name "Iryna Herzon" has been updated in Chapter 2 and also added to author contributions under 2.4 as well. The book and the chapter have been updated with the changes.

The updated version of this chapter can be found at
https://doi.org/10.1007/978-3-031-69856-9_2

M. Järvinen and H. Paulomäki (eds.), *Designing Renewable Energy Systems within Planetary Boundaries*, Green Energy and Technology,
https://doi.org/10.1007/978-3-031-69856-9_10

MIX
Papier aus verantwortungsvollen Quellen
Paper from responsible sources
FSC® C105338

If you have any concerns about our products, you can contact us on
ProductSafety@springernature.com

In case Publisher is established outside the EU, the EU authorized representative is:
Springer Nature Customer Service Center GmbH
Europaplatz 3, 69115 Heidelberg, Germany

Printed by Libri Plureos GmbH
in Hamburg, Germany